BIOLOGICAL PHYSICS

Books in the
Key Papers in Physics Series
published by the
American Institute of Physics

Gallium Arsenide
edited by John S. Blakemore

NMR in Biomedicine: The Physical Basis
edited by Eiichi Fukushima

Best of Soviet Semiconductor
Physics and Technology 1987–1988
edited by Mikhail Levinshtein and Michael Shur

Nonclassical Effects in Quantum Optics
edited by Pierre Meystre and Daniel F. Walls

Piezoelectricity
edited by Carol Zwick Rosen, Basavaraj V. Hiremath,
and Robert E. Newnham

Biological Physics
edited by Eugenie V. Mielczarek, Elias Greenbaum,
and Robert S. Knox

BIOLOGICAL PHYSICS

Edited by

Eugenie V. Mielczarek, Elias Greenbaum, and Robert S. Knox

AIP

American Institute of Physics

New York

American Institute of Physics
335 East 45th Street
New York, NY 10017-3483

Library of Congress Cataloging-in-Publication Data

Biological physics / edited by Eugenie V. Mielczarek, Elias Greenbaum, and
  Robert S. Knox.
      p.   cm.—(Key papers in physics)
    Reprints of papers originally published in various journals before 1990.
    Includes bibliographical references.
    ISBN 0-88318-855-4
    1. Biophysics. I. Mielczarek, Eugenie V.   II. Greenbaum, Elias. III. Knox,
Robert S. (Robert Seiple).   IV. Series.
    QH505.B4665   1993
    574.19′1–dc20
                                                              93-3280
                                                            CIP

Table of Contents

Section IV: Information Generation and Transfer

Section V: Experimental Techniques

Section VI: Photosynthesis

Preface

Biological physics as a field of research is the study of the physical processes governing living systems.

This book provides a sampling of the field. Represented by publications in journals as diverse as *Physical Review* and *Scientific American*, the volume is intended for physicists unfamiliar with the field of biological physics and for chemists and biologists working in the field.

The papers demonstrate how experimental techniques and physical theory have been adapted to answer questions specific to the dynamics of life. Some of them contain original research results, while some are reviews written for the scientifically literate public. When perusing the papers this becomes obvious: many physicists trained in fields unrelated to biology, such as condensed matter and nuclear physics, have been motivated to apply their expertise to problems of biological function; physical theory and techniques are used to investigate living systems on every scale, from millimeter to subängstrøm, and from every aspect, from structure to energetics. Many of the authors of researches in this volume were motivated to tackle questions of biological function by their curiosity and ability to see parallels. Mechanics is used to decipher bacterial motility, spectroscopy, and crystallography to pin down the structure and function of oxygenating hemoglobin, and femtosecond pulsed lasers to study the primary quantum event in photosynthesis. Ultimately all functions of living systems are determined by molecular dynamics. For example, biological materials are forms of condensed matter in which long-range forces are responsible for the conformational motions and the complex spectacular chemistry which governs the functioning of living matter.

Often after the application of established physical techniques and theory to biological problems, new technologies and theories have originated. For example, NMR, EPR, X-ray, and optical spectroscopy were originally developed by physicists in a totally nonbiological context. Yet no modern hospital can diagnose or determine medical protocol without these techniques.

Biological physics is an interdisciplinary field. The boundaries of an interdisciplinary field are amorphous. With a few exceptions, we decided to focus primarily on molecular function. Thus, we eliminated biomechanics, the study of forces and torques applied to muscles and skeleton, animal studies, which use physical techniques such as radioisotope counting; and a substantial portion of physical chemistry.

Because this is the first reprint volume that surveys the field of biological physics we had to limit our selection to papers published before 1990. We made very little attempt to balance the number of papers in each category. The papers were selected and classified; the number of papers chosen was determined by the desired size of the volume. We wish to thank the authors and publishers of the papers for their permission to reprint them.*

Several research problems cross categorical boundaries. It will come as no surprise to physicists that, of the 40 papers selected, one-fourth of these involve the understanding of photosynthesis. A formidable array of experimental and theoretical techniques have elucidated our planet's ability to cycle CO_2. The papers on the functioning of hemoglobin describe our current understanding of the other half of this cycle. Neural functioning and protein dynamics also span categorical boundaries.

We have included papers of all recipients of the prize in biological physics awarded by the American Physical Society through 1989. In chronological order the prize winners are Roderick Clayton and George Feher, Paul Lauterbur, Howard Berg, Edward Purcell, John Hopfield, Hartmut Michel and Johann Diesenhofer, and Britton Chance.

We hope that the variety of problems, depth of theory, and sophistication of experimental techniques provides an introduction to the understanding of the physical functioning of living systems and stimulates the curiosity of other scientists.

Eugenie V. Mielczarek
Elias Greenbaum
Robert S. Knox

---

*In compiling the articles chosen for inclusion in this volume, the editors conducted a search of the literature and solicited suggestions from colleagues. No collection of this size can accommodate everyone's taste; we are prepared to hear that all the wrong papers have ended up on the cutting room floor. If readers will let us know which papers might have been included, we will consider the desirability of a second volume.

Section I
Infrastructure

Contents

Introduction

An infrastructure is a network or assembly of structures essential to the operation of a whole system. In biological systems certain well-known classes of biomolecules comprise it: proteins, nucleic acids, lipids, porphyrins, carotenoids, collagen, and smaller energy-rich molecules such as adenosine triphosphate (ATP).

Lipids in the form of bilayer membranes have the important function of delimiting regions of space in order to confine reactants. Proteins in myriad forms act as catalysts for the reactions. Nuclei acids carry the information necessary for replication and thereby growth. Porphyrins and carotenoids, usually in conjunction with special proteins, carry electrons and electronic excitation energy. Collagen provides mechanical structure.

The articles in Sec. I provide examples of physical studies on certain primary biomolecules and their superstructures. The emphasis is on what is probably the most essential, the biological membrane, and on certain important proteins. Other parts of the infrastructure will be treated in later sections dealing with specific biological functions of cells and subcellular assemblies.

Our reprint collection opens with a description of the infrastructure, especially proteins, painted in broad strokes by Frauenfelder. The membrane and its functions are described by Chance *et al.*, and its behavior as a condensed matter system with thermodynamic phases is treated by Nagle and Scott.

The marvel of the protein as an object of physical study begins with the tendency for a polypeptide to undergo rapid folding into a unique three-dimensional structure. Two rather general papers dealing with the structure problem (Scheraga; and Zimm and Bragg) are followed by two dealing with the myoglobin molecule. Perutz concentrates on the connection between its structure and function, while Brill *et al.* examine its vibrational properties. The pigment-protein interactions in the photosynthetic reaction-center protein are described by Michel *et al.* This paper, as well as the following one by Hogan and Austin, our only DNA entry, touch on the important fact that interactions between the elements of infrastructure are frequently central factors in the structure-function relationship.

Physics and Biology/ HANS FRAUENFELDER

1. INTRODUCTION

Ideally we would like to understand the behavior of a complex biological system, such as a "simple" organism like a microbe, in terms of the behavior of the constituent atoms. Such organisms consist of the order of 10^{20} atoms. These are not simple many-body systems, but are built from subunits that are arranged in a hierarchy as shown in Fig. 1. Where does life begin in this chain? There is no unambiguous answer; the complexity of the systems increases as we move up from atoms towards organisms. Unique characteristics of life begin with the *biomolecules*. Some of these perform essentially the same function when isolated as they do when incorporated in the living system. I will therefore mainly discuss biomolecules.[1,2]

Two types of biomolecules dominate: *nucleic acids* and *proteins*. The first store and transmit information and direct the construction of proteins; the second are the machines that perform the functions necessary for life. The essential features are shown in Fig. 2. (In reality the functions are not so neatly separated between nucleic acids and proteins, but the complications are unimportant here.)

The information is stored in the form of "three-letter words" on a very long linear unbranched deoxyribonucleic acid (DNA) molecule. The information on the DNA is read and transcribed onto a ribonucleic acid (RNA) molecule and transported to a ribosome, where the protein assembly takes place. The protein is also built as a linear chain, but the building blocks of nucleic acids and

Reprinted from *Physics in a Technological World* (AIP, New York, 1988), pp. 255–267; © American Institute of Physics.

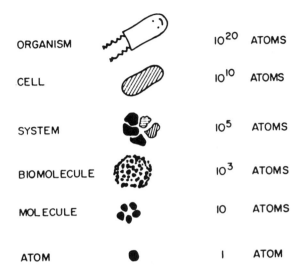

FIGURE 1 From atoms to organisms.

proteins differ. Nucleic acids are built from four different nucleotides; proteins from twenty different amino acids. The RNA instructs the ribosome in which order to connect the nucleic acids to form the primary sequence of the protein. The instruction involves the translation from the DNA and RNA language (three-letter words formed from an alphabet of four letters) to the protein

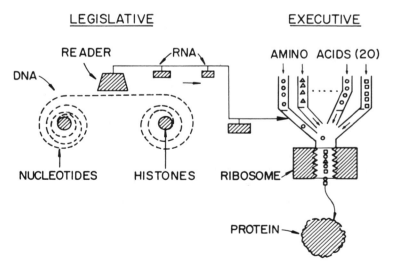

FIGURE 2 *Biomolecules.* Nucleic acids store and transmit information and direct the construction of proteins. The information is stored on DNA and transported to the factory (ribosome) by RNA.

language (consisting of an alphabet of twenty amino acids). When the primary chain emerges from the ribosome it folds into the functionally active three-dimensional structure.

A special class, globular proteins, fold into nearly close-packed structures. They consist typically of about 200 amino acids and have a diameter of a few nanometers (Fig. 3). Many of their physical properties are determined by their unique construction: Along the backbone, the bonds are covalent and hence so strong that they cannot be broken by thermal fluctuations. The cross connections that hold the protein together are hydrogen bonds, disulfide bridges, and Van der Waals forces; these are "weak" and can be broken by thermal fluctuations. This asymmetry in cohesion leads to unique characteristics that are

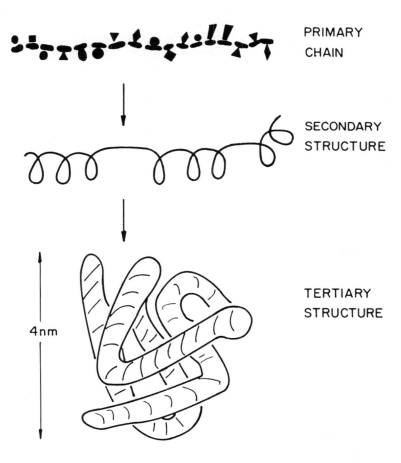

PRIMARY
CHAIN

SECONDARY
STRUCTURE

TERTIARY
STRUCTURE

4 nm

FIGURE 3   The covalently linked primary chain of amino acids (backbone) folds into the working tertiary protein structure. The globular protein is surrounded by a hydration shell and functions only in its presence.

important for the function of proteins: In contrast to solids, proteins are flexible and coupled to the solvent. Their size is such that they are at the border between classical and quantum systems. The proteins that exist now are not randomly constructed, they are the result of more than 3 Gy of R&D (research and development). Their ground state is structurally not unique: Most of the side chains of the amino acids can assume a number (n) of positions without changing the total energy of the protein appreciably. For a protein with N amino acids we consequently can expect a total number of approximately isoenergetic states of the order of n^N. With $n = 2$, $N = 200$, the number of possible ground states is of the order of 10^{60}. Proteins are highly degenerate systems.

The construction of proteins leads to the realization that *biological numbers* are far larger than astronomical numbers. Consider a protein with $N = 200$. With 20 different amino acids, the total number of possible proteins is 20^{200}. Of course, not all of the possible combinations fold and form useful systems. Nevertheless if we try to construct one copy of each possible arrangement, and then fill the universe with these copies, we need of the order of 10^{100} universes to complete the task. The human system contains about 10^5 different proteins. Each protein can occur in many minor modifications. So far, only a small number of proteins has been well studied—and "well" here means at most at the level where solid state physics was in 1920 and nuclear physics in 1930. Biomolecules, at the lowest biological level shown in Fig. 1, already show an exceedingly rich diversity and complexity and we can expect that their exploration will continue for a very long time.

Loosely speaking, physical studies of biological systems can be divided into *biophysics* and *biological physics*. In biophysics, physics is the servant and physical tools are used to study biological problems. Very few of the tools of classical and modern physics, from specific heat measurements to synchrotron radiation, are not used. In biological physics, on the other hand, the biosystem is considered just like any other physical system and the research aims at finding new phenomena, concepts, and laws. In practice, the border between the two fields is not sharp. In the following, I will try to present some examples of both.

2. STRUCTURE

Exploring a biological system without knowing its spatial structure is like being lost in the mountains at night without a map. While the coarse structure of organisms is often accessible to the knife and the naked eye, the exploration becomes more difficult as we descend the ladder of Fig. 1. Every tool of the physicist has had impact: Microscope, electron microscope, electron, x-ray, and neutron diffraction, EXAFS, and NMR all have added greatly to the knowledge of biological structures. Here only one approach, x-ray diffraction, will be sketched because it yields a rich variety of information.

A. X-Ray Diffraction Studies of Proteins

In principle the electron distribution in a protein can be determined without a protein single crystal. Assume that monochromatic x rays are scattered by a single oriented protein molecule. The scattering intensity $I(\theta)$ at the scattering angle θ is then proportional to $|f|^2$, the absolute square of the scattering amplitude f; f in turn is proportional to the Fourier transform of the charge density. If $I(\theta)$ is measured for a number of orientations of the protein molecule, the charge distribution can be found by Fourier transformation. The principle is the same as in the structure determination of particles and nuclei. One hurdle in all these structure determinations is the phase problem: In order to invert directly, the scattering amplitude f is required, but the experiment yields only the absolute square. The scattering phase must consequently be determined in a separate experiment.[3,4]

In practice, the single protein molecule is replaced by a protein single crystal which can be oriented and which yields a larger scattering intensity and suffers less radiation damage. The scattering intensity then is no longer a smooth function of the scattering angle; the interference from the many essentially identical protein molecules in the crystal leads to the discrete Laue–Bragg pattern. The positions of the spots in the diffraction pattern depend on the wavelength of the monochromatic x rays and on the lattice parameters of the single crystal; the intensities of the spots are determined by the electron distribution within each unit cell. The protein structure, i.e., the charge distribution or the arrangement of the nonhydrogen atoms in the protein, is consequently found from the spot intensities.

Proteins are rather complicated systems and it is difficult or even impossible to visualize their structure from the data in numerical form. For many years elaborate models built with great care from sticks and balls served as guides. They had many disadvantages; if the model was space-filling, the interior could not easily be examined; if the atoms were represented by dots, the impression was misleading. Computer graphics now gives beautiful representations which permit enlargement of important details, "walks" inside proteins, and examination of the effect of small changes in composition.[5]

B. Information from X-Ray Diffraction

X-ray diffraction experiments are capable of yielding far more than just the average or static structure of proteins. We sketch here some of the additional information that can be extracted from scattering experiments.

(i) *Dynamic structure.* In the simplest treatment of x-ray diffraction from a single crystal it is assumed that all equivalent atoms are at the correct periodic position. In this ideal situation, constructive interference is maximized and the Laue–Bragg spots possess maximum intensities. If the atoms are spread out over a linear dimension $\langle x^2 \rangle^{1/2}$, the constructive interference is reduced and the contribution of a particular atom to the total intensity of a Laue–Bragg spot is multiplied by the Debye–Waller factor[6]

$$T = \exp\left\{ -8\pi^2 \langle x_i^2 \rangle \sin^2\theta / \lambda^2 \right\}$$

where $\langle x_i^2 \rangle$ is the mean-square displacement (msd) of the atom i and λ the wavelength of the x rays. If the x-ray diffraction data are taken with sufficient accuracy, they yield both the average position and the msd of all nonhydrogen atoms in a protein.[7]

(ii) *Thermal Expansion Coefficient.* If the average positions of all nonhydrogen atoms in a protein have been determined at a number of temperatures, a linear expansion coefficient can be determined between any two atoms.[8] In a liquid or an isotropic solid, the coefficient is a scalar; in a protein, it depends on the positions of the atoms and becomes a tensor field. Because the expansion coefficient is related to the fluctuations in volume and energy, such a measurement can yield information about dynamic aspects of proteins.

(iii) *Time evolution.* In some cases, a protein reaction can be started by a rapid process, for instance initiated by a laser flash. In general, the structure of the protein will be changed by the reaction. In principle, the evolving structure can be studied by fast x-ray diffraction.

C. Structure: The Future

Even though structure studies have already yielded rich information about proteins and nucleic acids, it is likely that the future results will far surpass what is now known. This prognosis is based on a number of current developments: Synchrotron radiation sources are beginning to be widely available; their high intensities, widely variable wavelengths, and pulse structure make them ideal sources for biomolecular studies. Area detectors are speeding up data collection and make digital data acquisition feasible. Adequate computer power allows a rapid evaluation of the data. Data gathering is further speeded up by the use of the original Laue technique: Most of the work performed so far has been done with monochromatic x rays, yielding well-defined Laue–Bragg spots. It turns out that the Laue technique, where the incident x-ray beam is *not* monochromatic, still produces resolvable spots, but increases the speed enormously. These various improvements together make structure studies of biomolecules far more powerful than before and may well lead to another revolution in structural biology.

Additional insight into biomolecular phenomena is coming from the combination of x-ray diffraction with cryogenics and with high pressure. Since pressure and temperature are the two most important thermodynamic variables, knowledge of the protein structure and of the msd of the atoms participating in the dynamics as functions of the two variables adds greatly to the understanding of biological processes. Such studies, also, are only at a beginning.

3. COMPLEXITY

Traditionally, physicists study simple systems and attempt to extract generally valid laws from the simplest system that exhibits certain characteristic properties. The two prototypes of simple systems are the perfect gas (complete disorder) and the ideal solid (perfect order). We can assign complexity 0 to both of these systems, because neither can be used to store or transmit information. Studies of such simple systems have contributed greatly to physics. Many interesting systems, from proteins to brains, are neither fully ordered nor completely disordered; they are truly complex.[9] While some physicists have always been attracted to complex systems (turbulence!), most have stayed away from them. Within the past few decades, however, studies of disordered systems have increased markedly and complexity has become an acceptable concept even to physicists. The following examples show that complexity in proteins may connect these systems to glasses and spin glasses, the disordered systems that are better known to many physicists.

A. Nonexponential Time Dependence

We usually assume that the time dependence of a physical process is, to a first approximation, either sinusoidal or exponential. A large class of phenomena, however, deviate profoundly from either of these simple forms. These processes, sometimes called "endless", were first observed and described by W. Weber in 1835.[10,11] They typically extend over many orders of magnitude in time and can often be described by a power law or a "stretched exponential". Such processes have been measured in mechanical creep, discharge of capacitors, dielectric relaxation, and many other processes, and they always imply complexity.[12,13]

In proteins, nonexponential time dependence characterizes a number of processes. The best studied example is the rebinding of a small molecule such as dioxygen (O_2) to myoglobin.[14] Myoglobin (Mb) contains 153 amino acids, has a molecular weight of about 18 000 dalton, and has dimensions of about $3 \times 3 \times 4$ nm^3.[1] It contains a small organic molecule, protoheme, with an iron atom at its center. Mb acts as oxygen (O_2) carrier; O_2 binds reversibly at the iron atom: $Mb + O_2 \leftrightarrow MbO_2$. The binding process can be studied by photodissociation. Starting with MbO_2, the bond between the heme iron and the dioxygen molecule is broken by a laser flash and the subsequent rebinding is monitored in the visible[14] or the infrared.[15] It turns out that rebinding is not exponential in time, but approximately follows a power law,

$$N(t) = (1 + t/t_0)^{-n}$$

where $N(t)$ is the survival probability, i.e., the fraction of Mb molecules that have not rebound an O_2 at time t after photodissociation, and where t_0 and n are two temperature-dependent parameters. Similar nonexponential time developments are found in all heme proteins that have been studied.[16]

A simple explanation for the observed time dependence is based on the assumption that binding is governed by a barrier of height H at the heme iron and that each protein molecule possesses a different H. Denote with $g(H)dH$ the probability of having a protein with barrier between H and $H + dH$ and assume further that the rate coefficient $k(H)$ for binding is connected to H by an Arrhenius relation, $k(H) = A \exp\{-H/RT\}$. The survival probability then is given by

$$N(t) = \int dH \, g(H) \exp[-k(H)t] \, .$$

With a suitably determined probability distribution $g(H)$, the experimental data can be reproduced well between about 40 and 160 K. Below about 40 K, quantum-mechanical tunneling becomes important and the Arrhenius relation is no longer valid. Above about 160 K, competing processes obscure the direct rebinding.

B. Conformational Substates

Why do different myoglobin molecules possess different barrier heights at low temperatures? In the introduction we noted that a given primary sequence may not always fold into the same final globular structure. We now use this idea to introduce the concept of conformational substates (CS): We assume that a given protein in a given state (for instance MbCO) can assume a large number of structurally slightly different substates as sketched in Fig. 4. The energy surface of a protein consequently is postulated to be not smooth, but to contain a very large number of valleys, separated by energy mountains. The different CS or valleys then can possess different barrier heights H. At low temperatures, say below about 180 K, a protein remains frozen into a particular CS with corresponding barrier height H, and rebinding after photodissociation is nonexponential in time. At high temperatures, say 300 K, a given protein fluctuates rapidly from CS to CS and rebinding can be exponential in time.[14]

Corroborating evidence for CS comes from the Debye–Waller factor of myoglobin.[17,18] If all proteins were identical, and if no lattice imperfections existed in the protein single crystal, the msds of all atoms would be given by the vibrational contribution. If CS exist and have structural meaning, then the same atom should occupy slightly different positions in different protein molecules, leading to msd larger than the vibrational ones. Figure 5 shows the msd for Mb at two different temperatures. As expected for CS, the msd depends strongly on the positions of the atoms and is much larger than predicted by a purely vibrational contribution.

C. Proteins, Glasses, and All That

The existence of a highly structured energy landscape with many deep and approximately isoenergetic valleys may well be one significant characteristic of many (all?) complex systems. Highly degenerate ground states have been pos-

CONFORMATIONAL SUBSTATES

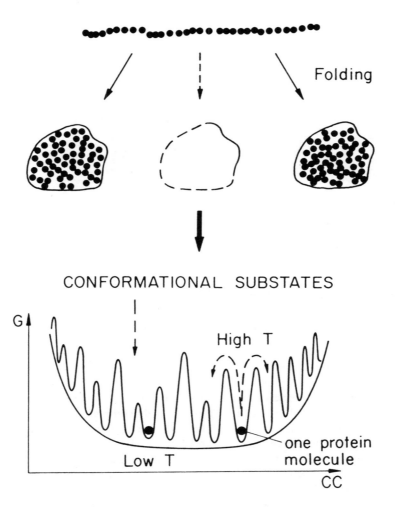

FIGURE 4 *Conformational substates.* A given primary sequence does not lead to a unique tertiary structure, but to a very large number of related, but in detail different, protein structures. These can be represented as points in a conformational space where each CS corresponds to a valley in the energy landscape.

tulated to explain the characteristics of spin glasses,[19,20] glasses,[21] and neural networks,[22–24] evolutionary biology,[25,26] optimization problems,[27] and learning.[28] All of these fields are at a beginning and many problems are unsolved or not even clearly recognized yet. In the case of the conformational substates of proteins, some questions can be posed: Is there just one tier of substates or do the

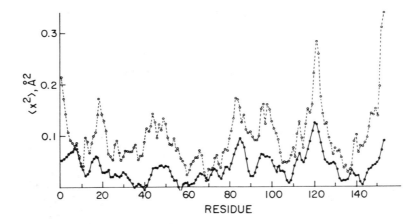

FIGURE 5   Mean-square displacements $\langle x^2 \rangle$ of the backbone atoms of metMb as a function of the amino acid number. The values are averages over the backbone atoms of each amino acid; the amino acid (residue) number labels the amino acids along the primary sequence. (After Ref. 18.)

CS fall into a hierarchical arrangement?[29] Is the organization of the CS ultrametric?[20] How are the substates, defined in the energy space, related to the structure and function of the protein?

4. FUNCTION

Biological systems perform functions. Even the simplest function executed at the lowest biological level of Fig. 1 is extremely complex when considered in detail. One goal of the work at this level is a "physics of function," a description of the function in terms of the static and dynamic structure of the biosystem. We are far from a full solution of this problem even in the simplest case, but work done so far already indicates interesting and promising relations and connections. Some of the emerging concepts are sketched here.

A. Equilibrium Fluctuations and "Fims"

Consider again myoglobin as example. Its function is dioxygen storage and it therefore must have at least two *states*, MbO_2 and Mb. In each of these states it possesses a very large number of conformational substates. The existence of states and substates leads to two types of motions, equilibrium fluctuations (EF) and functionally important motions, or fims. In a given state at high temperature, a protein fluctuates from CS to CS. These EF are governed by the general principles of equilibrium thermodynamics: A small system such as a protein does not have sharp values of internal energy, entropy, and volume.

These quantities fluctuate about their mean values.[30] The protein function, the transition from Mb to MbO_2 or its inverse, is performed through fims. Fims are nonequilibrium processes.

B. The Fluctuation–Dissipation Relation

EF and fims are often related. Studies of EF can consequently help elucidate the function. The first connection between an equilibrium and a nonequilibrium property was found by Einstein in his celebrated relation of the diffusion coefficient D to the friction coefficient f.[31] Later Nyquist related the voltage fluctuations across a resistor to the resistance.[32] General relations between equilibrium fluctuations and dissipative processes were formulated by Onsager,[33] Callen and Welton,[34] and many others.[35-37] It is likely that fluctuation–dissipation relations will be central for studying and understanding biological phenomena.

C. Proteinquakes

In any biomolecular reaction, the structures of the initial and final states are usually different. During and after the reaction, the biomolecule must rearrange its structure. Consider myoglobin again as an example. When it binds O_2, a stress is established at the active center, the heme group. Upon photodissociation, the stress is relieved and the strain energy is dissipated in the form of waves and through the propagation of a deformation. The process is similar to an earthquake and we call it a "proteinquake".[29] A proteinquake can happen because of the special structure of proteins as shown in Fig. 3: While the released energy is not sufficient to break the covalent bonds that link the backbone, the cross connections can be broken and rearranged, thus leading to a slightly different structure. Proteinquakes can be monitored by following changes in spectroscopic markers and, just as with earthquakes, we expect that they can give information about dynamic features.

D. Theories

So far, the discussion has been based predominantly on experimental observations. Ultimately there may be a fundamental theory that describes much of biology and biological processes, based on physical principles. We are far from such a theory. Nevertheless, progress has been rapid in the description and understanding of the phenomena on the lowest biological level: nucleic acids and proteins. The most successful direction so far has been the application of molecular dynamics, pioneered by Karplus and his collaborators.[38,39] One main limitation with this technique has been the restricted time range that could be covered. Most existing computations extend only to about 100 psec. Typical biologically important processes are much slower. New techniques promise to overcome the time limitations.[39,40]

5. OUTLOOK

Despite the vast amount of observational material already in existence, biophysics and biological physics are only at a beginning. We can expect that physics will continue to interact strongly with biology. Actually, the connection also includes chemistry and mathematics. New tools that become available in physics will continue to be applied to biological problems. We can expect that the flow of information will not be one way; biological systems will provide new information on many old and new parts of physics, from reaction theory and transport phenomena to complexity, cooperativity, and nonlinear processes.

ACKNOWLEDGMENTS

This contribution was written while the author was a Senior U.S. Scientist Awardee of the Alexander von Humboldt Foundation. The work was supported in part by U.S. National Institutes of Health grant GM 18051, National Science Foundation grant DMB 82-09616, and Office of Naval Research grant N00014-86-K-00270.

REFERENCES

1. L. Stryer, *Biochemistry* (Freeman, San Francisco, 1981).
2. H. Frauenfelder, Helv. Phys. Acta **57**, 165 (1984).
3. T. L. Blundell and L. N. Johnson, *Protein Crystallography* (Academic, New York, 1976).
4. J. D. Dunitz, *X-Ray Analysis and the Structure of Organic Molecules* (Cornell University Press, Ithaca, NY, 1979).
5. R. Langridge, T. E. Ferrin, I. D. Kuntz, and M. L. Connolly, Science **211**, 661 (1981).
6. B. T. M. Willis and A. W. Pryor, *Thermal Vibrations in Crystallography* (Cambridge University Press, Cambridge, 1975).
7. G. A. Petsko and D. Ringe, Ann. Rev. Biophys. Bioeng. **13**, 331 (1984).
8. H. Frauenfelder, H. Hartmann, M. Karplus, I. D. Kuntz, Jr., J. Kuriyan, F. Parak, G. A. Petsko, D. Ringe, R. F. Tilton, M. L. Connolly, and N. Max, Biochemistry **26**, 254 (1987).
9. B. A. Huberman and T. Hogg, Physics **22D**, 376 (1986).
10. W. Weber, Ann. Phys. Chem. (Poggendorf) **34**, 247 (1835).
11. J. T. Bendler, J. Stat. Phys. **36**, 625 (1984).
12. E. W. Montroll and J. T. Bendler, J. Stat. Phys. **34**, 129 (1984).
13. J. Klafter and M. E. Shlesinger, Proc. Nat. Acad. Sci. USA **83**, 848 (1986).
14. R. H. Austin, K. W. Beeson, L. Eisenstein, H. Frauenfelder, and I. C. Gunsalus, Biochemistry **14**, 5355 (1975).
15. A. Ansari, J. Berendzen, D. Braunstein, B. R. Cowen, H. Frauenfelder, M. K. Hong, I. E. T. Iben, J. B. Johnson, P. Ormos, T. B. Sauke, R. Scholl, A. Schulte, P. J. Steinbach, J. Vittitow, and R. D. Young, Biophys. Chem. **26**, 337 (1987).
16. H. Frauenfelder, F. Parak, and R. D. Young, Ann. Rev. Biophys. Biophys. Chem. (1988).
17. H. Frauenfelder, G. A. Petsko, and D. Tsernoglou, Nature **280**, 558 (1979).
18. H. Hartmann, F. Parak, W. Steigemann, G. A. Petsko, D. Ringe Ponzi, and H. Frauenfelder, Proc. Natl. Acad. Sci. USA **79**, 4967 (1982).

19. M. Mezard, G. Parisi, N. Sourlas, G. Toulouse, and M. Virasoro, Phys. Rev. Lett. **52**, 1156 (1984).
20. R. Rammal, G. Toulouse, and M. A. Virasoro, Rev. Mod. Phys. **58**, 765 (1986).
21. J. Jäckle, Rep. Prog. Phys. **49**, 171 (1986).
22. J. J. Hopfield, Proc. Natl. Acad. Sci. USA **79**, 2554 (1982).
23. J. J. Hopfield and D. W. Tank, Science **233**, 625 (1986).
24. *Neural Networks for Computing*, edited by J. S. Denker, Am. Inst. Phys. Conf. Proc. **151** (1986).
25. M. Eigen, in *Emerging Syntheses in Science*, edited by D. Pines (Santa Fe Institute, Santa Fe, NM, 1985).
26. P. W. Anderson, Proc. Natl. Acad. Sci. USA **80**, 3386 (1983).
27. G. Baskaran, Y. Fu, and P. W. Anderson, J. Stat. Phys. **45**, 1 (1986).
28. G. Toulouse, S. Dehaene, and J. P. Changeux, Proc. Natl. Acad. Sci. USA **83**, 1695 (1986).
29. A. Ansari, J. Berendzen, S. F. Bowne, H. Frauenfelder, I. E. T. Iben, T. B. Sauke, E. Shyamsunder, and R. D. Young, Proc. Natl. Acad. Sci. USA **82**, 5000 (1985).
30. A. Cooper, Proc. Natl. Acad. Sci. USA **73**, 2740 (1976).
31. A. Einstein, Ann. Phys. **17**, 549 (1905).
32. H. Nyquist, Phys. Rev. **32**, 110 (1928).
33. L. Onsager, Phys. Rev. **37**, 405 (1931).
34. H. B. Callen and T. B. Welton, Phys. Rev. **83**, 34 (1951).
35. L. Onsager and S. Machlup, Phys. Rev. **91**, 1505 (1953).
36. R. Kubo, Rep. Progr. Phys. **29**, 255 (1966).
37. P. Hanggi, Helv. Phys. Acta **51**, 202 (1979).
38. M. Karplus and J. A. McCammon, CRC Crit. Rev. Biochem. **9**, 293 (1981).
39. J. A. McCammon and S. C. Harvey, *Dynamics of Proteins and Nucleic Acids* (Cambridge University Press, Cambridge, 1987).
40. R. Elber and M. Karplus, Science **235**, 318 (1987).

Biological membranes

Britton Chance, Paul Mueller, Don De Vault and L. Powers

For solid-state physicists and engineers the "ultimate in miniaturization" would be to produce devices with structures that are about 8 or 10 nm across—about a tenth of the smallest scale that can currently be produced. (See PHYSICS TODAY, November 1979, page 25.) Biological systems, however, have, in a sense, solved the problems associated with such small microstructures. The fundamental unit of many cell functions, the lipid bilayer membrane (figure 1), is 4 nm thick; in regions where the membrane carries proteins it may be as much as 10 nm thick. Other elements of the cell, such as the microtubules that provide its structural framework, have similar dimensions.

These biological microstructures carry out a remarkable variety of functions, mechanical, chemical and electronic. In this article we will concentrate on the flow of electrons along and through the lipid bilayer membranes. Specifically we will consider some of the processes involved in respiration and photosynthesis.

Membranes

Much of the physiological and biochemical activity takes place not in the bulk cytoplasm of the cell but in membranes that are distributed throughout the cell and its organelles as well as forming their boundaries. These membranes consist of bilayers of fatty-acid esters (lipids) whose long hydrocarbon chains are repelled by the aqueous environment of the cell but are attracted to each other. The polar ends of the molecules (often consisting of phosphate esters) face the cytoplasm. In many cases protein molecules ("mem-

brane proteins") are associated with these bilayers, often forming integral parts of the membrane. The electron micrographs of figure 2 clearly show many such membranes, some of which have a grainy appearance because they are associated with particularly large proteins. Figure 3 shows the structure of a typical membrane with embedded membrane proteins.

The membrane itself is impermeable to water and an excellent insulator. Many of the proteins in the membrane provide pores for active or passive transport of biologically important atoms and molecules through the membrane, thus maintaining or establishing electrical or chemical gradients. A nerve impulse, for example, consists of a propagating wave of electrical depolarization that derives its energy from a transmembrane ionic concentration gradient maintained by molecular ion pumps fueled by high-energy phosphate compounds such as adenosine triphosphate (ATP).

The membrane proteins have molecular weights between 100 and 250 kilodaltons, diameters between 4 and 10 nm, and are often composed of several subunits. They can be packed densely up to 10^{12} molecules per cm^2 or distributed sparsely, mixed together or organized in patches. Figure 4 shows examples of two such arrangements. The proteins are embedded in the membrane, and their amino acids that are in the hydrocarbon interior of the membrane are highly hydrophobic and make close contact with the lipid chains.

In most biological systems each distinct function is carried out by a separate molecule either alone or as part of a larger complex. In a few systems the

same molecule can perform more than one function. For example, in the mitochondrion of the body cell one molecule serves both to transfer electrons to oxygen and to transfer protons across the membrane. (Strictly speaking, what is transferred is not protons but positive aqueous hydrogen ions; biochemists colloquially call these "protons.")

The electron-transfer systems we will discuss are the primary elements in a complex chain of energy-conversion mechanisms that, starting with a redox compound or a photon as a source, translate protons, generating an electrical potential gradient, then convert this gradient to a high-energy compound such as ATP, and finally utilize the ATP in many metabolic reactions and the maintenance of ionic concentration gradients.

Mitochondria

In eukaryotic cells (that is, cells with nuclei, as distinguished from bacteria and some other primitive organisms) respiration—the transfer of electrons from electron donors to reduce oxygen to water—takes place in mitochondria (on the left in figure 2). These are prominent organelles, about 2 microns in diameter, within the cells of every tissue and organ, separated from the rest of the cell by a membrane and containing many folded membranes as well. They appear to multiply independently of the remainder of the cell

Vesicles formed in water from lipid-bilayer membranes. The membranes were produced by hydration of lipids at low ionic strength. These vesicles are large enough for direct electrical measurements.   Figure 1

Reprinted from *Physics Today* **33**, 32–38 (October 1980); © American Institute of Physics.

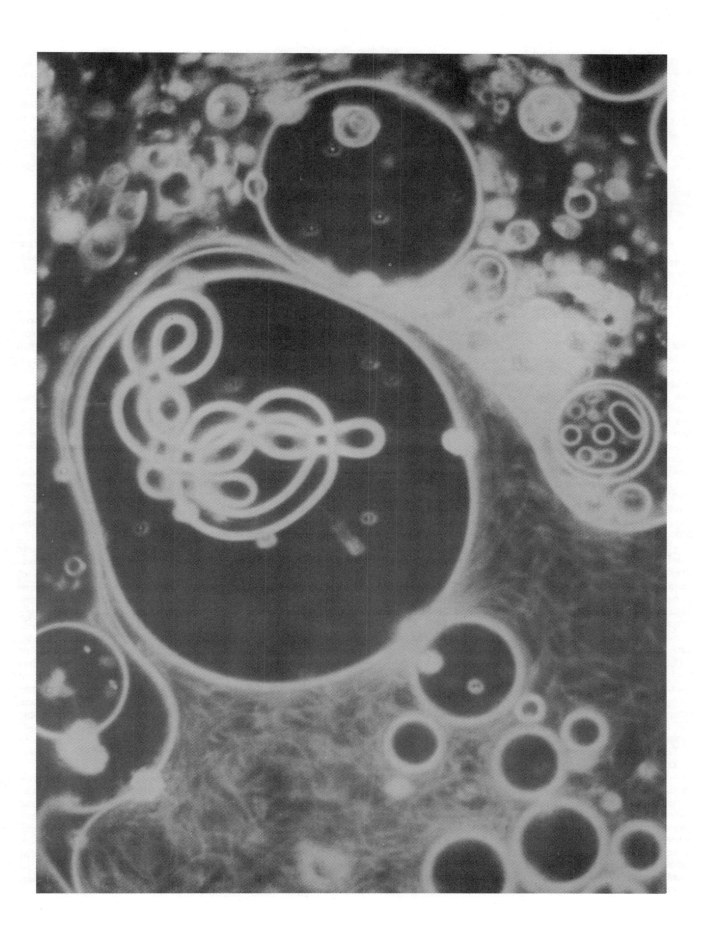

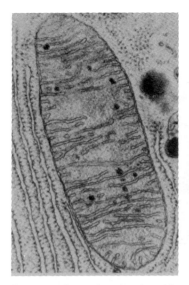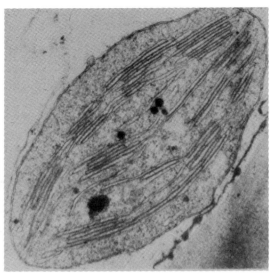

Cross sections of a mitochondrion and a chloroplast. These small, intracellular organelles are specialized to carry out oxidative phosphorylation or photosynthesis, respectively. Their membranes contain the molecular electron and ion-transfer components that are the subject of this article. The electron micrographs show areas of about 2 microns by 4 microns; they were taken by K. Porter and J. Antanavage.   Figure 2

and carry some genetic information independent of the cell's (apparently with some differences in genetic code). It is widely believed that mitochondria are derived from once-free bacteria (similar in characteristics to *Paracoccus denitrificans*) that entered into an intimate symbiotic relationship with what then became the eukaryotic cell, increasing the efficiency of oxidative metabolism by over thirty fold.

The mitochondrion supplies "low cost" energy for the cell by burning foodstuffs in the presence of molecular oxygen transported to the cell by the blood, and utilizing the energy stored in the resulting transmembrane proton gradient for the formation of ATP. It does this in a series of steps involving an assembly of electron transfer components called "the respiratory chain" (figure 5). The electron carriers of the respiratory chain are arranged in three pools that consist of several membrane proteins whose redox potentials are roughly the same. Electrons are shuttled between pools by smaller, mobile components such as ubiquinone, which moves by diffusing through the membrane lipid, and cytochrome *c*, which diffuses along the surface of the membrane.

The components of each pool are near thermodynamic equilibrium.[1] The energy that is given up by the electrons as they drop from pool to pool is used to form molecules of ATP from ADP (adenosine diphosphate). Coupled to the electron flow is a proton flow, which—in at least one case—takes place through a transmembrane pump. Electrons and protons thus pass from the reduced form of nicotinamide adenosine dinucleotide (NADH)

to oxygen, reducing oxygen to water and giving up energy to ATP in the process. A second electron and proton can enter the sequence midway, derived from the oxidation of succinic acid to fumaric acid.

The substrates of the respiratory chain, NADH and succinate, are derived from earlier steps in the metabolic process. Glucose, for example is converted to CO_2 and H_2O in several steps, each of which releases some of the energy stored in the molecule. First the glucose molecule is split into two molecules of lactic acid ("glycolysis"); next, molecules of CO_2 are sequentially split off from the lactic acid (the oxygen required is derived from water molecules) and the hydrogen is transferred to NAD and the respiratory chain (the "Krebs cycle"); finally the hydrogen is transferred to oxygen molecules to form water in the respiratory chain.

The overall chemical reaction of the respiratory chain is

$$4e^- + O_2 + 4H^+ \rightarrow 2H_2O$$

The electron flow along the chain (equivalently, the reaction rate) is quite rapid: Each carrier in a chain carries about 100–300 electrons per second.

Many of the molecules in the respiratory chain carry metal atoms, often found encased in porphyrin rings—as, iron is in hemoglobin. For example, the cytochromes (named for their intense pigmentation) contain porphyrin rings that carry iron atoms whose valence state can change from ferrous (Fe^{++}) to ferric (Fe^{+++}); iron atoms are also found in some proteins held by sulfur atoms in "iron–sulfur" centers. Other proteins contain copper, which transfers electrons by shifting between

its cuprous (Cu^+) and cupric (Cu^{++}) states.

The protein components of these molecules, which are often quite large, appear to function to provide the metal atoms with the appropriate dielectric and chemical environment. The excess charge associated with the addition or removal of electrons from the metal atoms can, in some, of these cases, be stabilized by being spread around a large region of the molecule. As we shall see later, the proteins can also serve to adjust the energy levels of electron carriers so that the electrons can tunnel from one to the other.

Cytochrome oxidase

To illustrate some of the properties of the molecules in the respiratory chain, let us consider the cytochrome-oxidase complex, the last member in the chain, and the molecular complex that reacts directly with the molecular oxygen delivered to the cell by the circulating blood.

Cytochrome oxidase is a very large protein that spans the mitochondrial membrane. It extends about 5 nm beyond the membrane into the cytoplasm; it extends very little into the mitochondrion. (The membrane itself is about 4 nm thick.) The complex has seven subunits and a total mass of about 120 kilodaltons.

Two of the subunits (cytochromes *a* and a_3) contain iron atoms and two others (including one of the largest units) contain copper. The metal atoms are paired, as they often are in such complexes, but, interestingly, they are paired in heterogeneous iron–copper pairs instead of the more common homogeneous pairs.

The metal atoms are bound to their protein environment in a highly "covalent" way and with a high level of coordination so that electrons transferred to them can readily tunnel between atoms. Recent x-ray absorption studies have shown that the copper that is spin-paired to the iron of cytochrome a_3 has a local environment (that is, charge density) similar to that found in the "blue copper proteins" of some bacteria, while the other copper has a more covalent environment, possibly held by nitrogen and sulfur atoms of the protein. The iron–copper pair that involves cytochrome *a* appears to function as an acceptor for the electrons from cytochrome *c*.

The electrons removed from cytochrome *c* are then transferred to the other iron–copper pair, and by them to oxygen. The iron atom in cytochrome a_3 is the initial binding site for molecular oxygen. This atom appears to be so closely held to the copper atom that electron paramagnetic resonance as well as some portions of absorption bands are suppressed. Apparently the local structure permits the iron and copper atoms (as Fe^{++} and Cu^+) to bind the oxygen molecule and transfer electrons to it; once the oxygen is reduced, the metal atoms bind the product in a peroxide bridge.[3] Protons from the surrounding medium serve to form hydrogen peroxide, still held by the iron–copper complex (now Fe^{+++} and Cu^{++}). Two more electrons from the rest of the oxidase complex then serve to form two OH^- ions, which are released into the surrounding water.

The iron–copper complex reacts avidly with oxygen. In fact, all these reaction are observed readily in the frozen state, for example at $-125\,°C$, where the large molecules are essentially immobile and electrons must tunnel between them. At higher temperatures the reactions proceed rapidly, with oxygen passing through its four possible reduction states in microseconds, stopping momentarily at the two-electron reduced state to form the peroxide.

The structure of the surrounding proteins appears to be designed to store electrons and to deliver them to the bound oxygen molecule. Although the structure of cytochrome oxidase—as for other large molecular complexes—is not yet fully clarified, some of the important features appear to be shared with smaller, better-understood molecules such as cytochrome c and hemoglobin. One example of such a common feature is an iron (or sometimes other metal) atom bound into the porphyrin ring of a heme group. The ring contains four nitrogen atoms coupled with a system of conjugated double bonds that permits an extra charge to be delocalized over a fairly large area—effectively increasing the capacitance of the system—and helps to match the energy levels of the electron donors and acceptors.

Photosynthesis

In the biosphere the ultimate source of energy is, of course, sunlight, which is converted to chemical energy by photosynthesis. In fact, we now believe that all the oxygen in the atmosphere (on which all respiration depends) was produced by primitive photosynthetic organisms over several millions of years.

Plant cells contain special organelles, (at the right in figure 2), called chloroplasts, to perform photosynthesis. These, like mitochondria, are apparently derived from independent photosynthetic bacteria rather like the modern blue-green algae. In photosynthesis, electrons, instead of being released to a low-energy electron sink (molecular oxygen), are recycled. Their energy is raised in a special molecule, chlorophyll, by photons.

The photosynthetic apparatus in bacteria (such as the blue-green algae)

appears to be more primitive than that found in eukaryotes. The central part is a "reaction center:" a large molecular complex (with a mass of about 90 kilodaltons) that spans the chromatophore membrane of the bacteria. The reaction centers are fairly densely packed in the membrane, as figure 3a shows.

The reaction centers contain several subunits:[4]

▶ two bacterio-chlorophyll molecules that probably serve to collect the photons at the reaction center

▶ a bound pair of bacterio-chlorophyll molecules

▶ a pair of bacterio-pheophytin molecules (essentially chlorophylls without the magnesium atom at the center of the porphyrin ring)

▶ two quinone molecules and an iron atom.

Their exact location and orientation in the reaction center are not known.

When a photon is absorbed by the reaction center its energy is transferred to the bound pair of bacterio-chlorophyll molecules. The excited pair is then oxidized, losing an electron to the bacterio-pheophytin molecules; these, in turn, transfer the electron to the first of the quinone molecules, which passes it to the second. Apparently the chlorophyll dimer is near the center of the membrane while the quinones are near one of the surfaces.[5] The rapid transfer of charge to the quinones thus gives rise to a transmembrane potential.

From the quinones the electron is

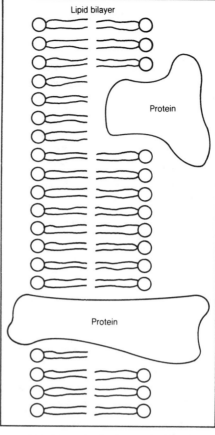

Lipid bilayer containing membrane proteins. The proteins can serve as channels through the membrane for selected ions, as active trans-membrane pumps for specific molecules, as catalysts in chemical reactions, or as structural units in the membrane. Figure 3

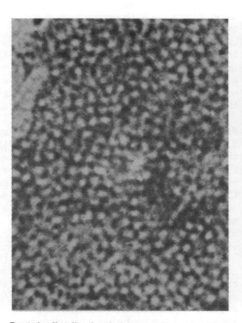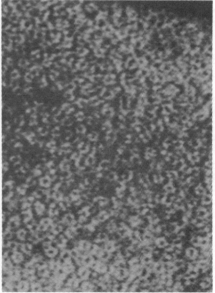

Protein distribution in biomembranes. The electron micrograph at left shows photosynthetic reaction centers of the bacterium *Rhodopseudomonas spheroides* in a reconstituted membrane. The micrograph at right shows densely packed acetylocholine receptors in the membranes of the electric organ of the torpedo, an electric fish. These receptors are ionic channels that are opened by the binding of acetylcholine. They appear in this face view as rings composed of five or six subunits with a central opening, presumably the channel entrance. The electron micrographs were made by J. Antanavage. Figure 4 transferred to another complex—very much like that found in the respiratory chain—that uses the electron to reduce cytochrome c. The reduced cytochrome c in turn donates its electron to the reaction center where it returns to the bacterio-chlorophyll, thus completing the cycle. Because the binding site for the cytochrome c is, apparently, on the opposite side from the quinone molecules, the return of the electron stabilizes the charge separation. The quantum efficiency of the system is nearly 100%.

On its return path, the electron is associated with a proton, so that in addition to the cyclic electron flow there is a net transfer of protons from one side to the other. The proton gradient in turn drives other chemical reactions, such as the synthesis of ATP.

Electron tunnelling

As we have seen, many of the most fundamental metabolic processes involve the transfer of electrons between macromolecules—often between metal atoms held by the macromolecules. In ordinary chemical reactions, such a transfer would take place via a temporary chemical bond. In the case of the macromolecules, the distance between electron-carriers remains large enough (5–30 Å) that electrons must tunnel through the potential barrier separating the molecules.

This tunnelling process appears to be the central feature of the cellular microelectronics. Thermal agitation is constantly changing the parameters of these systems, and the electrons jump from one molecule to another when the conditions are right. Evolution appears to have designed the electron-carrying molecules and the membranes in which they are embedded so as to enhance the right conditions for jumping or to control the jumping.

The probability of tunnelling through a potential barrier is a function both of its height and width. The transfer of electrons is made easier when the molecules are close together, as they can be when one of the molecules is small (cytochrome c or the quinones, for example), or when the molecules have configurations that bring their carriers close together. The height of the barrier is determined by the electronic state of the molecules. The tunnelling is also affected by the relation of the energy levels on the two sides of the barrier: it is enhanced when the levels match up.

The potential energy of the electron and of the carriers is greatly influenced by the random fluctuations of the electrical fields produced by thermally agitated polar groups in the surrounding molecules. The reverse is also true: the electron jump exerts a sudden change of electrical forces on the sur-

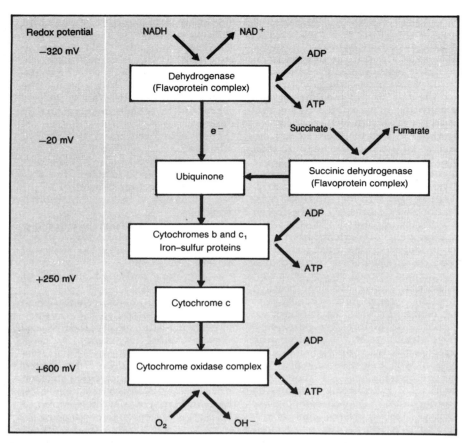

Respiratory chain. The diagram shows the last stages in the conversion of sugars to CO_2 and water, as it occurs in mitochondria: the transfer of electrons (and hydrogen ions) to oxygen to form water. In the process several protons are transferred across the membrane; these can then return via the ATPase complex, generating molecules of ATP, which serves as an energy source for other biochemical processes.   Figure 5

rounding molecules. Thus, the electronic motion is coupled to the nuclear (atomic) motion. The energy required to adjust the nuclear equilibrium positions from "electron on donor" to "electron on acceptor" (without, however, moving the electron) is called the "reorganizational energy." It is, for the sorts of systems we are considering, on the order of an electron volt.

In figure 7 we show the potential energy of an electron for several configurations of the nuclei. For curve A the nuclear configuration is such that the electron is most stable at the donor, in configuration B it is most stable at the acceptor. The curves represent the equilibria prevailing before the electron transfer (A) and after (B). A net amount of energy ΔE is released in the transfer.

If the thermal energy is much smaller than the reorganizational energy, E_R, the nuclei rarely assume configuration B spontaneously. However, they can assume intermediate configurations, such as C, for which they only require an energy E^* on the order of the thermal energy. In this configuration the electron can tunnel through the potential barrier from the donor to the acceptor and the molecules can relax to configuration B, releasing[6] an energy

$E^* + \Delta E$. The reaction rate in this case is given by the usual sort of expression:

$$K(T) = Ke^{-E^*/kT}$$

where K depends on the probability that the electron will tunnel once the nuclear configuration is right. The exponential factor gives the probability that the nuclei will reach the proper configuration by thermal agitation. At low temperatures the tunnelling may still be possible, even in the absence of thermal agitation, if the wavefunctions of the nuclear states A and B both extend as far as configuration C; in that case nuclear tunnelling can occur without thermal activation, to be followed by electron tunnelling.

Experiments have shown that some rates of electron transfer in photosynthesis are temperature independent. In these cases, then, it appears[7] that low-frequency vibrational modes are not strongly coupled to the electron-transfer process.[8] If so, this is an achievement of "biological microelectronics" that has probably not yet been duplicated in ordinary chemical or artificial systems.

Control of electron transfer

Both photosynthesis and respiration are controlled by the metabolic needs of

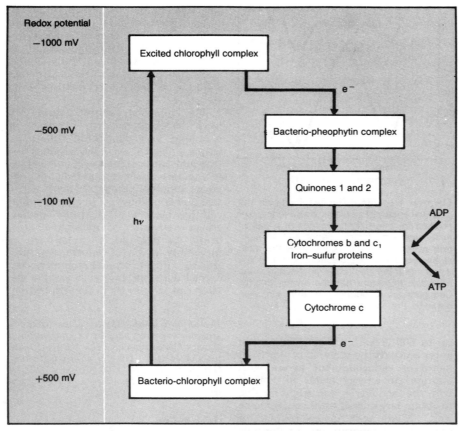

Redox potential

−1000 mV

−500 mV

−100 mV

+500 mV

Excited chlorophyll complex

e⁻

Bacterio-pheophytin complex

Quinones 1 and 2

ADP

Cytochromes b and c₁
Iron–sulfur proteins

ATP

Cytochrome c

e⁻

hν

Bacterio-chlorophyll complex

Photosynthesis. The diagram shows the first stages in the conversion of sunlight to chemical energy, as it occurs in photosynthetic bacteria. The chlorophyll and pheophytin molecules, together with some others, make up the bacterial "reaction center," found in the plasma membrane of these bacteria. One molecule of ATP per photon is formed when protons transferred across the membrane return via the ATPase complex. Figure 6

the cell—whose survival, in fact, depends on that control. Respiratory control, for example, turns off electron transport when the energy needs of the cell are met, thus preventing waste of foodstuffs. Clearly, there is a feedback system that couples the energy demand of the cell back to the respiratory chain either in terms of chemical intermediates or in terms of trans-membrane potentials or ion gradients.

The rate at which the electrons are transferred by one of the enzyme systems of the sort we have described could be controlled by electric fields along the direction between the donor and acceptor. Such a field would, in effect, alter ΔE in figure 7. Because the activation energy is a function of ΔE, the field would affect the rate of electron transfer. In one model[6] of the transfer system, for example, in which the nuclei are assumed to behave as simple harmonic oscillators, the activation energy is

$$E^* = (E_R - \Delta E)^2/4E_R$$

where E_R is, again, the reorganizational energy. Where ΔE has the same value as E_R the activation energy vanishes (and, incidentally, the reaction rate becomes temperature-independent). The rate at which E^* varies

with ΔE also vanishes, and small variations in ΔE can thus not be used to control the kinetics of the reaction. For such systems other means of control must be found. On the other hand, if ΔE is small compared to E_R (the situation shown in figure 7), E^* varies linearly with ΔE and the reaction rate also depends on ΔE.

In more complex systems, of course, the control can take on a greater variety of forms. Just what the control mechanisms are in biological systems is difficult to determine: so many parameters affect their behavior that it is difficult to untangle their effects. What is needed are more precise probes of living systems and some simpler systems in which to study the phenomena.

Membrane reconstruction

Most natural membranes contain proteins acting in parallel and having many different functions. Moreover, mitochondria, chloroplasts, and bacteria are too small for reliable direct electrical measurements. For these and other reasons, the development of methods for the isolation of individual membrane proteins and their insertion into lipid bilayers has led to new insights into specific mechanisms and structures. Reconstituted membranes

are in the form of either small (300–1000 Å diameter) vesicles or planar bilayers that separate two aqueous phases large enough to allow direct electrical measurements. Experimenters have been able to insert a great variety of membrane proteins, including members of the mitochondrial electron-transport chain, into the membrane of small vesicles. These protein-carrying vesicles exhibit activities such as oxidation-coupled protein transfer and oxidative phosphorylation. Through such studies the central ideas of energy coupling and transformation in the "chemi-osmotic theory"[9] can be tested experimentally. According to this model the major function of systems such as the reaction centers or the respiratory chain is, as we have suggested, to produce a proton gradient across the membrane. This proton gradient then drives other enzyme systems, in particular the ATP-generating complex: every pair of protons that flows through the ATPase complex produces one molecule of ATP. In photosynthetic organisms the proton gradient can also be used to reduce NAD^+; the resulting NADH serves as a general reducing agent (for example, driving the Krebs cycle in reverse).

Lipids can be made to form monolayers on the surface of water: One places a drop of lipid dissolved in an organic solvent such as hexane on the surface of water; the drop spreads to form a film and after the solvent evaporates one has a layer of lipid on the water (hydrocarbons pointing up). One can coat such monolayers onto each side of a plate with a hole in it, forming a bilayer membrane across the hole.) Such bilayers are by now fairly well studied objects. They have a capacity of 0.5–0.7 microfarad/cm², a typical resistivity of 10^8ohm cm² and an elastic constant of 90–100 dynes/cm. The lipids move about freely in the plane of the membrane and occasionally flip around to the opposite side.

Lipids dispersed in water naturally form globules of lipids with their polar ends pointing out or vesicles enclosed by bilayer membranes. The size of the vesicles can be controlled by adjusting the ionic strength of the water. Figure 1 shows a collection of large vesicles formed in this way. Such a vesicle can be turned into planar membranes by attaching it to an opening in a partition and then breaking the vesicle open on one side with an electric shock.

Several groups have been able to produce protein-carrying membranes and to study their electrochemical properties. Among the enzyme systems studied in this way are cytochrome oxidase, bacterial reaction centers[10] (see PHYSICS TODAY, September, page 19), rhodopsin, acetylcholine receptor[11] and several other ionic channels derived from bacterial and mitochondrial membranes. Some of these channels have voltage-dependent conductances and display all the electrokinetics of electrically excitable membranes, including action potentials.[12] In many cases one can observe the opening and closing of individual channels; the statistics of this channel activity give clues about the gating mechanisms for the channels. In planar membranes containing cytochrome oxidase one can directly measure the currents and potentials associated with the transfer of electrons from cytochrome c to molecular oxygen.

Membranes containing reaction centers are, of course, particularly interesting for studies of photosynthesis. The photo-induced currents and potentials observed in bilayers containing reaction centers agree well with the electron transfer mechanism derived from spectroscopic data. Such membranes give currents in response to steady-state illumination that consist of an initial peak, a smaller steady-state current, and a reverse current on cessation of illumination. The action spectrum matches the absorption spectrum of isolated reaction centers.

With excess reduced cytochrome c added to one side of the membrane as an electron source and additional ubiquinones as acceptors in the bilayer, the integrated peak current represents the transfer of several electrons per reaction center from the cytochrome c to the ubiquinones. The integrated peak current is determined by the number of added quinones. The steady-state current results from the oxidation of the quinones by O_2, which maintains a steady supply of electron acceptors. The reverse current is due probably to transmembrane equilibration of the anionic semiquinone.

The total charge transferred, as measured from the current integral in response to a single turn-over laser flash, matches the number of reaction centers in the membrane as estimated from the reaction center concentration of the membrane-forming solution. The membranes contain approximately 2.5×10^9 reaction centers/cm^2. The amplitude of the flash-induced potential change is commensurate with the membrane capacity and the integral of the current due to a single flash. Each center can transfer about 5×10^9 electrons per second, and with about 2×10^{11} centers per cm^2 the membrane can reach current densities of about 150 A/cm^2 in response to brief intense flashes of light, but under physiological conditions the electron-transfer rates are much lower. However, a densely packed (6×10^{12} centers per cm^2) oriented monolayer of reaction centers could theoretically give transient currents of

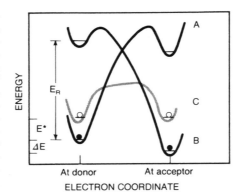

Electron transfer. The graph shows the potential energy of an electron as a function of position for three configurations of a pair of molecules—an electron donor and an electron acceptor. In configuration A the most stable state has the electron bound to the donor. In B the electron is most stable at the acceptor; this state has the lowest energy. Configuration C permits electron tunneling between the two sites.   Figure 7

up to 600 A/cm^2. If electrons could enter and exit the reaction center from metal or semiconductor layers with appropriate energy levels in contact with the monolayer, one might be able to obtain large steady-state currents as well. Many such layers coupled in series would generate large potentials and power.

Preliminary experiments with condensed monolayers of reaction centers sandwiched between two transparent evaporated metal layers have demonstrated the feasibility of this approach, and although no practical applications are foreseen, these experiments may be of use in the design and construction of energy-conversion devices.

Such studies illuminate the function of the various components of biological microstructures. The mechanisms for producing the functions, are, however, far from being fully understood. To a great extent, this is so because the structures of the molecular machines have not yet been resolved at the atomic level. In most cases, we have only a vague outline of the size and arrangement of the subunits at resolutions not much below 2 nm.

Membrane proteins do not readily produce crystals of high quality suitable for x-ray diffraction, precluding the use of that powerful tool for structural analysis. In some cases, however, experimenters have been able to produce fairly high-resolution maps. For bacterial rhodopsin (which acts as a light-driven proton pump) both the amino-acid sequence and the folding pattern through the membrane are known. At present, we have a fairly complete understanding of the correlation between structure and function down to the atomic level for only one protein, gramicidin, a simple cation-

selective channel.[13] But although detailed structural data are an indispensible prerequisite, they do not guarantee full mechanistic understanding. For example, we do not fully understand how hemoglobin works although we have a very clear picture of its structure.

The biological microstructures we have discussed perform a variety of chemical and electronic functions. Other biological systems, whose structure and function are as yet only vaguely understood, serve as gated channels, power supplies, pumps, receptors, effectors and transducers. In an individual cell they perform all the functions necessary to sustain life. Assembled into a brain, for example, they perform as incredibly powerful information processors. Physicists and engineers concerned with building ever smaller devices can look to them for inspiration.

* * *

The authors would like to acknowledge the generous research support of the National Science Foundation and the National Institutes of Health: USNIH grants GM27308, GM12202 and HL18708 (Chance); USNIH grant GM25256 (Mueller); and NSF grant RCM78-23194 (De Vault).

References

1. F. Wilson *et al.*, Ann. Rev. Biophys. Bioeng. **3**, 203 (1974).
2. T. Frey *et al.*, J. Biol. Chem. **253**, 4389 (1978).
3. B. Chance *et al.* in *Cytochrome Oxidase, Developments in Biochemistry* **5**, T. King, K. Okunuki, Y. Orii, B. Chance, eds., Elsevier, New York (1979), page 353.
4. W. Sistrom, R. Clayton, eds., *The Photosynthetic Bacteria*, Plenum, New York (1978).
5. J. Katz in *Light-Induced Charge Separation in Biology and Chemistry*, H. Gerischer, J. Katz, eds., Dahlem Konferenzen, Berlin (1979) page 331; J. Jackson, P. Dutton, Biochim. Biophys. Acta **325**, 102 (1973).
6. R. Marcus, J. Chem Phys. **43**, 679 (1965).
7. J. Hopfield, Proc. Nat. Acad. Sci. USA **71**, 3640 (1974); J. Jortner, J. Chem Phys. **64**, 4860 (1976).
8. A. Kuznetzoff, N. Sondergård, J. Ulstrup, Chem. Phys. **29**, 383 (1978); A. Sarai, Biochim. Biophys. Acta **589**, 71 (1980).
9. P. Mitchell, Biol. Rev. **41**, 445 (1966).
10. N. K. Packham, C. Packham, P. Mueller, D. M. Tiede, P. L. Dutton, FEBS lett. **110**, 101 (1980); M. Schonfeld, M. Montal, G. Feher, Proc. Nat. Acad. Sci. USA **76**, 6351 (1979).
11. H. Schindler, U. Quast, Proc. Nat. Acad. Sci. USA **77**, 305 (1980).
12. P. Mueller, *The Neurosciences, Fourth Study Program*, F. Schmitt, F. Worden, eds., MIT Press, Cambridge (1979), page 641.
13. D. Urry, Proc. Nat. Acad. Sci. USA **68**, 1907 (1971).

□

Biomembrane phase transitions

John F. Nagle and Hugh L. Scott

Along with DNA and proteins, cell membranes are among the principal organizational structures of living matter. Over the past decade the study of cell membranes has become a focus of effort ranging from the very biologically oriented to the very physically oriented. One property that is especially appropriate for physical studies is the phase transition that occurs in many cell membranes.

The best documented case is found in the cytoplasmic membrane of *Acholeplasma laidlawii,* which is a primitive organism with a large surface-to-volume ratio. After the cell is grown, at a particular temperature T_g, the membrane is extracted and calorimetric measurements made. These show a specific-heat anomaly about 20 celsius degrees broad centered near or slightly below T_g. When these cells are grown at different growth temperatures, the calorimetric anomaly follows the change in T_g.[1] (For a review, see reference 2.) This suggests that the phase transition is not just some physical phenomenon that happens to occur, but that it has real biological relevance.

There are many other examples of physically induced biological changes related to the membrane phase transition. Organisms that exist in cold environments have membrane components giving rise to reduced phase-transition temperatures. Other properties, such as transport, the response to anesthetics and immunological response, can be altered by ·changing the thermodynamic state of the membrane involved.

A molecular sketch of a biomembrane, according to the conventional wisdom, is shown in figure 1. Many important functions of cells are carried out by membrane-bound enzymes, which are proteins; they are indicated by the large objects in the figure. However, the primary structural component of many membranes is the lipid (fat) bilayer, consisting of the more or less regularly spaced molecules shown in figure 1.

Lipid bilayers

Figure 2 shows an enlarged view of space-filling molecular models of these lipid molecules, as they would appear in the top half of the bilayer in membranes. The lipid molecule is well suited to forming bilayer membranes in water because the hydrocarbon tails, like oil, do not mix well with water, and so are called "hydrophobic." But the charged head groups can reduce their electrostatic energy by associating intimately with water with its high dielectric constant, and so are called "hydrophilic." (Other well known molecules with both properties are soaps or surfactants, which have a single hydrocarbon tail. Apparently the presence of two tails on lipids leads to the formation of bilayers, whereas the single-tailed soaps tend to form small spherical structures in water, called "micelles.")

Experiments have conclusively demonstrated that the phase transition in *Acholeplasma laidlawii* is due exclusively to the lipid component of the membrane.[1,2] Therefore physical techniques that elucidate structure, and theory that deals with extended phases, are appropriate tools to study this phenomenon. Equally important, this means that phase-transition studies performed on "model membranes" made from highly ·purified and commercially available lipids can be expected to yield useful information. These model membranes are similar in their structural properties to biomembranes; furthermore, they can be much better characterized physically and chemically than living cell membranes, which have a complicated assortment of different lipids. The phase transitions are also much sharper with purified lipids of a single homogeneous type, and this makes quantitative measurement and theoretical interpretation much easier.

The simplest model membrane system is made simply by mixing lipid and water to form a dispersion. Unfortunately, the structures formed in this way are not *single* isolated bilayers but *multi*bilayers, as illustrated by the first sketch in figure 3. However, it appears now that the molecular interactions between adjacent bilayers are small compared to other interactions, so these are appropriate systems for phase-transition studies. These multibilayer systems are similar to lyotropic liquid-crystal systems of the smectic type. Liquid-crystal physicists, such as Peter Pershan of Harvard University, are studying these systems, using birefringence as well as quasielastic and Brillouin-scattering techniques developed for liquid crystals. By these means the elastic properties of lipid bilayers can be determined. These techniques and x-ray diffraction also show that, with low and varying water concentration, lipids have a rich variety of phase behavior in addition to the phase transition in the presence of excess water, which is of primary biological interest.[2]

The existence of various phases and the fact that lipids form bilayers in excess water whereas soaps form micelles leads us naturally to ask for a quantitative theory to explain the observed macromolecular structures. Such a theory is likely to be as difficult and remote as a quantitative theory to explain observed

Reprinted from *Physics Today* **31**, 38–47 (February 1978); © American Institute of Physics.

crystal structures in solid-state physics. However, great progress has been made in the study of non-structural phase transitions, such as magnetism, in the solid state by ignoring the question of how the crystal structure arose and accepting it as a given. A similar path can be followed with the phase transition of biological importance in lipids. In particular, the lipids form bilayer structures both above and below this transition, so this structure will be assumed in the further discussion below.

Disordered chains

The basic nature of the transition in multibilayer dispersions is revealed by calorimetry and x-ray diffraction.[2] The transition enthalpy (heat of transition) ΔH is very large, about 9 kcal/mole for phospholipids with chains of 16 carbon atoms. This gives an increase in entropy, $\Delta S = \Delta H/T$, of about 15 R, where R is the gas constant. If the new degrees of freedom were of a simple two-state spin type then, from $\Delta S \approx R \ln 2$, over 20 such degrees of freedom are activated at the transition. This immediately implicates the activation of internal molecular degrees of freedom in the lipid molecules, and the only portions with this many degrees of freedom and the appropriate regularity are the hydrocarbon-chain tails. X-ray studies show that the tails are quite regularly packed below the transition, with a fairly well-defined spacing of 4.8 Å between extended parallel chains. Above the transition, however, the diffraction pattern is diffuse and quite similar to those obtained from long-chain liquid hydrocarbons.

Thermodynamic evidence that the transitions in lipid bilayers mainly involve hydrocarbon-chain disordering comes from recent volume-versus-temperature measurements performed by one of us

(Nagle) and Allan Wilkinson. Figure 4 shows a plot of transition temperature versus $1/(n - \delta)$, where n is the hydrocarbon chain length; $\delta = 3$ was chosen to straighten the lines. In the limit $n \to \infty$, at which $1/(n - 3) \to 0$, the transition temperature extrapolates to the transition temperature of polyethylene. The volume change, also shown in this figure, extrapolates as $n \to \infty$ to two thirds of the

volume change in polyethylene. However, shorter hydrocarbon chains also have a "premelting transition" a few degrees below the melting transition with a volume change that extrapolates to about one third that of polyethylene. Thus, in the infinite-chain limit the transition in lipid bilayers corresponds to the transition from the premelted state of hydrocarbons to the melted state.

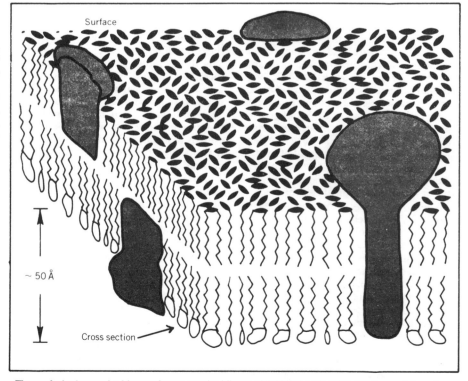

The main features of a biomembrane are the bilayer of lipid (fat) molecules and the proteins, shown in color. At low temperature the hydrocarbon tails of the lipid molecules appear as regular zigzag lines, as the cross section indicates. The tails are joined in pairs by a backbone, shown on the top surface. Omitted from the top but shown on the bottom layer are the head groups that are attached to the backbones of the lipid molecules.   Figure 1

In such systems the entropic force that drives the transition is rotation about any one of the carbon–carbon bonds in the hydrocarbon chains. Such rotations are internally impeded, as shown in figure 5. The *rotational isomeric* model, used extensively in polymer research, replaces the continuum of rotational angles by the *trans* angle (lowest in energy) and the two degenerate *gauche* angles, which have energies about 0.5 kcal/mole (*RT* when *T* = 250 K). Intermediate angles have much higher energies and are therefore ignored at biological temperatures. The model therefore consists of three states, called "rotamers," for each of the carbon–carbon bonds, one trans rotamer and two gauche rotamers. In the low-temperature phase most of the molecules are in the all-trans state, with the chains parallel to each other. If one tries to change a trans rotamer to a gauche rotamer independently of other rotamers, the free end of that hydrocarbon chain strikes another hydrocarbon chain, as illustrated in figure 2. The molecule on the left in this figure has one jog (one gauche rotation); the middle molecule has one kink (gauche–trans–gauche sequence) and one all-trans chain; the one on the right has two all-trans chains. The excluded-volume interaction therefore requires that the disordering of rotamers be a cooperative effect, which explains the existence of a sharp transition.

In addition to the rotamer energy and the excluded-volume-interaction energy, there is also a strong van der Waals energy, which pulls the hydrocarbon chains towards each other. The transition is accompanied by a volume change of about 4%, which does work against this van der Waals energy. Estimates by one of the authors[3] attribute about half of the measured heat of transition to this increase in van der Waals energy, a little less than half to the formation of gauche rotamers, and much smaller energy changes to the other interactions. Such estimates indicate that the high-temperature phase does not have as many gauche rotamers per carbon–carbon bond as do liquid hydrocarbons. This is reasonable, because the lipid bilayers have an additional constraint: Each hydrocarbon chain is pinned at one end to the water interface. Direct evidence that the chain melting in lipid bilayers is less extensive than in hydrocarbons and polyethylene is that the heat of transition and the volume change per CH_2 group are only about one third as large in the lipids (with 16 CH_2 groups per chain) as for the other systems.

Resonance studies

The disordered nature of the hydrocarbon chains in the high-temperature phase is verified by electron spin resonance,[4] nuclear resonance[5] and Raman spectroscopy. The resonance work also measures a directional order parameter of the sort used to define the order in liquid

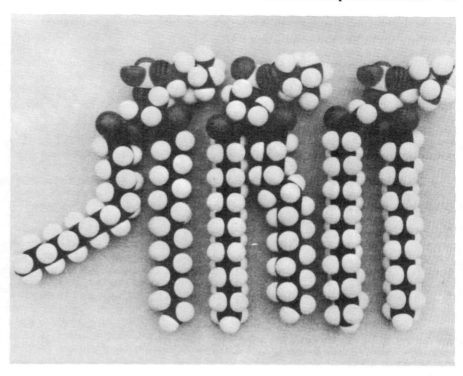

Three molecular models of the same lipid in different configurations. The lipid, dipalmitoyl phosphatidylcholine, consists of two hydrocarbon tails, $(CH_2)_{14}CH_3$, linked to a head group, $(CH_3)_3N^+(CH_2)_2PO_4^-$, by ester linkages and a glycerol backbone, $(OCO)_2(CH_2)_2CH$. The molecule on the left has a single jog (gauche rotation) in one tail, while the other tail is all trans. The middle molecule has a kink (gauche–trans–gauche sequence) in one tail. Figure 2

crystals. In the nmr case these order parameters are

$$S_n = \tfrac{1}{2} \langle 3 \cos^2\theta_n - 1 \rangle$$

where n denotes the position on the hydrocarbon chain from the head group and θ_n is the angular deviation of bond number n from its orientation in an all-trans chain normal to the bilayer surface. However, these order parameters should not be confused with those used in critical phenomena. In critical phenomena the order parameter is identically zero in the high-temperature disordered phase, whereas in bilayers the S_n's are nonzero in the high-temperature phase as well as in the low-temperature phase; they are therefore not the canonical Landau order parameters for the phase transition. The fact that the order parameters S_n are nonzero in both phases simply means that both phases are liquid crystals (of smectic type). Although the description "gel-to-liquid-crystal transition" is often used, the transition is really a polymer type of phase transition within a liquid-crystal phase. Nevertheless, the fundamental question as to the existence of a useful measurable order parameter in the Landau sense remains open. A macroscopic, phenomenological order parameter will be discussed below.

An extraordinary amount of resonance work has been done on lipid bilayers. One interesting discrepancy involves the measurement of the S_n by esr versus nmr. Measurements of S_n by esr, pioneered by

Harden McConnell's group at Stanford, require the use of spin probes; these are made by attaching a nitroxide free-radical group with an unpaired electron to the nth position on a hydrocarbon chain. Although the free radical is not excessively large, it does represent a local perturbation, which has been blamed for the difference between the esr and nmr results. Determinations of S_n by nmr use lipids that were synthesized with deuterium substituted in the nth position, and so have negligible perturbation. However, McConnell[4] suggested that the order parameters measured by esr should be different from those measured by nmr because the different hyperfine splittings set different time scales to the S_n averages, 10^{-7}–10^{-9} sec for esr and 10^{-5}–10^{-6} sec for nmr.

The picture of the molecular motions that is proposed by McConnell[4] and Sunney Chan and his co-workers at Cal Tech[5] is that, in addition to rotamer disorder in the high-temperature phase, entire molecules are tilted with respect to the normal to the bilayer and this tilt is measured by esr.[4] The tilting is necessarily cooperative, and in the low-temperature phase it is frozen in with long-range order, so it is seen by x-ray diffraction. However, in the high-temperature phase the bilayer is fluid enough that reorientation of the local tilt is fast on the nmr time scale (and is therefore averaged) but is not fast on the esr time scale. In contrast, rotamer motion is supposed to

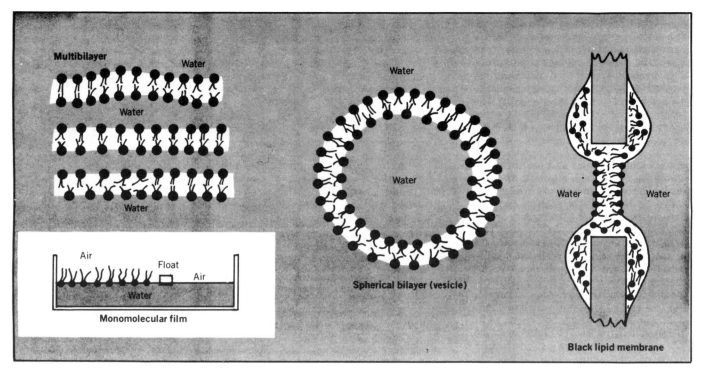

Four "model membranes." Parallel sheets of lipid bilayers separated by water form *multibilayers*; the sheets can be rippled. The application of ultrasound converts the multibilayers into *vesicles*, spherical bilayers about 300 Å in radius. Coated carefully on a hole in a plastic sheet and allowed to thin until it appears black, the lipid forms a *black lipid membrane*. The small wiggly lines indicate that some of the hydrocarbon solvent may still be present. A *monomolecular film* may be spread carefully on the surface of a water-filled trough. Such a film, which may be thermodynamically similar to half a bilayer, permits the measurement of the lateral pressure against the float shown.   Figure 3

be fast on both time scales. Thus, to compare the results one must first perform an ensemble average of the esr S_n.[4] This possible resolution of the discrepancy is attractive because it supports the contention that good quantitative measurements using different techniques on complicated biological systems do not need to be in direct competition but can be complementary and reveal unsuspected subtleties that further the understanding of the system.

In addition to multibilayer dispersions there are other model systems that are very useful. Ultrasonic agitation breaks up multibilayer dispersions and a large fraction of the material is then in the form of vesicles, each consisting of a single, roughly spherical bilayer enclosing an interior water volume rather like a prototypical cell, as illustrated in the second diagram in figure 3.

Figure 6, an electron micrograph of a sonicated sample of egg yolk lipid, shows two views of single-walled vesicles as well as small multilamellar structures. Unfortunately, the small diameter of the vesicles (about 300 Å) puts a packing strain on the lipid molecules that is different on the two sides of the bilayer. This is evidenced by a reduction in transition temperature by several degrees and the narrowing of nmr lines due to increased molecular freedom. Although these considerations of the packing strain mean that quantitative phase-transition studies must be interpreted with caution,[6]

these sonicated vesicles are very useful in biological studies because of the ease with which substances can be included inside the vesicles during sonication and then removed from the outside solution to form concentration differentials.

Another model system, which allows controlled access to both sides of the membrane, is that of the so-called "black lipid membranes." These are formed by application of a phospholipid–hexane (typically) solution to a small aperture to form the third structure shown in figure 3. It is a single bilayer, which is so thin that it is optically black. Black lipid membranes have been studied extensively and prepared by various techniques.[7] Some problems involving their use as model membranes include their fragile nature and the possibility that they contain some of the hydrocarbon solvent (such as hexane) as well as lipids.

Critical points in monolayers?

It can also be argued that a monomolecular film at an air–water interface, also shown in figure 3, is a model system for membranes.[2,3] One motivation for using monolayers as a model membrane system is that they provide an extra pair of experimentally controllable thermodynamic variables, namely the area per lipid molecule A and the lateral pressure π, which are not experimentally accessible in the other model systems. The intuitive idea that monolayers are model membrane systems is that a bilayer essentially con-

sists of two back-to-back monolayers with only weak interactions between them. Of course, this can only be true if water is prevented from contacting the hydrocarbon region in the single monolayers, as is the case. A slightly less obvious condition for a valid comparison concerns the free water–air interface behind the movable barrier in the figure. This pulls the barrier to the right with a force of 70 dynes/cm due to the surface tension of water, while the free hydrocarbon–air interface on the left pulls it to the left with a force estimated from the surface tension of bulk hydrocarbons to be about 20 dynes/cm. Therefore, to compare monolayers with bilayers, we must apply a lateral surface pressure of about 50 dynes/cm to the monolayer.[3] When this is done the measured transition temperatures for lecithin monolayers and bilayers are in agreement.

Monolayer isotherms, shown in figure 7, definitely show a transition for temperatures below some temperature T_c. Most interpretations of these isotherms are in terms of a two-phase coexistence region, also indicated in the figure. It is not known why the observed isotherms are not flat in the supposed two-phase region, although possibly the speed at which the experiments are done precludes true equilibrium in the transition region into the solid-like phase. In this regard it should be noted that Steve Hui at Roswell Park Memorial Institute, Buffalo, has observed domain structure in the

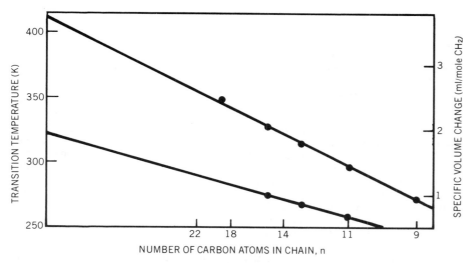

Transitions in lipid bilayers. The main transition temperatures are shown in black and the volume changes in color; the horizontal axis is linear in $1/(n-3)$.   Figure 4

low-temperature phase of homogeneous lipids.[8] The model system in this case is one of single bilayers on an electron-microscope grid, in a special hydrated chamber. This observance of domain structure as well as earlier monolayer work shows that the low-temperature phase becomes rigid and solid-like.

The conventional interpretation of the monolayer experiments[9] suggests a critical point, as shown in figure 7. For surface pressures above π_c or for temperatures above T_c there is no transition. The area change ΔA across the two-phase region in the figure behaves like a typical Landau order parameter, although it is a macroscopic, not a microscopic order parameter. The conjugate field is then the lateral surface pressure π. It is noteworthy that no indication of such a critical point appears when the ordinary three-dimensional pressure P and volume V are studied, either in lipids or polyethylene. This indicates that P and V in bulk lipids are not analogous to π and A. Estimates of the critical surface pressure[9] in 16-carbon phospholipids have placed it near 50 dynes/cm. If the analogy between monolayers and bilayers is taken seriously, this suggests that bilayers at the transition are near a critical point.

One experimental result that supports the contention that the phase transition in multibilayers is near critical is the permeability of the bilayer to sodium ions,[10] which shows an anomalous maximum at T_c. Although this permeability is low compared to its value in real membranes, which have sodium pumps, gated pores or special carrier molecules, it may be a useful probe of bilayer properties. In particular, the rate-limiting step in Na^+ appears to be the transfer of the ions from water through the head-group region and into the hydrocarbon region. This step would be greatly facilitated by critical fluctuations in the lateral packing of the lipids, which would open up small holes where the ions could enter.

Such fluctuations are large near a critical point, at which the lateral compressibility, $-A^{-1}(\partial A/\partial \pi)_T$, is high. As has also been recently suggested by Seb Doniach at Stanford, the anomalous peak in the Na^+ permeability therefore may be evidence for critical fluctuations. However, it must be emphasized that calorimetric and volumetric measurements on multibilayer dispersions show a transition that is only about a degree wide and the precise temperature variation changes from sample to sample. Thus, any critical region must be fairly narrow, at least for these measurements; the conventional presumption is that the phase transition in multibilayers is of first order.

At this point we can speculate on why real biomembranes might like to grow in the thermodynamic region of a phase transition. In order to grow, new material, such as lipids, cholesterol and protein, must be inserted into the membrane because there is no free edge to add on to—as there is in crystal growth. Such insertion is made easier if the lateral compressibility is high, because then local holes could be forced open with less expenditure of energy to accomodate new molecules. If one assumes that there is a growth enzyme already in the membrane, which forces open such holes, then this insertion would not require the vicinity of a critical point but merely a high lateral compressibility, such as occurs in monolayers in the two-phase region.[4] It must be cautioned that many membranes have transition temperatures that are considerably lower than the growth temperature, so a phase transition state is not necessary for these membranes. However, it appears that most membranes do require the disorder or fluidity associated with having the temperature above or within the lipid phase transition.

Lateral movement

In addition to permeability, another dynamical property of lipid bilayers that

is biologically relevant is the lateral motion of molecules within the plane of the membrane. The measure of this lateral mobility is the coefficient of lateral diffusion D_L, which is related to the mean-square displacement $\langle r^2 \rangle$ over a time t by the usual relation

$$\langle r^2 \rangle = 4 D_L t \qquad (1)$$

Some of the first quantitative measurements of D_L were made by McConnell and his co-workers at Stanford by analyzing the time dependence of the esr spectra that result when multibilayers are prepared with a high concentration of spin label in a localized region.[4] They find $D_L \approx 2 \times 10^{-8}$ cm²/sec.

Watt Webb and his co-workers at Cornell University have studied the lateral mobility of both lipids and proteins in multibilayers, black lipid membranes and biological membranes, using two novel optical techniques:[11] fluorescence correlation spectroscopy and fluorescence photobleaching recovery. The first technique has been used to measure D_L in black lipid membranes. After the model system is prepared with a small concentration of fluorescent molecules present, a laser is focussed through a microscope onto a small portion of the membrane and the resulting fluorescence intensity is monitored. Over a time τ this intensity fluctuates, due to variations in the concentration of the fluorescent probes in the illuminated region. The usual measure of such fluctuations is the time autocorrelation function

$$g(\tau) = [\langle I(t)\,I(t+\tau)\rangle - \langle I(t)\rangle^2]/\langle I(t)\rangle^2 \quad (2)$$

where $I(t)$ is the fluorescence intensity at time t. Analysis of the two-dimensional diffusion problem leads to the result

$$g(\tau) = \left[\langle N\rangle\left(1 + \frac{\tau}{\tau_c}\right)\right]^{-1} \quad (3)$$

where $\langle N\rangle$ is the average number of fluorescent molecules in the illuminated area πr^2 and $\tau_c = r^2/4D_L$ is the characteristic diffusion time for movement out of the illuminated area. From a plot of $g(\tau)$ versus τ, Webb and his co-workers find $D_L \gtrsim 10^{-7}$ cm²/sec in a variety of black lipid membranes prepared by different techniques and with varying compositions.

The fluorescence photobleaching recovery technique involves monitoring the recovery of fluorescence in a small area of a membrane illuminated by a focussed laser, after the area is irreversibly bleached by a high-intensity laser pulse. The rate of recovery of fluorescence in the area yields the diffusion coefficient through the characteristic time τ_c. From these measurements the Cornell workers find, for membranes of various mammalian cells such as rat myoblasts, $D \approx 2 \times 10^{-10}$ cm²/sec for the proteins and $D \approx 9 \times 10^{-9}$ cm²/sec for the lipid components.[11]

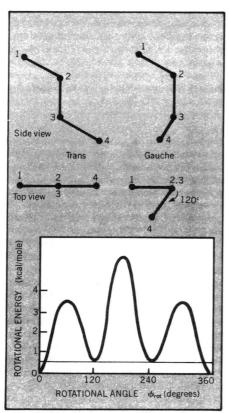

Rotational isomerism affects the shape of a hydrocarbon chain. In the upper diagrams, the left-hand chain is in the ground, trans, state; that on the right has undergone a gauche rotation about the bond connecting carbon atoms 2 and 3. The potential energy as a function of rotation angle shows three minima.   Figure 5

With the preceding values of the lateral mobility the time required for a molecule to travel around a cell of size 10 microns can be estimated. Using $D_L \approx 10^{-8}$ cm^2/sec and $\langle r^2 \rangle = (10$ microns$)^2$ gives $\tau_c \approx 25$ sec for lipids; using $D_{protein} \approx 2 \times 10^{-10}$ and the same $\langle r^2 \rangle$ gives $\tau_c \approx 20$ min for proteins. This is in good agreement with measured redistribution times for cell-surface antigens after cell fusion.[10] Thus it appears that lateral mobility is, for many cells, biologically operational. Measurements of diffusion constants for various membrane components involved in physiological processes have already shed new light on some of the molecular mechanisms of membrane function, and research continues along these lines in several laboratories. Understanding the immobilization of substantial fractions of various membrane proteins is a current concern.

The previous measurements of D_L were for the high-temperature fluid phase. Below T_c, measurements by Webb and his co-workers show that D_L is smaller by several orders of magnitude for multibilayers. This supports the idea that the low-temperature phase is much more rigid and solid-like. Curiously, however, D_L versus T for black lipid membranes does not show any phase transition. The

reason for this difference is imputed to be due to the solvent in the black lipid membranes, which implies that the two model systems are not strictly equivalent, at least in dynamic properties. Paul Fahey and Webb have recently found that D_L does change by several orders of magnitude for large solvent-free single bilayer vesicles formed by a new technique.

The various lipids mentioned in connection with figures 1 and 2 have phase transitions at different temperatures. Roughly, adding two CH$_2$ groups to each hydrocarbon chain in the membrane raises its T_c by about 15 degrees. Unsaturating the chains to form a single C=C double bond lowers T_c by about 45 degrees. Removing the CH$_3$ groups from the head regions raises T_c by about 20 degrees. Such perturbations are vitally important biologically. In particular, the way in which cells lower their membrane transition temperatures is to include more short-chain and unsaturated-chain lipids in the membrane composition. However, from a fundamental point of view transition-temperature changes of 10–20 degrees out of 300 K are indeed perturbations and do not indicate drastic qualitative differences.

Mixtures

Some important features of biomembrane phase transitions can not be mimicked by model systems composed of a single kind of lipid but require lipid mixtures. As is typical of all two-component mixtures there are upper and lower transition temperatures, T_{cu} and T_{cl} respectively. For temperatures between the transition points, the lipids A and B in each bilayer are laterally phase-separated into an A-rich (relatively fluid) phase and a B-rich (relatively solid) phase. As the membrane temperature is raised, the A-rich phase grows at the expense of the B-rich phase until there is only a single fluid phase at T_{cu}. As T is lowered to T_{cl} the system may condense into a single solid phase or, depending on the particular lipid mixture, a solution of two solid phases may form.[4]

With more than two lipid components there is, of course, the possibility of more than two coexisting phases. This lateral phase separation leads to another speculation as to why a cell might prefer to grow a membrane that is in a phase-transition region. There are many different membrane-bound proteins that carry out vital biological functions. These proteins may require different lipid environments to function. For example, some proteins may require the presence of lipids with head groups having a net charge, whereas other proteins may require neutral lipids. Clearly, it is possible to accomodate such different requirements in the phase-transition region of a multicomponent mixture of lipids where there are different phases with different lipid characteristics,

but this is not possible in a single phase.

Another kind of phase separation that takes place is that the compositions of lipids on the inside layer of the bilayer is often different from that of the outside layer. Differences in composition can be maintained easily because the flip-flop time for a molecule to go from one side of the bilayer to the other is quite long, as McConnell's group showed, using spin labelling.[4] Such a difference, which is called "membrane asymmetry," can be important in preferential binding, in orientation of membrane proteins and in the overall shape of the membrane.

In addition to the preceding speculation concerning the possible effect of lipids on the function of proteins there is esr evidence that the presence of proteins in bilayers affects the lipids that surround proteins lying entirely or almost entirely within the hydrocarbon region of lipid bilayers.[2,4] These studies suggest that each protein is surrounded by layers of "boundary lipid" with the lipids closest to the protein relatively immobile compared to lipids in the high-temperature phase. Such boundary lipids do not contribute to the enthalpy change of the phase transition.

The role played by cholesterol in membranes has been under active investigation for many years, at least in part because of its role in atherosclerosis. The influence of cholesterol upon the lipid phase transition is well known from calorimetric studies. As one adds cholesterol to multibilayers of a single-component lipid, the enthalpy change at the transition decreases, while the transition temperature does not change appreciably, until the transition disappears at 33–50% cholesterol.[2] The disappearance of the transition is generally thought to be due to the rigid nature of the steroid nucleus of the cholesterol molecule, which prevents the cooperative isomerization of the long hydrocarbon chains of the lipids. In any case cholesterol not only hardens the arteries, it also kills the phase transition!

Theoretical approaches

Biological phenomena are often too poorly characterized and involve too many variables to be amenable to the kind of theory developed for physical phenomena. However, the preceding discussion shows that the phase transition in lipid bilayers has been fairly well characterized, so it is not premature to develop theories involving statistical-mechanical calculations and theories of phase transitions; a growing number of theoreticians with backgrounds in physics have turned their attention to this phenomenon.

A proper treatment of transitions requires the calculations of the partition function

$$Z = \sum \exp(-E/kT) \qquad (4)$$

where the sum runs over all configurations

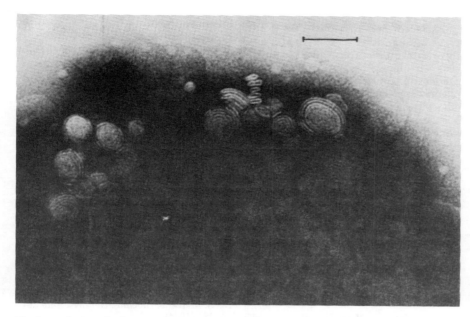

Electron micrograph of sonicated egg-yolk lecithin; the bar indicates a length of 0.1 micron. The bilayers appear bright in contrast to the negative stain of phosphotungstic acid. The larger objects are multilamellar structures. The circular light patches near the edge of the stained drop are interpreted as single-bilayer vesicles, which may be disks or spheres. Also visible are disk-shaped vesicles aggregated in stacks, known as "rouleaux." Figure 6

of the system and E is the energy of a configuration. In a system as complex as a lipid bilayer there are many contributions to E. From the preceding discussion we may write

$$E = E_{xvol} + E_{rot} + E_{vdW} + E_{other} \quad (5)$$

Here E_{xvol} is the strongly repulsive excluded-volume interaction that prevents atoms from occupying the same volume; E_{rot} is the rotamer energy for the hydrocarbon chains (this equals n_g times about 0.5 kcal/mole, where n_g is the number of gauche rotamers); E_{vdW} is the van der Waals attractive energy between the hydrocarbons (the scale of this energy term has been determined to be 1.84 kcal per mole of CH_2 from sublimation studies on hydrocarbons). Other contributions to

the energy, E_{other}, involve the *vibrational and kinetic energies,* which for each degree of freedom will have approximately the same classical value of $\frac{1}{2}kT$ in either phase; *head-group interactions,* including electrostatic interactions and possible weak hydrogen-bond interactions, either direct or mediated by bound water, which by the observed shifts in T_c can be assumed to be small; *interactions between the two layers that form the bilayer,* which appear from the agreement between monolayer and bilayer T_c's to be small, and *interfacial interactions* with the bulk water, which do not change much as long as the basic structure is that of the bilayer. Basic statistical calculations therefore concentrate on E_{xvol}, E_{rot} and E_{vdW}.

Even when E_{other} is neglected, the statistical-mechanical problem is formidable because of the excluded-volume interaction as well as the long-range van der Waals interaction, both of which make rigorous calculations intractable even for simple fluid systems. There are basically two alternatives for the theoretician to consider:

▶ Simplify the model, for example by reducing the dimensionality, so as to obtain a system for which equation 4 may be evaluated exactly.

▶ Retain the original model and evaluate equation 4 with approximate computational techniques.

These two approaches are complementary and, through study of a wide variety of theoretical models and careful comparison with experiment, a clear picture of the nature of the microscopic interactions that lead to the observed phase properties of bilayers should emerge.

The two authors of this article are involved in theoretical efforts using the two different approaches described above. Nagle simplified the basic model until it could be solved exactly by the Pfaffian-dimer technique in statistical mechanics.[3] In this model the hydrocarbon chains are infinitely long and restricted to a two-dimensional lattice, but the excluded-volume constraint is rigorously satisfied. With no free parameters the calculation gives the correct value for T_c for bilayers and the correct value of the critical point for monolayers. This critical point is of an unusual variety in statistical mechanics, and has been called a "$\frac{3}{2}$-order" critical point.[3]

Scott, on the other hand, has developed an approximation method for the treatment of the hard-core forces in monolayers, which is adapted from the Flory method in polymer statistics.[12] The calculation yields a classical critical point. In addition Scott is performing Monte Carlo simulations of the hydrocarbon region in a monolayer. These calculations emphasize the vital role the hard-core forces play in determining chain conformations, and they shed some additional light on the reason the esr and nmr parameters differ.

Stjepan Marčelja, a physicist at Zagreb University, has developed a mean-field approximation, with which he carefully treats the rotational isomerism.[13] His treatment of the intermolecular interactions is similar to the Maier–Saupe methods for liquid crystals. As is also the case for liquid crystals, the theory produces fairly good quantitative agreement with experiment, but the approximate treatment introduces parameters with values that are either undetermined or at variance with knowledge of intermolecular interactions. Bruce Hudson, Hans Andersen and their co-workers at Stanford have performed calculations that make use of mean-field and scaled-particle theory, and which concentrate on the

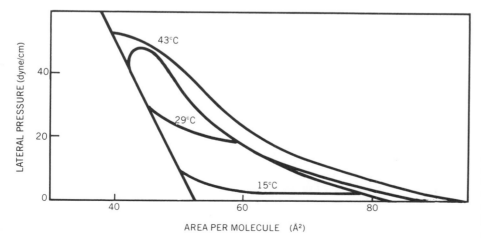

The lateral pressure of phospholipid monolayers as a function of their molecular area, taken at three temperatures. The colored line shows the postulated coexistence (two-phase) region. The data are from S. W. Hui and collaborators, reference 9. Figure 7

behavior of T_c with chain length and on two-component mixtures.[14]

Theoretical model building and statistical-mechanical calculations often focus and sharpen experimental investigation in traditional areas of physics, and the same is true in some of the biomembrane studies discussed here. For example, one of us (Nagle),[3] by analysis of the general energy equation, 5, pointed out that if E_{other} is small the relative volume change $\Delta V/V$ must be in the range 3–4%. In this case the experiments supported the theory. Other cases, yet to be fully supported or rejected by experiment, include the possible subcritical nature of the phase transition in bilayers mentioned earlier, and an hypothesis put forward by one of us (Scott) concerning reasons for the unusual shape of monolayer isotherms.

Prognosis

We have attempted to describe an area of research that is relevant to the biological sciences and also amenable to the types of quantitative experimental and theoretical studies that appeal to physicists. In biomembrane-related research there are of course messy problems that plague experimentalist and theoretician alike but, as we have shown, progress has been made. With continuing contributions from, and interactions between, physicists, biophysicists, chemists, biochemists and biologists, the next decade should see even more rapid advancement.

References

1. D. L. Melchior, J. M. Steim, Biochim. Biophys. Acta 466, 148 (1977).
2. D. Chapman, Quart. Rev. Biophys. 8, 185 (1975).
3. J. F. Nagle, J. Membrane Biol. 27, 233 (1976).
4. H. M. McConnell, in Spin Labeling (L. J. Berliner, ed.), Academic, New York (1976), chapter 13.
5. N. O. Petersen, S. I. Chan, Biochemistry 16, 2657 (1977).
6. C. H. A. Seiter, S. I. Chan, J. Am. Chem. Soc. 95, 7541 (1973).
7. H. T. Tien, Bilayer Lipid Membranes (BLM) Theory and Practice, Marcel Dekker, New York (1974).
8. S. W. Hui, D. F. Parsons, Science 184, 77 (1974).
9. S. W. Hui, M. Cowden, D. Papahadjopoulos, D. F. Parsons, Biochim. Biophys. Acta 382, 265 (1975).
10. D. Papahadjopoulos, K. Jacobson, S. Nir, T. Isac, Biochim. Biophys. Acta 311, 330 (1973).
11. W. W. Webb, in Electrical Phenomena at the Biological Membrane Level (E. Roux, ed.), Elsevier (1977), pages 119–156.
12. H. L. Scott, Biochim. Biophys. Acta 406, 329 (1975).
13. S. Marčelja, Biochim. Biophys. Acta 367, 162 (1974).
14. R. E. Jacobs, B. Hudson, H. C. Andersen, Biochemistry 16, 4349 (1977).   □

Influence of Interatomic Interactions on the Structure and Stability of Polypeptides and Proteins

HAROLD A. SCHERAGA, *Baker Laboratory of Chemistry, Cornell University, Ithaca, New York 14853*

Synopsis

Several examples are cited to demonstrate how conformational-energy calculations provide information about the manner in which interatomic interactions influence the structure and stability of polypeptides and proteins, and of intermolecular complexes involving these same species. These include the screw senses of α-helical polyamino acids, helix–coil transition parameters, properties of β-bends, and the structures of gramicidin S and of synthetic poly-(tripeptide) models of collagen. In all of these calculations, the multiple-minimum problem has been surmounted. A description is provided as to how the multiple-minimum problem is being approached in calculations on globular proteins. These computational methods have also been applied to protein structure refinement and to the calculation of structures of homologous proteins and of enzyme–substrate complexes.

INTRODUCTION

Conformational-energy calculations are proving to be very effective in elucidating how interatomic interactions dictate the stable conformations of polypeptides and proteins, and of intermolecular complexes involving these same species. This is demonstrated, in this article dedicated to Professor Shiro Akabori on his 80th birthday, by citing several selected results of computations carried out in our laboratory. A more extensive discussion of the contributions from many laboratories was provided in a review by Némethy and Scheraga.[1]

Stable conformations correspond to local minima of the free energy of the whole system (solutes plus solvent). Small polypeptides are (Boltzmann-averaged) ensembles over the local minima, whereas larger systems, such as fibrous and globular proteins, are presumably in their global minima in the native conformation, although consideration is given to the possibility (for which no evidence exists at the present time) that native proteins may be metastable.

A brief history of the start of our involvement in such computational methods (to take advantage of experimental information on distance constraints, including loop formation) was described by Wako and Scheraga.[2] Initially, the calculations were carried out with a hard-sphere potential,[3–6] but subsequently more complete empirical potential functions were used,[7–13] including the role of aqueous[14,15] and nonaqueous solvents.[16]

Reprinted from *Biopolymers* **20**, 1877–1899 (1981); © John Wiley & Sons, Inc.

Liquori and coworkers,[17] Ramachandran et al.,[18] and Brant and Flory[19] initiated similar studies at about the same time. In the ensuing years, much effort has been devoted to the acquisition of reliable potential functions, to the treatment of solvation of the functional groups of the polypeptide chain, to the underlying theory of the conformational properties of chain molecules, and to the application of computational methods to a variety of polypeptide and protein systems.

METHODOLOGY

Our currently used main potential function is ECEPP (Empirical Conformational Energy Program for Peptides),[10] an atom-centered potential parameterized on crystal and gas-phase structural and thermodynamic data, and recently adapted[20] for use with a high-speed array processor in conjunction with a minicomputer host. United-atom[21] and united-residue[22] versions of ECEPP have been developed to reduce computational time. Though ECEPP is adequate for most applications, current efforts are continuing to try to improve the form of the potential functions and the parameters therein; for example, an alternative potential, EPEN (Empirical Potential based on the Interactions of Electrons and Nuclei), has been developed[11,12] and improved.[13]

Most of the computations are carried out with (backbone and side-chain) dihedral angles as the independent variables, keeping the geometry (bond lengths and bond angles) fixed. In some cases, however, flexible geometry is used. For a protein of 100 amino acid residues, there are 500–600 variable dihedral angles. Some details are given in Ref. 23.

The free energy of hydration is taken care of by a hydration-shell model,[14,15] and efforts are being made to improve it by providing a better representation of the thermodynamic parameters for the hydration of the functional groups on a polypeptide chain[24-26]; for example, Monte Carlo[27,28] (R. H. Kincaid and H. A. Scheraga, work in progress) and molecular-dynamics[29] simulations are being carried out on water and aqueous solutions of small-molecule solutes. Also, a treatment has been provided to compute the entropy of association of proteins in aqueous solution.[30]

Theory and procedures have been developed for computing first and second derivatives of the conformational energy, for minimizing the conformational energy, and for evaluating statistical weights (including conformational entropy).[31-35] Stable configurations of the various systems under study are attained by direct minimization, by Monte Carlo or molecular-dynamic methods.

MULTIPLE-MINIMUM PROBLEM

Because of the existence of many minima in the empirical potential-energy function, for any number of independent degrees of freedom, minimization procedures lead only to the local minimum that is closest to the starting conformation (the multiple-minimum problem). Since efficient mathematical procedures for surmounting potential barriers (to move from one minimum to a lower one) do not exist, except for spaces of low dimensionality,[32,33,36-39] various strategies[40] have had to be developed to surmount the multiple-minimum problem. This has been accomplished[23] for small open-chain and cyclic peptides and for synthetic analogs of fibrous proteins but not yet for globular proteins. The solution has been achieved, in the former cases, because the number of variables is small, and a limited but sufficient number of starting conformations could be selected (for subsequent energy minimization) to assure the statistically complete enough coverage of conformational space to locate the global minimum. For globular proteins, however, the number of variables is too large for this approach to be computationally feasible, and other strategies are being devised.

COMPUTATIONS ON MODEL PEPTIDES AND POLYAMINO ACIDS

Terminally blocked single amino acid residues, X, i.e. N-acetyl-X-N'-methyl amides, and regular-repeating (i.e., helical) structures of homopolymers involve only two backbone variables (ϕ,ψ), for fixed ω, and the conformational energies can be represented as two-dimensional contour plots on (ϕ,ψ) diagrams. The multiple-minimum problem is thereby surmounted by visual inspection of the contour diagrams. (As indicated earlier, experimental methods such as nuclear magnetic resonance spectroscopy would detect a Boltzmann-averaged ensemble over the local minima.) Larger structures are built up from low-energy structures of smaller ones, and their energies minimized. For cyclic peptides, exact ring closure[41,42] is imposed as a constraint during minimization.

Properties of Helices and Coils

In this manner, the preferred helix senses of many α-helical homopolyamino acids (constrained to have the same conformation in each residue) have been computed,[43-46] with good agreement with experiment. In addition, an understanding has been gained as to how the side chain–backbone interactions influence the helix sense. For example, the right-handedness of α-helical poly(methyl-L-glutamate) and the left-handedness of α-helical poly(methyl-L-aspartate) are attributable to the different interactions in each polymer between the dipoles of the side-chain ester group and the nearest backbone amide group.[44] As another example, whereas poly-(benzyl-L-aspartate) forms a left-handed α-helix, the p-Cl-benzyl derivative is right-handed but the m-Cl-benzyl derivative is left-handed. Figure 1 illustrates the lowest energy left- and right-handed α-helical structures of the m-Cl-benzyl derivative, with the energy of the left-handed form being lower than that of the right-handed form. The side-chain C-Cl dipole and

Left - handed   Right - handed

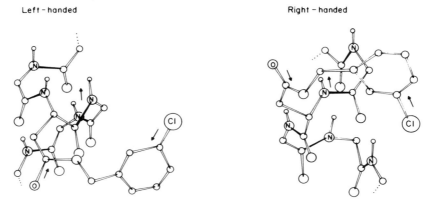

Fig. 1. Orientation of the side chains of the lowest energy left- and right-handed α-helices (Ref. 46). The solid arrows represent the directions of the C-Cl, ester, and amide dipoles, respectively. [Reprinted with permission from *J. Am. Chem. Soc.* **92**, 1109 (1970). Copyright 1970 American Chemical Society.]

the nearest backbone amide dipole are antiparallel in the left-handed form but parallel in the right-handed form.[46] This electrostatic contribution to the total energy is the main factor stabilizing the left-handed helix. This prediction has been confirmed by experiment.[47]

Interactions involving the side chains can be expressed in another way, viz., in terms of the temperature dependence of the Zimm and Bragg helix stability constant, s.[48] The thermodynamic parameters of the thermally induced helix–coil (order–disorder) transition of polyamino acids in water have been evaluated by treating the helix by the small-vibration approximation and the coil by the single-residue nearest-neighbor approximation.[49–52] Figure 2 provides a comparison of calculated[52] and experimental[53,54] curves for the temperature dependence of s for poly(L-valine) in water. It can be seen that the calculations match the observed increase in s with increasing temperature, reflecting primarily the characteristic

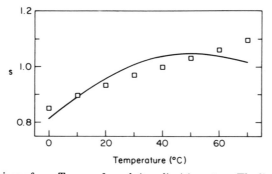

Fig. 2. Comparison of s vs T curves for poly(L-valine) in water. The line is a calculated one (Ref. 52), and the squares are the experimental results (Ref. 53). [Reprinted with permission from *Macromolecules* **7**, 459 (1974). Copyright 1974 American Chemical Society.]

increase[55] in hydrophobic bond strength (involving the valine side chain) in this temperature range.

Similarly, using a Debye-Hückel screening potential for treating charged groups,[56] it has been possible to account for the isothermal pH-induced helix–coil transition of poly(L-lysine) in aqueous salt solution. The transition between two ordered forms of poly(L-proline), the isothermal solvent-induced polyproline I ↔ II interconversion between *cis*- and *trans*-peptide forms, respectively, has also been treated[16,57-59] (see Fig. 3).

Interactions in Crystals

The stable crystal arrangements of several homopolymers, including the thermally induced conversion between α- and ω-helical forms of polyamino acids, have been computed.[60,61] These involve intermolecular, crystal-packing degrees of freedom, as well as the internal ones, but the number

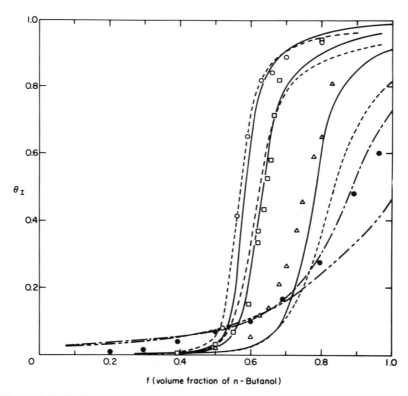

f (volume fraction of n - Butanol)

Fig. 3. Calculated transition curves (Ref. 16) of poly(L-proline) in *n*-butyl alcohol:benzyl alcohol at 70°C. The fraction of form I helix, θ_I, is plotted against the volume fraction of *n*-butyl alcohol. The experimental points (Ref. 59) are shown for degrees of polymerization of 217 (○), 90 (□), 33 (△) and 14 (●). The various solid and dashed curves are based on different assumptions made in the computations (Ref. 16). [Reprinted with permission from *Macromolecules* **8,** 516 (1975). Copyright 1975 American Chemical Society.]

of degrees of freedom is still small enough to surmount the multiple-minimum problem. Whereas poly(p-Cl-benzyl-L-aspartate) exists as a right-handed α-helix in solution, it is possible to convert it to an ω-helix in the crystal. Figure 4, based on computations of Fu et al.,[61] illustrates how favorable interchain interactions enable the backbone to adopt the ω-helical form in the crystal. X-ray fiber-diffraction studies have demonstrated the presence of the ω-helix in crystals of this polymer.[62]

β-Bends

A peptide structural feature to which much attention has been devoted is the β-bend, first examined by Venkatachalam.[63] The conformational stabilities and characteristics of several bends[64–70] and multiple-bends,[71] involving a small number of degrees of freedom, have been examined by these methods, and a differential-geometrical analysis of bend structures has also been provided.[72] The differential-geometrical analysis (illustrated in Fig. 5) demonstrates that there is a structural continuum of bends, ranging from those that resemble four-C^α segments of α_R-helices through essentially flat bends to those that resemble four-C^α segments of α_L-heli-

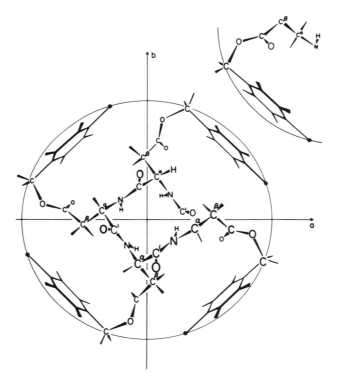

Fig. 4. View of the crystalline ω-helical form of poly(p-Cl-benzyl-L-aspartate) (Ref. 61). [Reprinted with permission from *Macromolecules* **7**, 468 (1974). Copyright 1974 American Chemical Society.]

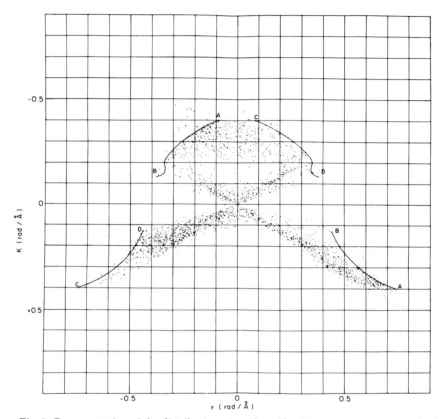

Fig. 5. Representation of the distribution of residues (dots) in a curvature-torsion (κ, τ) plane for a sample of 22 proteins (Ref. 72). The two curves AB are equivalent conformationally, as are the two curves CD. [Reprinted with permission from *Macromolecules* **13,** 1440 (1980). Copyright 1980 American Chemical Society.]

ces[72] (this continuity is indicated in Fig. 5 by the conformational equivalence of the two AB curves and the two CD curves, respectively).

Recently, several cyclic peptides, viz., cyclo(L-Ala-Gly-ε-aminocaproyl),[73–75] cyclo(L-Ala-L-Ala-ε-aminocaproyl),[76] and cyclo(L-Ala-D-Ala-ε-aminocaproyl),[76] have been synthesized as models for β-bends, and the spectroscopic and theoretical conformational properties of such structures have been elucidated. Figure 6 shows the structure of the cyclic L-Ala-Gly peptide. The corresponding L-Ala-L-Ala and L-Ala-D-Ala molecules exist in solution and in the solid state primarily as type I (and III) and type II bend conformations, respectively.[76] The theoretical[73] and experimental[75,76] analysis of the L-Ala-Gly molecule has shown that it exists in solution predominantly as a type II bend, with a type I (and III) bend present in a small amount, depending on temperature and solvent. (The CD data for the L-Ala-L-Ala and L-Ala-D-Ala compounds[76] led to an estimate of the

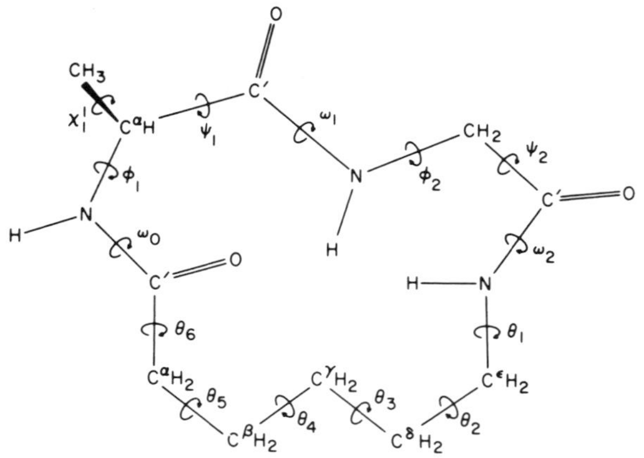

Fig. 6. Structural formula of cyclo(L-Ala-Gly-ε-aminocaproyl), showing the dihedral angles allowed to vary in calculations of the conformational energy (Ref. 73). (Reprinted with permission from *Macromolecules* 14, in press. Copyright 1981 American Chemical Society.)

relative amounts of type I and II bends from the circular dichroism data for the L-Ala-Gly compound.[74]) It should be noted that type I and type III bends do not constitute two distinct classes of bends in terms of distributions of dihedral angles, but they are subsets of a continuous distribution[70] (see Fig. 7). They differ from each other in terms of the orientation of the third peptide group in the bend. On the basis of the orientation of the same peptide group, two subclasses of type II bends can also be distinguished.[70]

Gramicidin S

As an example of computations on a cyclic peptide,[77] we cite the cyclic decapeptide gramicidin S. An initial 10,541 conformations were examined, with C_2 symmetry imposed in the computations, to assure the attainment of the global minimum (Figs. 8 and 9). A differential-geometrical analysis[78] of the computed structure[77] indicated that it was in good agreement[79] (including the prediction of a hydrogen bond between the side-chain N^δ of ornithine and the backbone CO of phenylalanine) with a subsequently determined[80] x-ray crystal structure of a hydrated gramicidin S–urea complex. It was noted[79] that, in the crystal, gramicidin S is dimerized, forming an intermolecular antiparallel four-stranded β-sheet (see Fig. 10).

FIBROUS PROTEINS

Collagen is an example of a fibrous protein which involves interchain association to form a triple-stranded coiled-coil structure. The geometrical criteria for formation of coiled-coil structures of polypeptide chains have

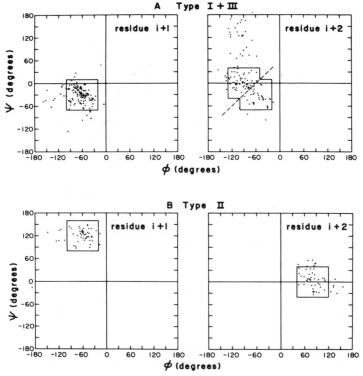

Fig. 7. Distribution of dihedral angles for the (middle two) residues $i + 1$ and $i + 2$ in bends (A) of types I and III and (B) of type II, in 23 proteins of known structure (Ref. 70).   The squares indicate ranges of ±40° around the "standard" values of (ϕ,ψ) in each type of bend.   The dashed line indicates the boundary between type I and type III bends for residue $i + 2$.   [Reprinted with permission from *Biochem. Biophys. Res. Commun.* **95**, 320 (1980).   Copyright 1980 Academic Press.]

been developed,[81,82] and conformational-energy calculations have been carried out on several synthetic poly(tripeptide) analogs, poly(Gly-*X*-*Y*), of collagen.[83-86]   Because of the regularity conditions imposed on each tripeptide in the computations, the number of degrees of freedom was small, so that, again, the multiple-minimum problem was surmounted by an adequate coverage of conformational space.   The computations indicated that poly(Gly-Pro-Pro), poly(Gly-Pro-Hypro), and poly(Gly-Pro-Ala) formed stable triple-stranded coiled-coil collagenlike structures, whereas poly(Gly-Ala-Pro) did not, in agreement with experiment.   Figure 11 illustrates the calculated lowest energy (collagenlike) structure[83] of poly(Gly-Pro-Pro).   The computations also provided an explanation for the association of the chains [in contrast to the single-chain structures of α-helical forms of, e.g., poly(γ-benzyl-L-glutamate)]; viz., the resulting interchain interactions in collagen lower the energy of a tripeptide unit

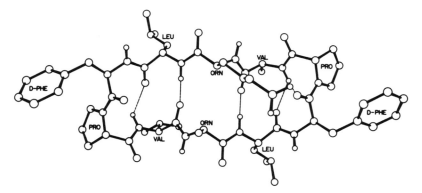

Fig. 8. Calculated conformation of gramicidin S of lowest energy (Ref. 77), looking down the C_2 symmetry axis. For clarity, one ornithine and one valine side chain have been deleted. Dashed lines denote hydrogen bonds. The Cartesian coordinates of this calculated structure are given in Table VIII of Ref. 77. [Reprinted with permission from *Macromolecules* **8,** 750 (1975). Copyright 1975 American Chemical Society.]

below that in the nonassociated single chain. Furthermore, the interchain interactions induce a slight conformational change in going from the low-energy form of the single chain to the lower-energy form of the triple-stranded complex.

After completion of the calculations[83] on poly(Gly-Pro-Pro), it was learned that Okuyama et al.[87] had carried out a single-crystal x-ray structure analysis of (Pro-Pro-Gly)$_{10}$. Our structure is in agreement with theirs, with an rms deviation of 0.3 Å for all (non-hydrogen) atoms, based on a comparison of the x-ray coordinates (kindly provided us by Professor M. Kakudo) and our computed ones.

GLOBULAR PROTEINS

The strategies involving energy minimization for a globular protein are based on the prior use of approximate methods to reach the correct potential energy well, after which minimization of the complete energy function can be carried out to reach the minimum.[1,88,89] While Monte Carlo and molecular-dynamics methods can surmount energy barriers, and indeed have been applied to small cyclic[90] and open-chain[91] peptides, these methods have not yet been applied successfully to protein folding.

Time to Fold a Globular Protein

Aside from problems due to the slow rate of convergence, Monte Carlo and molecular-dynamics methods explore conformational space on a time scale equivalent to that of picoseconds, whereas globular proteins take much longer to fold (anywhere from nanoseconds to milliseconds). Were it not for the dominant role of short-range interactions,[89,92–94] which direct folding

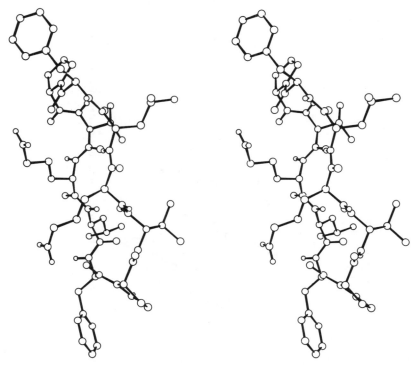

Fig. 9. A stereoscopic illustration of the calculated structure of Fig. 8 (Ref. 77). [Reprinted with permission from *Macromolecules* **8,** 750 (1975). Copyright 1975 American Chemical Society.]

along a limited number of pathways to the native structure, protein folding would be much slower; i.e., the protein does not explore the whole conformational space in the folding process.

Prediction of Values of (ϕ_i, ψ_i)

A variety of short-range approximate procedures (reviewed elsewhere[1]) are used to predict the conformational states of the residues of a protein. Recently, a nearest-neighbor Ising model has been proposed,[92] to predict the actual values of the backbone dihedral angles (ϕ_i, ψ_i) rather than the conformational states. A test of this model on bovine pancreatic trypsin inhibitor, applied only to the nonhelical and nonbend regions (to which the nearest-neighbor Ising model is applicable), indicated that its accuracy is 60%, compared to 47% for the best prediction method in the literature, to 73% for a "perfect" prediction [which is less than 100% because the usually defined conformational regions are so large that a perfect prediction of the region does not always lead to good values of (ϕ_i, ψ_i)], and to 15% for a random prediction. The extension of this method to longer-range interactions may improve the predictability in the helical and bend regions.

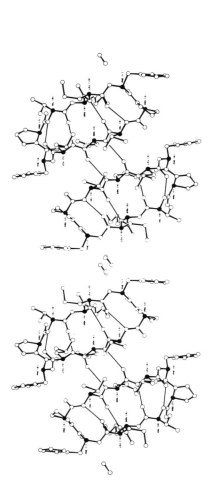

Fig. 10. A stereo ORTEP diagram (Ref. 79) of the x-ray structure (Ref. 80) of a pair of gramicidin S molecules and the associated urea molecules.

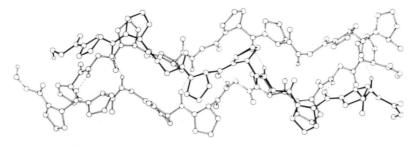

Fig. 11. Triple-stranded coiled-coil complex of poly(Gly-Pro-Pro) of lowest energy (Ref. 83). Each chain has an $N_1H_1\cdots O_2C_2$ hydrogen bond, indicated by a dashed line; i.e., the first NH of a tripeptide of a given chain is hydrogen-bonded to the CO of the second residue of the corresponding tripeptide on the chain that is counterclockwise from the given one, when looking from the N- to the C-terminus.

Possible Folding Mechanisms: Nucleation

These results on the prediction of values of (ϕ_i, ψ_i), and related ones from immunochemical experiments,[93,94] provide strong support for the dominance of short-range interactions. In other words, the residues of a socalled denatured protein are practically in their native state because of short-range interactions; e.g., the equilibrium constant between the unfolded and folded forms of reduced ribonuclease[94] is ∼0.06, a very large number. Thus, small fluctuations in the conformations of the residues in the unfolded form can bring about the formation of nuclei[95] (possibly involving proline isomerization[96]), which can then induce the subsequent rapid folding of the rest of the molecule. This mechanism accounts for the relative rapidity of protein folding mentioned above.

Several views of nucleation have been developed, all of them being essentially different aspects of fundamentally the same mechanism. In one view,[95] folding starts in segments of the chain with the highest density of nonpolar residues that can form hydrophobic bonds,[55,97] as illustrated in Fig. 12. In another view,[98] initial folding nuclei are those (including α-helices[99]) that appear near the diagonal in the protein contact map, as illustrated[100] in Fig. 13. A differential-geometric analysis of protein structures (S. Rackovsky and H. A. Scheraga, *Macromolecules*, in press) provides a unifying framework for the proposed nucleation mechanisms and shows that there is relatively low selectivity in nucleation on a scale of four C^α's, and that only certain combinations of four-C^α structures occur on a scale of five C^α's; further, a higher degree of selectivity in nucleation appears on the five-C^α length scale than on the four-C^α scale. Figure 14 is a schematic representation of folding of a protein which involves, as an example, two nucleation sites. From kinetics studies of the pathway of folding of ribonuclease A, we have shown that the nuclei can form either early or late in the folding process, depending on the conditions used to induce folding (Y. Konishi, T. Ooi, and H. A. Scheraga, unpublished).

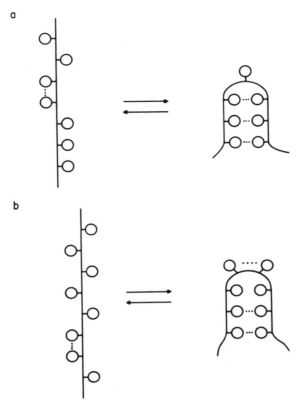

Fig. 12. Schematic representation of nucleation step (Ref. 95) in which a hydrophobic pocket is formed from an ensemble of unfolded species [which may or may not contain neighbor–neighbor hydrophobic bonds (Refs. 55, 97) as indicated by the species on the left side of the equilibria in (a) and (b)]. In (a), only one side chain is involved in the "bend," and it does not participate in hydrophobic bonds with nearby residues. In (b), two side chains (which themselves can form a neighbor–neighbor hydrophobic bond) are involved in the "bend," and the pocket is shown as an imperfect one in that one pair of side chains is not sufficiently nonpolar to form a hydrophobic bond. Hydrophobic bonds are indicated by dotted lines. For pockets of types (a) and (b), at least 5 and 4 residues, respectively, are assumed to be required, so that the pockets are large enough for the bends to be stereochemically feasible. Pockets of both types are included in the search algorithm to locate the nucleation site (Ref. 95). [Reprinted with permission from *Macromolecules* 11, 819 (1978). Copyright 1978 American Chemical Society.]

Longer-Range Interactions

With short-range interactions properly accounted for, and the approximations extended to four residues on each side to take medium-range interactions into account,[101] it is still necessary to introduce long-range interactions into the model (H. Meirovitch and H. A. Scheraga, unpublished) before beginning energy minimization; otherwise, minimization is trapped in a nonnative local minimum. This can be accomplished by dividing the conformational space of a protein into classes characterized by different

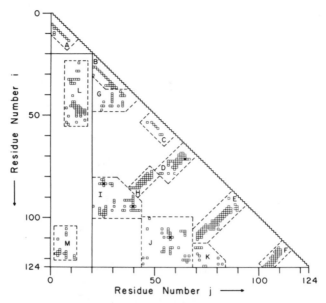

Fig. 13. Contact map of ribonuclease S (Ref. 100). Each point of the map represents the presence (square) or absence (no marking) of a contact between two amino acid residues i and j. Contacts between residues are omitted from the figure whenever $|i - j| \leq 4$. The pairs of half-cystine residues forming the disulfide bridges are denoted by black squares. Contact regions (A–M) are bounded by dashed lines. Those contact regions near the diagonal are possible nucleation sites.

spatial geometric arrangements of the loops formed by disulfide bonds (illustrated for a hypothetical protein, with two disulfide bonds, in Fig. 15) and then randomly selecting one or more members of each class for subsequent energy minimization. Because of the limited number of classes,

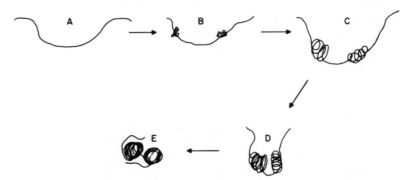

Fig. 14. Schematic representation of stages of folding in a protein with, say, two nucleation sites (Ref. 89). A represents the ensemble of unfolded species; B illustrates formation of two nucleation sites; C shows the growth of the nucleated regions; D represents the association of two contact regions or domains; and E indicates the final stages of folding. [Reprinted, by permission of the publisher, from *Protein Folding*, Jaenicke, Ed. (1980), p. 261. Copyright 1980 by Elsevier North Holland, Inc.]

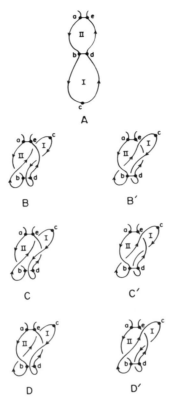

Fig. 15. Schematic representation of the seven possible geometric arrangements of the loops in a hypothetical protein having two disulfide bonds (H. Meirovitch and H. A. Scheraga, unpublished.)

and the availability of an array processor, such an approach is feasible. The selection of a starting conformation from the class of the native structure is a necessary but not a sufficient condition to attain the native structure by energy minimization.   It is thus necessary to introduce other constraints, e.g., placement of the residues in their correct local conformational state enables energy minimization to bring the resulting structure closer to the native one (H. Meirovitch and H. A. Scheraga, unpublished).   Other constraints, such as the radial distribution of hydrophobic and hydrophilic residues,[102,103] or the location of nucleation sites,[95] can be incorporated into this procedure.   Use can also be made of empirical information on the packing of α- and β-structures[104–108] and on the natural right-handed twists of β-sheets.[109]   Short- and medium-range constraints can also be introduced by distance-geometry methods.[110–114]   (If distance-geometry methods are used by themselves, i.e., without being coupled to an energy-minimization algorithm, then a very large number of distances must be known accurately to obtain a relatively error-free structure.[2])   The closeness of a computed structure to the experimental one (in a test of the procedure on a protein of known structure) can be assessed by comparison of triangle contact maps (such as that of Fig. 13) and by use of the conformational distance function of a recently developed differential-geometric representation of polypeptide chains.[72]

APPLICATION TO PROTEIN STRUCTURE REFINEMENT

Conformational-energy calculations have been useful in potential-energy constrained refinements of x-ray crystal structures of proteins. For reasons discussed elsewhere,[115] our approach has been to fix bond lengths and bond angles at standard values in such refinements and to vary only dihedral angles during energy minimization. Thus, for example, a sterically unhindered structure of lysozyme has been obtained by energy minimization.[116]

Recently, a potential-energy constrained real-space refinement procedure has been developed to use low- rather than high-resolution x-ray data to obtain a high-resolution structure (Ref. 117 and S. Fitzwater and H. A. Scheraga, unpublished). It was applied to the 2.5-Å resolution data of bovine pancreatic trypsin inhibitor and yielded a reasonable approximation to the structure that had been obtained[118] earlier by use of 1.5-Å resolution data.

CALCULATION OF STRUCTURES OF HOMOLOGOUS PROTEINS

This methodology is also applicable to the determination of structures of proteins that are homologous to other proteins of known structure. It has been applied to compute the structure of α-lactalbumin[119] from that of lysozyme and those of three snake venom inhibitors[115] from that of bovine pancreatic trypsin inhibitor.

ENZYME SUBSTRATE INTERACTIONS

These computational methods have also been applied to enzyme–substrate interactions (M. R. Pincus and H. A. Scheraga, unpublished), e.g., to those involving α-chymotrypsin[120] and lysozyme[121–124] (M. R. Pincus & H. A. Scheraga, *Biochemistry*, in press). For illustrative purposes, we consider lysozyme, which cleaves polymers and copolymers of *N*-acetylglucosamine (NAG) and *N*-acetylmuramic acid (NAM).[125] The active site can accommodate a hexasaccharide, and cleavage takes place between the fourth and fifth residues of this substrate. The lactic acid side chain of NAM would naturally be expected to lead to a different set of interactions compared to NAG, and both NAG and NAM would be expected to differ from glucose in their binding affinities to the active site. We have examined the binding of various size oligomers of NAG and NAM, from the monomer to the hexamer, to the active site of lysozyme[121–124] (M. R. Pincus and H.

A. Scheraga, *Biochemistry*, in press).   This is essentially a docking problem, where, in the last analysis, a flexible substrate is allowed to approach a flexible enzyme (exploring the whole conformational space of the substrate and of the active site of the enzyme) to find the most stable binding disposition.

Several low-energy structures were found, e.g., a right-side complex with a distorted (half-chair) conformation for the D residue of $(NAG)_6$ and a left-side undistorted complex.[124]   (The terms "right" and "left" refer to the positions of the E and F residues of the substrate in the active-site cleft of the enzyme, as viewed in Fig. 16.)   The left-side complex is the most stable one found for $(NAG)_6$; its F site includes Arg 45, Asn 46, and Thr 47, and residue D is near the surface of the active site, away from Glu 35 and Asp 52.   This position of the D ring is compatible with experimental results of Schindler et al.[126]   Actually, two right-side complexes were found.   One has an undistorted D residue, which is close to Glu 35 and Asp 52, with the F site including Phe 34 and Arg 114 and nonoptimal contacts in the E and F sites.   The other is similar to this right-side complex but has a distorted (half-chair) D ring and good contacts in the E and F sites; it is similar to one obtained earlier by model building.[125]   While the left-side complex appears to be the most stable one for binding, the other two may play a role in recognition, since three productive binding modes have been observed in stopped-flow and relaxation studies of the binding of $(NAG)_6$ to lysozyme.[127]

The N-acetyl group of NAG provides a considerable fraction of the binding energy and may account for the preference of the enzyme for binding $(NAG)_6$ compared to the corresponding glucose hexamer.   A very

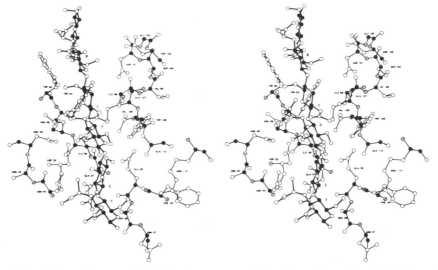

Fig. 16.   Stereo ORTEP view of the lowest energy right-side conformation for binding of $(NAG\text{-}NAM)_3$ to lysozyme (M. R. Pincus and H. A. Scheraga, *Biochemistry*, in press).

important lesson learned from this and other studies is that it is practically impossible to pinpoint one or two specific interactions as being the most significant for the recognition process.   Instead, the binding energy that leads to recognition and specificity is a sum over small contributions from many nonbonded interactions.

These calculations have been extended to the binding of NAG/NAM copolymers to the active site of lysozyme (M. R. Pincus and H. A. Scheraga, *Biochemistry*, in press).   As with the homopolymer (NAG)$_6$, the hexasaccharide (NAG-NAM)$_2$–(NAG)$_2$ binds preferentially on the left side of the active-site cleft.   The alternating copolymer (NAG-NAM)$_3$, however, binds with its F-site residue preferentially on the right side of the active-site cleft (see Fig. 16).   The lactic acid side chain prevents good binding to the F site on the left side.

Recently, an x-ray structure of the complex with the trisaccharide NAM-NAG-NAM has been reported.[128]   The computed and observed complexes appear to be similar, showing binding in sites B, C, and D (Fig. 17), and movements of Trp 62 and Trp 63 toward the substrate and Trp 108 away from it, upon binding.   Further, both methods show that the ring NH of Trp 63 forms a hydrogen bond to the C$=$O of NAG in site C and that the D ring is near the surface of the active-site cleft.

CONCLUDING REMARKS

A variety of conformational problems in peptide and protein chemistry have been treated by the computational methods described here.   Reasonable potential functions are available, and current research should lead

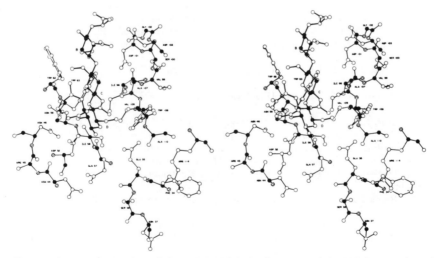

Fig. 17. Stereo ORTEP view of the energy-minimized x-ray structure of the complex of NAM-NAG-NAM with lysozyme (M. R. Pincus and H. A. Scheraga, *Biochemistry*, in press).

to an improved treatment of hydration effects.   The multiple-minimum problem has been surmounted for various types of structures, except globular proteins, but current approaches, involving the use of various constraints, may lead to a solution of this problem.   The agreement between computed and experimental structures and properties, e.g., helix screw sense, helix–coil transition parameters, gramicidin S, poly(Gly-Pro-Pro), is encouraging and demonstrates the power of empirical potential-energy methods in conformational analysis.   It remains to be seen how well the computed structure of α-lactalbumin will compare with experiment, whether the computed structures of lysozyme–hexasaccharide complexes will be confirmed experimentally, and whether the multiple-minimum problem can indeed be surmounted for a globular protein.   One of the most important results from the computational approach is the insight that it has provided as to how interatomic interactions influence the structure and stability of peptides and proteins.

This research has been supported by research grants from the National Science Foundation (PCM 79-20279, PCM 77-09104, and PCM 79-18336) and from the National Institute of General Medical Sciences of the National Institutes of Health, U.S. Public Health Service (GM-14312 and GM-25138).

References

1. Némethy, G. & Scheraga, H. A. (1977) *Q. Rev. Biophys.* **10,** 239–352.
2. Wako, H. & Scheraga, H. A. (1981) *Macromolecules*, in press.
3. Némethy, G. & Scheraga, H. A. (1965) *Biopolymers* **3,** 155–184.
4. Scheraga, H. A., Leach, S. J., Scott, R. A. & Némethy, G. (1965) *Discuss. Faraday Soc.* **40,** 268–277.
5. Leach, S. J., Némethy, G. & Scheraga, H. A. (1966) *Biopolymers* **4,** 369–407.
6. Leach, S. J., Némethy, G. & Scheraga, H. A. (1966) *Biopolymers* **4,** 887–904.
7. Scott, R. A. & Scheraga, H. A. (1966) *Biopolymers* **4,** 237–238.
8. Scott, R. A. & Scheraga, H. A. (1966) *J. Chem. Phys.* **44,** 3054–3069.
9. Gibson, K. D. & Scheraga, H. A. (1966) *Biopolymers* **4,** 709–712.
10. Momany, F. A., McGuire, R. F., Burgess, A. W. & Scheraga, H. A. (1975) *J. Phys. Chem.* **79,** 2361–2381.
11. Shipman, L. L., Burgess, A. W. & Scheraga, H. A. (1975) *Proc. Natl. Acad. Sci. USA* **72,** 543–547.
12. Burgess, A. W., Shipman, L. L. & Scheraga, H. A. (1975) *Proc. Natl. Acad. Sci. USA* **72,** 854–858.
13. Snir, J., Nemenoff, R. A. & Scheraga, H. A. (1978) *J. Phys. Chem.* **82,** 2497–2530.
14. Gibson, K. D. & Scheraga, H. A. (1967) *Proc. Natl. Acad. Sci. USA* **58,** 420–427.
15. Hodes, Z. I., Némethy, G. & Scheraga, H. A. (1979) *Biopolymers* **18,** 1565–1610.
16. Tanaka, S. & Scheraga, H. A. (1975) *Macromolecules* **8,** 516–521.
17. De Santis, P., Giglio, E., Liquori, A. M. & Ripamonti, A. (1963) *J. Polym. Sci.* **A1,** 1383–1404.
18. Ramachandran, G. N., Ramakrishnan, C. & Sasisekharan, V. (1963) *J. Mol. Biol.* **7,** 95–99.
19. Brant, D. A. & Flory, P. J. (1965) *J. Am. Chem. Soc.* **87,** 2791–2800.
20. Pottle, C., Pottle, M. S., Tuttle, R. W., Kinch, R. J. & Scheraga, H. A. (1980) *J. Comput. Chem.* **1,** 46–58.
21. Dunfield, L. G., Burgess, A. W. & Scheraga, H. A. (1978) *J. Phys. Chem.* **82,** 2609–2616.
22. Pincus, M. R. & Scheraga, H. A. (1977) *J. Phys. Chem.* **81,** 1579–1583.

23. Scheraga, H. A. (1977), in *Peptides: Proceedings of the Fifth American Peptide Symposium*, Goodman, M. & Meienhofer, J., Eds., Wiley, New York, pp. 246–256.

24. Scheraga, H. A. (1979) *Acc. Chem. Res.* **12**, 7–14.

25. Paterson, Y., Némethy, G. & Scheraga, H. A. (1981) *Ann. NY Acad. Sci.* **367**, 132–150.

26. Némethy, G., Peer, W. J. & Scheraga, H. A. (1981) *Annu. Rev. Biophys. Bioeng.* **10**, 459–497.

27. Owicki, J. C. & Scheraga, H. A. (1977) *J. Am. Chem. Soc.* **99**, 7403–7418.

28. Owicki, J. C. & Scheraga, H. A. (1978) *J. Phys. Chem.* **82**, 1257–1264.

29. Rapaport, D. C. & Scheraga, H. A. (1981) *Chem. Phys. Lett.* **78**, 491–494.

30. Steinberg, I. Z. & Scheraga, H. A. (1963) *J. Biol. Chem.* **238**, 172–181.

31. Gibson, K. D. & Scheraga, H. A. (1969) *Physiol. Chem. Phys.* **1**, 109–126.

32. Gibson, K. D. & Scheraga, H. A. (1969) *Proc. Natl. Acad. Sci. USA* **63**, 9–15.

33. Gibson, K. D. & Scheraga, H. A. (1969) *Proc. Natl. Acad. Sci. USA* **63**, 242–245.

34. Gō, N. & Scheraga, H. A. (1969) *J. Chem. Phys.* **51**, 4751–4767.

35. Gō, N. & Scheraga, H. A. (1976) *Macromolecules* **9**, 535–542.

36. Gibson, K. D. & Scheraga, H. A. (1970) *Comput. Biomed. Res.* **3**, 375–384.

37. Crippen, G. M. & Scheraga, H. A. (1969) *Proc. Natl. Acad. Sci. USA* **64**, 42–49.

38. Crippen, G. M. & Scheraga, H. A. (1971) *Arch. Biochem. Biophys.* **144**, 453–466.

39. Crippen, G. M. & Scheraga, H. A. (1973) *J. Comput. Phys.* **12**, 491–497.

40. Simon, I., Némethy, G. & Scheraga, H. A. (1978) *Macromolecules* **11**, 797–804.

41. Gō, N. & Scheraga, H. A. (1970) *Macromolecules* **3**, 178–187.

42. Gō, N. & Scheraga, H. A. (1973) *Macromolecules* **6**, 273–281.

43. Scott, R. A. & Scheraga, H. A. (1966) *J. Chem. Phys.* **45**, 2091–2101.

44. Ooi, T., Scott, R. A., Vanderkooi, G. & Scheraga, H. A. (1967) *J. Chem. Phys.* **46**, 4410–4426.

45. Yan, J. F., Vanderkooi, G. & Scheraga, H. A. (1968) *J. Chem. Phys.* **49**, 2713–2726.

46. Yan, J. F., Momany, F. A. & Scheraga, H. A. (1970) *J. Am. Chem. Soc.* **92**, 1109–1115.

47. Erenrich, E. H., Andreatta, R. H. & Scheraga, H. A. (1970) *J. Am. Chem. Soc.* **92**, 1116–1119.

48. Zimm, B. H. & Bragg, J. K. (1959) *J. Chem. Phys.* **31**, 526–535.

49. Gō, N., Gō, M. & Scheraga, H. A. (1968) *Proc. Natl. Acad. Sci. USA* **59**, 1030–1037.

50. Gō, M., Gō, N. & Scheraga, H. A. (1970) *J. Chem. Phys.* **52**, 2060–2079.

51. Gō, M., Gō, N. & Scheraga, H. A. (1971) *J. Chem. Phys.* **54**, 4489–4503.

52. Gō, M., Hesselink, F. T., Gō, N. & Scheraga, H. A. (1974) *Macromolecules* **7**, 459–467.

53. Alter, J. E., Andreatta, R. H., Taylor, G. T. & Scheraga, H. A. (1973) *Macromolecules* **6**, 564–570.

54. Chang, M. C., Fredrickson, R. A., Powers, S. P. & Scheraga, H. A. (1981) *Macromolecules*, in press.

55. Némethy, G. & Scheraga, H. A. (1962) *J. Phys. Chem.* **66**, 1773–1789.

56. Hesselink, F. T., Ooi, T. & Scheraga, H. A. (1973) *Macromolecules* **6**, 541–552.

57. Tanaka, S. & Scheraga, H. A. (1975) *Macromolecules* **8**, 494–503.

58. Tanaka, S. & Scheraga, H. A. (1975) *Macromolecules* **8**, 504–516.

59. Ganser, V., Engel, J., Winklmair, D. & Krause, G. (1970) *Biopolymers* **9**, 329–352.

60. McGuire, R. F., Vanderkooi, G., Momany, F. A., Ingwall, R. T., Crippen, G. M., Lotan, N., Tuttle, R. W., Kashuba, K. L. & Scheraga, H. A. (1971) *Macromolecules* **4**, 112–124.

61. Fu, Y. C., McGuire, R. F. & Scheraga, H. A. (1974) *Macromolecules* **7**, 468–480.

62. Takeda, Y., Iitaka, Y. & Tsuboi, M. (1970) *J. Mol. Biol.* **51**, 101–113.

63. Venkatachalam, C. M. (1968) *Biopolymers* **6**, 1425–1436.

64. Lewis, P. N., Momany, F. A. & Scheraga, H. A. (1971) *Proc. Natl. Acad. Sci. USA* **68**, 2293–2297.

65. Lewis, P. N., Momany, F. A. & Scheraga, H. A. (1973) *Biochim. Biophys. Acta* **303**, 211–229.

66. Nishikawa, K., Momany, F. A. & Scheraga, H. A. (1974) *Macromolecules* 7, 797–806.
67. Zimmerman, S. S., Shipman, L. L. & Scheraga, H. A. (1977) *J. Phys. Chem.* 81, 614–622.
68. Zimmerman, S. S. & Scheraga, H. A. (1977) *Biopolymers* 16, 811–843.
69. Zimmerman, S. S. & Scheraga, H. A. (1978) *Biopolymers* 17, 1849–1869.
70. Némethy, G. & Scheraga, H. A. (1980) *Biochem. Biophys. Res. Commun.* 95, 320–327.
71. Isogai, Y., Némethy, G., Rackovsky, S., Leach, S. J. & Scheraga, H. A. (1980) *Biopolymers* 19, 1183–1210.
72. Rackovsky, S. & Scheraga, H. A. (1980) *Macromolecules* 13, 1440–1453.
73. Némethy, G., McQuie, J. R., Pottle, M. S. & Scheraga, H. A. (1981) *Macromolecules*, in press.
74. Deslauriers, R., Evans, D. J., Leach, S. J., Meinwald, Y., Minasian, E., Némethy, G., Rae, I. D., Scheraga, H. A., Somorjai, R. L., Stimson, E. R., van Nispen, J. W. & Woody, R. W. (1981) *Macromolecules*, in press.
75. Maxfield, F. R., Bandekar, J., Krimm, S., Evans, D. J., Leach, S. J., Némethy, G. & Scheraga, H. A. (1981) *Macromolecules*, in press.
76. Bandekar, J., Evans, D. J., Krimm, S., Leach, S. J., Lee, S., McQuie, J. R., Minasian, E., Némethy, G., Pottle, M. S., Scheraga, H. A., Stimson, E. R. & Woody, R. W. (1981) *Macromolecules*, in press.
77. Dygert, M., Gō, N. & Scheraga, H. A. (1975) *Macromolecules* 8, 750–761.
78. Rackovsky, S. & Scheraga, H. A. (1978) *Macromolecules* 11, 1168–1174.
79. Rackovsky, S. & Scheraga, H. A. (1980) *Proc. Natl. Acad. Sci. USA* 77, 6965–6967.
80. Hull, S. E., Karlsson, R., Main, P., Woolfson, M. M. & Dodson, E. J. (1978) *Nature* 275, 206–207.
81. Crick, F. H. C. (1953) *Acta Crystallogr.* 6, 685–689.
82. Nishikawa, K. & Scheraga, H. A. (1976) *Macromolecules* 9, 395–407.
83. Miller, M. H. & Scheraga, H. A. (1976) *J. Polym. Sci., Polym. Symp.* 54, 171–200.
84. Miller, M. H., Némethy, G. & Scheraga, H. A. (1980) *Macromolecules* 13, 470–478.
85. Miller, M. H., Némethy, G. & Scheraga, H. A. (1980) *Macromolecules* 13, 910–913.
86. Némethy, G., Miller, M. H. & Scheraga, H. A. (1980) *Macromolecules* 13, 914–919.
87. Okuyama, K., Tanaka, N., Ashida, T. & Kakudo, M. (1976) *Bull. Chem. Soc. Jpn.* 49, 1805–1810.
88. Scheraga, H. A. (1971) *Chem. Rev.* 71, 195–217.
89. Scheraga, H. A. (1980) in *Protein Folding*, Jaenicke, R., Ed., Elsevier, Amsterdam, pp. 261–288.
90. Gō, N. & Scheraga, H. A. (1978) *Macromolecules* 11, 552–559.
91. Rapaport, D. C. & Scheraga, H. A. (1981) *Macromolecules*, in press.
92. Dunfield, L. G. & Scheraga, H. A. (1980) *Macromolecules* 13, 1415–1428.
93. Chavez, L. G. & Scheraga, H. A. (1980) *Biochemistry* 19, 996–1004.
94. Chavez, L. G. & Scheraga, H. A. (1980) *Biochemistry* 19, 1005–1012.
95. Matheson, R. R. & Scheraga, H. A. (1978) *Macromolecules* 11, 819–829.
96. Brandts, J. F., Halvorson, H. R. & Brennan, M. (1975) *Biochemistry* 14, 4953–4963.
97. Poland, D. C. & Scheraga, H. A. (1965) *Biopolymers* 3, 315–334.
98. Tanaka, S. & Scheraga, H. A. (1977) *Macromolecules* 10, 291–304.
99. Finkelstein, A. V. & Ptitsyn, O. B. (1976) *J. Mol. Biol.* 103, 15–24.
100. Némethy, G. & Scheraga, H. A. (1979) *Proc. Natl. Acad. Sci. USA* 76, 6050–6054.
101. Ponnuswamy, P. K., Warme, P. K. & Scheraga, H. A. (1973) *Proc. Natl. Acad. Sci. USA* 70, 830–833.
102. Meirovitch, H., Rackovsky, S. & Scheraga, H. A. (1980) *Macromolecules* 13, 1398–1405.
103. Meirovitch, H. & Scheraga, H. A. (1980) *Macromolecules* 13, 1406–1414.
104. Levitt, M. & Chothia, C. (1976) *Nature* 261, 552–558.

105. Chothia, C., Levitt, M. & Richardson, D. (1977) *Proc. Natl. Acad. Sci. USA* **74,** 4130–4134.
106. Sternberg, M. J. E. & Thornton, J. M. (1977) *J. Mol. Biol.* **115,** 1–17.
107. Richmond, T. J. & Richards, F. M. (1978) *J. Mol. Biol.* **119,** 537–555.
108. Cohen, F. E., Sternberg, M. J. E. & Taylor, W. R. (1980) *Nature* **285,** 378–382.
109. Chothia, C. (1973) *J. Mol. Biol.* **75,** 295–302.
110. Crippen, G. M. (1975) *J. Comput. Phys.* **18,** 224–231.
111. Havel, T. F., Crippen, G. M. & Kuntz, I. D. (1979) *Biopolymers* **18,** 73–81.
112. Kuntz, I. D., Crippen, G. M. & Kollmann, P. A. (1979) *Biopolymers* **18,** 939–957.
113. Ycas, M., Goel, N. S. & Jacobsen, J. W. (1978) *J. Theor. Biol.* **72,** 443–457.
114. Goel, N. S. & Ycas, M. (1979) *J. Theor. Biol.* **77,** 253–305.
115. Swenson, M. K., Burgess, A. W. & Scheraga, H. A. (1978) in *Frontiers in Physico-chemical Biology*, Pullman, B., Ed., Academic Press, pp. 115–142.
116. Warme, P. K. & Scheraga, H. A. (1974) *Biochemistry* **13,** 757–767.
117. Fitzwater, S. & Scheraga, H. A. (1980) *Acta Crystallogr., Sect. A* **36,** 211–219.
118. Deisenhofer, J. & Steigemann, W. (1975) *Acta Crystallogr., Sect. B* **31,** 238–250.
119. Warme, P. K., Momany, F. A., Rumball, S. V., Tuttle, R. W. & Scheraga, H. A. (1974) *Biochemistry* **13,** 768–782.
120. Platzer, K. E. B., Momany, F. A. & Scheraga, H. A. (1972) *Int. J. Pept. Protein Res.* **4,** 187–219.
121. Pincus, M. R., Burgess, A. W. & Scheraga, H. A. (1976) *Biopolymers* **15,** 2485–2521.
122. Pincus, M. R., Zimmerman, S. S. & Scheraga, H. A. (1976) *Proc. Natl. Acad. Sci. USA* **73,** 4261–4265.
123. Pincus, M. R., Zimmerman, S. S. & Scheraga, H. A. (1977) *Proc. Natl. Acad. Sci. USA* **74,** 2629–2633.
124. Pincus, M. R. & Scheraga, H. A. (1979) *Macromolecules* **12,** 633–644.
125. Imoto, T., Johnson, L. N., North, A. C. T., Phillips, D. C. & Rupley, J. A. (1972) in *The Enzymes*, 3rd ed., Vol. 7, Boyer, P. D., Ed., Academic Press, New York, pp. 665–868.
126. Schindler, M., Assaf, Y., Sharon, N. & Chipman, D. M. (1977) *Biochemistry* **16,** 423–431.
127. Holler, E., Rupley, J. A. & Hess, G. P. (1975) *Biochemistry* **14,** 2377–2385.
128. Kelly, J. A., Sielecki, A. R., Sykes, B. D., James, M. N. G. & Phillips, D. C. (1979) *Nature* **282,** 875–878.

Received February 18, 1981
Accepted March 3, 1981

Author's Note:

For an updated view of this, see my article in *Chemica Scripta,* 29a, 3–13 (1989).

Theory of the Phase Transition between Helix and Random Coil in Polypeptide Chains

B. H. Zimm and J. K. Bragg

General Electric Research Laboratory, Schenectady, New York

(Received March 5, 1959)

The transition between the helical and randomly coiled forms of a polypeptide chain is discussed by reference to a simple model that allows bonding only between each group and the third preceding. Two principal parameters are introduced, a statistical parameter that is essentially an equilibrium constant for the bonding of segments to a portion of the chain that is already in helical form, and a special correction factor for the initiation of a helix. A third parameter which specifies the minimum number of segments in a random section between two helical portions has only a minor effect on the results. The partition function for this model is handled in two alternative ways, either as a summation suitable for short chains, or in terms of the eigenvalues and eigenvectors of a characteristic matrix; the latter is more suitable for long chains. A transition from the random to the helical form is encountered as either the bonding parameter or the chain length is increased. The critical value of the bonding parameter is unity for long chains, while the sharpness of the transition depends on the initiation parameter.

Depending on the values of the bonding parameter and the chain length, one of the following configurations dominates: random coils, single helices with occasional disorder at the ends, and for longer chains, helices occasionally broken by random sections. In rather narrow transition regions, mixtures of these forms may be found. A diagram is given that displays the relationships of these forms.

The theory is compared with published data on polybenzyl-glutamate with fair agreement.

I. INTRODUCTION

DOTY, Blout, and co-workers[1] have recently found that polypeptide chains in solution can be reversibly converted from the randomly coiled form to the α helix of Pauling *et al.*[2] The transformation is remarkably sharp. A change of a few degrees in temperature or a few percent in solvent composition is sufficient to complete it, and it seems fully to merit the term "phase transition" that has been applied to it.[1(b)]

The polypeptide chain consists of amide groups

connected by intermediatry carbon atoms. In the α helix, the hydrogen atom of each amide group forms a hydrogen bond with the oxygen atom of the third preceding amide group. We shall refer to an amide group plus one adjacent carbon as a *segment* of the chain.

The amide group is a rigid planar structure. However, according to Pauling *et al.*[2] there is some freedom of rotation about the bonds to the adjacent carbon atoms. Therefore, if the hydrogen bonds are broken, the chain can assume the randomly coiled configurations usual to chain polymers.

Since there can be hardly any doubt that the transformation would occur in a single, isolated chain, we have the novelty of a rather sharp transition in a one-dimensional system. Transformations in other such systems, such as the one-dimensional ferromagnet,[3,4] are quite diffuse. Furthermore, it has been shown[5] that different macroscopic phases cannot coexist in a one-dimensional system.

The transition is of obvious importance to the full understanding of the formation and stability of proteins. The construction of a theory should therefore be interesting from several points of view.

This paper presents a simple model of the chain that facilitates calculation of the dependence of the partition function on the hydrogen bonding. This model gives a rather sharp transition from the random to the helical form as the strength of the hydrogen bonds is increased beyond a critical value, in agreement with the experimental observations.

The sharpness of the transition is due to the following consequence of the model. The formation of the first turn of the helix is difficult because of a large reduction of entropy. Once formed, however, this turn acts as a nucleus to which further turns can add by hydrogen bonding. Thus this transformation has the property of nucleation characteristic of other sharp transitions.

Associated with the tendency to nucleate is a property that might be called a boundary tension. That is, such faults in the helical structure as exist tend to consist of a number of missing hydrogen bonds at adjacent segments, rather than of missing bonds distributed at random. Further, disorder is propagated inward from the ends of the helix, in a way similar to the inward propagation of disorder from the surface of a crystal lattice.

[1] (a) Doty, Holtzer, Bradbury, and Blout, J. Am. Chem. Soc. **76**, 4493 (1954); (b) P. Doty and J. T. Yang, *ibid.* **78**, 498 (1956); (c) Doty, Bradbury, and Holtzer, *ibid.* **78**, 947 (1956); (d) E. R. Blout and A. Asadourian, *ibid.* **78**, 955 (1956); (e) P. Doty and R. D. Lundberg, *ibid.* **78**, 4810 (1956); (f) P. Doty and K. Iso (private communication).

[2] Pauling, Corey, and Branson, Proc. Natl. Acad. Sci. U. S. **37**, 205, 241 (1951).

[3] E. Ising, Z. Physik **31**, 253 (1925).

[4] H. A. Kramers and G. H. Wannier, Phys. Rev. **60**, 252, 263 (1941).

[5] L. Landau and E. Lifshitz, *Statistical Physics* (Oxford University Press, New York, 1938), p. 232.

Reprinted from *Journal of Chemical Physics* **31**, 526–535 (1959); © American Institute of Physics.

Schellman[6] has presented for this transformation a very simple theory, in which he considers special effects at the ends of the helix, equivalent to our nucleation, but ignores the possibility of alternation between helix and coil in the middle of the chain. Further, the transition appears in the theory as perfectly sharp, although dependent upon chain length. Our theory, described below, differs in that it gives a diffuse transition but confirms the dependence on chain length. In addition, it yields a convenient description of the alternation of helical and coiling regions which is important under some circumstances with long chains.

The treatment in this paper differs in another way from that of Schellman. The latter uses the heat and the entropy of adding a segment to the helix as his basic parameters. We prefer to employ two statistical parameters, one for the nucleation of the helix and one for its further growth. While both methods are equally correct, the expression of the results appears to be somewhat more direct in terms of the statistical parameters.

II. THE MODEL

This section presents a simple model of the chain that is intended to represent the significant physical features of the system, and at the same time is amenable to evaluation by simple means. Specifically, the model distinguishes between the contribution of a bonded segment and of an unbonded segment to the partition function, and additionally considers the influence of the state of neighboring segments on these contributions. To describe the model in detail, we first have to establish a notation for configurations of the chain.

It is convenient to base the description of the chain on the helical configuration. We assume that a given state of the chain can be completely described by the state of the oxygen atoms alone; i.e., by a statement as to whether or not each is bonded to the hydrogen of the third preceding segment. This amounts to assuming that bonding of a segment, if it occurs at all, is always to the third preceding segment. The state of a chain of n segments can then be described by a sequence of $n-3$ symbols, each of which can have one of two values. We establish the convention that the *first* three segments are always unbonded. This amounts to selecting as the "beginning" of the chain that end of the helix that has three unbonded oxygen atoms. If the digit 1 represents a bonded oxygen atom and 0 an unbonded atom, then a state of the chain is described by a sequence such as

$$000111000011\cdots.$$

Since our object is the writing down of a partition function, we must now make some specific assumptions about the statistical weights to be attached to particular states. Our concern lies primarily with the thermodynamics associated with the transition from

random coil to helix; hence, it is not necessary to describe the quantum states or phase space of the individual segments in detail, as long as the relative weights of the random and helical forms are correctly represented. The following simple set of assumptions about the relative weights appears to be adequate.

The statistical weight* of a given state of the chain is assumed to be the product of the following factors:

(1) The quantity unity for every 0 that appears (unbonded segment).
(2) The quantity s for every 1 that follows a 1 (bonded segment).
(3) The quantity σs for every 1 that follows μ or more 0's (boundary between bonded and unbonded regions).
(4) The quantity 0 for every 1 that follows a number of 0's less than μ.

The effect of assumption (4) is that sequences of less than μ zeros do not appear. For the α helix, μ is usually considered to be about three.[6] The meaning of the first three weights is as follows. The factor unity is arbitrarily assigned as the statistical weight of a segment when it is not bonded into the helix. The factor s measures the contribution to the partition function of a bonded segment relative to that of an unbonded segment. This factor contains a decrease in statistical weight owing to restriction of freedom of motion, but is enhanced by the Boltzmann factor resulting from the bond energy. Finally, an abnormally large decrease in statistical weight is assumed to be caused by the first bond after μ or more unbonded segments since such a bond decreases the freedom of the segments intervening between the bonding oxygen and hydrogen, as well as restricting the freedom of the bonding segment itself. Since the same Boltzmann factor is involved, this contribution to the partition function is written σs, where σ is less than unity.

These assumptions constitute a highly simplified representation of the problem. The formalism is capable of dealing with more detailed assumptions without undue difficulty, but our present knowledge is too incomplete to justify a more refined model. For example, one might introduce a set of σ's, $\sigma(k)$, to give the decrease in statistical weight due to a bond following k unbonded segments. The plot of $\sigma(k)$ versus k would be expected to look like the curve shown in Fig. 1. The assumption of a single value of σ, and the ban on sequences of less than μ 0's, is the approximation shown by the dotted line.

There are two nonrigid bonds in each segment.[2] If the degree of restriction of the phase space of each on entering the helix is r, then we should have s proportional to r^2. Similarly, if the formation of a bond

---

[6] J. A. Schellman, Compt. rend. trav. lab. Carlsberg, Ser. chim. **29**, No. 15 (1955).

* By *statistical weight* we mean the factor that a segment contributes to the partition function, including, if appropriate, a Boltzmann factor; it is not just the number of quantum states, in contrast to one popular usage.

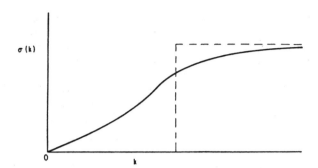

Fig. 1. The weighting factor $\sigma(k)$ for the initiation of a helix after k unbonded segments (schematic). The approximation used in this paper is represented by the dashed line.

restricts to the same degree the segments intervening between the oxygen and hydrogen atoms of the bond, we should have σs proportional to r^6 with the same constant, since there are six nonrigid bonds per turn. Then we have $\sigma = r^4$. Since r can hardly be greater than about $\frac{1}{3}$, σ must be of the order of 10^{-2} or less. The result is that the first turn of the helix can only be formed with difficulty.

We neglect several other possible effects that might have to be considered in a complete treatment. For example, one might introduce a correlation between the statistical weights for hydrogen bonding in one turn of the helix with the presence or absence of bonds in the preceding turn. This has been considered by Hill,[7] but without the correlation between successive bonds in the same turn of the helix on which we base the present paper. In our view the interactions between successive turns is likely to be of secondary importance compared to the interactions within a turn. We also neglect the possibility of hydrogen bonding to other than the segments characteristic of the alpha helix. This phenomenon would not be expected to occur except when the alpha helix was unstable, but under these conditions very few hydrogen bonds would form anyway; therefore we do not believe that the phenomenon is ever of major importance. Further, we assume that only helices of one sense, right- or left-handed, can form with a given chain; this seems to be in accord with experiment for all polypeptides bearing

side groups, where the side group interaction is apparently strong enough to establish a preference for one handedness over the other. Finally, we make no explicit reference to interactions between the side groups; to a considerable extent the effects of these interactions can be included in the parameter s.

III. MATHEMATICAL TREATMENT

Direct Derivation of the Partition Function and the Probability of Bonding

A formal representation of the partition function Q for a chain of n segments may be obtained from the above assumptions by direct enumeration of the number of ways of arranging a given number of zeros and ones in a chain always starting with three zeros. For example, with μ taken as unity, we have obtained the formula

$$Q = 1 + \sum_{l}^{(n-2)/2} \sigma^l \sum_{k}^{n-l-2} \frac{(k-1)!(n-k-2)!s^k}{l!(l-1)!(k-l)!(n-k-l-2)!}, \quad (1)$$

where $(n-2)/2$ is the largest integer less than $(n-2)/2$. Though this formula does not appear to be attractive for calculation in general, it is useful when the product of $n\sigma$ is small and s is appreciably greater than unity, since then only the first term of the summation over l is important. It will be shown that for rather short chains these are just the conditions under which the helix is formed. Physically, this corresponds to conditions under which only one helical section (unbroken sequence of 1's) would be expected.

It is easy to show that the expression $(d \ln Q/d \ln s)$ is the average number of hydrogen bonds formed in the chain at a given value of s, since the number of hydrogen bonds in any state is equal to the power of s in the corresponding term in the partition function. We define θ as the fraction of possible hydrogen bonds formed,

$$\theta = \frac{1}{(n-3)} \frac{d \ln Q}{d \ln s}. \quad (2)$$

Then, keeping only the first term of the summation over l, we get from Eq. (1),

$$\theta = \frac{\sum_{1}^{n-3} k(n-k-2)s^k}{(n-3)\left[1/\sigma + \sum_{1}^{n-3} k(n-k-2)s^k\right]}, \quad (3a)$$

$$= \frac{(n-3)(s-1) - 2 + [(n-3)(s-1)+2s]s^{-n+2}}{(n-3)(s-1)\{1+(s-1)^2s^{-n+1}/\sigma - [(n-3)(s-1)+s]s^{-n+2}\}}. \quad (3b)$$

These formulas are useful for calculation for small values of n. Since they are valid only when there is one unbroken helical sequence in the chain, they are inde-

pendent of the parameter μ which specifies the minimum possible number segments involved in a break in the helix. It appears, however, that any attempt to use all the terms of Eq. (1) would lead to very complicated expressions for large n. Fortunately, other

[7] T. L. Hill, J. Polymer Sci. 23, 549 (1957).

methods are available, and these will be discussed in the following section.

The Matrix Method

There are several well-known methods for evaluating complicated partition functions, *e.g.*, the method of the maximum term and the method of steepest descents, but one method, the matrix method[4,7–9] is particularly well adapted to this chain problem, and to it we shall confine our attention. While the method as applied to infinite chains has been adequately described in the above references, we have found it desirable to extend its scope to include finite chains. For this reason we give a brief discussion of the method.

Formally the method is capable of taking into account interactions between distant segments as well as nearest neighbors. Our formal description will be given for the general case. Much of the later development is in terms of nearest neighbor interactions only ($\mu=1$), but a model which includes certain longer-range effects is also discussed ($\mu=3$).

If the physical situation requires the inclusion of effects between segments whose positions in the chain differ by the integer μ, the matrix method requires the state description of the chain to be in terms of the "joint configuration" of μ successive segments. Following Kramers and Wannier, we use an indexing of states based on binary numbers. To illustrate, let $\mu=3$. The various configurations of a group of three segments are described by sequences of 0's and 1's such as 110. The sequence can be interpreted as a binary number, and we define a single index for the configurations as the value of this binary number. Thus,

State	Index
000	0
001	1
010	2
011	3
· · ·	· · ·

and so on. There are 2^μ possible states for a group of μ segments.

The matrix method involves operations on a *statistical weight vector* $\mathbf{a}_i$. This column vector has 2^μ components, one for each joint configuration of the segments $i-\mu+1$, $i-\mu+2\cdots$, i. Each component, $\mathbf{a}_{i,l}$, is the statistical weight of the lth joint configuration of the segments $i-\mu+1$ through i, including the contributions to the statistical weight of all compatible configurations of the preceding $i-\mu$ segments. The partition function of the chain is just the sum of all the components of $\mathbf{a}_n$, where n is the number of segments in the entire chain.

[8] E. W. Montroll, J. Chem. Phys. **9**, 706 (1941); G. F. Newell and E. W. Montroll, Phys. Rev. **25**, 159 (1953).
[9] E. N. Lasettre and J. P. Howe, J. Chem. Phys. **9**, 747 (1941).

The vectors $\mathbf{a}_1$, $\mathbf{a}_2$, and $\mathbf{a}_3$ are taken to be

$$\mathbf{a}_1=\mathbf{a}_2=\mathbf{a}_3=(1, 0, 0, \cdots, 0), \qquad \mu<4, \qquad (4)$$

since the first three segments cannot bond to preceding hydrogens. If μ is equal to or greater than four, further consideration may be necessary to establish $\mathbf{a}_4, \cdots, \mathbf{a}_\mu$. The succeeding discussion is for $\mu<4$, but the necessary modification for $\mu\geq4$ is formally a minor one.

As long as i is greater than the larger of 3 and μ, the vectors $\mathbf{a}_i$ can be generated by the use of a $2^\mu\times2^\mu$ matrix operator $\mathbf{M}$,

$$\mathbf{a}_i{}^\dagger=\mathbf{M}\mathbf{a}_{i-1}{}^\dagger, \qquad (5)$$

where the symbol † indicates the transposed or column vector. The matrix embodies the physical assumptions of the problem. The element M_{kl} is the factor to be multiplied to the statistical weight upon adding the ith segment, if the segments i through $i-\mu+1$ form joint configuration k while the segments $i-1$ through $i-\mu$ are in joint configuration l. (For μ greater than 1, the matrix will have at most the fraction $2^{1-\mu}$ of its elements nonzero.)

The vector $\mathbf{a}_n$ is given by

$$\mathbf{a}_n{}^\dagger=\mathbf{M}^{n-3}\mathbf{a}_3{}^\dagger, \qquad (6)$$

and the partition function is

$$Q=\omega\mathbf{M}^{n-3}\alpha^\dagger \qquad (7a)$$

$$\alpha=(1, 0, 0, \cdots, 0), \qquad (7b)$$

$$\omega=(1, 1, 1, \cdots, 1). \qquad (7c)$$

Calculations based on Eqs. (7) are relatively easy. If the matrix $\mathbf{M}$ can be diagonalized,

$$\mathbf{\Lambda}=\mathbf{T}^{-1}\mathbf{M}\mathbf{T}, \qquad (8)$$

the diagonal matrix can be easily raised to the required power; the elements of the diagonal matrix $\mathbf{\Lambda}^k$ are the kth powers of the elements of $\mathbf{\Lambda}$. When k is large the kth power of the largest element of $\mathbf{\Lambda}$ is so much greater than the others that it alone needs to be considered; this is the classical case discussed in the references.[4,8,9]

The diagonal elements of $\mathbf{\Lambda}$ are the eigenvalues of $\mathbf{M}$. Corresponding to each eigenvalue are two eigenvectors, a row vector and a column vector, since $\mathbf{M}$ is not symmetrical. The row vector is the eigenvector for $\mathbf{M}$ operating to the left and the column vector for $\mathbf{M}$ operating to the right. The column eigenvectors constitute the columns of $\mathbf{T}$ and the row vectors the rows of $\mathbf{T}^{-1}$.

The matrix $\mathbf{M}$ is unsymmetrical, and there are certain unsymmetrical matrices that cannot be diagonalized. The matrices encountered in our work can in general be diagonalized, although in the limiting case where σ is zero a matrix is formed that cannot be. This limiting case is that of a perfectly sharp transition. However, this limit can be evaluated after the entire computation has been done for finite σ.

Average States of Individual Segments

The model makes it easy to obtain approximations to the state of any given segment of the chain. The statistical weight to be attached to a given joint configuration of the segments $(i-\mu+1)$ through i is given by the sum of the statistical weights of all configurations of the entire chain consistent with the given joint configuration. Now the vector,

$$\mathbf{a}_i{}^\dagger = \mathbf{M}^{i-3}\boldsymbol{\alpha}^\dagger, \qquad (9)$$

has components which are the aggregate statistical weights of the possible joint configurations of segments $i-\mu+1$ through i, taking account also of the preceding part of the chain. In a similar way the row vector,

$$\mathbf{b}_i = \boldsymbol{\omega}\mathbf{M}^{n-i}, \qquad (10)$$

has components which are the aggregate statistical weights provided to each joint configuration by the states available to the remainder of the chain.

The definitions of $\mathbf{a}_i$ and $\mathbf{b}_i$ are such that

$$\mathbf{b}_i \cdot \mathbf{a}_i{}^\dagger = Q, \qquad (11)$$

since Q is the sum of all statistical weights. The probability that segments i through $i-\mu+1$ are in joint configuration l is therefore

$$p_i(l) = (1/Q)b_{i,l}a_{i,l}. \qquad (12)$$

Several special cases of Eq. (12) are of interest. First let us consider the state of a segment near the middle of a long chain. Then, by neglecting large powers of all eigenvalues of $\mathbf{M}$ relative to the same powers of the largest, λ_0, we may express $\mathbf{a}_i$ and $\mathbf{b}_i$ in terms of the principal eigenvectors only:

$$\mathbf{b}_i = \lambda_0{}^{n-i}\left(\sum_{k=0}^{\rho-1} T_{k0}\right)(T_{00}{}^{-1}, T_{01}{}^{-1}, T_{02}{}^{-1}, \cdots, T_{0\rho-1}{}^{-1}),$$

$$\mathbf{a}_i = \lambda_0{}^{i-3}T_{00}{}^{-1}(T_{00}, T_{10}, T_{20}, \cdots, T_{\rho-1,0}), \qquad (13)$$

where $\rho=2^\mu$. Equation (12) then gives the simple expression,

$$p_i(l) = \frac{b_{i,l}a_{i,l}}{\displaystyle\sum_{l=0}^{\rho-1}{}'b_{i,l}a_{i,l}} = T_{0l}{}^{-1}T_{l0}. \qquad (14)$$

If the segment of interest is near the end of a chain, the approximation involving the largest eigenvalue can be applied to the $\mathbf{a}$ vector but not the $\mathbf{b}$ vector. This leads to an expression

$$p_i(l) = \frac{b_{i,l}T_{l0}}{\displaystyle\sum_{l=0}^{\rho-1}{}'b_{i,l}T_{l0}}, \quad i\approx n, \qquad (15)$$

for the probability that the joint configuration is state l. Simplifying the expression further depends on the form of the matrix $\mathbf{M}$; later an approximate form will be given.

It is also of interest to discuss the occurrence of breaks in the helix. Such a break (unbonded section) must be bounded by the configurations 10 and 01, i.e., it is described by a sequence such as $\cdots 100001\cdots$. According to Eq. (14) the probability of the sequence 01 at the $(i-1)$th and ith positions near the middle of the chain, which is equal to the probability of the sequence 10, is

$$P_i(01) = \sum_{l=1}^{\rho-3}{}''p_i(l) = \sum_{l=1}^{\rho-3}{}''T_{0l}{}^{-1}T_{l0}, \qquad (16)$$

where the double-primed sum includes only every fourth term, $l=1, 5, 9, \cdots \rho-3$. This formula applies when $\mu\geq 2$. The special case $\mu=1$ will be discussed at a later point.

The Form of the Operator M

The matrix $\mathbf{M}$ is of order $\rho\times\rho$, where $\rho=2^\mu$. Only certain elements can be nonzero. The assumptions listed above give a matrix of the form illustrated below for the case of $\mu=3$:

$$\mathbf{M}=\begin{pmatrix} 1 & 0 & 0 & 0 & 1 & 0 & 0 & 0 \\ \sigma s & 0 & 0 & 0 & 0 & 0 & 0 & 0 \\ 0 & 1 & 0 & 0 & 0 & 1 & 0 & 0 \\ 0 & s & 0 & 0 & 0 & s & 0 & 0 \\ 0 & 0 & 1 & 0 & 0 & 0 & 1 & 0 \\ 0 & 0 & 0 & 0 & 0 & 0 & 0 & 0 \\ 0 & 0 & 0 & 1 & 0 & 0 & 0 & 1 \\ 0 & 0 & 0 & s & 0 & 0 & 0 & s \end{pmatrix}. \qquad (17)$$

It can be shown that the characteristic equation of the matrix,

$$|M-I\lambda|=0, \qquad (18)$$

where $\mathbf{I}$ is the unit matrix, can be reduced in this case to

$$\lambda^{\mu-1}(1-\lambda)(s-\lambda)=\sigma s. \qquad (19)$$

Furthermore, the trace is $1+s$, independent of μ.

Since σ is small, approximations to the roots of $\mathbf{M}$ are 1 and s with $\mu-1$ very small roots, provided $\sigma^{1/\mu}$ is less than unity. Therefore the nature of the partition function, which depends mainly on the large eigenvalues, will be to a large extent independent of μ. This being the case, we shall illustrate in most detail the case $\mu=1$. In a later section we shall give some of the results of the case $\mu=3$. In the case $\mu=1$ the steps described in the foregoing are especially simple.

The matrix $\mathbf{M}$, for $\mu=1$, is

$$\mathbf{M}=\begin{pmatrix} 1 & 1 \\ \sigma s & s \end{pmatrix}. \qquad (20)$$

The characteristic equation is

$$(1-\lambda)(s-\lambda) = \sigma s, \qquad (21)$$

and the roots are

$$\lambda = \tfrac{1}{2}\{1 + s \pm [(1-s)^2 + 4\sigma s]^{\frac{1}{2}}\}. \qquad (22)$$

We designate the larger of these by λ_0, the smaller by λ_1. The transformation that diagonalizes $\mathbf{M}$ is

$$\mathbf{T} = \begin{pmatrix} 1 & 1 \\ \lambda_0 - 1 & \lambda_1 - 1 \end{pmatrix}, \qquad (23)$$

$$\mathbf{T}^{-1} = \frac{1}{\lambda_1 - \lambda_0} \begin{pmatrix} \lambda_1 - 1 & -1 \\ -\lambda_0 + 1 & 1 \end{pmatrix}. \qquad (24)$$

The expression for the partition function, Eq. (7a), is

$$Q = (1,1)\mathbf{M}^{n-3}(1,0)^{\dagger} = (1,1)\mathbf{T}\mathbf{\Lambda}^{n-3}\mathbf{T}^{-1}(1,0)^{\dagger}, \qquad (25)$$

which becomes, using Eq. (22) and the fact that $\lambda_0 + \lambda_1 = 1 + s$,

$$Q = \frac{\lambda_0^{n-2}(\lambda_0 - s) + \lambda_1^{n-2}(s - \lambda_1)}{\lambda_0 - \lambda_1}. \qquad (26)$$

According to Eq. (2), the average number of hydrogen bonds is

$$\theta = [s/(n-3)]\{[(n-2)\lambda_0'/\lambda_0 + (\lambda_0' - 1)/(\lambda_0 - s)]\lambda_0^{n-2}(\lambda_0 - s)$$

$$+ [(n-2)\lambda_1'/\lambda_1 + (1 - \lambda_1')/(s - \lambda_1)]\lambda_1^{n-2}(s - \lambda_1)\}/[\lambda_0^{n-2}(\lambda_0 - s) + \lambda_1^{n-2}(s - \lambda_1)] - \left(\frac{s}{n-3}\right)\left(\frac{\lambda_0' - \lambda_1'}{\lambda_0 - \lambda_1}\right), \qquad (27)$$

where the prime denotes differentiation by s.

In order to discuss the state of a particular bond, say one in the middle or near an end of the chain, we need the vectors $\mathbf{a}_i$ and $\mathbf{b}_i$. From Eqs. (13), (23), and (24) these are

$$\mathbf{a}_i = [\lambda_0^{i-3}(\lambda_0 - s)/(\lambda_0 - \lambda_1)](1, \lambda_0 - 1), \qquad (28)$$

$$\mathbf{b}_i = [\lambda_0^{n-i+1}/(\lambda_0 - \lambda_1)](\lambda_0 - s, 1) \qquad (29)$$

to the approximation involved in Eq. (13). Equation (14) gives directly the probability of the state l (0 or 1) at the ith segment near the center of the chain as

$$p_i(l) = T_{0l}^{-1} T_{l0}. \qquad (30)$$

The state of a segment near the end of a long chain is approximately given by Eq. (15). This can be simplified for large s in the present case by actually computing successive powers of the matrix (20). If a small number of these are calculated, and terms involving σ are dropped, it may be seen that the vector $\mathbf{b}_i$, Eq. (10), is approximately

$$\mathbf{b}_i = (1, 1 + s + s^2 + \cdots + s^{n-i}). \qquad (31)$$

Then Eq. (15) yields

$$p_{n-i}(0) = T_{00}/[T_{00} + T_{10}(1 + s + \cdots + s^{n-i})] \qquad (32a)$$

$$= 1/[1 + (\lambda_0 - 1)(1 + s + \cdots + s^{n-i})]. \qquad (32b)$$

But for large s, $\lambda_0 \approx s$, so that we have

$$p_{n-i}(0) = s^{-(n-i+1)}, \quad s > 1. \qquad (33)$$

The end segment has the probability s^{-1} of being unbonded, the next segment the probability s^{-2}, and so on. The mean number of unbonded segments at one end, obtained by summing over the above probabilities, is $1/(s-1)$. These results are valid only for chains long enough so that the two ends do not influence each other. A formula for short chains is given in the next section.

The general formula for computing breaks in the chain, Eq. (16), cannot be used when $\mu = 1$. However, it is easy to show in analogy to Eq. (14) that the probability of a sequence 01 near the middle of the chain, $P_i(01)$, is

$$P_i(01) = T_{00}^{-1} M_{01} T_{10}/\lambda_0, \quad \mu = 1. \qquad (33a)$$

To illustrate these formulas, approximation may be made to the roots of Eq. (21). Since $\sigma \ll 1$, we have the results shown in Table I. Then the probability of an unbonded segment, $p_i(0)$, near the center of the chain is, from Eq. (14),

$$\begin{cases} 1 - \sigma s/(s-1)^2, & s < 1, \\ \tfrac{1}{2}, & s = 1, \\ \sigma s/(s-1)^2, & s > 1. \end{cases} \qquad (34)$$

From Eq. (33a) the probability of a change from bonded to unbonded region, $P_i(01)$, at any given segment near the middle is

$$\begin{cases} \sigma s/(1-s), & s < 1, \\ (\sigma)^{\frac{1}{2}}/2, & s = 1, \\ \sigma/(s-1), & s > 1. \end{cases} \qquad (35)$$

The probability that the last segment of a *short* chain is bonded is of interest in connection with the polymerization kinetics. We can find the formula from Eq. (10) with Eqs. (8), (23), and (24),

$$p_n(1) = \frac{(\lambda_0^{n-3} - \lambda_1^{n-3})(s - \lambda_1)(\lambda_0 - s)}{\lambda_0^{n-2}(\lambda_0 - s) + \lambda_1^{n-2}(s - \lambda_1)}. \qquad (36)$$

TABLE I.

	$s < 1$	$s \approx 1$	$s > 1$
λ_0	$1 + \sigma s/(1-s)$	$(1+s)/2 + \sqrt{\sigma}$	$s + \sigma s/(s-1)$
λ_1	$s - \sigma s/(1-s)$	$(1+s)/2 - \sqrt{\sigma}$	$1 - \sigma s/(s-1)$

This reduces to a simple form if we use the above approximations to the eigenvalues and assume that ns^{-n} is much less than unity:

$$p_n(1) = \frac{\sigma s(s-1)(s^{n-3}-1)}{\sigma s^{n-1}+(s-1)^2}, \quad s>1. \tag{37}$$

When n is large this agrees with Eq. (33).

We note in passing that the above approximate eigenvalues substituted into Eq. (27) for θ yield Eq. (3b) if the first power only of σ is retained.

The Case of $\mu = 3$

The case of $\mu = 3$ is of special interest because it has been assumed[6] that this corresponds to real polypeptide chains. The right-hand and left-hand eigenvectors are respectively:

$$(1, \sigma s/\lambda, \sigma s/\lambda^2, \sigma s^2/\lambda^2, \lambda-1, 0, s(\lambda-1), s^2(\lambda-1))^\dagger, \tag{38a}$$

$$\frac{\lambda(\lambda-s)}{4\lambda^2-3\lambda s-3\lambda+2s}\left(1, \frac{\lambda-1}{\sigma s}, \frac{1}{\lambda^2}, \frac{\lambda-1}{\sigma s}, \frac{1}{\lambda}, \frac{\lambda-1}{\sigma s}, \frac{1}{\lambda^2}, \frac{\lambda-1}{\sigma s}\right). \tag{38b}$$

In these formulas the eigenvalue, λ_0 or λ_1, corresponding to the desired eigenvector is to be inserted for λ. These eigenvalues are the two largest roots of the secular equation, Eq. (19). For $s>1$ we have the approximations,

$$\lambda_0 = s + \sigma/s(s-1), \tag{39a}$$

$$\lambda_1 = 1 - \sigma s/(s-1). \tag{39b}$$

We omit other formulas for the eigenvalues since this case differs from that of $\mu = 1$ only when s is large.

IV. RESULTS AND DISCUSSION

The discussion of this problem is somewhat complicated by the uncertainty regarding the value of μ, the parameter that represents the minimum number of hydrogen bonds that can be broken in one sequence. The formulas are generally simplest when μ is assumed to be unity, although some larger value, perhaps three, is more realistic. Fortunately, we find that many of the interesting results are practically independent of the value chosen. We shall therefore give the discussion in terms of μ equal to unity, except where we consider breaks in the sequence of hydrogen bonds in long helices.

The Transition

The first noteworthy result is the existence at large n of a transition which becomes sharper as σ is decreased. For very large n the partition function is dominated by the largest eigenvalue, λ_0, raised to the $(n-3)$ power. The fraction of hydrogen bonds is then given approximately by

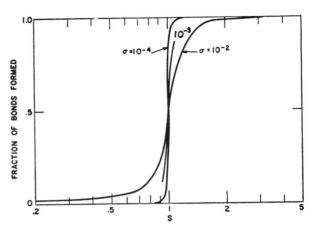

FIG. 2. Fraction of intersegment hydrogen bonds θ as a function of the equilibrium constant s for various values of the initiation parameter σ.

$$\theta = d\ln\lambda_0/d\ln s. \tag{40}$$

The results from this formula are shown in Fig. 2 for the case of $\mu = 1$ and various values of σ.

In this case when σ is unity there is no interaction between the states of successive segments, and s is just the equilibrium constant for the formation of the hydrogen bonds; the fraction of hydrogen bonds then shows a gradual rise with increasing s according to the formula

$$\theta = s/(1+s). \tag{41}$$

Quite different behavior appears at the other extreme when σ approaches zero; in this case there is an almost sharp transition at $s=1$ corresponding to the intersection of the two branches of λ_0, 1, and s. In view of the form of Eq. (21) and since λ is very nearly unity at the transition, the shape of the transition curve is not perceptibly dependent on the parameter μ which specifies the minimum number of hydrogen bonds that can be broken at one place. The value of unity is thus a critical value of s at which long chains go substantially into the helical form.

Critical Size

In a corresponding fashion there is also a critical value of the size n at which substantial helix formation appears for any given value of σ and for s greater than unity. From Eq. (3b), it appears that this value is approximately that at which

$$(s-1)^2 s^{-n+1} = \sigma. \tag{42}$$

The actual behavior of θ as a function of n is shown in Fig. 3 for several values of s and σ. These results were calculated from Eqs. (3a, b) as well as Eq. (27); a few values were also calculated by the corresponding formulas for the case of $\mu = 2$, but the differences were insignificant. Thus the critical size effect is also independent of μ, at least at small n.

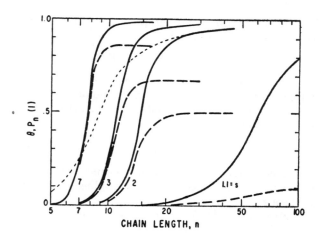

FIG. 3. Fraction of intersegment hydrogen bonds Θ and probability that the last segment be hydrogen-bonded, $p_n(1)$, Eq. (37), as functions of the number of segments n at various values of s and σ. While Eq. (37) specifically refers to the case where $\mu = 1$, the results would not be significantly different for other values of μ. Solid lines, Θ at $\sigma = 10^{-4}$ and the indicated values of s; thin dashed curve, Θ for $s = 2$ and $\sigma = 10^{-2}$; heavy dashed curves, $p_n(1)$ at $\sigma = 10^{-4}$ and the indicated values of s.

The critical size effect offers the most definitive method of determining the two important parameters s and σ from experimental data on the fraction of hydrogen bonds. If a sequence of polymers of different chain lengths is available, the data may be compared to theoretical curves for different s and σ until the best fit is found.

Equilibrium Constants

Equilibrium constants can be defined for various processes involving the helix-coil transition. For example, a sort of an equilibrium constant is the ratio of the number of hydrogen-bonded segments to the number of unbonded segments, which is $\theta/(1-\theta)$. Another ratio accessible to direct measurement (by optical rotatory power) is the ratio of the number of segments in helical form to those in random form; this is $[(n-3)\theta+3]/(n-3)(1-\theta)$. The ratio of the number of molecules with any amount of helical content whatever to those without is another equilibrium constant, and is equal to $Q-1$.

Even the parameter s can be thought of as the equilibrium constant for a certain process, that of incorporating into a helical section the first adjoining segment from a long random section, since the ratio of the aggregate of the statistical weights of those states with a helical section of, say, $k+1$ segments in length to the aggregate of the weights of those states with a section of k segments is practically s, if the adjacent random section is sufficiently greater than the minimum length μ. By a familiar thermodynamic relation we then have

$$d \ln s/dT = \Delta H/RT^2, \qquad (43)$$

where T is the absolute temperature, and ΔH is the enthalpy change on converting one segment from the random to the helical form under the conditions described in the foregoing.

Temperature Dependence and the Heat of Helix Formation

Experimental data are available in some cases for the variation of θ with temperature, but the values of s and σ cannot be obtained from these curves unless data are available for different chain lengths. For example, data of Doty and Yang[1(b)] and Doty and Iso[1(f)] for polybenzyl-L-glutamate are shown in Fig. 4 together with the theoretical curves for two values of σ. In each case the relation between s and the temperature has been adjusted to give the best fit to the experimental curves. It is evident that a small change in $d \ln s/dT$ would be sufficient to make either value of σ satisfactory for either value of n alone, but with the two together the choice of $\sigma = 2 \times 10^{-4}$ with $d \ln s/dT = 0.00614$ is clearly preferable.

From Eq. (43) we immediately calculate that ΔH is $+990$ cal/mole. This heat, it should be remembered, includes the heat of desorption of solvent from the random-form segment when the latter is transformed to helix. The positive sign of ΔH, corresponding to heat adsorbed on helix formation, would be unintelligible otherwise.

In fitting the curves to the data we have assumed that σ does not depend on the temperature because of

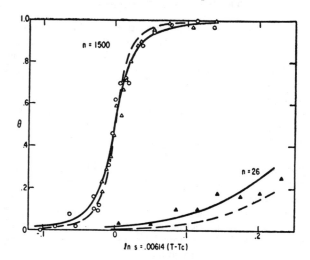

FIG. 4. Comparison of theoretical curves of fraction of segments intramolecularly hydrogen-bonded Θ with observations of Doty and Yang[1(b)] and Doty and Iso[1(f)] on poly-γ-benzyl-L-glutamate. The fraction intramolecularly bonded was assumed to be a linear function of the optical rotation and $\ln s$ was assumed to be linear in the temperature T; T_c is the temperature at which Θ is 0.5. Circles, Doty and Yang; triangles, Doty and Iso. Solid lines correspond to $\sigma = 2 \times 10^{-4}$; dashed lines, $\sigma = 1 \times 10^{-4}$. Doty and Yang's measurements on a sample of degree of polymerization, n, of 84 have been omitted because there is some doubt about the molecular-weight distribution of this sample (private communication from Professor Doty).

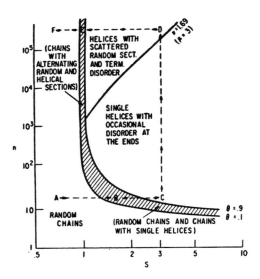

FIG. 5. The $n-s$ plane, calculated for $\sigma=10^{-4}$, and showing the characteristics of the chains in the various regions. The two contours of constant Θ are the (arbitrarily chosen) boundaries of the transition region. The line of constant v corresponds to Eq. (45). The dotted line is the circuit described in the text.

the interpretation given earlier in terms of a ratio of available phase space; this ratio should not change much with temperature. We might hazard a guess that σ should be likewise independent of solvent, depending only on the constitution of the polymer. Data are not yet available to test this point.

The Dominant Configurations

We turn now to the description of the dominant configurations of the chains under various conditions. The situation is epitomized in Fig. 5. For small values of n or s the chains are in the random, unbonded configuration. At larger values of n and s the helical configurations dominate, but in different ways in the sectors of moderate n and large s or large n and moderate s; in the former sector each chain contains only one unbroken helical section, in the latter, several.

For the case of $\mu=1$ we have already seen how the probability of a break in the helix depends upon σ and s, Eq. (35). The case of $\mu=3$, corresponding to no less than three consecutive segments being unbonded at once, is more realistic, however, when both n and s are large. Let us define v as the average number of unbroken helical sections per molecule; by analogy with Eq. (2) this is given by

$$v=(d\ln Q/d\ln\sigma).\qquad(44)$$

With the eigenvalues and eigenfunctions of Eqs. (38) and (39) we can calculate easily the necessary terms of Q and obtain v by differentiation. The result is

$$v=\sigma[n/s^{2}(s-1)+0(1)],\qquad(45)$$

where $0(1)$ stands for terms of order unity. Likewise,

by the use of Eq. (15), the probability of finding a particular segment unbonded is

$$P(0)=\sigma(3s-2)/s^{2}(s-1)^{2}.\qquad(46)$$

Since v is approximately the average number of sequences of unbonded segments, the ratio of $nP(0)$ to v is the average number of unbonded segments in one sequence; this is

$$(3s-2)/(s-1).\qquad(47)$$

The average number of unbonded segments at a break in the helix is thus three or more, depending only on the value of s.

Returning to Fig. 5, we see that the two sectors of the region of helices are separated from each other by a line of constant v. The exact value of v is arbitrary; we have selected $v=1+\ln2$ for purposes of illustration, since along this line half the chains contain one unbroken helical section.

Let us proceed in sequence through the five distinct regions of the diagram to become acquainted with their characteristics, following the circuit indicated by the dotted line. We begin in the region of random chains at point A where n and s are small. If we maintain chain length n and increase the equilibrium constant s, we soon enter the transition region B, where chains containing helices start to appear in the ensemble. The critical value of s for a given n is the one that satisfies Eq. (42). Equation (42) implies that the aggregate statistical weights of the states containing helices are approximately equal to those of the random states. This has an interesting consequence; since θ is near one-half, about half of the chains of the ensemble must be nearly completely in the helical form while half are still in the random form. At any given time the individual chains "make a choice" between the two extreme forms; mixed forms are not favored at small n.

In the region beyond the transition C, most of the chains are in the helical form, except at the ends, where the sizable fraction of random configurations indicated by Eqs. (33) and (37) remain. The end effect depends only on s; therefore the fraction of bonds θ depends almost entirely on s alone in this region (compare curves for $s=2$, Fig. 3).

The end effect is still present in the same way in the next region of the diagram D at large n and large s, but disorder also appears in the middle of the chains as we increase n. This disorder takes the form of short sequences of broken bonds, as we have already seen.

When we decrease s the amount of disorder increases in all its manifestations: the length of the breaks, Eq. (47); the number of independent helical sections, Eq. (45); and the probability of segments unbonded at the ends, Eq. (33). Eventually we enter another branch of the transition region near $s=1$, point E, as shown in Fig. 5. A characteristic of this branch is the fact that single chains contain substantial sections in both helical and random forms in contrast to the situation encountered in the lower branch of the transition region. The average combined length of a sequence of bonded segments followed by a sequence of unbonded segments is just the reciprocal of $P(01)$, which is given by Eq. (35) with sufficient accuracy when s is unity. At the midpoint of the transition the average length of a helical sequence is half of the reciprocal of $P(01)$, or $\sigma^{-\frac{1}{2}}$. The magnitude of this number is noteworthy. By comparison, if the bonds were arranged at random, the average length of a bonded sequence would be two. When we decrease s further the relative lengths of the random and helical sections disproportionate rapidly, until finally the chains become almost purely random in configuration, and we return to the region of random chains, point F.

Kinetics of Polymerization

The kinetics of polymerization have been found to show different rate constants when the polymer is in the helical or random forms. According to Doty and Lundeberg,[1(e)] the addition of monomer to the helical form occurs several times faster than the addition to the random form in dioxane solution. Presumably the rate depends upon the condition of the nth segment of the helix. For this reason we have plotted $p_n(1)$, Eq. (37), in Fig. 3. Here also the critical size is important, but the limiting value of $p_n(1)$ at large n is never quite as large as the limiting value of θ. Doty and Lundberg found that polybenzylglutamate in dioxane solution showed a rather sharp transition in rate of addition of monomer at about $n=8$; this would correspond to $s=5$ if σ is assumed to be 10^{-4} as is suggested by the apparent sharpness of the transition and in accordance with the results cited in the foregoing in the section on temperature dependence.

Relation to Other Work

Subsequent to Schellman's original publication[6] and more or less simultaneously with each other, a number of workers have been developing theories of the helix-random coil transition. Several preliminary accounts have already appeared[10-13] and others, in addition to the present paper, are now being published.[14-17] Insofar as we have been able to ascertain, there is substantial agreement about the results, although considerable divergence in the methodology and emphasis. Our justification for adding one more report on the topic is the fact that we alone seem to have made extensive use of the matrix method, which allows the simplest treatment on a unified basis of the various phenomena of interest for various modifications of the basic model.

It remains to clarify the relation of these results to the well known demonstration that a one-dimensional system cannot show a sharp phase transition in the usual sense. To be exact, the usual demonstration, as given for example by Landau and Lifshitz,[5] states that a sharp transition cannot occur unless the boundary tension between the two phases is infinite, since otherwise the two phases will always mix with each other to an appreciable extent. An infinite boundary tension corresponds in our treatment to σ equaling zero which is the only circumstance under which we find the transition to be sharp. Thus there is no contradiction. In fact, the prediction that the two phases will mix with each other when the boundary tension is finite is in complete accord with our result that a long chain near the transition point consists of alternating helical and random sections. It is this alternation of short sections of each phase that is characteristic of a one-dimensional system and that causes the transition to be diffuse.

V. ACKNOWLEDGMENT

We wish to acknowledge gratefully the help of Miss Ann Warner in performing some and checking others of the calculations.

[10] L. Peller, thesis, Princeton University (1957).
[11] J. H. Gibbs and E. A. DiMarzio, J. Chem. Phys. **28**, 1247 (1958).
[12] B. H. Zimm and J. K. Bragg, J. Chem. Phys. **28**, 1246 (1958).
[13] Rice, Wada, and Geiduschek, Discussions Faraday Soc. **25**, 130 (1958).
[14] J. H. Gibbs and E. A. DiMarzio, J. Chem. Phys. **30**, 271 (1959).
[15] J. A. Schellman (private communication).
[16] L. Peller (private communication).
[17] T. L. Hill, J. Chem. Phys. **30**, 383 (1959).

Density of Low-Energy Vibrational States in a Protein Solution

A. S. Brill, F. G. Fiamingo,[a] D. A. Hampton,[b] P. D. Levin,[c] and R. Thorkildsen[d]

Department of Physics, University of Virginia, Charlottesville, Virginia 22901

(Received 10 December 1984)

Electron paramagnetic resonance measurements on the aquo complex of sperm whale skeletal myoglobin in solution at $T < 4$ K show that, at phonon energies around 20 cm^{-1}, the density of vibrational states is that of a three-dimensional system.

The result reported here would not have been looked for four years ago. Then in 1980, Stapleton *et al.*[1] interpreted the unusual temperature dependence of electron spin-lattice relaxation in low-spin ferric heme proteins in terms of a fractional dimension, now called the fracton[2] or spectral[3] dimension ($\tilde{d}$), for the space characterizing the density of vibrational states responsible for the Raman process. This seminal idea led to computation of the fractal dimension ($\bar{d}$) of protein structures from the x-ray crystallographic coordinates, and correlations between $\bar{d}$ and polypeptide structure and function are being investigated.[4] The values of $\tilde{d}$ observed for low-spin ferric proteins are in the range 0.8 to 1.7.

Alexander and Orbach[2] critically addressed the relation between $\bar{d}$ (a structural parameter) and $\tilde{d}$ (which quantifies the vibrational eigenvalue density of states), demonstrating that these two parameters are not necessarily the same. With regard to polymers, they suggested that the role of the medium could not be neglected, and that it would certainly not be unexpected to find $\tilde{d} = d$, the Euclidean dimensionality of the entire system, for a structure of fractional $\bar{d}$ intimately embedded in a solvent. Alexander and Orbach also pointed out that crossover to $\tilde{d} = d$ ("normal behavior") occurs at low frequencies where the vibrational length scale is greater than the size of the finite fractal object.

The measurements reported below evolved from experiments started a number of years ago with the goal of quantitatively characterizing the four-level model[5] of the high-spin ferric ion in complexes of myoglobin, hemoglobin, and other heme proteins for which the limited set of states shown in Fig. 1 is adequate. This model generates the tetragonal and rhombic splitting parameters D and E in the spin Hamiltonian

$$\mathcal{H}_{spin} = D(S_z^2 - \tfrac{1}{3}S^2) + E(S_x^2 - S_y^2) \tag{1}$$

and, with these, the principal values of the effective g tensor. The calculation of D from

$$D = \frac{\lambda^2}{5}\left[\frac{1}{\Delta_1} - \frac{1}{\Delta_2}\right] \tag{2}$$

requires knowledge of the spin-orbit coupling constant, λ, which is reduced from the free-ion value (~ 420 cm^{-1}) by electron delocalization. Spin-orbit coupling mixes $^4A_2$ and ^4E into $^6A_1$, depleting the latter by

$$\eta_{s.o.}^2 = \frac{\lambda^2}{5}\left[\frac{1}{\Delta_1^2} + \frac{1}{\Delta_1\Delta_2} + \frac{1}{\Delta_2^2}\right], \tag{3}$$

and the in-plane g values are closely approximated by[6,7]

$$g_{1,2} = 6.01 \pm 24\left(\frac{E}{D}\right) - 18.7\left(\frac{E}{D}\right)^2 - 12\eta_{s.o.}^2. \tag{4}$$

In simulation of the X-band EPR spectra of heme proteins in solution, it is necessary to include distributions in E/D and $\eta_{s.o.}$.[8,9] We use the method of Scholes, Isaacson, and Feher[10] to determine D, and, with data acquisition under microcomputer control, observe effects of a distribution in D.[11,12] The large linewidths observed in far-infrared spectra of related hemeprotein complexes have also been attributed to distribution in D.[13] Distributions in D, E, and $\eta_{s.o.}$ arise from distributions in the energy differences Δ_1, Δ_2, and γ. These spreads, in turn, are associated with the ensemble of protein conformations frozen in as the temperature is lowered to the cryogenic region of the measurements.[14,15] The value of λ is taken as fixed because the covalency of the coordination complex is not expected to vary significantly among the members of the ensemble. An upper limit to the width of the distribution in electron delocalization onto a ligand can be obtained from the widths of the corresponding electron-nuclear double resonance signals; for example, with the field along the heme normal of aquo myoglobin, the histidine ϵ-nitrogen signals have halfwidths (0.15 MHz) at half height of only 1.3% of the ligand hyperfine interaction (11.5 MHz).[16] The

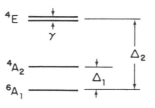

FIG. 1. Four-level model of high-spin ferric heme.

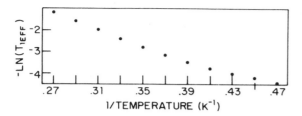

FIG. 2. Temperature dependence of $T_{1\text{eff}}$ (in microseconds), aquo sperm whale skeletal myoglobin, Expt. WMAB83.

sites in myoglobin are nearly axial and, for the properties discussed below, γ and E are taken as zero. In this case, the parameters needed are λ, the mean values Δ_1^0 and Δ_2^0, and the rms widths σ_1 and σ_2 of the distributions (taken Gaussian) in Δ_1 and Δ_2.

The measurement of D is based upon the temperature dependence of spin-lattice relaxation in a region where the Orbach process dominates the rate.[10] For a site with first excited state at zero-field splitting energy $2D$, the relaxation time T_1 is given (with F a frequency factor) by

$$1/T_1 = FD^{\tilde{d}}\exp(-2D/k_B T). \qquad (5)$$

$D^{\tilde{d}-1}$ is proportional to the density of effective vibrational states (e.g., for vibrations propagating in a three-dimensional space, $\tilde{d}=3$). Because of the Raman-process observations of Stapleton and co-workers of $0.8 < \tilde{d} < 1.7$ in low-spin ferric proteins, we have taken $\tilde{d}$ as a parameter to be obtained from the data. For a system in which all sites have the same D, $\tilde{d}$ could only be obtained in the Orbach region if an absolute formula for relaxation rate were available. However, when there is a distribution in D, the resulting spread in $1/T_1$ over the ensemble of sites is influenced by $\tilde{d}$, and $\tilde{d}$ can be obtained by comparison of data with simulations as below.

With a distribution in relaxation rates, the time course of recovery from a saturating power pulse is not a single exponential, and, in the presence of some drift, the slowest relaxing sites introduce uncertainty in the base line (asymptote for full recovery). In order to reduce this problem and provide uniform conditions for simulations, we have adopted a standard experimental timing protocol.[12] The total time of observation is adjusted iteratively until it is $8T_{1\text{eff}}$, where $T_{1\text{eff}}$ characterizes the least-squares single-exponential fit to the observed part of the recovery curve. The aperture time of the boxcar integrator is adjusted in parallel to be a fixed fraction ($\frac{1}{5}$) of $T_{1\text{eff}}$. In the course of monitoring recovery, the aperture is moved this width 40 times, at intervals long compared with the response of the EPR output filter. Data are obtained at increments of 0.02 K^{-1} in inverse temperature and provide, as a function of temperature, $T_{1\text{eff}}$ (Fig. 2) and the differ-

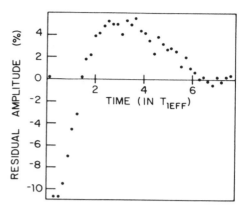

FIG. 3. Residuals between an experimental saturation recovery curve and its least-squares-fit single exponential, Expt. WMAB83, $T = 2.17$ K.

ence between the experimental recovery curve and least-squares-fit single exponential (Fig. 3). The parameter which best quantifies the latter residual curves is the trough amplitude, $A_{\min}$; its temperature dependence for two sets of data is shown in Fig. 4(a).

With the nonlinear least-squares Marquardt algorithm, data sets of the kind shown in Fig. 2 are fitted very well ($\chi^2 \sim 10^{-3}$) by the function

$$1/T_{1\text{eff}} = a\exp(-2D_{\text{eff}}/kT) + bT. \qquad (6)$$

The second term accounts for the appearance of curvature in the logarithmic plot as the temperature decreases. At temperatures below 2.7 K (0.37 K^{-1}), the behavior of $A_{\min}$ indicates that bT cannot arise solely from a nondistributed direct process.[12] The value of D_{eff} obtained by a matching of Eq. (6) to the data is $7.9(3) \pm 4\%$ cm^{-1}. The relation of D_{eff} to the central value $D_0 = (\lambda^2/5)(1/\Delta_1^0 - 1/\Delta_2^0)$ was investigated by simulations of saturation recovery. This calculation also provides: (1) the relation between $\langle\eta_{\text{s.o.}}\rangle$ and $\eta_{\text{s.o.}}^0$ [calculated from Eq. (3) with Δ_1^0, Δ_2^0]; (2) the rms deviation $\delta\eta_{\text{s.o.}}$ of $\eta_{\text{s.o.}}$ from $\langle\eta_{\text{s.o.}}\rangle$; (3) the temperature dependence of $A_{\min}$. First Δ_1 and Δ_2 were given independent Gaussian distributions of rms widths σ_1 and σ_2 about central values Δ_1^0 and Δ_2^0 to generate, with Eq. (2), distributions $P(D)$. Calculations were made for λ within the range 320–380 cm^{-1}. D was restricted to positive values and the Gaussians were truncated at 4σ, so that

$$\sigma_1 + \sigma_2 < (\Delta_2^0 - \Delta_1^0)/4. \qquad (7)$$

The values of $P(D)$ were binned into 100 equal intervals in D from 0 to $2D_0$ with the last bin also receiving the few entries with $D > 2D_0$. Recovery curves were then synthesized as the sum of 100 contributions, each characterized by bin central energy D_i and weight $P(D_i)$. Also included was a rate term of the (direct process) form BT in parallel with the Orbach process,

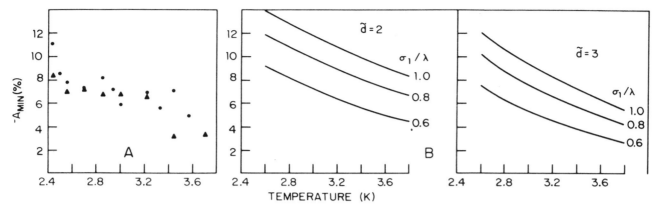

FIG. 4. Temperature dependence of A_{min}. (a) Experiments WMA883, circles, and WMAB83, triangles. (b) Calculated for $D_0 = 7.93$ cm^{-1}, $\eta_{s.o.}^0 = 0.094$, $\sigma_2 = 0$, $\lambda = 340$ cm^{-1}, $B/F = 1 \times 10^{-8}$ cm$^{-\tilde{d}}$K^{-1}.

with relative amounts assigned so that the onset of effectiveness of BT fell between the low end and middle of the temperature range used experimentally. In a manner exactly analogous to the standard experimental protocol, values of T_{1eff} and A_{min} were obtained from the simulated recovery curves, and the temperature dependence of $(T_{1eff})_{simulated}$ was fitted to Eq. (6). For $\Delta_1^0 \approx 2000$ cm^{-1}, $\Delta_2^0 \approx 6000$ cm^{-1}, 200 cm^{-1} $< \sigma_1 < 400$ cm^{-1}, and $1 \leq \tilde{d} \leq 3$, the results of these calculations can be summarized as follows: (1) The effect of the spread in Δ_2 permitted by Eq. (7) is at least an order of magnitude smaller than the effect of the spread in Δ_1; (2) D_{eff} differs from D_0 by no more than $\pm 4\%$; (3) $\langle \eta_{s.o.} \rangle$ does not exceed $\eta_{s.o.}^0$ by more than 5%, Fig. 5; (4) $\delta \eta_{s.o.}$ is linear in σ_1 to within 30%, Fig. 5; (5) $|A_{min}|$ increases monotonically with decreasing temperature and, for a given temperature, the value of A_{min} depends strongly upon $\tilde{d}$ and σ_1/λ. This behavior, shown in Fig. 4(b), can be understood as follows.[12] The residual function arises from the distribution in recovery rates. The faster-than-average rates

express themselves early, and A_{min} is the greatest amount of actual signal recovery above that of the best single-exponential fit to the data. The distribution in rates is the convolution of the distribution in D with the rate function, Eq. (5). The latter has a maximum at $D = \tilde{d}k_BT/2$ which, for $T < 4$ K, is much smaller than D_0 (about 8 cm^{-1}) at the peak of the distribution. It follows that the fractional spread in rates, upon which the amplitude of the residual curve depends, is proportional to the logarithmic derivative of Eq. (5) evaluated at the center of the D distribution, namely

$$\left[\left[\frac{1}{T_1} \right]^{-1} \frac{d}{dD} \left(\frac{1}{T_1} \right) \right]_{D=D_0} = \frac{\tilde{d}}{D_0} - \frac{2}{k_BT}. \tag{8}$$

This factor, Eq. (8), gets less negative, and hence the fractional spread in rates decreases, with increasing $\tilde{d}$.

Simulations of EPR spectra place $\langle \eta_{s.o.} \rangle$ and $\delta \eta_{s.o.}$ for the aquo complex of sperm whale myoglobin in the ranges

$$0.094 \leq \langle \eta_{s.o.} \rangle \leq 0.097, \quad 0.010 \leq \delta \eta_{s.o.} \leq 0.013.$$

Points (2)–(4) above permit one to take as excellent approximations $D_0 = 7.9$ cm^{-1}, $\eta_{s.o.}^0 = 0.094$, and $0.010 \leq \delta \eta_{s.o.} \leq 0.013$. Then, with D_0 and $\eta_{s.o.}^0$, one can calculate Δ_1^0, Δ_2^0 as functions of λ; $\delta \eta_{s.o.}$ as a function of λ and σ_1; and A_{min} versus temperature as a function of λ, σ_1, and $\tilde{d}$. Examples are shown in Figs. 5 and 4(b). Comparison of the experimentally allowed ranges of $\langle \eta_{s.o.} \rangle$, $\delta \eta_{s.o.}$, and A_{min} with the results of the calculations constrains the input parameters λ, σ_1, and $\tilde{d}$. For example, the $\lambda = 340$ cm^{-1} plot of $\delta \eta_{s.o.}$ of Fig. 5 shows that, for sperm whale myoglobin, σ_1/λ is restricted to the range 0.70 to 0.875. Thus, in Fig. 4(b), the $\sigma_1/\lambda = 0.60$ functions cannot be considered for this protein complex, and $\tilde{d} = 2$ is clearly ruled out because most of the experimental A_{min} values lie below even the $\sigma_1/\lambda = 0.60$ curve. In this manner one

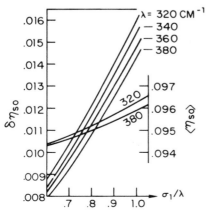

FIG. 5. $\langle \eta_{s.o.} \rangle$ and $\delta \eta_{s.o.}$ as functions of σ_1/λ for $D_0 = 7.93$ cm^{-1}, $\eta_{s.o.}^0 = 0.094$, $\sigma_2 = 0$.

excludes $\tilde{d} \lesssim 2.5$ (for all λ) and $\lambda = 380$ cm^{-1} (for all $\tilde{d}$), and constrains σ_1 to 220 cm$^{-1} < \sigma_1 < 270$ cm^{-1}. For the $\sigma_1/\lambda = 0.8$ case of Fig. 4(b), the rms width of the distribution in D is 1.81 cm^{-1}.

We conclude that, in the energy region associated with high-spin ferric myoglobin spin-lattice relaxation through the Orbach process, $\tilde{d}$ is close or equal to 3. The reason for this can be seen by considering the acoustic wavelength of phonons with energy in the neighborhood of $2D_0$. Taking the speed of sound in the protein to be 2×10^5 cm/s, one finds the vibrational wavelength to be about 40 Å, comparable with the overall size $(2\overline{R})$ of the myoglobin molecule (roughly $25 \times 35 \times 45$ Å^3). The Orbach process in any paramagnetic center with a zero-field splitting of 20 cm^{-1} or less, bound to any protein of the size of myoglobin or smaller, should exhibit $\tilde{d} \approx 3$. The Raman process, depending as it does upon phonons of shorter wavelength, has an associated spectral density which is more sensitive to structural properties of the protein. On this basis the experimental difference in $\tilde{d}$ between low- and high-spin ferric proteins arises from a difference in the region of energy of the phonons producing the measured effect rather than from the difference in binding of the metal ions to the macromolecules.

Thus, the value of $\tilde{d}$ which would have been assumed four years ago, and is supported by a wavelength versus size argument, is found experimentally. In addition to demonstrating a novel determination of the dimensionality of vibrational space, the result reported here enables one to take $\tilde{d} = 3$ and proceed to utilize the methods outlined above to quantify the low-lying energies and their widths in other high-spin ferric heme protein complexes for which $\overline{R}D < 2 \times 10^{-6}$.

We thank P. G. Debrunner and H. J. Stapleton for helpful discussions. The support of the National Science Foundation (currently PCM81-04377) and the U.S. Public Health Service (T32 GM07294) is acknowledged with thanks.

(a)Current address: Department of Physiological Chemistry, Ohio State University, College of Medicine, Columbus, Ohio 43210.

(b)Current address: Formative Technologies, Pittsburgh, Pa. 15213.

(c)Current address: Biology Department, Brookhaven National Laboratory, Upton, N.Y. 11973.

(d)Current address: AT&T Bell Laboratories, Murray Hill, N.J. 07974.

[1]H. J. Stapleton, J. P. Allen, C. P. Flynn, D. G. Stinson, and S. R. Kurtz, Phys. Rev. Lett. **45**, 1456 (1980).

[2]S. Alexander and R. Orbach, J. Phys. (Paris), Lett. **43**, L625 (1982).

[3]R. Rammal and G. Toulouse, J. Phys. (Paris), Lett. **44**, L13 (1983).

[4]J. P. Allen, J. T. Colvin, D. G. Stinson, C. P. Flynn, and H. J. Stapleton, Biophys. J. **38**, 299 (1982).

[5]J. S. Griffith, Proc. Roy. Soc. (London), Ser. A **235**, 23 (1956).

[6]C. P. Scholes, J. Chem. Phys. **52**, 4890 (1970).

[7]F. G. Fiamingo, Ph.D. thesis, University of Virginia, 1980 (unpublished).

[8]A. S. Brill, F. G. Fiamingo, and D. A. Hampton, in *Frontiers of Biological Energetics,* edited by P. L. Dutton *et al.* (Academic, New York, 1978), Vol. 2, p. 1025.

[9]C. E. Schulz, R. Rutter, J. T. Sage, P. G. Debrunner, and L. P. Hager, Biochemistry **23**, 4743 (1984).

[10]C. P. Scholes, R. A. Isaacson, and G. Feher, Biochim. Biophys. Acta **244**, 206 (1971).

[11]R. Thorkildsen, Ph.D. thesis, University of Virginia, 1981 (unpublished).

[12]P. D. Levin, Ph.D. thesis, University of Virginia, 1984 (unpublished).

[13]P. M. Champion and A. J. Sievers, J. Chem. Phys. **72**, 1569 (1980).

[14]A. S. Brill, in *Tunneling in Biological Systems,* edited by B. Chance *et al.* (Academic, New York, 1979), p. 561.

[15]P. G. Debrunner and H. Frauenfelder, Annu. Rev. Phys. Chem. **33**, 283 (1982).

[16]C. P. Scholes, R. A. Isaacson, and G. Feher, Biochim. Biophys. Acta **263**, 448 (1972).

Authors' Note:

Additional information about the analysis of the electron paramagnetic resonance pulse saturation recovery kinetics can be found in P. D. Levin and A. S. Brill, *J. Phys. Chem.* **92**, 5103 (1988).

Pigment − protein interactions in the photosynthetic reaction centre from *Rhodopseudomonas viridis*

H.Michel[1], O.Epp[2] and J.Deisenhofer[2]

Max-Planck-Institut für Biochemie, [1]Abteilung Membranbiochemie, and [2]Abteilung Strukturforschung II, D-8033 Martinsried, FRG

Communicated by H.Michel

An X-ray structure analysis of the photosynthetic reaction centre from the purple bacterium *Rhodopseudomonas viridis* provides structural details of the pigment-binding sites. The photosynthetic pigments are found in rather hydrophobic environments provided by the subunits L and M. In addition to apolar interactions, the bacteriochlorophylls of the primary electron donor ('special pair') and the bacteriopheophytins, but not the accessory bacteriochlorophylls, form hydrogen bonds with amino acid side chains of these protein subunits. The two branches of pigments which originate at the primary electron donor, and which mark possible electron pathways across the photosynthetic membrane, are in different environments and show different hydrogen bonding with the protein: this may help to understand why only one branch of pigments is active in the light-driven electron transfer. The primary electron acceptor, a menaquinone (Q_A), is in a pocket formed by the M subunit and interacts with it by hydrophobic contacts and hydrogen bonds. Competitive inhibitors of the secondary quinone Q_B (*o*-phenanthroline, the herbicide terbutryn) are bound into a pocket provided by the L subunit. Apart from numerous van der Waals interactions they also form hydrogen bonds to the protein.

Key words: chlorophyll/electron transfer/herbicide/quinone/X-ray structure analysis

Introduction

Photosynthetic reaction centres are complexes comprised of pigments and integral membrane proteins which catalyse the light-driven electron transfer across the photosynthetic membranes. The reaction centres from the purple photosynthetic bacteria are well characterised (for reviews, see Feher and Okamura, 1978; Hoff, 1982; Parson, 1982). Most of them contain three protein subunits which are called H (heavy), M (medium) and L (light) subunits according to their apparent mol. wts as determined by SDS−polyacrylamide gel electrophoresis. However, the sequence work has shown that the H subunits are the smallest ones (Youvan *et al.*, 1984; Michel *et al.*, 1985, 1986). The reaction centres from several purple photosynthetic bacteria, e.g. that from *Rhodopseudomonas viridis* contain a tightly bound cytochrome molecule which re-reduces the photo-oxidised primary electron donor.

The reaction centre from *Rps. viridis* contains four bacteriochlorophyll *b*s, two bacteriopheophytin *b*s, one menaquinone, one non-heme iron and one ubiquinone. The reaction centres from most of the other purple photosynthetic bacteria contain bacteriochlorophyll *a* instead of bacteriochlorophyll *b*, bacteriopheophytin *a* instead of bacteriophytin *b*, and the menaquinone is replaced by another ubiquinone. Two of the bacteriochlorophyll molecules

form the primary electron donor ('special pair'). The existence of such a special pair had been postulated from *Rhodobacter sphaeroides* (formerly called *Rps. spheroides*, see Imhoff *et al.*, 1984) on the basis of e.p.r. experiments (Norris *et al.*, 1971). one of the two 'accessory' bacteriochlorophyll molecules may be the very first electron acceptor, but this point is controversial (see Kirmaier *et al.*, 1985; Martin *et al.*, 1986; Shuvalov and Duysens, 1986). The electron is transferred to one of the bacteriopheophytins, then to the tightly bound primary quinone, Q_A (the menaquinone in *Rps. viridis*, Shopes and Wraight, 1985) and finally to the loosely bound secondary quinine, Q_B.

The reaction centres from *Rps. viridis* and *Rb. sphaeroides* could be crystallised (Michel, 1982; Allen and Feher, 1984; Chang *et al.*, 1985). The crystallised reaction centres are photochemically active (Zinth *et al.*, 1983; Allen and Feher, 1984; Gast and Norris, 1984).

The crystallographic analysis of the reaction centre crystals from *Rps. viridis* at 3 Å resolution provided a complete picture of pigments and protein subunits (Deisenhofer *et al.*, 1984, 1985). A striking result was the symmetric arrangement of the pigments. A local 2-fold rotation axis runs between the monomers constituting the special pair near the periplasmic side of the membrane, and through the non-heme iron near the cytoplasmic side of the membrane. Since the tetrapyrrole rings of the accessory bacteriochlorophylls and of the two bacteriopheophytins are also related by the local dyad, two structurally equivalent branches are formed that could be used for electron transfer across the membrane. However, the phytyl side chains do not obey the symmetry and, in addition, the symmetry is broken by the presence of only one quinone at the right-hand side branch (Figure 4 in Deisenhofer *et al.*, 1984). The asymmetry is also evident from the spectroscopic inequivalence of the two bacteriopheophytins. The bacteriopheophytin absorbing light at longer wavelengths is involved in the electron transfer (Vermeglio and Paillotin, 1982). Comparison of absorbance spectra of crystals taken with polarised light (Zinth *et al.*, 1983) and the coordinates of the pigments obtained by X-ray structure analysis shows that the bacteriopheophytin absorbing light at the longer wavelengths is the one close to the quinone (Zinth *et al.*, 1985; Knapp *et al.*, 1985). The quinone must be the primary quinone, a menaquinone, since the secondary quinone is lost during isolation and crystallisation (Shopes and Wraight, 1985; Gast *et al.*, 1985). These observations establish that the right-hand side branch with the quinone is used for light-driven electron transfer across the membrane.

The fold of the L and M subunits is very similar and they bind the photosynthetic pigments in a symmetric manner. Each of these subunits contains five membrane-spanning helices; these helices, and parts of the polypeptide chains connecting them, obey the local 2-fold symmetry that relates the photosynthetic pigments. The accessory bacteriochlorophyll and the bacteriopheophytin of the right-hand side branch, which are involved in the light-driven electron transfer, are more closely associated with the subunit L than with M. However, the binding site for the ring system of the menaquinone is made up exclusively by the connection of the transmembrane helices D and E from the M subunit. The binding site for the secondary quinone has also been established by soaking quinones and competitive inhibitors into the crystals and subsequent difference Fourier analysis. These compounds bind into an empty pocket of the protein which is provided by the connection of the transmembrane helices D and E of the L subunit.

In parallel, the genes coding for the subunits were isolated and sequenced (Michel *et al.*, 1985, 1986). The derived amino acid sequences were incorporated into the atomic model of the reaction centre. Now a detailed analysis of the pigment−protein interaction becomes possible. Since only one out of two structurally very similar branches is used for light-driven electron transfer, the subunits L and M provide not only a scaffold for the pigments, but pigment−protein interaction determines which branch is used for light-driven electron transfer. Other points of interest are the binding sites for the quinone molecules, the unusual coordination of the non-heme iron and the binding sites for the competitive inhibitors of the secondary quinone. Several of these inhibitors are used as herbicides since they displace the secondary quinones in photosystem II reaction centres of chloroplasts, thus blocking light-driven electron transfer (Pfister *et al.*, 1981). However, their binding sites and their modes of binding are not known. Here we describe and discuss the binding sites of the bacteriochlorophylls, bacteriopheophytins, quinones, the non-heme iron and the competitive quinone inhibitors, based on an atomic model refined at 2.9 Å resolution.

Results and Discussion

Interactions between special pair bacteriochlorophylls and the protein environment

As described recently (Deisenhofer *et al.*, 1984, 1985) the bacteriochlorophylls constituting the special pair are found between the L and M subunits in a primarily hydrophobic environment. Their nearly parallel macrocycles are perpendicular to the plane of the membrane; their pyrrole rings I are stacked on top of each other, such that lines connecting the nitrogens of pyrrole rings I and III enclose an angle of ∼ 144°. The central Mg ions are liganded to histidine L173 or M200, which are located close to the periplasmic ends of the membrane-spanning helices D of the L and M subunits. (The bacteriochlorophylls constituting the special pair will be designated BC_{LP} and BC_{MP}, where L and M refer to the ligation of the Mg ions to the L or M subunits,

and P indicates pair. Similarly the accessory bacteriochlorophylls will be designated BC_{LA} and BC_{MA}.) The acetyl groups at the pyrrole rings I point towards the central Mg ions of the other bacteriochlorophyll in the pair. Crystallographic refinement shows clearly that the carbonyl groups are approximately parallel to the macrocycles, and that they are too far (∼ 3.5 Å) from the central Mg ions of the other bacteriochlorophyll to act as ligands. Thus, the Mg ions may be regarded as five-coordinated, in agreement with recent resonance Raman data (Lutz and Robert, 1985). Both acetyl groups are hydrogen bonded to amino acid side chains as is shown in Figure 1; the acetyl group of BC_{LP} to histidine L168 and the actyl group of BC_{MP} to tyrosine M195. The keto carbonyl group of ring V of BC_{LP} forms an additional hydrogen bond to threonine L248. There may be also a hydrogen bond between the hydroxyl group of serine M203 and the ester carbonyl group at ring V of BC_{MP}. However, hydrogen bonds between the serine and two protein carbonyl groups are equally possible: the model is not accurate enough to decide between these possibilities. Another point of interest is that a considerable number of aromatic side chains form part of the special pair binding site; these are the side chains of phenylalanines M154, M194, L160, L181, L241 of tyrosines L162, M195 and M208, and of tryptophans L167, M183 and M199. Tyrosine L162 has its aromatic ring perpendicular to the macrocycles of the special pair bacteriochlorophylls and is located halfway between the special pair and the nearest heme group of the cytochrome subunit. It is conserved between *Rps. viridis*, *Rb. capsulatus* and *Rb. sphaeroides*, and may play a role in re-reducing the photo-oxidised special pair.

The distribution of polar and apolar protein side groups in the immediate vicinity of the special pair bacteriochlorophylls is clearly asymmetric. The differences in the environment could also lead to a differently distorted geometry of the special pair bacteriochlorophylls. The combined effects may cause preferred electron release on the BC_{LP} side of the special pair in the excited state, and initiate the asymmetric electron transport across the photosynthetic membrane.

Interestingly, tyrosine M195 is replaced by a phenylalanine in *Rb. sphaeroides* (for sequence comparisons see Michel *et al.*, 1986), threonine L248 by a methionine, and only histidine L168 and the non-hydrogen-bonded tyrosine M208 are conserved. These changes of amino acids may influence the electron distribution at the special pair. E.p.r. studies on the photo-oxidised primary electron donor of both species show that the unpaired

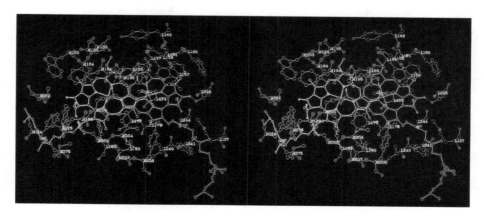

Fig. 1. (Stereo pair): special pair and surrounding protein residues: residue numbers indicated at C^α positions. Empty circles: carbons; half-filled small circles: oxygens; filled circles: nitrogens and other atoms. Colours: light blue: BC_{LP}; yellow: BC_{MP}; orange: residues from L-subunit; blue: residues from M-subunit; purple: possible hydrogen bonds. Figures 1−5 were produced by a computer program written by A.M.Lesk and K.D.Hardman (1985).

electron is shared symmetrically between both special pair bacteriochlorophylls in *Rb. sphaeroides* but not in *Rps. viridis* (Thornber *et al.*, 1977).

Interactions between the accessory bacteriochlorophylls and the protein environment

The accessory bacteriochlorophylls are in van der Waals contact with the special pair bacteriochlorophylls on one side and the bacteriopheophytin on the other side. Protein residues from subunits L and M contribute to their binding pockets (see Figure 2); in addition, parts of them are also accessible from the membrane environment. The Mg atoms of both bacteriochlorophylls are liganded to histidine residues (BC$_{LA}$:L153, BC$_{MA}$:M180) in the helical part of the segments which connect the membrane-spanning helices C and D of subunits L and M. There are a considerable number of van der Waals interactions but hydrogen bonds between accessory bacteriochlorophylls and the surrounding amino acid residues are not formed. The aromatic side chain of tyrosine M208 is in contact with BC$_{LA}$, BC$_{LP}$ and BC$_{MP}$; the symmetry-related phenylalanine L181 is in contact with BC$_{MA}$, BC$_{MP}$ and BC$_{LP}$. The refined electron density map shows an elongated feature in the environment of BC$_{MA}$ which is as yet unexplained. It may represent a lipid molecule; more likely it represents part of the carotenoid dihydroneurosporene which was found in the crystals by chemical analysis (I.Sinning, H.Michel, C.H.Eugster, unpublished).

Interactions between the bacteriopheophytins and the protein environment

Each bacteriopheophytin, named BP$_L$ or BP$_M$ according to the predominant interaction with either the L or M subunit, is found in a pocket formed mainly by non-polar residues from both subunits. BP$_L$ (BP$_M$) is located near the centres of the membrane-spanning helices C and E of the L subunit (M subunit), its macrocycle is parallel to the transmembrane helix D of the M subunit (L subunit). The bacteriopheophytins are partially accessible from the membrane space. BP$_L$ and BP$_M$ appear to be hydrogen bonded to the side chains of tryptophan L100 and M127, respectively via the ester carbonyl groups of ring V (see Figure 3). The hydrogen bond between BP$_M$ and tryptophan M127 seems to be significantly shorter (~ 2.7 Å) than the hydrogen bond between BP$_L$ and tryptophan L100 (~ 3.2 Å). The aromatic rings of three phenylalanines (L97, L121 and L241) are in contact with BP$_L$, whereas only one (M148) is found near BP$_M$. The most important difference in the sites of BP$_L$ and BP$_M$ is the presence of one glutamic acid residue (L104) near ring V of BP$_L$. This glutamic acid residue is in a hydrophobic environment, and does not have a counter-charge in its neighbourhood. However, it is at the correct distance to the keto carbonyl group of ring V of BP$_L$ to form a hydrogen bond, and it is therefore most likely protonated. A negatively charged carboxyl group would electrostatically prevent the formation of the bacteriopheophytin anion during the light-driven electron transfer.

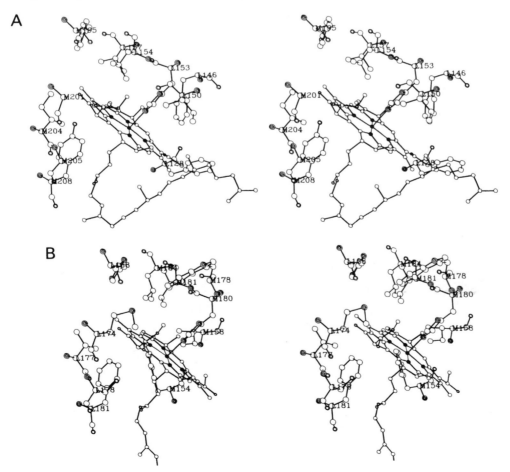

Fig. 2. (Stereo pairs): accessory bacteriochlorophylls and surrounding protein residues. (**A**) B$_{LA}$; (**B**) BC$_{MA}$. Atom types as in Figure 1.

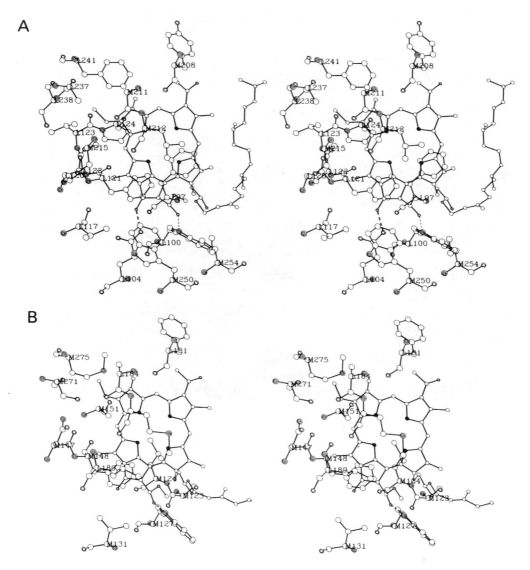

Fig. 3. (Stereo pair): bacteriopheophytins and surrounding protein residues. (**A**) BP$_L$; (**B**) BP$_M$. Possible hydrogen bonds are indicated with dashed lines. Atom types as in Figure 1.

This glutamic acid residue is conserved between bacterial L subunits, and even the D1 proteins of photosystem II, which has been proposed to be the equivalent of the L subunit in the photosystem II reaction centre (Michel and Deisenhofer, 1986; Hearst, 1986). In particular, glutamic acid L104 may contribute to the spectroscopic red shift of BP$_L$ compared with BP$_M$.

Another interesting amino acid is tryptophan M250 which forms part of the binding site of the primary quinone, Q$_A$ (see below). Its π-electron system 'bridges' the π-electron systems of BP$_L$ and Q$_A$ and may be of crucial importance for light-driven electron transfer. In a position equivalent by local symmetry phenylalanine L216 forms part of the binding site of the secondary quinone Q$_B$, but due to its smaller size it cannot serve as a 'bridge' between BP$_M$ and Q$_B$. Tryptophan M250 is also conserved between bacterial M subunits and the D 2 proteins of photosystem II reaction centres, which have been proposed to be structurally and functionally equivalent (Michel and Deisenhofer, 1986; Hearst, 1986).

The binding sites of the primary quinone, the non-heme iron and the competitive inhibitors

As reported previously (Deisenhofer *et al.*, 1984, 1985) the primary quinone Q$_A$ which is menaquinone 9 in *Rps. viridis* (Shopes and Wraight, 1985; Gast *et al.*, 1985) is located close to BP$_L$. Its head group is bound exclusively to residues of the connection of the transmembrane helices D and E of the M subunit. Its isoprenoid side chain is mainly in contact with BP$_L$ and the L subunit. An important part of the binding site for the quinone head group is formed by tryptophan M250 whose indole ring is nearly parallel to the quinone ring system (see Figure 4). The distance between both ring systems is ~3.2 Å. The oxygen atoms of Q$_A$ appear to be hydrogen bonded to the peptide nitrogen of alanine M258, and to an imidazole nitrogen of histidine M217. Other parts of quinone-binding site are formed by the side chains of alanine M216, valine M263 and threonine M220.

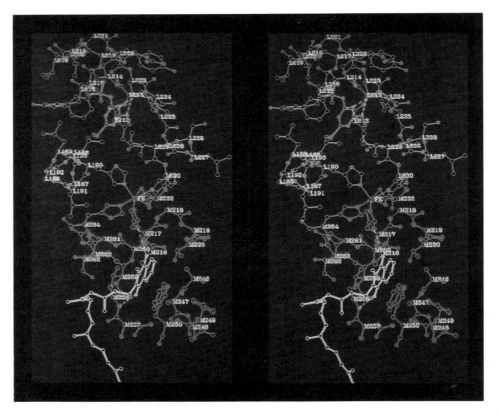

Fig. 4. (Stereo pair): menaquinone (yellow), non-heme-iron (red), terbutryn (light blue) and protein residues forming binding pockets. Possible hydrogen bonds are indicated in purple with dashed lines. Atom types as in Figure 1.

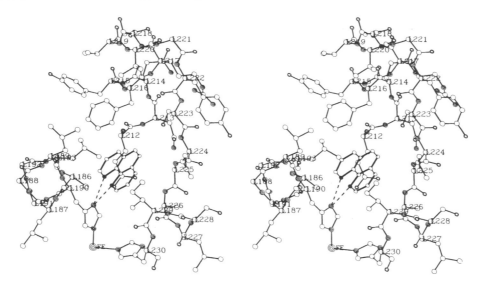

Fig. 5. (Stereo pair): *o*-phenanthroline and protein residues forming the binding pocket. Atom types as in Figure 1.

In agreement with recent e.p.r. (Dismukes *et al.*, 1984) and EXAFS data (Bunker *et al.*, 1982) Q_A is not directly bound to the ferrous non-heme iron atom. We find that five protein side chains, the histidines L190, L230, M217, M264 with the N_ϵ atoms, and glutamic acid M232 with the carboxylate group participate in binding the iron atom. The carboxylate group seems to act as a bidentate ligand. The distorted octahedral environment found for the iron atom is in good agreement with EXAFS

data (Bunker *et al.*, 1982; Eisenberger *et al.*, 1982).

The reaction centres loose the secondary quinone, Q_B, which is ubiquinone 9 in *Rps. viridis*, during isolation and crystallisation (Shopes and Wraight, 1985; Gast *et al.*, 1985). As described briefly (Deisenhofer *et al.*, 1985), quinones with shorter side chains can be reconstituted into the crystallised reaction centres. After data collection and analysis of the data by the difference Fourier technique, eight possible binding sites were found for ubiquinone 1 which was due to the high ubiquinone concentration (2 mM) which had to be used for technical reasons. Similar experiments with *o*-phenanthroline and terbutryn [2-thiomethyl-4-ethylamino-6-*t*-butylamino-*s*-triazin] which are thought to be competitive inhibitors for Q_B (Pfister *et al.*, 1981) yielded three binding sites for *o*-phenanthroline and one for terbutryn. The unique terbutryn-binding site was common to all three compounds. It is related to the Q_A-binding site by the local 2-fold symmetry which relates the L and M subunits, and the photosynthetic pigments. In addition it is the only binding site of ubiquinone-1 which is compatible with spectroscopic data (Dismukes *et al.*, 1984; Butler *et al.*, 1984). Unfortunately, the electron density for ubiquonine-1 was not well-defined enough to position and to orient the ubiquinone-1 uniquely. However, this was possible with terbutryn and *o*-phenanthroline (Figures 4 and 5). *o*-Phenanthroline binds close to the non-heme iron. Its two nitrogen atoms form a shared hydrogen bond with the imidazole nitrogen of histidine L190. Additionally it is in close contact with isoleucine L229 and leucine L193.

The binding site for terbutryn is further away from the non-heme iron atom. A hydrogen bond is possible between the peptide nitrogen of isoleucine L224 and N3 of the *s*-triazine ring system. A second hydrogen bond with the ethylamino nitrogen of terbutryn as hydrogen bond donor and the side chain oxygen of serine L223 as acceptor can be observed. The hydroxyl group of serine L223 also acts as a hydrogen bond donor to the side chain oxygen of asparagine L213. Extensive contacts are formed with valine L220, isoleucine L229 and phenylalanine L216. An interesting side chain is glutamic acid L212 which also participates in the formation of the terbutryn-binding site. It could be involved in donating protons to the reduced ubiquinone.

As shown in *Rb. sphaeroides*, resistance towards the herbicide terbutryn can be caused by a mutation which changes isoleucine L229 into methionine. The somewhat larger side chain of methionine may interfere with binding of terbutryn; specific chemical effects of the methionine sulphur atom may also be involved. Inspection of the herbicide-binding site and the ubiquinone-binding site shows that in this mutant also the ubiquinone-binding site must be influenced. This expectation is in agreement with the observed altered electron flow from Q_A to Q_B (Schenck *et al.*, 1986).

Resistance against the same type of herbicide is found in chloroplasts of several *Chlamydomonas* mutants (Erickson *et al.*, 1985). Mutation of phenylalanine 255 to tyrosine or of serine 264 to alanine or glycine in the D1 proteins causes herbicide resistance. Phenylalanine 255 in D1 corresponds to phenylalanine L216, which is in contact with terbutryn in the *Rps. viridis* reaction centre. However, it is not evident from our data why this mutation results in herbicide resistance. Serine 264 to D2 is very close to serine L223 in the alignment b of Michel *et al.* (1986). Thus the mutation of serine 264 to glycine or alanine may abolish one of the two possible hydrogen bonds between terbutryn and the protein subunit, and decrease the binding of terbutryn by several orders of magnitude.

Materials and methods

The initial atomic model of the reaction centre (Deisenhofer *et al.*, 1984, 1985) was built according to an electron density map at 3 Å resolution calculated with phases from multiple isomorphous replacement. Complete sequence information on subunits H, L and M (Michel *et al.*, 1985, 1986), and partial sequence information on the cytochrome subunit (Weyer, Michel and Lottspeich, unpublished) was incorporated into the model. Subsequently, crystallographic refinement of this model, using the method of Jack and Levitt (1978) and an interactive graphics display system driven by the program FRODO (Jones, 1978), was carried out

at a resolution of 2.9 Å. During this process the crystallographic R-value of the model could be reduced to 0.231 for 54341 unique reflections (R = $\Sigma \mid \mid F_{obs} \mid - \mid F_{calc} \mid \mid / \Sigma \mid F_{obs} \mid$, where F_{obs} and F_{calc} are observed and calculated structure factor amplitudes, respectively). The refined model of the 'native' reaction centre was then used as a starting point for model interpretation of difference Fourier maps, and for crystallographic refinement at 2.9 Å resolution of the complexes of the reaction centre with terbutryn and *o*-phenanthroline (Deisenhofer *et al.*, 1985); R-values of 0.229 (43 022 unique reflections) for the terbutryn complex, and of 0.238 (38 663 unique reflections) for the *o*-phenanthroline complex were obtained. The limited resolution available did not allow reliable determination of all atomic positions in the reaction centre. This especially applies to peptide orientations and small side chains, e.g. serine. A detailed description of crystallographic refinement of reaction centre models will be given elsewhere.

Amino acid residues were counted as neighbours of pigment molecules if interatomic distances of <4 Å were calculated from the refined atomic coordinates; residues neighbouring exclusively phytyl chains or the isoprenoid chain of the menaquinone were omitted. A donor−acceptor distance of <3.4 Å, and suitable geometry were used as criteria for assuming a hydrogen bond. The rather large cut-off distance for hydrogen bonds is necessary because of the expected inaccuracy of some of the atomic coordinates.

Acknowledgements

We thank Professors D.Oesterhelt and R.Huber for support and discussions; Ciba-Geigy AG, Basel, for performing an X-ray structure analysis of terbutryn and providing the coordinates, Dr Pfister (Ciba-Geigy AG) for repeated gifts of terbutryn, and the Deutsche Forschungsgemeinschaft (SFB 143) and the Max-Planck-Gesellschaft for financial help.

References

Allen,J. and Feher,G. (1984) *Proc. Natl. Acad. Sci. USA*, **81**, 4795−4799.
Breton,J. (1985) *Biochim. Biophys. Acta*, **810**, 235−245.
Bunker,G., Stern,E.A., Blankenship,R.E. and Parson,W.W. (1982) *Biophys. J.*, **37**, 539−551.
Butler,W.F., Calvo,R., Fredkin,D.R., Isaacson,R.A., Okamura,M.Y. and Feher,G. (1984) *Biophys. J.*, **45**, 947−973.
Chang,C.H., Schiffer,M., Tiede,D., Smith,U. and Norris,J. (1985) *J. Mol. Biol.*, **186**, 201−203.
Deisenhofer,J., Epp,O., Miki,K., Huber,R. and Michel,H. (1984) *J. Mol. Biol.*, **180**, 385−398.
Deisenhofer,J., Epp,O., Miki,K., Huber,R. and Michel,H. (1985) *Nature*, **318**, 681−624.
Dismukes,G.C., Frank,H.A., Friesner,R. and Sauer,K. (1984) *Biochim. Biophys. Acta*, **764**, 253−271.
Eisenberger,P., Okamura,M.Y. and Feher,G. (1982) *Biophys. J.*, **37**, 523−538.
Erickson,J.M., Rahire,M., Rochaix,J.D. and Mets,L. (1985) *Science*, **228**, 204−207.
Feher,G. and Okamura,M.Y. (1978) In Clayton,R.K. and Sistrom,W.R. (eds), *The Photosynthetic Bacteria*. Plenum Press, NY, pp. 349−386.
Gast,P. and Norris,J.R. (1984) *FEBS Lett.*, **177**, 277−280.
Gast,P., Michalski,T.J., Hunt,J.E. and Norris,J.R. (1985) *FEBS Lett.*, **179**, 325−328.
Hearst,J. (1986) In Staehelin,A.C. and Arntzen,C.J. (eds), *Encyclopedia of Plant Physiology: Photosynthesis III*. Springer, Berlin, Vol. 19, pp. 382−389.
Hoff,A.J. (1982) In Fong,F.K. (ed.), *Molecular Biology, Biochemistry, and Biophysics*. Springer, Berlin, Vol. 35, pp. 80−151, 322−326.
Imhoff,J.F., Trüper,H.G. and Pfenning,N. (1984) *Int. J. Syst. Bacteriol.*, **34**, 340−343.
Jack,A. and Levitt,M. (1978) *Acta Crystallogr.*, **A34**, 931−935.
Jones,T.A. (1978) *J. Appl. Crystallogr.*, **11**, 268−272.
Kirmaier,C., Holten,D. and Parson,W.W. (1985) *FEBS Lett.*, **185**, 76−82.
Knapp,E.W., Fischer,S.F., Zinth,W., Sander,M., Kaiser,W., Deisenhofer,J. and Michel,H. (1985) *Proc. Natl. Acad. Sci. USA*, **82**, 8463−8467.
Lesk,A.M. and Hardman,K. (1985) *Methods Enzymol.*, **115**, 381−390.
Lutz,M. and Robert,B. (1985) In Michel-Beyerle,M.E. (ed.), *Antennas and Reaction Centers of Photosynthetic Bacteria*. Springer, Berlin, pp. 138−145.
Martin,J.-L., Breton,J., Hoff,A.J., Migus,A. and Antonetty,A. (1986) *Proc. Natl. Acad. Sci. USA*, **83**, 957−961.
Michel,H. (1982) *J. Mol. Biol.*, **158**, 567−572.
Michel,H. and Deisenhofer,J. (1986) In Staehelin,A.C. and Arntzen,C.J. (eds), *Encyclopedia of Plant Physiology: Photosynthesis III*. Springer, Berlin, Vol. 19, pp. 371−381.
Michel,H., Weyer,K.A., Gruenberg,H. and Lottspeich,F. (1985) *EMBO J.*, **4**, 1667−1672.
Michel,H., Weyer,K.A., Gruenberg,H., Dunger,I., Oesterhelt,D. and Lottspeich,F. (1986) *EMBO J.*, **5**, 1149−1158.

Norris,J.R., Uphaus,R.A., Crespi,H.L. and Katz,J.J. (1971) *Proc. Natl. Acad. Sci. USA,* **68**, 625−628.

Parson,W.W. (1982) *Annu. Rev. Biophys. Bioeng.,* **11**, 57−80.

Pfister,K., Steinback,K.E., Gardner,G. and Arntzen,C.J. (1981) *Proc. Natl. Acad. Sci. USA,* **78**, 981−985.

Schenck,C.C., Sistrom,W.R., Bunzow,J.R., Rambousek,E.L. and Capaldi,R.A. (1986) *Biochemistry,* in press.

Shopes,R.J. and Wraight,C.A. (1985) *Biochim. Biophys. Acta,* **806**, 348−356.

Shuvalov,V.A. and Duysens,L.N.M. (1986) *Proc. Natl. Acad. Sci. USA,* **83**, 1690−1694.

Thornber,J.P., Dutton,P.L., Fajer,J., Forman,A., Holten,D., Olsen,J.M., Parson,W.S., Prince,R.C., Tiede,D.M. and Windsor,T.W. (1977) In Hall,D.O., Coombs,J. and Goodwin,T.W., *Proceedings of the Fourth International Congress on Photosynthesis.* The Biochemical Society, London, pp. 55−70

Vermeglio,A. and Paillotin,G. (1982) *Biochem. Biophys. Acta,* **681**, 32−40.

Youvan,D.C, Bylina,E.J., Alberti,M., Begusch,H. and Hearst,J.E. (1984) *Cell,* **37**, 949−957.

Zinth,W., Kaiser,W. and Michel,H. (1983) *Biochim. Biophys. Acta,* **723**, 128−131.

Zinth,W., Sander,M., Dobler,J., Kaiser,W. and Michel,H. (1985) In Michel-Beyerle,M.E. (ed.), *Antennas and Reaction Centers of Photosynthetic Bacteria.* Springer, Berlin, pp. 97−102.

Received on 27 June 1986

THE HEMOGLOBIN MOLECULE

by M. F. Perutz

In 1937, a year after I entered the University of Cambridge as a graduate student, I chose the X-ray analysis of hemoglobin, the oxygen-bearing protein of the blood, as the subject of my research. Fortunately the examiners of my doctoral thesis did not insist on a determination of the structure, otherwise I should have had to remain a graduate student for 23 years. In fact, the complete solution of the problem, down to the location of each atom in this giant molecule, is still outstanding, but the structure has now been mapped in enough detail to reveal the intricate three-dimensional folding of each of its four component chains of amino acid units, and the positions of the four pigment groups that carry the oxygen-combining sites.

The folding of the four chains in hemoglobin turns out to be closely similar to that of the single chain of myoglobin, an oxygen-bearing protein in muscle whose structure has been elucidated in atomic detail by my colleague John C. Kendrew and his collaborators. Correlation of the structure of the two proteins allows us to specify quite accurately, by purely physical methods, where each amino acid unit in hemoglobin lies with respect to the twists and turns of its chains.

Physical methods alone, however, do not yet permit us to decide which of the 20 different kinds of amino acid units occupies any particular site. This knowledge has been supplied by chemical analysis; workers in the U.S. and in Germany have determined the sequence of the 140-odd amino acid units along each of the hemoglobin chains. The combined results of the two different methods of approach now provide an accurate picture of many facets of the hemoglobin molecule.

In its behavior hemoglobin does not resemble an oxygen tank so much as a molecular lung. Two of its four chains shift back and forth, so that the gap between them becomes narrower when oxygen molecules are bound to the hemoglobin, and wider when the oxygen is released. Evidence that the chemical activities of hemoglobin and other proteins are accompanied by structural changes had been discovered before, but this is the first time that the nature of such a change has been directly demonstrated. Hemoglobin's change of shape makes me think of it as a breathing molecule, but paradoxically it expands, not when oxygen is taken up but when it is released.

When I began my postgraduate work in 1936 I was influenced by three inspiring teachers. Sir Frederick Gowland Hopkins, who had received a Nobel prize in 1929 for discovering the growth-stimulating effect of vitamins, drew our attention to the central role played by enzymes in catalyzing chemical reactions in the living cell. The few enzymes isolated at that time had all proved to be proteins. David Keilin, the discoverer of several of the enzymes that catalyze the processes of respiration, told us how the chemical affinities and catalytic properties of iron atoms were altered when the iron combined with different proteins. J. D. Bernal, the X-ray crystallographer, was my research supervisor. He and Dorothy Crowfoot Hodgkin had taken the first X-ray diffraction pictures of crystals of protein a year or two before I arrived, and they had discovered that protein molecules, in spite of their large size, have highly ordered structures. The wealth of sharp X-ray diffraction spots produced by a single crystal of an enzyme such as pepsin could be explained only if every one, or almost every one, of the 5,000 atoms in the pepsin molecule occupied a definite position that was repeated in every one of the myriad of pepsin molecules packed in the crystal. The notion is commonplace now, but it caused a sensation at a time when proteins were still widely regarded as "colloids" of indefinite structure.

In the late 1930's the importance of the nucleic acids had yet to be discovered; according to everything I had learned the "secret of life" appeared to be concealed in the structure of proteins. Of all the methods available in chemistry and physics, X-ray crystallography seemed to offer the only chance, albeit an extremely remote one, of determining that structure.

The number of crystalline proteins then available was probably not more than a dozen, and hemoglobin was an obvious candidate for study because of its supreme physiological importance, its ample supply and the ease with which it could be crystallized. All the same, when I chose the X-ray analysis of hemoglobin as the subject of my Ph.D. thesis, my fellow students regarded me with a pitying smile. The most complex organic substance whose structure had yet been determined by X-ray analysis was the molecule of the dye phthalocyanin, which contains 58 atoms. How could I hope to locate the thousands of atoms in the molecule of hemoglobin?

The Function of Hemoglobin

Hemoglobin is the main component of the red blood cells, which carry oxygen from the lungs through the arteries to the tissues and help to carry carbon dioxide through the veins back to the lungs. A single red blood cell contains about 280 million molecules of hemoglobin. Each molecule has 64,500 times the weight of a hydrogen atom and is made up of about 10,000 atoms of hydrogen, carbon, nitrogen, oxygen and sulfur, plus four atoms of iron, which are more important than all the rest. Each iron atom lies at the center of the group of atoms that form the pigment called heme, which gives blood its red color and its ability to combine with oxygen. Each heme group is enfolded in one of the four chains of amino acid units that collectively constitute the protein part of the molecule, which is called globin. The four chains of globin consist of two identical pairs. The members of one pair are known as alpha chains and those of the other as beta chains. Together the four chains contain a total of 574 amino acid units.

In the absence of an oxygen carrier a liter of arterial blood at body temperature could dissolve and transport no more than three milliliters of oxygen. The presence of hemoglobin increases this quantity 70 times. Without hemoglobin large animals could not get enough oxygen to exist. Similarly, hemoglobin is responsible for carrying more than 90 percent of the carbon dioxide transported by venous blood.

Each of the four atoms of iron in the hemoglobin molecule can take up one molecule (two atoms) of oxygen. The reaction is reversible in the sense that oxygen is taken up where it is plentiful, as in the lungs, and released where it is scarce, as in the tissues. The reaction is accompanied by a change in color: hemoglobin containing oxygen, known as oxyhemoglobin, makes arterial blood look scarlet; reduced, or oxygen-free, hemoglobin makes venous blood look purple. The term "reduced" for the oxygen-free form is really a misnomer because "reduced" means to the chemist that electrons have been added to an atom or a group of atoms. Actually, as James B. Conant of Harvard University demonstrated in 1923, the iron atoms in both reduced hemoglobin and oxyhemoglobin are in the same electronic condition: the divalent, or ferrous, state. They become oxidized to the trivalent, or ferric, state if hemoglobin is treated with a ferricyanide or removed from the red cells and exposed to the air for a considerable time; oxidation also occurs in certain blood diseases. Under these conditions hemoglobin turns brown and is known as methemoglobin, or ferrihemoglobin.

Ferrous iron acquires its capacity for binding molecular oxygen only through its combination with heme and globin. Heme alone will not bind oxygen, but the specific chemical environment of the globin makes the combina-

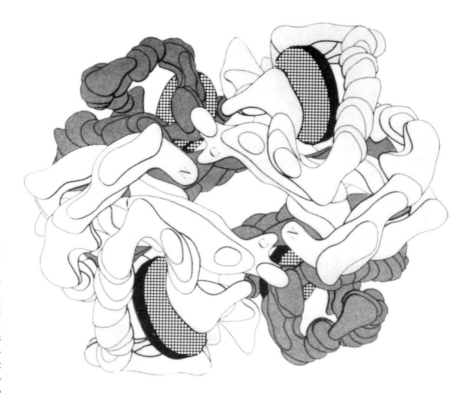

HEMOGLOBIN MOLECULE, as deduced from X-ray diffraction studies, is shown from above (*top*) and side (*bottom*). The drawings follow the representation scheme used in three-dimensional models built by the author and his co-workers. The irregular blocks represent electron-density patterns at various levels in the hemoglobin molecule. The molecule is built up from four subunits: two identical alpha chains (*light blocks*) and two identical beta chains (*dark blocks*). The letter "N" in the top view identifies the amino ends of the two alpha chains; the letter "C" identifies the carboxyl ends. Each chain enfolds a heme group (*checkered disk*), the iron-containing structure that binds oxygen to the molecule.

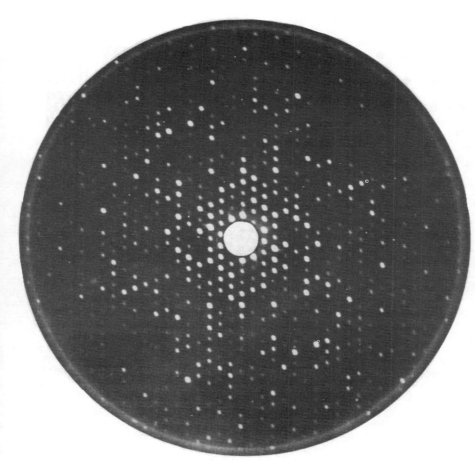

X-RAY DIFFRACTION PATTERN was made from a single crystal of hemoglobin that was rotated during the photographic exposure. Electrons grouped around the centers of the atoms in the crystal scatter the incident X rays, producing a symmetrical array of spots. Spots that are equidistant from the center and opposite each other have the same density.

tion possible. In association with other proteins, such as those of the enzymes peroxidase and catalase, the same heme group can exhibit quite different chemical characteristics.

The function of the globin, however, goes further. It enables the four iron atoms within each molecule to interact in a physiologically advantageous manner. The combination of any three of the iron atoms with oxygen accelerates the combination with oxygen of the fourth; similarly, the release of oxygen by three of the iron atoms makes the fourth cast off its oxygen faster. By tending to make each hemoglobin molecule carry either four molecules of oxygen or none, this interaction ensures efficient oxygen transport.

I have mentioned that hemoglobin also plays an important part in bearing carbon dioxide from the tissues back to the lungs. This gas is not borne by the iron atoms, and only part of it is bound directly to the globin; most of it is taken up by the red cells and the noncellular fluid of the blood in the form of bicarbonate. The transport of bicarbonate is facilitated by the disappearance of

an acid group from hemoglobin for each molecule of oxygen discharged. The reappearance of the acid group when oxygen is taken up again in the lungs sets in motion a series of chemical reactions that leads to the discharge of carbon dioxide. Conversely, the presence of bicarbonate and lactic acid in the tissues accelerates the liberation of oxygen.

Breathing seems so simple, yet it appears as if this elementary manifestation of life owes its existence to the interplay of many kinds of atoms in a giant molecule of vast complexity. Elucidating the structure of the molecule should tell us not only what the molecule looks like but also how it works.

The Principles of X-Ray Analysis

The X-ray study of proteins is sometimes regarded as an abstruse subject comprehensible only to specialists, but the basic ideas underlying our work are so simple that some physicists find them boring. Crystals of hemoglobin and other proteins contain much water and, like living tissues, they tend to lose their regularly ordered structure on dry-

ing. To preserve this order during X-ray analysis crystals are mounted wet in small glass capillaries. A single crystal is then illuminated by a narrow beam of X rays that are essentially all of one wavelength. If the crystal is kept stationary, a photographic film placed behind it will often exhibit a pattern of spots lying on ellipses, but if the crystal is rotated in certain ways, the spots can be made to appear at the corners of a regular lattice that is related to the arrangement of the molecules in the crystal [see illustration at left]. Moreover, each spot has a characteristic intensity that is determined in part by the arrangement of atoms inside the molecules. The reason for the different intensities is best explained in the words of W. L. Bragg, who founded X-ray analysis in 1913—the year after Max von Laue had discovered that X rays are diffracted by crystals—and who later succeeded Lord Rutherford as Cavendish Professor of Physics at Cambridge:

"It is well known that the form of the lines ruled on a [diffraction] grating has an influence on the relative intensity of the spectra which it yields. Some spectra may be enhanced, or reduced, in intensity as compared with others. Indeed, gratings are sometimes ruled in such a way that most of the energy is thrown into those spectra which it is most desirable to examine. The form of the line on the grating does not influence the positions of the spectra, which depend on the number of lines to the centimetre, but the individual lines scatter more light in some directions than others, and this enhances the spectra which lie in those directions.

"The structure of the group of atoms which composes the unit of the crystal grating influences the strength of the various reflexions in exactly the same way. The rays are diffracted by the electrons grouped around the centre of each atom. In some directions the atoms conspire to give a strong scattered beam, in others their effects almost annul each other by interference. The exact arrangement of the atoms is to be deduced by comparing the strength of the reflexions from different faces and in different orders."

Thus there should be a way of reversing the process of diffraction, of proceeding backward from the diffraction pattern to an image of the arrangement of atoms in the crystal. Such an image can actually be produced, somewhat laboriously, as follows. It will be noted that spots on opposite sides of the center of an X-ray picture have the same

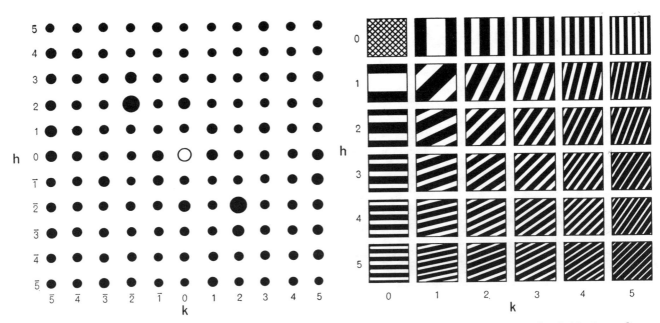

INTERPRETATION OF X-RAY IMAGE can be done with a special optical device to generate a set of diffraction fringes (*right*) from the spots in an X-ray image (*left*). Each pair of symmetrically related spots produces a unique set of fringes. Thus the spots in-dexed 2,$\bar{2}$ and $\bar{2}$,2 yield the fringes indexed 2,2. A two-dimensional image of the atomic structure of a crystal can be generated by printing each set of fringes on the same sheet of photographic paper. But the phase problem (*below*) must be solved first.

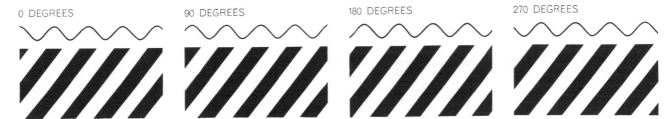

PHASE PROBLEM arises because the spots in an X-ray image do not indicate how the fringes are related in phase to an arbitrarily chosen common origin. Here four identical sets of fringes are related by different phases to the point of origin at the top left corner. The phase marks the distance of the wave crest from the origin, measured in degrees. One wavelength is 360 degrees.

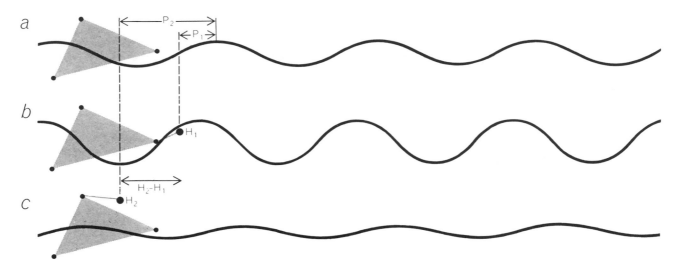

HEAVY-ATOM REPLACEMENT METHOD provides information about phases by changing the intensities of the X-ray diffraction pattern. In *a* a highly oversimplified protein (a triangle of three atoms) scatters a sinusoidal wave that represents the amplitude and phase of a single set of fringes. In *b* and *c*, after heavy atoms H_1 and H_2 are attached to the protein in different positions, the wave is changed in amplitude and phase. The heavy atoms can serve as points of common origin for measuring the magnitude of the phases (P_1 and P_2) of waves scattered by the unaltered protein. The distance between H_1 and H_2 must be accurately known.

degree of intensity. With the aid of a simple optical device each symmetrically related pair of spots can be made to generate a set of diffraction fringes, with an amplitude proportional to the square root of the intensity of the spots. The device, which was invented by Bragg and later developed by H. Lipson and C. A. Taylor at the Manchester College of Science and Technology, consists of a point source of monochromatic light, a pair of plane-convex lenses and a microscope. The pair of spots in the diffraction pattern is represented by a pair of holes in a black mask that is placed between the two lenses. If the point source is placed at the focus of one of the lenses, the waves of parallel light emerging from the two holes will interfere with one another at the focus of the second lens, and their interference pattern, or diffraction pattern, can be observed or photographed through the microscope.

Imagine that each pair of symmetrically related spots in the X-ray picture is in turn represented by a pair of holes in a mask, and that its diffraction fringes are photographed. Each set of fringes will then be at right angles to the line joining the two holes, and the distance between the fringes will be inversely proportional to the distance between the holes. If the spots are numbered from the center along two mutually perpendicular lines by the indices h and k, the relation between any pair of spots and its corresponding set of fringes would be as shown in the top illustration on the preceding page.

The Phase Problem

An image of the atomic structure of the crystal can be generated by printing each set of fringes in turn on the same sheet of photographic paper, or by superposing all the fringes and making a print of the light transmitted through them. At this point, however, a fatal complication arises. In order to obtain the right image one would have to place each set of fringes correctly with respect to some arbitrarily chosen common origin [see middle illustration on preceding page]. At this origin the amplitude of any particular set of fringes may show a crest or trough or some intermediate value. The distance of the wave crest from the origin is called the phase. It is almost true to say that by superposing sets of fringes of given amplitude one can generate an infinite number of different images, depending on the choice of phase for each set of fringes. By itself the X-ray picture tells us only about the amplitudes and nothing about the phases of the fringes to be generated by each pair of spots, which means that half the information needed for the production of the image is missing.

The missing information makes the diffraction pattern of a crystal like a hieroglyphic without a key. Having spent years hopefully measuring the intensities of several thousand spots in the diffraction pattern of hemoglobin, I found myself in the tantalizing position of an explorer with a collection of tablets engraved in an unknown script. For some time Bragg and I tried to develop methods for deciphering the phases, but with only limited success. The solution finally came in 1953, when I discovered that a method that had been developed by crystallographers for solving the phase problem in simpler structures could also be applied to proteins.

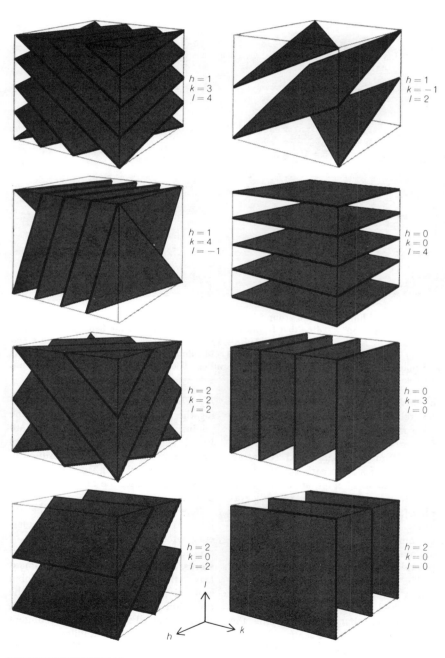

$h = 1$
$k = 3$
$l = 4$

$h = 1$
$k = -1$
$l = 2$

$h = 1$
$k = 4$
$l = -1$

$h = 0$
$k = 0$
$l = 4$

$h = 2$
$k = 2$
$l = 2$

$h = 0$
$k = 3$
$l = 0$

$h = 2$
$k = 0$
$l = 2$

$h = 2$
$k = 0$
$l = 0$

THREE-DIMENSIONAL FRINGES are needed to build up an image of protein molecules. For this purpose many different X-ray diffraction images are prepared and symmetrically related pairs of spots are indexed in three dimensions: h, k and l and $\bar{h}$, $\bar{k}$ and $\bar{l}$. Each pair of spots yields a three-dimensional fringe like those shown here. Fringes from thousands of spots must be superposed in proper phase to build up an image of the molecule.

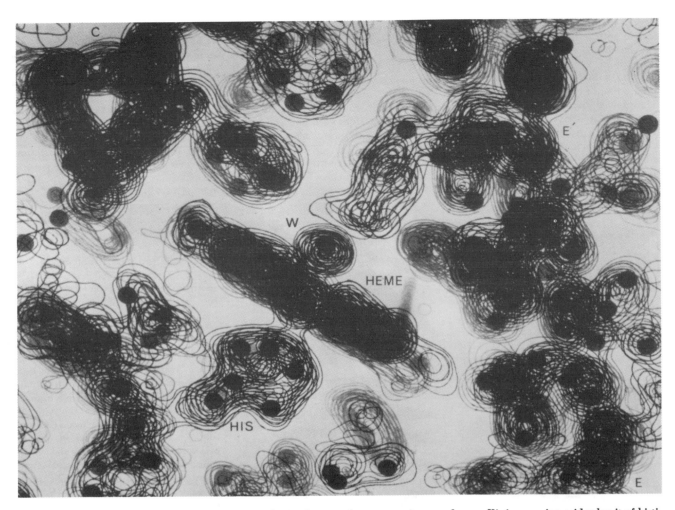

CONTOUR MAPS, drawn on stacked sheets of clear plastic, show a portion of the myoglobin molecule as revealed by superposition of three-dimensional fringe patterns. The maps were made by John C. Kendrew and his associates at the University of Cambridge. Myoglobin is very similar to the beta chain of hemoglobin. The heme group is seen edge on. *His* is an amino acid subunit of histidine that is attached to the iron atom of the heme group. *W* is a water molecule linked to the iron atom. The region between *E* and *E'* represents amino acid subunits arranged in an alpha helix. *C* is an alpha helix seen end on. The black dots mark atomic positions.

In this method the molecule of the compound under study is modified slightly by attaching heavy atoms such as those of mercury to definite positions in its structure. The presence of a heavy atom produces marked changes in the intensities of the diffraction pattern, and this makes it possible to gather information about the phases. From the difference in amplitude in the absence or presence of a heavy atom, the distance of the wave crest from the heavy atom can be determined for each set of fringes. Thus with the heavy atom serving as a common origin the magnitude of the phase can be measured. The bottom illustration on page 67 shows how the phase of a single set of fringes, represented by a sinusoidal wave that is supposedly scattered by the oversimplified protein molecule, can be measured from the increase in amplitude produced by the heavy atom H_1.

Unfortunately this still leaves an am-

biguity of sign; the experiment does not tell us whether the phase is to be measured from the heavy atom in the forward or the backward direction. If n is the number of diffracted spots, an ambiguity of sign in each set of fringes would lead to 2^n alternative images of the structure. The Dutch crystallographer J. M. Bijvoet had pointed out some years earlier in another context that the ambiguity could be resolved by examining the diffraction pattern from a second heavy-atom compound.

The bottom illustration on page 67 shows that the heavy atom H_2, which is attached to the protein in a position different from that of H_1, diminishes the amplitude of the wave scattered by the protein. The degree of attenuation allows us to measure the distance of the wave crest from H_2. It can now be seen that the wave crest must be in front of H_1; otherwise its distance from H_1 could not be reconciled with its distance from

H_2. The final answer depends on knowing the length and direction of the line joining H_2 to H_1. These quantities are best calculated by a method that does not easily lend itself to exposition in nonmathematical language. It was devised by my colleague Michael G. Rossmann.

The heavy-atom method can be applied to hemoglobin by attaching mercury atoms to the sulfur atoms of the amino acid cysteine. The method works, however, only if this attachment leaves the structure of the hemoglobin molecules and their arrangement in the crystal unaltered. When I first tried it, I was not at all sure that these stringent demands would be fulfilled, and as I developed my first X-ray photograph of mercury hemoglobin my mood alternated between sanguine hopes of immediate success and desperate forebodings of all the possible causes of failure. When the diffraction spots ap-

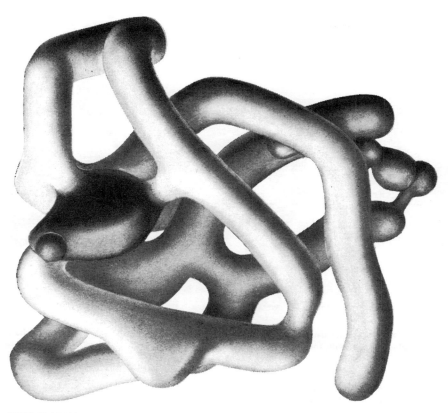

MYOGLOBIN MOLECULE, as first reconstructed at low resolution by Kendrew and his co-workers in 1957, had this rather repulsive visceral appearance. The sausage-like knot marks the path of the amino acid chain of the molecule. The dark disklike shape (here placed at an incorrect angle) is the heme group. A more detailed and more correct view of myoglobin, as seen from the other side, appears at bottom right on the opposite page.

compounds of the protein, each with heavy atoms attached to different positions in the molecule. Then the results have to be corrected by various geometric factors before they are finally used to build up an image through the superposition of tens of thousands of fringes. In the final calculation tens of millions of numbers may have to be added or subtracted. Such a task would have been quite impossible before the advent of high-speed computers, and we have been fortunate in that the development of computers has kept pace with the expanding needs of our X-ray analyses.

While I battled with technical difficulties of various sorts, my colleague John Kendrew successfully applied the heavy-atom method to myoglobin, a protein closely related to hemoglobin [see "The Three-dimensional Structure of a Protein Molecule," by John C. Kendrew; SCIENTIFIC AMERICAN, December, 1961]. Myoglobin is simpler than hemoglobin because it consists of only one chain of amino acid units and one heme group, which binds a single molecule of oxygen. The complex interaction phenomena involved in hemoglobin's dual function as a carrier of oxygen and of carbon dioxide do not occur in myoglobin, which acts simply as an oxygen store.

Together with Howard M. Dintzis and G. Bodo, Kendrew was brilliantly successful in managing to prepare as many as five different crystalline heavy-atom compounds of myoglobin, which meant that the phases of the diffraction spots could be established very accurately. He also pioneered the use of high-speed computers in X-ray analysis. In 1957 he and his colleagues obtained the first three-dimensional representation of myoglobin [see illustration on this page].

It was a triumph, and yet it brought a tinge of disappointment. Could the search for ultimate truth really have revealed so hideous and visceral-looking an object? Was the nugget of gold a lump of lead? Fortunately, like many other things in nature, myoglobin gains in beauty the closer you look at it. As Kendrew and his colleagues increased the resolution of their X-ray analysis in the years that followed, some of the intrinsic reasons for the molecule's strange shape began to reveal themselves. This shape was found to be not a freak but a fundamental pattern of nature, probably common to myoglobins and hemoglobins throughout the vertebrate kingdom.

In the summer of 1959, nearly 22 years after I had taken the first X-ray

peared in exactly the same position as in the mercury-free protein but with slightly altered intensities, just as I had hoped, I rushed off to Bragg's room in jubilant excitement, expecting that the structure of hemoglobin and of many other proteins would soon be determined. Bragg shared my excitement, and luckily neither of us anticipated the formidable technical difficulties that were to hold us up for another five years.

Resolution of the Image

Having solved the phase problem, at least in principle, we were confronted with the task of building up a structural image from our X-ray data. In simpler structures atomic positions can often be found from representations of the structure projected on two mutually perpendicular planes, but in proteins a three-dimensional image is essential. This can be attained by making use of the three-dimensional nature of the diffraction pattern. The X-ray diffraction pattern on page 66 can be regarded as a section through a sphere that is filled with layer after layer of diffraction spots. Each pair of spots can be made to generate a set of three-dimensional fringes like the ones shown on page 68. When their phases have been measured, they can be superposed by calculation to build up a three-dimensional image of the protein. The final image is represented by a series of sections through the molecule, rather like a set of microtome sections through a piece of tissue, only on a scale 1,000 times smaller [see illustration on preceding page].

The resolution of the image is roughly equal to the shortest wavelength of the fringes used in building it up. This means that the resolution increases with the number of diffracted spots included in the calculation. If the image is built up from part of the diffraction pattern only, the resolution is impaired.

In the X-ray diffraction patterns of protein crystals the number of spots runs into tens of thousands. In order to determine the phase of each spot accurately, its intensity (or blackness) must be measured accurately several times over: in the diffraction pattern from a crystal of the pure protein and in the patterns from crystals of several

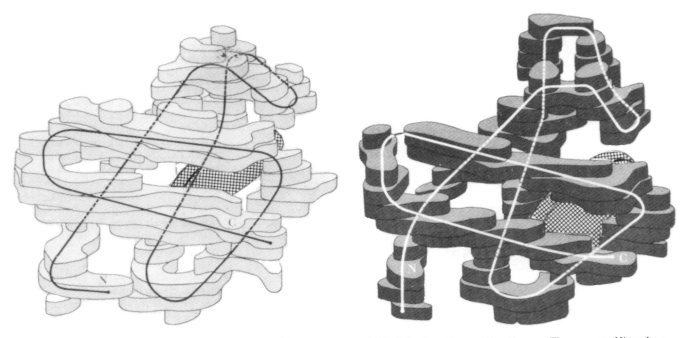

HEMOGLOBIN CHAINS, alpha at left and beta at right, are redrawn from models built by the author and his colleagues. The superposed lines show the course of the central chain. A heme group (*checkered*) is partly visible, tucked in the back of each model.

pictures of hemoglobin, its structure emerged at last. Michael Rossmann, Ann F. Cullis, Hilary Muirhead, Tony C. T. North and I were able to prepare a three-dimensional electron-density map of hemoglobin at a resolution of 5.5 angstrom units, about the same as that obtained for the first structure of myoglobin two years earlier. This resolution is sufficient to reveal the shape of the chain forming the backbone of a protein molecule but not to show the position of individual amino acids.

As soon as the numbers printed by the computer had been plotted on contour maps we realized that each of the four chains of hemoglobin had a shape closely resembling that of the single chain of myoglobin. The beta chain and myoglobin look like identical twins, and the alpha chains differ from them merely by a shortcut across one small loop [*see illustration below*].

Kendrew's myoglobin had been extracted from the muscle of the sperm whale; the hemoglobin we used came

from the blood of horses. More recent observations indicate that the myoglobins of the seal and the horse, and the hemoglobins of man and cattle, all have the same structure. It seems as though the apparently haphazard and irregular folding of the chain is a pattern specifically devised for holding a heme group in place and for enabling it to carry oxygen.

What is it that makes the chain take up this strange configuration? The extension of Kendrew's analysis to a high-

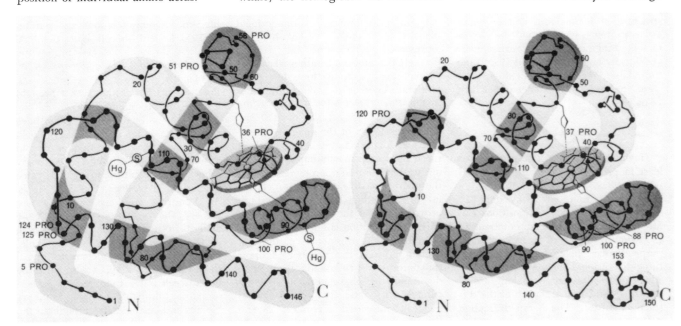

BETA CHAIN AND MYOGLOBIN appear at left and right. Every 10th amino acid subunit is marked, as are proline subunits (*gray*), which often coincide with turns in the chain. Balls marked "Hg" show where mercury atoms can be attached to sulfur atoms (*S*).

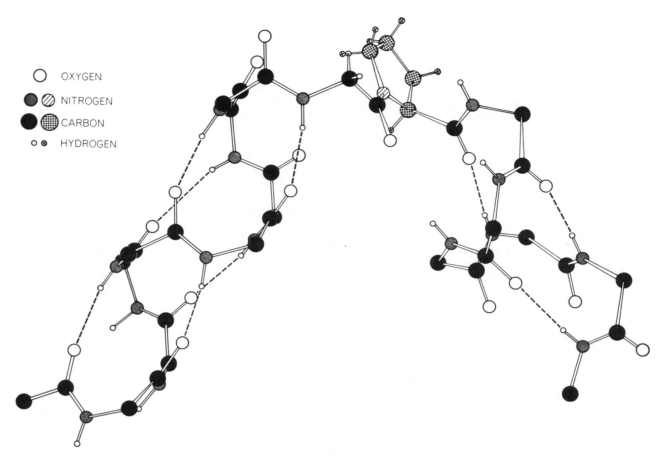

CORNER IN HEMOGLOBIN MOLECULE occurs where a subunit of the amino acid proline (*checkered disks*) falls between two helical regions in the beta chain. The chain is shown bare; all hydrogen atoms and amino acid side branches, except for proline, are removed.

OXYGEN
NITROGEN
CARBON
HYDROGEN

er resolution shows that the chain of myoglobin consists of a succession of helical segments interrupted by corners and irregular regions. The helical segments have the geometry of the alpha helix predicted in 1951 by Linus Pauling and Robert B. Corey of the California Institute of Technology. The heme group lies embedded in a fold of the chain, so that only its two acid groups protrude at the surface and are in contact with the surrounding water. Its iron atom is linked to a nitrogen atom of the amino acid histidine.

I have recently built models of the alpha and beta chains of hemoglobin and found that they follow an atomic pattern very similar to that of myoglobin. If two protein chains look the same, one would expect them to have much the same composition. In the language of protein chemistry this implies that in the myoglobins and hemoglobins of all vertebrates the 20 different kinds of amino acid should be present in about the same proportion and arranged in similar sequence.

Enough chemical analyses have been done by now to test whether or not this is true. Starting at the Rockefeller Institute and continuing in our laboratory, Allen B. Edmundson has determined the sequence of amino acid units in the molecule of sperm-whale myoglobin. The sequences of the alpha and beta chains of adult human hemoglobin have been analyzed independently by Gerhardt Braunitzer and his colleagues at the Max Planck Institute for Biochemistry in Munich, and by William H. Konigsberg, Robert J. Hill and their associates at the Rockefeller Institute. Fetal hemoglobin, a variant of the human adult form, contains a chain known as gamma, which is closely related to the beta chain. Its complete sequence has been analyzed by Walter A. Schroeder and his colleagues at the California Institute of Technology. The sequences of several other species of hemoglobin and that of human myoglobin have been partially elucidated.

The sequence of amino acid units in proteins is genetically determined, and changes arise as a result of mutation. Sickle-cell anemia, for instance, is an inherited disease due to a mutation in one of the hemoglobin genes. The mutation causes the replacement of a single amino acid unit in each of the beta chains. (The glutamic acid unit normally present at position No. 6 is replaced by a valine unit.) On the molecular scale evolution is thought to involve a succession of such mutations, altering the structure of protein molecules one amino acid unit at a time. Consequently when the hemoglobins of different species are compared, we should expect the sequences in man and apes, which are close together on the evolutionary scale, to be very similar, and those of mammals and fishes, say, to differ more widely. Broadly speaking, this is what is found. What was quite unexpected was the degree of chemical diversity among the amino acid sequences of proteins of similar three-dimensional structure and closely related function. Comparison of the known hemoglobin and myoglobin sequences shows only 15 positions—no more than one in 10—where the same amino acid unit is present in all species. In all the other positions one or more replacements have occurred in the course of evolution.

What mechanism makes these diverse

chains fold up in exactly the same way? Does a template force them to take up this configuration, like a mold that forces a car body into shape? Apart from the topological improbability of such a template, all the genetic and physicochemical evidence speaks against it, suggesting instead that the chain folds up spontaneously to assume one specific structure as the most stable of all possible alternatives.

Possible Folding Mechanisms

What is it, then, that makes one particular configuration more stable than all others? The only generalization to emerge so far, mainly from the work of Kendrew, Herman C. Watson and myself, concerns the distribution of the so-called polar and nonpolar amino acid units between the surface and the interior of the molecule.

Some of the amino acids, such as glutamic acid and lysine, have side groups of atoms with positive or negative electric charge, which strongly attract the surrounding water. Amino acid side groups such as glutamine or tyrosine, although electrically neutral as a whole, contain atoms of nitrogen or oxygen in which positive and negative charges are sufficiently separated to form dipoles; these also attract water, but not so strongly as the charged groups do. The attraction is due to a separation of charges in the water molecule itself, making it dipolar. By attaching themselves to electrically charged groups, or to other dipolar groups, the water molecules minimize the strength of the electric fields surrounding these groups and stabilize the entire structure by lowering the quantity known as free energy. The side groups of amino acids such

as leucine and phenylalanine, on the other hand, consist only of carbon and hydrogen atoms. Being electrically neutral and only very weakly dipolar, these groups repel water as wax does. The reason for the repulsion is strange and intriguing. Such hydrocarbon groups, as they are called, tend to disturb the haphazard arrangement of the liquid water molecules around them, making it ordered as it is in ice. The increase in order makes the system less stable; in physical terms it leads to a reduction of the quantity known as entropy, which is the measure of the disorder in a system. Thus it is the water molecules' anarchic distaste for the orderly regimentation imposed on them by the hydrocarbon side groups that forces these side groups to turn away from water and to stick to one another.

Our models have taught us that most electrically charged or dipolar side groups lie at the surface of the protein molecule, in contact with water. Nonpolar side groups, in general, are either confined to the interior of the molecule or so wedged into crevices on its surface as to have the least contact with water. In the language of physics, the distribution of side groups is of the kind leading to the lowest free energy and the highest entropy of the protein molecules and the water around them. (There is a reduction of entropy due to the orderly folding of the protein chain itself, which makes the system less stable, but this is balanced, at moderate temperatures, by the stabilizing contributions of the other effects just described.) It is too early to say whether these are the only generalizations to be made about the forces that stabilize one particular configuration of the protein chain in preference to all others.

At least one amino acid is known to be a misfit in an alpha helix, forcing the chain to turn a corner wherever the unit occurs. This is proline [*see illustration on opposite page*]. There is, however, only one corner in all the hemoglobins and myoglobins where a proline is always found in the same position: position No. 36 in the beta chain and No. 37 in the myoglobin chain [*see bottom illustration on page 71*]. At other corners the appearance of prolines is haphazard and changes from species to species. Elkan R. Blout of the Harvard Medical School finds that certain amino acids such as valine or threonine, if present in large numbers, inhibit the formation of alpha helices, but these do not seem to have a decisive influence in myoglobin and hemoglobin.

Since it is easier to determine the sequence of amino acid units in proteins than to unravel their three-dimensional structure by X rays, it would be useful to be able to predict the structure from the sequence. In principle enough is probably known about the forces between atoms and about the way they tend to arrange themselves to make such predictions feasible. In practice the enormous number of different ways in which a long chain can be twisted still makes the problem one of baffling complexity.

Assembling the Four Chains

If hemoglobin consisted of four identical chains, a crystallographer would expect them to lie at the corners of a regular tetrahedron. In such an arrangement each chain can be brought into congruence with any of its three neighbors by a rotation of 180 degrees about one of three mutually perpendicular

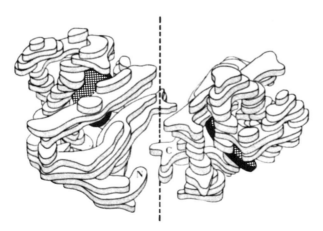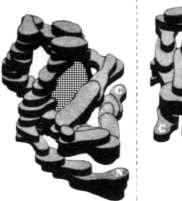

FOUR CHAINS OF HEMOGLOBIN are arranged in symmetrical fashion. Two alpha chains (*left*) and two beta chains (*right*) face each other across an axis of symmetry (*broken vertical lines*). In the assembled molecule the two alpha chains are inverted over the two beta chains and nested down between them. When arranged in this manner, the four chains lie at the corners of a tetrahedron.

axes of symmetry. Since the alpha and beta chains are chemically different, such perfect symmetry is unattainable, but the actual arrangement comes very close to it. As a first step in the assembly of the molecule two alpha chains are placed near a twofold symmetry axis, so that a rotation of 180 degrees brings one chain into congruence with its partner [*see illustration on preceding page*].

Next the same is done with the two beta chains. One pair, say the alpha chains, is then inverted and placed over the top of the other pair so that the four chains lie at the corners of a tetrahedron. A true twofold symmetry axis now passes vertically through the molecule, and "pseudo-axes" in two directions perpendicular to the first relate the alpha to the beta chains. Thus the arrangement is tetrahedral, but because of the chemical differences between the alpha and beta chains the tetrahedron is not quite regular.

The result is an almost spherical molecule whose exact dimensions are $64 \times 55 \times 50$ angstrom units. It is astonishing to find that four objects as irregular as the alpha and beta chains can fit together so neatly. On formal grounds one would expect a hole to pass through the center of the molecule because chains of amino acid units, being asymmetrical, cannot cross any symmetry axis. Such a hole is in fact found [*see top illustration on page 65*].

The most unexpected feature of the oxyhemoglobin molecule is the way the four heme groups are arranged. On the basis of their chemical interaction one would have expected them to lie close together. Instead each heme group lies in a separate pocket on the surface of the molecule, apparently unaware of the existence of its partners. Seen at the present resolution, therefore, the structure fails to explain one of the most important physiological properties of hemoglobin.

In 1937 Felix Haurowitz, then at the German University of Prague, discov-

ered an important clue to the molecular explanation of hemoglobin's physiological action. He put a suspension of needle-shaped oxyhemoglobin crystals away in the refrigerator. When he took the suspension out some weeks later, the oxygen had been used up by bacterial infection and the scarlet needles had been replaced by hexagonal plates of purple reduced hemoglobin. While Haurowitz observed the crystals under the microscope, oxygen penetrated between the slide and the cover slip, causing the purple plates to dissolve and the scarlet needles of hemoglobin to re-form. This transformation convinced Haurowitz that the reaction of hemoglobin with oxygen must be accompanied by a change in the structure of the hemoglobin molecule. In myoglobin, on the other hand, no evidence for such a change has been detected.

Haurowitz' observation and the enigma posed by the structure of oxyhemoglobin caused me to persuade a graduate student, Hilary Muirhead, to attempt an X-ray analysis at low resolution of the reduced form. For technical reasons human rather than horse hemoglobin was used at first, but we have now found that the reduced hemoglobins of man and the horse have very similar structures, so that the species does not matter here.

Unlike me, Miss Muirhead succeeded in solving the structure of her protein in time for her Ph.D. thesis. When we examined her first electron-density maps, we looked for two kinds of structural change: alterations in the folding of the individual chains and displacements of the chains with respect to each other. We could detect no changes in folding large enough to be sure that they were not due to experimental error. We did discover, however, that a striking displacement of the beta chains had taken place. The gap between them had widened and they had been shifted sideways, increasing the distance between their respective iron atoms from 33.4 to 40.3 angstrom units [*see illustration on page 76*]. The arrangement of the two alpha chains had remained unaltered, as far as we could judge, and the distance between the iron atoms in the beta chains and their nearest neighbors in the alpha chains had also remained the same. It looked as though the two beta chains had slid apart, losing contact with each other and somewhat changing their points of contact with the alpha chains.

F. J. W. Roughton and others at the University of Cambridge suggest that the change to the oxygenated form of

RESIDUE NUMBER	HEMOGLOBIN			MYOGLOBIN
	ALPHA	BETA	GAMMA	
81	MET	LEU	LEU	HIS
82	PRO	LYS	LYS	GLU
83	ASN	GLY	GLY	ALA
84	ALA	THR	THR	GLU
85	LEU	PHE	PHE	LEU
86	SER	ALA	ALA	LYS
87	ALA	THR	GLN	PRO
88	LEU	LEU	LEU	LEU
89	SER	SER	SER	ALA
90	ASP	GLU	GLU	GLN
91	LEU	LEU	LEU	SER
92	HIS	HIS	HIS	HIS
93	ALA	CYS	CYS	ALA
94	HIS	ASP	ASN	THR
95	LYS	LYS	LYS	LYS
96	LEU	LEU	LEU	HIS
97	ARG	HIS	HIS	LYS
98	VAL	VAL	VAL	ILEU
99	ASP	ASP	ASP	PRO
100	PRO	PRO	PRO	ILEU
101	VAL	GLU	GLU	LYS
102	ASP	ASN	ASN	TYR

ALA ALANINE	GLY GLYCINE	PRO PROLINE
ARG ARGININE	HIS HISTIDINE	SER SERINE
ASN ASPARAGINE	ILEU ISOLEUCINE	THR THREONINE
ASP ASPARTIC ACID	LEU LEUCINE	TYR TYROSINE
CYS CYSTEINE	LYS LYSINE	VAL VALINE
GLN GLUTAMINE	MET METHIONINE	
GLU GLUTAMIC ACID	PHE PHENYLALANINE	

AMINO ACID SEQUENCES are shown for corresponding stretches of the alpha and beta chains of hemoglobin from human adults, the gamma chain that replaces the beta chain in fetal human hemoglobin and sperm-whale myoglobin. Colored bars show where the same amino acid units are found either in all four chains or in the first three. Site numbers for the alpha chain and myoglobin are adjusted slightly because they contain a different number of amino acid subunits overall than do the beta and gamma chains. Over their full length of more than 140 subunits the four chains have only 20 amino acid subunits in common.

hemoglobin takes place after three of the four iron atoms have combined with oxygen. When the change has occurred, the rate of combination of the fourth iron atom with oxygen is speeded up several hundred times. Nothing is known as yet about the atomic mechanism that sets off the displacement of the beta chains, but there is one interesting observation that allows us at least to be sure that the interaction of the iron atoms and the change of structure do not take place unless alpha and beta chains are both present.

Certain anemia patients suffer from a shortage of alpha chains; the beta chains, robbed of their usual partners, group themselves into independent assemblages of four chains. These are known as hemoglobin *H* and resemble normal hemoglobin in many of their properties. Reinhold Benesch and Ruth E. Benesch of the Columbia University College of Physicians and Surgeons have discovered, however, that the four iron atoms in hemoglobin *H* do not interact, which led them to predict that the combination of hemoglobin *H* with oxygen should not be accompanied by a change of structure. Using crystals grown by Helen M. Ranney of the Albert Einstein College of Medicine, Lelio Mazzarella and I verified this prediction. Oxygenated and reduced hemoglobin *H* both resemble normal human reduced hemoglobin in the arrangement of the four chains.

The rearrangement of the beta chains must be set in motion by a series of atomic displacements starting at or near the iron atoms when they combine with oxygen. Our X-ray analysis has not yet reached the resolution needed to discern these, and it seems that a deeper understanding of this intriguing phenomenon may have to wait until we succeed in working out the structures of reduced hemoglobin and oxyhemoglobin at atomic resolution.

Allosteric Enzymes

There are many analogies between the chemical activities of hemoglobin and those of enzymes catalyzing chemical reactions in living cells. These analogies lead one to expect that some enzymes may undergo changes of structure on coming into contact with the substances whose reactions they catalyze. One can imagine that the active sites of these enzymes are moving mechanisms rather than static surfaces magically endowed with catalytic properties.

Indirect and tentative evidence suggests that changes of structure involv-

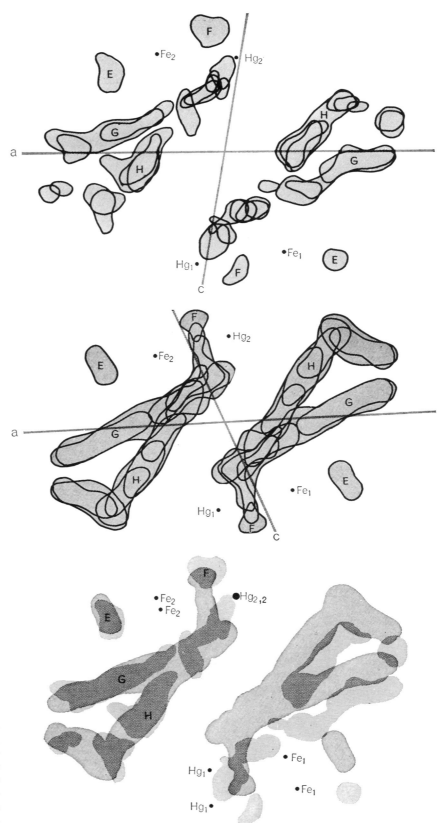

MOVEMENT OF HEMOGLOBIN CHAINS was discovered by comparing portions of the two beta chains in "reduced" (oxygen-free) human hemoglobin (*top*) with the same portions of horse hemoglobin containing oxygen (*middle*). The bottom illustration shows the outlines of the top and middle pictures superposed so that the mercury atoms (*Hg₂*) and helical regions (*E, F, G, H*) of the two chains at left coincide. The iron atoms (*Fe₂*) do not quite match. The chains at right are now seen to be shifted with respect to each other.

ing a rearrangement of subunits like that of the alpha and beta chains of hemoglobin do indeed occur and that they may form the basis of a control mechanism known as feedback inhibition. This is a piece of jargon that biochemistry has borrowed from electrical engineering, meaning nothing more complicated than that you stop being hungry when you have had enough to eat.

Constituents of living matter such as amino acids are built up from simpler substances in a series of small steps, each step being catalyzed by an enzyme that exists specifically for that purpose. Thus a whole series of different enzymes may be needed to make one amino acid. Such a series of enzymes appears to have built-in devices for ensuring the right balance of supply and demand. For example, in the colon bacillus the amino acid isoleucine is made from the amino acid threonine in several steps. The first enzyme in the series has an affinity for threonine: it catalyzes the removal of an amino group from it. H. Edwin Umbarger of the Long Island Biological Association in Cold Spring Harbor, N.Y., discovered that the action of the enzyme is inhibited by isoleucine, the end product of the last enzyme in the series. Jean-Pierre Changeux of the Pasteur Institute later showed that isoleucine acts not, as one might have expected, by blocking the site on the enzyme molecule that would otherwise combine with threonine but probably by combining with a different site on the molecule.

The two sites on the molecule must therefore interact, and Jacques Monod, Changeux and François Jacob have suggested that this is brought about by a rearrangement of subunits similar to that which accompanies the reaction of hemoglobin with oxygen. The enzyme is thought to exist in two alternative structural states: a reactive one when the supply of isoleucine has run out and an unreactive one when the supply exceeds demand. The discoverers have coined the name "allosteric" for enzymes of this kind.

The molecules of the enzymes suspected of having allosteric properties are all large ones, as one would expect them to be if they are made up of several subunits. This makes their X-ray analysis difficult. It may not be too hard to find out, however, whether or not a change of structure occurs, even if it takes a long time to unravel it in detail. In the meantime hemoglobin will serve as a useful model for the behavior of more complex enzyme systems.

Importance of DNA stiffness in protein–DNA binding specificity

M. E. Hogan* & R. H. Austin†

* Department of Molecular Biology and † Department of Physics, Princeton University, Princeton, New Jersey 08544, USA

From the first high-resolution structure of a repressor bound specifically to its DNA recognition sequence[1] it has been shown that the phage 434 repressor protein binds as a dimer to the helix. Tight, local interactions are made at the ends of the binding site, causing the central four base pairs (bp) to become bent and overtwisted. The centre of the operator is not in contact with protein but repressor binding affinity can be reduced at least 50-fold in response to a sequence change there[2]. This observation might be explained should the structure of the intervening DNA segment vary with its sequence, or if DNA at the centre of the operator resists the torsional and bending deformation necessary for complex formation in a sequence dependent fashion. We have considered the second hypothesis by demonstrating that DNA stiffness is sequence dependent. A method is formulated for calculating the stiffness of any particular DNA sequence, and we show that this predicted relationship between sequence and stiffness can explain the repressor binding data in a quantitative manner. We propose that the elastic properties of DNA may be of general importance to an understanding of protein–DNA binding specificity.

The fact that DNA behaves as a stiff elastic rod has been recognized since Peterlin's early analysis of light scattering data[3]. This rigidity is expressed as the persistence length P of the molecule, which is related to the Young's modulus, E (ref. 4):

$$P = EI/kT \qquad (1)$$

where I is the surface moment of inertia for a right cylinder, k is Boltzmann's constant and T is the absolute temperature; E is related to the stress which develops when the long axis of a rod is strained. The intrinsic shear modulus, G, relates the restoring stress when a lateral or torsional strain is applied to the material. The Young's and shear moduli are related by Poisson's ratio μ, which is also intrinsic:

$$G = E/2(1+\mu) \qquad (2)$$

where μ is close to 0.5 for polymeric materials[5,6].

Recently, it has been found that E and G are not constant for DNA, but are a function of bp composition[7-10] which is not surprising as the stacking and the hydrogen bonding stabilizing DNA are strongest for G·C bps. The available experiments do not measure the intrinsic moduli directly, but instead the extrinsic bending constant B and the torsional constant C, given by;

$$C = IG/L \qquad (3)$$

and

$$B = EIL \qquad (4)$$

where L is the length of the molecule under strain and I is the area moment of inertia.

In Table 1 we present measured values for the Young's modulus of DNA as a function of base composition. Although the modulus E is the fundamental parameter, physical techniques measure EI. To extract E from those data, we have calculated I, using the crystallographic radius for B-DNA (11 Å), although it will be shown that the significant conclusions to be drawn from these analyses are not dependent on the value chosen. It is clear that the available data are not comprehensive, and the theories used to calculate EI from the data not perfect, especially in analysis of anisotropy and quenching data. Consequently, when the individual methods are applied to similar random sequence DNA samples, calculated values for E can vary by as much as a factor of three. Nevertheless, there is a good qualitative agreement among the various methods which we have emphasized by normalizing the outcome of each, so

Reprinted from *Nature* **329**, 263–266 (1987); © Macmillan Magazines, Ltd.

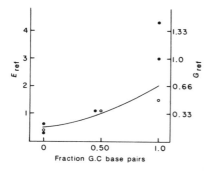

Fig. 1 Summary of the relation between stiffness and DNA sequence. The normalized Young's moduli E catalogued in Table 1 have been plotted against the mole fraction of G·C bps in the helix being studied. The corresponding shear moduli (G_{ref}) have been calculated from the Young's modulus, as described in the text. Solid line, compositional dependence predicted by equations (9) and (10) for random DNA sequences, using the consensus stiffness parameters in Table 2. ●, Measurements on random sequences or non-alternating synthetic helices, such as poly(dA)·poly(dT); ○, measurements on DNA sequences with a simple alternating sequence, such as poly (dA-dT) · poly(dA-dT) or poly (dG-dC) · poly(dG-dC).

that for a random sequence DNA (45% G + C) the Young's modulus is adjusted to a traditional value $(1.1 \times 10^9 \text{ dyn cm}^{-2})$, corresponding to a persistence length of 175 bps at 300 K. Although this normalization is not strictly valid, the arguments presented below show this first order correction provides a useful insight into the properties of DNA.

The shear modulus of DNA, G, is also presented in Table I, calculated from the Young's modulus using equation (2) and assuming that μ is 0.5, which has previously given close agreement with G as measured for DNA[7]. The relation does not hold for the homopolymer poly(dA) · poly(dT) (ref. 7), but this discrepancy could result from the atypical B-helix structure assumed by poly(A) sequences[11]. Because of this ambiguity, and because the Young's modulus of DNA has been measured by several methods whereas the shear modulus has not, our opinion is that, at present, the sequence dependence of DNA stiffness is best evaluated from calculations using the Young's modulus (as measured by light scattering, pulsed birefringence, triplet anisotropy decay and triplet quenching).

A conservative interpretation of the available data suggests that, at the extremes of base composition, E, and hence the bending and torsional stiffness of DNA, can vary by at least a factor of four (Fig. 1; Table 1).

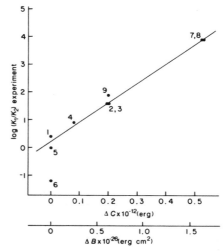

Fig. 2 Sequence dependence of 434 repressor binding: theory compared with experiment. The measured change in repressor binding affinity (from ref. 2) is compared to the change in DNA stiffness calculated from theory. See Table 2 for the numerical values. Experimental data are presented as log (K_1/K_2) plotted against ΔC or ΔB. Individual pairs (data against theory for each operator sequence) are identified by the numbering system in Table 2. Because C and B are proportional, their correlation with experiment has been presented together, by appropriate labelling of the abscissa. Solid line, linear least-squares analysis, assuming that the error in log(K_1/K_2) is ±0.2, as ascertained by the scatter of the binding data. A good linear correlation of that kind suggests that the measured binding constant change is determined by DNA stiffness effects.

Ethidium bromide fluorescence anisotropy experiments have surprisingly failed to detect sequence-dependent effects on the shear modulus[6,12]. The discrepancy between the ethidium bromide data and other physical measurements of DNA stiffness is a source of continuing controversy, but could be related to the fact that short-lived fluorescence decay measurements are most sensitive to high-frequency (local) motions of DNA, and for that reason are susceptible to changes in the DNA molecule which occur near the dye probe binding site. High-frequency DNA motions may be crucial to an understanding of its stability and function, but we will limit our present discussion of stiffness variation to results verified by several techniques.

To understand protein binding specificity in terms of DNA stiffness variation, it is first necessary to consider the work required to deform a helix. The mechanics of elastic media

Method Sample	Triplet anisotropy $E(E_{ref})$	Birefringence $E(E_{ref})$	Light scatter $E(E_{ref})$	Triplet quenching $E(E_{ref})$	Consensus values E	G
poly(dG) · poly(dC)	3.0 (2.4)			1.33 (4.3)	2	0.66
poly(dG-dC) · poly(dG-dC)		1.5 (1.5)			2	0.66
poly(dA-dC) · poly(dG-dT)	1.4 (1.1)				1	0.33
Random	1.4 (1.1)	1.1 (1.1)	1.1 (1.1)	0.34 (1.1)	1	0.33
poly(dA) · poly(dT)	0.8 (0.6)			0.10 (0.32)	0.5	0.167
poly(dA-dT) · poly(dA-dT)		0.36 (0.36)			0.5	0.167

Table 1 DNA flexibility measurements

These data are a compilation of data available in the literature. In each instance the Young's modulus E or shear modulus G has been presented in units of 10^9 dyne cm^{-2}. For each method which has been listed, a random sequence DNA was also measured. To compare the sequence dependence of those data more directly, we have normalized measured values E such that for each method, random sequence DNA is assigned the value 1.1×10^9. Those normalized moduli are presented in the parenthesis as E_{ref}. Triplet anisotropy[7] pulsed electric birefringence decay[8] light scattering[8] and triplet lifetime quenching[10] measurements are cited. For the birefringence and light scattering data, the original authors have calculated a persistence length for DNA. That value has been converted to a Young's modulus using equation (1) in the text and $I = 2.3 \times 10^{-28}$ cm^4. Consensus values for the Young's moduli E are an evaluation of the available literature and may be refined in the context of future experiments. Consensus values for G have been calculated using equation (2).

specifies that if a length L of DNA, with Young's modulus E and area moment of inertia I is strained by bending into an arc of radius S or twisting through an angle ϕ (in radians) the work done is;

$$U(\phi) = -IG\phi^2/2L = -C\phi^2/2 \qquad (5)$$

and

$$U(S) = -EIL/2S^2 = -B/2S^2 \qquad (6)$$

This linear elastic model may be too simple, because it overlooks the possibility that DNA bending and twisting are directional[13] and assumes that DNA can be modelled as a continuum rod. But, given that the same simplification is built into theories used to interpret the hydrodynamic and scattering data (see Table 1), the above equations are adequate approximations.

To complete our model, it is necessary to describe how the contribution of individual bps sums to produce a net bending constant B and torsional constant C. DNA flexibility as measured experimentally does not involve a disruption of bps but is a property of the duplex state, so sequence-dependent stiffness variation must be a function for the vertical (base-stacking) interaction between the bps. To a first approximation then, the stiffness of a DNA segment should vary as the frequency of nearest neighbours in the helix. Thus, a helix can be modelled as a heterogeneous elastic rod with E and G varying with the type of nearest neighbours. Equilibrium statics says that for a heterogeneous rod composed of N equally spaces subunits, the net torsional constant is:

$$1/C = 1/C_1 + 1/C_2 + \ldots 1/C_n \qquad (7)$$

and the bending constant is:

$$1/B = 1/N^2(1/B_1 + 1/B_2 + \ldots 1/B_n) \qquad (8)$$

where the subscript n refers to the individual elements of the rod. If, as the data suggest (Fig. 1), DNA stiffness is dominated by the nearest-neighbour interactions and is relatively insensitive to bp inversion we can usefully write the above equations;

$$1/C = \frac{L}{I}(f_{aa}/G_{aa} + f_{ag}/G_{ag} + f_{gg}/G_{gg}) \qquad (9)$$

and

$$1/B = \frac{1}{IL}(f_{aa}/E_{aa} + f_{ag}/E_{ag} + f_{gg}/E_{gg}) \qquad (10)$$

where L is the overall length of the helix segment in cm, G_{aa} (for example) is the shear modulus between two adjacent A·T bps and f_{aa} is the frequency of AT nearest neighbours, normalized to unity. The value for each element of E or G has been assigned by reference to the consensus values listed in Table 1. Thus, for any DNA segment

$$f_{aa} + f_{ag} + f_{gg} = 1 \qquad (11)$$

Using the above formalism, we can evaluate the contribution of DNA stiffness variation to the binding specificity of the 434 repressor. In Table 2 we have catalogued the three base-stacking interactions at the centre of the operator for each of the synthetic binding sites studied by Koudelka et al.[2] Based upon those nearest neighbour frequencies, we have used equation (9) to calculate the bending and torsional spring constant for the region (Table 2). Next, as a first approximation, we propose that when DNA sequence is changed at the centre of the operator (where there are no protein contacts), the resulting change in repressor affinity is due exclusively to a change in the binding free energy which must be spent to bend or twist the central DNA segment. This free energy is stored as bending and torsional strain, as described by equations (5) and (6). If two particular operator sequences labelled 1 and 2 have torsional strain energies $U_1(\phi)$ and $U_2(\phi)$ stored for a fixed angular strain ϕ, then the dissociation constant change which results from the stiffness difference can be expressed as the ratio K_1/K_2:

$$K_1/K_2 = \exp[2(U_2(\phi) - U_1(\phi))/kT] \qquad (12)$$

Further, by taking the log of both sides of the above equation and substituting in equation (5) we find:

$$\log(\{K_1/K_2\}) = \phi^2/kT \times (C_2 - C_1) \qquad (13)$$

Similarly, in terms of bending energetics:

$$\log(K_1/K_2) = 1/kTS^2 \times (B_2 - B_1) \qquad (14)$$

Note that in equations (13) and (14) we assume that the twist angle ϕ and the bending radius S each represent one degree of freedom and that the energy of mechanical deformation has a negligible entropy contribution.

The calculated torsion and bending constants are presented for each synthetic operator in Table 2, along with the binding data measured by Koudelka et al.[2]. The correlation between theory and experiment is presented in Fig. 2, which shows that the stiffness model accounts well for the sequence dependence

Table 2 A calculated relation between 434 repressor binding and stiffness

Operator number	Name	Central sequence	f_{aa}	f_{ag}	f_{gg}	C $\times 10^{-12}$	B $\times 10^{-26}$	K_1/K_2 exper	$\log\{K_1/K_2\}$ exper
0	14	ATAT	1	0	0	0.37	1.15	1	0
1	6T	TTAA	1	0	0	0.37	1.15	1.5	0.4
2	6C	CTAG	1/3	2/3	0	0.57	1.8	5	1.6
3	6G	GTAC	1/3	2/3	0	0.57	1.8	5	1.6
4	6G/2	GTAT	2/3	1/3	0	0.45	1.4	2.5	0.9
5	7A	AATT	1	0	0	0.37	1.15	1	0
6	7A/2	AAAT	1	0	0	0.37	1.15	0.3	-1.2
7	7C	ACGT	0	2/3	1/3	0.9	2.8	50	3.9
8	7G	AGCT	0	2/3	1/3	0.9	2.8	50	3.9
9	7G/2	AGAT	1/3	2/3	0	0.57	1.8	7	1.9

The binding data of Koudelka et al.[2] have been summarized here. The central DNA sequence of each 14-bp long operator has been specified (positions 6-9). All other nucleotides are unchanged relative to operator 14, which is: ACAATATATATTGT. Binding data (K_1/K_2 experimental) are presented as the repressor dissociation constant, relative to that measured with the unmodified binding site (fragment 14). For example, 50 signifies that in ref. 2, 434 repressor binding to operator 7G was found to be 50 times weaker than to unmodified operator fragment 14. f_{aa}, f_{ag}, f_{gg} refer to the frequency of nearest neighbour associations at the centre of each synthetic operator. C is the calculated torsional spring constant for that central region in units of 10^{-12} dyne, derived by assuming $G_{aa} = 0.167 \times 10^9$ dyne cm^{-2}; $G_{ag} = 0.33 \times 10^9$ dyne cm^{-2}; $G_{gg} = 0.66 \times 10^9$ dyne cm^{-2}. B is the calculated bending constant for the central region, in units of 10^{-26} dyne cm^2, derived by assuming $E_{aa} = 0.5 \times 10^9$ dyne cm^{-2}; $E_{ag} = 1.0 \times 10^9$ dyne cm^{-2}; $E_{gg} = 2.0 \times 10^9$ dyne cm^{-2}. See text for details.

of the data. It is interesting that the one sequence permutation which is not adequately fitted by these stiffness calculations is 7A/2 (number 6 in Fig. 2). That manipulation creates a three-nucleotide oligo-(A) sequence at the centre of the synthetic operator (Table 2). Such oligo-(A) segments have been described as the source of permanent helix bends[11,14,15]. Our simple elastic analysis has probably failed to account for such a secondary structural effect in operator 6.

Although Anderson et al. point out that the centre 4 bp of the 434 repressor complex is substantially bent and overtwisted[1], the absolute change in twist or curvature cannot yet be determined as the structure of the uncomplexed binding site has not been measured. In the absence of this information, it is useful to quantify the relationship between theory and experiment in Fig. 2 in the context of two limiting models. If it is presumed that the sequence dependence of repressor binding in ref. 2 is determined only by torsional stiffness differences, the slope of Fig. 2 specifies that, within the central 4 bp, 32° of twist change will account for all the measured affinity change among the 10 synthetic binding sites. Also, in terms of bending, a 33 Å radius of curvature would account for all the measured binding constant difference (excluding fragment 6).

From the Anderson et al. data[1] it is clear that the bend and twist change occurring on binding 434 repressor is about half that specified by the two limiting models. Consequently, when bending and torsional flexibility are both considered, it is likely that the simple stiffness formalism which we have presented can account for the sequence dependence of 434 repressor binding to its operator sequence, both qualitatively and quantitatively.

Additional experimentation is required to understand the rules relating DNA sequence to DNA flexibility, but the data at hand suggest that the sequence dependence of DNA stiffness may be significant whenever a protein (or other ligand) must bend or twist DNA to form its bound complex. Within the limits of the model, the formalism which we have presented can be used to assess the contribution of bending or twisting energetics to this class of site-specific binding interaction.

This work benefited from discussion with Professor Gerry Manning and Professor Robert Hopkins, who has independently developed a similar model for DNA rigidity as a function of bp composition. R.H.A. acknowledges support from the Office of Naval Research, and M.E.H. from the NSF and the National Cancer Institute.

Received 15 June; accepted 3 August 1987.

1. Anderson, J. E., Ptashne, M. & Harrison, S. C. Nature 326, 846–852 (1987).
2. Koudelka, G. B., Harrison, S. B. & Ptashne, M. Nature 326, 886–888 (1987).
3. Peterlin, A. Nature 171, 259–262 (1953).
4. Landau, L. & Lifshitz, E. M. in Statistical Physics (Addison-Wesley, Reading, 1958).
5. Barkley, M. D. & Zimm, B. H. J. chem. Phys. 70, 2991–3007 (1979).
6. Millar, D. P., Robbins, R. J. & Zuwail, A. H. J. chem. Phys. 76, 2080–2094 (1982).
7. Hogan, M., LeGrange, J. & Austin, B. Nature 304, 752–754 (1983).
8. Thomas, T. J. & Bloomfield, V. A. Nucleic Acids Res. 11, 1919–1931 (1983).
9. Chen, H. H., Rau, D. C. & Charney, E. J. biomolec. struct. Dynam. 2, 709–719 (1985).
10. Berkoff, B., Hogan, M., LeGrange, J. & Austin, R. Biopolymers 25, 307–316 (1986).
11. Wu, H. M. & Crothers, D. M. Nature 308, 509–513 (1984).
12. Fujimoto, B. S., Shibata, J. H., Schurr, R. L. & Schurr, J. M. Biopolymers 24, 1009–1022 (1985).
13. Fratini, A. V., Kopka, M. L., Drew, H. R. & Dickerson, R. E. J. biol. Chem. 257, 14686–14707 (1982).
14. Ulanovsky, L., Bodner, M., Trifonov, E. N. & Choder, M. Proc. natn. Acad. Sci. U.S.A. 83, 862–866 (1986).
15. Hagermann, P. Nature 321, 449–450 (1986).

Section II
Cells

Contents

Introduction

The cell, consisting primarily of an enclosing membrane, a nucleus, the cytoplasm, and various organelles such as chloroplasts and mitochondria, is the building block of macroscopic biology. While the cell is generally the object of purely biological investigation, the articles selected for this section reveal an arena for fascinating physical investigation.

Berg and Anderson describe the "motor" that drives the flagella of a cell. Purcell then discusses the remarkable hydrodynamic hurdles that unicellular life must overcome. Berg and Purcell examine the statistical physics associated with the chemistry underlying the motional processes. Some cells have evolved a mechanism to sense the Earth's magnetic field; Ofer *et al.* discuss the motional dynamics of the tiny magnets involved in this process.

Bacteria Swim by Rotating their Flagellar Filaments

IT is widely agreed that bacteria swim by moving their flagella, but how this motion is generated remains obscure[1,2]. A flagellum has a helical filament, a proximal hook, and components at its base associated with the cell wall and the cytoplasmic membrane. If there are several flagella per cell, the filaments tend to form bundles and to move in unison. When viewed by high-speed cinematography, the bundles show a screw-like motion. It is commonly believed that each filament propagates a helical wave[3]. We will show here that existing evidence favours a model in which each filament rotates.

The idea that the flagellar filaments might rotate relative to the cell body as rigid or semi-rigid helices has appeared intermittently in the literature, but a convincing case for such a model has not been made. Rigid rotation was noted as a possibility by Stocker in 1956[4]. Doetsch argued for it in 1966[5,6], but he abandoned the idea in 1969[2,7] in favour of one in which the flagella "wobble" and only appear to rotate. Recently, Mussill and Jarosch[8] photographed preparations of *Spirillum volutans* squeezed between slides in which the flagellar bundles remain "motionless" while the wave movement appears in the cell body. They conclude that "the body must rotate around the point of insertion of the flagellar bundle". Rotation of this kind is assumed in a hydromechanical analysis of *Spirillum* by Chwang *et al.*[9]. Since the flagellar filaments of *Spirillum* arise separately[10], an alternative possibility is that each filament rotates individually. This alternative, which we favour, makes sense for peritrichously flagellated bacteria as well. The flagellar bundles of these cells cannot rotate as a unit, because their filaments arise at widely distant points on the surface of the cell.

It is possible to envisage a biological rotary motor consistent with electron micrographs of structures at the base of a flagellum. In their model of the basal end of a flagellum extracted from *Escherichia coli*, DePamphilis and Adler[11] show a rod extending from the end of the proximal hook and enclosed by rings of diameter 22.5 nm. Suppose that the ring associated with the cytoplasmic membrane, the M ring[12], is rigidly coupled to the rod yet free to rotate in the membrane, and that the rod is able to rotate freely within the other components through which it passes. Rotation of the flagellum would then be possible if the periphery of the M ring was linked to the cell wall by cross bridges of the kind found in skeletal muscle.

Such a rotary motor would be capable of delivering the power required to drive a flagellum. Coakley and Holwill[13] estimate the power dissipation of an isolated filament running at 50 Hz to be about 4×10^{-10} erg s^{-1}. If the cross bridges step along the periphery of the M ring with a step length of 8 nm (ref. 14), the ring will rotate once about every nine steps, and the filament can be driven at 50 Hz if each cross bridge steps at a rate of about 450 s^{-1}, which is a reasonable figure[15]. Huxley and Simmons[14] judge the external work which can be done per cycle of attachment and detachment of a cross bridge to be about 3×10^{-13} erg. Therefore, three cross bridges stepping at the above rate can generate the necessary power. If there are three cross bridges, the average force which each must exert is about 3.8×10^{-7} dyne, a value only slightly larger than the isometric force exerted by a cross bridge in the experiments of Huxley and Simmons[14]. There is room for five cross bridges along the periphery of the M ring, if their spacing is similar to that in skeletal muscle[15].

Although cross bridges of the kind found in skeletal muscle have not been seen at the basal end of a flagellum, they may have been washed away in the procedure of DePamphilis and Adler[16] or obscured in thin-sectioned material by components of the cell wall. Of course, the construction of the motor may be entirely different. We have made the above calculations only to demonstrate the plausibility of a biological rotary motor. Indeed, there appears to be as much, if not more, structural evidence for this model than for one in which the flagellar filaments propagate helical waves[1,2], especially if the energy for the external work which each element of the filament must do is provided by a driving mechanism at the base[3,17].

The model in which the filaments rotate individually may not be argued against on the basis of the greater power required to rotate a helical filament than to propagate a helical wave. Coakley and Holwill[13] find that most of the power dissipation is due to the lateral displacement of the fluid, which is the same in both models. For an isolated filament the power dissipation for rotation is only 0.5% larger than the power dissipation for wave propagation; for a bundle of 200 filaments separated by aqueous fluid of mean thickness equal to the filament radius, the figure is 50%. The latter increase in dissipation may be unrealistically large, since there is no reason *a priori* to suppose that the filament-to-filament separation is as small as the filament radius. Nevertheless, the dissipation per filament for the bundle is relatively small, because the filaments work together in displacing the fluid laterally. For the bundle of 200 filaments rotating individually and in phase, it is only 1.5% of the dissipation for the isolated filament[13].

One cannot distinguish the various models by the fact that the body of a freely swimming cell rotates about an axis parallel to the direction of motion in a sense opposite to the rotation (or the apparent rotation) of the flagellar bundle. As discussed by Taylor[18], the torque responsible for the rotation of the body is generated, in the main, by the lateral displacements. This effect is an essential component in more recent theories[13,19,20].

One can, however, distinguish the models by experiments sensitive to the rotation of filaments relative to one another. This occurs when the filaments rotate individually (Fig. 1) but not when they propagate helical waves, "wobble"[7], or rotate as a group[8]. If the filaments rotate individually and two are linked together, it is likely that the entire bundle will stop. The linked filaments will stop, because they can no longer rotate relative to one another; the others will stop, because their lateral motion will be blocked by the first two.

Cross-linking experiments have been done, and the results clearly favour individual rotation. Greenbury and Moore[21] find that the peritrichous bacterium *Salmonella typhimurium* remains motile when as many as 10^5 univalent flagellar antibodies are adsorbed per cell, a mass of antibody four times

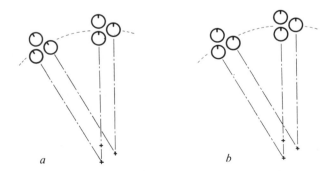

Fig. 1 A schematic cross-sectional view of a bundle of three flagellar filaments rotating individually (*a*) or propagating helical waves (*b*) at two successive instants of time. The section is in a plane normal to the helical axes (+), and the motion is in the frame of reference of the body of the cell. Each filament is marked at a fixed point. In (*a*) the filaments rotate relative to one another; in (*b*) they do not. The thrusts generated are the same.

Reprinted from *Nature* 245, 380–382 (1973); © Macmillan Magazines, Ltd.

that of the mass of the flagella. About half of the cells are immobilized when about 200 bivalent antibodies are adsorbed per cell. They conclude that the bivalent antibody acts by linking the filaments together. An alternative interpretation is that flagellin molecules loaded with univalent antibodies still function, but molecules in the same filament cross linked together do not. This possibility is eliminated in an experiment by DiPierro and Doetsch[22], who compare the effect of univalent and bivalent flagellar antibodies on *E. coli*, which is peritrichous, and on *Pseudomonas fluorescens*, which has only one filament. In the presence of univalent flagellar antibodies both kinds of bacteria remain motile. In the presence of bivalent antibodies *E. coli* stops, but *P. fluorescens* remains motile until the cells are linked together. "Very few non-motile single cells were observed in the *Pseudomonas* culture being immobilized, whereas with the *Escherichia* culture, large numbers of *single* non-motile organisms were observed, and in both cases this condition existed *before* the appearance of microscopically visible clumps"[22].

Results of experiments with bacteriophages that attack flagella provide additional evidence for the individual rotation model. Meynell[23] notes that rapidly swimming *Salmonella abortus-equi* stop abruptly on swimming into a drop of χ phage at high titre, even after the phage has been inactivated by radiation or over-centrifugation. The adsorption of the phage did not cause any morphological changes in the flagella, as far as could be seen by electron microscopy. Raimondo *et al.*[24] conclude that the adsorption of one PBS1 phage to one flagellum of *Bacillus subtilis* is sufficient to render the entire complement of flagella non-functional. They obtain the same results with phage ghosts treated with DNase, and conclude that "the functioning of all the flagella of *B. subtilis* is under the control of one 'motor'". These results can be explained if the filaments rotate. If the tail fibres of PBS1 wrap tightly enough around even a single filament, the bundle will jam, since the protruding body of the phage will not be able to rotate past the adjacent filaments.

The individual rotation model suggests a mechanism by which flagellotropic phages are able to infect cells. The phage χ attaches to the filament with one tail fibre and then travels to the base of the flagellum where it injects its DNA; if the filament is inactive, the phage fails to reach the site of injection[25]. Assuming that the binding of the phage to the filament is not completely rigid, so that some relative motion is possible, it is conceivable that the phage moves down the filament like a nut on a bolt, the grooves between the helical rows of flagellin molecules serving as the threads. If this notion is correct, the phage will move to the base of the flagellum if the filament is rotating and the "threads" are intact; motion of the cell body is not required. There is a mutant of *Salmonella* which is completely non-motile because its filaments are straight[26]. The defect is in the flagellin, not in the components at the flagellar base. The flagellin molecules are still arranged in helical rows[27]. As we would predict, the mutant is fully sensitive to χ (ref. 26). There is a similar mutant of *B. subtilis* which remains sensitive to PBS1 (ref. 28).

Although bundle formation by peritrichously flagellated bacteria is not well understood, the forces responsible are thought to be hydrodynamic. If the viscosity of the medium is increased slightly, for example by the addition of methylcellulose, the bundles scatter more light[4]. The bacteria also swim more rapidly[29]. We find in tracking experiments[30] that they swim less erratically. These effects can all be explained, if, as argued by Stocker[4], the increased viscous stress causes more filaments to join bundles. If hydrodynamic forces, in fact, cause bundle formation, the observation that the mutant of *Salmonella* possessing straight flagella forms bundles[26] provides further evidence for the rotation of the filaments,

since in this case interactions due to lateral displacements are absent. We suggest that the bending of the filaments around the sides of the cell is facilitated by the proximal hook, which serves as a flexible coupling or universal joint.

The hydrodynamic basis for the synchronization of filaments within a bundle is more firmly established. Flagella of adjacent spermatozoa[31] or flagellar bundles of adjacent bacteria[4] tend to move in phase. Taylor[31] calculated the forces acting to synchronize parallel undulating sheets and found these forces to be large for sheet spacings smaller than the wavelength of the undulations. Although explicit calculations of this kind have not been made for thin filaments, the work of Coakley and Holwill cited earlier[13] indicates a large reduction in power dissipation for synchronous rotation within a bundle.

Work by Silverman and Simon[32] on "polyhook" mutants of *E. coli* which provides strong support for the rotation model has recently come to our attention. When cells which have abnormally long proximal hooks but no filaments are treated with anti-hook antibody, they form pairs which counter-rotate. This result can be explained if the proximal hooks are driven by rotary motors and the antibodies link a single hook from one cell to one or more hooks from another.

If, as suggested by existing evidence, bacterial flagella rotate, the structures at the base of the flagellum deserve more attention than they have received thus far.

This work was supported by grants from the National Science Foundation and the US Atomic Energy Commission.

HOWARD C. BERG

Department of Molecular, Cellular and
Developmental Biology,
University of Colorado,
Boulder, Colorado 80302

ROBERT A. ANDERSON

Sandia Laboratories,
Albuquerque, New Mexico 87115

Received July 2; revised August 23, 1973.

1 Smith, R. W., and Koffler, H., *Adv. microbiol. Physiol.*, **6**, 219 (1971).
2 Doetsch, R. N., *CRC Crit. Rev. Microbiol.*, **1**, 73 (1971).
3 Lowy, J., and Spencer, M., *Symp. Soc. exp. Biol.*, **22**, 215 (1968).
4 Stocker, B. A. D., *Symp. Soc. gen. Microbiol.*, **6**, 19 (1956).
5 Doetsch, R. N., *J. theor. Biol.*, **11**, 411 (1966).
6 Doetsch, R. N., and Hageage, G. J., *Biol. Rev.*, **43**, 317 (1968).
7 Vaituzis, Z., and Doetsch, R. N., *J. Bact.*, **100**, 512 (1969).
8 Mussill, M., and Jarosch, R., *Protoplasma*, **75**, 465 (1972).
9 Chwang, A. T., Wu, T. Y., and Winet, H., *Biophys. J.*, **12**, 1549 (1972).
10 Murray, R. G. E., and Birch-Andersen, A., *Can. J. Microbiol.*, **9**, 393 (1963).
11 DePamphilis, M. L., and Adler, J., *J. Bact.*, **105**, 384 (1971).
12 DePamphilis, M. L., and Adler, J., *J. Bact.*, **105**, 396 (1971).
13 Coakley, C. J., and Holwill, M. E. J., *J. theor. Biol.*, **35**, 525 (1972).
14 Huxley, A. F., and Simmons, R. M., *Nature*, **233**, 533 (1971).
15 Huxley, H. E., *Science, N.Y.*, **164**, 1356 (1969).
16 DePamphilis, M. L., and Adler, J., *J. Bact.*, **105**, 376 (1971).
17 Klug, A., *Symp. Int. Soc. Cell Biol.*, **6**, 1 (1967).
18 Taylor, G., *Proc. R. Soc.*, A **211**, 225 (1952).
19 Chwang, A. T., and Wu, T. Y., *Proc. R. Soc.*, B **178**, 327 (1971).
20 Schreiner, K. E., *J. Biomechanics*, **4**, 73 (1971).
21 Greenbury, C. L., and Moore, D. H., *Immunology*, **11**, 617 (1966).
22 DiPierro, J. M., and Doetsch, R. N., *Can. J. Microbiol.*, **14**, 487 (1968).
23 Meynell, E. W., *J. gen. Microbiol.*, **25**, 253 (1961).
24 Raimondo, L. M., Lundh, N. P., and Martinez, R. J., *J. Virol.*, **2**, 256 (1968).
25 Schade, S. Z., Adler, J., and Ris, H., *J. Virol.*, **1**, 599 (1967).
26 Iino, T., and Mitani, M., *J. gen. Microbiol.*, **49**, 81 (1967).
27 O'Brien, E. J., and Bennett, P. M., *J. molec. Biol.*, **70**, 133 (1972).
28 Martinez, R. J., Ichiki, A. T., Lundh, N. P., and Tronick, S. R., *J. molec. Biol.*, **34**, 559 (1968).
29 Shoesmith, J. G., *J. gen. Microbiol.*, **22**, 528 (1960).
30 Berg, H. C., and Brown, D. A., *Nature*, **239**, 500 (1972).
31 Taylor, G., *Proc. R. Soc.*, A **209**, 447 (1951).
32 Silverman, M. R., and Simon, M. I., *J. Bact.*, **112**, 986 (1972).

PHYSICS OF CHEMORECEPTION

HOWARD C. BERG AND EDWARD M. PURCELL, *Department of Molecular, Cellular, and Developmental Biology, University of Colorado, Boulder, Colorado 80309 and the Department of Physics, Harvard University, Cambridge, Massachusetts 02138 U.S.A.*

ABSTRACT   Statistical fluctuations limit the precision with which a microorganism can, in a given time T, determine the concentration of a chemoattractant in the surrounding medium. The best a cell can do is to monitor continually the state of occupation of receptors distributed over its surface. For nearly optimum performance only a small fraction of the surface need be specifically adsorbing. The probability that a molecule that has collided with the cell will find a receptor is $Ns/(Ns + \pi a)$, if N receptors, each with a binding site of radius s, are evenly distributed over a cell of radius a. There is ample room for many independent systems of specific receptors. The adsorption rate for molecules of moderate size cannot be significantly enhanced by motion of the cell or by stirring of the medium by the cell. The least fractional error attainable in the determination of a concentration $\bar{c}$ is approximately $(T\bar{c}aD)^{-1/2}$, where D is the diffusion constant of the attractant. The number of specific receptors needed to attain such precision is about a/s. Data on bacteriophage adsorption, bacterial chemotaxis, and chemotaxis in a cellular slime mold are evaluated. The chemotactic sensitivity of *Escherichia coli* approaches that of the cell of optimum design.

INTRODUCTION

In the world of a cell as small as a bacterium, transport of molecules is effected by diffusion, rather than bulk flow; movement is resisted by viscosity, not inertia; the energy of thermal fluctuation, kT, is large enough to perturb the cell's motion. In these circumstances, what are the physical limitations on the cell's ability to sense and respond to changes in its environment? What, for example, is the smallest change in concentration of a chemical attractant that a bacterium could be expected to measure reliably in a given time? We begin our analysis by reviewing some relevant features of diffusive transport and low Reynolds number mechanics. This will lead to certain conclusions about selective acquisition of material by a cell and how this acquisition may be influenced by the cell's movement. We then develop a theory of the signal-to-noise relation for measurement of concentration by a cell with specific receptors, discuss its implications for chemotactic behavior, and compare theory with experiment.

DIFFUSIVE INTAKE

Consider a spherical cell of radius a immersed in an unbounded medium. The medium contains in low concentration some molecules of species X with diffusion

Reprinted from *Biophysical Journal* **20**, 193–219 (1977); © Biophysical Society.

constant D. The local concentration of X will be denoted by c and expressed in molecules per cubic centimeter (1 M = 6×10^{20} cm^{-3}). The spatial and temporal variation of c is governed by the diffusion equation

$$D \nabla^2 c = \partial c / \partial t. \tag{1}$$

Suppose the cell is a perfect sink for the molecules X, sequestering or otherwise immobilizing every X molecule that reaches its surface. Then in the steady state the current J of molecules to the cell, in molecules per second, is given by

$$J = 4 \pi a D c_\infty, \tag{2}$$

where c_∞ is the concentration far from the cell, assumed to be maintained at a steady value.[1]

For our purposes it will be useful and instructive to relate this formula to an electrical analogue. Comparing the·time-independent diffusion equation $\nabla^2 c = 0$ with Laplace's equation for the electrostatic potential ϕ in charge-free space, $\nabla^2 \phi = 0$, we observe that the diffusive current density $\mathbf{F} = -D \operatorname{grad} c$ is the analogue of the electric field vector $\mathbf{E} = -\operatorname{grad} \phi$. The total diffusive current J entering a closed surface S is given by $J = \int_S \mathbf{F} \cdot d\mathbf{s}$, whereas the total electric charge Q on a surface is given by $Q = 1/4\pi \int_S \mathbf{E} \cdot d\mathbf{s}$. Because the cell is a perfect sink, c must be zero at its surface, which therefore corresponds to a surface at constant potential. We see that Eq. 2 is equivalent to the statement that the static charge Q on a spherical conductor in vacuum is $\phi_\infty a$, if ϕ_∞ is the difference in potential between the conductor and points far away. And in general, the steady-state diffusive current to a totally absorbing body of any shape and size can be written as

$$J = 4 \pi C D c_\infty, \tag{3}$$

where C is the electrical capacitance (in cgs units of centimeters) of an isolated conductor of that size and shape.

Solutions are available for the capacitances of a variety of conductors. As an example, the capacitance of an isolated thin conducting disk of radius b, $2b/\pi$ in cgs units (1), provides us with the diffusive current to both sides of a disk-like sink: $J = 8bDc_\infty$. The same result provides us with the current through a circular aperture of radius b in a thin membrane which separates regions of concentration c_1 and c_2: $J_{1,2} = 4bD(c_1 - c_2)$.

Another simple result, perhaps not quite so familiar, is easily obtained by way of the electrical analogy. Consider again the completely absorbing sphere of radius a in an unbounded medium. Let a molecule X be released at a point in the medium a distance r from the center of the sphere. What is the probability P_c that the molecule will eventually be captured by the sphere? The exact answer, which we shall make use of

---

[1] If a steady state has been established, the right-hand side of Eq. 1 is zero. Then a solution with spherical symmetry and with $c = 0$ at $r = a$ is $c = c_\infty (1 - a/r)$. The density of inward current, $D dc/dr$, is c_∞ / r^2, giving for the total inward current the result in Eq. 2.

later, is

$$P_c = a/r. \tag{4}$$

It can be found by considering the capacitance of concentric spherical shells.[2] Notice that the result has nothing to do with the solid angle subtended by the absorber as seen from the point of release; with increasing distance, P_c goes down not as $1/r^2$, but only as $1/r$.

The electrical analogy does not extend to time-dependent diffusion. The relaxation of a charge distribution within a homogeneous poor conductor, which might be thought to resemble superficially the relaxation by diffusion of a spatially varying concentration, is governed by a first-order equation and is characterized by a size-independent time constant. In contrast, the characteristic time for a change brought about by diffusion in a region of size a is a^2/D.

Absorption by Specific Receptors

Let us apply some of this to a cell that carries on its surface specific receptors for species X. Each receptor has a binding site that we shall idealize as a circular patch of radius s. Suppose N receptors are distributed more or less uniformly over the surface of the cell. The cell's radius is again a; the fraction of its surface occupied by binding sites is $Ns^2/4a^2$. Any X molecule that touches a binding site is immediately (or within a time short compared to the interval between arrivals) captured and transported through the cell wall, clearing the site for its next catch. The surface of the cell between these absorbent patches is impermeable to and does not bind X molecules. Under these rules, what is the total current of X molecules assimilated by the cell, in a medium of X concentration c_∞?

For $N = 1$, and if $s \ll a$, the current is the same as that to one side of the disk-like sink already mentioned, and is given by

$$J_1 = 4Dsc_\infty. \tag{5}$$

With only a few widely separated receptors the total current J will be almost N times as great. But as N increases, the receptors will begin to interfere with one another, the presence of one sink depressing the concentration in the vicinity of a neighboring sink, and vice versa. If N is so large that the surface is entirely covered by absorbent patches, the whole cell becomes a spherical sink, and the current, which is the largest current a cell of that size, however equipped, could collect by diffusion, is given by Eq. 2 as

$$J_{max} = 4\pi aDc_\infty. \tag{6}$$

---

[2]To derive this, consider the diffusive current from a continuous source distributed uniformly over a shell of radius r. The probability sought is the ratio of the inward current flowing to the sphere of radius a to the total current, inward plus outward. The electrical equivalent is a spherical capacitor with the inner conductor at zero potential, the same as the potential at infinity. The ratio of the inward diffusive current to the total current is the same as the ratio of the charge on the inner surface of the outer conductor to the total charge on that conductor, namely a/r.

We can find the current for any number N of receptors by solving the following analogous problem in electrostatics: Over the surface of an insulating sphere of radius a are evenly distributed N conducting disks of radius s, connected together by infinitesimal wires so as to form a single conductor. The insulating sphere is itself impermeable to electric field, which is to say that its dielectric constant is zero, requiring the electric field just outside the surface of the dielectric to have vanishing normal component. What is the capacitance of this object? The answer, derived in Appendix A, is

$$C = Nsa/(Ns + \pi a). \tag{7}$$

The only assumption made in the derivation of this equation is that the distance between neighboring disks is large compared with the size of a disk, or equivalently, that the fraction of the sphere's surface covered by disks is small. This condition will be satisfied in all our applications of Eq. 7. Translating Eq. 7 into a formula for the cell's intake of X molecules by diffusion, we find

$$J = 4\pi D c_\infty Nsa/(Ns + \pi a) = J_{max} Ns/(Ns + \pi a). \tag{8}$$

For large N the intake approaches that of the completely absorbing cell, as it ought to. But it can become *almost* that large before more than a small fraction of the cell's surface is occupied by absorbent patches. A reasonably generous allotment of area for one patch might be a few hundred square angstroms. Let us take the patch radius s equal to 10 Å and the cell radius a equal to 1 μm. According to Eq. 8, the intake is half of J_{max} for $N = \pi a/s = 3,100$. The receptor patches then occupy somewhat less than $1/1,000$ of the cell's surface. The distance between neighboring patches is about 60 times the patch radius. It is important that the receptor patches be well dispersed. If they were combined into a single absorbent patch of the same total area, the current would be severely reduced, from $J_{max}/2$ down to $J_{max}/(3,100)^{1/2}$. If the same number of receptors were distributed randomly over the surface of the cell, rather than uniformly, as assumed in the derivation of J/J_{max}, the current would be only slightly smaller than $J_{max}/2$. A numerical calculation comparing uniformly distributed with randomly distributed receptors showed that the difference in current, for the same number of receptors N, does not exceed a few percent if N is larger than 50.

Qualitatively the outcome reflects the fact that a diffusing molecule that has bumped against the surface of the cell is by that very circumstance destined to wander around in that vicinity for a time, most likely hitting the cell many times before it wanders away for good. Insight is gained by developing this idea quantitatively. Fig. 1 shows the path of a diffusing molecule that has touched the cell's surface at a sequence of points $A, B, \ldots F$, none of which happened to lie in a receptor patch. This hypothetical path is unrealistic in one respect: the total number of distinct encounters with the surface in a finite interval after one encounter is really very large—in the limit of continuous diffusion, infinite. But clearly, that does not give the molecule an infinite number of independent tries at hitting an absorbent patch. Two contacts such as C and D, close together compared to the dimension s of a patch, must count as only one try. The effective number of independent tries must be something like the number of

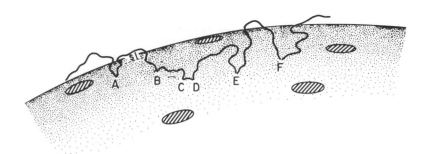

FIGURE 1   The path of a diffusing molecule that has touched the surface of a cell of radius a at a sequence of points $A, B, \ldots F$. The receptor patches, shown shaded, are of radius s. A and B constitute independent tries at hitting a patch, but C and D do not. Note between A and B the excursion of distance s perpendicular to the surface of the sphere.

path segments whose ends on the cell surface are separated by a distance greater than s. Such a path segment is likely to include an excursion of similar magnitude perpendicular to the surface. Let us therefore assume, for a rough calculation, that extension out to a distance s from the surface will serve as the necessary and sufficient condition for the ends of a segment to be at least a distance s apart.

How many such excursions are to be expected after a molecule has once touched the cell? The probability P_s that a molecule now located a distance s from the surface of a sphere of radius a will hit the surface of the sphere at least once before escaping to infinity is precisely equivalent to the "capture probability" P_c given by Eq. 4, which we now rewrite as

$$P_s = a/(a + s). \tag{9}$$

The probability that a molecule now at $r = a + s$ will execute exactly n excursions to the surface, separated by reappearances at $r = a + s$ and followed by diffusion to infinity, is $P_s^n(1 - P_s)$. It follows that the average number of such excursions is

$$\bar{n} = \sum_{n=0}^{\infty} nP_s^n(1 - P_s) = P_s/(1 - P_s) = a/s. \tag{10}$$

The probability of not hitting a receptor patch in a single random encounter is $\beta = 1 - (Ns^2/4a^2)$. If the contacts we have just enumerated can be taken as independent tries, the probability that a molecule starting at $r = a + s$ survives all subsequent contacts until it escapes to infinity is

$$P_{esc} = \sum_{n=0}^{\infty} \beta^n P_s^n(1 - P_s) = (1 - P_s)/(1 - \beta P_s). \tag{11}$$

Eq. 11 reduces to

$$P_{esc} = 4a/(4a + Ns). \tag{12}$$

Since $1 - P_{esc}$ is the fraction of all arriving molecules that ultimately are captured, we have for the resulting current

$$J/J_{max} = Ns/(4a + Ns), \qquad (13)$$

to be compared with our exact formula, Eq. 8. The close numerical agreement is fortuitous. But this does show clearly how the remarkable effectiveness of dispersed receptors arises from the multiplicity of encounters of a single diffusing molecule with the surface of the cell. The number $\bar{n}$ would be about 1,000 in our earlier example.

Some general conclusions can now be drawn. The number of receptors a cell can usefully employ is not much larger than the ratio of cell diameter to patch diameter; more receptors than that fail to increase the intake much. Receptor patches of adequate number cover only a small fraction of the surface of the cell. Hundreds of such receptor systems can be accommodated, each capable of collecting its particular molecular species almost as effectively as if the entire surface of the cell were dedicated to that single task. Other constraints aside, the best arrangement of receptors of a given type is maximum dispersal, with different receptor systems thoroughly intermingled.

Two-Stage Capture

Adam and Delbrück (2) considered a two-stage capture process involving adsorption followed by diffusion of the adsorbed molecule over the surface of the cell. Suppose that an *X* molecule that touches the cell at any point becomes attached, but so weakly that it can migrate by two-dimensional diffusion until it either desorbs or encounters an *X*-receptor. This will increase the rate at which molecules are captured by receptors by an amount that depends on the coefficient of surface diffusion, D', and the mean time of residence on the surface before desorption, $\bar{t}_r$. Of course, a molecule just desorbed has a very good chance of diffusing back to the surface and being readsorbed, so the total time available to a particular molecule for random exploration of the cell surface will be many times $\bar{t}_r$. But that time and $\bar{t}_r$ itself do not need to be involved explicitly in the result, as we shall see.

Following Adam and Delbrück, we consider first the mean time $\bar{t}_c$ between adsorption on the cell's surface and capture, by a receptor, of a molecule that never desorbs. This time $\bar{t}_c$ is to be averaged over all possible starting positions, that is, over the whole surface of the cell. An approximate formula for $\bar{t}_c$, agreeing closely with the one given by Adam and Delbrück, is

$$\bar{t}_c = (1.1\,a^2/ND')\ln(1.2\,a^2/Ns^2). \qquad (14)$$

As before, *N* is the number of *X* receptors on the cell, *a* is the radius of the cell, and *s* is the radius of the binding site. The receptors are assumed to be uniformly distributed; the binding sites take up only a small fraction of the cell's surface ($Ns^2 \ll 4a^2$). Eq. 14 has been adapted from an exact formula that can be derived for the case of an absorber in the center of a disk with an impermeable perimeter, as explained in Appendix B. The result was checked for a square lattice of absorbers in an independent computation by a relaxation method, and also by a Monte Carlo calculation. The latter confirmed that the distribution of times-to-capture is exponential. That simplifies our problem, for it implies that, averaged over all positions, the probability for capture within an interval dt is $dt/\bar{t}_c$, independent of the starting time; the mean rate at which X molecules are being captured by receptors at any time depends only on the number of X molecules on the cell's surface at that time. Denote that number by m and its time average by $\bar{m}$. Let J' be the average current absorbed by the cell by way of the two-stage process. Then

$$J' = \bar{m}/\bar{t}_c. \tag{15}$$

As long as J' is small compared with $J_{\max} = 4\pi a D c_\infty$, the number $\bar{m}$ will be close to its equilibrium value for the given concentration of X molecules in the medium. In that case we expect $\bar{m}$ to be given approximately by

$$\bar{m} \simeq 4\pi a^2 d c_\infty \exp{(E_A/kT)}, \tag{16}$$

where E_A is the energy of adsorption and d is a distance of molecular size. The factor $4\pi a^2 d$ is roughly the volume accessible to a molecule adsorbed but still free to move over the surface. When Eqs. 14–16 are combined, we find

$$J' \simeq 4\pi N D' d c_\infty \exp{(E_A/kT)}/\ln{(a^2/Ns^2)}. \tag{17}$$

Let us compare this with the current that would be collected without the aid of surface diffusion (Eq. 8), which for small $J/J_{\max}$ is

$$J = 4NDsc_\infty. \tag{18}$$

The two-step process will be dominant if J' exceeds J, that is, if

$$(\pi d/s)(D'/D)\exp{(E_A/kT)} > \ln{(a^2/Ns^2)}. \tag{19}$$

In order of magnitude, the logarithm will by typically around 10, the factor $\pi d/s$ roughly unity. Suppose D'/D is as large as 0.1. Then the two-stage process will be important if E_A is greater than $kT\ln{(100)}$, about 3 kcal/mol.

There may well be a number of systems that rely on the two-stage process. On the other hand, as we have shown, it is not necessary to invoke a two-stage process to explain efficient collection by a cell with many receptors. Nor is the advantage of a two-stage process—when Eq. 19 is satisfied and such an advantage exists—to be attributed simply to a reduction from a three-dimensional to a two-dimensional diffusion process, as Adam and Delbrück implied. They noted that the logarithmic factor in Eq. 14, which they called the "tracking factor," is a measure of the difficulty of "finding the target" in a two-dimensional random walk, and that the corresponding factor for three-dimensional diffusion is much larger, being of order (space size/ target size), rather than the logarithm of that ratio. But target finding in three dimensions, that is, finding the cell itself, is required as a first step in both the two-stage and the one-stage processes. Any X molecule that arrives at the surface of the cell has already overcome the difficulty of that three-dimensional search and is now assured a period of diffusion in close proximity to the surface, which it will probe at a large number of points. Indeed, one could regard this motion close to the cell wall as quasi-two-dimensional diffusion—not quite as effective, to be sure, as an equally rapid diffusion on the surface itself, but by no means as poor a substitute as the contrast of the two- and three-dimensional tracking factors might suggest.

These results have been derived for receptors with binding sites that are perfect sinks. It is not obvious how they apply in chemotaxis, where we are concerned with the time-average occupation of receptors that bind molecules of attractant temporarily, subsequently releasing them to the medium. It will turn out, however, that the formula for the current absorbed by receptors with binding sites that are perfect sinks is precisely what we shall need to analyze the fluctuations in the occupation of receptors in the more general case.

EFFECT ON INTAKE OF ACTIVE MOVEMENT

It is natural to ask whether a cell cannot, by some movement of its own, increase its intake of X molecules. Before addressing that question, we remind the reader that in the environment of the microorganism the mechanics of the medium is dominated by viscosity. The ratio of inertial forces to viscous forces is expressed in the Reynolds number:

$$R = Lv\rho/\eta = Lv/\nu, \tag{20}$$

where L and v are a length and a velocity typical of the motion under consideration, η is the viscosity of the fluid, and ρ is its density. The kinematic viscosity ν is defined as η/ρ. The smallest kinematic viscosity we need to consider is that of water, about 10^{-2} cm^2/s, and the largest velocities we shall encounter are well below 10^{-2} cm/s. Then, even if L is as large as 10^{-3} cm, we have $R \leq 10^{-3}$. In most cases the Reynolds number will be smaller by one or two orders of magnitude. Inertia is utterly negligible in all the processes we shall consider. To emphasize that, we may remark that if a bacterium the size of *E. coli*, swimming in water at top speed, about 30 μm/s, were suddenly to stop moving its flagella, it would coast less than 10^{-9} cm!

Stirring

Consider first what might be called local stirring. Let the organism be equipped with suitable active appendages with which to manipulate the fluid in its vicinity. Can it thereby significantly increase the rate at which molecules reach its receptors? This has been suggested as a possible major function of flagella (3, 4). Consider the following rather loose dimensional argument: Transport by stirring is characterized by some velocity V_s, the speed of the appendage, and by a length L, its distance of travel, which determine a characteristic time $t_s = L/V_s$. On the other hand, movement of molecules over a distance L by diffusion alone is characterized by a time $t_D = L^2/D$. Roughly speaking, stirring will be more effective than diffusion only if $t_s < t_D$, which is to say, only if

$$V_s > D/L. \tag{21}$$

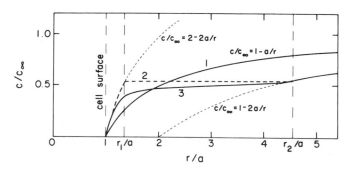

FIGURE 2   Relative concentration in the vicinity of a spherical absorber, for three cases: (1) No stirring; the current absorbed is $J_0 = 4\pi a D c_\infty$. (2) Volume between r_1 and r_2 stirred infinitely rapidly, fluid stationary elsewhere; current absorbed is $2J_0$. (3) Finite stirring speed; region inside r_1 still dominated by diffusion; current absorbed is $2J_0$.

For local stirring the distance L cannot be much larger than the size of the organism itself. With $L = 1$ μm and $D = 10^{-5}$ cm²/s, Eq. 21 calls for speeds of order 10^3 μm/s. That is faster than any motions bacterial cells exhibit. However, for larger organisms lower speeds suffice. In fact, the feasibility of effective local stirring depends on size even more strongly than Eq. 21 might suggest. That becomes clear when we reckon the cost in energy dissipated in viscous friction, energy which the organism itself is obliged to supply if it is doing the stirring.

Consider a spherical cell of radius a that is a perfect sink for molecules X. Without stirring the cell collects a current of X molecules given by Eq. 2: $J_0 = 4\pi a D c_\infty$. The concentration in the neighborhood of the cell is given by $c = c_\infty (1 - a/r)$, shown as curve 1 in Fig. 2. Now let us introduce local stirring with the aim of doubling the cell's steady intake. The object is to transfer fluid from a distant region of relatively high concentration to a place much closer to the cell, thereby increasing the concentration gradient near the absorbing surface. Of course, the depleted parcels of fluid must be carried back again—some more or less complicated pattern of circulation must be maintained by means we need not specify.

An idealized limiting case of this process could be described as follows. Let the stirred volume extend from a sphere of radius $r_1 > a$ out to a sphere of radius r_2, and let the stirring be so vigorous as to keep the concentration uniform at all times throughout this region. Elsewhere, that is, both for $a < r < r_1$ and $r > r_2$, the fluid is stationary and transport of X molecules is effected by diffusion only. If the current is to be twice J_0, r_1 and r_2 are related as follows:

$$r_1 = 2ar_2/(r_2 + 2a). \qquad (22)$$

The dependence of relative concentration on distance is shown by curve 2 in Fig. 2.

This case demanded unlimited rapidity of stirring, and hence unlimited expenditure of energy. We should expect that in a practical case the concentration $c(r)$, which now must be understood as a mean concentration at any given r, will behave more like curve 3 in Fig. 2, rising somewhat with increasing r through the stirred volume. The boundary at $r = r_1$, though no longer sharp, still locates the essential transition from diffusion-dominated transport to convective transport. We can express this by the condition that at $r = r_1$ half the current density be due to diffusion and half to fluid motion, that is, by the condition $D\,dc/dr = cV_r/4$. Here V_r is the maximum radial velocity of flow at r_1. The factor $\frac{1}{4}$ derives from the assumption of a sinusoidal pattern of radial velocities over the spherical surface, together with an assumption that fluid parcels having the greatest outward velocity are fully depleted. Since the convective current at r_1, $4\pi r_1^2\, c(r_1)\, V_r/4$, is equal to half of the total current, $4\pi\, aDc_\infty$, and $c(r_1)$ is approximately $2c_\infty(1 - a/r_1)$, we get

$$V_r = 2aD/r_1(r_1 - a). \qquad (23)$$

This is a specific example of the general relation expressed by Eq. 21.

Whatever the pattern of circulation, the region from $r = a$ out to at least $r = r_1 + (r_1 - a)$ must contain velocity gradients and, in particular, rates of shearing deformation, as large as $V_r/(r_1 - a)$. The square of the rate of shearing deformation determines the local rate of energy dissipation in a viscous fluid (ref. 5, p. 54). Thus, Eq. 23 implies a lower bound on the rate at which energy must be expended in stirring. The mean rate of energy dissipation per unit volume of fluid is approximately

$$\tfrac{1}{2}\eta[V_r/(r_1 - a)]^2 = 2\eta a^2 D^2/r_1^2(r_1 - a)^4. \qquad (24)$$

The volume involved is approximately $8\pi r_1^2(r_1 - a)$, so the total stirring power expended in this region is $16\pi\eta a^2 D^2/(r_1 - a)^3$. We neglect the dissipation in the larger portion of the stirred volume that extends from $2r_1 - a$ out to r_2, since the velocity gradients there can be much smaller. Using Eq. 22 to express the result in terms of r_2, the outer limit of the stirred volume, we find as a lower bound on the stirring power, S, required to double the cell's intake

$$S = \frac{16\pi\eta D^2}{a}\left(\frac{r_2 + 2a}{r_2 - 2a}\right)^3. \qquad (25)$$

The energy cost per unit volume of cell is

$$\frac{S}{(4\pi/3)a^3} \geq \frac{12\eta D^2}{a^4}\left(\frac{r_2 + 2a}{r_2 - 2a}\right)^3. \qquad (26)$$

For example, if $r_2 = 6a$, which would call for a rather extensive stirring apparatus, the specific power demand would be at least $100\ \eta D^2/a^4$. For $\eta = 10^{-2}$ P, $D = 10^{-5}$cm^2/s, and $a = 1\ \mu$m, this amounts to 0.1 W/cm^3, more than 10^4 times the specific power demand required to propel the sphere at a speed $v_0 = 30\ \mu$m/s.[3]

For an organism as small as a bacterium in a medium like water, the cost of increas-

---

[3]The force required to propel a sphere at speed v_0 is $6\pi\eta av_0$ (Stokes' law). The power dissipated is $6\pi\eta av_0^2$. Division by the volume of the sphere gives the specific power demand: $9\eta v_0^2/2a^2$.

ing the intake by local stirring would appear to be prohibitive. The prospect is somewhat more favorable if the viscosity of the medium is high. Since D varies as $1/\eta$, the product ηD^2 also varies as $1/\eta$. Stirring may also be more useful if the molecule in question is so large that its diffusion constant is small. Note, however, that for any molecule whose configuration is such that it can be enclosed in a sphere of radius R, the product ηD^2 cannot be smaller than $(kT/6\pi R)^2/\eta$.

The most striking aspect of Eq. 26 is the strong dependence of the specific energy demand on cell radius. If the cell in our previous example had had a radius of 10 μm, the stirring power required would have been reduced to 10^{-5} W/cm^3. Local stirring for the purpose of increasing intake changes from a hopelessly futile to a possibly useful activity somewhere in the range of cell size between a few microns and a few tens of microns. We emphasize that this conclusion does not depend on the details of the stirring mechanism. It should be noted, as well, that the largest possible gain in intake, even if all other constraints could be ignored, would be by the factor r_2/a, because the current is limited ultimately to what can diffuse into the stirred region.

Stirring of the fluid on a larger scale by some external agent can in principle increase the current absorbed by the cell, but to be effective, it must convey fresh solution into the region of low concentration around the cell faster than diffusion into the absorber depletes it, that is, in a time short compared to a^2/D. What is required is a continuous shearing deformation with a transverse velocity gradient greater than D/a^2. It is shown in another article[4] that if a suspension of spheres of radius a is stirred vigorously enough to double the rate at which diffusing material is absorbed, the mechanical power expended in agitating the fluid must be approximately 500 $\eta D^2/a^4$ ergs/cm^3 per s. For the same values of η, D, and a as we assumed in our previous example, the external stirring power required is 0.5 W/cm^3 of fluid stirred. Here again we find strong dependence on the size of the absorbing particles.

Swimming

Can a cell in a medium of uniform concentration increase its material intake by swimming? One is tempted to suppose that a moving cell might scoop up the X molecules that lie in its path (3) or move suddenly to a region in which the local concentration is c_∞ (4). This is not the case. The molecules in front of the cell are carried out of its way along with the fluid it must push aside to move. The cell carries with it a layer of liquid that is practically stationary in its frame of reference. Every molecule that reaches the surface of the cell must cross this layer by diffusion. The controlling relations are essentially the same as those involved in stirring, of which swimming could be viewed as a special case. But here the question can be formulated and answered more precisely.

Let a spherical cell of radius a be propelled at constant velocity v_0 through a fluid containing at concentration c_∞ molecules for which the cell is a perfect sink. By adopt-

---

[4]Purcell, E. M. 1978. The effect of fluid motions on the absorption of molecules by suspended particles. *J. Fluid Mech.* In press.

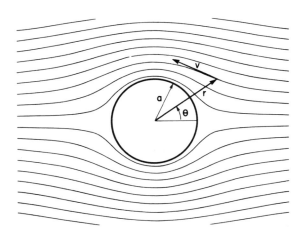

FIGURE 3   A sphere moving at constant velocity through a viscous fluid.   Flow lines show the direction of flow in a frame of reference fixed to the sphere.   In that frame the components of fluid velocity are given by Eq. 27.

ing the frame of reference of the cell and polar coordinates, as shown in Fig. 3, the flow around the sphere is the Stokes' velocity field described (ref. 5, p. 65) by

$$v_r = -v_0 \cos \theta (1 - 3a/2r + a^3/2r^3),$$
$$v_\theta = v_0 \sin \theta (1 - 3a/4r - a^3/4r^3). \tag{27}$$

To find the current to the cell, the equation to be solved is

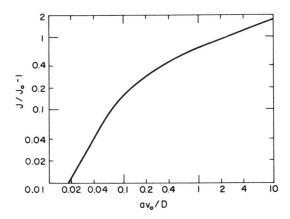

FIGURE 4   The increase in current to a spherical absorber resulting from motion at speed v_0. J and J_0 are the currents collected by moving and stationary spheres, respectively.   The curve is a log-log plot of the fractional increase, $J/J_0 - 1$, as a function of the dimensionless velocity parameter av_0/D, where a is the radius of the sphere and D the diffusion constant of the molecules absorbed.   The fractional increase $J/J_0 - 1$ is   proportional to v_0^2 for $v_0 \ll D/a$ and to $v_0^{1/3}$ for $v_0 \gg D/a$.

$$D \nabla^2 c - \mathbf{v} \cdot \operatorname{grad} c = 0, \tag{28}$$

with the components of $\mathbf{v}$ given by Eq. 27. An approximate solution has been obtained by a relaxation method. Thanks to the axial symmetry, the three-dimensional problem can be reduced to diffusion with drift on an appropriately modified two-dimensional grid. The result for $J/J_0 - 1$, the fractional increase in the current collected by the moving sphere compared with that collected by a stationary sphere in the same medium, is plotted in Fig. 4 against the dimensionless velocity parameter av_0/D. For $av_0/D \ll 1$, the increase in current, $J-J_0$, is proportional to $(av_0/D)^2$. That was to be expected. The increase cannot depend on the direction of motion and can hardly have a singularity at $v_0 = 0$; therefore, it must depend on even powers of v_0 in the neighborhood of $v_0 = 0$. There is, to be sure, an increase in current to the forward half of the sphere that is linear in v_0; our solution shows a fractional increase of approximately $1.5 \, av_0/D$. There is a corresponding linear decrease in current to the rear half of the sphere, leaving the total current with an initial quadratic rise. But our major concern is with values of av_0/D large enough to make an appreciable difference in $J-J_0$. Here the current increases much more slowly with v_0. One can show that in the high-velocity limit J/J_0 is proportional to $(av_0/D)^{1/3}$. For values of av_0/D as large as 10, our numerical solution exhibits that behavior.[5]

For a bacterium in pursuit of a typical nutrient, $a = 1 \, \mu$m, $v_0 = 30 \, \mu$m/s, $D = 10^{-5} \, $cm^2/s, and $av_0/D = 0.03$. The increase in intake is only 2.5% (Fig. 4). The speed required to double the intake is enormous, about 3 mm/s. The situation is somewhat more favorable if the diffusion constant D is very small. For example, if $a = 1 \, \mu$m and $D = 5 \times 10^{-8} \, $cm^2/s, the intake is doubled at a speed of only $15 \, \mu$m/s.

At the speed $v_0 = 2.5D/a$, the speed required to double the intake, the specific power demand is $28\eta D^2/a^4$; see footnote 3. This is to be compared with the result from our earlier example of stirring, $100 \, \eta D^2/a^4$. The mechanical efficiency of flagellar propulsion is at best a few percent (8), so the actual power requirement lies well above our estimated lower bound for stirring in general.

We conclude that in a uniform medium motility cannot significantly increase the cell's acquisition of material. At least that is true for a cell as small as a bacterium and for molecules of moderate size. In a nonuniform medium, on the other hand, a motile organism can seek out, as in chemotaxis, more favorable regions. The search must involve the detection of changes in concentration, a comparison of concentrations at different places or different times.

---

[5]Approximate analytical treatments of this problem have been published, for example, by Friedlander (6) and Acrivos and Taylor (7). Their results are expressed in terms of the Nusselt number, in our notation $2J/J_0$, and the Peclet number, in our notation $2av_0/D$. While agreeing that for large values of v_0 J/J_0 is proportional to $v_0^{1/3}$, the papers cited predict a linear dependence of $J-J_0$ on v_0 for small values of v_0. We do not understand how that dependence can arise.

MEASUREMENT OF CONCENTRATION

The Perfect Instrument

We want to see how well a microorganism can determine the concentration of molecules X in its vicinity. We shall assume that the organism derives its information from the state of occupation of specific receptors, and we shall study presently the capability of such a measuring procedure. But it will be instructive to consider first a hypothetical instrument, a device that can instantly count all the X molecules in some fixed volume V of medium. The expected count is $V\bar{c}$, where $\bar{c}$ is the "true" concentration of X molecules, the mean over a very long time. At the dilutions with which we are concerned there is no doubt that the X molecules diffuse independently. If the concentration is inferred from the result of a single count, a fractional error of magnitude $\Delta c/\bar{c} \approx (V\bar{c})^{-1/2}$ is probable. However, given a sufficient length of time to make the determination, we could by repeating the count reduce the uncertainty—provided that we wait long enough between counts to insure that the next population counted is statistically independent of the previous one. The waiting time required is approximately the time it takes a molecule to diffuse out of the sample volume, roughly $v^{2/3}/D$. That is, we can make about $TD/V^{2/3}$ independent counts in the alloted time T, which will reduce the rms fractional error in the determination of $\bar{c}$ to something like

$$\Delta c/\bar{c} \approx (TD/V^{2/3})^{-1/2}(V\bar{c})^{-1/2} = (TD\bar{c})^{-1/2}V^{-1/6}. \tag{29}$$

If the sample space is a spherical volume of radius a, this gives

$$\Delta c/\bar{c} \approx (1.61\,TD\bar{c}a)^{-1/2}. \tag{30}$$

If $a = 1\ \mu\text{m}$ and $D = 10^{-5}\ \text{cm}^2/\text{s}$, a concentration of 10^{-6} M ($\bar{c} = 6 \times 10^{14}\ \text{cm}^{-3}$) could be measured with 1% uncertainty in something like 0.01 s.

Numerical factors of order unity would appear in such relations if they were formulated precisely. A "perfect" instrument whose performance can be described precisely is a counter that registers at every instant the exact number m of X molecules that are at that moment inside a spherical region of radius a. The function $\dot{m}(t)$, the instrument's output, contains all the information about the ambient concentration $\bar{c}$ that can be obtained without looking outside that sphere. If we are given the output for time T, starting at t_1, the best estimate of $\bar{c}$ is $\bar{c} = (3/4\pi a^3)m_T$, where m_T is the average of $m(t)$ over the time of observation:

$$m_T = \frac{1}{T}\int_{t_1}^{t_1+T} m(t)\,dt. \tag{31}$$

The question now becomes, if we repeat this operation many times, starting at widely separated times t_k, what is the expected fluctuation in the values of m_T? We need to compute the mean square deviation of m_T, $\langle m_T^2 \rangle - \langle m_T \rangle^2$. The brackets indicate an average over a large number of independent runs, each of duration T. The average of the m_T's, $<m_T>$, is of course just $(4\pi/3)a^3\bar{c}$.

A useful tool is the autocorrelation function of $m(t)$, defined by

$$G(\tau) = <m(t)m(t + \tau)>, \tag{32}$$

in which the average indicated by the brackets is over an unlimited time. $G(\tau)$ is an even function of τ: $G(\tau) = G(-\tau)$. From the definition of m_T it follows that

$$m_T^2 = \frac{1}{T^2}\int_{t_1}^{t_1+T} dt' \int_{t_1}^{t_1+T} m(t)m(t')\, dt, \tag{33}$$

from which by introducing the autocorrelation function $G(\tau)$ we obtain

$$<m_T^2> = \frac{1}{T^2}\int_0^T dt' \int_0^T G(t' - t)\, dt, \tag{34}$$

reducing our problem to the determination of $G(\tau)$.

To find $G(\tau)$, consider a large number N of X molecules confined to a spherical volume of radius $R \gg a$ within which lies our spherical sample volume of radius a. Let $w_j(t)$ be the function which is 1 if molecule j is inside the smaller sphere at time t and 0 if it is not. Then the correlation function of $m(t)$ can be written as follows:

$$<m(t)m(t + \tau)> = \left\langle \sum_{j=1}^N w_j(t)w_j(t + \tau) \right\rangle + \left\langle \sum_{j\neq1}^N \sum_{i=1}^N w_j(t)w_i(t + \tau) \right\rangle \tag{35}$$

Clearly $<w_j> = a^3/R^3$, and since w_j and w_i are independent, the average of the double sum, in which there are $N(N - 1)$ terms, is $N(N - 1)a^6/R^6$, or $(Na^3/R^3)^2$ for large N. In the single sum the average of one of the N terms is $(a^3/R^3)u(\tau)$, where $u(\tau)$ is the probability that if a certain molecule is found inside the sample volume at time t, it will be found inside it at the later time $t + \tau$, this probability having been averaged over a uniform distribution of initial positions throughout the spherical volume $r < a$. We now have

$$<m(t)m(t + \tau)> = (Na^3/R^3)u(\tau) + (Na^3/R^3)^2. \tag{36}$$

But Na^3/R^3 is just $<m>$, so the correlation function is

$$G(\tau) = <m>u(\tau) + <m>^2. \tag{37}$$

Actually we shall not need $u(\tau)$ itself but only the characteristic time τ_0 defined by

$$\tau_0 = \int_0^\infty u(\tau)\, d\tau, \tag{38}$$

which is easier to calculate. The value of τ_0, derived in Appendix C, is $\frac{2}{5}a^2/D$.

It is an appropriate measure of the time for the contents of the sample volume to be renewed by diffusion in and out.

Returning to Eq. 34, let us consider observation times T much longer than τ_0. In that case, remembering that $G(-\tau) = G(\tau)$, the integral becomes

$$<m_T^2> = (1/T^2) \int_0^T dt'(T<m>^2 + 2\tau_0<m>)$$

$$= <m_T>^2 + (2\tau_0/T)<m_T>. \quad (39)$$

This gives us an exact formula for the mean square fluctuation in m_T:

$$<\Delta m_T^2> = <m_T^2> - <m_T>^2 = (2\tau_0/T)<m_T> = (4a^2/5DT)\overline{m}, \quad (40)$$

from which we obtain the rms fractional error in concentration $\bar{c}$ in one such measurement:

$$\Delta c_{rms}/\bar{c} = (5\pi TD\bar{c}a/3)^{-1/2}. \quad (41)$$

The rms error is smaller by a factor 0.55 than the estimate of Eq. 30.

This result for the perfect instrument will provide us with a convenient standard of comparison. For instance, any procedure capable of determining the concentration with an rms error of 1% in an observation time T may be said to be equivalent to a perfect instrument sampling a spherical volume of radius about $2{,}000/TD\bar{c}$.

A Single Receptor

We assume that a receptor has a binding site capable of binding one, but only one, X molecule. We shall describe the history of this site by a function $p(t)$ that has the value 1 when the site is occupied and 0 when it is empty. The time average occupation $\bar{p}$ is determined by a single dissociation constant K, the concentration in moles per liter for which $\bar{p} = 0.5$. Let us denote by $c_{1/2}$ the same concentration in molecules per cubic centimeter. In equilibrium at concentration c the expected average occupancy is

$$\bar{p} = c/(c + c_{1/2}). \quad (42)$$

After a molecule has become attached to a binding site there is a constant probability, per unit time, that it will be released. Let the probability of detachment in an interval dt be dt/τ_b. Then τ_b is the average time a molecule stays bound to a site. As before, we shall describe the binding site as a circular patch of radius s. If the patch were a sink, the current to it would be $4Dsc$, as in Eq. 5. Suppose that a molecule that arrives at a vacant binding site sticks with probability α. If it doesn't stick on its first contact, it may very soon bump into the site again—and again. If these encounters occur within a time interval short compared to τ_b, their result is equivalent merely to a larger value of α. As we have no independent definition of the patch radius s, we may as well absorb the effective α into s, writing for the probability that a vacant patch becomes occupied during dt simply $4Dsc\, dt$. Since $\bar{p}$ is the probability that a receptor patch is occupied and $1 - \bar{p}$ the probability that it is empty, in the steady state the following relation must hold:

$$\bar{p}/\tau_b = 4(1 - \bar{p})\,Dsc. \tag{43}$$

In particular, since $\bar{p} = \frac{1}{2}$ for $c = c_{1/2}$,

$$\tau_b = (4Dsc_{1/2})^{-1}. \tag{44}$$

Thus for $K = 10^{-6}$M, $D = 10^{-5}$cm^2/s, and $s = 10$ Å, we would have $\tau_b = 4 \times 10^{-4}$ s.

If the only information about the ambient concentration c is the function $p(t)$ for one receptor recorded for a time T, the best use that can be made of it is to form the average,

$$p_T = (1/T) \int_{t_1}^{t_1+T} p(t)\,dt, \tag{45}$$

take that as an estimate of $\bar{p}$, and use Eq. 42 to derive c:

$$c/c_{1/2} = p_T/(1 - p_T). \tag{46}$$

To compute the uncertainty in such a determination of c we proceed exactly as we did above with the measurement of m_T. All we need is the correlation function for $p(t)$, for which we shall use the same symbol, $G(\tau)$:

$$G(\tau) = \langle p(t)p(t + \tau)\rangle. \tag{47}$$

Consider data from a large number n of pairs of observations, one at t, the other at $t + \tau$, with random values of t but always the same value of τ. Segregate those pairs in which the first observation found $p = 1$ and ignore the others. If n is very large, there will be about $n\bar{p}$ such pairs. Of these, according to the definition of $G(\tau)$, the number with $p(t + \tau)$ also equal to 1 will be $nG(\tau)$. These "1,1" pairs are the only ones for which $p(t)p(t + \tau) \neq 0$. Now consider the result of shifting the time of the second observation from $t + \tau$ to $t + \tau + d\tau$. Some of the $nG(\tau)$ 1,1 pairs will become 1,0 pairs; $nG\,d\tau/\tau_b$ of them will do so. Some of the 1,0 pairs, of which there were $n\bar{p} - nG$, will become 1,1 pairs; the number doing so will be $n(\bar{p} - G)[\bar{p}/(1 - \bar{p})]d\tau/\tau_b$. We should now have $nG(\tau + d\tau)$ 1,1 pairs, which requires that

$$dG = -G\,d\tau/\tau_b + (\bar{p} - G)[\bar{p}/(1 - \bar{p})]\,d\tau/\tau_b. \tag{48}$$

Integrating and requiring that $G(0) = \bar{p}$, we obtain

$$G(\tau) = \bar{p}^2 + \bar{p}(1 - \bar{p})\exp[-|\tau|/(1 - \bar{p})\tau_b]. \tag{49}$$

We now use Eq. 34 to calculate $\langle p_T^2\rangle$, assuming $T \gg \tau_b$, with the result

$$\langle p_T^2\rangle - \langle p_T\rangle^2 = (2/T)\bar{p}(1 - \bar{p})^2\tau_b. \tag{50}$$

For the rms error in $\bar{c}$ inferred from such a measurement we get

$$\Delta c_{rms}/\bar{c} = (2\tau_b/T\bar{p})^{1/2}. \qquad (51)$$

This result can be expressed in a surprisingly simple and illuminating form. Using Eqs. 42 and 44, we can transform Eq. 51 into

$$\Delta c_{rms}/\bar{c} = (\nu/2)^{-1/2}, \qquad (52)$$

where

$$\nu = 4 D s \bar{c}(1 - \bar{p})T. \qquad (53)$$

The current $4 D s \bar{c}$ is the rate at which molecules would be captured by the receptor patch if it were a perfect sink. It is therefore the rate at which molecules arrive at the receptor if we only count "new" molecules, those that have not been there before. Since the probability that the patch is already occupied when any molecule arrives is $\bar{p}$, the rate at which new molecules are captured by the patch is $4 D s \bar{c}(1 - \bar{p})$. Hence the number ν is just the total number of new molecules that have occupied the receptor patch during the observation period T. We see that the fractional error in the determination of $\bar{c}$ depends on this number ν and nothing else! Evidently, once a particular molecule has occupied a receptor patch, subsequent visits by the same molecule contribute no information whatever about the concentration in the medium. Indeed, that ought to be true if the molecules are diffusing independently. Such diffusion is a Markov process, which is to say that the probability of a future configuration is determined completely by the present configuration, regardless of the past. It follows that we can draw from the molecule's future behavior no inference about the past that is not already implied by its present position. This observation is the key to the generalization of our result to include cells with many receptors.

It is interesting to compare the performance of one receptor, as a "$\bar{c}$-measuring" device, with the perfect instrument described earlier. Using Eq. 41, we find that the single receptor is equivalent to a perfect instrument with a spherical sampling volume of radius $(6/5\pi)(1 - \bar{p})s$, a radius approximating that of the single receptor patch. With $\bar{p} = \frac{1}{2}$, $s = 10$ Å, and $D = 10^{-5}$ cm²/s, a concentration of 10^{-6} M could be measured with 1% uncertainty in about 17 s.

A System of Many Receptors

As we turn toward our ultimate goal, assessing the performance of a cell with many receptors, a formidable complication looms ahead. The "signal" from which $\bar{c}$ is to be determined is now the total instantaneous occupation of the N receptors on the cell, that is, the function

$$P(t) = \sum_{j=1}^{N} p_j(t). \qquad (54)$$

Now the histories of the occupation of two receptors on the same cell, especially of two receptors relatively close to one another, are not statistically independent. A molecule just released at receptor j is necessarily favorably situated to wander into receptor k. Put another way, we should expect the two fluctuations, $p_j(t) - \bar{p}$ and $p_k(t) - \bar{p}$, to exhibit some positive cross-correlation. The effect would depend on the distance between the receptors and would be extremely difficult to handle rigorously. Fortunately, this threatening complication vanishes when we realize that, as we have just learned in the case of the single receptor, we need be concerned only with the capture of new molecules. Here "new" molecules are those that have not previously occupied *any* receptor on the cell. For once a particular molecule has occupied a receptor anywhere on the cell, its subsequent history is statistically determined and can convey no further information about the ambient concentration. The current of new molecules to a receptor patch is equal to the current that would reach that patch if all the receptor patches on the cell were perfect absorbers. The capture rate for the receptor patch is this current times $(1 - \bar{p})$. Therefore, the rate at which new molecules are captured by the receptor system is just $J(1 - \bar{p})$, where J is given by Eq. 8 and $\bar{p}$ is given, as before, by Eq. 42. This does not mean that the cell is obliged to identify the new molecules and avoid counting the others. What we are asserting is that the statistical error in the cell's inferred value of $\bar{c}$, given the receptor occupation history $P(t)$, will be the same as if new molecules only had been recorded.

The capture of a new molecule by one receptor and the capture of a new molecule (necessarily a different molecule) by another receptor are statistically independent events. With respect to such events, the history of N receptors observed for time T is statistically indistinguishable from that of a single receptor observed for time NT. It follows that the probable error in a value of $\bar{c}$ inferred from this information will be the same as Eq. 52 predicts for a single receptor if $\nu = TJ(1 - \bar{p})$. Thus we arrive rather suddenly at our final result:

$$\Delta c_{\text{rms}}/\bar{c} = [\tfrac{1}{2}TJ(1 - \bar{p})]^{-1/2} = [2\pi TD\bar{c}Nsa(1 - \bar{p})/(Ns + \pi a)]^{-1/2}. \quad (55)$$

If the number of receptors, N, is such as to make $J/J_{\text{max}} = \tfrac{1}{2}$, namely $N = \pi a/s$, this becomes

$$\Delta c_{\text{rms}}/\bar{c} = [\pi TDca(1 - \bar{p})]^{-1/2}. \quad (56)$$

On comparing this with Eq. 41, we find that the equivalent sample volume for the perfect instrument would be a sphere of radius $\tfrac{3}{5}(1 - \bar{p})a$.

APPLICATIONS

Bacteriophage Adsorption

Eq. 8 provides a solution to a classic problem in bacteriophage adsorption (see ref. 9). Why is the initial rate of adsorption of phage to a bacterium so close to the diffusion-limited rate for a perfectly adsorbing cell, given that the receptor binding sites cover only a small fraction of the surface? As we have seen, the answer lies in the large number of independent tries that each diffusing particle has at hitting a binding site once it has bumped into the surface of the cell. This is a statistical property of diffusion per se; it is true regardless of the structure of the diffusing particle. Schwartz (9) found a hyperbolic dependence of the initial rate of adsorption of bacteriophage λ on the number of λ-receptors in samples of *E. coli* grown under different cultural conditions (his Fig. 2a). A least-squares fit (10) of Eq. 8 to this data gives J/c_∞ = $(2.14 \pm 0.10) \times 10^{-10}$ cm^3/s and $\pi a/s$ = 483 $\pm$ 79. With $a = 8 \times 10^{-5}$ cm, we find $s = 52$ Å. Recall that s is an effective radius; it depends on the size of the binding site and of the phage and on the probability that a phage, having arrived at a binding site, is adsorbed. Adsorption occurs at the half-maximum rate with fewer than 500 receptors per cell, when only 0.5% of the surface is specifically adsorbing. The value for J/c_∞ determined by the data is larger than the value computed from Eq. 2, 5.03×10^{-11} cm^3/s, by a factor of 4. The diffusion constant for bacteriophage λ is quite small, 5×10^{-8} cm^2/s; in this case, the specific energy demand for stirring (Eq. 26) is not prohibitive, and the bacterium could double the adsorption rate by swimming (Fig. 3, $av_0/D \approx 3$). But we do not know whether Schwartz's bacteria were motile. A systematic error in the measurement of the number of phage or bacteria could explain the discrepancy.

Bacterial Chemotaxis

Studies of chemotaxis are most advanced for the enteric bacteria *Escherichia coli* and *Salmonella typhimurium* (for reviews, see refs. 11–14). These cells execute a three-dimensional random walk (15). They swim steadily along smooth trajectories (run), move briefly in a highly erratic manner (tumble or twiddle), and then run in new directions. They sense concentrations of attractants or repellents as a function of time (16, 17) and bias their random walk accordingly. Runs that carry a cell to higher concentrations of an attractant or to lower concentrations of a repellent are extended. The available evidence is consistent with a model in which a bacterium measures the difference in the fraction of receptors bound in successive intervals of time (17–19), i.e., in which the response is proportional to $d\bar{p}/dt$. The random walk can be biased most effectively if the measurements are made in a time interval short compared to the mean run length (ref. 17, Fig. 3); information gathered during a run is of little value once the cell has chosen a new direction at random. The time available for gradient determination and chemotactic response could not, in any case, exceed the time, τ_{rot}, which characterizes the Brownian rotation of the cell. There is no way, even in principle, for a bacterium to preserve an orientation reference frame for a time much longer than τ_{rot}—unless, of course, it could use some external clue such as the direction of illumination. In the case of *E. coli*, τ_{rot} is typically a few seconds, somewhat longer than the length of a run (15), so the run length remains the controlling limit on gradient measurement and response time.

Let the period of time devoted to each measurement of $\bar{p}$ be T. The difference between two successive measurements will be significant if that difference is larger than the standard deviation in the difference, i.e., if

$$(T/\bar{c})\partial\bar{c}/\partial t > \sqrt{2}\,\Delta c_{\mathrm{rms}}/\bar{c}. \tag{57}$$

This inequality places a condition on T that can be found by substitution of Eq. 55:

$$T > \left[\pi a D\left(\frac{Ns}{Ns + \pi a}\right)\left(\frac{\bar{c}c_{1/2}}{\bar{c} + c_{1/2}}\right)\left(\frac{1}{\bar{c}}\frac{\partial\bar{c}}{\partial t}\right)^2\right]^{-1/3}. \tag{58}$$

If the temporal gradient is generated by the movement of the cell at velocity v through a spatial gradient $\partial\bar{c}/\partial x$, $(1/\bar{c})\partial\bar{c}/\partial t = (v/\bar{c})\partial\bar{c}/\partial x$. The time required to complete the temporal comparison is $2T$.

Working with enzymatically generated temporal gradients of the attractant L-glutamate, Brown and Berg (ref. 17, Fig. 1) found that the mean run length doubled (increased from 0.67 to 1.34 s) for $d\bar{p}/dt = 1.05 \times 10^{-3}$ s^{-1}, i.e., for $\bar{c} = 1.61$ mM, $(1/\bar{c})\partial\bar{c}/\partial t = 4.35 \times 10^{-3}$ s^{-1}, and $c_{1/2} = 2.3$ mM. If $a = 0.8$ μm, $D = 9 \times 10^{-6}$ cm^2/s, and $Ns/(Ns + \pi a) = 0.5$, we find $2T > 0.087$ s. The time required to detect a temporal gradient of $\frac{1}{10}$ the magnitude would be $10^{2/3} = 4.64$ times longer, or about 0.4 s.

Working with defined spatial gradients of the attractant L-serine, Dahlquist et al. (18) found that a trajectory of length 10 μm was doubled by a gradient of decay length about 1.4 cm. With $c_{1/2} = 1.0$ mM, $(1/\bar{c})\partial\bar{c}/\partial x = 0.7$ cm^{-1}, $v = 15$ μm/s, $a = 0.8$ μm, $D = 10^{-5}$ cm^2/s, and $Ns/(Ns + \pi a) = 0.5$, we find $2T \geq 0.27$ s. A gradient $\frac{1}{10}$ as steep could be detected in about 1.2 s.

Using the capillary assay and attractants detected by the aspartate and galactose chemoreceptors, Mesibov et al. (19) found threshold responses for DL-α-methylaspartate and D-galactose at concentrations (the initial concentration in the capillary) $\bar{c}_0 \simeq 4 \times 10^{-7}$ M and 2.5×10^{-8} M, respectively, with $c_{1/2} \simeq 1.3 \times 10^{-4}$ M and 6×10^{-7} M, respectively. These experiments are difficult to analyze, because the response is complex and the gradient near the mouth of the capillary is hard to define. Using Adler's interpolation (ref. 20, Fig. 5), we estimate $\bar{c} = 10^{-2}\bar{c}_0$ and $(1/\bar{c})\partial\bar{c}/\partial x = 80$ cm^{-1}. As before, we assume $Ns/(Ns + \pi a) = 0.5$. With $a = 0.8$ μm, $D = 10^{-5}$ cm^2/s, and $v = 15$ μm/s, we find for α-methylaspartate $2T > 0.6$ s, for galactose $2T > 1.4$ s.

Taken together, these results imply that *E. coli* and *S. typhimurium* are able to make temporal comparisons of concentrations in time intervals of about 1 s. If times much longer than this were required, cells in spatial gradients could not effectively bias their random walks; for *E. coli,* even 5 s would be prohibitive (17). Thus, the design of the chemotaxis machinery appears to be nearly optimum. This implies, for example, that the receptors are dispersed widely over the surface of the cell, rather than concentrated at the base of each flagellum, and that essentially every capture of a molecule by a receptor contributes to the signal controlling the direction of rotation of the flagella.

It is of interest to ask whether, in principle, a bacterium could navigate by comparing the concentration at the front to that at the back, that is, by a strictly spatial mechanism. In this case we require

$$(a/\bar{c})\partial\bar{c}/\partial x \ > \ \sqrt{2}\,\Delta c_{rms}/\bar{c}, \tag{59}$$

where Δc_{rms} is the standard deviation in the measurement made by half of the cell. A condition on T is obtained by substituting Eq. 55 and using for J the current to one half of the cell:

$$T \ > \ \left[\frac{1}{2}\,\pi a^3 D \left(\frac{Ns}{Ns \ + \ \pi a}\right)\left(\frac{\bar{c}c_{1/2}}{\bar{c} \ + \ c_{1/2}}\right)\left(\frac{1}{\bar{c}}\,\frac{\partial\bar{c}}{\partial x}\right)^2\right]^{-1}. \tag{60}$$

For the experiment of Dahlquist et al. (18), we find $T > 1.7$ s. The comparison probably could be made in less time, since the length of the cell is roughly twice its diameter. Thus, on the basis of this analysis a mechanism involving spatial comparisons remains feasible.

But there is a much more serious problem: motion of the cell will generate an apparent spatial gradient. As noted in our discussion of swimming, the flux to a perfectly absorbing sphere of radius a moving at velocity v_0 in a solution of uniform concentration is greater in the front than in the back by a factor of about $1 + 3av_0/D$. As shown in Appendix D, the flux to a stationary perfectly absorbing sphere of radius a in a spatial gradient of decay length L is greater in the front than in the back by a factor $1 + a/L$. Thus, the moving sphere finds itself in an apparent spatial gradient of decay length $D/3v$. In the experiment of Dahlquist et al. (18), this decay length is of order 2×10^{-3} cm: the apparent gradient is 600 times steeper than the real gradient! The problem remains severe but is less dramatic when we realize that the cell takes up only a few percent of the molecules that reach its surface (ref. 4, Table 3); molecules not absorbed by the front half of the cell could still be counted by a receptor system operating independently at the back.

Slime Mold Chemotaxis

The spatial mechanism is much more effective if the cell is large. Mato et al. (21), working with the cellular slime mold *Dictyostelium discoideum,* observed a threshold response for cyclic AMP (with cells 0.6 cm away from a point source of the attractant) when $\bar{c} = 4.3 \times 10^{-9}$ M, $\partial\bar{c}/\partial x = 3.6 \times 10^{-8}$ M/cm, and $c_{1/2} \simeq 10^{-8}$ M. Assuming $a = 5\ \mu$m, $D = 5 \times 10^{-6}$ cm^2/s, and $Ns/(Ns + \pi a) = 0.5$, we find, using Eq. 60, $T > 16$ s. This interval is short compared to the duration of the wave of cyclic AMP diffusing past the cells. If the cells absorb (or destroy) an appreciable fraction of the cyclic AMP, then an apparent gradient of decay length 0.8 cm would be generated were they to crawl through a solution of uniform concentration at about 0.2 μm/s, their usual speed during a chemotactic response. This is comparable to the decay length of the threshold gradient in the experiment of Mato et al. (1.2 cm).

How long would it take the slime mold to sense the spatial gradient by making temporal comparisons? If the crawl velocity is 0.2 μm/s, the spatial gradient of Mato et al. could be detected in about 17 s (Eq. 58). A pseudopod of one-fourth the radius moving twice as fast would require a similar time. This analysis does not allow us to rule out a temporal mechanism. But note that whatever the mechanism, the measurement must be made over an appreciable period of time.

We thank Steven M. Block and Francis D. Carlson for comments on the manuscript.

Computations were performed with facilities supported by National Science Foundation Grant PCM 74-23522. Other aspects of the work were supported by National Science Foundation Grants BMS 75-05848 and PCM 77-08543.

Received for publication 13 May 1977.

REFERENCES

1. SMYTHE, W. R. 1950. Static and Dynamic Electricity. McGraw-Hill Book Company, New York. 2nd edition. 114, where C is given in mks units.
2. ADAM, G., and M. DELBRÜCK. 1968. Reduction of dimensionality in biological diffusion processes. *In* Structural Chemistry and Molecular Biology. A. Rich and N. Davidson, editors. W.H. Freeman & Company, Publishers, San Francisco, Calif. 198–215.
3. CARLSON, F. D. 1962. A theory of the survival value of motility. *In* Spermatozoan Motility. D.W. Bishop, editor. American Association for the Advancement of Science, Washington, D.C. 137–146.
4. KOCH, A. L. 1971. The adaptive responses of *Escherichia coli* to a feast and famine existence. *Adv. Microb. Physiol.* 6:147–217.
5. LANDAU, L.D., and E. M. LIFSCHITZ. 1959. Fluid Mechanics. Pergamon Press, Ltd., Oxford, Great Britain.
6. FRIEDLANDER, S. K. 1957. Mass and heat transfer to single spheres and cylinders at low Reynolds numbers. *Am. Inst. Chem. Eng. J.* 3:43–48.
7. ACRIVOS, A., and T. D. TAYLOR. 1962. Heat and mass transfer from single spheres in Stokes flow. *Phys. Fluids.* 5:387–394.
8. PURCELL, E. M. 1976. Life at low Reynolds number. *In* Physics and Our World: A Symposium in Honor of Victor F. Weisskopf. AIP Conference Proceedings, No. 28. K. Huang, editor. American Institute of Physics, New York. 49–64. Reprinted: 1977. *Am. J. Phys.* 45:3–11.
9. SCHWARTZ, M. 1976. The adsorption of coliphage Lambda to its host: Effect of variations in the surface density of receptor and in phage-receptor affinity. *J. Mol. Biol.* 103: 521–536.
10. WILKINSON, G. N. 1961. Statistical estimations in enzyme kinetics. Biochem. J. 80:324–332.
11. ADLER, J. 1975. Chemotaxis in bacteria. *Annu. Rev. Biochem.* 44:341–356.
12. BERG, H. C. 1975. Bacterial behavior. *Nature (Lond.).* 254:389–392.
13. BERG, H. C. 1975. Chemotaxis in bacteria. *Annu. Rev. Biophys. Bioeng.* 4:119–136.
14. KOSHLAND, D. E., JR. 1977. A response regulator model in a simple sensory system. *Science (Wash. D.C.).* 196:1055–1063.
15. BERG, H. C., and D. A. BROWN. 1972. Chemotaxis in *Escherichia coli* analysed by three-dimensional tracking. *Nature (Lond.).* 239:500–504.
16. MACNAB, R., and D. E. KOSHLAND, JR. 1972. The gradient-sensing mechanism in bacterial chemotaxis. *Proc. Natl. Acad. Sci. U.S.A.* 69:2509–2512.
17. BROWN, D. A., and H. C. BERG. 1974. Temporal stimulation of chemotaxis in *Escherichia coli*. *Proc. Natl. Acad. Sci. U.S.A.* 71:1388–1392.
18. DAHLQUIST, F. W., R. A. ELWELL, and P. S. LOVELY. 1976. Studies of bacterial chemotaxis in defined concentration gradients. A model for chemotaxis toward L-serine. *J. Supramol. Struc.* 4:329–342.
19. MESIBOV, R., G. W. ORDAL, and J. ADLER. 1973. The range of attractant concentrations for bacterial chemotaxis and the threshold and size of response over this range. *J. Gen. Physiol.* 62:203–223.
20. ADLER, J. 1974. A method for measuring chemotaxis and use of the method to determine optimum conditions for chemotaxis by *Escherichia coli*. *J. Gen. Microbiol.* 74:77–91.
21. MATO, J. M., A. LOSADA, V. NANJUNDIAH, and T. M. KONIJN. 1975. Signal input for a chemotactic response in the cellular slime mold *Dictyostelium discoideum*. *Proc. Natl. Acad. Sci. U.S.A.* 72:4991–4993.

APPENDIX A

The Capacitance of N *Conducting Disks of Radius* s *on an Insulating Sphere of Radius* a *and Dielectric Constant Zero*

We first treat the N disks as a system of independent conductors with different charges q_j and potentials ϕ_j, $j = 1, 2, \ldots N$. The charges and potentials are connected by linear relations involving the so-called potential coefficients, h_{jk}:

$$\phi_j = \sum_k h_{jk} q_k. \qquad (A1)$$

Put zero charge on all but the kth disk. Then $\phi_k = h_{kk} q_k$. The presence of the $N - 1$ uncharged disks can affect ϕ_k only through dipole or higher moments induced in them by the field of the lone charge q_k. That is a second-order effect which can be neglected if, as we shall now assume, the disk radius s is small compared to the distance between neighboring disks, approximately $a(4\pi/N)^{1/2}$. If the other disks were not there, the capacitance of a single disk on the insulating sphere would be just half that of an isolated conducting disk, or s/π. So we must have, to first order in the ratio $Ns^2/4\pi a^2$, $h_{kk} = \pi/s$. As the disks are all equivalent, this holds for all $k = 1, \ldots N$. Now the potential of one of the uncharged disks is

$$\phi_j = h_{jk} q_k. \qquad (A2)$$

Consider the sum

$$[1/(N - 1)] \sum_{j \neq k} \phi_j = [q_k/(N - 1)] \sum_{j \neq k} h_{jk}. \qquad (A3)$$

If N is large, this is essentially the average over the spherical surface of the potential due to the single charge q_k. If the sphere were empty space with dielectric constant unity instead of zero, that average would be simply q_k/a, because the average of a harmonic function over a sphere is equal to its value at the center of the sphere. In our case, however, the electric field outside the sphere is not that of a single charge q_k, but rather that plus the field of a set of "image charges" within the sphere, the combined field being such as to satisfy the boundary condition of zero normal component at the spherical surface. So we should add to q_k/a the contribution of the image-charge distribution to the average potential. But that contribution is zero, since the net charge of the image distribution is zero. (If a sphere contains no net charge, the average of the potential over the surface is that due to external charges alone.) Hence we are left with

$$[q_k/(N - 1)] \sum_{j \neq k} h_{jk} = q_k/a. \qquad (A4)$$

We have tacitly assumed that N is large, so we can write

$$\sum_{j \neq k} h_{jk} = N/a. \qquad (A5)$$

Referring back to Eq. A1, if we now put the same charge q on every disk so that the total charge is Nq, the common potential ϕ will be

$$\phi = \phi_k = h_{kk} q + \sum_{j \neq k} q h_{jk} = \pi q/s + Nq/a, \qquad (A6)$$

from which we obtain the capacitance of the combination:

$$C = Nq/\phi = Nsa/(Ns + \pi a).$$ (A7)

APPENDIX B

The Mean Time to Capture in Two-Dimensional Diffusion

In a space of two dimensions a particle is released at time $t = 0$ at the point x, y. The particle then diffuses at a rate determined by a two-dimensional diffusion constant D' until it eventually blunders into the boundary of an absorber, where, at time t, it is captured. Let this be repeated very many times, starting always at the same point x, y. Let W denote the mean of all the observed times-to-capture for this starting point. To find the equation satisfied by the function $W(x, y)$, picture the diffusion as a random walk on a square lattice, with step length δ and step time Δt. Consider a particle now at the lattice point x, y, from which the mean time to capture is $W(x, y)$. One step-time later this particle will be with equal probability at one of the four lattice points $x \pm \delta, y \pm \delta$. It must therefore be true that

$$W(x, y) = \Delta t + \tfrac{1}{4}[W(x + \delta, y) + W(x - \delta, y)$$
$$+ W(x, y + \delta) + W(x, y - \delta)].$$ (B1)

If we now shrink the step-length and step-time so as to approach the continuous diffusion limit with $D' = \delta^2/4\Delta t$, Eq. B1 becomes

$$D' \nabla^2 W + 1 = 0.$$ (B2)

In electrical terms, this is just Poisson's equation for a region of uniform charge density, with W the potential and $1/4\pi D'$ the charge density. As boundary conditions we require $W = 0$ on all absorbing boundaries. On a line of symmetry the normal component of grad W must vanish. The same condition holds at an impermeable, nonabsorbing boundary.

A number of cases are now almost trivially soluble, including the case of diffusion in an annular region treated by Adam and Delbrück (2): a circular absorber of radius s (a in their notation) centered within an impermeable boundary of radius b. All we need is the electrical potential $\phi(r)$ between concentric cylinders, the space being filled with uniform space charge opposite in sign and equal in total amount to the charge on the inner cylinder (thus insuring zero gradient at $r = b$). In this way we find

$$W(r) = (2b^2 \ln r - 2b^2 \ln s - r^2 + s^2)/4D'.$$ (B3)

We seek the mean of W over all starting points in the annular space, the quantity Adam and Delbrück call $\tau^{(2)}$ and we call $\bar{t}_c$. In this case

$$\bar{t}_c = \frac{1}{\pi(b^2 - s^2)} \int_s^b 2\pi r W(r) dr = \frac{b^4}{2D'(b^2 - s^2)} \ln \frac{b}{s} - \frac{3b^2 - s^2}{8D'}.$$ (B4)

For our application we might have defined t_c as the average of W over the whole region $r < b$, including the absorber, but the distinction is unimportant if $b \gg s$, which is the case of interest. In this limit Eq. B4 reduces to

$$\bar{t}_c = (b^2/2D')[\ln(b/s) - \tfrac{3}{4}].$$ (B5)

Adam and Delbrück's approximate result for this case, obtained by rather laborious means, is in our notation

$$\bar{t}_c \simeq (0.5b^2/D')[\ln(b/s) - 0.5], \qquad (B6)$$

in excellent agreement with Eq. B5 as far as the logarithmic term.

Our actual problem was concerned with an unbounded array of evenly spaced absorbers. Let these be circular patches of radius s on a square lattice of spacing $d \gg s$. We need only solve Eq. B2 in the unit cell, a square of edge b with the absorber at its center. W is zero at the absorber and its normal derivative vanishes on the unit cell boundary. An approximation that ought to be adequate for $b \gg s$ is obtained simply by taking over Eq. B5 for a circular boundary and setting $b^2 = (4/\pi)d^2$. Another approach is to subdivide the unit cell with a grid and solve Eq. B2 numerically. Such a treatment has been carried out by a relaxation method for meshes as fine as $b/40$ and for absorbers consisting of a square set of mesh points ranging in size from a single point to 7^2 points. Both of these approaches gave results adequately represented by Eq. 14, as did a Monte Carlo study of random walks on a square grid with a central sink. The Monte Carlo results included information about the distribution of times-to-capture, in the form of the first four moments of the distribution of all path lengths to capture with all starting points equally weighted. The ratios of the observed moments corresponded closely to those for an exponential distribution, i.e., a distribution in which the probability of a time-to-capture greater than t_c is proportional to $\exp(-t_c/\bar{t}_c)$.

The derivation of Eq. B2 generalizes easily to spaces of higher or lower dimensionality. Eq. B2 remains unchanged, it being understood that in ν dimensions ∇^2 is the ν-dimensional Laplacian and D' the ν-dimensional diffusion constant. For a spherical absorber of radius a in a spherical vessel of radius b we find

$$W(r) = (2b^3/a - 2b^3/r + a^2 - r^2)/6D, \qquad (B7)$$

which leads to the mean time to capture

$$\bar{t}_c = \frac{b^6}{3Da(b^3 - a^3)} \left(1 - \frac{9}{5}\frac{a}{b} + \frac{a^3}{b^3} - \frac{1}{5}\frac{a^6}{b^6}\right). \qquad (B8)$$

This reduces to $b^3/3aD$ in the limit $b \gg a$. Thus, the three-dimensional "tracking factor" of Adam and Delbrück is $b/3a$.

APPENDIX C

Calculation of τ_0 Defined by Eq. 38.

Let a sphere of radius a contain at time $t = 0$ unit amount of a uniformly distributed diffusing substance. The function $u(\tau)$ is defined as the fraction of the original material that remains inside the sphere at $t = \tau$. If a concentrated source of unit amount were released at any point P at time zero, the concentration in the neighborhood of some other point Q at a later time would be

$$f(\rho, \tau) = (4\pi D\tau)^{-3/2} \exp(-\rho^2/4D\tau) \qquad (C1)$$

where ρ is the distance between P and Q. We can use this to express $u(\tau)$ as follows:

$$u(\tau) = \int dv \int (3/4\pi a^3)f(|r - r'|, \tau) dv', \qquad (C2)$$

both integrals to be extended throughout the volume of the sphere.

We want to calculate τ_0 defined by

$$\tau_0 = \int_0^\infty u(\tau)\,d\tau. \qquad (C3)$$

Substitute from Eqs. C1 and C2 into Eq. C3 and carry out the integration over τ first:

$$\int_0^\infty f(\rho,\tau)\,d\tau = (1/4D\rho\pi^{3/2})\int_0^\infty (e^{-x}\,dx/\sqrt{x}) = 1/4\pi D\rho. \qquad (C4)$$

Then

$$\tau_0 = (1/4\pi D)\int dv \int (3/4\pi a^3)(dv'/|r - r'|). \qquad (C5)$$

The integral over v' can be recognized as the potential at r within a uniform spherical charge distribution of total charge unity, which is $3/2a - r^2/2a^3$. The integration over v now yields

$$\tau_0 = \tfrac{2}{5}a^2/D. \qquad (C6)$$

APPENDIX D

Diffusive Current to Two Halves of a Stationary, Perfectly Absorbing Sphere in a Uniform Gradient

Let a sphere of radius a be immersed in a uniform gradient $\partial c/\partial z = c_0/L$, with $L \gg a$. The solution to the equation $\nabla^2 c = 0$, with boundary conditions $c = 0$ at $r = a$ and $\partial c/\partial z \to c_0/L$ at $r \to \infty$, is

$$c = c_0(1 - a/r) + (c_0 a^3/L)(z/r^3 - z/a^3). \qquad (D1)$$

The current density at the surface of the sphere is

$$D\frac{\partial c}{\partial r}\bigg|_{r=a} = Dc_0(1/a + \cos\theta/L), \qquad (D2)$$

where θ is the angle measured from the $+z$-axis (spherical coordinates). The current to the forward half of the sphere, J_+, is obtained by integrating Eq. D2 over the surface from $\theta = 0$ to $\theta = \pi/2$; the current to the rear half of the sphere, J_-, is obtained by integrating Eq. D2 over the surface from $\theta = \pi/2$ to $\theta = \pi$. We find

$$J_\pm = 2\pi a D c_0(1 \pm a/2L), \qquad (D3)$$

and

$$J_+/J_- = 1 + a/L. \qquad (D4)$$

Authors' Note:

In Eq. 20 and on the following line, ν in the numerator should be v, for velocity.

On p. 206, par. 1, line 15, $v^{2/3}$ should be $V^{2/3}$.

In Eq. 35, the lower index in the first summation in the second bracket should read $j \neq i$.

There is an error in Eq. D1 that propagates back to the main text. In Eq. D1, the terms in the final parentheses should appear in the reverse order: $z/a^3 - z/r^3$. In Eq. D2, the final term should be $3\cos\theta/L$. In Eq. D3, the final term should be $3a/2L$. In Eq. D4, the final term should be $3a/L$. On p. 214, par. 2, line 7, read $1 + 3a/L$; in line 8, read D/v_0; in line 9, read 6×10^{-3} cm and 200 times steeper. On p. 214, par. 3, line 8, read 2.4 cm.

Life at low Reynolds number

E. M. Purcell
Lyman Laboratory, Harvard University, Cambridge, Massachusetts 02138
(Received 12 June 1976)

Editor's note: This is a reprint (slightly edited) of a paper of the same title that appeared in the book *Physics and Our World: A Symposium in Honor of Victor F. Weisskopf,* published by the American Institute of Physics (1976). The personal tone of the original talk has been preserved in the paper, which was itself a slightly edited transcript of a tape. The figures reproduce transparencies used in the talk. The demonstration involved a tall rectangular transparent vessel of corn syrup, projected by an overhead projector turned on its side. Some essential hand waving could not be reproduced.

This is a talk that I would not, I'm afraid, have the nerve to give under any other circumstances. It's a story I've been saving up to tell Viki. Like so many of you here, I've enjoyed from time to time the wonderful experience of exploring with Viki some part of physics, or anything to which we can apply physics. We wander around strictly as amateurs equipped only with some elementary physics, and in the end, it turns out, we improve our understanding of the elementary physics even if we don't throw much light on the other subjects. Now this is that kind of a subject, but I have still another reason for wanting to, as it were, needle Viki with it, because I'm going to talk for a while about viscosity. Viscosity in a liquid will be the dominant theme here and you know Viki's program of explaining everything, including the heights of mountains, with the elementary constants. The viscosity of a liquid is a very tough nut to crack, as he well knows, because when the stuff is cooled by merely 40 degrees, its viscosity can change by a factor of a million. I was really amazed by fluid viscosity in the early days of NMR, when it turned out that glycerine was just what we needed to explore the behavior of spin relaxation. And yet if you were a little bug inside the glycerine, looking around, you wouldn't see much change in your surroundings as the glycerine cooled. Viki will say that he can at least predict the *logarithm* of the viscosity. And that, of course, is correct because the reason viscosity changes is that it's got one of these activation energy things and what he can predict is the order of magnitude of the exponent. But it's more mysterious than that, Viki, because if you look at the Chemical Rubber Handbook table you will find that there is almost no liquid with viscosity much lower than that of water. The viscosities have a big range *but they stop at the same place.* I don't understand that. That's what I'm leaving for him.[1]

Now, I'm going to talk about a world which, as physicists, we almost never think about. The physicist hears about viscosity in high school when he's repeating Millikan's oil drop experiment and he never hears about it again, at least not in what I teach. And Reynolds's number, of course, is something for the engineers. And the *low* Reynolds number regime most engineers aren't even interested in—except possibly chemical engineers, in connection with fluidized beds, a fascinating topic I heard about from a chemical engineering friend at MIT. But I want to take you into the world of very low Reynolds number—a world which is inhabited by the overwhelming majority of the organisms in this room. This world is quite different from the one that we have developed our intuitions in.

I might say what got me into this. To introduce something that will come later, I'm going to talk partly about how microorganisms swim. That will not, however, turn out to be the only important question about them. I got into this through the work of a former colleague of mine at Harvard, Howard Berg. Berg got his Ph.D. with Norman Ramsey, working on a hydrogen maser, and then he went back into biology which had been his early love, and into cellular physiology. He is now at the University of Colorado at Boulder, and has recently participated in what seems to me one of the most astonishing discoveries about the questions we're going to talk about. So it was partly Howard's work, tracking *E. coli* and finding out this strange thing about them, that got me thinking about this elementary physics stuff.

Well, here we go. In Fig. 1, you see an object which is moving through a fluid with velocity v. It has dimension a. In Stokes's law, the object is a sphere, but here it's anything; η and ρ are the viscosity and density of the fluid. The ratio of the inertial forces to the viscous forces, as Osborne Reynolds pointed out slightly less than a hundred years ago, is given by $av\rho/\eta$ or av/ν, where ν is called the *kinematic* viscosity. It's easier to remember its dimensions: for water, $\nu \approx 10^{-2}$ cm^2/sec. The ratio is called the Reynolds number and when that number is small the viscous forces dominate. Now there is an easy way, which I didn't realize at first, to see who should be interested in small Reynolds numbers. If you take the viscosity η and square it and divide by the density, you get a force (Fig. 2). No other dimensions come in at all. η^2/ρ is a force. For water, since $\eta \approx 10^{-2}$ and $\rho \approx 1$, $\eta^2/\rho \approx 10^{-4}$ dyn. That is a force that will tow *anything,* large or small, with a Reynolds number of order of magnitude 1. In other words, if you want to tow a submarine with Reynolds number 1 (or strictly speaking, $1/6\pi$ if it's a spherical submarine) tow it with 10^{-4} dyn. So it's clear in this case that you're interested in small Reynolds number if you're interested in *small forces* in an absolute sense. The only other people who are interested in low Reynolds number, although they usually don't have to invoke it, are the geophysicists. The Earth's mantle is supposed to have a viscosity of 10^{21} P. If you now work out η^2/ρ, the force is 10^{41} dyn. That is more than 10^9 times the gravitational force that half the Earth exerts on the other half! So the conclusion is, of course, that in the flow of the mantle of the Earth the Reynolds number is *very* small indeed.

Now consider things that move through a liquid (Fig. 3). The Reynolds number for a man swimming in water might be 10^4, if we put in reasonable dimensions. For a goldfish or a tiny guppy it might get down to 10^2. For the animals that we're going to be talking about, as we'll see in a mo-

Reprinted from *American Journal of Physics* **45**, 3–11 (1977); © American Association of Physics Teachers.

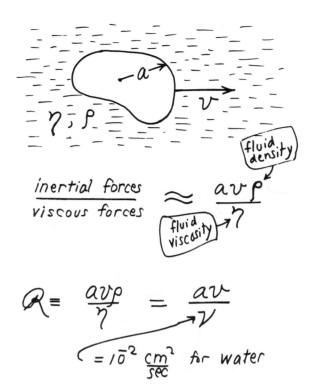

$$\frac{\text{inertial forces}}{\text{viscous forces}} \approx \frac{a v \rho}{\eta}$$

(fluid density) → ρ

(fluid viscosity) → η

$$\mathcal{R} \equiv \frac{a v \rho}{\eta} = \frac{a v}{\nu}$$

$$\nu = 10^{-2} \frac{cm^2}{sec} \text{ for water}$$

Figure 1.

ment, it's about 10^{-4} or 10^{-5}. For these animals inertia is totally irrelevant. *We* know that $F = ma$, but they could scarcely care less. I'll show you a picture of the real animals in a bit but we are going to be talking about objects which are the order of a micron in size (Fig. 4). That's a micron scale, not a suture, in the animal in Fig. 4. In water where the kinematic viscosity is 10^{-2} cm/sec these things move around with a typical speed of 30 μm/sec. If I have to push that animal to move it, and suddenly I stop pushing, how

$$\frac{\eta^2}{\rho} = force$$

for water; $\frac{\eta^2}{\rho} = 10^{-4}$ dynes

This force will tow anything, large or small, at $\mathcal{R} \approx 1$

Earth's mantle has $\eta \approx 10^{21}$

$\frac{\eta^2}{\rho} = 10^{41}$ dynes

$\mathcal{R} \lll 1$

Figure 2.

Figure 3.

far will it coast before it slows down? The answer is, about 0.1 Å. And it takes it about 0.6 μsec to slow down. I think this makes it clear what low Reynolds number means. Inertia plays no role whatsoever. If you are at very low Reynolds number, what you are doing at the moment is entirely determined by the forces that are exerted on you *at that moment,* and by nothing in the past.[2]

It helps to imagine under what conditions a man would be swimming at, say, the same Reynolds number as his own sperm. Well, you put him in a swimming pool that is full of molasses, and then you forbid him to move any part of his body faster than 1 cm/min. Now imagine yourself in that condition: you're under the swimming pool in molasses, and now you can only move like the hands of a clock. If under those ground rules you are able to move a few meters in a couple of weeks, you may qualify as a low Reynolds number swimmer.

I want to talk about swimming at low Reynolds number in a very general way. What does it mean to swim? Well, it means simply that you are in some liquid and are allowed to deform your body in some manner. That's all you can do.

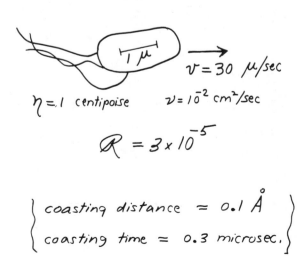

$v = 30 \ \mu/sec$

$\eta = 1$ centipoise $\nu = 10^{-2} \ cm^2/sec$

$$\mathcal{R} = 3 \times 10^{-5}$$

$$\left\{ \begin{array}{l} coasting \ distance = 0.1 \ \mathring{A} \\ coasting \ time = 0.3 \ microsec. \end{array} \right\}$$

Figure 4.

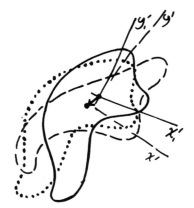

Figure 5.

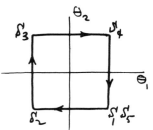

Figure 7.

Move it around and move it back. Of course, you choose some kind of cyclic deformation because you want to keep swimming, and it doesn't do any good to use a motion that goes to zero asymptotically. You have to keep moving. So, in general, we are interested in cyclic deformations of a body on which there are no external torques or forces except those exerted by the surrounding fluid. In Fig. 5, there is an object which has a shape shown by the solid line; it changes its shape to the dashed contour and then it changes back. When it finally gets back to its original shape, the dotted contour, it has moved over and rotated a little. It has been swimming. When it executed the cycle, a displacement resulted. If it repeats the cycle, it will, of course, effect the same displacement, and in two dimensions we'd see it progressing around a circle. In three dimensions its most general trajectory is a helix consisting of little kinks, each of which is the result of one cycle of shape change.

There is a very funny thing about motion at low Reynolds number, which is the following. One special kind of swimming motion is what I call a reciprocal motion. That is to say, I change my body into a certain shape and then I go back to the original shape by going through the sequence in reverse. At low Reynolds number, everything reverses just fine. Time, in fact, makes no difference—only config-

Navier - Stokes:

$$-\nabla p + \eta \nabla^2 \vec{v} = \cancel{\rho \frac{\partial \vec{v}}{\partial t}} + \cancel{\rho(\vec{v} \cdot \nabla)\vec{v}}$$

If $\mathcal{R} \ll 1$:

Time doesn't matter. The pattern of motion is the same, whether slow or fast, whether forward or backward in time.

The Scallop Theorem

Figure 6.

uration. If I change quickly or slowly, the pattern of motion is exactly the same. If you take the Navier–Stokes equation and throw away the inertia terms, all you have left is $\nabla^2 v = p/\eta$, where p is the pressure (Fig. 6). So, if the animal tries to swim by a reciprocal motion, it *can't go anywhere*. Fast or slow, it exactly retraces its trajectory and it's back where it started. A good example of that is a scallop. You know, a scallop opens its shell slowly and closes its shell fast, squirting out water. The moral of this is that the scallop at low Reynolds number is no good. It can't swim because it only has one hinge, and if you have only one degree of freedom in configuration space, you are bound to make a reciprocal motion. There is nothing else you can do. The simplest animal that can swim that way is an animal with two hinges. I don't know whether one exists but Fig. 7 shows a hypothetical one. This animal is like a boat with a rudder at both front and back, and nothing else. This animal can swim. All it has to do is go through the sequence to configurations shown, returning to the original one at S_5. Its configuration space, of course, is two dimensional with coordinates θ_1, θ_2. The animal is going around a loop in that configuration space, and that enables it to swim. In fact, I worked this one out just for fun and you can prove from symmetry that it goes along the direction shown in the figure. As an exercise for the student, what is it that distinguishes that direction?

You can invent other animals that have no trouble swimming. We had better be able to invent them, since we know they exist. One you might think of first as a physicist, is a torus. I don't know whether there is a toroidal animal, but whatever other physiological problems it might face, it clearly could swim at low Reynolds number (Fig. 8). Another animal might consist of two cells which were stuck together and were able to roll on one another by having

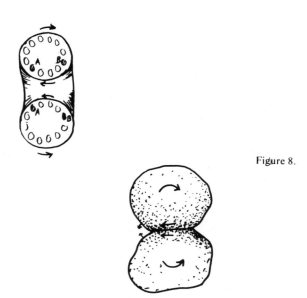

Figure 8.

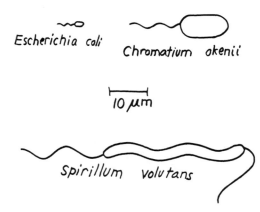

Figure 10.

some kind of attraction here while releasing there. That thing will "roll" along. I described it once as a combination caterpillar tractor and bicycle built for two, but that isn't the way it really works. In the animal kingdom, there are at least two other more common solutions to the problem of swimming at low Reynolds number (Fig. 9). One might be called the flexible oar. You see, you can't row a boat at low Reynolds number in molasses—if you are submerged—because the stiff oars are just reciprocating things. But if the oar is flexible, that's not true, because then the oar bends one way during the first half of the stroke and the other during the second half. That's sufficient to elude the theorem that got the scallop. Another method, and the one we'll mainly be talking about, is what I call a corkscrew. If you keep turning it, that, of course, is not a reciprocal change in configuration space and that will propel you. At this point, I wish I could persuade you that the direction in which this helical drive will move is *not* obvious. Put your-

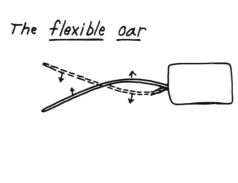

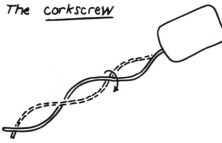

Figure 9.

self back in that swimming pool under molasses and move around very, very slowly. Your intuitions about *pushing water backwards* are irrelevant. That's not what counts. Now, unfortunately, it turns out that the thing does move the way your naive, untutored, and actually incorrect argument would indicate, but that's just a pedagogical misfortune we are always running into.

Well, lets look at some real animals (Fig. 10). This figure I've taken from a paper of Howard Berg that he sent me. Here are three real swimmers. The one we're going to be talking about most is the famous animal, *Escherichia coli,* at A, which is a very tiny thing. Then there are two larger animals. I've copied down their Latin names and they may be old friends to some of you here. This thing (*S. volutans*) swims by waving its body as well as its tail and roughly speaking, a spiral wave runs down that tail. The bacterium *E. coli* on the left is about 2 μm long. The tail is the part that we are interested in. That's the flagellum. Some *E. coli* cells have them coming out the sides; and they may have several, but when they have several they tend to bundle together. Some cells are nonmotile and don't have flagella. They live perfectly well, so swimming is not an absolute necessity for this particular animal, but the one in the figure does swim. The flagellum is only about 130 Å in diameter. It is much thinner than the cilium which is another very important kind of propulsive machinery. There is a beautiful article on cilia in this month's *Scientific American.*[3] Cilia are about 2000 Å in diameter, with a rather elaborate apparatus inside. There's not room for such apparatus inside this flagellum.

For a long time there has been interest in how the flagellum works. Classic work in this field was done around 1951, as I'm sure some of you will remember, by Sir Geoffrey Taylor, the famous fluid dynamicist of Cambridge. One time I heard him give a fascinating lecture at the National Academy. Out of his pocket at the lecture he pulled his working model, a cylindrical body with a helical tail driven by a rubber-band motor inside the body. He had tested it in glycerine. In order to make the tail he hadn't just done the simple thing of having a turning corkscrew, because at that time nearly everyone had persuaded themselves that the tail doesn't rotate, it waves. Because, after all, to rotate you'd have to have a rotary joint back at the animal. So he had sheathed the turning helix with rubber tubing anchored to the body. The body had a keel. I remember Sir Geoffrey Taylor saying in his lecture that he was embarrassed that he hadn't put the keel on it first and he'd had to find out that

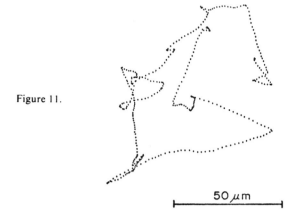

Figure 11.

50 μm

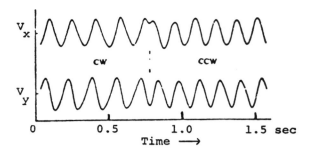

Figure 12.

he needed it. There has since been a vast literature on this subject, only a small part of which I'm familiar with. But at that time G. I. Taylor's paper in the *Proceedings of the Royal Society* could conclude with just three references: H. Lamb, *Hydrodynamics;* G. I. Taylor (his previous paper); G. N. Watson, *Bessel Functions.* That is called getting in on the ground floor.

To come now to modern times, I want to show a picture of these animals swimming or tracking. This is the work of Howard Berg, and I'll first describe what he did. He started building the apparatus when he was at Harvard. He was interested in studying not the actual mechanics of swimming at all but a much more interesting question, namely, why these things swim and where they swim. In particular, he wanted to study chemotaxis in *E. coli*—seeing how they behave in gradients of nutrients and things like that. So he built a little machine which would track a single bacterium in *x, y, z* coordinates—just lock onto it optically and track it. He was able then to track one of these bacteria while it was behaving in its normal manner, possibly subject to the influence of gradients of one thing or another. A great advantage of working with a thing like *E. coli* is that there are so many mutant strains that have been well studied that you can use different mutants for different things. The next picture (Fig. 11) is one of his tracks. It shows a projection on a plane of the track of one bacterium. The little dots are about 0.1 sec apart so that it was actually running along one of the legs for a second or two and the speed is typically 20–40 μm/sec. Notice that it swims for a while and then stops and goes off in some other direction. We'll see later what that might suggest. A year ago, Howard Berg went out on a limb and wrote a paper in *Nature*[4] in which he argued that, on the basis of available evidence, *E. coli* must swim by *rotating* their flagella, not by waving them. Within the year a very elegant, crucial experiment by Silverman and Simon at UC–San Diego showed that this fact is the case.[5,6] Their experiment involved a mutant strain of *E. coli* bacteria which don't make flagella at all but only make something called the proximal hook to which the flagella would have been attached. They found that with antihook antibodies they could cause these things to glue together. And once in a while one of the bacteria would have its hook glued to the microscope slide, in which case the whole body rotated at constant angular velocity. And when two hooks glued together, the two bodies counter-rotated, as you would expect. It's a beautiful technique. Howard was ready with his tracker and the next picture[7] (Fig. 12) shows his tracker

following the end of one of these tethered *E. coli* cells which is stuck to the microscope slide by antibody at the place where the flagellum should have been. Plotted here are the two velocity components V_x and V_y. The two velocity components are 90° out of phase. The point being tracked is going in a circle. In the middle of the figure, you see a 90° phase change in one component, a reversal of rotation. They can rotate hundreds of revolutions at constant speed and then turn around and rotate the other way. Evidently the animal actually has a rotary joint, and has a motor inside that's able to drive a flagellum in one direction or the other, a most remarkable piece of machinery.

I got interested in the way a rotating corkscrew can propel something. Let's consider propulsion in one direction only, parallel to the axis of the helix. The helix can translate and it can rotate; you can apply a force to it and a torque. It has a velocity v and an angular velocity Ω. And now remember, at low Reynolds number everything is linear. When everything is linear, you expect to see matrices come in. Force and torque must be related by matrices with constant coefficients, to linear and angular velocity. I call this little 2 × 2 matrix the propulsion matrix (Fig. 13). If I knew its elements A, B, C, D, I could then find out how good this rotating helix is for propelling anything.

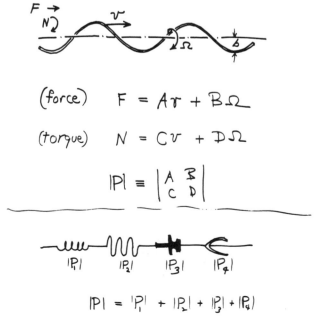

(force) $F = Av + B\Omega$

(torque) $N = Cv + D\Omega$

$$|P| \equiv \begin{vmatrix} A & B \\ C & D \end{vmatrix}$$

$|P_1|$ $|P_2|$ $|P_3|$ $|P_4|$

$$|P| = |P_1| + |P_2| + |P_3| + |P_4|$$

Figure 13.

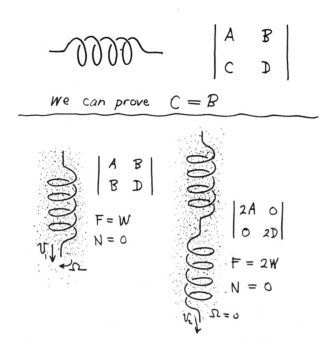

We can prove $C = B$

Figure 14.

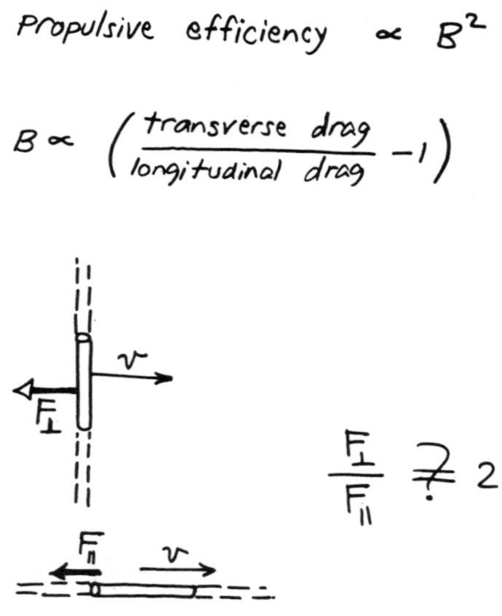

$$\text{Propulsive efficiency} \propto B^2$$

$$B \propto \left(\frac{\text{transverse drag}}{\text{longitudinal drag}} - 1 \right)$$

$$\frac{F_\perp}{F_{\parallel}} \overset{?}{=} 2$$

Figure 15.

are supposed to differ in a certain limit by a factor of 2. But for the models I've tested that factor is more like 1.5. Since it's that factor minus 1 that counts, that's very bad for efficiency. We thought that if you want something to rotate more while sinking, it would be better not to use a round wire. Something like a slinky ought to be better. I made one and measured its off diagonal elements. Surprise, surprise, it was no better at all! I don't really understand that, because the fluid mechanics of these two situations is not at all simple. In each case there is a logarithmic divergence that you have to worry about, and the two are somewhat different in character. So that theoretical ratio of two I referred to is probably not even right.

When you put all this in and calculate the efficiency, you find that it's really rather low even when the various parameters of the model are optimized. For a sphere which is driven by one of these helical propellers (Fig. 16), I will

Well, let's try to go on by making some assumptions. If two corkscrews or other devices on the same shaft are far enough from one another so that their velocity patterns don't interact, their propulsive matrices just add. If you allow me that assumption, then there is a very nice way, which I don't have time to explain, of proving that the propulsion matrix must be symmetrical (Fig. 14). So actually the motion is described by only three constants, not four, and they are very easily measured. All you have to do is make a model of this thing and drop in a fluid at you are interested in or not, because these constants are independent of that. And so I did that and that's my one demonstration. I thought this series of talks ought to have one experiment and there it is. We're looking through a tank not of glycerine but of corn syrup, which is cheaper, quite uniform, and has a viscosity of about 50 P or 5000 times the viscosity of water. The nice part of this is you can just lick the experimental material off your fingers.

Motion at low Reynolds number is very majestic, slow, and regular. You'll notice that the model is actually rotating but rather little. If that were a corkscrew moving through a cork of course, the pattern in projection wouldn't change. It's very very far from that, it's *slipping,* so that it sinks by several wavelengths while it's turning around once. If the matrix were diagonal, the thing would not rotate at all. So all you have to do is just see how much it turns as it sinks and you have got a handle on the off-diagonal element. A nice way to determine the other elements is to run two of these animals, one of which is a spiral and the other is two spirals, in series, of opposite handedness. The matrices add and with two spirals of opposite handedness, the propulsion matrix must be diagonal (Fig. 14). That's not going to rotate; it better not.

The propulsive efficiency is more or less proportional to the square of the off-diagonal element of the matrix. The off-diagonal element depnds on the difference between the drag on a wire moving perpendicular to its length and the drag on a wire moving parallel to its length (Fig. 15). These

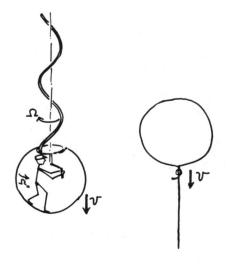

PROPULSIVE EFFICIENCY $\approx$ 1 %

Figure 16.

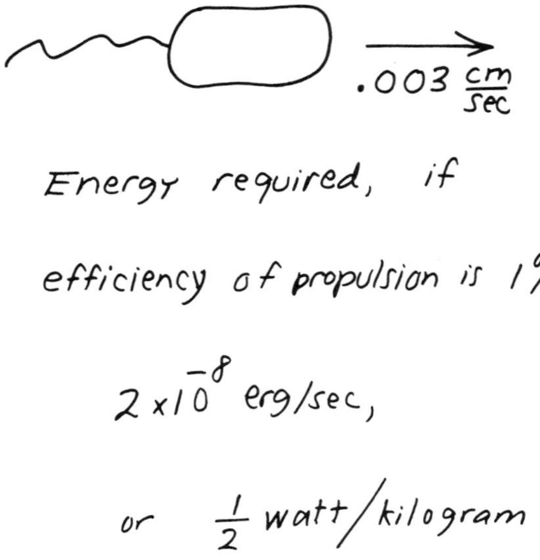

Energy required, if

efficiency of propulsion is 1%:

$$2 \times 10^{-8} \text{ erg/sec},$$

or $\frac{1}{2}$ watt/kilogram

Figure 17.

define the efficiency as the ratio of the work that I would have to do just to pull that thing along to what the man inside it turning the crank has to do. And that turns out to be about 1%. I worried about that result for a while and tried to get Howard interested in it. He didn't pay much attention to it, and he shouldn't have, because it turns out that efficiency is really not the primary problem of the animal's motion. We'll see that when we look at the energy requirement. How much power does it take to run one of these things with a 1% efficient propulsion system, at this speed in these conditions? We can work it out very easily. Going 30 μm/sec, at 1% efficiency will cost us about 2×10^{-8} ergs/sec at the motor. On a per weight basis, that's a 0.5 W/kg, which is really not very much. Just moving things around in our transportation system, we use energy at 30 or 40 times that rate. This bug runs 24 h a day and only uses 0.5 W/kg. That's a small fraction of its metabolism and its energy budget. Unlike us, they do not squander their energy budget just moving themselves around. So they don't care whether they have a 1% efficient flagellum or a 2% efficient flagellum. It doesn't really make that much difference. They're driving a Datsun in Saudi Arabia.

So the interesting question is not how they swim. Turn anything—if it isn't perfectly symmetrical, you'll swim. If the efficiency is only 1%, who cares. A better way to say it is that the bug can collect, by diffusion through the surrounding medium, enough energetic molecules to keep moving when the concentration of those molecules is 10^{-9} M. I've now introduced the word diffusion. Diffusion is important because of another very peculiar feature of the world at low Reynolds number, and that is, stirring isn't any good. The bug's problem is not its energy supply; its problem is its environment. At low Reynolds number you can't shake off your environment. If you move, you take it along; it only gradually falls behind. We can use elementary physics to look at this in a very simple way. The time for transporting anything a distance l by stirring, is about l divided by the stirring speed v. Whereas, for transport by diffusion, it's l^2 divided by D, the diffusion constant. The ratio of those two times is a measure of the effectiveness of stirring versus that of diffusion for any given distance and diffusion constant.

I'm sure this ratio has someone's name but I don't know the literature and I don't know whose number that's called. Call it S for *stirring number*.[8] It's just lv/D. You'll notice by the way that the Reynolds number was lv/ν. ν is the kinematic viscosity in cm²/sec, and D is the diffusion constant in cm²/sec, for whatever it is that we are interested in following—let us say a nutrient molecule in water. Now, in water the diffusion constant is pretty much the same for every reasonably sized molecule, something like 10^{-5} cm²/sec. In the size domain that we're interested in, of micron distances, we find that the stirring number S is 10^{-2}, for the velocities that we are talking about (Fig. 18). In other words, this bug can't do anything by stirring its local surroundings. It might as well wait for things to diffuse, either in or out. The transport of wastes away from the animal and food to the animal is entirely controlled locally by diffusion. You can thrash around a lot, but the fellow who just sits there quietly waiting for stuff to diffuse will collect just as much.

At one time I thought that the reason the thing swims is that if it swims it can get more stuff, because the medium is full of molecules the bug would like to have. All my instincts as a physicist say you should move if you want to scoop that stuff up. You can easily solve the problem of diffusion in the velocity field represented by the Stokes flow around a sphere—for instance, by a relaxation method. I did so and found out how fast the cell would have to go to increase its food supply. The food supply if it just sits there is $4\pi aND$ molecules/sec, where a is the cell's radius (Fig. 19) and N is the concentration of nutrient molecules. To increase its food supply by 10% it would have to move at a speed of 700 μm/sec, which is 20 times as fast as it can swim. The increased intake varies like the square root of the bug's velocity so the swimming does no good at all in that respect. But what it can do is find places where the food is better or more abundant. That is, it does not move like a cow

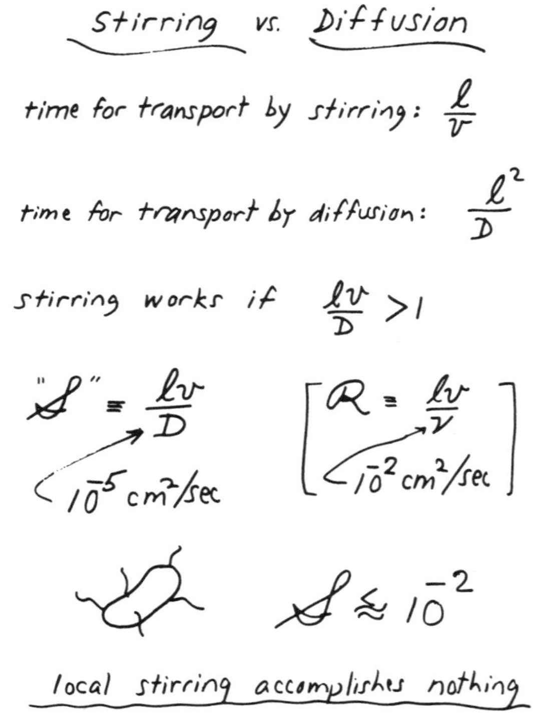

Figure 18.

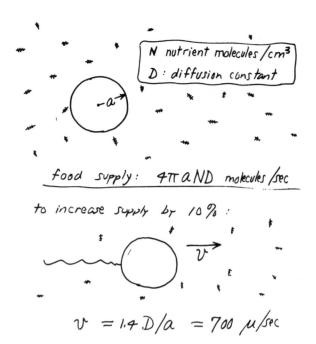

N nutrient molecules/cm³
D : diffusion constant

food supply: $4\pi a N D$ molecules/sec

to increase supply by 10% :

$$v = 1.4 \, D/a = 700 \; \mu/sec$$

Figure 19.

that is grazing a pasture—it moves to find *greener pastures*. And how far does it have to move? Well, it has to move far enough to outrun diffusion. We said before that stirring wouldn't do any good locally, compared to diffusion. But suppose it wants to run over there to see whether there is more over there. Then it must outrun diffusion, and how do you do that? Well, you go that magic distance, D/v. So the rule is then, to outswim diffusion you have to go a distance which is equal to or greater than this number we had in our S constant. For typical D and v, you have to go about 30 μm and that's just about what the swimming bacteria were doing. If you don't swim that far, you haven't gone any-where, because it's only on that scale that you could find a

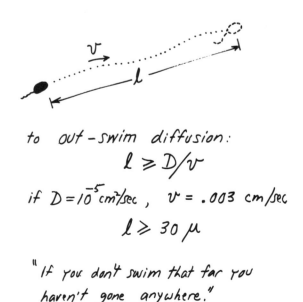

to out-swim diffusion:

$$\ell \geqslant D/v$$

if $D = 10^{-5} cm^2/sec$, $v = .003 \; cm/sec$

$$\ell \geqslant 30 \; \mu$$

" If you don't swim that far you haven't gone anywhere."

Figure 20.

difference in your environment with respect to molecules of diffusion constant D (Fig. 20).

Let's go back and look at one of those sections from Berg's track (Fig. 11). You'll see that there are some little trips, but otherwise you might ask why did it go clear over here and stop. Why did it go back? Well, my suggestion is, and I'd like to put this forward very tentatively, that the reason it does is because it's trying to outrun diffusion. Otherwise, it might as well sit still, as indeed do the mutants who don't have flagella. Now there is still another thing that I put forward with even more hesitation because I haven't tried this out on Howard yet. When he did his chemotaxis experiments, he found a very interesting behavior. If these things are put in a medium where there is a gradient of something that they like, they gradually work their way upstream. But if you look at how they do it and ask what rules they are using, what the algorithm is to use the current language, for finding your way upstream, it turns out that it's very simple. The algorithm is: if things are getting better, don't stop so soon. If, in other words, you plot, as Berg has done in some of his papers, the distribution of path lengths between runs and the little stops that he calls "twiddles," the distribution of path lengths if they are going *up* the gradient gets longer. That's a very simple rule for working your way to where things are better. If they're going down the gradient, though, they don't get shorter. And that seems a little puzzling. Why, if things are getting worse, don't they change sooner? My suggestion is that there is no point in stopping sooner. There is a sort of bedrock length which outruns diffusion and is useful for sampling the medium. Shorter paths would be a ridiculous way to sample. It may be something like that, but as I say, I don't know. The res-idue of education that I got from this is partly this stuff about simple fluid mechanics, partly the realization that the mechanism of propulsion is really not very important except, of course, for the physiology of that very myserious motor, which physicists aren't competent even to conjecture about.

I come back for a moment to Osborne Reynolds. That was a very great man. He was a professor of engineering, actually. He was the one who not only invented Reynolds number, but he was also the one who showed what turbu-lence amounts to and that there is instability in flow, and all that. He is also the one who solved the problem of how you lubricate a bearing, which is a very subtle problem that I recommend to anyone who hasn't looked into it. But I discovered just recently in reading in his collected works that toward the end of his life, in 1903, he published a very long paper on the details of the *submechanical universe,* and he had a complete theory which involved small particles of diameter 10^{-18} cm. It gets very nutty from there on. It's a mechanical model, the particles interact with one another and fill all space. But I thought that, incongruous as it may have seemed to put this kind of stuff in between our studies of the submechanical universe today, I believe that Osborne Reynolds would not have found that incongruous, and I'm quite positive that Viki doesn't.

[1](1976 footnote) As no one will be surprised to hear, Professor Weisskopf has recently shown me how this can be explained. I hope he will com-municate it to AJP readers.

[2](1976 footnote) In that world, Aristotle's mechanics is *correct!* See A. Franklin, Am. J. Phys. **44**, 527–528 (1976).

[3]P. Satir, Sci. Am. **231**, 45 (October 1974).

[4]H. C. Berg and R. A. Anderson, Nature **245**, 380 (1973).

[5]M. Silverman and M. Simon, Nature **249**, 73 (1974).

[6]S. H. Larson, R. W. Reader, E. N. Kort, W-W. Tso, and J. Adler, Nature **249**, 74 (1974).

[7]H. C. Berg, Nature **249**, 77 (1974).

[8]I've recently discovered that its official name is the *Sherwood number*, so S is appropriate, after all! There is a list of all the dimensionless ratios that have acquired names—an astonishingly long list—in the *McGraw-Hill Encyclopedia of Science and Technology* (1971).

BIBLIOGRAPHY OF RECENT REVIEW ARTICLES

H. C. Berg, Nature **254**, 389 (1975).

H. C. Berg, Ann. Rev. Biophys. Biolog. **4**, 119 (1975).

J. Adler, Ann. Rev. Biochem. **44**, 341 (1975).

H. C. Berg, Sci. Am. **233**, 36 (August 1975).

MAGNETOSOME DYNAMICS IN MAGNETOTACTIC BACTERIA

S. Ofer†, I. Nowik, and E. R. Bauminger
Racah Institute of Physics, Hebrew University, Jerusalem, Israel

G. C. Papaefthymiou and R. B. Frankel
Francis Bitter National Magnet Laboratory, Massachusetts Institute of Technology, Cambridge, Massachusetts 02139

R. P. Blakemore
Department of Microbiology, University of New Hampshire, Durham, New Hampshire 03424

ABSTRACT   Diffusive motions of the magnetosomes (enveloped Fe_3O_4 particles) in the magnetotactic bacterium *Aquaspirillum magnetotacticum* result in a very broad-line Mössbauer spectrum ($\Gamma \sim 100$ mm/s) above freezing temperatures. The line width increases with increasing temperature. The data are analyzed using a bounded diffusion model to yield the rotational and translational motions of the magnetosomes as well as the effective viscosity of the material surrounding the magnetosomes. The results are $\langle \theta^2 \rangle^{1/2} < 1.5°$ and $\langle x^2 \rangle^{1/2} < 8.4$ Å for the rotational and translational motions, respectively, implying that the particles are fixed in whole cells. The effective viscosity is 10 cP at 295 K and increases with decreasing temperature. Additional Fe^{3+} material in the cell is shown to be associated with the magnetosomes. Fe^{2+} material in the cell appears to be associated with the cell envelope.

INTRODUCTION

Iron accounts for 2% of the dry weight of the magnetotactic bacterium *Aquaspirillum magnetotacticum* (1). Most of the iron (80–90%) is present in the form of intracytoplasmic, enveloped, 40–50 nm wide particles of Fe_3O_4 (2, 3). Cells also contain ferrous iron and hydrous-ferric-oxide (ferrihydrite) (4). The enveloped Fe_3O_4 particles, which are termed magnetosomes (3), are arranged in a chain that longitudinally traverses the cell in close proximity to the inner surface of the cytoplasmic membrane (3). The number of magnetosomes in the chain is variable, depending upon the culture conditions, but typically averages 20. Magnetosomes are enveloped by electron-transparent and electron-dense layers and each is separated from those adjacent to it by 10 nm regions containing cytoplasmic material free of ribosomes or other particulate elements (3). The chemical composition of the distinctive region surrounding bacterial magnetite grains is unknown but may be important in their formation. Magnetite particles extracted from cells by brief sonication retain an envelope although their interparticle separation is <50% of that separating particles in chains within intact cells (3).

The magnetosomes impart a magnetic dipole moment to the cell, parallel to the axis of motility (2). According to the passive orientation hypothesis (5), the cell is oriented as it swims in the geomagnetic field by the torque exerted on the magnetic dipole moment by the field. The fact that the entire bacterium is oriented implies that the motions of the individual particles relative to each other are small and that the effective viscosity of the magnetosome surroundings in the cell is high compared with the viscosity of water; otherwise, the magnetic dipole moment of the cell could change its orientation with respect to the axis of motility. However, until now there has been no way to determine the motions of the magnetosomes in the cells.

In this paper we report that Mössbauer spectroscopy of whole cells above the freezing point of water can be used to measure the motions of the magnetosomes and to determine the effective viscosity of the magnetosome surroundings. The determination is based on the fact that in cells at ambient temperatures the magnetosomes will undergo small diffusive displacements, the magnitude of which are related to the viscosity of their surroundings. It has been previously shown that for iron containing colloidal particles introduced into a viscous fluid, the Mössbauer line width is inversely proportional to the viscosity of the fluid (6). Furthermore, it has been demonstrated recently that the Mössbauer effect (ME) on iron-containing proteins in cells can be used to determine the viscosity of the iron containing environment (7).

In the present case, Mössbauer measurements were performed between 90 and 295 K on packed cells *A. magnetotacticum*, which were cultured in chemically

---

†Deceased.

Reprinted from *Biophysical Journal* 46, 57–64 (1984); © Biophysical Society.

defined media containing ferric quinate enriched in Fe^{57} (the Mössbauer sensitive isotope). The spectra were analyzed using a bounded diffusion model derived below. We find that rotational and translational motions of the individual particles are small ($\langle\theta^2\rangle^{1/2} < 1.5°$; $\langle x^2\rangle^{1/2} < 8.4$ Å) and that the effective viscosity of the cytoplasm of the magnetotactic bacteria is ~15 times greater than the viscosity of water.

In addition, the results give information about the location of the other iron-containing materials, precursors to Fe_3O_4 precipitation (4), in the cells. The hydrous-ferric-oxide is associated with the magnetosomes, whereas the Fe^{2+} is localized elsewhere in the cell, possibly in the cell envelope.

METHODS

Organism and Culture Conditions

A. magnetotacticum strain MS-1 was used throughout (8). Cells enriched in Fe-57 were cultured microaerobically on a chemically defined medium as described previously (1). The medium contained tartaric and succinic acids as carbon sources, ferric quinate as the principal iron source and sodium nitrate, ammonium sulfate, or a combination as the nitrogen source. Cells were grown in 10-liter glass carboys to late logarithmic or early stationary phase (10–14 d; 2.0×10^8 cells/ml, at 30°C). They were harvested by continuous-flow centrifugation in an electrically driven centrifuge equipped with water cooling. The cells were washed twice at 5°C in 10–20 ml of cold 50 mM potassium phosphate buffer (pH 6.9) and centrifuged into a pellet with the consistency of paste. The wet cell material was packed into 1 ml plastic Mössbauer absorber cells with tight-fitting covers. These Mössbauer samples were either frozen and subsequently analyzed, or, in one case, analyzed immediately without freezing. The experimental spectra were found to be independent of whether the cells were frozen at the end or the beginning of the series of measurements, and except for measurements close to the freezing point, showed no temperature hysteresis.

$^{57}FeCl_3$ was prepared by first dissolving 80 mg Fe_2O_3, isotopically enriched to 90% ^{57}Fe, in 4.5 ml analytical grade concentrated HCl. Excess HCl was then boiled off to concentrate the resulting $FeCl_3$ to a sample volume of 1 ml. The concentrate was diluted with 5–10 ml distilled H_2O and again concentrated by boiling. This procedure was repeated two more times and the $^{57}FeCl_3$ was finally taken to near dryness. The salt was then dissolved in 100 ml distilled H_2O and combined with 0.19 *g D*-quinic acid (Sigma Chemical Co., St. Louis, MO) to chelate the metal. 20 ml of this 0.01 M stock of ferric quinate, sterilized by autoclaving, was aseptically added to each 10 liters of medium prior to inoculation.

Mössbauer Spectroscopy

Mössbauer measurements were performed using a conventional, constant acceleration spectrometer, with velocities extending up to ± 100 mm/s. A 100 mCi source of Co^{57} in Rh was maintained at room temperature. The absorber consisting of packed cells was held in a cryostat in helium vapor. Measurements between 90 and 295 K were carried out with the temperature stabilized to within 0.1 K. The spectra were least-squares fitted by computer, assuming Lorentzian line shapes.

EXPERIMENTAL RESULTS

The Mössbauer spectrum of whole packed cells at $T < 265$ K consisted primarily of the spectrum due to Fe_3O_4 (Fig. 1) (4). The spectrum consists of sharp, magnetic hyperfine lines between approximately +8 and −8 mm/s. Six lines

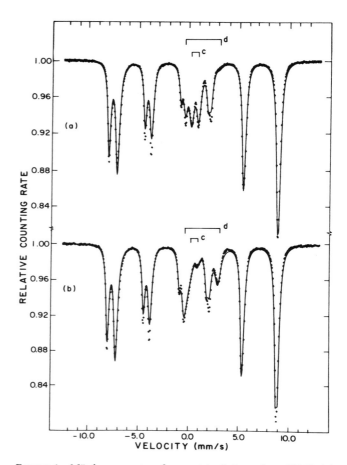

FIGURE 1 Mössbauer spectra of magnetotactic bacteria at 200 K. (*a*) The spectrum of a sample frozen immediately after harvesting the cells. (*b*) The spectrum obtained in a sample which was held above 285 K for a few days before freezing. Lines at position *c* correspond to an Fe^{3+} quadrupole doublet. Lines at position *d* correspond to an Fe^{2+} doublet. The solid lines are least squares computer fits to the experimental spectra.

are due to Fe^{3+} in tetrahedral sites and six lines are due to Fe^{3+} and Fe^{2+} in octahedral sites. The two highest velocity lines of the tetrahedral and octahedral site spectra fortuitously superpose giving 10 resolvable lines. There is no discernible absorption for velocities greater than ~8 mm/s and less then −8 mm/s. In samples kept frozen after collection of the bacteria, there were two additional quadrupole doublets (Fig. 1 *a*). One doublet, with intensity corresponding to ~13% of the total iron, had parameters typical of Fe^{3+} (isomer shift relative to iron metal $\delta = 0.42 \pm 0.02$ mm/s; quadrupole splitting $\Delta E_Q = 0.63 \pm 0.02$ mm/s). The other doublet, with intensity corresponding to <2% of total iron, had parameters typical of Fe^{2+} ($\delta = 1.12 \pm 0.02$ mm/s; $\Delta E_Q = 2.88 \pm 0.02$ mm/s). These doublets have been ascribed to precursors in the biomineralization of Fe_3O_4 in the bacteria (4). When the cells were held above 280 K in the sealed container for several days and refrozen, the Fe_3O_4 spectrum was unchanged, but the intensities of the two doublets had reversed, indicating that the Fe^{3+} material in the cell had been reduced to Fe^{2+} (Fig. 1 *b*).

The Mössbauer spectrum of the whole cells at $T > 275$ K was dramatically different from that of the frozen cells ($T < 265$ K) (Fig. 2). At 275 K it consisted primarily of a broad line of width $\Gamma = 72 \pm 1$ mm/s. The width of the broad line increased with increasing temperature to $\Gamma = 139$ mm/s at $T = 295$ K (Fig. 3). However, the total spectral intensity was temperature independent and equal to the total spectral intensity of the sharp line spectrum of the frozen cells (Fig. 4). Some hysteresis in the solid-liquid transition was noted in spectra obtained at 270 K. If the sample temperature had been increased from 265 K, the sharp-line spectrum was observed. However, if the sample temperature had been decreased from 275 K the broad line

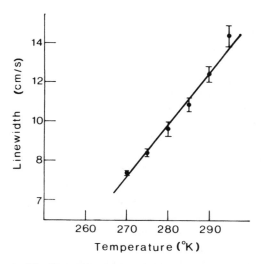

FIGURE 3   The line width of the broad line in the Mössbauer spectra plotted as function of temperature. Note units in centimeters per second on the ordinate.

spectrum was obtained. For $T > 275$ K, computer analysis showed that the intensity of the sharp-line Fe_3O_4 spectrum superposed on the broad line spectrum was $<0.2\%$.

The temperature dependence of the additional quadrupole doublet depended on whether the iron was primarily Fe^{3+} or Fe^{2+}. When the additional iron was Fe^{3+}, as evidenced by the parameters of the doublet in the $T = 265$ K spectrum, there was no residual doublet superposed on the broad-line spectrum at $T \geq 275$ K. However, when the additional iron was primarily Fe^{2+}, the low intensity, sharp line Fe^{2+} doublet remained superposed on the broad-line spectrum (Fig. 2 b).

The dramatic change of the spectral shape from a normal sharp-line spectrum between -8 and $+8$ mm/s to an extremely broad-line spectrum with width of ~ 100 mm/s cannot be produced by the onset of diffusive motions of the whole bacteria. Such diffusive motions should have the same effect on the shape of the spectra corresponding to all iron within the cell. The fact that the Fe^{2+} spectrum above 270 K is narrow and well defined proves that the effect of the diffusive motions of the whole bacteria on the shape of the broad spectrum is extremely small. The

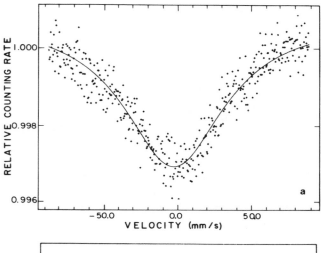

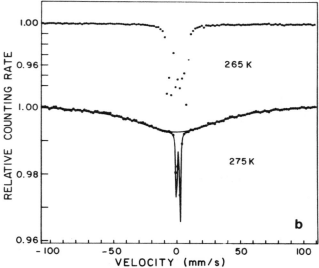

FIGURE 2   (a) Mössbauer spectrum at 275 K of the sample corresponding to the spectrum in Fig. 1 a. The width of the broad line is the same as in b (275 K). There is no superposed sharp-line spectrum due to Fe^{3+}. (b) Mössbauer spectra obtained at 265 K and 275 K of the sample corresponding to the spectrum in Fig. 1 b. The spectrum at 265 K is very similar to that at 200 K, but is shown here on an extended velocity scale. Only the experimental points are shown in this spectrum. The solid line in the 275 K spectrum is a least squares computer fit to the experimental points, consisting of a wide line of width (72 ± 1) mm/s and a well-defined doublet of small relative intensity, corresponding to Fe^{2+}.

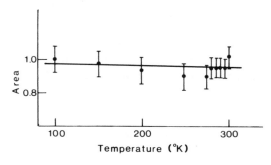

FIGURE 4   The total area of the Mössbauer spectrum as function of temperature.

broadening produced by the whole cell diffusive motions is expected to be <0.5 mm/s because of the large size of the cells (~3 μm) and the large effective viscosity in the packed cell sample.

The striking spectral change at 270 K can be explained by the onset of diffusive motions of the Fe_3O_4 particles in the bacteria as they are warmed through the solid-liquid phase transition of the cytoplasmic fluid at 270 K. Evidence for this comes from the fact that for freeze-dried cells (2, 4) the sharp-line spectrum persists at 300 K and the broad-line spectrum is never observed. Below we present an analysis of the broad-line spectra based on an extension of the "bounded diffusion" model previously developed for iron-containing proteins in the whole cells (7, 9, 10). The theory is presented below. From the analysis we derive the diffusion constant D of the magnetosomes, the effective viscosity η of the magnetosome environment, and a limit for the mean-squared translational displacement $\langle x^2 \rangle$ and rotational displacement $\langle \theta^2 \rangle$ of the magnetosomes as a function of temperature.

THEORY

The Extended Bounded Diffusion Model

A full calculation of the Mössbauer absorption spectra of overdamped harmonically bound particles in translational Brownian motion is given in references 9 and 10. We derive here an extension of this model for rigid spherical particles of radius R participating in both translational and rotational diffusive motions. As we do not know the exact restoring moments acting on the magnetosomes in the bacteria, we estimate the maximum contribution of the rotational diffusion to the line width. The maximum contribution will be obtained in the case of the free rotational diffusion in a viscous medium neglecting restoring moments.

Pure Translation Diffusion. The Brownian translational motion in one dimension of a particle of mass m which is bound to a center by a harmonic force $-mw^2x$, damped by a frictional force $-m\beta\,dx/dt$, and acted upon by random forces $F(t)$ has been treated in a classic paper by Uhlenbeck and Ornstein (11). In that paper, the classical self-correlation function $G(x, x_0, t)$ of the motion is derived in terms of two parameters, the diffusion constant D given by $k_BT/m\beta$ and the ratio between the harmonic and frictional force constants $\alpha = w^2/\beta$. Equipartition of energy gives

$$\langle x^2 \rangle = \frac{k_B T}{mw^2} = D/\alpha. \qquad (1)$$

The pure translational Brownian motion in three dimensions of a particle in the overdamped case ($w \ll \beta$), will

yield a Mössbauer spectrum (10) given by

$$I(\omega) = \frac{1}{2\pi} \int_{-\infty}^{\infty} dt$$

$$\exp\left[-i(\omega - \omega_o)t - \frac{\Gamma}{2}|t| - \frac{k^2D}{\alpha}(1 - e^{-\alpha|t|}) \right] \quad (2)$$

where Γ is the natural line width of the γ-ray (0.1 mm/s) and k is the wave number of the γ-ray. Eq. 2 can also be expressed as the sum of a narrow line with natural line width and an infinite sum of Lorentzian lines (12). Expanding $\exp(k^2 \langle x^2 \rangle e^{-\alpha t})$ in series yields

$$I(\omega) = \exp(-k^2 \langle x^2 \rangle) \frac{\Gamma/2\pi}{(\Gamma/2)^2 + (\omega - \omega_o)^2}$$

$$+ \sum_{n-1}^{\infty} \frac{1}{\pi} \exp(-k^2 \langle x^2 \rangle) \frac{(k^2 \langle x^2 \rangle)^n}{n!}$$

$$\cdot \frac{\Gamma/2 + n\alpha}{(\Gamma/2 + n\alpha)^2 + (\omega - \omega_o)^2} . \qquad (3)$$

The relative intensity of the narrow line is given by $e^{-k^2\langle x^2 \rangle}$. The sum of the broad Lorentzian lines can be approximated by a single Lorentzian line, the width of which (in units of α) is shown in Fig. 5 as a function of $k^2 \langle x^2 \rangle$. For $(k^2 \langle x^2 \rangle) < 0.1$, the intensity of the wide line will be <10%. For $(k^2 \langle x^2 \rangle) > 4$, where the relative intensity of the narrow line is negligible, the half-line width of the wide line is approximately $k^2\alpha \langle x^2 \rangle = k^2D$, where D is the

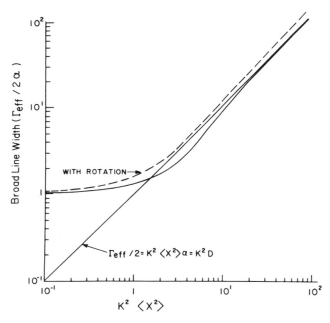

FIGURE 5 The theoretical width of the wide line (in units 2α) as function of $k^2\langle x^2 \rangle$. The solid curve line corresponds to pure bound translational diffusion. The dashed line corresponds to a combination of bound translational diffusion with free rotational diffusion. The straight line corresponds to a line width of $2k^2D$ (width obtained in the case of free diffusion).

translational diffusion constant $D = k_B T / 6\pi R \eta$. R is the radius of the spherical particles and η is the viscosity of the medium. This width is also obtained in the case of free translational diffusion (13). The total width in energy units is $2\hbar k^2 D$ ergs.

Rotational Diffusion of a Sphere Neglecting Restoring Moments.

Free rotational diffusion (with no translational diffusion and no restoring moments) also leads to a spectrum which is a sum of Lorentzian lines (14). The contribution to the Mössbauer spectrum from a rotating nucleus at distance r from the center of the sphere is

$$I_r(\omega) = j_0^2(kr) \frac{\Gamma/2\pi}{(\Gamma/2)^2 + (\omega - \omega_o)^2}$$
$$+ \sum_{l=1}^{\infty} \frac{(2l+1)j_l^2(kr)(\Gamma/2 + l(l+1)D_r)}{(\Gamma/2 + l(l+1)D_r)^2 + (\omega - \omega_o)^2}. \quad (4)$$

Here $j_l(kr)$ are the spherical Bessel functions and D_r is the rotational diffusion constant $D_r = k_B T / 8\pi R^3 \eta$. For a sphere of radius R,

$$D_r = 0.75 \cdot D / R^2. \quad (5)$$

The Mössbauer spectrum obtained from the whole rotating sphere will be given by

$$I(\omega) = \frac{3}{R^3} \int_0^R I_r(\omega) r^2 dr. \quad (6)$$

The spectrum includes a line with natural line width whose relative intensity is given by

$$\frac{3}{R^3} \int_0^R j_0^2(kr) r^2 dr = \frac{3}{2} \left(kR - \frac{1}{2} \sin 2kR \right) \Big/ (kR)^3. \quad (7)$$

In our case, where $R \sim 200$ Å, and $k = 7.3$ Å^{-1} (for the 14.4 keV γ-ray of Fe^{57}) the relative intensity of the narrow line is $<10^{-6}$. The sum of the broad lines can be approximated by a single broad Lorentzian lines.

Combination of Translational Diffusion and Free Rotational Diffusion.

For particles in a sphere, participating in both bound-translational and free-rotational diffusive motions (without restoring moments), the two forms of motion are uncorrelated and the "intermediate scattering function" (15) can be expressed as the product of the translational diffusion function

$$F_{tr}(k, t) = \{\exp[-k^2 \langle x^2 \rangle (1 - e^{-\alpha t})]\} \quad (8)$$

and the rotational diffusion function (14)

$$F_{rot}(r, k, t) = \sum_{l=0}^{\infty} (2l+1) j_l^2(kr) \exp[-l(l+1)D_r t]. \quad (8a)$$

The average spectrum for all nuclei in the sphere is now given by

$$I(\omega) = \frac{3}{R^3} \int_0^R r^2 dr \frac{1}{2\pi} \int_{-\infty}^{\infty}$$
$$\cdot \exp\left[-i(\omega - \omega_o)t - \frac{\Gamma}{2}|t|\right]$$
$$\cdot F_{tr}(k, t) F_{rot}(r, k, t) dt. \quad (9)$$

$I(\omega)$ can be expressed as a double infinite sum of Lorentzian lines;

$$I(\omega) = \sum_{n=0}^{\infty} \sum_{l=0}^{\infty} \frac{A_{nl}\Gamma_{nl}/2\pi}{(\Gamma_{nl}/2)^2 + (\omega - \omega_o)^2} \quad (10)$$

where $\Gamma_{nl}/2 = (\Gamma/2) + n\alpha + l(l+1)D_r$ and

$$A_{nl} = \exp(-k^2 \langle x^2 \rangle) \frac{(k^2 \langle x^2 \rangle)^n}{n!} (2l+1) \frac{3}{R^3} \int_0^R j_l^2(kr) r^2 dr.$$

The value of the last integral is given by

$$\int_0^R j_l^2(kr) r^2 dr = (R^3/2)[j_l^2(kR) + j_{l-1}^2(kR)$$
$$- (2l+1) j_l(kR) j_{l-1}(kR)/kR]. \quad (11)$$

For a sphere of radius R, D_r is given in Eq. 5, and the values of $\Gamma_{nl}/2$ are given by

$$\Gamma_{nl}/2 = \Gamma/2 + \alpha[n + 0.75l(l+1)k^2 \langle x^2 \rangle / k^2 R^2]. \quad (12)$$

The Mössbauer spectrum given by Eq. 10 is composed of a relatively narrow subspectrum corresponding to $n = 0$ and a broad spectrum corresponding to $n \geq 1$. The relative intensity of the $n = 0$ subspectrum is $\exp(-k^2 \langle x^2 \rangle)$. We approximate each subspectrum by an effective Lorentzian line with a width given by the harmonic average of the Lorentzian lines of the subspectrum.

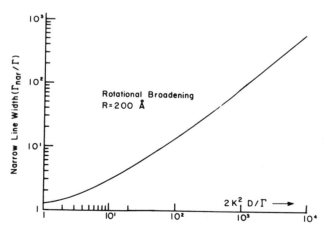

FIGURE 6   The width of the $n = 0$ subspectrum Γ_{nar}, in units of Γ (the natural width of the Mössbauer absorption line, ~ 0.1 mm/s) as function of $2k^2 D/\Gamma$ for spheres of a radius of 200 Å participating in bounded translational diffusive motions and in free (no restoring moments) rotational diffusive motions (see Eq. 13).

The width of the $n = 0$ subspectrum is given by

$$
\begin{aligned}
\frac{1}{\Gamma_{nar}} &= \exp(k^2 \langle x^2 \rangle) \sum_{l-0}^{\infty} \frac{A_{0l}}{\Gamma_{0l}} \\
&= \frac{\exp(k^2 \langle x^2 \rangle)}{\Gamma} \sum_{l-0}^{\infty} \\
&\cdot \frac{A_{0l}}{1 + 0.75l(l+1)\dfrac{2k^2 D}{\Gamma}}.
\end{aligned} \qquad (13)
$$

This width Γ_{nar} is a function of R and $2k^2 D/\Gamma$. Assuming that $R = 200$ Å, the values of Γ_{nar} as a function of $2k^2 D/\Gamma$ were calculated and are shown in Fig. 6.

Similarly, the effective width of the broad line corresponding to $n \geq 1$ is given by

$$
\frac{1}{\Gamma_{eff}} = \frac{\exp(k^2 \langle x^2 \rangle)}{\exp(k^2 \langle x^2 \rangle) - 1} \sum_{n-1}^{\infty} \sum_{l-0}^{\infty} \frac{A_{nl}}{\Gamma_{nl}}. \qquad (14)
$$

The values of $\Gamma_{eff}/2\alpha$ for $R = 200$ Å were calculated as a function of $k^2 \langle x^2 \rangle$ and are shown in Fig. 5 (dashed line). When $(k^2 \langle x^2 \rangle) > 4$, the relative intensity of the narrow line is negligible. The rotational diffusive motions contribute to the width of the broad line, and the total width of the broad line is larger than $2k^2 D$. The width of the broad line depends on the radius of the spheres participating in the diffusive motions. In Fig. 7 the width of the broad line in units of $2k^2 D$ is plotted as a function of $k^2 R^2$, [for $(k^2 \langle x^2 \rangle) > 4$]. For $k^2 R^2 > 10$, the ratio $\Gamma_{eff}/2k^2 D$ reaches a saturation value of 1.271. For $R = 200$ Å, $k^2 R^2 = 2 \cdot 10^6$ and the width of the broad line is $2 \cdot 1.271 \, k^2 D$.

The calculations lead to the conclusion that for a sphere with $R = 200$ Å, and $(k^2 \langle x^2 \rangle) > 4$, the spectrum consists only of a broad line, the width of which is given by $2\sigma k^2 D$, where $1 < \sigma < 1.27$. σ is closer to 1.0 when the rotational diffusive motions are small and may be neglected. σ is close to 1.27 when the rotational diffusion may be regarded as "free" (restoring moments may be neglected).

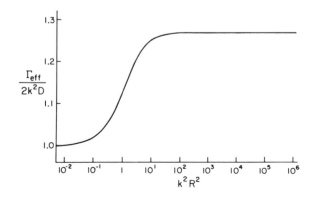

FIGURE 7 The width of the broad line ($n \geq 1$) in units of $2k^2 D$ as function of $k^2 R^2$ for a sphere of radius R participating in both translational and free rotational diffusion, assuming that $k^2(\langle x^2 \rangle) > 4$.

ANALYSIS OF EXPERIMENTAL RESULTS AND DISCUSSION

We analyze the experimental results using the extended bounded diffusion model described above, assuming that the magnetic particles are spheres of radius 230 Å. We do not make any assumptions about the strength of the restoring forces and moments. We treat the motions in the overdamped limit, since according to the theory, the experimentally observed broad Lorentzian lines are only obtained in the overdamped limit. We carry out the analysis for two extreme possibilities: (a) pure translational diffusion, without rotational diffusion; and (b) translational diffusion with free rotational diffusion (neglecting restoring moments).

(a) For pure translational diffusion, the relative intensity of the narrow line is given by $\exp(-k^2 \langle x^2 \rangle)$. In the experimental spectra above 275 K, this intensity is $>0.2\%$, thus yielding $(k^2 \langle x^2 \rangle) > 6$. We see from the graph in Fig. 5, corresponding to pure translation motions, that for $(k^2 \langle x^2 \rangle) > 6$, $\Gamma_{eff} = (0.9 \pm 0.1) \cdot 2 k^2 D$.

(b) In the case of translational diffusion with free rotational diffusion, one could try the assumption that the observed broad line is the line corresponding to $n = 0$ (Eq. 13) and that the line corresponding to $n > 1$ (Eq. 14) is weak and very broad and is, therefore, not observed experimentally. The meaning of such an assumption is that the particle is very strongly bound as far as translational motions are concerned, but is free to rotate in the viscous medium, which is extremely unlikely. The observed broad spectrum must, therefore, include both the $n = 0$ and the $n \geq 1$ lines. From Fig. 6 it follows that one of the following two relations always holds: (a) $\Gamma_{nar} < 80 \, \Gamma$; or, (b) $\Gamma_{nar} < 10^{-1} \cdot 2k^2 D$. According to Fig. 3, the experimental width W is larger than $800 \, \Gamma$. From Fig. 5 it follows that $W > 2k^2 D$. The conclusion is, therefore, that always $\Gamma_{nar} < 1/10 W$. As the experimental spectra do not contain a narrow spectrum to within 10% in amplitude, the relative area of the narrow spectrum is <0.01, and the value of $k^2 \langle x^2 \rangle$ is, therefore, larger than 4.6. Using the graph corresponding to free rotation in Fig. 5, for values of $k^2 \langle x^2 \rangle$ larger than 4.6, we obtain the relation $\Gamma_{eff} = (1.17 \pm 0.10) 2k^2 D$.

From the expressions obtained for the width of the broad spectrum in the two extreme cases (pure translational diffusion and translational diffusion with free rotational diffusion), we conclude that the expression for the total width of the spectra in the case of a combination of translational diffusion with any rotational diffusion is $\Gamma_{eff} = (1.03 \pm 0.23) 2k^2 D$. Using this expression for Γ_{eff}, the diffusion constants of the magnetic particles are calculated from the experimental values of the line widths given in Fig. 3. D changes between $(50 \pm 12) \cdot 10^{-10}$ cm^2/s at 270 K to $(96 \pm 22) \cdot 10^{-10}$ cm^2/s at 295 K. The effective viscosity η of the medium surrounding the magnetic particles is calculated using the formula $D = kT/6\pi R\eta$, taking

R as 230 Å (6). The value of η as a function of temperature is shown in Fig. 8. η changes from 17 cP at 270 K to 10 cP at 295 K. We estimate that the errors in the absolute values of η are ~30%. The errors on the relative values of η as a function of temperature are ~5%. The viscosity of water as a function of temperature is also shown in Fig. 8. The effective viscosity of the medium surrounding the magnetosomes is thus ~15 times greater than the viscosity of water but its temperature dependence is quite similar to that of water.

As seen from Fig. 4, the areas of the absorption spectra above 270 K are equal, to within 5%, to the areas below 270 K. This proves that the spectra observed experimentally above 270 K are the whole spectra at these temperatures. The magnetic particles do not participate in motions additional to those which are responsible for the spectra observed in the velocity range between -150 mm/s and $+150$ mm/s.

The cytoplasm of a bacterium is probably not a homogeneous medium. The effective viscosities determined in the present work are the average viscosities of the materials surrounding the magnetic particles. A comparison of the present result with those obtained previously by measuring rotational correlation times for tempone in the cytoplasm of *E. coli* bacteria (16) shows that the effective viscosity in the magnetic bacteria is larger by a factor of 2 than in *E. coli*.

We now estimate the size of the displacements of a magnetic particle from its equilibrium position, as a result of its diffusive motions. We have already concluded from the experimental values of the upper limit of the relative intensity of the narrow spectrum that $(k^2 \langle x^2 \rangle) > 4.6$. Using the value $k^2 = 53 \cdot 10^{16}$ cm^{-2}, we find that $(\langle x^2 \rangle) > 0.12$ Å^2. ($\langle x^2 \rangle$ is the mean square translational deviation in the x direction; $\langle r^2 \rangle$, the total mean square translational deviation, is equal to $3\langle x^2 \rangle$; $\langle r^2 \rangle$ is, therefore, larger than 0.36 Å^2.) If we only take into account restoring forces and restoring moments produced by the magnetic interac-

tions, then the energy of a magnetic particle is approximately given by

$$E \approx (-4M^2/d^3)(1 - 0.5\,\theta^2 - 1.5\,x^2/d^2) \qquad (15)$$

where M is the magnetic moment of each magnetic particle, d is the distance between the centers of adjacent particles, x is the displacement of the particle in a direction perpendicular to the chain of magnetic particles, and θ is the rotation around an axis perpendicular to the chain. Eq. 15 is obtained by treating each magnetic particle as a magnetic dipole located at the center of the particle and having a magnetic moment M. In our case $M = 4 \cdot 10^{-14}$ emu (particle dimension is 220 Å and the saturation magnetization of bulk Fe$_3$O$_4$ is 480 G/cm^3) and $d = 500$ Å. Thus, $m = 12\,M^2/d^5 = 6$ erg/cm^2. Equipartition of energy gives $\langle x^2 \rangle = k_B T/m = 70$ Å^2. If additional restoring forces exist then $\langle x^2 \rangle < 70$ Å^2. The conclusion is, therefore, that 0.12 Å$^2 < (\langle x^2 \rangle) < 70$ Å^2 or 0.35 Å $< [\langle x^2 \rangle]^{1/2} < 8.4$ Å.

From Eq. 15 the restoring moment constant C is equal to $4M^2/d^3 = 0.5 \cdot 10^{-10}$ erg. Equipartition of energy gives $\langle \theta^2 \rangle < k_B T/C$. Thus $\langle \theta^2 \rangle < 8.4 \cdot 10^{-4}$ or $[\langle \theta^2 \rangle]^{1/2} < 0.029$ rad, which means that diffusive rotations are $< 1.5°$. The upper limit for the displacements due to rotational diffusion is $(a/2)[\langle \theta^2 \rangle]^{1/2} = 6$ Å ($a = 420$ Å is the dimension of the particle). The small value of $\langle x^2 \rangle^{1/2}$ and $\langle \theta^2 \rangle^{1/2}$ imply that the particles are fixed relative to each other in the cell.

The fact that the additional Fe^{3+} quadrupole doublet in the spectrum broadened together with the Fe$_3$O$_4$ lines is consistent with previous cell fractionation studies (4). The Fe^{3+} in the cell has been characterized as a hydrous-ferric-oxide precursor to Fe$_3$O$_4$ precipitation that is physically associated with the magnetosomes. Hence this material would participate in the diffusive motions of the magnetosomes and a broad line spectrum is expected. Because of the relatively low intensity of the quadrupole doublet, its wide-line spectrum is buried in the Fe$_3$O$_4$ wide-line spectrum.

The fact that the sharp-line Fe^{2+} spectrum remains even when the Fe$_3$O$_4$ lines have broadened shows that the Fe^{2+} material is not associated with the magnetosomes. (Fe^{2+} in the cells is thought to result from reduction of chelated Fe^{3+} which the cells take up from the external medium [4]. The Fe^{2+} is subsequently reoxidized and deposited as hydrous-ferric-oxide.) As noted above, in wet, packed cells held anaerobically above freezing temperature, degradative processes reduce the hydrous-ferric-oxide to Fe^{2+}. If the Fe^{2+} remained associated with the magnetosomes, or if the Fe^{2+} was dissolved in the cytoplasm, diffusive motion would broaden the sharp-line spectrum at $T > 275$ K, contrary to experiment. This suggests that the Fe^{2+} is not associated either with the magnetosomes or with the cytoplasm in the cells. The Fe^{2+} is very probably associated with the cell wall. Metal ions are bound by the cell wall (17) and peptidoglycan (18) in some gram-positive bacte-

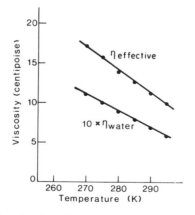

FIGURE 8   The viscosity of the medium surrounding the magnetic particles as function of temperature. The viscosity of water is shown in the graph on an extended scale.

ria. On the other hand, heavy metals are accumulated intracellularly in gram-negative species (19). In gram-negative *A. magnetotacticum*, ferrous iron could be transiently associated with the cell envelope during its conversion from the iron quinate complex outside the cell to ferric iron and ultimately Fe_3O_4 within the cell.

We thank Dr. F. F. Torres de Araujo for participation in the early phases of this study and Dr. S. G. Cohen for discussion of the results.

Drs. Ofer, Nowik, and Bauminger were partially supported by the Stiftung Volkswagenwerk. Drs. Papaefthymiou and Frankel were partially supported by the Office of Naval Research. Dr. Blakemore was supported by the Office of Naval Research and the National Science Foundation. The Francis Bitter National Magnet Laboratory is supported by the National Science Foundation.

Received for publication 15 September 1983 and in final form 5 March 1984.

REFERENCES

1. Blakemore, R. P., D. Maratea, and R. S. Wolfe. 1979. Isolation and pure culture of a freshwater magnetic spirillum in chemically defined medium. *J. Bacteriol.* 140:720–729.

2. Frankel, R. B., R. P. Blakemore, and R. S. Wolfe. 1979. Magnetite in freshwater magnetotactic bacteria. *Science (Wash. DC).* 203:1355–1356.

3. Balkwill, D. L., D. Maratea, and R. P. Blakemore. 1980. Ultrastructure of a magnetotactic spirillum. *J. Bacteriol.* 141:1399–1408.

4. Frankel, R. B., G. C. Papaefthymiou, R. P. Blakemore, and W. O'Brien. 1983. Fe_3O_4 precipitation in magnetotactic bacteria. *Biochim. Biophys. Acta.* 763:147–159.

5. Frankel, R. B., and R. P. Blakemore. 1980. Navigational compass in magnetic bacteria. *J. Magn. Mater.* 15–18:1562–1564.

6. Craig, P. P., and N. Sutin. 1963. Mössbauer effect in liquids: influence of diffusion broadening. *Phys. Rev. Lett.* 11:460–462.

7. Bauminger, E. R., S. G. Cohen, S. Ofer, and U. Bachrach. 1982. Study of storage iron in cultured chick embryo fibroblasts and rat glimoa cells, using Mössbauer spectroscopy. *Biochim. Biophys. Acta.* 720:133–140.

8. Maratea, D., and R. P. Blakemore. 1981. *Aquaspirillum magnetotacticum* sp. nov., a magnetic spirillum. *Int. J. Syst. Bacteriol.* 31:452–455.

9. Bauminger, E. R., S. G. Cohen, I. Nowik, S. Ofer, and J. Yariv. 1983. Dynamics of heme iron in crystals of metmyoglobin and deoxymyoglobin. *Proc. Natl. Acad. Sci. USA.* 80:736–740.

10. Nowik, I., S. G. Cohen, E. R. Bauminger, and S. Ofer. 1983. Mössbauer absorption in overdamped harmonically bound particles in Brownian motion. *Phys. Rev. Lett.* 50:1528–1530.

11. Uhlenbeck, G. E., and L. S. Ornstein. 1930. On the theory of the Brownian motion. *Phys. Rev.* 36:823–841.

12. Parak, F., E. W. Knapp, and D. Kucheida. 1982. Protein dynamics. Mössbauer spectroscopy on deoxymyoglobin crystals. *J. Mol. Biol.* 161:177–194.

13. Singwi, K. S., and A. Sjolander. 1960. Resonance absorption of nuclear gamma rays and the dynamics of atomic motions. *Phys. Rev.* 120:1093–1102.

14. Springer, T. 1972. Quasielastic neutron scattering for the investigation of diffusive motions in solids and liquids. *In* Springer Tracts in Modern Physics. G. Höhle, editor. Springer-Verlag, Berlin. 63:67.

15. van Hove, L., 1954. Correlations in space and time and Born approximation scattering in systems of interacting particles. *Phys. Rev.* 95:249–262.

16. Keith, A. D., and W. Snipes. 1974. Viscosity of cellular protoplasm. *Science (Wash. DC).* 183:666–668.

17. Beveridge, T. J., C. W. Forsberg, and R. J. Doyle. 1982. Major sites of metal binding in *Bacillus licheniformis* walls. *J. Bacteriol.* 150:1438–1448.

18. Beveridge, T. J., and R. G. E. Murray. 1976. Uptake and retention of metals by cell walls of *Bacillus subtilis*. *J. Bacteriol.* 127:1502–1518.

19. Strandberg, G. W., S. E. Shumate, III, and J. R. Parrott, Jr. 1981. Microbial cells as biosorbants for heavy metals: accumulation of uranium by *Saccharomyces cerevesiae* and *Pseudomonas aeruginosa*. *Appl. Environ. Microbiol.* 41:237–245.

Section III
Energetics

Contents

Introduction

Life is applied bioenergetics. The physicist's understanding of bioenergetics has proceeded at both the microscopic and macroscopic levels. The most fundamental process involved in energy conversion in biological systems is the electron transfer reaction which results in the creation of separated electrostatic charges. All biological systems have learned the trick of capturing and conserving the energy contained in separated charges rather than allowing them to merely recombine in a wasteful back reaction. In his paper on electron transfer between two fixed sites in biological molecules, Hopfield develops a theory of electron transfer based on quantum mechanical tunneling. Two temperature ranges are considered by him. At low temperatures, a temperature-independent range which proceeds by tunneling is presented. At higher temperatures, vibronic coupling in the individual molecules produces an activation energy. In photosynthetic systems, electron transfer is driven by photon absorption. The history of the photon after initial absorption and prior to initiating charge transfer is of fundamental importance. In his lucid translation of Förster's classic 1948 *Ann. Phys.* article, Knox relates the quantum mechanical treatment of the transfer of electronic excitation energy between similar molecules in solution. An application of exciton migration, random walk theory, and trapping in photosynthetic antenna pigment beds is developed by Pearlstein. Elegant picosecond electrical measurements on excitation trapping and charge stabilization in a photosynthetic bacterium are described by Deprez, Trissl, and Breton.

Hemeproteins are among the most studied biomolecules using physical techniques. Perutz presents a grandmaster's account of how hemoglobin functions in respiratory gas exchange. Mössbauer spectroscopy is an analytical tool that has found important applications in elucidating the electronic and chemical environment of the iron site in hemeproteins. Using Mössbauer spectroscopy, Keller and Debrunner analyze the temperature dependence of the effective mean-square displacement of the active iron center in oxymyoglobin, recognizing three contributions to the motion of this site. Elaborating on this concept Nowik *et al.* apply to macromolecular systems calculations of the spectral shapes of Mössbauer absorption spectra for harmonically bound nuclei in Brownian motion, enabling one to decode the conformational motion of iron proteins.

The decreased deformability of the sickled red cell produced by gelation of hemoglobin S is the fundamental cause of sickle cell disease. Eaton and Hofrichter review developments in the relation between hemoglobin S gelation and sickle cell disease. Hemoglobin is able to perform its function because it is circulated throughout the body by the pumping action of the heart. By discussing the relationship of the electrocardiogram to charge distribution in the heart, Hobbie provides an interesting example of how elementary concepts in electrostatics can be used to understand qualitative aspects of this important chemical diagnostic tool.

The last paper in this section models the dependence of membrane potential and calcium concentration in beta islet cells. The model reveals, for a nondriven excitable membrane, the possibility for an aperiodic energetic response which can be characterized as chaotic behavior.

Electron Transfer Between Biological Molecules by Thermally Activated Tunneling

(photosynthesis/oxidative phosphorylation/cytochromes)

J. J. HOPFIELD

Department of Physics, Princeton University, Princeton, New Jersey 08540; and Bell Laboratories, Murray Hill, New Jersey 07974

Contributed by J. J. Hopfield, June 24, 1974

ABSTRACT   A theory of electron transfer between two fixed sites by tunneling is developed. Vibronic coupling in the individual molecules produces an activation energy to transfer at high temperatures, and temperature-independent tunneling (when energetically allowed) at low temperature. The model is compared with known results on electron transfer in *Chromatium* and in *Rhodopseudomonas spheroides*. It quantitatively interprets these results, with parameters whose scale is verified by comparison with optical absorption spectra. According to this description, the separation between linking sites for electron transfer is 8–10 Å in *Chromatium*, far smaller than earlier estimates.

The transfer of an electron from one molecule to another is an essential part of oxidative phosphorylation and photosynthesis. The reversible oxidation of the heme of cytochrome c by cytochrome oxidase is a specific example of such a process, one of several electron transfers in the sequence of reactions resulting in oxidative phosphorylation. The overall effectiveness of such processes as photosynthesis or oxidative phosphorylation depends both on there being a large electron transfer rate for desired transfers and a small rate for inappropriate transfers. A particular cytochrome (or iron–sulfur protein) seems to have, as its sole chemical function, the ability to exchange electrons with two other molecules A and B, which (apparently) cannot directly exchange electrons. The absence of direct exchange may be due either to spatial localization or stereochemical constraints.

Since electron-transfer proteins play a specific chemical role, one should be able to explain in quantitative physical terms how the observed functional properties are related to aspects of molecular structure. There are two major obstacles to attempting such an explanation at present. First, very little is known about the relative geometry of the donor and acceptor during the electron transfer process. Second, even when a geometry is known or surmised, the mechanism of electron transfer is unsure. A wide variety of transfer descriptions have been utilized for particular systems. Hodges, Holwerda, and Gray (1) have described the electron transfer between cytochrome c and Fe(EDTA) in terms of the "outer sphere electron transfer" of solution electrochemistry (2, 3). In the Winfield mechanism (4) of electron transfer in cytochrome c, the electron is visualized as being passed along a chain of binding sites with a thermally activated transfer between these sites. Thermal activation of an electron to a "conduction band" and, thence, free motion to a second site has been suggested (5). Quantum mechanical tunneling of the electron between two sites has also been invoked (5, 6).

In this paper, we bypass the first problem by assuming the electron to be transferred between two sites in fixed geometry.

We bypass possible Winfield-like complications, and assume that there are no other electron states available at low enough energies to be thermally accessible. Section I shows that transfer between two fixed sites, in suitable approximation, is mathematically isomorphic with the conceptually simpler problem of excitation transfer by the Förster (7, 8) (dipolar) mechanism. In Section II, the simplest possible model of the coupling of electronic states to molecular thermal motions is developed and used to calculate the temperature-dependent electron transfer rate. The model is compared with experimental results in Section III.

I. The two–site tunneling description of electron transfer

We consider the problem of the transfer of an electron between two sites a and b, with the electron initially in a wave function φ_a localized around site a. The final state will have the electron in φ_b, localized around b. φ_a and φ_b weakly overlap, as sketched in Fig. 1. Because of the overlap between these wave functions, there is a matrix element T_{ab} of the Hamiltonian between these two one-particle states. The meaning of T_{ab} can be seen from the special case of sites a and b being equivalent, in which case the overlap generates a splitting $2T_{ab}$ between the bonding and anti-bonding states $(\varphi_a \pm \varphi_b)/\sqrt{2}$. The smallness of T_{ab} results from the exponential decrease of wave functions in the barrier penetration region between the sites. The transfer process thus involves tunneling, and T_{ab} will be called the *tunneling matrix element*.

To develop a simple analog with excitation transfer, consider two atoms a and b. Let the product $\psi_a'\psi_b$ be the wave function for the electrons on a *and* b when atom a is excited and b is in its ground state, and $\psi_a\psi_b'$ be the wave function for all electrons when atom b is excited and a is in its ground state. Define for convenience $\Psi_a = \psi_a'\psi_b$ and $\Psi_b = \psi_a\psi_b'$. There is an excitation transfer matrix element U_{ab} between states Ψ_a and Ψ_b. In the Förster (7, 8) description, excitation transfer arises from the coupling of the transition dipoles on atoms a and b, and is proportional to each transition dipole and to the inverse cube of their separation. If atoms a and b are identical, the symmetric and anti-symmetric *excitation* states are split by $2U_{ab}$. Thus, while the mechanisms of generating U_{ab} and T_{ab} are totally different, the identifications

$$U_{ab} \to T_{ab}, \qquad \Psi_a \to \varphi_a, \qquad \Psi_b \to \varphi_b$$

makes the mathematical descriptions of transfer identical.

In the calculation of the rate of excitation transfer inter- and intramolecular vibrations of the atoms play an essential role by giving an energy width to states that would otherwise

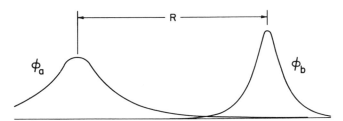

FIG. 1. The wave functions φ_a and φ_b, well separated by a distance R, with exponentially decaying tails overlapping weakly. Irrelevant detail near the wave function centers is omitted.

be infinitely sharp. The single atom state characterizing an excitation on atom a is given a spectral shape $S_a(E)$ characterized by a weighted optical emission spectrum at energy E of the transition $\psi_a' \to \psi_a$, with an appropriate normalization. This emission spectrum includes all effects of the interaction of the electronic excitation on atom a with its environment. Similarly, the excitation of b is characterized by its absorption spectrum $S_b'(E)$, including all effects of the motions of atoms in the transition $\psi_b \to \psi_b'$. The rate of excitation transfer from a to b by the Förster mechanism can then be written (7, 8)

$$W_{ab} = (2\pi/\hbar)|U_{ab}|^2 \int_{-\infty}^{\infty} S_a(E)S_b'(E)dE. \qquad [1]$$

While U_{ab} is essentially temperature-independent, W_{ab} and W_{ba} are both temperature-dependent due to the temperature dependences of the spectra involved in the overlap integral.

A precise parallel exists for the transfer of an electron from site a to b, including the effects of vibronic coupling. The analog to the emission spectrum $S_a(E)$ of $\psi_a' \to \psi_a$ is the electron removal spectral distribution $D_a(E)$. In the presence of the coupling between the electronic state φ_a and the nuclear motions, the removal of an electron (which can be thought of as being destroyed or transferred to a fictional state of zero energy) is characterized by a distribution of energies $D_a(E)$. $D_a(E)$ is broad for exactly the same reasons of atomic position readjustment that make $S_a(E)$ broad. Similarly, there is an electron insertion spectrum $D_b'(E)$ that describes the distribution of energy changes that result from the insertion of an electron (from the fictional state at zero energy) into electronic state φ_b. The rate of electron transfer can then be written in exact analogy to Eq. 1 as

$$W_{ab} = (2\pi/\hbar)|T_{ab}|^2 \int_{-\infty}^{\infty} D_a(E)D_b'(E)dE. \qquad [2]$$

This equation can also be directly calculated from the usual quantum mechanical expression for first-order transition rates. W_{ab} is in this problem always due to tunneling, although the usual temperature dependence of $D_a(E)$ and $D_b'(E)$ will make this tunneling rate temperature-dependent.

Eq. 2, with $D(E)$ approximated by a high temperature form of Eq. 4, represents a special case of the general theory (9) of electron transfer in solution electrochemistry. The suppositions (a) of fixed, well-separated sites, (b) of independent atomic motions interacting with the electron at each site, and (c) no important effect of atomic motions on $|T_{ab}|$ are particularly appropriate to transfer between distant sites embedded in a more or less rigid matrix. These approximations are relevant to many cases of transfer in biological

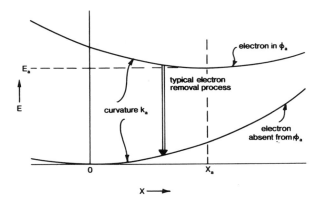

FIG. 2. The description of an electron removal process by a configuration-coordinate diagram. The two curves represent the total energy of the system as a function of the coordinate for the two states with and without the electron. $D_a(E)$ is the thermal probability distribution of the vertical separation between these states.

systems, and lack some of the complexity of the electrochemical problem. At the same time, Eq. 2 will permit extensions beyond Gaussian and high-temperature spectral functions. The treatment of our simpler problem is modeled on the usual description of tunneling through an insulating barrier between two metals (10).

II. Results from a symmetric model of D(E)

We next describe the simplest available model of $D(E)$, which can be based on the analog to a symmetric configuration-coordinate description (11) of optical emission spectra. Fig. 2 gives the essence of the physical description of this configuration-coordinate description. The upper curve describes the energy as a function of a vibrational coordinate x with the electron in state φ_a, and the lower curve the energy as a function of the same coordinate in the absence of an electron in φ_a. In the *symmetric model*, the curvature k_a of the ground and excited states are the same, and there is no entropy of electron transfer. At temperature T, the classical probability distribution of being at x if the electron is present is

$$P(x) = \left(\frac{k_a}{2\pi\kappa T}\right)^{1/2} \exp(-k_a(x - x_a)^2/2\kappa T). \qquad [3]$$

The electronic removal is a vertical transition between the two energy curves. Given $P(x)$, the energy distribution $D_a(E)$ is

$$D_a(E) = \left(\frac{1}{2\pi\kappa T k_a x_a^2}\right)^{1/2} \exp\left(-\frac{(E - E_a + \frac{1}{2}k_a x_a^2)^2}{2\kappa T \cdot k_a x_a^2}\right). \qquad [4]$$

Eq. 4 is based on a classical probability distribution, valid when κT is greater than the vibrational energies $\hbar\omega = \kappa T_a$ of the relevant vibrational coordinates. The single most important effect of quantum mechanical corrections is to product a zero-point width (11) to the distribution Eq. 3. The modification of Eq. 4 valid (with restrictions—see next paragraph) to lower temperatures is

$$D_a(E) = (1/2\pi\sigma_a^2)^{1/2} \exp{-((E - E_a + \frac{1}{2}k_a x_a^2)^2/2\sigma_a^2)} \qquad [5]$$

$$\sigma_a^2 = k_a x_a^2(\kappa T_a/2) \coth T_a/2T$$

which reduces to Eq. 4 at high temperature.

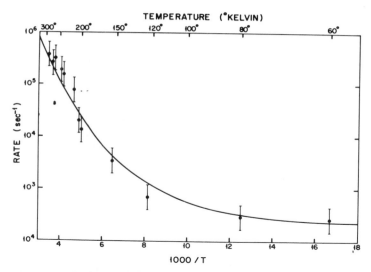

Fig. 3. The rate of photolysis-initiated transfer of an electron from cytochrome in *Chromatium* as a function of temperature. The experimental points are from deVault and Chance (5). The solid line is a plot of Eq. 8 with $T_a = T_b = 350°K$, $E_a - E_b = 0.05$ eV, $\frac{1}{2}k_ax_a^2 = \frac{1}{2}k_bx_b^2 = 0.5$ eV, and $|T_{ab}| = 4 \times 10^{-4}$ eV.

The form of $D_b'(E)$ follows from an exactly parallel model, except that $P(x)$ is replaced by

$$(k_b/2\pi\kappa T)^{1/2} \exp(-k_bx^2/2\kappa T) \quad [6]$$

and the transition is in the opposite direction, whence

$$D_b'(E) = (1/2\pi\sigma_b^2)^{1/2} \exp(-(E + E_b - \frac{1}{2}k_bx_b^2)^2/2\sigma_b^2) \quad [7]$$

$$\sigma_b^2 = k_bx_b^2(\kappa T_b/2) \coth T_b/2T.$$

From Eqs. 2, 5, and 7

$$W_{ab} =$$
$$2\pi/\hbar|T_{ab}|^2(1/2\pi\sigma^2)^{1/2} \exp(-(E_a - E_b - \Delta)^2/2\sigma^2) \quad [8]$$

where

$$\sigma^2 = (k_ax_a^2/2)\kappa T_a \coth T_a/2T + (k_bx_b^2/2)\kappa T_b \coth T_b/2T$$

and

$$\Delta = \frac{1}{2}k_ax_a^2 + \frac{1}{2}k_bx_b^2 \qquad W_{ba} = W_{ab}\exp(-(E_a - E_b)/\kappa T).$$

At high temperatures, quantum effects are unimportant, and the approximations involved in calculating Eq. 8 are valid. The process appears thermally activated, for at high temperatures, $\sigma^2 \propto T$. At low temperatures, the use of this expression is limited at best to the case $E_a > E_b + \kappa T_{a \text{ or } b}$, with further possible restrictions depending on the size of σ. Eq. 8 is a strong coupling result, also requiring $\frac{1}{2}k_ax_a^2/\kappa T_a \gg 1$.

III. The scale of parameters

The general scale of parameters for electron transfer can be established by making a fit to an appropriate experiment. The transfer of an electron from a cytochrome to fill a hole made available by a flash of light (the earliest stages of photosynthesis) has been studied in *Chromatium* by deVault and Chance (5). This transfer seems likely to come close to the idealized problem the theory describes. It has simple kinetics, has been studied over a wide range of temperatures, persists to very low temperatures, and does not appear closely coupled

to water and its phases. Figure 3 shows the experimental electron transfer rate and an approximate fit to the experimental data by use of Eq. 8. In our ignorance of the two sites of transfer, it is pointless to differentiate between the two sites, a and b, as far as vibronic parameters are concerned. We pick $k_ax_a^2 = k_bx_b^2$ and $T_a = T_b$. Eq. 8, thus reduced, contains four effective parameters, namely T_a, $k_ax_a^2$, $|T_{ab}|$, and $(E_a - E_b)$.

The characteristic temperature T_a is $350° \pm 70°$. (About $150°$ marks the turning point of the data between two regions of temperature behavior, and the characteristic turning point involves $T_a/2$ in Eq. 5.) The other three parameters are not uniquely determined. Fortunately, $E_a - E_b$ is well limited by usual constraints on electron transfer. E_a is greater than E_b, for the electron transfer takes place even at zero temperatures. But successive steps in electron transfer chains normally have their standard redox potentials within about 0.05 V (unless energy is being usefully extracted in the step). $(E_a - E_b)$ occurs only in the exponential of Eq. 8. If it is given a "typical" value of 0.05 V (1.18 kcal), then Δ is so large in order to fit the data that the values of both Δ and T_{ab} are insensitive to whether $(E_a - E_b)$ is in error by a factor of five. We thus obtain $\frac{1}{2}k_ax_a^2 = 0.5 \pm 0.1$ eV (11.5 kcal) and $|T_{ab}| = 4 \times 10^{-4 \pm 0.6}$ eV (9.6 cal). The value of $\frac{1}{2}k_ax_a^2$ is sharply constrained by the high temperature activation energy. $|T_{ab}|$ is much less definitely determined because it occurs only as a prefactor.

The magnitude of T_a and $\frac{1}{2}k_ax_a^2$ can be directly checked on a semiquantitative basis by comparing the parameters just determined, appropriate to *adding* (or *removing*) an electron to a cytochrome, to the parameters relevant to optically *exciting* a similar heme electron without removing it. The general considerations that were used to generate the shape of $D_a(E)$ in Section II are identical with those of the configuration-coordinate description of the broadening of optical spectral lines. The only important difference between optical excitation of electrons and the removal of electrons is that the removal of an electron is a somewhat larger perturbation, so the *effective* $\frac{1}{2}k_ax_a^2$ for an optical transition of the heme is expected to be comparable to, but smaller than, that for electron removal. Similar vibrations will be involved, so T_a should be essentially the same for the two cases. The full width at half maximum for an optical transition with a distortion parameter $(1/2 k_a x_a^2)_{optical}$ is

$$\text{width}(T) = 2.34[(\kappa T_a)(1/2 k_a x_a^2)_{optical} \coth (T_a/2T)]^{1/2}.$$

If we use the value of $1/2 k_a x_a^2$ deduced from tunneling, the predicted linewidth as a function of temperature is 0.290 eV at $100°$K and below, 0.337 eV at $200°$K, and 0.388 eV at $300°$K. For a typical Soret transition at 430 nm, the corresponding full widths at half maximum are 43, 50, and 58 nm, respectively. For comparison, the full width at half maximum of the Soret band of typical six-coordinated iron in oxyhemoglobin at room temperature (12) is about 32 nm, and it sharpens (13) about 10% on going to $210°$K. As anticipated, the scale of the temperature variation is similar to that interpreted from the electron transfer data. The optical linewidth is similar in scale but somewhat smaller than the electron transfer linewidth, also as expected. The optical transition has been calculated for a typical Soret $\pi-\pi^*$ transition, while the electron to be added is placed in an "iron" orbital. However, the nuclear magnetic resonance spectra (14) show that the highest energy orbital on ferric iron, the orbital from which the electron is being transferred, is widely delocalized on the heme, and the $\pi-\pi^*$ optical transition is strongly mixed with iron d-states. Since the orbitals involved in electron transfer and in the optical excitation are similar mixtures, there is no need to distinguish between them in their general properties.

The magnitude of the tunneling matrix element can be used to construct an approximate distance between the donor and acceptor. T_{ab} will fall approximately exponentially with separation R, with a characteristic length determined by the barrier height. An upper limits to the barrier height is about half the $\sigma \rightarrow \sigma^*$ band-gap of the surrounding material, yielding a height of about 3 eV. That hemes in hemoglobin are not readily photooxidized in the Soret band suggests that the barrier is not unusually small. A 2-eV barrier height is taken as a reasonable estimate. For two carbon atoms in a π-bonding configuration, then

$$T_{ab} \approx 2.7 \exp(-0.72 R)$$

where T_{ab} is in electron volts and R in Ångstroms. The prefactor is evaluated by getting the correct π resonance integral (15) at a normal bond length.* If two large aromatic groups of N_a and N_b atoms are in contact through one "edge" atom on each, T_{ab} will be multiplied by a normalization factor $(N_a N_b)^{-1/2}$, and R then measures the separation between the edge atoms. Based on such an edge-to-edge contact between two such π-systems with $N_a \approx N_b \approx 20$, the T_{ab} of 4×10^{-4} eV corresponds to a separation of 8.0 Å between the two atoms through which the transfer takes place.

The scale of parameters shows that within a fixed geometry, the rate of electron transfer will be greatly enhanced in transfer from an excited state. In the example just examined, the rate of electron transfer W_{ab} at low temperatures is 260 sec^{-1}. In an excited state higher by 1.0 eV instead of by 0.05 eV but with no other factor changed, the argument of the exponential in Eq. 8 would vanish. For the excited state, the

---

* We have used a resonance integral of 1.0 eV as a compromise between various views.

rate of transfer would be almost temperature independent, and enhanced from the low-temperature ground state result (above) by a factor of 1.7×10^7. In addition, the excited state is less well bound, and the tunneling barrier appears less high. With the parameters previously used, the tunneling barrier would now be only 1 eV high, and the matrix element T_{ab} would be raised a factor of 6 from its ground-state value. The total increase in the rate of transfer from the excited state compared to the ground state is a factor of 6×10^8, with the dominant factor arising from the change in the exponential of Eq. 8, which eliminates the usual Stokes shift suppression of the transfer rate. This general effect in Marcus theory has recently been noted (16).

Experiments perhaps related to his calculation have been carried out on the bacterium *Rhodopseudomonas spheroides* and on photosynthetic reaction centers taken from them. The rate of electron transfer from an *excited* state of molecule a to molecule b is $\geq 1.4 \times 10^{11}$ sec^{-1} (17). The rate of electron transfer from b back to the *ground* state of a below $80°$K is 30 sec^{-1} (6). This enormous difference in rates may be an indication of the strong effect of the energy difference on the transfer rate. [The failure of the reaction b $\rightarrow$ a to be thermally activated (18) can be accommodated within the general tunneling framework, but involves details too complicated to treat here.]

IV. Discussion

Our conclusions on the nature and range of the electron transfer process are totally different from the interpretations that have previously been used in analyzing the same experiments. We believe that previous errors of principle and of emphasis are responsible for the divergent conclusions. These models and their problems are summarized for comparison.

(a) *"Low Temperature" Tunneling Description.* These calculations (Eq. 1 of ref. 5 or Eq. 4 of ref. 6) are based on the penetration of a square barrier by a particle that is *otherwise free*. In order to make a comparison with a problem of transfer between two localized sites, an effective collision frequency was introduced in an *ad hoc* fashion. It was guessed that this frequency factor should be constant and about 10^{15} sec^{-1}. No discussion of the physics of how that frequency factor came about was given (5, 6). The approach omits the Franck–Condon factors that must always be present. The present paper is equivalent to constructing a detailed quantum mechanical description of this frequency factor. Because of vibronic coupling and Stokes shifts, this frequency factor is many orders of magnitude smaller than previously assumed, and is temperature dependent. The estimates of 30 Å for the transfer distance (5, 6), therefore, lack a legitimate theoretical basis.

(b) *"Barrier Fluctuation" Description of Temperature-Dependent Tunneling.* Two descriptions of temperature-dependent transfer rates have been given (5). In these descriptions, the effective distance an electron must tunnel is modulated by thermal fluctuations, and tunneling is easier at high temperatures. One of these descriptions was rejected by its authors on the basis of unreasonable parameters required, while the other seemed to its authors reasonable. Both descriptions make the same error as in (a) of completely failing to come to terms with the frequency factor.

Extensive studies have been made of tunneling through an insulating barrier between two metals. In this case, the tunneling current is closely related to Eq. **2**, with the functions $D_a(E)$ and $D_b'(E)$ replaced by their appropriate counterparts in metals. The tunneling matrix element is generally not appreciably temperature dependent. Though the analog is not exact, it suggests that the temperature dependence of the barrier is not the most likely source of temperature dependence.

(c) *Transfer by Thermal Excitation to a Free Electron State.* The idea of this transfer description (5) is that the 3.3 kcal (0.14 eV) activation energy observed for electron transfer from cytochrome c represents the binding energy of an electron in cytochrome c with respect to the conduction band of the surrounding material, and that thermally freed electrons react rapidly with the hole generated by the photon initiating the transient process. The cytochrome then acts like a donor in silicon. This description is unfortunately not internally consistent. Donors of a depth of only 0.14 eV will be largely ionized (18) at room temperature, with their electrons in the conduction band (unless the cytochrome c concentration is greater than 1 mM). This model has also been used to suggest that the barrier height for tunneling is only 0.14 eV (with concomitant huge distances possible in electron transfer), an interpretation we believe to be erroneous.

The experiments in *Chromatium* on which these various interpretations of electron transfer have been based are all consistent with a single model containing four parameters, of which one is insensitively involved, and of which two others can be semiquantitatively verified in optical absorption studies. Enough information is available to evaluate the tunneling parameter T_{ab} of the theory and to estimate the separation between the linking sites on the donor and acceptor as 8 Å. (A barrier height of 1 eV would have increased this estimate by 2.5 Å.) This distance is so much smaller than previous estimates (5, 6) (30–80 Å) for such transfers that, if correct, it must profoundly affect the view of the structural requirements for electron transfer.

A relatively short range of electron transfer is probably imperative to the operation of electron transport molecules. They seem to be used to exchange electrons with another molecule of similar standard redox potential. If they could also exchange electrons with molecules with a very different redox potential, this would short-circuit the useful paths of the oxidative phosphorylation or photosynthetic electron transport chains. The transfer mechanism described here has a propensity toward such short-circuits. It was shown in Section III that, other things being equal, electron transfer between levels differing by one volt in redox is about 10^9

times faster at room temperature than electron transfer between equivalent levels. These short-circuiting transfers must be prevented (i.e., other things kept from being equal, as by preventing approach) by structural and stereochemical considerations. When the range of electron transfer is too great, it would be impossible to prevent these short-circuiting transfers. (This rapid transfer between levels of considerably different redox potentials may, however, be functionally useful in photosynthesis for separating electrons and holes in spite of its free energy cost.)

This description of electron transfer by tunneling can provide a framework for interpreting the function of structural features of electron transport molecules. To apply the model more generally and quantitatively it will be necessary to extend descriptions of $D(E)$ to include the case in which there is an entropy change on electron transfer, a feature not included in the present description.

I thank R. G. Shulman, J. D. McElroy, and J. D. E. McIntyre for discussions of the problem. The work at Princeton was supported in part by NSF Grant GH40474.

1. Hodges, H. L., Holwerda, R. A. & Gray, H. B. (1974) *J. Amer. Chem. Soc.* **96**, 3132–3137.
2. Marcus, R. A. (1964) *Annu. Rev. Phys. Chem.* **15**, 155–196.
3. Newton, W. T. (1968) *J. Chem. Educ.* **45**, 571–575.
4. Takano, T., Swanson, R., Kallai, O. B. & Dickerson, R. E. (1971) *Cold Spring Harbor Symp. Quant. Biol.* **36**, 397–404.
5. DeVault, D. & Chance, B. (1966) *Biophys. J.* **6**, 825–847.
6. McElroy, J. D., Mauzerall, D. C. & Feher, G. (1974) *Biochim. Biophys. Acta* **333**, 261–278.
7. Förster, T. (1946) *Naturwissenschaften* **33**, 166–182.
8. Dexter, D. L. (1953) *J. Chem. Phys.* **21**, 836–850.
9. Levich, V. G. (1965) "Present state of the theory of oxidation-reduction in solution," in *Advances in Electrochemistry and Electrochemical Engineering*, eds. Delahay, P. & Tobias, C. W. (Interscience, New York), Vol. 4, pp. 249–372.
10. Schrieffer, J. R. (1964) in *Superconductivity* (W. A. Benjamin, Inc., New York), pp. 78–80.
11. Klick, C. C. & Schulman, J. H. (1957) *Solid State Phys.* **5**, 97–172.
12. Antonini, E. & Brunori, M. (1971) in *Hemoglobin and Myoglobin in their Interaction with Ligands* (American Elsevier, New York), p. 18.
13. Treu, J. (1973) "The Magnetic Circular Dichroism of Hemoglobin," Ph.D. Dissertation, Princeton University.
14. Shulman, R. G., Glarum, S. H. & Karplus, M. (1971) *J. Mol. Biol.* **57**, 93–115.
15. Salem, L. (1966) in *The Molecular Orbital Theory of Conjugated Systems* (W. A. Benjamin, Inc., New York), pp. 109, 138, and 143.
16. Efrima, S. & Bixon, M. (1974) *Chem. Phys. Lett.* **25**, 34–47.
17. Netzel, T. L., Rentzepis, P. M. & Leigh, J. (1973) *Science* **182**, 238–241.
18. Parson, W. W. (1967) *Biochim. Biophys. Acta* **131**, 154–172.

Intermolecular Energy Migration and Fluorescence*

by Th. Förster

Abstract

A quantum-mechanical treatment of the transfer of electronic excitation energy between similar molecules in solution is given, in an expansion of the earlier theories of J. and F. Perrin and of the classical-physical considerations of the present author. The critical molecular separation below which transfer occurs during the excitation lifetime is shown to be calculable from the absorption and fluorescence spectra and the excitation lifetime of the molecule. For fluorescein and chlorophyll a there result values of 50 and 80 A, which correspond to the average molecular separations in solutions of 3.2×10^{-3} and 7.7×10^{-4} moles/liter, respectively. For the regions above and below the critical concentration formulas are given for the calculation of energy migration away from the primary molecule, and these are in good agreement with existing measurements of concentration depolarization of fluorescence. Application of the theory to analogous energy transfer problems in molecular crystals and in the photosynthetic apparatus of plants is discussed.

§ 1. Observations on Energy Migration

It is known that optically excited atoms and molecules are capable of giving up their electronic excitation energy to other atoms and molecules during collisions.[1] In many cases this event seems to be a one-electron transition, such as in the quenching of fluorescence of pigment solutions by impurities according to a mechanism proposed by Weiss.[2,3] In other cases a transfer of energy occurs without the transfer of matter of any kind, such as, for example, in the case of sensitized fluorescence in vapors (Franck and Cario[4,5,6]).

In the case of excited materials of higher density, *e.g.*, in absorbing crystals, a migration of excitation energy over many molecules and to correspondingly large distances is often observed. Of course, in crystals with appreciable electronic conductivity one expects energy transfer through electronic motion, as, for example, in the inorganic crystal phosphors (Riehl[7]) and in the silver halide crystals of photographic emulsions (Gurney and Mott[8]). But in crystals without electronic conductivity energy migration is also observed. Thus the quenching of anthracene crystal fluorescence by traces of naphthacene[9,10,10a,11] can be understood only as a pure energy (not electron) migration from the absorbing anthracene molecules to the naphthacene molecules. Analogous effects in the Scheibe[12] pseudoisocyanine-type dye polymers are also to be thus explained. Of more general interest are processes in the photosynthetic apparatus of plants and in chromosomes, in which an energy migration is likewise discussed (Gaffron and Wohl;[13] Möglich, Rompe, and Timofeev-Ressovsky[14]).

Particularly clear conclusions can be reached in the case of solutions of fluorescent material, because of the smaller density of material capable of being excited. In dye solutions transfer of excitation energy takes place through the inactive solvent to molecules up to perhaps 50A away. The proof of this comes from measurements of the polarization of the fluorescence in viscous solvents.

If a dilute solution of a dye in glycerine is irradiated with polarized light, the fluorescent radiation is also strongly polarized (Weigert[15]). This effect, also observed in other viscous solvents, can be understood if we assume that the polarized excitation creates an anisotropic distribution of excited molecules, and that, if the molecular orientation relaxation time is longer than the excitation lifetime (about 10^{-8} sec), this anisotropy remains throughout the radiative process.

---

* A translation from the German of the article by Th. Förster originally appearing in *Annalen der Physik*, ser. 6, vol. 2, 55-75 (1948). See Acknowledgements.

Reprinted from *Annalen der Physik* **2**, 55–75 (1948); © American Institute of Physics.

It was observed by Gaviola and Pringsheim[16] as well as by Weigert and Käppler[17] that the fluorescence polarization of dilute viscous dye solutions dropped sharply with increasing concentration. This effect, called concentration depolarization, has been investigated at various times since, largely in the case of fluorescein. The most reliable measurements appear to be those of Feofilov and Sveshnikov.[18] They are carried out in water-free glycerin and include a measurement of the concentration dependence of the fluorescence yield. Figure 1 shows the results of the fluorescein measurements. Plotted there are the ratio p/p_0, where p is the degree of polarization and p_0 is the asymptotic value of p at vanishing concentration,[19] and the corresponding ratio η/η_0 where η is the fluorescence yield, as a function of concentration on a logarithmic scale. What interests us here is the fact that the degree of polarization, as shown in Fig. 1, has already dropped to half its maximum value in a 2.3×10^{-3} molar solution. For the dyes eosin and rhodamine G, which show a similar pattern, the half-value concentrations are 2.8 and 1.5×10^{-3} mol/l., respectively.

It is inconceivable that the relaxation time for molecular orientation decreases when these critical concentrations are exceeded. The excitation lifetimes of the molecules are equally unlikely to be lengthened; rather, they are shortened, in accordance with the concentration quenching to be discussed later. One can understand the concentration depolarization, therefore, only by assuming that molecules other than those primarily excited take over the radiation process. A trivial possibility in this connection would be reabsorption of the primary fluorescence and the subsequent occurrence of secondary fluorescence. Gaviola and Pringsheim[16] already noted, however, that this did not suffice even to approximate the cause of the strong depolarization. Because of the Stokes shift of the fluorescence spectrum from the absorption spectrum, only a tiny part of the primary fluorescence would be reabsorbed and cause secondary fluorescence. In Fig. 1 the small part of the depolarization attributable to this effect has already been considered as a correction and only that larger part that cannot be understood in terms of reabsorption is shown.

J. Perrin was the first to note[20,21] that in addition to radiation and reabsorption, a transfer of energy *(transfert d'activation)* could also take place through direct electrodynamic interaction between the primarily excited molecule and its neighbors. He presented a theory of such processes based on classical physics, and F. Perrin[22] later gave a corresponding quantum mechanical theory, the latter leaning on Kallmann and London's theory[23] of excitation energy transfer between various atoms in the gas phase. The classical and quantum mechanical treatments lead to identical results, so the essence of the matter is the model chosen. If one conceives of the dye molecules as having precisely defined electron oscillator frequencies, then energy transfers during the lifetime of the excited state are predicted to occur when the intermolecular distance is below a critical value of the order of $\lambda/2\pi$, where λ is the wavelength of

the radiation of the oscillator (J. Perrin). Upon consideration of level widths due to collision with the solvent molecules, one arrives at a critical molecular separation of the order of $(\lambda/\pi) \sqrt[3]{\bar{t}/\tau}$ (F. Perrin), where $\bar{t}$ is the average time between two such collisions and τ is the average excitation lifetime. With F. Perrin's assumed values $\bar{t} = 10^{-14}$ to 10^{-13} sec one obtains, *e.g.*, for fluorescein ($\lambda \sim 5000$ A), a critical distance of 150-250 A. The range of concentration corresponding to these distances (2×10^{-5} to 1×10^{-4} mol/l.) lies considerably below that in which the depolarization actually occurs (*cf.* Fig. 1). F. Perrin has already conjectured that the source of this discrepancy might be the difference in absorption and fluorescence frequencies giving rise to the Stokes shift, which is not considered in this model.

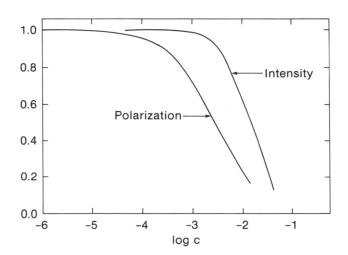

Fig. 1. Polarization and intensity of the fluorescence of fluorescein in glycerine (after Feofilov and Svenshnikov, ref. 18). The concentration is in soles per liter. The values of polarization and intensity at infinite dilution are set equal to 1.

In view of the great interest that now surrounds the problem reviewed above, it appears desirable to extend the Perrin theory in the light of our present knowledge of the structure of dye molecules. The author has given a preliminary classical treatment.[24,25] The energy migration is related to quantities such as the absorption and fluorescence spectra and the excitation lifetime all of which can be obtained exclusively from experiment. The solvent molecule collision data (accessible only as approximate estimates) do not appear, in contrast with F. Perrin's theory, so a more stringent test of the present theory can be made. This theory will now be given a quantum mechanical basis and will be developed further.

§ 2. Formal Treatment of Energy Transfer

Consider a single absorbing molecule in a solvent that is transparent in the spectral region concerned. If this molecule is excited by irradiation at time $t = 0$, then its excitation probability $\rho(t)$ decays from its initial value

$\rho(0) = 1$ through radiation, and, in general, through radiationless processes besides. And indeed

$$\frac{d\rho(t)}{dt} = -\frac{1}{\tau}\rho(t), \qquad \rho(t) = e^{-t/\tau},$$

where τ is the average lifetime of the exponentially decaying excited state. This is given in terms of the radiation probability and the rates of concurrent radiationless quenching processes as follows:

$$\frac{1}{\tau} = \frac{1}{\tau_S} + L.$$

Here τ_S is the natural lifetime of the excited state in the absence of all quenching processes and L is the number of the latter per unit time. The ratio of $1/\tau_S$ to L determines the quantum yield η of fluorescence, whose value can range from 0 to 1. Its value is given by the overall radiation probability and therefore

$$\eta = \frac{1}{\tau_S}\int_0^\infty \rho(t)dt = \frac{\tau}{\tau_S}. \qquad (1)$$

If through any quenching process η is diminished (relative to its value η_0 at infinite dilution), then according to Eq. (1) a proportional decrease in the lifetime occurs.

Consider now a number of similar absorbing molecules in the solution. If there exists the possibility of transfer of excitation energy from one molecule k to another molecule l, then there will be a decay of the excitation probability of the first molecule at a rate proportional to ρ_k and an increase of the same magnitude in that of the second molecule. Considering these processes as occurring between all pairs of molecules one obtains

$$\frac{d\rho_k(t)}{dt} = \sum_l F_{kl}[\rho_l(t) - \rho_k(t)] - \frac{1}{\tau}\rho_k(t), \quad F_{lk} = F_{kl} \quad (2)$$

Here, F_{kl} is the number of transfers $k \to l$ per unit time that would obtain if ρ_k were held constant. Equation (2) implies that the decay of excitation is accompanied by an equilibration of the excitation energy among the individual molecules. The ansatz differs essentially from one such as that of energy transfer by fluorescence reabsorption, since in the latter case radiation and transfer would be successive rather than simultaneous processes.

For the sum of the excitation probabilities of the individual molecules, Eq. (2) gives

$$\Sigma_k\rho_k(t) = \Sigma_k\rho_k(0)e^{-t/\tau}.$$

The decay of excitation in the whole system is therefore not changed by the transfer processes, so Eq. (1) also remains valid.

It is to be expected that the F_{kl} decrease rapidly with increasing distance, so a transfer will occur during the excitation lifetime only below a critical distance. If a single unexcited molecule 2 is present within this critical distance from an originally excited molecule 1 [initial

conditions $\rho_1(0) = 1$, $\rho_2(0) = 0$], then the solutions of Eqs. (2) run as follows:

$$\rho_1(t) = \frac{1}{2}(1 + e^{-2Ft})e^{-t/\tau},$$

$$\rho_2 = \frac{1}{2}(1 - e^{-2Ft})e^{-t/\tau} \quad (F = F_{12}).$$

The total radiation probabilities of the two molecules are

$$\eta_{1,2} = \frac{1}{\tau_S}\int_0^\infty \rho_{1,2}(t)dt$$

or

$$\eta_1 = \eta\frac{1 + \tau F}{1 + 2\tau F}, \qquad \eta_2 = \eta\frac{\tau F}{1 + 2\tau F}. \qquad (3)$$

For $\tau F \ll 1$ (little interaction during the lifetime) the primary molecule alone radiates; for $\tau F \gg 1$ (strong interaction) both radiate with equal probability.

§ 3. Mechanism of Energy Transfer

The quantities F_{kl} are given by the probability for the transition from a state in which molecule k is excited and molecule l is unexcited into another state with the roles of k and l interchanged. The stationary molecular vibrational states of the two molecules in these initial and final states will be described by the quantum numbers v'_k, v_l and v_k, v'_l, respectively. Their corresponding contributions to the energy are w'_k, w_l and w_k, w'_l respectively. (The quantities associated with the excited molecule are denoted by a prime.)

It will be assumed that the energy transfer is due to the interaction between the electronic systems of the two molecules, as in the Perrin theory, and that it proceeds substantially more slowly than the nuclear motion. Then for a quantum-mechanical treatment it suffices to write the eigenfunctions of the molecules as functions of the electronic coordinates alone, their dependence on the nuclear coordinates being suppressed. If $\vec{r}_k$ denotes collectively the spatial coordinates of the electrons of molecule k, then these eigenfunctions in the ground and excited states are, respectively,

$$\phi(w_k, \vec{r}_k), \quad \phi'(w'_k, \vec{r}_k).$$

The parameters w_k and w'_k are used here in place of the corresponding quantum numbers v_k and v'_k to describe the stationary molecular vibration states. Since the surrounding solvent must be considered in the description of the molecule, the energy parameters represent continuous variables.

If W_0 denotes the energy of the purely electronic excitation on either of the two molecules, the energy given up by molecule k during the transfer process described above is equal to

$$W_0 + w'_k - w_k.$$

Correspondingly, the energy taken up by molecule l is

$$W_0 + w'_l - w_l.$$

The arithmetic mean of these two values will be denoted by

$$W = W_0 + \frac{1}{2}(w_k' - w_k + w_l' - w_l). \qquad (4)$$

During a transfer the total energy changes by an amount

$$\Delta W = w_k + w_l' - w_k' - w_l. \qquad (5)$$

In terms of the new quantities introduced here the vibrational energies of the final state may be written as follows:

$$w_k = w_k' + W_0 - W + \frac{1}{2}\Delta W,$$
$$w_l' = w_l + W - W_0 + \frac{1}{2}\Delta W. \qquad (6)$$

Next let us assume certain definite initial values of w_k' and w_l. Then the parameters w_k and w_l' of the final state have no definite values; in fact, transitions occur to a continuous manifold of separate states. According to a formula given by Dirac,[30] the transition probability (defined here as the total number of transitions per unit time for each molecule found in the initial state) is given by

$$\frac{2\pi}{\hbar} \int |u_{kl}(w_k', w_l; w_k, w_l')|^2 \cdot D \cdot dW, \quad \text{with } \Delta W = 0.$$

Here $\hbar$ is Planck's constant divided by 2π and u_{kl} is the matrix element of the interaction, which is regarded as a purely electronic interaction $u(\vec{r}_k, \vec{r}_l)$:

$$u_{kl}(w_k', w_l; w_k, w_l') =$$
$$\iint \phi_k'(w_k', \vec{r}_k)^* \phi_l(w_l, \vec{r}_l)^* u(\vec{r}_k, \vec{r}_l)$$
$$\times \phi_k(w_k, \vec{r}_k)\phi_l'(w_l', \vec{r}_l)d\vec{r}_k d\vec{r}_l. \qquad (7)$$

$D = \partial(v_k, v_l')/\partial(W, \Delta W)$ is the functional determinant of the coordinate transformation from the quantum numbers v_k, v_l' to the energy quantities W, ΔW. The restriction $\Delta W = 0$ expresses conservation of energy during the transition, which is not hypothesized elsewhere. One has

$$D = \frac{\partial(v_k, v_l')}{\partial(W, \Delta W)} = \frac{dv_k}{dw_k} \cdot \frac{dv_l'}{dw_l'} \cdot \frac{\partial(w_k, w_l')}{\partial(W, \Delta W)}. \qquad (8)$$

The differential quotients appearing here represent the density of states at the energy concerned. It is convenient to make them unity by a suitable normalization of the final state eigenfunctions $\phi(w_k, \vec{r}_k)$ and $\phi'(w_l', \vec{r}_l)$, but to retain the usual normalization for the initial state eigenfunctions $\phi'(w_k', \vec{r}_k)$ and $\phi(w_l, \vec{r}_l)$. We have then

$$\int_{w_k=w}^{w_k=w+1} \int |\phi(w_k, \vec{r}_k)|^2 d\vec{r}_k dw_k = 1,$$

$$\int |\phi'(w_k', \vec{r}_k)|^2 d\vec{r}_k = 1.$$

For the functional determinant of the right side of Eq. (8) one finds the value 1 using Eq. (6). Therefore D is also equal to 1 and the transition probability becomes

$$\frac{2\pi}{\hbar} \int |u_{kl}(w_k', w_l; w_k' + W_0 - W, w_l + W - W_0)|^2 dW.$$

(Note that because of the normalization chosen, the matrix element u_{kl} is not an energy, but a dimensionless quantity.)

Now it will be assumed that in the initial state the molecular vibration energy has no definite value, but rather that it changes rapidly during the excitation transfer. The frequencies of occurrence of individual values are then given by two distribution functions $g(w_l)$ and $g'(w_k')$, normalized to 1 on an energy scale, for a molecule in the ground and excited state, respectively. It can be assumed that these correspond to thermal equilibrium.

Thus the quantity defined in §2 becomes

$$F_{kl} = \frac{2\pi}{\hbar} \int_{W=0}^{\infty} \int_{w_l=0}^{\infty} \int_{w_k'=0}^{\infty} g'(w_k') \cdot g(w_l)$$
$$\times |u_{kl}(w_k', w_l; W_0 - W + w_k', W - W_0 + w_l)|^2$$
$$\times dw_k' dw_l dW. \qquad (9)$$

The interaction energy $u(\vec{r}_k, \vec{r}_l)$ is given by the Coulomb interaction of the moving electron charges, if the molecular separation is large compared with the intramolecular atomic separations. It is therefore

$$u(\vec{r}_k, \vec{r}_l) = -\frac{e^2}{n^2 |\vec{r}_k - \vec{r}_l|},$$

wherein the square of the solvent refractive index n serves as the dielectric constant. At molecular separations that are large also compared with the molecules' dimensions,[31] this energy reduces in the well known way to the interaction energy of two dipoles. For the matrix element one obtains, accordingly,

$$u_{kl}(w_k', w_k; w_k, w_l') =$$
$$\frac{1}{n^2 R_{kl}^5}\{R_{kl}^2[\vec{M}_k^*(w_k, w_k') \cdot \vec{M}_l(w_l, w_l')]$$
$$- 3[\vec{M}_k^*(w_k, w_k') \cdot \vec{R}_{kl}][\vec{M}_l(w_l, w_l') \cdot \vec{R}_{kl}]\}. \qquad (10)$$

Here $\vec{R}_{kl}$ is the vector between the molecular centers of gravity and R_{kl} is its absolute value.[31a] The matrix element of the transition moment, $\vec{M}_k$, is defined by

$$\vec{M}_k(w_k, w_k') = -e \int \phi_k(w_k, \vec{r}_k)^* \vec{r}_k \phi_k'(w_k', \vec{r}_k)d\vec{r}_k.$$

The corresponding molecular center of gravity is chosen as the origin of electron position vectors $\vec{r}_k$. Of the two arguments in the transition moment matrix element, the first denotes the nuclear vibrational energy in the electronic ground state, the second that in the electronic excited state.

According to Eq. (10) the interaction energy of the molecules depends on their relative orientation as well as their separation. In the case of rigidly oriented molecules in viscous solvents, the only case in which energy transfer alone is observable as fluorescence depolarization, the strict application of the theory requires calculation for every individual orientation and subsequent averaging. In less viscous solvents, when energy transfer takes place more slowly than rotational Brownian motion, F_{kl} is to be averaged over all orientations of both molecules. Since the orientation parameters are independent of the integration variables, the averaging may be carried out under the integral. An easy calculation yields[31b]

$$\overline{|u_{kl}(w_k', w_l; w_k, w_l')|^2} = \frac{2}{3n^4 R_{kl}^6} M_k(w_k, w_k')^2 \cdot M_l(w_l, w_l')^2 \tag{10'}$$

where M is the absolute value of the transition moment $\vec{M}$; because of the identity of the two molecules, the index k or l may be dropped. Because of the normalization chosen for the eigenfunctions $\phi(w_k)$ and $\phi'(w_l')$, $M_k(w_k, w_k')^2 dw_k$ is the sum of the squares of the moments for those transitions between an electronic excited state having a fixed vibrational energy w_k' and ground electronic states with vibrational energies in the region w_k to $w_k + dw_k$. In a similar way $M_l(w_l, w_l')^2 dw_l'$ relates to the region dw_l' of the excited state. According the Eqs. (9) and (10') the transition probability averaged over orientations is therefore

$$\overline{F_{kl}} = \frac{4\pi}{3\hbar n^4 R_{kl}^6}$$
$$\times \int_{W=0}^{\infty} \left[\int_{w_k'=0}^{\infty} g'(w_k') M(W_0 - W + w_k', w_k')^2 dw_k' \right.$$
$$\times \left. \int_{w_l=0}^{\infty} g(w_l) M(w_l, W - W_0 + w_l)^2 dw_l \right] \cdot dW. \tag{9'}$$

§ 4. Connection with Spectra

The two integrals in the brackets [Eq. (9')] have a direct connection with the absorption and fluorescence spectra of the molecules. That is, the probability of emission of a photon in the interval W, $W + dW$ from an excited state with vibrational energy w' is given by the Einstein spontaneous transition probability to the states between $W_0 - W + w'$ and $W_0 - W - dW + w'$ of the ground electronic state; it will be denoted by $A(W_0 - W + w', w')dW$. The connection with the matrix element of the transition moment is given by[32]

$$A(W_0 = W + w', w') = \frac{4}{3} \frac{nW^3}{\hbar^4 c^3} |M(W_0 - W + w', w')|^2.$$

For a thermal equilibrium distribution of the vibrational motion in the excited state the average is

$$A(W) = \overline{A(W_0 - W + w', w')}$$
$$= \frac{4}{3} \frac{nW^3}{\hbar^4 c^3}$$
$$\times \int_{w'=0}^{\infty} g'(w') |M(W_0 - W + w', w')|^2 dw'. \tag{11}$$

$A(W)dW$ is the number of quanta emitted per unit time per excited molecule in the energy range $W, W+dW$ and gives the fluorescence intensity in the corresponding frequency interval. The integral occuring here is the same one that appears first in the brackets on the right side of Eq. (9'). Consequently the latter may be related to the fluorescence spectrum.

In a similar way the second integral may be related to the absorption spectrum. Suppose there exists radiation with an energy density $\sigma(W)dW$ in the photon energy region between W and $W + dW$. This radiation causes absorptive transitions from the ground electronic state with vibrational energy w to the excited state with vibrational energy between $W - W_0 + w$ and $W + dW - W_0 + w$. The transition probability is $\sigma(W) \cdot B(w, W - W_0 + w)dW$, wherein there is the following connection with the transition moment matrix elements:

$$B(w, W - W_0 + w) = \frac{4\pi^2}{3n^2\hbar} |M(w, W - W_0 + w)|^2.$$

For a thermal equilibrium distribution in the ground state the average is

$$B(W) = \overline{B(w, W - W_0 + w)}$$
$$= \frac{4\pi^2}{3n^2\hbar} \int_{w=0}^{\infty} g(w) |M(w, W - W_0 + w)|^2 dw. \tag{12}$$

$B(W)$ can be expressed in terms of $\epsilon(W)$, the molar decadic extinction coefficient of the absorber, as follows:

$$B(W) = \frac{c(\ln 10)}{nN'W} \epsilon(W). \tag{13}$$

Here $N' = 6.02 \times 10^{20}$ is the number of molecules per millimole.[33] The second integral of Eq. (9') may be related to the absorption spectrum through Eqs. (12) and (13).

Since the absorption spectrum can be measured more easily and precisely than the fluorescence spectrum, it is convenient to relate the first integral to the absorption spectrum too. In molecules of the kind we are considering, in which the electronic excitation energy is distributed over a large number of atoms, the vibrational degrees of freedom of the excited state are little different from those of the ground state. Therefore the transition moment is an approximately symmetric function of the vibrational energy in the ground and excited states:

$$M(w, w') \sim M(w', w).$$

(In contrast to our earlier usage, w and w' are now two arbitrarily fixed energies, which may belong to either the ground or excited state. The order is as given before, however; the argument appearing first corresponds to the ground state and the second to the excited state).

Because of the approximate correspondence of the vibrational degrees of freedom of the two states, it is also the case that

$$g(w) \sim g'(w).$$

As a consequence of this symmetry it follows from Eqs. (11) and (12) that

$$B(W) \sim \frac{\pi^2 \hbar^3 c^3}{n^3(2W_0 - W)^3} A(2W_0 - W),$$

whereby a correspondence between the fluorescence spectrum and the absorption spectrum reflected about $W = W_0$ is established. The approximate mirror symmetry of the intensity pattern in absorption and fluorescence spectra is well known and is addressed by Levschin's[34] law of "mirror correspondence," which, quantitatively, is somewhat differently formulated. With this symmetry relation and Eqs. (11)-(13), Eq. (9') finally yields[33a]

$$\begin{aligned} F_{kl} &= \frac{3(\ln 10)^2 \hbar c^2}{4\pi^3 n^2 (N')^2 R_{kl}^6} \\ &\times \int_{W=0}^{\infty} \frac{\epsilon(W)\epsilon(2W_0 - W)}{W(2W_0 - W)} dW \qquad (14) \\ &\sim \frac{3\hbar c^2 J(W_0)}{4\pi^3 n^2 (N')^2 W_0^2 R_{kl}^6} \end{aligned}$$

with

$$J(W_0) = (\ln 10)^2 \int_{W=0}^{\infty} \epsilon(W)\epsilon(2W_0 - W) dW. \qquad (15)$$

Here use is made of the fact that the absorption and fluorescence spectra of similar molecules, and consequently also the functions $\epsilon(W)$ and $\epsilon(2W_0 - W)$, overlap each other only in a narrow region on both side of W_0, so that [in the denominator of Eq. (14)] one can set $W \sim W_0$. One may write this expression for F_{kl} conveniently in the form

$$F_{kl} = \frac{1}{\tau_0} \left(\frac{R_0}{R_{kl}} \right)^6 \qquad (16)$$

with

$$R_0 = \sqrt[6]{\frac{3\hbar \tau_0 c^2 J(W_0)}{4\pi^3 n^2 (N')^2 W_0^2}}. \qquad (17)$$

According to Eq. (16) and the definition of F_{kl}, R_0 is the critical molecular separation, mentioned earlier, below which energy transfer sets in during the excitation lifetime. For the evaluation of Eq. (17) it is convenient to express the molar extinction coefficient as a function of the frequency $\nu = W/h$ or the wavenumber $\bar\nu = \nu/c$ of the photons, and also W_0 in terms of the corresponding frequency ν_0 or wavenumber $\bar\nu_0$ of the purely electronic transition. $J(W_0)$ is then converted into integrals $J(\nu_0)$ or $J(\bar\nu_0)$ that are formulated in analogy with Eq. (15):

$$R_0 = \sqrt[6]{\frac{3\tau_0 c^2 J(\nu_0)}{8\pi^4 n^2 (N')^2 \nu_0^2}} = \sqrt[6]{\frac{3\tau_0 c^2 J(\bar\nu_0)}{8\pi^4 n^2 (N')^2 \bar\nu_0^2}}. \qquad (17')$$

Approximate values of R_0 are obtained most readily if the overlap of the absorption and fluorescence spectra are written in terms of the quantity

$$\frac{1}{\Delta\nu} = \frac{\int \epsilon(\nu)\epsilon(2\nu_0 - \nu) d\nu}{[\int \epsilon(\nu) d\nu]^2}$$

and the following well-known but admittedly approximate relation is used:

$$\frac{1}{\tau_S} \sim \frac{8\pi(\ln 10)n^2 \nu_0^2}{N'c^2} \int \epsilon(\nu) d\nu$$

(cf., e.g., F. Perrin[22]). (All integrals are to be extended over the whole relevant absorption region.) One then obtains

$$\begin{aligned} R_0 &\sim \frac{\lambda_0}{2\pi n} \sqrt[6]{\frac{3}{8} \frac{\tau_0}{\tau_S^2 \Delta\nu}} \\ &= \frac{\lambda_0}{2\pi n} \sqrt[6]{\frac{3}{8} \frac{\nu_0^2}{\tau_0 \Delta\nu}}. \end{aligned} \qquad (17'')$$

By dropping the sixth root these expressions would yield the (such too large) values of J. Perrin's theory. If there were complete overlapping of the absorption and fluorescence spectra $[\epsilon(2\nu_0 - \nu) \sim \epsilon(\nu)]$, $\Delta\nu$ would become roughly equal to their width at half maximum. If this width were determined through the chopping of the emitted radiation as a consequence of phase-changing collisions with solvent molecules in time intervals of average length $\bar{t}$, then for $\eta_0 = 1$ and $n = 1$ this F. Perrin-like formula results:

$$R_0 \sim \frac{\lambda_0}{2\pi} \sqrt[6]{\frac{\bar{t}}{\tau}}.$$

In fact, the absorption and fluorescence spectra of similar molecules are far from completely overlapping. $1/\Delta\nu$ is therefore only about $1/10$ of the reciprocal half width and in dyes this is about $1/20$ of the absorption frequency, from which for a dye absorbing at 6000A we have $\Delta\nu = 5 \times 10^{15} \text{sec}^{-1}$. With $\tau_0 \sim 10^{-8}$ sec and $\nu_0 = 1$, $n = 1$, we obtain a value $R_0 \sim 70$A, smaller than the value resulting from J. and F. Perrin's theories and in better agreement with observations.

Equation (17') is the most suitable starting point for precise calculations of R_0. For fluorescein in water $n = 1.34$ and $\tau_0 = 5.07 \times 10^{-9}$ sec, the latter according to recent direct fluorometric measurements by Szymonowski.[34] The spectra used for the calculation of $J(\nu_0)$ are reproduced in Figure 2. $\epsilon(\nu)$ is the absorption spectrum with a maximum at 20,400cm^{-1} measured in an aqueous solution 10^{-5} molar in fluorescein and 0.01 normal in NaOH. Only a few points of the fluorescence spectrum were measured to fix the maximum at 19,400 cm^{-1}. The arithmetic average of the maxima gives the value $\bar\nu_0 = 19,900$cm-1 with which the reflected curve $\epsilon(2\bar\nu_0 - \bar\nu)$ is calculated. Through graphical integration over the product function also shown in Fig. 2, $J(\nu_0)$ is found to be 7.0×10^{12}cm^3 (mMol)$^{-2}$. With these values Eq. (17') yields

$$R_0 = 50\text{A}.$$

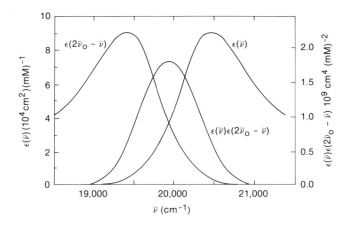

Fig. 2. Absorption curve $\epsilon(\bar{\nu})$, the reflected curve $\epsilon(2\bar{\nu}_0 - \bar{\nu})$, and their product curve for fluorescein in water (10^{-5}M in solution with 0.01N NaOH).

For chlorophyll a, the main constituent of natural chlorophyll, $\tau_0 \sim 3 \times 10^{-8}$ sec (calculated by F. Perrin[36] from the viscosity dependence of the fluorescence polarization). According to the photographic spectra of Dhéré and coworkers[37] the absorption and fluorescence maxima in ether appear to lie almost in the same place ($\bar{\nu}_{abs} = 15130 \text{cm}^{-1}$, $\bar{\nu}_{fl} = 15100 \text{cm}^{-1}$). From this information and the absorption spectrum measured quantitatively by Sprecher von Bernegg[38] one obtains $J(\bar{\nu}_0) = 1.1 \times 10^{13} \text{cm}^3 \text{(mMol)}^{-2}$. With $n = 1.35$ for ethyl ether it follows that

$$R_0 = 80\text{A (chlorophyll).}^{38a}$$

Therefore, energy transfer between chlorophyll molecules is to be expected at distances still larger than those between dye molecules of the fluorescein type. This is effected by the small breadth of the absorption maximum (half width 500cm^{-1}) and the strong overlapping of the absorption and fluorescence spectra.

It is quite possible that absorption and fluorescence maxima actually lie farther apart than it appears in photographic spectra. Decreasing plate sensitivity toward long wavelengths simulates a situation in which the absorption maximum appears at lower wavenumbers and the fluorescence maximum at higher wavenumbers. Indeed, the true absorption maximum according to Sprecher von Bernegg[38] lies at 15200cm^{-1}, thus 70cm^{-1} higher than the photographically determined maximum. Under the assumption of an equally large shift of the fluorescence maximum in the opposite direction there results a supposedly true position of 15025cm^{-1} and $\bar{\nu}_0 = 15100\text{cm}^{-1}$. But even here a strong overlapping of the spectra remains, with $J(\bar{\nu}_0 \sim 0.90 \times 10^{13}\text{cm}^3 \text{(mMol)}^{-2}$ and $R_0 = 77\text{A}$ (or 57.5A^{38a}).

Quinine must be considered as a representative of one other class of fluorescent compounds. According to

F. Perrin the fluorescence lifetime of its ions in acid solution is of the same order of magnitude (4×10^{-8}sec) as that of chlorophyll. The absorption maximum is, however, considerably lower and broader[39] [$\epsilon_{max} = 7000\text{cm}^2 \text{(mMol)}^{-1}$, half width 5000cm^{-1}]. Because of the small overlapping and the width of the spectra $J(\bar{\nu})$ is only of the order of magnitude of $10^8\text{cm}^3 \text{(mMol)}^{-2}$ and R_0 is therefore a factor of 10 smaller than in the case of chlorophyll. Energy migration is possible, therefore, only in very dense aggregates of the molecule, such as in crystals. For anthracene, esculin, and other fluorescent compounds of this kind the same is likely to hold.

The materials considered above have high fluorescence yields. Of two molecules with the same spectrum but different fluorescence yields η, the weaker fluorescing also has a shorter lifetime and therefore according to Eqs. (17) and (17') the smaller value of R_0. Energy transfer over large distances between two molecules is possible only when the fluorescence yield is high.

§ 5. Energy Transfer in Solution

Knowing the distance dependence of the coupling term F_{kl} contained in Eq. (16) (inversely proportional to the sixth power), we can calculate the effects of energy transfer in the case of an assumed statistical distribution of dye molecules in solution. It suffices to consider only the nearest neighbor of the excited molecule at low concentrations. If one molecule is at the origin and its nearest neighbor is at some distance between R and $R + dR$, then there is an empty volume $v = 4\pi R^3/3$ surrounding the first molecule and the second is somewhere within a volume dv ($= 4\pi R^2 dR$). According to elementary formulas of statistics, in a uniform distribution over a volume V the probability that an arbitrary one of a very large number N' of molecules lies in dv, and is nearest to the origin, is

$$w(v)dv = \lim_{N' \to \infty} \frac{N'dv}{V}\left(1 - \frac{v}{V}\right)^{N'-1} = e^{-\xi}d\xi \quad (18)$$

with

$$\xi = \frac{N'v}{V} = N'cv. \quad (19)$$

When N' is again taken to be equal to the number of molecules per millimole, V is the molar volume measured in liters and c is the concentration in the usual units of moles/liter. (Confusion with the speed of light, which was also denoted by c is earlier intermediate calculations, hardly needs to be feared.)

Under the assumed limitation of transfer to the nearest neighbors of excited molecules one obtains the predicted radiation probability from the primary molecules by averaging the radiation probability calculated by Eq. (3) over the distribution of molecular separations given by Eq. (18). The quantity τF appearing in Eq. (3) is, according to Eqs. (16), (1), and (19),

$$\tau F = \frac{\tau}{\tau_0}\left(\frac{R_0}{R}\right)^6 = \frac{\eta}{\eta_0}R_0^6\left(\frac{4\pi}{3v}\right)^2 = \frac{\eta}{\eta_0}R_0^6\left(\frac{4\pi N'c}{3\xi}\right)^2.$$

In terms of a critical concentration defined by

$$c_0 = \frac{3}{4\pi} \frac{1}{N' R_0^3} = \left(\frac{7.35 \times 10^{-8}}{R_0} \right)^3, \qquad (20)$$

this becomes

$$\tau F = \frac{\eta}{\eta_0} \left(\frac{c}{c_0} \right)^2 \frac{1}{\xi^2}. \qquad (21)$$

Moreover, a dimensionless concentration parameter will be introduced:

$$\gamma = \sqrt{\frac{\eta}{\eta_0}} \cdot \frac{c}{c_0}. \qquad (22)$$

Then according to Eqs. (3) and (21) the fraction of energy radiated from the primary molecule, averaged over all distances, is given by[42]

$$\begin{aligned}
\frac{\eta_1}{\eta} &= \int_0^\infty \frac{1 + \tau F}{1 + 2\tau F} e^{-\xi} d\xi \\
&= 1 - \frac{\gamma}{\sqrt{2}} [\sin \gamma\sqrt{2} \cdot \mathrm{Ci}\gamma\sqrt{2} + \cos \gamma\sqrt{2} \qquad (23) \\
&\quad \cdot (\frac{\pi}{2} - \mathrm{Si}\gamma\sqrt{2})].
\end{aligned}$$

Si and Ci are the sine integral and cosine integral functions, respectively. The energy radiated according to Eq. (23) from the primary molecule as a function of the concentration parameter γ is shown in Fig. 3. For $\gamma = 1$ (corresponding to the critical concentration when the quantum yield equals that at infinite dilution), $\eta_1/\eta = 0.645$, so the photon is more likely to be emitted from a neighboring molecule than from molecule 1. Above the concentration c_0, considerable energy migration from the primary molecule therefore occurs. Using Eq. (20) the following values of the critical concentration may be inferred from the critical distances calculated in § 4:

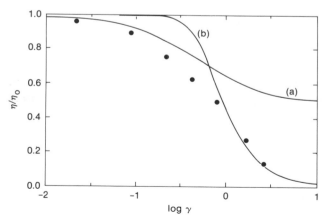

Fig. 3. Contribution of the primarily excited molecules to the radiation as a function of the dimensionless parameter γ [Eq. (22)). Curve a [Eq. (23)) holds for low concentrations, curve b [Eq. (29)] for high. The points are obtained by interpreting the polarization measurements on fluorescein in glycerine according to § 6.

Fluorescein (in ethanol) ....$c_0 = 3.2 \times 10^{-3}$ Mol/l
Chlorophyll (in ethyl ether) $c_0 = 7.7 \times 10^{-4}$ Mol/l

On the other hand, the values for quinine and similar compounds lie several orders of magnitude higher.

Because it is based on consideration of only the nearest neighbor molecules, Eq. (23) gives correct values for the primary molecule's radiation probability only at concentrations below the critical concentration. At higher concentrations more distant molecules must also be considered. On the other hand, in this case it is to be expected that the statistical fluctuations of the molecular separations that were considered in the derivation of Eq. (23) are less important, and that carrying out the calculations for an average molecular distribution leads to an approximately correct result.

It will therefore be assumed that the molecular centers of gravity form a regular lattice of some kind. The position vector of molecule k is therefore taken to be

$$\vec{R}_k = \vec{k} R_g,$$

where R_g is the lattice constant and $\vec{k}$ is a dimensionless vector whose components are the indices of the lattice point, a rational triad. If further

$$\vec{k} - \vec{l} = \vec{s}_{kl},$$

then according to Eq. (16)

$$F_{kl} = \frac{1}{\tau_0} \left(\frac{R_0}{R_g} \right)^6 \frac{1}{|\vec{s}_{kl}|^6}. \qquad (24)$$

Now let a transition to a continuum be made, in which ρ_k, defined only on the lattice points, is interpolated through the use of a continuous and differentiable function $\rho(\vec{R}, t)$. With appropriate normalization this function represents the average spatial distribution of the excitation probability. Expansion in a Taylor series yields

$$\rho_l = \rho_k - R_g(\vec{s}_{kl} \cdot \vec{\nabla}\rho) + \frac{1}{6} R_g^2 |\vec{s}_{kl}|^2 \vec{\nabla}^2 \rho + \ldots . \qquad (25)$$

Second-order terms not invariant under rotations and terms of third and higher order have been omitted from Eq. (25). Upon summation over all points of the regular lattice chosen for the calculation, Eq. (2) must acquire a form that is independent of the direction of the crystal axes, with the help of Eqs. (24) and (25). Therefore all the non-rotationally-invariant terms of first and second order in R_g drop out and there remains the partial differential equation[42a]

$$\frac{\partial \rho}{\partial t} = D\vec{\nabla}^2 \rho - \frac{1}{\tau}\rho. \qquad (26)$$

where we have used the abbreviation

$$D = \frac{R_0^6}{6\tau_0 R_g^4} \sum_l \frac{1}{|\vec{s}_{kl}|^4}.$$

For z molecules per unit cell there are z/R_g^3 molecules per unit volume. Since on the other hand this number is equal to $N'c$ in a solution of concentration c, it follows that

$$R_g = \sqrt[3]{\frac{z}{N'c}}$$

and

$$D = \frac{s}{\tau_0} R_0^6 (N'c)^{4/3} \qquad (27)$$

where the dimensionless constant s is defined by

$$s = \frac{1}{6z^{4/3}} \sum_l \frac{1}{|\vec{s}_{kl}|^4}. \qquad (28)$$

This sum is to be extended over all lattice points. Its evaluation yields, in the following cases:

Simple cubic lattice .... $s = 2.76$
Face-centered cubic lattice .... $s = 2.66$
Body-centered cubic lattice .... $s = 2.66$
Diamond lattice .... $s = 3.02$

Naturally, these lattice types are unlikely to be realized as possible configurations of the molecules statistically distributed in solution, but at any rate they approximate the statistical distribution better than do less uniform distributions. Therefore, and because of the small variation in the numerical values of s, it appears justified to adopt Eq. (27), so far derived only for special regular lattices with an average value of s, also for the statistical molecular distribution of molecules in the solution. For further calculations the value $s = 2.81$ will be used.

In the framework of the continuum theory carried out here, Eq. (26) represents the decay of the excitation probability combined with its spatial equilibration. The term involving the decay can be eliminated by the substitution

$$\rho(\vec{R}, t) = \sigma(\vec{R}, t)e^{-t/\tau},$$

whereupon Eq. (26) is converted into

$$\frac{\partial \sigma}{\partial t} = D\vec{\nabla}^2 \sigma. \qquad (26')$$

This is identical to the well-known differential equation of heat conduction and diffusion. This means that the excitation probability spreads out among the molecules in solution according to the same formal that which govern these processes. Therefore the excitation energy diffuses over distances that are proportional to the square root of the time, as in the Einstein equation for the average displacement in the case of material particles. The quantity D has the dimensions of area/time, like the diffusion coefficient or the thermal conductivity, and is therefore directly comparable with these quantities.[43] According to Eq. (27) D is concentration dependent, and indeed the value corresponding to the critical concentration of Eq. (20) is

$$D_0 = \frac{9s}{16\pi^2} \frac{1}{\tau_0(N'c_0)^{2/3}} \sim \frac{0.159}{\tau_0(N'c_0)^{2/3}}.$$

For example, with the data for fluorescein $D_0 = 2.0 \times 10^{-5} \text{cm}^2\text{sec}^{-1}$. Even in water solution the molecular diffusion coefficient is smaller than this value (for the similar dye rhodamine B it has been measured[44] as $2.8 \times 10^{-6} \text{cm}^2\text{sec}^{-1}$), and the diffusion coefficients are naturally still smaller in viscous or, particularly, "solid" solvents such as glycerine or sugar, respectively. It is therefore understandable that for fluorescent dyes the energy migration considered here outweighs any convective migration due to the diffusion of excited molecules. On the contrary, for fluorescent materials of the quinine type, because of the higher critical concentration D_0 is substantially smaller and will be exceeded by the material diffusion constant in solvents of normal viscosity, so that there the convective transport of energy prevails.

The migration of the excitation energy away from the primary molecule will be represented by the well known source function from heat conduction theory:

$$\sigma(\vec{R}, t) = \left(\frac{1}{\sqrt{4\pi Dt}}\right)^3 e^{-R^2/4Dt} \qquad (|\vec{R}| = R).$$

The excitation probability $\sigma_1(t)$ of the primary molecule located at $R = 0$ is obtained by integration over its lattice cell, which may be replaced here by a sphere of equal volume whose radius is

$$R_\kappa = \sqrt[3]{\frac{3}{4\pi N'c}}.$$

Therefore

$$\sigma_1(t) = 4\pi \int_0^{R_\kappa} \sigma(\vec{R}, t)R^2 dR$$

$$= \frac{4}{\sqrt{\pi}} \int_0^{\xi_\kappa} e^{-\xi^2} \xi^2 d\xi$$

$$= \phi(\xi_\kappa) - \frac{2}{\sqrt{\pi}} \xi_\kappa e^{-\xi_\kappa^2},$$

Here $\phi(\xi_\kappa) = \frac{2}{\sqrt{\pi}} \int_0^{\xi_\kappa} e^{-\xi^2} d\xi$ is the Gauss error function and[44a]

$$\xi_\kappa = \frac{R_\kappa}{2\sqrt{Dt}} = \frac{\kappa}{2\gamma}\sqrt{\frac{\tau}{t}}.$$

In addition to the dimensionless concentration γ already defined in Eq. (22) the following dimensionless constant has been introduced:[44a]

$$\kappa = \frac{1}{\sqrt{s}}(4\pi/3)^{2/3} = 1.55.$$

The radiation probability of the primary molecule is therefore obtained by calculating

$$\eta_1 = \frac{1}{\tau_S} \int_0^\infty \rho_1(t)dt$$

$$= \frac{1}{\tau_S} \int_0^\infty \sigma_1(t)e^{-t/\tau}dt$$

$$= \frac{1}{\tau_S} \int_0^\infty [\phi(\xi_\kappa) - \frac{2}{\sqrt{\pi}} \xi_\kappa e^{-\xi_\kappa}]e^{-t/\tau}dt.$$

From Laplace transformation formulas (see, *e.g.*, Magnus-Oberhettinger, ref. 42, p. 128) and with the help of Eq. (1), one obtains

$$\frac{\eta_1}{\eta} = 1 - (1 + \frac{\kappa}{\gamma})e^{-\kappa/\gamma}. \qquad (29)$$

Because statistical fluctuations in the molecular separations have not been considered, Eq. (29) holds only for such concentrations at which energy migration over several molecules occurs ($\gamma \gg 1$). For $\gamma \sim 1$ energy migration is greater than it would be according to either of the two formulas. In Fig. 3 the dependence of the two functions on γ is shown.

§ 6. Concentration Depolarization

In the energy transfer calculation of § 3 the interaction energy of two molecules was replaced by its average u_{kl} over all orientations at a fixed distance. Strictly speaking, therefore, the theory holds only for low-viscosity solvents in which Brownian rotational motion of the fluorescent molecules is fast in comparison with the energy transfer. But a study of energy transfer based on concentration depolarization is possible only in viscous or solid solutions, since only there can polarized fluorescence be observed without perturbations from other processes. There the molecules are rigidly oriented, and energy transfer proceeds faster or slower than it would according to our theory, depending on whether the interaction energy lies above or below the average. Since in the case of concurrent processes the faster ones are favored, it is to be assumed that in general energy transfer goes somewhat faster than calculated here, but the deviation from the theory should hardly be considerable.

An exact treatment of concentration depolarization would require a vectorial theory, in which the individual components of the molecular transition moments are considered. Here we are interested less in the concentration depolarization as such than in its application to the study of energy transfer. A vectorial theory will therefore be relinquished and the scalar theory formulated above will be used for an approximate calculation of the concentration depolarization. This is possible because the radiation is maximally polarized when emitted from the set of primary molecules and considerably less polarized when emitted from the others. For simplicity it will be assumed that in the latter case it is unpolarized, which leads to an error partially compensated by the more rapid energy transfer between rigidly oriented molecules. Upon additively combining the degree of polarization from the contributions of the individual molecules, it is found that $p/p_{max} = \eta_1/\eta$, where p_{max} is the maximum degree of polarization in dilute solution and the other quantities have their previous meanings. A more exact calculation done by summing the contributions of individual molecules to the polarization components of the light[45] yields for polarized excitation and observation in the same or opposite direction

$$\frac{p}{p_{max}} = \frac{6\eta_1/\eta}{5 + \eta_1/\eta} \qquad (30)$$

For unpolarized excitation and observation in a perpendicular direction one obtains in the same way

$$\frac{p}{p_{max}} = \frac{9\eta_1/\eta}{10 - \eta_1/\eta}. \qquad (30')$$

In our approximation, the concentration dependence of η_1/η therefore yields that of the degree of polarization. When comparing with experiment it is to be noted, of course, that the dimensionless variable γ defined in Eq. (22) is proportional to the concentration only in the case of constant fluorescence yield. According to Fig. 1 this is the case at the onset of concentration depolarization, but when it has increased appreciably the yield drops. Although this so-called concentration quenching is closely connected with the process of energy migration and a theoretical treatment is equally accessible,[46] the connection between γ and the dye concentration will be represented here by the empirical data of Fig. 1. The critical concentration calculated in § 5 for a water solution, $c_0 = 3.2 \times 10^{-3}$ Mol/l, has been used since the corresponding data for glycerine are not at hand.

The contribution of the primary molecules to the radiation, calculated from Eq. (30) using the polarization data of Feofilov and Sveshnikov,[18] is reduced to a γ scale in the manner described above and compared with the theoretical results in Fig. 3. The points thus obtained agree well at low concentrations with curve *a* [Eq. (23)] and at high concentrations with curve *b* [Eq. (20)]. Thus the concentration dependences expressed by Eqs. (23) and (29), as well as the method of calculating the critical concentration c_0, are confirmed.

The polarization of the dyes eosin and rhodamine B show concentration dependence similar to that of fluorescein. As already noted, the calculated and measured critical concentrations are of the same order of magnitude. The polarization of trypaflavin, likewise measured by Feofilov and Sveshnikov, shows irregular behavior in that it rises again with increasing concentration after passing through a minimum at $p/p_{max} \sim 0.45$. This dye shows the further peculiarity that its concentration quenching sets in at considerably lower concentrations than in the case of the other dyes. Consequently the change in shape of the polarization curve due to the conversion from the concentration to the γ scale is especially great here. If one carries out the transformation on the basis of empirical quenching data, the minimum vanishes on the γ scale and there remains a monotonic dependence as in the case of the other dyes. The uniqueness of this dye, therefore, is due only to the quenching (whose explanation cannot be discussed here), which sets in at low concentrations. The polarization minimum occurs because above a certain concentration the reduction of the excitation lifetime because of quenching outweighs the faster energy transfer, so that the energy transfer from the primary molecule decreases again after passing through a maximum.

§ 7.   Energy Migration at High Concentrations

Above the critical concentration c_0 the fraction of energy remaining on the primary molecule is given by Eq. (29). At very high concentration this goes over into

$$\frac{\eta_1}{\eta}\frac{\kappa^2}{2\gamma^2} = \frac{\kappa^2}{2}\frac{\eta_0}{\eta}\frac{c_0^2}{c^2}. \qquad \left(\frac{\kappa^2}{2} \sim 1.20\right) \qquad (29')$$

In the case of constant fluorescence yield ($\eta = \eta_0$) the fraction of energy radiating from the primary molecules therefore drops off as the inverse square of the concentration. Whereas in this kind of transfer process none of the other radiating molecules has any larger rate of radiation than that of the primary molecule, their number increases as the square of the concentration. At one hundred times the critical concentration this is something like 10^4 molecules (always assuming $\eta = \eta_0$).

Now as we have already noted, quenching occurs in fluorescent dye solutions at higher concentrations, the fluorescence yield rapidly dropping. Through the concomitant shortening of the excitation lifetime this can, as in the case of trypaflavin, make the [relative] radiation probability of the primary molecule rise again. Concentration quenching therefore prevents energy transfer of the kind we are considering to a large number of molecules.

Generally concentration quenching is in no way forced to be tied to energy transfer. Moreover, as we will discuss below, it is by all means possible to have cases in which no quenching occurs at the highest concentrations. Here the broader assumptions of the theory will be examined to ascertain their limitations at high concentrations.

At concentrations so high that the molecular separations are comparable with the extent of the electronic systems of the molecules, the interaction energy can no longer be represented as that of two dipoles [Eq. (10)]. In this case energy transfer occurs by the same mechanism as in dilute solutions, but the quantitative formulation of the theory is no longer valid.

A further limitation is due to the assumption of thermal equilibrium in the vibrational motion. If the theory predicts a transfer time of the order of magnitude of the equilibrium time, it is no longer valid. This time interval is that of a few molecular collisions in the liquid, and therefore of the order of 10^{-13} sec. At yet stronger interactions the energy can finally transfer in times shorter than a molecular vibration period (3×10^{-14} sec). An excited molecule therefore gives up the energy it received before it has returned to the nuclear configuration of the unexcited state. In this case the structure of the absorption spectrum is determined, not (as in the other case) by the coupling between electron and nuclear motion, but rather by the coupling of the electron systems of neighboring molecules with one another. The excitation process itself is essentially shared with the neighboring molecules, and it is more correct to attribute the excitation energy to the whole system of molecules than to individual molecules. Theoretically this is the case treated by Frenkel,[47,48] Peierls,[49] and Franck and Teller,[50] and it will not be discussed further here. It appears to be realized in Scheibe dye polymers.

As noted before, energy migration in molecular solutions of fluorescent dyes is limited by concentration quenching. E.g., in the case of fluorescein in glycerine the fluorescence yield is about 20% of the maximum when the concentration is about ten times the critical concentration. Consequently, according to Eq. (29') energy migration should occur over about 100 molecules but it appears to do so over about 20, which also agrees with the depolarization measurements. In the region below this concentration, where alone the theory is of interest, its assumptions seem to be valid. The average intermolecular distance is still large compared with molecular dimensions. Since for an excitation lifetime of about 10^{-8} sec the duration of single transfers is no less than about 10^{-10} sec, the assumption of thermal equilibrium seems to be a good one.

The concentration quenching, which annihilates the excitation energy in highly concentrated dye solutions, appears to be by no means necessarily associated with energy transfer. For example, organic compounds such as anthracene are capable of fluorescing in the crystalline state, while at the same time high-concentration energy migration effects are seen distinctly (e.g., in anthracene with naphthacene impurities, according to the observations of Winterstein and Schön,[9] Bowen,[10] and Weigert[11]). On the other hand, as it has been demonstrated amply,[24] concentration quenching can be traced to a more or less strong association with the presence of dimer molecules that annihilate the incoming radiation or the migrating excitation energy. It is understandable that such dimer molecules might occur in solution but not in ordered crystals.

According to this conception of the process, concentration quenching should be absent in quasi-crystalline structures in which the absorbing molecules are fixed in definite places. Our present conception of the structure of the photosynthetic apparatus of plants (cf., e.g., Frey-Wissling[51]), and of biological material in general, leads us to assume that the chlorophyll molecules are embedded in some kind of regular way in their carrier substance, which excludes association into dimers. In fact the native chlorophyll reradiates at least a part of the energy not used in photosynthesis. Its local concentration amounts to about 0.1 Mol/l, or one hundred times critical, so that according to Eq. (29') energy migration over about 10 molecules is possible if the quantitative predictions of the theory are still valid in this extreme case. A discussion of the assimilation process from this standpoint has been carried out in another publication.[52]

Niedernjesa, Kr. Göttingen
(Received by the editor 5 May 1947)

Acknowledgements

Thanks are due to many colleagues for pointing out errors of various kinds in the informal editions of the translation. The translator is especially indebted to Prof. Aadne Ore, whose keen eye located several dozen.

This translation was prepared in early 1973 in order to make more readily available Förster's original elegant treatment of excitation transfer and diffusion. Indeed, his own excellent reviews in English tended to eclipse the original article and some of its unique features. Prof. Förster received copies and wrote, approving both its content and the plan to make it available, just weeks before his untimely death in May 1974. The translation is dedicated to the memory of this excpetional scientist and gentleman.

> – Robert S. Knox, Department of Physics and Astronomy, University of Rochester, Rochester, NY 14627.

References

1. This is the case of so-called collisions of the second kind.
2. J. Weiss, Nature **141**, 248 (1938).
3. J. Weiss, Trans. Faraday Soc. **35**, 48 (1939).
4. J. Franck, Z. Physik **9**, 859 (19-22).
5. G. Cario, Z. Physik **10**, 185 (1922).
6. G. Cario and J. Franck, Z. Physik **17**, 202 (1923).
7. N. Riehl, Naturwiss. **28**, 601 (1946).
8. R. W. Gurney and N. F. Mott, Proc. Roy. Soc. (London) **164**, 151 (1938).
9. A. Winterstein and K. Schön, Naturwiss. **22**, 237 (1934).
10. E. J. Bowen, Nature **142**, 1081 (1938).
10a. E. J. Bowen, Nature **159**, 706 (1947).
11. F. Weigert, Trans. Faraday Soc. (London) **36**, 1033 (1940).
12. G. Scheibe, A. Schöntag, and F. Katheder, Naturwiss. **27**, 499 (1939).
13. H. Gaffron and K. Wohl, Naturwiss. **24**, 81, 103 (1936).
14. F. Möglich, R. Rompe, and H. W. Timofeev-Ressovsky, Naturwiss. **30**, 409 (1941).
15. F. Weigert, Verh. deutsch. physik. Ges. **23**, 100 (1920).
16. E. Gaviola and P. Pringsheim, Z. Physik **24**, 24 (1924).
17. F. Weigert and G. Käppler, Z. Physik **25**, 99 (1924).
18. P. P. Feofilov and B. J. Sveshnikov, J. Phys. USSR **3**, 493 (1940).
19. This asymptotic value, here, is 40%.
20. J. Perrin, 2me Conseil de Chimie Solvay, Bruxelles (1924); Gauther-Villars Paris (1925), p. 322. [The years on these two references probably should be 1925 and 1926.]
21. J. Perrin, C. R. Acad. Sci. (Paris) **184**, 1097 (1927).
22. F. Perrin, Ann. Chim. Physique **17**, 283 (1932).
23. H. Kallmann and F. London, Z. Physik. Chem. (B) **2**, 207 (1928).
24. Th. Förster, Naturwiss. **33**, 166 (1946).
25. Meanwhile the author has learned that Vavilov has considered the problem at hand fully in several earlier publications.[26-29] Insofar as it can be seen from those references that are accessible, his treatment is only in partial agreement with the present one in method and in results. Publication of this paper therefore seems justifiable to the author.
26. S. I. Vavilov and P. P. Feofilov, C. R. USSR **34**, 220 (1942).
27. S. I. Vavilov, J. Phys. USSR **7**, 141 (1943); Z. R. USSR **42**, 331 and **45**, 47 (1944).
28. S. I. Vavilov, C. R. USSR **42**, 331 (1944).
29. S. I. Vavilov, C. R. USSR **45**, 47 (1944).
30. P. A. M. Dirac, Proc. Roy. Soc. (London) **A144**, 259 (1927).
31. In the case of large organic molecules the distinction between this condition and the preceding one is essential.
31a. Translator's note: vector notation in current use has been introduced.
31b. Corrections to the original equation are due to Prof. A. Ore.
32. The form of this relation differs from the usual one only in that the energy of the photons instead of their frequency appears, and in that it relates to a continuous spectrum and a medium of refractive index n.
33. This constant appears herein place of Avogadro's number because the usual chemical concentration unit Mol/liter or mMol/cm^3 has been adopted.
33a. Translator's note: starting at Eq. (14) and ending at the expression for $1/\tau_S$, several missing subscripts "zero" have been added to the original text.
34. V. L. Levschin, Z. Physik **72**, 368 (1931).
35. W. Szymonowski, Z. Physik **95**, 440 (1935).
36. F. Perrin, Ann. Chim. Physique **12**, 169 (1929).
37. Ch. Dhéré, C. R. Acad. sci. (Paris) **158**, 64 (1914).
38. A. Sprecher von Bernegg, E. Heierle, and F. Almasy, Biochem. Z. **283**, 45 (1935).
38a. Translator's note: More recent measurements show that the fluorescence lifetime of chlorophyll a under these conditions is 5×10^{-9} sec, six times smaller, so Förster's estimate of R_0 must be reduced by a factor of $\sqrt[6]{6}$ to about 60 A. For a discussion, see L. N. M. Duysens, Prog. Biophys. **14**, 1 (1964), p. 48.
39. Absorption spectrum: Fischer,[40] *cf.* also Landolt-Bornstein tables, II. Suppl. Vol., p. 673; fluorescence spectrum: Kortüm and Finkh.[41]
40. H. Fischer, thesis, Zurich (1925).
41. G. Kortüm and B. Finkh, Spectrochim. Acta **2**, 138 (1944).
42. The integration is done by using a simple transformation given by W. Magnus and F. Oberhettinger, *Formeln und Sätze für die speziellen Funktionen der Mathematischen Physik* (Springer, Berlin, 1943), p. 124.
42a. Translator's note: To conform with current usage, and thereby to emphasize the historical importance

of this paper, we have chosen to use D instead of Λ for exciton diffusion constants. There is little likelihood that this D will become confused with the functional determinant of section 3. Furthermore, we use vector notation for the gradient and Laplacian operators.

43. The analogy between an excited state and a material particle, which Frenkel underscored by the introduction of the term "exciton," is also brought out here. D can be interpreted as the diffusion coefficient of such an exciton.

44. S. I. Vavilov, Z. Physik **31**, 750 (1925).

44a. Translator's note: minor corrections to the original equations have been made (the orignal ξ_κ contained an extra factor $2\eta_0/\eta$ and κ contained an extra $\sqrt{1/2}$). In η_1 τ_0 has been replaced by τ_S.

45. *Cf.* the calculations of F. Perrin.[36]

46. The path toward this has been pointed out in an earlier publication (Förster[24]).

47. J. Frenkel, Phys. Rev. **37**, 17, 1276 (1931).

48. J. Frenkel, Phys. Z. Sowjet. **9**, 158 (1936).

49. R. Peierls, Ann. Physik (5) **13**, 905 (1932).

50. J. Franck and E. Teller, J. Chem. Phys. **6**, 861 (1938).

51. A. Frey-Wissling, Protoplasms **29**, 279 (1938).

52. Th. Förster, Z. Naturforsch. **2b**, (1947), to appear. Translator's note: this was probably meant to refer to the subsequent article "Experimentelle und theoretische Untersuchung des Zwischenmolekularen Übergangs von Elektronenanregungsenergie," Z. Naturforsch. **4a**, 321 (1949).

EXCITON MIGRATION AND TRAPPING IN PHOTOSYNTHESIS

Robert M. Pearlstein

Battelle Memorial Institute, Columbus Laboratories, Columbus, OH 43201, USA

(*Received* 23 *October* 1981; *accepted* 26 *January* 1982)

Abstract—Mobile electronic excited states, excitons, undergo random walks through the antenna chlorophyll arrays of photosynthetic organisms. The time interval from exciton creation, by photon absorption, until its first arrival at a reaction center (RC) is called the "first passage time" (FPT) of the random walk. A theory of exciton migration and trapping presented here predicts that the exciton lifetime, as measured from chlorophyll fluorescence decay in chromatophores or P700 complexes, is a linear function of the fractional number of quanta absorbed directly by the antenna, not by the RC. The slope of this line is the FPT, and its intercept is the exciton's lifetime as limited only by photoconversion at the RC. This photoconversion-limited lifetime is simply related to the *in situ* photoconversion rate constant via two parameters, each of which is experimentally accessible. It is also possible to obtain values of individual Förster rate constants, at least approximately, from measurements of exciton lifetime as functions of temperature and excitation wavelength. This new theory, based on lattice random walk models, receives some support from fluorescence measurements done on *Rhodopseudomonas sphaeroides* R26 chromatophores. In its present form the theory is only applicable to one-antenna-component systems, like *Rp. sphaeroides* R26 or *Rhodospirillum rubrum* chromatophores or P700 complexes, but should be readily extendible to multi-antenna-component systems including whole chloroplasts.

I. INTRODUCTION

In photosynthetic organisms, almost all absorbed photons create mobile electronic excited singlet (S_1)* states in the arrays of antenna chlorophyll (Chl) or bacteriochlorophyll (Bchl) (Knox, 1975, 1977; Sauer, 1975, 1978; Pearlstein, 1982). These mobile states, called "excitons", are believed to migrate to the photochemically active reaction centers (RC's) by random walks over the antenna, with excitation energy sequentially "hopping" from Chl to Chl (Knox, 1975; Pearlstein, 1982). Exciton migration is thus the usual precursor of photochemistry in photosynthesis.

Despite its seeming universality and importance in photosynthetic light harvesting, exciton migration and trapping (i.e. conversion to a charge-separated state) at the RC is not completely understood. This is not for a lack of theoretical treatments. Beginning with the work by Bay and Pearlstein (1963), many theories of the kinetics of exciton migration and trapping have been put forward (see Knox, 1975, 1977; Pearlstein, 1982, for reviews), but their practical value for the interpretation of experimental results has been quite limited. Either these theories involve numerical calculations that give little insight into the functional dependence of experimental quantities on exciton parameters (Robinson, 1967; Altman *et al.*, 1979; Shipman, 1980), or they analytically derive equations some of whose parameters appear to be experimentally

inaccessible (Pearlstein, 1967; Montroll, 1969; Hemenger *et al.*, 1972).

In this paper, I describe an *analytical* theory (i.e. one in which parametric dependences are explicit) of the kinetics of exciton migration and trapping, together with prescriptions for relating theoretical parameters to experimentally accessible quantities. The theory is based on that of Hemenger *et al.* (1972), but extends it in three ways. First, as noted, is a closer coupling to experiment. Second, the results are described in a way that is immediately more useful to experimentalists. Third, the single most important equation, which expresses exciton lifetime in terms of energy transfer rate constants and other parameters, is generalized in a way that is particularly useful in photosynthesis research. The generalized equation predicts a linear dependence of exciton lifetime on that fraction of absorbed quanta that is not absorbed directly by the RC. Previous theoretical treatments consider only a single, fixed partitioning of initial excitation between antenna and RC, and thus overlook the possibility of an exciton-lifetime dependence on that fraction. This new prediction allows a straightforward experimental test of the theory. If validated for a particular photosynthetic system, the theory can then immediately be used to derive such interesting quantities as the *in situ* rate of S_1 conversion at the RC, and the "first-passage-time" of the exciton random walk (see below), directly from experimental data on that system.

The interesting phenomena of exciton collisions and mutual annihilations, both singlet–singlet (Breton and Geacintov, 1980) and singlet–triplet (Monger and Parson, 1977), are *not* covered here. The present

---

*Abbreviations: Bchl, bacteriochlorophyll; Chl, chlorophyll; FPT, first passage time; HPL theory, theory due to Hemenger *et al.* (1972); RC, reaction center; S_0, S_1 and S_n, singlet states.

theory is thus applicable only in the limit of low exciting light intensity.

Most photosynthetic organisms have two or more antenna pigment types (*Rhodospirillum rubum* and *Rhodopseudomonas sphaeroides* R26 are exceptions), which differ either chemically or as a presumed result of differing physical interactions with their lipoprotein environments (Thornber *et al.*, 1979; Cogdell and Thornber, 1980). For each organism, it is useful to distinguish that class of antenna pigments, always a Chl or a Bchl, whose $S_0 \rightarrow S_1$ transition energy lies lowest. Excitons within that class, called here the *proximal* antenna, migrate directly to RC's, hence are the main topic of this paper. Excitations of other classes, loosely referred to here as "accessory pigments" whether Chl or not, are not considered unless noted otherwise.

In Section II, I review general theoretical concepts, including Förster rate constants, master equation, kinetic model of antenna–RC coupling, and random walk aspects. Section III explains and extends specifically the theory of Hemenger *et al.* (1972). Section IV describes how parameters that appear in the theoretical expressions of Section III can be derived from experimental quantities, largely without *ad hoc* assumptions. In Section V, the theory is applied to the analysis of fluorescence experiments. The final Section proposes new experimental tests, and suggests possibly useful extensions of the theory.

II. THEORETICAL CONCEPTS

The excitons are assumed to be of Frenkel-type, i.e. no charge separation occurs in the antenna (cf. Mauzerall, 1982). Possible coherent, or wavelike, aspects of the exciton motion are ignored. Although substantial coherent motion would probably alter the exciton trapping kinetics (Pearlstein, 1968; Hemenger *et al.*, 1974), its effects are thought to be slight in photosynthetic antennae (Knox, 1977; Pearlstein, 1982). In the absence of coherence, the exciton motion can be described purely in terms of singlet excitation hopping from Chl to Chl.

The hopping motion is induced by the interactions of the $S_0 \rightarrow S_1$ transition dipole moments of the Chl's*. The rate constant F_{ij} for intermolecular excitation transfer between antenna Chl molecules i and j is given by Förster's formula (Förster, 1965)

$$F_{ij} = C\kappa_{ij}^2 I/(\tau_0 n^4 R_{ij}^6) \qquad (1)$$

In Eq. 1, κ_{ij}^2 is a factor that depends on the mutual angular geometry of the two transition dipoles, R_{ij} is the distance between them, n is the refractive index of the medium (assumed constant

---

*If some of the Chl's are closely juxtaposed, as in the same protein complex, it may be thought essential to use transition monopoles (Weiss, 1972) in any calculation. However, to a good approximation this appears to be unnecessary (Pearlstein and Hemenger, 1978).

over the antenna), and τ_0 is the Chl radiative lifetime. The overlap integral, I, is

$$I = \int \mathscr{F}(v)\epsilon(v) \, dv/v^4 \qquad (2)$$

where $\epsilon(v)$ is the antenna Chl molar extinction coefficient on a wave number scale, and $\mathscr{F}(v)$ is the corresponding fluorescence intensity normalized so that $\int \mathscr{F}(v) \, dv = 1$. For R_{ij} in nm and τ_0 in s, F_{ij} is in Hz if the constant, C is 8.785×10^{17}.

In terms of the rate constants F_{ij}, one can write down a set of coupled rate equations (collectively called a "master" equation; see Knox, 1975, 1977; Pearlstein, 1982) for the probabilities as functions of time that each antenna Chl molecule is in the S_1 state (i.e. the exciton "resides" on that Chl). If the S_1 state did not decay, i.e. had an infinite lifetime, the master equation would be

$$\dot{\rho}_i = \sum_j F_{ij}(\rho_j - \rho_i) \qquad (3)$$

In Eq. 3, $\rho_i(t)$ is the probability that the ith antenna Chl molecule is in the S_1 state at time t, $\dot{\rho}_i = d\rho_i/dt$, and the sum is taken over all antenna Chl's (cf. Section III). If each antenna Chl molecule has the same decay channels for the S_1 state's energy (e.g. radiation, intersystem crossing, internal conversion), with K_A = sum of rate constants for all such channels, the master equation becomes

$$\dot{\rho}_i = \sum_j F_{ij}(\rho_j - \rho_i) - K_A\rho_i. \qquad (4)$$

Both Eqs. 3 and 4 describe the time-dependent redistribution of the excitation probabilities from some initial condition. According to these equations, the exciton motion is very much like that of a diffusing or randomly walking particle. Interpretations of certain quantities in terms of random walk concepts are given below.

Neither Eq. 3 nor Eq. 4 explicitly includes the effects of RC's on the exciton motion. Hemenger *et al.* (1972) introduced a master equation that does this in a quite general way. In the next Section, I discuss their equation and its treatment. Here, I describe their kinetic model of antenna–RC coupling in photosynthetic terms. As Fig. 1 shows, the coupling involves three types of rate constants in addition to those of Eq. 4. Two of these, the trapping and de-trapping rate constants F_{Ti} and F_{Di}, are given by Förster's formula, Eq. 1, with parameters appropriate to RC Chl (as singlet excitation acceptor for F_{Ti}, or as singlet excitation donor for F_{Di}). The third rate constant, k_p, is that for conversion of the RC singlet to the radical pair state. (Strictly speaking, there is a fourth rate constant, K_{RC}. Usually, however, $K_{RC} \sim K_A$, and need not be considered separately.) Without k_p, excitation is more likely to be found at the RC if $F_{Ti} > F_{Di}$ for all neighbors, but is not permanently trapped there unless all of the F_{Di}'s are negligible. In general, it is the large

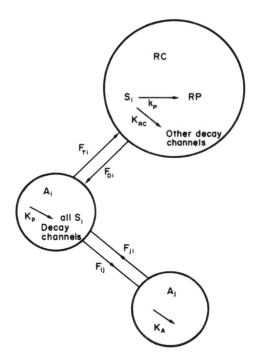

Figure 1. Schematic diagram showing the kinetic coupling of antenna and RC for exciton transport and trapping. For simplicity, only one RC, one nearest-neighboring antenna Chl (A_i), and one more distant antenna Chl (A_j) are shown. Rate processes are denoted by arrows with labelled rate constants. Those inside circles are for decay of the Chl singlet state, with k_p the rate constant for photochemical conversion to the radical-pair (RP) state. $F_{ij} = F_{ji}$ is the reversible Förster rate constant for excitation transfer between antenna Chl's i and j, F_{Ti} is the Förster rate constant for transfer from a nearest-neighboring antenna Chl to the RC (trapping rate constant), and F_{Di} is the Förster rate constant for transfer from the RC back to that same neighboring antenna Chl (detrapping rate constant).

magnitude of k_p that "traps" the exciton at the RC by providing an especially rapid decay channel for the Chl S_1 state only at that site. How the three RC kinetic parameters are included in a master equation is discussed in Section III.

A particularly interesting quantity, proportional to the instantaneous Chl fluorescence intensity, is defined by

$$P(t) \equiv \sum_i \rho_i(t) \qquad (5)$$

$P(t)$ is the probability that the exciton still exists, i.e. has not yet decayed, at time t; $P(0) = 1$. Because the antenna rate constants are reversible ($F_{ij} = F_{ji}$), it follows from summing Eq. 3 on i that $P(t) = 1$ for all time, i.e. with no decay channels there is no decay. Similarly, from Eq. 4 one has $P(t) = \exp(-K_A t)$, i.e. a set of decay channels that is the same for all the Chl's leads to purely exponential decay at a rate independent of all exciton parameters. In the presence of RC's with large k_p's, however, it is not as easy to write down the analytic form of $P(t)$, although in many cases it is well approximated as a single exponential

(Pearlstein, 1982). Situations in which $P(t)$ deviates from monoexponentiality are described in Section III in the context of regular lattices. A cause of nonexponentiality in any type of array is exciton–exciton annihilation which requires high exciting-light intensities and introduces second-order exciton decay kinetics (see Breton and Geacintov, 1980, for a review). At low exciton densities, it is known theoretically that $P(t)$ would be nonexponential in a random array of pigment molecules, for example solute molecules in a solvent, provided the number of excitation donor molecules is much less than the number of acceptors (Förster, 1949; Wolber and Hudson, 1979). However, the shape of $P(t)$ is unknown for a random array in which the number of donors is much greater than the number of acceptors, the situation in a photosynthetic system where the donors are the antenna Chl's and the acceptors are the RC's. These sources of nonexponentiality are not considered further here.

The quantity $P(t)$ contains less information than the set of quantities $\rho_i(t)$, but it is in principle easier to calculate and is also readily accessible experimentally. An even easier quantity to calculate, although still less informative, is the lowest moment of $P(t)$

$$M_0 \equiv \int_0^\infty P(t)\,dt \qquad (6)$$

M_0, which obviously has units of time, is of interest because it is a measure of the exciton lifetime. If $P(t)$ is exponential, M_0 is just the "$1/e$" decay time. If the photochemical channel is the most important decay route for the exciton, the case of greatest interest, M_0 is the mean duration of the exciton random walk regardless of the shape of $P(t)$. For exponential $P(t)$

$$M_0^{-1}(K_A) = M_0^{-1}(0) + K_A \qquad (7)$$

The quantity $M_0(0)$ is the mean duration of an exciton random walk that is terminated only by trapping.

In random walk terminology (Montroll, 1964), the interval from exciton creation until it first reaches the RC is called the "first passage time" (FPT). It may happen that the exciton walks away again, i.e. is not trapped on its first "visit", and so must make multiple visits. However, it is not necessary to know the number of visits in order to determine $M_0(0)$. The formalism of Hemenger et al. (1972) gives expressions for both $M_0(0)$ and FPT without any assumptions regarding number of visits which, in any case, is not a readily accessible quantity experimentally.

III. LATTICE THEORY

In the remainder of this paper, I assume that the RC's and the individual antenna Chl molecules occupy the sites of a regular lattice, not necessarily a simple one. There are several justifications of this assumption. First, there is experimental evidence for two-dimensional crystalline arrays of membrane proteins in thylakoids (Park and Biggins, 1964) and of antenna Bchl a–protein in green photosynthetic bacteria (Staehelin *et al.*, 1980). Second, it can be argued that regular arrays of this sort have value for the organism. They ensure that excitation can leave the vicinity of a closed (oxidized) RC as rapidly as it approaches an open one; certain types of irregularities may not provide this advantage. (This argument does not hold for accessory pigments, which are not considered here.) Finally, as I show here, lattice theory makes definite predictions that are readily tested experimentally.

The first part of this section sets forth the main conclusions of Hemenger *et al.* (HPL theory) in terms of the notation used in Section II. The second part describes a mathematically straightforward, but novel, extension of HPL theory that predicts a linear dependence of $M_0(0)$, defined in Eq. 7, on the fraction of absorbed quanta that is absorbed directly by RC's.

(A) HPL theory

The HPL master equation (Hemenger *et al.*, 1972), which generalizes Eq. 3 to include exciton trapping by the RC, is

$$\dot{\rho}_i = \sum_j F_{ij}(\rho_j - \rho_i) + \sum_j S_{ij}\rho_j \qquad (8)$$

Here, F_{ij} is defined as in Eq. 3, except when lattice site i or j is occupied by a RC; in that case, F_{ij} is to be calculated as if both sites were occupied by antenna Chl's. The rate constants S_{ij} are defined to have the following properties: If $i \neq j$, $F_{ij} + S_{ij}$ is the actual Förster rate constant (not necessarily reversible) from site j to site i, even if one of the sites is occupied by a RC. The quantity

$$\left(-S_{ii} + \sum_{j \neq 1} F_{ij} \right)$$

is the total rate constant with which excitation leaves site i (antenna or RC). This somewhat artificial separation of rate constants between the F_{ij} and the S_{ij} is merely a bookkeeping device that simplifies analysis of Eq. 8.

For a lattice of N sites with a single RC at the origin of coordinates ($i = 0$), retaining only terms corresponding to interactions of the RC with its nearest-neighboring antenna Chl's (likewise for antenna Chl–antenna Chl interactions), HPL show that

$$S_{ij} = (F_T - F_A)(\delta_{i0} - \delta_{jm}) + (F_D - F_A)\delta_{im}\delta_{j0}$$
$$- [k_p + q(F_D - F_A)]\delta_{i0}\delta_{j0} \qquad (9)$$

In Eq. 9, F_A is the reversible Förster rate constant between nearest-neighboring antenna Chl's, F_T and F_D are, respectively, the trapping and de-trapping

---

*HPL define the dimensionless Laplace transform of a function of time, $y(t)$, by

$$\mathscr{L}\{y\} \equiv F_A \int_0^\infty e^{-sF_A t} y(t)\,dt$$

The function $y(t)$ can be, for example, $\rho_i(t)$, or the Green's function.

rate constants, and k_p is the photoconversion rate constant. For simplicity, isotropic exciton migration and trapping has been assumed, but this condition can be relaxed (Hemenger *et al.*, 1972). Other symbols used in Eq. 9 include q, the lattice coordination number (i.e. the number of nearest neighbors of any site); the Kronecker δ's ($\delta_{ij} = 1$ if $i = j$, $= 0$ otherwise); and the subscript m, which designates any site that is a nearest neighbor of $i = 0$. The restriction to nearest-neighbor interactions is not a serious one because of the inverse sixth power dependence on intermolecular distance of the Förster rate constants—Eq. 1. (See also the discussion below for the special case $F_D = F_T = F_A$.) Equations 8 and 9 as written do not include the rate constant K_A of Eq. 4. However, as discussed in Section IV, the effect of K_A either can be included via Eq. 7 or is likely to be negligible.

The two most interesting quantities that can be obtained from the solution of Eq. 8 are $P(t)$, defined by Eq. 5, and $\rho_0(t)$, the time-dependent probability that the exciton resides on the RC itself. In fact, these two are quite generally related by

$$P(t) = 1 - k_p \int_0^t \rho_0(u)\,du \qquad (10)$$

so that it is only necessary to solve Eq. 8 for $\rho_0(t)$ to obtain both. HPL give a formal solution of Eq. 8, with the S_{ij} defined by Eq. 9, for $\rho_0(t)$ in terms of lattice Green's functions. In principle, the Green's functions, which are the solutions of a simpler master equation (Hemenger *et al.*, 1972), are known for a variety of lattices. In practice, for two- and three-dimensional lattices, simple analytic expressions for them are not available. For this reason, HPL do not provide explicit formulae for $\rho_0(t)$ or $P(t)$. However, as HPL show, both M_0, the lowest moment of $P(t)$ from Eq. 6, and the extent to which $P(t)$ deviates from exponentiality, can nonetheless be analytically derived. The reason is that both M_0 and a measure of exponentiality depend only on the behavior of the Laplace-transformed Green's functions near $s = 0$, where s is the transform variable*. For small s, the transformed Green's function, $g_{ij}(s) \simeq -(1/Ns) + f_{ij}$, where the f_{ij}'s are calculable constants that depend on N and lattice structure. In terms of these constants

$$M_0 = [1 + (F_D/F_T)(N - 1)]k_p^{-1}$$
$$+ \left\{ [1 - \rho_0(0)](N - 1)[(qF_T)^{-1} - (qF_A)^{-1}] \right.$$
$$\left. + \left[-f_{00} + \sum_i f_{0i}\rho_i(0) \right] NF_A^{-1} \right\} \qquad (11)$$

For the special case that $F_D = F_T = F_A$, Eq. 11 reduces to

$$M_0 = Nk_p^{-1} + \left[-f_{00} + \sum_i f_{0i}\rho_i(0) \right] NF_A^{-1} \qquad (12)$$

It is easy to see from Eq. 12 that exciton trapping and photoconversion can be described as an additive sequence of events. The term in F_A^{-1} is the time required for the initial random walk to the RC, the first-passage-time; the term in k_p^{-1} is the time required for photoconversion at the RC. In what is called the "diffusion-limited" case, values of parameters are such that the F_A-term dominates; the reverse is true in the "trap-limited" case. In diffusion-limited exciton migration and trapping the random walk terminates essentially on the first visit to the RC*. It is important to note that *both* terms scale with N ($N - 1$ = the antenna/RC ratio, i.e. photosynthetic unit size). The at-first-sight surprising (although rigorous) result that M_0 scales with N, even when the exciton starts at the RC itself, and even in the diffusion limit, can be understood heuristically as follows. If the rate constant k_p is not infinite, the exciton has a finite probability to escape the RC before photoconversion. Once it escapes, it executes a random walk whose average length before returning to the RC is exactly N steps, regardless of lattice geometry or random-walk step-time (Montroll, 1964). Thus, if the exciton initially starts at the RC, it spends on the average only $(1/N)$th of its time there, where it is quenched with (photoconversion) rate-constant k_p. As a result, its lifetime is just $(k_p/N)^{-1}$, the value given by Eq. 12. If the exciton initially starts somewhere other than the RC, its first-passage-time to the RC must be added to the total time required for quenching (in the sense of Eq. 6), which is the result expressed by Eq. 12. Similar remarks can be made about Eq. 11, where the first-passage-time is given by the terms in braces.

Perhaps the most interesting remark to be made about Eq. 12 is that HPL obtain exactly the same formal expression for M_0 (in the special case $F_D = F_T = F_A$) when the restriction to nearest-neighbor interactions among antenna chlorophylls is eliminated. Although the values of the f_{oi}'s differ somewhat without the nearest-neighbor restriction, dependence of M_0 on parameters is unchanged. This reinforces the assertion made earlier that the nearest-neighbor restriction is not a serious limitation of the theory.

The quantity $\rho_i(0)$ in Eq. 11 is the probability that the exciton is on site i at $t = 0$. (Experimentally, establishing any initial condition requires time and wavelength selectivity; see Section IV.) HPL considered explicitly two different initial conditions, each of which is physically realizable in principle. The "uniform" initial condition is one in which each site is equally likely to be excited initially, i.e. $\rho_i(0) = 1/N$ independent of i. In this case, because $\Sigma_i f_{oi} = 0$, Eq. 11 reduces to

$$M_0 = [1 + (F_D/F_T)(N - 1)]k_p^{-1}$$
$$+ [(qF_T)^{-1} - (qF_A)^{-1}][(N - 1)^2/N]$$
$$+ \alpha N F_A^{-1} \qquad (13)$$

where $\alpha \equiv -f_{00}$ ($f_{00} < 0$). Under the "impurity" initial condition (so-called because the RC can be viewed as an impurity in an otherwise perfect lattice of antenna Chl's), $\rho_i(0) = \delta_{io}$, i.e. only the RC is excited initially. In this case, Eq. 11 reduces to

$$M_0 = [1 + (F_D/F_T)(N - 1)]k_p^{-1} \qquad (14)$$

This result shows that if the exciton starts at the trap, its lifetime is exactly that found in the purely trap-limited case (with arbitrary initial condition), independent of antenna parameters; the first-passage-time is necessarily zero. In Part (B) of this Section, I generalize the results expressed by Eqs. 13 and 14 in a way that lends them much greater usefulness experimentally.

Up to this point, I have described a lattice of N sites with a single RC at the origin. HPL note that this is completely equivalent to an infinite lattice that has one RC per N antenna Chl's, provided that the RC's occur periodically, i.e. form a "superlattice". [Experimentally, this corresponds to the equivalence of "puddle" and "lake" models of the photosynthetic unit (see Robinson, 1967) at vanishing light intensity.] HPL also show that for values of N ($\gtrsim 30$) relevant to photosynthetic organisms, corrections to the foregoing M_0-formulae for randomly rather than periodically placed RC's are negligible. The shape of $P(t)$ is similarly insensitive to the relative placement of RC's.

Calculating the detailed shape of $P(t)$ is more difficult than calculating M_0 (Hemenger *et al.*, 1972). For hopping transfer and quenching on two- or three-dimensional lattices, with uniform initial excitation (all exciton starting sites equally likely), $P(t)$ is well-approximated as a single exponential, $\exp(-\gamma_0 t)$, where $\gamma_0 \simeq M_0^{-1}$. This single (lowest or zero) characteristic diffusion mode is not necessarily so dominant with other initial conditions. For example, it breaks down completely if initially only the RC itself is excited, provided that the exciton quenching is diffusion limited. Even with uniform initial excitation, higher characteristic diffusion modes can contribute significantly if there is long-range quenching or slow back transfer from the RC. (By "long-range quenching" is meant a situation in which exciton hopping into the RC from its non-nearest neighbors competes with nearest neighbor hops between antenna Chls. "Slow back transfer" signifies $F_T \gg F_D \sim k_p \sim \gamma_0$.) Neither of these conditions, i.e. long-range quenching

---

*An expression for the average number of visits in terms of HPL parameters can be developed as follows: from Eq. 11, the time required for all visits after the first one is $[1 + (F_D/F_T)(N - 1)]k_p^{-1}$. The average number of steps in a random walk that begins *and* ends at the RC is exactly N (Montroll, 1964), and the time per step of the walk is $(qF_A)^{-1}$ (Bay and Pearlstein, 1963; Hemenger *et al.*, 1972). Thus, the time between consecutive visits is $N(qF_A)^{-1}$, and the *total* number of visits is $1 + [1 + (F_D/F_T)(N - 1)](qF_A/Nk_p)$. For the special case, $F_D/F_T = 1$, this becomes simply, $1 + (qF_A/k_p)$; in this case, if the stepping rate-constant equals the photoconversion rate-constant, the number of visits is precisely two.

or slow back transfer, is easy to create in photosynthetic systems. For lattice models of the Chl arrays in these, the rule is therefore zero-mode dominance of hopping exciton migration and quenching (with uniform initial excitation).

Zero-mode dominance can also break down in lattices where the exciton hopping is critically anisotropic (Hemenger *et al.*, 1972). Critical anisotropy is even less likely to occur in photosynthetic Chl "lattices" than, say, long-range quenching, but anisotropy generally is quite possible. The effects of anisotropy, including critical anisotropy, on the kinetics of incoherent exciton quenching are discussed in Hemenger *et al.* (1972), and in Pearlstein (1982).

(B) Extension of HPL theory: a more general initial condition useful for photosynthetic studies

The two initial conditions considered by HPL are the uniform condition (all sites, individual antenna Chl's and RC, equally likely to be excited at $t = 0$), which leads to Eq. 13, and the impurity condition (only the RC excited at $t = 0$), which leads to Eq. 14. In general, the relative probability of exciting antenna or RC initially is a function of experimental conditions, e.g. type of sample, temperature and wavelength of exciting light. It is useful, therefore, to generalize the HPL results to deal with an arbitrary excitation situation. It turns out both that it is straightforward to do so, and that the results have unexpected predictive value.

One begins the generalization from Eq. 11. At $t = 0$, the probability that the RC is excited is $\rho_0(0) \equiv \rho_{RC}$. Because there are $N - 1$ proximal antenna sites per RC, each of which is assumed here to have an identical absorption spectrum, the probability that *each* antenna Chl is excited at $t = 0$ is $\rho_i(0) = (1 - \rho_{RC})/(N - 1)$ for all $i \neq 0$ (the RC site). Thus, in Eq. 11, one can substitute

$$\sum_i f_{0i}\rho_i(0) = f_{00}\rho_{RC} + \left(\frac{1 - \rho_{RC}}{N - 1}\right)\sum_{i \neq 0} f_{0i}$$

Taking advantage of the fact that

$$\sum_i f_{0i} = 0$$

(i.e. the sum on *all* i, including $i = 0$), one has that

$$\sum_{i \neq 0} f_{0i} = -f_{00},$$

so that

$$\sum_i f_{0i}\rho_i(0) = f_{00}(N\rho_{RC} - 1)/(N - 1)$$

and Eq. 11 reduces to

$$\begin{aligned} M_0 = &[1 + (F_D/F_T)(N - 1)]k_p^{-1} \\ &+ (qF_A)^{-1}\{(F_A/F_T) \\ &+ [(q\alpha N^2)/(N - 1)^2] - 1\} \\ &\times (N - 1)(1 - \rho_{RC}) \end{aligned} \quad (15)$$

where $\alpha \equiv -f_{00}$. Equation 15, which is the desired result, contains Eqs. 13 and 14 as special cases, which may be seen by substituting $\rho_{RC} = 1/N$ to yield Eq. 13, or $\rho_{RC} = 1$ to yield Eq. 14.

Equation 15 makes the following immediately useful prediction: M_0 is linear in $1 - \rho_{RC}$. In other words, the lifetime of the exciton as limited by quenching at the RC is a linear function of that fraction of absorbed quanta that are not absorbed directly by the RC. The slope of the line is exactly the first-passage-time of the exciton's random walk to the RC. Its intercept is the photoconversion-limited exciton lifetime. In the next section, I show how the *in situ* photoconversion rate constant, k_p, and other useful parameters can be readily derived from experimental data that is reduced with the aid of Eq. 15, for any photosynthetic system in which the linear relation of M_0 and $1 - \rho_{RC}$ holds.

IV. EXPERIMENTAL ACCESSIBILITY OF PARAMETERS

In order to test Eq. 15, one has to experimentally determine M_0 as a function of $1 - \rho_{RC}$. To do this requires wavelength and time selectivity of both exciting light and detection of sample response. At each excitation wavelength the fractional optical density, ρ_{RC}, due to RC absorbance must be determined, a procedure that by present techniques is straightforward for a number of photosynthetic systems, including purple bacterial chromatophores (Aagaard and Sistrom, 1972; Monger *et al.*, 1976) and P700 particles (Olson and Thornber, 1979; Mullet *et al.*, 1980). The excitation must be at wavelengths long enough not to excite any accessory pigments (as defined in the Introduction), otherwise the analysis becomes more complicated than described here, with additional parameters to be taken into account. In purple bacterial RC's, the bacteriopheophytins may have to be viewed as accessory pigments in a similar sense, i.e. they may transfer excitation to the proximal antenna with reasonable efficiency.

With regard to sample response, one may detect either the absorbance loss (bleaching) of the longest-wavelength RC band, or fluorescence decay of antenna plus RC (equivalently, decay of $S_1 \rightarrow S_n$ absorbance; see Holten *et al.*, 1980); the two quantities are connected by Eq. 10. Here I describe the analysis in terms of fluorescence. In many cases, the RC contribution to fluorescence even at early times is negligible, so that the fluorescence may be detected at any convenient wavelength, or broadband with due care to avoid overlap with absorption bands. If fluorescence coming directly from the RC is not negligible, it is essential to detect with sufficient bandwidth to encompass the emissions of both sources.

The measurements are most reliably performed with a pulsed excitation source having pulse widths of at most a few tens of ps, and a detector having a similar time resolution. (The more stably and reliably the pulses can be produced and detected, i.e. the less "jitter" in the system, the less stringent become the resolution requirements, unless for example it is desired to study the detail of a nonexponential decay at short times.) In practice this means a ps laser with some degree of wavelength tunability and a streak camera detector. In analyzing the data, it may be necessary to deconvolute the system response, integrate the deconvoluted fluorescence decay curve, and/or correct for K_A (Eq. 7). Deconvolution is required unless the pulse width is much less than the lifetime of the shortest-lived component of the decay, M_0 for an exponential decay. Integration, Eq. 6, is required if the decay is nonexponential. (Because the decay lifetime determined by integration agrees with that from a straight-line fit to a semi-log plot only for an exponential decay, such a comparison is a sensitive way to test for small deviations from exponentiality.) A K_A-correction is probably not necessary for a sample in which the RC's are ready to perform optimal photochemistry. Under circumstances where such a correction is necessary, K_A may be difficult to estimate reliably. For purple bacteria, the Bchl fluorescence lifetime in reconstituted "chromatophores" that are devoid of RC's (Heathcote and Clayton, 1977), or in isolated antenna Bchl–protein complexes (Bolt, 1980), may provide an approximate value of K_A^{-1}.

Once M_0 and $1 - \rho_{RC}$ have been experimentally determined over a range of values of the latter for a given sample, a plot of M_0 vs. $1 - \rho_{RC}$ provides an immediate test of the theory, which predicts that the plot is a straight line. If so, as noted at the end of the previous section, the slope and intercept, respectively, measure the first-passage-time of the exciton random walk and the photoconversion-limited exciton lifetime. The latter is proportional to k_p^{-1}, which should equal the lifetime of the Chl excited singlet state in isolated RC's (Holten *et al.*, 1980), provided there is no effect of the environmental change. Thus, the *in situ* (i.e. in the membrane) lifetime of the Chl singlet in the RC can be established if the constant of proportionality is determined. From Eq. 15, that constant is $1 + (F_D/F_T)(N - 1)$. As already noted, $N - 1$ is the photosynthetic unit size which is known or determinable for many photosynthetic systems. The Förster rate constants F_D and F_T are, individually, not so readily determinable experimentally. However, from Eq. 1

$$F_D/F_T = I_D/I_T \qquad (16)$$

i.e. the ratio of Förster constants equals the ratio of overlap integrals because the geometric factors cancel. The overlap integrals are themselves straightforwardly calculated, at least for purple bacteria, from Eq. 2 with steady-state absorption and fluorescence spectral data: I_D is the overlap of the (isolated) RC fluorescence spectrum with the antenna absorption, I_T is the overlap of antenna fluorescence with RC absorption.

As Eq. 16 shows, a *ratio* of Förster rate constants can be experimentally accessible even when the individual rate constants are not. Other accessible ratios of Förster rate constants occur if I_j ($j = A$, D or T) is a function of some experimentally controllable variable, V, and the geometric factors in Eq. 1 are not, for then

$$F_j(V_1)/F_j(V_2) = I_j(V_1)/I_j(V_2) \qquad (17)$$

where V_1 and V_2 are two different values of V. The case, $V = T$, the absolute temperature, is of great potential interest. It is clear at least for *Rp. sphaeroides* R26 that overlap integrals do vary as a function of T. Inspection of the RC absorption and fluorescence spectra and the corresponding antenna spectra shows that the overlap integrals I_D and I_A change greatly with temperature at least between 77 and 300 K (Bolt, 1980). At 77 K both the de-trapping rate and random-walk stepping rate must become much smaller.

This raises the possibility of evaluating the individual terms in the expression for the first-passage-time of the exciton's random walk to the RC, τ_{FPT}. From Eq. 15, this is given by the slope of M_0 vs. $1 - \rho_{RC}$, i.e. (neglecting corrections of order $1/N$)

$$\tau_{FPT} = \frac{N}{qF_T} + (\alpha - q^{-1})\frac{N}{F_A} \qquad (18)$$

Making use of Eq. 17, one may write Eq. 18 in the form

$$\tau_{FPT}(T_i) = \frac{N}{qF_T(T_0)}\left[\frac{I_T(T_0)}{I_T(T_i)}\right]$$
$$+ (\alpha - q^{-1})\frac{N}{F_A(T_0)}\left[\frac{I_A(T_0)}{I_A(T_i)}\right] \qquad (19)$$

where T_0 is some chosen reference temperature, and T_i is some other temperature. (Strictly speaking the temperature dependence of the radiative lifetime, τ_0, should be checked. It is a trivial extension to include any temperature variation of τ_0 in this calculation.) Then for any two temperatures, T_i and T_j, and reference temperature T_0, it follows from Eq. 19 that

$$\frac{N}{qF_T(T_0)} = \frac{b_j\tau_i - b_i\tau_j}{a_ib_j - a_jb_i} \qquad (20)$$

where

$\tau_i = \tau_{FPT}(T_i)$
$a_i = I_T(T_0)/I_T(T_i)$
$b_i = I_A(T_0)/I_A(T_i)$

similarly for the j-subscripted quantities. Similarly

$$\frac{(\alpha - q^{-1})N}{F_A(T_0)} = \frac{a_j\tau_i - a_i\tau_j}{a_jb_i - a_ib_j} \qquad (21)$$

All of the quantities on the right-hand sides of Eqs. 20 and 21 are experimentally accessible and temperature dependent. Those on the left-hand sides are, for fixed T_0, temperature independent. Thus, one must first check whether each of the combinations of experimental quantities, $(b_j\tau_i - b_i\tau_j)/(a_ib_j - a_jb_i)$, and $(a_j\tau_i - a_i\tau_j)/(a_jb_i - a_ib_j)$, are indeed constant over some temperature range. If either is not constant, it indicates a temperature variation of an orientation (κ^2) or a distance (R^6) factor in Eq. 1 for F_T or F_A. The temperature dependences of the combination-expressions are therefore direct indicators of any temperature variations of the energy transfer geometry.

If, over some temperature range, the right-hand expressions in Eqs. 20 and 21 are constant, they would then yield the values of $(qF_T)^{-1}$ and $(\alpha - q^{-1})F_A^{-1}$ for those temperatures. Additional information would be needed to obtain the Förster rates and the lattice parameters separately. However, for three-dimensional lattices the expression $\alpha - q^{-1} \simeq 1.3$ within a few percent; fortuitously, for $N = 60$ (the value for $Rp.$ $sphaeroides$ R26) the same is true for two-dimensional lattices (Pearlstein, 1982). (This is so even though q itself can vary from 3 to 12.) Thus, it is possible to obtain an estimate of the mean antenna Förster constant, F_A, with a total uncertainty that is probably not much greater than experimental error. Evaluating q and F_T would be more difficult, but by setting $q = 6$, F_T can be estimated to within a factor of two.

It must be noted that M_0 from Eq. 15 depends on the state of the RC's mainly through k_p, and to a lesser extent through other parameters, particularly F_D and F_T. It would be interesting to compare M_0's for the two limiting cases, all RC's reduced (i.e. ready for photooxidation), and all RC's oxidized. One would expect the slope, τ_{FPT}, to be relatively unaffected while the intercept, which is proportional to k_p^{-1}, should increase markedly.

V. ANALYSIS OF FLUORESCENCE EXPERIMENTS

It is probably fair to say that no experiment performed to date meets all of the criteria (time resolution, several long-wavelength excitations, low exciting-pulse energy, etc.) necessary to test rigorously the prediction of Eq. 15 that M_0 is linear in $1 - \rho_{RC}$. However, two fluorescence experiments performed with $Rp.$ $sphaeroides$ R26 chromatophores at 300 K provide data that is interesting to re-examine in light of the considerations in this paper. One of these, by Wang and Clayton (1971), measured the absolute fluorescence quantum yield, ϕ, at each of two excitation wavelengths, 810 and 850 nm. At low exciting-light intensity (virtually all RC's prepared to undergo photooxidation), they found $\phi_{810} = 0.015$ and $\phi_{850} = 0.022$. They state that ρ_{RC} (810 nm) = 0.5, and ρ_{RC} (850 nm) $\simeq$ 0. If one uses the well-known relation, $\tau = \phi\tau_0$, where τ is the actual fluorescence lifetime* and τ_0 is the radiative lifetime [$\sim$18 ns for Bchl a in

*In the same integrated sense as M_0.

purple bacteria, according to Zankel et $al.$ (1968)], one has $\tau_{810} = 270$ ps and $\tau_{850} = 400$ ps. Fitting a straight line to these two points, one finds

$$M_0 = 140\,\text{ps} + 260\,(1 - \rho_{RC})\,\text{ps} \qquad (22)$$

Thus, from the data of Wang and Clayton, one might infer that $\tau_{FPT} = 260$ ps, and the photoconversion-limited time = 140 ps. It is known that for $Rp.$ $sphaeroides$ R26, $F_D/F_T \lesssim 1$ (at 300 K), $N = 60$ (Campillo et $al.$, 1977), and $k_p^{-1} \lesssim 4$ ps (Holten et $al.$, 1980). With the specific values, $F_D/F_T = 0.5$, $N = 60$, and $k_p^{-1} = 4$ ps, Eq. 15 yields a photoconversion-limited time = 120 ps. However, because of the many uncertainties, this apparent agreement may be purely fortuitous. For example, the value of k_p^{-1} is quite uncertain, a straight line can always be fitted to just two points, no correction for K_A has been considered, and it may be that excitation at 810 nm yields a slightly different de-trapping rate "constant" initially than does excitation at longer wavelengths (because de-trapping from the RC is possibly fast enough to compete with intra-RC singlet energy transfer). Moreover, if this analysis is correct, it would imply that for a uniform initial condition ($\rho_{RC} \simeq 0$), $M_0/\tau_{FPT} \lesssim 2$. This is rather close to the diffusion limit, contrary to prevalent belief for this organism (Pearlstein, 1982), and would therefore predict noticeable deviation from exponentiality of fluorescence decay (not observable in the Wang–Clayton experiment) for $\rho_{RC} \gtrsim 0.5$.

On the other hand, the foregoing analysis of the Wang–Clayton experiment receives some support from the more recent data of Campillo et $al.$ (1977), who excited chromatophores only at 530 nm with 20-ps laser pulses of suitably low intensity and rep rate. According to Campillo et $al.$ $\rho_{RC} = 0.19$ under their experimental conditions. Equation 22 predicts that for $\rho_{RC} = 0.19$, $M_0 = 350$ ps, which compares well with the lifetime, 300 ± 50 ps, measured by Campillo et $al.$ Crudely put, the latter measurement adds a third point to the straight-line fit of Wang and Clayton's data. Again, because of the uncertainties involved (e.g. that part of the 530-nm excitation absorbed directly by the RC's in this carotenoidless mutant excites primarily the bacteriopheophytins, which may have distinct de-trapping kinetics), even the two experiments together do not constitute a real test of present predictions. Nonetheless, theory and experiment so far appear to be consistent.

VI. DISCUSSION

The theory of exciton migration and trapping presented here makes several predictions that are either new altogether, or at least are new to photosynthesis. Chief among the former is the linearity of the Chl or Bchl excited-singlet-state lifetime with the fraction of absorbed excitation quanta that are absorbed directly by the antenna, provided that only the RC or the proximal antenna is initially excited. The slope of the line is the mean first-passage-time of the exciton's random walk to the RC, and the intercept is the exciton's lifetime as limited only by photoconversion at the RC in the presence of detrapping. This photoconversion-limited lifetime is simply related to the *in situ* photoconversion rate constant via two parameters each of which is experimentally accessible. Values of individual Förster rate constants possibly can be obtained, at least approximately, from measurements of excited-state lifetimes as functions of both temperature and excitation wavelength. Perhaps the greatest benefit of being able to measure both slope (first-passage-time) and intercept (photoconversion-limited lifetime) is that the former is more sensitive to changes of geometry that affect energy transfer than the latter, while the reverse is true for changes of the state of the RC (see end of Section IV).

Among the second category of predictions is the interesting one that the excited-state decay deviates from exponentiality if the exciton is created at the RC and its trapping is diffusion-limited, i.e. the mean first-passage-time (averaged over all starting sites in RC and proximal antenna) is much greater than the photoconversion-limited lifetime. Indications that this phenomenon may occur to some extent in *Rp. sphaeroides* R26 are noted in Section V, where it is also pointed out that none of the new predictions has yet received a rigorous experimental test. Such a test, at the University of Rochester's low-jitter ps laser facility (Mourou and Knox, 1980; Nordlund and Knox, 1981), is currently planned.

If the major new theoretical predictions are verified for long-wavelength excitation of such simple organisms as *Rs. rubrum* or *Rp. sphaeroides* R26, it will be worthwhile to extend the theoretical concepts in several ways. It should not be too difficult to generalize the theory to deal with shorter wavelength excitation of the one-antenna-component organisms, and perhaps also to model exciton migration and trapping for excitations created in accessory pigment of multi-antenna-component organisms. In addition, it will be of interest to explore the theoretical limits of validity of the linear relationship between exciton lifetime and antenna-absorption-fraction. Although, in the derivation of Eq. 15, that relationship appears to be a specific consequence of the properties of lattice Green's functions, it is an intuitively simple and appealing one that may apply under less restrictive assumptions.

In any event, irrespective of any skepticism of the assumptions underlying it, the current theory has the particular virtue of making several, correlated, readily testable predictions. It is thus a useful step toward a comprehensive theory of exciton migration and trapping in photosynthesis.

REFERENCES

Aagaard, J. and W. R. Sistrom (1972) *Photochem. Photobiol.* **15,** 209–225.

Altman, J. A., G. S. Beddard and G. Porter (1979) *Ciba Foundation Symp.* **61,** 191–197. Excerpta Medica, Amsterdam.

Bay, Z. and R. M. Pearlstein (1963) *Proc. Natl. Acad. Sci. USA* **50,** 1071–1078.

Bolt, J. (1980) Dissertation, University of California, Berkley, CA.

Breton, J. and N. E. Geacintov (1980) *Biochim. Biophys. Acta* **594,** 1–32.

Campillo, A. J., R. C. Hyer, T. G. Monger, W. W. Parson and S. L. Shapiro (1977) *Proc. Natl. Acad. Sci. USA* **74,** 1997–2001.

Cogdell, R. J. and J. P. Thornber (1980) *FEBS Lett.* **122,** 1–8.

Förster, Th. (1949) *Z. Naturforsch.* **4a,** 321–327.

Förster, Th. (1965) In *Modern Quantum Chemistry, Part III* (Edited by O. Sinanoglu), pp. 93–137. Academic Press, New York.

Heathcote, P. and R. K. Clayton (1977) *Biochim. Biophys. Acta* **459,** 506–515.

Hemenger, R. P., K. Lakatos-Lindenberg and R. M. Pearlstein (1974) *J. Chem. Phys.* **60,** 3271–3277.

Hemenger, R. P., R. M. Pearlstein and K. Lakatos-Lindenberg (1972) *J. Math. Phys.* **13,** 1056–1063.

Holten, D., C. Hoganson, M. W. Windsor, C. C. Schenck, W. W. Parson, A. Migus, R. L., Fork and C. V. Shank (1980) *Biochim. Biophys. Acta* **592,** 461–477.

Knox, R. S. (1975) In *Bioenergetics of Photosynthesis* (Edited by Govindjee), pp. 183–221. Academic Press, New York.

Knox, R. S. (1977) In *Topics in Photosynthesis, Vol. 2* (Edited by J. Barber), pp. 55–97. Elsevier, Amsterdam.

Mauzerall, D. (1982) *Israel. J. Chem.* In press.

Monger, T. G., R. J. Cogdell and W. W. Parson (1976) *Biochim. Biophys. Acta* **449,** 136–153.

Monger, T. G. and W. W. Parson (1977) *Biochim. Biophys. Acta* **460,** 393–407.

Montroll, E. W. (1964) *Proc. Symp. Appl. Math.* **16,** 193–220.

Montroll, E. W. (1969) *J. Math. Phys.* **10,** 753–765.

Mourou, G. and W. Knox (1980) *Appl. Phys. Lett.* **36,** 623–626.

Mullet, J. E., J. J. Burke and C. J. Arntzen (1980) *Plant Physiol.* **65,** 814–822.

Nordlund, T. M. and W. H. Knox (1981) *Biophys. J.* **36,** 193–201.

Olson, J. M. and J. P. Thornber (1979) In *Membrane Proteins in Energy Transduction* (Edited by R. A. Capaldi), pp. 279–340. Marcel Dekker, New York.

Park, R. B. and J. Biggins (1964) *Science* **144,** 1009–1010.

Pearlstein, R. M. (1967) *Brookhaven Symp. Biol.* **19,** 8–14.

Pearlstein, R. M. (1968) *Photochem. Photobiol.* **8,** 341–347.

Pearlstein, R. M. (1982) In *Photosynthesis: Energy Conversion by Plants and Bacteria* (Edited by Govindjee). Academic Press, New York. In press.

Pearlstein, R. M. and R. P. Hemenger (1978) *Proc. Natl. Acad. Sci. USA* **75,** 4920–4924.

Robinson, G. W. (1967) *Brookhaven Symp. Biol.* **19,** 16–45.

Sauer, K. (1975) In *Bioenergetics of Photosynthesis* (Edited by Govindjee), pp. 115–181. Academic Press, New York.

Sauer, K. (1978) *Acc. Chem. Res.* **11,** 257–264.

Shipman, L. L. (1980) *Photochem. Photobiol.* **31,** 157–167.

Staehelin, L. A., J. R. Golecki and G. Drews (1980) *Biochim. Biophys. Acta* **589,** 30–45.

Thornber, J. P., J. P. Markwell and S. Reinman (1979) *Photochem. Photobiol.* **29,** 1205–1216.

Wang, R. T. and R. K. Clayton (1971) *Photochem. Photobiol.* **13,** 215–224.

Weiss, C., Jr. (1972) *J. Mol. Spectrosc.* **44**, 37–80.

Wolber, P. K. and B. S. Hudson (1979) *Biophys. J.* **28**, 197–210.

Zankel, K. L., D. W. Reed and R. K. Clayton (1968) *Proc. Natl. Acad. Sci. USA* **61**, 1243–1249.

Excitation trapping and primary charge stabilization in *Rhodopseudomonas viridis* cells, measured electrically with picosecond resolution

(photosynthesis/electron transfer/light gradient/capacitative metal electrodes)

J. DEPREZ[†], H. W. TRISSL[‡], AND J. BRETON[†]

†Service de Biophysique, Département de Biologie, Centre d'Etudes Nucléaire de Saclay, 91191 Gif-sur-Yvette Cedex, France; and ‡Schwerpunkt Biophysik, Universität Osnabrück, Barbarastrasse 11, D-4500 Osnabrück, Federal Republic of Germany

Communicated by George Feher, November 4, 1985

ABSTRACT   The transmembrane primary charge separation in the photosynthetic bacterium *Rhodopseudomonas viridis* was monitored by electric measurements of the light-gradient type [Trissl, H. W. & Kunze, U. (1985) *Biochim. Biophys. Acta* 806, 136–144]. Excitation of whole cells with 30-ps laser pulses at either 532 nm or 1064 nm gave rise to a biphasic increase of the photovoltage. The fast phase, contributing about 50% of the total, rose with an exponential time constant ≤ 40 ps and was independent of the redox state of the quinone electron acceptor. It is assigned to the migration of the excitation energy in the antenna and its subsequent trapping by the reaction center, monitored by the ultrafast charge separation between the primary electron donor and the bacteriopheophytin intermediary acceptor. The slower phase (125 ± 50 ps) only occurred when the quinone was oxidized and disappeared when it was reduced (either chemically or photochemically). It is assigned to the forward electron transfer from the bacteriopheophytin to the quinone. The relative amplitudes of these two electrogenic steps demonstrate that the bacteriopheophytin intermediary acceptor is located halfway between the primary donor and the quinone.

The primary redox reactions in photosynthesis occur in a chlorophyll–protein complex called the reaction center (RC). Each RC is surrounded by a large number of other chlorophyll–protein complexes that function as an antenna. All these complexes are incorporated in the membranes of closed vesicles. The RC itself is inserted in an asymmetric way so that it carries electrons across the membrane, thereby creating a transmembrane potential.

Upon photon absorption by an antenna pigment, the singlet excitation energy migrates towards the RC, where it creates the excited state of the primary donor, P*. In the case of the photosynthetic bacterium *Rhodopseudomonas viridis*, the rate of this process is not known. However, in other purple bacteria (*R. sphaeroides* and *Rhodospirillum rubrum*), a time constant of 50–100 ps has been deduced by analysis of fluorescence decay kinetics (1–3).

Our knowledge of the primary photochemistry in bacterial photosynthesis stems from picosecond absorption measurements on isolated RCs lacking the antenna systems (4–6). After P* formation, the charge-separated state between P and the bacteriopheophytin electron acceptor H (P^+H^-) appears in less than 20 ps in the case of *R. viridis* RCs (5) and in 2.8 ± 0.2 ps in the case of *R. sphaeroides* (6). The subsequent electron transfer to a quinone, Q, requires 200 ± 50 ps in both species (4, 5). In these bacteria, the electron transfer from the primary to the secondary quinone takes place in $\approx 100 \ \mu$s (7). Further, in *R. viridis* the reduction of P^+ by the associated cytochrome *c* complex occurs with a 270-ns relaxation time (5). Thus, by varying the preillumination conditions (a saturating preflash at an adjustable time before a picosecond measuring flash, or continuous illumination in the presence of a reducing agent) one can achieve different redox states of the RC (PHQ^- or PH^-Q^-, respectively) (5, 7, 8).

The molecular organization of the pigments in the RC of *R. viridis* has been analyzed recently by x-ray crystallography (9). In the crystal structure, H is located approximately halfway between P and Q. The distance between P and Q is about 25 Å. Thus, during the sequential electron transfer from P to H and further to Q, an increase of the spatial separation of electric charges and, therefore, an increase of the strength of the transmembrane electric field must occur.

In a previous study, the relative positions of P, H, and Q in RCs of *R. sphaeroides* were determined with electrodes by comparing the magnitude of the electric field in the states $P^+H^-Q^-$ or P^+HQ^- (10). The two states were distinguished by the different kinetics of their back reactions. Assuming a linear array of the electron carriers in the protein, these results were interpreted in terms of effective distances. The time resolution in these experiments, 2 ns, was not sufficient to resolve any of the forward electron-transfer reactions. However, technical improvements of the light-gradient experiment yielded higher time resolution, as reported for plant chloroplasts and purple membranes (11, 12).

We have applied these faster techniques to whole cells of *R. viridis* and succeeded in separating kinetically the two steps of the electrogenic forward reaction. Since the initial charge separation between P and H is extremely rapid, the first phase could be used as a measure of the mean trapping time of excitons created in the antenna. Using excitation at 532 nm or 1064 nm with 30-ps laser pulses, the electric experiments allowed us to estimate a trapping time constant ≤ 40 ps and a time constant for the forward reaction of electron transfer from H to Q of 125 ± 50 ps.

MATERIALS AND METHODS

R. viridis cells were grown as described (13). For measurements, they were resuspended in Tris·HCl buffer (20 mM, pH 8.0), containing the electron mediator phenazine methosulfate (10 μM), at an optical density of 50 cm^{-1} at 532 nm. When needed, 100 mM sodium dithionite (in 0.1 M Tris·HCl, pH 9.0) was added just prior to the measurements. Purple membranes from *Halobacterium halobium* were obtained as described (11).

The electric measurements of the light-gradient type were carried out essentially as described (12). However, the measuring cell was reduced in size and built as a coaxial cell

Abbreviations: RC, reaction center; P, primary donor; H, intermediate electron acceptor; Q, quinone acceptor.

Reprinted from *Proceedings of The National Academy of Sciences* **83**, 1699–1703 (1986); © The National Academy of Sciences.

in an SMA-type micro coaxial connector. The bottom electrode was a Pt disk of 3 mm diameter and the top electrode was a Pt mesh (wire diameter 60 μm) of the same diameter. The spacing between the two electrodes was 0.1 mm.

The photovoltage signals were amplified by two cascaded 50-Ω amplifiers (Nucletude, Orsay, France), each having a specified bandwidth of 10–6000 MHz and a gain of 20 decibels. The signals were recorded as single sweeps on a 4-GHz oscilloscope (TSN 660.2, Thomson, Malakoff, France). The sensitivity of the oscilloscope was 70 mV per division on the direct-access mode. The trace was digitized by a Vidicon camera (TSN 1150, Thomson) and subsequently stored in a Hewlett–Packard HP87 computer, which was also used for signal-averaging and for controlling experimental parameters. The 10–90% response time of the whole detection device was found to be 100 ps, using a pulse generator with a rise time < 25 ps (S50, Tektronix, Beaverton, OR).

The excitation source was a mode-locked Nd-YAG laser (YG 402, Quantel, Orsay, France) delivering two simultaneous 30-ps-wide pulses at 1064 nm and 532 nm. One of these pulses was used to excite the sample while the other was sent to an Auston photocell (14), which triggered the oscilloscope. An optical delay of ≈85 ns was introduced between the two pulses in order to compensate for the internal delay of the oscilloscope. The pulse energy was measured using a RjP 7200 (Laser Precision, Yorkville, NY) instrument equipped with two detectors.

For experiments where saturating preflashes were needed, a dye laser (CMX 4, Chromatronix, Mountain View, CA) delivering 500-ns pulses at 600 nm was available. All flashes were sent to the coaxial cell via a bifurcated fiber-optic light guide.

RESULTS

The time course of the photovoltage elicited by a single 30-ps flash at 532 nm from whole cells of *R. viridis* is biphasic (Fig. 1, trace a). The fast phase of the rise is close to the instrumental limitation. We assign these two phases to the electric equivalents of the charge separations between P and H and between H and Q, respectively. This assignment is supported by the observation that upon chemical reduction of Q by dithionite prior to the flash, only the fast phase is present (Fig. 1, trace b). The 50% reduction in amplitude of this signal compared to that of trace a is an indication that only a partial electron transfer occurs in the RCs. If the sample is illuminated continuously so that the state PH$^-$Q$^-$ accumulates, then a 30-ps flash does not evoke any photovoltage (trace c).

The reduction by dithionite causes physicochemical changes in the measuring cell that might influence the quantitative evaluation of the two phases. Therefore, Q was differently reduced by a saturating preflash. Due to the known 270-ns kinetics of reduction of P$^+$ by cytochrome c-type donors (5), the RCs are expected to be in the state PHQ$^-$ before the picosecond flash. If the preflash was given 10 μs prior to the measuring 30-ps flash, the kinetics and amplitude of the photovoltage (Fig. 1, trace d) were indistinguishable from those shown in trace b. This result confirms our interpretation of the experiment with chemically reduced RCs (trace b).

At 532 nm, in addition to absorption by antenna pigments, some direct absorption by the RC pigments themselves occurs. Therefore the experiments were repeated with 30-ps pulses at 1064 nm, where only antenna pigments absorb. The kinetic traces were indistinguishable from the ones repre-

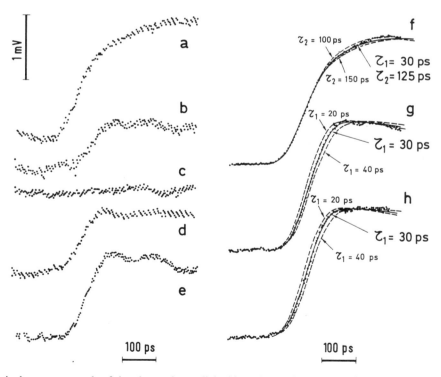

FIG. 1.   Traces a–d: single-sweep records of the photovoltage elicited by a 30-ps, 532-nm laser flash at an energy of 2 × 10^{14} photons per cm^2 on *R. viridis* whole cells in the presence of 10 μM phenazine methosulfate. Cells were either dark-adapted for 20 s (trace a), dark-adapted with the addition of 0.1 M Na$_2$S$_2$O$_4$ (trace b), continuously illuminated in the presence of 0.1 M Na$_2$S$_2$O$_4$ (trace c), or exposed to a saturating preillumination flash (600 nm, 500 ns wide) 10 μs before excitation. Trace e: single-sweep record of the photovoltage from purple membranes electrically preoriented (500 V/cm for 5 s). A baseline signal, recorded in the dark, has been subtracted from all the signal traces. Traces f–h: normalized averages of 15 traces corresponding to the individual sweeps depicted by traces a, b, and d, respectively. The continuous and broken curves are the calculated responses of the measuring equipment to assumed mono- or biexponential charge separations with the indicated time constants (τ). The amplitudes of the two components in trace f are equal (see text). The concentration of the sample corresponded to (0.8–1.2) × 10^{17} RCs per ml.

sented in Fig. 1, traces a, b, and d. This observation demonstrates that our measurements obtained upon 532-nm excitation are not affected by partial direct absorption by the RC pigments.

In order to compare the initial phase observed for *R. viridis* with the fastest reported electrogenic event from a biological object (15), purple membranes were preoriented electrically in the coaxial cell (11). Excitation of this sample with a 30-ps laser pulse at 532 nm elicited the photovoltage kinetics shown in Fig. 1, trace e. The rise of this trace is as fast as the one obtained with *R. viridis* cells in which Q was reduced (traces b and d). After deconvolution, trace e leads to a time constant of 30 ± 10 ps, in agreement with a previous report (15).

In order to improve the signal-to-noise ratio, a shift program to recenter the traces on top of each other was applied. In Fig. 1, traces f, g, and h are the kinetic traces obtained by averaging 15 individual curves corresponding to the conditions used to record traces a, b, and d, respectively. The continuous curves in traces f, g, and h are calculated by assuming monophasic (traces g and h) or biphasic (trace f) exponential kinetics of the displacement current in the cell (see *Appendix*). The time constants indicated in Fig. 1 result from the best fits of calculated curves to the measured photovoltage signals.

The 10–90% rise time observed in dark-adapted *R. viridis* cells (Fig. 1, trace f) was found to be 180 ± 15 ps. It was independent of the excitation energy over the range 5×10^{13} to 2×10^{15} photons per cm^2 (Fig. 2a, ●). The faster phase, as indicated by the 10–50% rise time, also was independent of the excitation energy (Fig. 2a, ○). The amplitude of the photovoltage (Fig. 2b) increased approximately linearly up to an energy of 3×10^{14} photons per cm^2, which corresponds to approximately 1 photon absorbed per 3 RCs, and then leveled off. This saturation is probably due to the decrease of anisotropy of the light-gradient effect, which is expected upon closure of most of the RCs (16–18). The small deviation from linearity observed below 3×10^{14} photons per cm^2 might indicate exciton-annihilation processes.

As mentioned in the Introduction, the state PHQ$^-$, which is present ≈1 μs after the preflash, converts in ≈100 μs to state PHQ. Thus, measurements of the photovoltage at various delays probe the different ratios of these states, offering the possibility to assess the consistency of the given assignments. The dependence of the amplitude and of the 10–90% rise time of the photovoltage are shown in Fig. 3 a and b, respectively. Both the amplitude and the rise time increase with increasing delay times. The two horizontal broken lines in a and in b result from measurements performed in the dark-adapted state in the presence (lower lines) or absence (upper lines) of dithionite. The photovoltage amplitude and the 10–90% rise time obtained at the shortest delay time (5 μs) are identical to the respective values obtained in the presence of dithionite.

The decay of the photovoltage signal observed with dark-adapted cells is shown on an extended time scale in Fig. 4a. This decay (3.0 ± 0.5 ns time constant) is due to the discharge of the cell capacitance in the 50-Ω preamplifier (11, 12). The faster decay (2 ± 0.5 ns) observed when Q is reduced prior to the picosecond flash (Fig. 4b) is due to a depolarization in the dielectric of the capacitor convoluted with the electronic decay. The depolarization is assigned to the back reaction between P$^+$ and H$^-$ which accelerates the decay kinetics of the photovoltage. A deconvolution procedure (*Appendix*) yields a time constant of 10 ± 3 ns for the depolarization, in agreement with the literature value for the back reaction (5).

DISCUSSION

Flash-induced photovoltages arising from the light gradient in a chloroplast suspension were first reported in 1973 (16, 17). The signals have been ascribed to the primary charge separation in the RCs. The time resolution of this technique has recently been improved by the use of metal electrodes forming a planar capacitor, in which the suspension of photosynthetic membranes constitutes the dielectric medium (12). At least one of the electrodes is semitransparent so that light propagation perpendicular to the plane of the electrode can establish an electric anisotropy, as described in the literature (16–18).

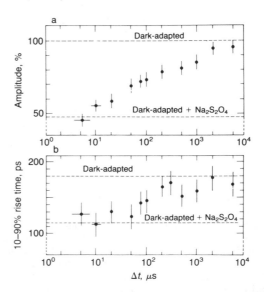

FIG. 3. Dependence of the amplitude (*a*) and of the 10–90% rise time (*b*) of the photovoltage measured after preillumination of *R. viridis* cells on the delay Δt between the preillumination flash (600 nm, 500 ns) and the 30-ps, 532-nm measuring pulse. The amplitude of the signal is normalized to the amplitude measured in the same excitation condition from dark-adapted cells. The two horizontal broken lines in *a* and *b* indicate the mean value obtained in the dark-adapted state in the presence (lower line) or in the absence (upper line) of 0.1 M Na$_2$S$_2$O$_4$. Other conditions were as in Fig. 1.

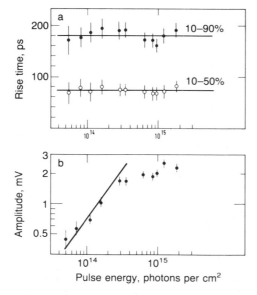

FIG. 2. Dependence of the 10–90% and 10–50% rise times (*a*; ● and ○, respectively) and of the amplitude (*b*) of the photovoltage from dark-adapted *R. viridis* cells on the energy of the 30-ps, 532-nm excitation pulse. Experimental conditions were as described for Fig. 1.

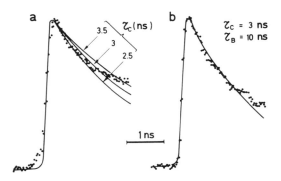

FIG. 4. Normalized averages of 15 traces corresponding to dark-adapted bacteria (*a*) and to bacteria given a saturating preilluminating flash 10 μs before excitation (*b*). The continuous curves in *a* depict the calculated decay of the measuring cell (time constant τ_C). Continuous curve in *b* takes into account a 10-ns time constant τ_B for the charge-recombination process (see *Appendix*).

In the present study, we have achieved a real-time resolution of 100 ps, which allows us to resolve one step of the forward electron transfer in whole cells of *R. viridis*. This is apparent as a slow-rising phase of the photovoltage in Fig. 1, traces a and f. Assuming a single-exponential time constant for this phase, the deconvolution analysis gives a time constant of 125 ± 50 ps, with a relative amplitude of 50 ± 5%. The disappearance of this phase upon photochemical (Fig. 1, trace d) or chemical (Fig. 1, trace b) reduction allows us to ascribe it to the electron transfer from H to Q. The significance of the slightly faster kinetics of this transfer in intact cells observed here compared to the 230-ps value reported for isolated RCs (5) is not clear.

The rapidly rising phase (Fig. 1, traces a and f), which is not time-resolved, must be ascribed to an electrogenic event occurring before the electron is transferred from H to Q. According to our present knowledge, there are only two primary photosynthetic processes that take place on this time scale. One is the charge separation between P and H, and the other is the migration of the excitation energy toward the RC where trapping occurs. In contrast to the charge-separation step, this last process is expected to be nonelectrogenic. However, the latter will manifest itself as a photovoltage by feeding the initial charge-separation step in the RCs. If the rate constants of these two processes were comparable, both processes could contribute to the kinetics of the fast rise. However, picosecond spectroscopic measurements have shown that the state P^+H^- is created in <20 ps in isolated *R. viridis* RCs (5) and in 2.8 ps in RCs of *R. sphaeroides* (6). Therefore, the fast phase in our measurements should be assigned to the migration of energy in the antenna.

According to our deconvolution procedure, the molecular process giving rise to the photovoltage in Fig. 1, traces g and h and to the fast phase in Fig. 1, trace f has an exponential time constant of 30 ± 10 ps. The same constant is also calculated for the rise of the photovoltage in purple membranes (Fig. 1, trace e). This value must be considered as an upper estimate, since it represents the instrumental time resolution. Thus, our data show that the time constant for the trapping of excitons in *R. viridis* is <40 ps even at the lowest excitation energy applied, which corresponds to approximately 1 exciton per 20 RCs. To our knowledge, this is the fastest reported trapping time in an intact photosynthetic organism.

The common experimental approach to measure trapping times is fluorescence decay analysis. This assumes that the fluorescence decay time of the antenna is intimately connected with the mean trapping time. There are no lifetime data available on the fluorescence of antenna pigments in *R.*

viridis. However, in the case of *R. sphaeroides* and *Rhodospirillum rubrum* with open RCs, fluorescence lifetimes of 60 ± 10 ps have been reported (2, 3). The lifetime increases to ≈200 ps upon saturation of photochemistry by background illumination (2, 3). It would thus be of interest to analyze the kinetics of the electrogenic events in these species as a function of the state of the RCs and compare them to the fluorescence decay kinetics.

In earlier photoelectric measurements, it was shown that the amplitude of the open-circuit photovoltages is a measure of the relative positions of electron carriers in RCs (10). Comparisons of the initial photovoltage amplitudes from interfacial layers of *R. sphaeroides* chromatophores, in which Q was either oxidized or reduced, showed that H was slightly closer to Q (one-third) than to P (two-thirds) (10). In these experiments, the RCs in the reduced state ($P^+H^-Q^-$) were characterized by the back-reaction kinetics of 15 ns. In the experiments reported here, the two steps of charge separation were distinguished by their forward reactions. The calculated amplitudes of the two rising phases, in the case of *R. viridis*, are about equal (Fig. 1, traces a and f), indicating that H lies about halfway between P and Q. This is further demonstrated by the halved amplitude of the total photovoltage when Q is reduced (Fig. 1, traces b and d). The location of H in the middle between P and Q, which follows from our photoelectric study, is in agreement with the results of recent x-ray diffraction studies on the crystallized RC of the same bacterium (9). This agreement means that the implicit assumption made for the distance estimation—namely, a homogeneous dielectric between the different electron carriers—is reasonable.

A comparable yield of the charge separation when the first quinone is in the reduced or oxidized state is another prerequisite for the estimation of relative distances from photovoltage amplitudes. The agreement on the location of H derived from the photoelectric data and from the x-ray results indicates that the trapping yield is nearly independent of the redox state of Q. Alternatively, the photoelectric measurements on the position of H with respect to P and Q in the present study on *R. viridis* and in a recent one on *R. sphaeroides* (10) show that, with fully connected antenna systems, the thermodynamic equilibrium between the states P^*HQ^- and $P^+H^-Q^-$ strongly favors the charge-separated state. This may not be the case in photosystem II of green plants, although this is still an open question (19).

The improved time-resolution of photoelectric measurements down to 100 ps in this study has made possible an investigation of the dynamics of exciton migration and trapping by the electrogenicity connected with the first step of primary photosynthetic charge-separation. It is significant that, after picosecond excitation of the antenna system, only those excitons that are successfully utilized for photosynthesis contribute to the photovoltage. This new approach to the very primary photosynthetic processes is completely different from the widely applied approach of fluorescence decay analysis, which probes only those excitons that are not utilized for photosynthesis. This complicates considerably the interpretation of fluorescence measurements. For example, the fluorescence decay kinetics of dark-adapted photosynthetic organisms can correspond to either the migration of excitons in the antenna before formation of P^* or the disappearance of recombination luminescence upon charge stabilization on Q (20, 21). The lengthening of the fluorescence lifetime observed upon photooxidation of the RCs in *R. sphaeroides* (2, 3) can be assigned to either an increased distance of migration of the excitons on their way to an open RC or to a specific deactivation process occurring in the vicinity of closed RCs. In contrast, if exciton capture by the RC (trapping) is measured by the dielectric polarization connected with the first electron-transfer reaction, the results can be interpreted much more directly.

The present time-resolution is limited by the frequency of the oscilloscope available for this study. It is conceivable that with a faster oscilloscope, the time resolution of such photoelectric measurements can be increased further. Thus, for the study of primary photosynthetic events, fast photoelectric measurements turn out to be an alternative to measurements of absorption changes and fluorescence lifetimes.

APPENDIX

The kinetics of the photovoltage traces in Fig. 1 were analyzed by assuming the consecutive reactions

$$\text{Antenna*-PHQ} \xrightarrow{k_1} \text{P}^+\text{H}^-\text{Q} \xrightarrow{k_2} \text{P}^+\text{HQ}^-, \quad [1]$$

where k_1 and k_2 are the rate constants. The time-dependent concentrations of the states $\text{P}^+\text{H}^-\text{Q}$ and P^+HQ^- will be denoted $n_1(t)$ and $n_2(t)$, respectively. The expressions for $n_1(t)$ and $n_2(t)$ can be found in textbooks.

The formation of the states $\text{P}^+\text{H}^-\text{Q}$ and P^+HQ^- is connected with a dielectric polarization of the medium in a capacitor. The corresponding diplacement current is given by

$$I_0(t) = K_1\left(\frac{dn_1}{dt}\right) + K_2\left(\frac{dn_2}{dt}\right), \quad [2]$$

where K_1 and K_2 are electrogenicity parameters that represent the change of the dipole strength associated with the charge-transfer steps $\text{PHQ} \rightarrow \text{P}^+\text{H}^-\text{Q}$ and $\text{PHQ} \rightarrow \text{P}^+\text{HQ}^-$, respectively.

Introducing $n_1(t)$ and $n_2(t)$ in Eq. 2 yields

$$I_0(t) = \text{constant}$$
$$\times \left[\frac{k_1}{k_1 - k_2}(k_1 - Kk_2)e^{-k_1 t} + k_2(K-1)e^{-k_2 t} \right] \quad [3]$$

where $K \equiv K_2/K_1$.

In the special case of *R. viridis* cells prepared in the state PHQ^- prior to the picosecond flash, the displacement current follows from Eq. 3 by taking $K_2 = 0$ and $k_2 = 10^{-8}\,\text{s}^{-1}$ which describes the charge recombination between P^+ and H^- (5).

Supposing a linear system, the measured output signal $V(t)$ is determined by the convolution of the displacement current $I_0(t)$ with the weighting functions of the subsystems:

$$V(t) = L(t) \otimes I_0(t) \otimes C(t) \otimes D(t). \quad [4]$$

$L(t)$ is the time distribution of the laser pulse described by a gaussian function with half-width 30 ps. $D(t)$ is the weighting function of the detection device (amplifiers and oscilloscope) described by a gaussian function with half-width 100 ps. $C(t)$ is the weighting function of the measuring cell which integrates the displacement current $I_0(t)$. It is described by $C(t) = \text{constant} \times e^{-t/R\Gamma}$, where Γ is the capacitance of the coaxial cell and R is the input impedance of the preamplifier. The decay observed in the nanosecond time scale with dark-adapted cells reflects only the electric characteristic of the measuring cell and permits (Fig. 4a) evaluation of the time constant $\tau_c = R\Gamma$ to 3 ± 0.5 ns. The faster decay observed in

the case of preilluminated or dithionite-treated cells (Fig. 4b) was fitted using a 10-ns time constant for the charge-recombination process (5).

The deconvolution of curves g and h in Fig. 1, taking into account this back reaction, yields the rate constant k_1, whereas the parameter K is deduced from the relative amplitude of the photovoltage (Fig. 1, curves a and b or d). The value of k_2 is then obtained from the deconvolution of curve f, using the values of k_1 and K determined above.

The convolution calculations were made with an integrating step-width of 3 ps (digital resolution on the faster time scale). The iteration steps were equal to 10 ps and 25 ps for the time constants associated with k_1 and k_2. The fitting procedure minimized the χ^2 parameter.

Curves f, g, and h (Fig. 1) depict the experimental data points and the calculated curves corresponding to the best fit and to the fits obtained with the next lower and higher values of the parameters of our iterative calculation.

We are greatly indebted to A. Dobek for his contributions during the initial development of this technique and to R. Burgei and P. Bouyer (Département de Physique Nucléaire, Moyennes Energies, Centre d'Etudes Nucléaire de Saclay) for discussions and for their generous loan of the fast oscilloscope. We also acknowledge active support from W. Junge and stimulating discussions with W. Leibl, A. W. Rutherford, and W. W. Parson.

1. Campillo, A. J., Hyer, R. C., Monger, T. G., Parson, W. W. & Shapiro, S. L. (1977) *Proc. Natl. Acad. Sci. USA* **74**, 1997–2001.
2. Sebban, P. & Moya, I. (1983) *Biochim. Biophys. Acta* **722**, 436–442.
3. Borisov, A. Yu., Freiberg, A. M., Godik, V. I., Rebane, K. K. & Timpman, K. E. (1985) *Biochim. Biophys. Acta* **807**, 221–229.
4. Kirmaier, C., Holten, D. & Parson, W. W. (1985) *FEBS Lett.* **185**, 76–82.
5. Holten, D., Windsor, M. W., Parson, W. W. & Thornber, J. P. (1978) *Biochim. Biophys. Acta* **501**, 112–126.
6. Martin, J.-L., Breton, J., Hoff, A. J., Migus, A. & Antonetti, A. (1986) *Proc. Natl. Acad. Sci. USA* **83**, 957–961.
7. Carithers, R. P. & Parson, W. W. (1975) *Biochim. Biophys. Acta* **387**, 194–211.
8. Prince, R. C., Tiede, D. M., Thornber, J. P. & Dutton, P. L. (1977) *Biochim. Biophys. Acta* **462**, 467–490.
9. Deisenhofer, J., Epp, O., Miki, K., Huber, R. & Michel, H. (1984) *J. Mol. Biol.* **180**, 385–398.
10. Trissl, H. W. (1983) *Proc. Natl. Acad. Sci. USA* **80**, 7173–7177.
11. Trissl, H. W. (1985) *Biochim. Biophys. Acta* **806**, 124–135.
12. Trissl, H. W. & Kunze, U. (1985) *Biochim. Biophys. Acta* **806**, 136–144.
13. Cohen-Bazire, G., Sistrom, W. R. & Stanier, R. Y. (1957) *J. Cell. Comp. Physiol.* **49**, 25–68.
14. Auston, D. H. (1977) in *Topics in Applied Physics*, ed. Shapiro, S. L. (Springer, Berlin), Vol. 18, pp. 186–190.
15. Groma, G. I., Szabo, G. & Varo, Gy. (1984) *Nature (London)* **308**, 557–558.
16. Witt, H. T. & Zickler, A. (1973) *FEBS Lett.* **37**, 307–310.
17. Fowler, C. F. & Kok, B. (1974) *Biochim. Biophys. Acta* **357**, 308–318.
18. Trissl, H. W., Kunze, U. & Junge, W. (1982) *Biochim. Biophys. Acta* **682**, 364–377.
19. Van Grondelle, R. (1985) *Biochim. Biophys. Acta* **811**, 147–195.
20. Godik, V. I. & Borisov, A. Yu. (1979) *Biochim. Biophys. Acta* **548**, 296–308.
21. Breton, J. (1983) *FEBS Lett.* **159**, 1–5.

Hemoglobin Structure and Respiratory Transport

by M. F. Perutz

Why grasse is greene, or why our
blood is red,
Are mysteries which none have reach'd
unto.
In this low forme, poore soule, what
wilt thou doe?
—JOHN DONNE,
"Of the Progress of the Soule"

When I was a student, I wanted to solve a great problem in biochemistry. One day I set out from Vienna, my home town, to find the Great Sage at Cambridge. He taught me that the riddle of life was hidden in the structure of proteins, and that X-ray crystallography was the only method capable of solving it. The Sage was John Desmond Bernal, who had just discovered the rich X-ray-diffraction patterns given by crystalline proteins. We really did call him Sage, because he knew everything, and I became his disciple.

In 1937 I chose hemoglobin as the protein whose structure I wanted to solve, but the structure proved so much more complex than any solved before that it eluded me for more than 20 years. First fulfillment of the Sage's promise came in 1959, when Ann F. Cullis, Hilary Muirhead, Michael G. Rossmann, Tony C. T. North and I first unraveled the architecture of the hemoglobin molecule in outline [see "The Hemoglobin Molecule," by M. F. Perutz; SCIENTIFIC AMERICAN Offprint 196]. We felt like explorers who have discovered a new continent, but it was not the end of the voyage, because our much-admired model did not reveal its inner workings: it provided no hint about the molecular mechanism of respiratory transport. Why not? Well-intentioned colleagues were quick to suggest that our hard-won structure was merely an artifact of crystallization and might be quite different from the structure of hemoglobin in its living environment, which is the red blood cell.

Hemoglobin is the vital protein that conveys oxygen from the lungs to the tissues and facilitates the return of carbon dioxide from the tissues back to the lungs. These functions and their subtle interplay also make hemoglobin one of the most interesting proteins to study. Like all proteins, it is made of the small organic molecules called amino acids, strung together in a linear sequence called a polypeptide chain. The amino acids are of 20 different kinds and their sequence in the chain is genetically determined. A hemoglobin molecule is made up of four polypeptide chains, two alpha chains of 141 amino acid residues each and two beta chains of 146 residues each. The alpha and beta chains have different sequences of amino acids but fold up to form similar three-dimensional structures. Each chain harbors one heme, which gives blood its red color. The heme consists of a ring of carbon, nitrogen and hydrogen atoms called porphyrin, with an atom of iron, like a jewel, at its center. A single polypeptide chain combined with a single heme is called a subunit of hemoglobin or a monomer of the molecule. In the complete molecule four subunits are closely joined, as in a three-dimensional jigsaw puzzle, to form a tetramer.

Hemoglobin Function

In red muscle there is another protein, called myoglobin, similar in constitution and structure to a beta subunit of hemoglobin but made up of only one polypeptide chain and one heme. Myoglobin combines with the oxygen released by red cells, stores it and transports it to the subcellular organelles called mitochondria, where the oxygen generates chemical energy by the combustion of glucose to carbon dioxide and water. Myoglobin was the first protein whose three-dimensional structure was determined; the structure was

solved by my colleague John C. Kendrew and his collaborators.

Myoglobin is the simpler of the two molecules. This protein, with its 2,500 atoms of carbon, nitrogen, oxygen, hydrogen and sulfur, exists for the sole purpose of allowing its single atom of iron to form a loose chemical bond with a molecule of oxygen (O_2). Why does nature go to so much trouble to accomplish what is apparently such a simple task? Like most compounds of iron, heme by itself combines with oxygen so firmly that the bond, once formed, is hard to break. This happens because an iron atom can exist in two states of valency: ferrous iron, carrying two positive charges, as in iron sulfate, which anemic people are told to eat, and ferric iron, carrying three positive charges, as in iron oxide, or rust. Normally, ferrous heme reacts with oxygen irreversibly to yield ferric heme, but when ferrous heme is embedded in the folds of the globin chain, it is protected so that its reaction with oxygen is reversible. The effect of the globin on the chemistry of the heme has been explained only recently with the discovery that the irreversible oxidation of heme proceeds by way of an intermediate compound in which an oxygen molecule forms a bridge between the iron atoms of two hemes. In myoglobin and hemoglobin the folds of the polypeptide chain prevent the formation of such a bridge by isolating each heme in a separate pocket. Moreover, in the protein the iron is linked to a nitrogen atom of the amino acid histidine, which donates negative charge that enables the iron to form a loose bond with oxygen.

An oxygen-free solution of myoglobin or hemoglobin is purple like venous blood; when oxygen is bubbled through such a solution, it turns scarlet like arterial blood. If these proteins are to act as oxygen carriers, then hemoglobin must be capable of taking up oxygen in the

Reprinted from *Scientific American* **239**, 92–125 (1978); © Scientific American, Inc.

lungs, where it is plentiful, and giving it up to myoglobin in the capillaries of muscle, where it is less plentiful; myoglobin in turn must pass the oxygen on to the mitochondria, where it is still scarcer.

A simple experiment shows that myoglobin and hemoglobin can accomplish this exchange because there is an equilibrium between free oxygen and oxygen bound to heme iron. Suppose a solution of myoglobin is placed in a vessel constructed so that a large volume of gas can be mixed with it and so that its color can also be measured through a spectroscope. Without oxygen only the purple color of deoxymyoglobin is observed. If a little oxygen is injected, some of the oxygen combines with some of the deoxymyoglobin to form oxygen bound to heme iron.

myoglobin, which is scarlet. The spectroscope measures the proportion of oxymyoglobin in the solution. The injection of oxygen and the spectroscopic measurements are repeated until all the myoglobin has turned scarlet. The results are plotted on a graph with the partial pressure of oxygen on the horizontal axis and the percentage of oxymyoglobin on the vertical axis. The graph has

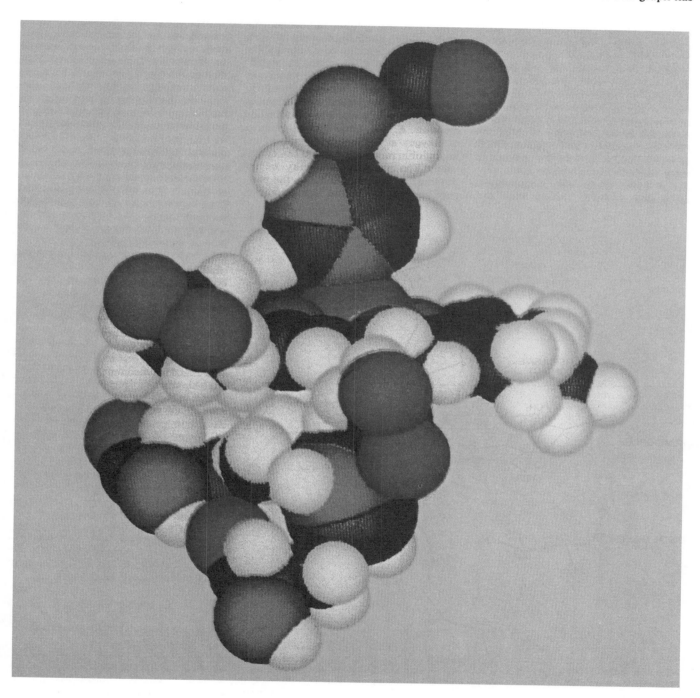

HEME GROUP is the active center of the hemoglobin molecule, the binding site for oxygen. The heme is a flat ring, called a porphyrin, with an iron atom at its center; it is seen here edge on and extending horizontally across the middle of the illustration. Three of the 16 amino acid residues of the globin that are in contact with the heme are also shown. In this computer-generated image each atom is represented by a sphere into which no other atom can penetrate unless the atoms are chemically bonded; where two atoms are bonded the spheres overlap. Carbon atoms are black, nitrogen atoms blue, oxygen atoms red, hydrogen atoms white and the iron atom is rust-colored. The model shows the deoxygenated heme; oxygen binds to the lower side of the iron atom. The picture was generated by Richard J. Feldmann and Thomas K. Porter of the National Institutes of Health from atomic coordinates determined by Giulio Fermi of the Medical Research Council Laboratory of Molecular Biology at Cambridge in England. A key to the structure is provided on page 4.

the shape of a rectangular hyperbola: it is steep at the start, when all the myoglobin molecules are free, and it flattens out at the end, when free myoglobin molecules have become so scarce that only a high pressure of oxygen can saturate them.

To understand this equilibrium one must visualize its dynamics. Under the influence of heat the molecules in the solution and in the gas are whizzing around erratically and are constantly colliding. Oxygen molecules are entering and leaving the solution, forming bonds with myoglobin molecules and breaking away from them. The number of iron-oxygen bonds that break in one second is proportional to the number of oxymyoglobin molecules. The number of bonds that form in one second is proportional to the frequency of collisions between myoglobin and oxygen, which is determined in turn by the product of their concentrations. When more oxygen is added to the gas, more oxygen molecules dissolve, collide with and

bind to myoglobin; this raises the number of oxymyoglobin molecules present and therefore also the number of iron-oxygen bonds liable to break, until the number of myoglobin molecules combining with oxygen in one second becomes equal to the number that lose their oxygen in one second. When that happens, a chemical equilibrium has been established.

The equilibrium is best represented by a graph in which the logarithm of the ratio of oxymyoglobin molecules (Y) to deoxymyoglobin molecules ($1 - Y$) is plotted against the logarithm of the partial pressure of oxygen. The hyperbola now becomes a straight line at 45 degrees to the axes. The intercept of the line with the horizontal axis drawn at $Y/(1 - Y) = 1$ gives the equilibrium constant K. This is the partial pressure of oxygen at which exactly half of the myoglobin molecules have taken up oxygen. The greater the affinity of the protein for oxygen, the lower the pressure needed to achieve half-saturation

and the smaller the equilibrium constant. The 45-degree slope remains unchanged, but lower oxygen affinity shifts the line to the right and higher affinity shifts it to the left.

If the same experiment is done with blood or with a solution of hemoglobin, an entirely different result is obtained. The curve rises gently at first, then steepens and finally flattens out as it approaches the myoglobin curve. This strange sigmoid shape signifies that oxygen-free molecules (deoxyhemoglobin) are reluctant to take up the first oxygen molecule but that their appetite for oxygen grows with the eating. Conversely, the loss of oxygen by some of the hemes lowers the oxygen affinity of the remainder. The distribution of oxygen among the hemoglobin molecules in a solution therefore follows the biblical parable of the rich and the poor: "For unto every one that hath shall be given, and he shall have abundance: but from him that hath not shall be taken away even that which he hath." This phenomenon suggests there is some kind of communication between the hemes in each molecule, and physiologists have therefore called it heme-heme interaction.

A better picture of the underlying mechanism of heme-heme interaction is obtained in a logarithmic graph. The equilibrium curve then begins with a straight line at 45 degrees to the axes, because at first oxygen molecules are so scarce that only one heme in each hemoglobin molecule has a chance of catching one of them, and all the hemes therefore react independently, as in myoglobin. As more oxygen flows in, the four hemes in each molecule begin to interact and the curve steepens. The tangent to its maximum slope is known as Hill's coefficient (n), after the physiologist A. V. Hill, who first attempted a mathematical analysis of the oxygen equilibrium. The normal value of Hill's coefficient is about 3; without heme-heme interaction it becomes unity. The curve ends with another line at 45 degrees to the axes because oxygen has now become so abundant that only the last heme in each molecule is likely to be free, and all the hemes in the solution react independently once more.

Cooperative Effects

Hill's coefficient and the oxygen affinity of hemoglobin depend on the concentration of several chemical factors in the red blood cell: protons (hydrogen atoms without electrons, whose concentration can be measured as pH), carbon dioxide (CO_2), chloride ions (Cl^-) and a compound of glyceric acid and phosphate called 2,3-diphosphoglycerate (DPG). Increasing the concentration of any of these factors shifts the oxygen equilibrium curve to the right, toward lower oxy-

CHEMICAL STRUCTURE of the heme group and surrounding amino acids is shown by a skeleton of lines connecting the centers of atoms. The only chemical bond between the heme and the protein that engulfs it is the link between the iron atom and the amino acid at the top, called the proximal histidine; the two amino acids at the bottom (the distal histidine and the distal valine) touch the heme but are not bonded to it. The proximal histidine is the principal path for communication between the heme and the rest of the molecule. In the deoxy state shown the iron protrudes above the porphyrin and may be hindered from returning to a centered position by repulsion between one corner of the proximal histidine and one of the porphyrin nitrogen atoms. Key was constructed with aid of a computer by R. Diamond of Cambridge.

gen affinity, and makes it more sigmoid. Increased temperature also shifts the curve to the right, but it makes it less sigmoid. Strangely, none of these factors, with the exception of temperature, influences the oxygen equilibrium curve of myoglobin, even though the chemistry and structure of myoglobin are related closely to those of the individual chains of hemoglobin.

What is the purpose of these extraordinary effects? Why is it not good enough for the red cell to contain a simple oxygen carrier such as myoglobin? Such a carrier would not allow enough of the oxygen in the red cell to be unloaded to the tissues, nor would it allow enough carbon dioxide to be carried to the lungs by the blood plasma. The partial pressure of oxygen in the lungs is about 100 millimeters of mercury, which is sufficient to saturate hemoglobin with oxygen whether the equilibrium curve is sigmoid or hyperbolic. In venous blood the pressure is about 35 millimeters of mercury; if the curve were hyperbolic, less than 10 percent of the oxygen carried would be released at that pressure, so that a man would asphyxiate even if he breathed normally.

The more pronounced the sigmoid shape of the equilibrium curve is, the greater the fraction of oxygen that can be released. Several factors conspire to that purpose. Oxidation of nutrients by the tissues liberates lactic acid and carbonic acid; these acids in turn liberate protons, which shift the curve to the right, toward lower oxygen affinity, and make it more sigmoid. Another important regulator of the oxygen affinity is DPG. The number of DPG molecules in the red cell is about the same as the number of hemoglobin molecules, 280 million, and probably remains fairly constant during circulation; a shortage of oxygen, however, causes more DPG to be made, which helps to release more oxygen. With a typical sigmoid curve nearly half of the oxygen carried can be released to the tissues. The human fetus has a hemoglobin with the same alpha chains as the hemoglobin of the human adult but different beta chains, resulting in a lower affinity for DPG. This gives fetal hemoglobin a higher oxygen affinity and facilitates the transfer of oxygen from the maternal circulation to the fetal circulation.

Carbon monoxide (CO) combines with the heme iron at the same site as oxygen, but its affinity for that site is 150 times greater; carbon monoxide therefore displaces oxygen, which explains why it is so toxic. In heavy smokers up to 20 percent of the oxygen combining sites can be blocked by carbon monoxide, so that less oxygen is carried by the blood. In addition carbon monoxide has an even more sinister effect. The combination of one of the four hemes in any hemoglobin molecule with carbon monoxide raises the oxygen affinity of the remaining three hemes by heme-heme interaction. The oxygen equilibrium curve is therefore shifted to the left, which diminishes the fraction of the oxygen carried that can be released to the tissues.

If protons lower the affinity of hemoglobin for oxygen, then the laws of action and reaction demand that oxygen lower the affinity of hemoglobin for protons. Liberation of oxygen causes hemoglobin to combine with protons and vice versa; about two protons are taken up for every four molecules of oxygen released, and two protons are liberated again when four molecules of oxygen are taken up. This reciprocal action is known as the Bohr effect and is the key to the mechanism of carbon dioxide transport. The carbon dioxide released by respiring tissues is too insoluble to be transported as such, but it can be rendered more soluble by combining with water to form a bicarbonate ion and a proton. The chemical reaction is written

$$CO_2 + H_2O \rightarrow HCO_3^- + H^+$$

In the absence of hemoglobin this reaction would soon be brought to a halt by the excess of protons produced, like a fire going out when the chimney is blocked. Deoxyhemoglobin acts as a buffer, mopping up the protons and tipping the balance toward the formation of soluble bicarbonate. In the lungs the process is reversed. There, as oxygen binds to hemoglobin, protons are cast off, driving carbon dioxide out of solution so that it can be exhaled. The reaction between carbon dioxide and water is catalyzed by carbonic anhydrase, an enzyme in the red cells. The enzyme speeds up the reaction to a rate of about half a million molecules per second, one of the fastest of all known biological reactions.

There is a second but less important mechanism for transporting carbon dioxide. The gas binds more readily to deoxyhemoglobin than it does to oxyhemoglobin, so that it tends to be taken up when oxygen is liberated and cast off when oxygen is bound. The two mechanisms of carbon dioxide transport are antagonistic: for each molecule of carbon dioxide bound to deoxyhemoglobin either one or two protons are released, which oppose the conversion of other molecules of carbon dioxide to bicarbonate. Positively charged protons entering the red cell draw negatively

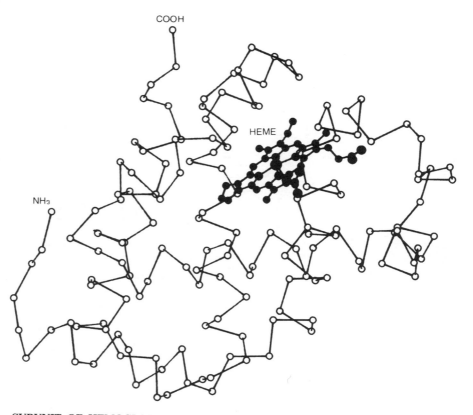

SUBUNIT OF HEMOGLOBIN consists of a heme group (*black*) enfolded in a polypeptide chain. The polypeptide is a linear sequence of amino acid residues, each of which is represented here by a single dot, marking the position of the central (alpha) carbon atom. The chain begins with an amino group (NH$_3$) and ends with a carboxyl group (COOH). Most of the polypeptide is coiled up to form helical segments but there are also nonhelical regions. The computer-generated diagram of a horse-hemoglobin subunit was prepared by Feldmann and Porter.

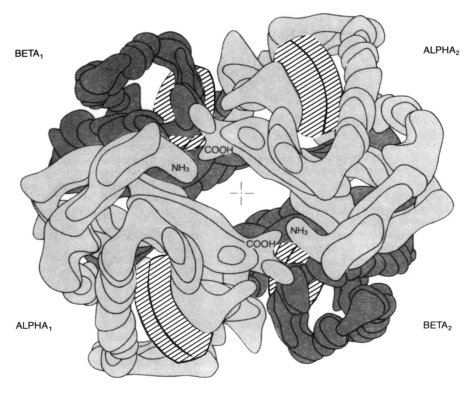

BETA₁   ALPHA₂

COOH
NH₃

NH₃
COOH

ALPHA₁   BETA₂

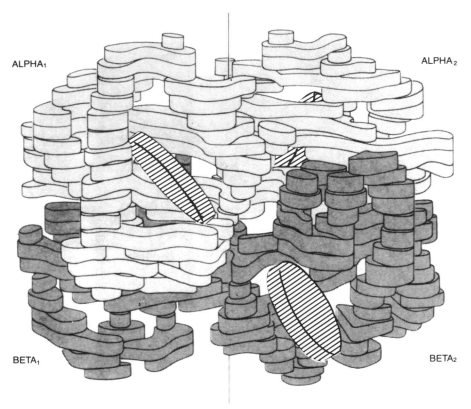

ALPHA₁   ALPHA₂

BETA₁   BETA₂

COMPLETE MOLECULE of hemoglobin is made up of four subunits, each of which consists of one polypeptide chain and one heme. There are two kinds of subunit, designated alpha (*white*) and beta (*gray*), which have different sequences of amino acid residues but similar three-dimensional structures. The beta chain also has one short extra helix. The four subunits, seen here in two views, are arranged at the vertexes of a tetrahedron around an axis of two-fold symmetry. Each heme (*lines*) lies in a separate pocket at the surface of the molecule.

charged chloride ions in with them, and these ions too are bound more readily by deoxyhemoglobin than by oxyhemoglobin. DPG is synthesized in the red cell itself and cannot leak out through the cell membrane. It is strongly bound by deoxyhemoglobin and only very weakly bound by oxyhemoglobin.

Heme-heme interaction and the interplay between oxygen and the other four ligands are known collectively as the cooperative effects of hemoglobin. Their discovery by a succession of able physiologists and biochemists took more than half a century and aroused many controversies. In 1938 Felix Haurowitz of the Charles University in Prague made another vital observation. He discovered that deoxyhemoglobin and oxyhemoglobin form different crystals, as though they were different chemical substances, which implied that hemoglobin is not an oxygen tank but a molecular lung because it changes its structure every time it takes up oxygen or releases it.

Theory of Allostery

The discovery of an interaction among the four hemes made it obvious that they must be touching, but in science what is obvious is not necessarily true. When the structure of hemoglobin was finally solved, the hemes were found to lie in isolated pockets on the surface of the subunits. Without contact between them how could one of them sense whether the others had combined with oxygen? And how could as heterogeneous a collection of chemical agents as protons, chloride ions, carbon dioxide and diphosphoglycerate influence the oxygen equilibrium curve in a similar way? It did not seem plausible that any of them could bind directly to the hemes, that all of them could bind at any other common site, although there again it turned out we were wrong. To add to the mystery, none of these agents affected the oxygen equilibrium of myoglobin or of isolated subunits of hemoglobin. We now know that all the cooperative effects disappear if the hemoglobin molecule is merely split in half, but this vital clue was missed. Like Agatha Christie, nature kept it to the last to make the story more exciting.

There are two ways out of an impasse in science: to experiment or to think. By temperament, perhaps, I experimented, whereas Jacques Monod thought. In the end our paths converged.

Monod's scientific life had been devoted to finding out what regulates the growth of bacteria. The key to this problem appeared to be regulation of the synthesis and catalytic activity of enzymes. Monod and François Jacob had discovered that the activity of certain enzymes is controlled by switching their synthesis on and off at the gene: they and others then found a second mode of regulation that appeared to operate switches on the enzymes themselves.

In 1965 Monod and Jean-Pierre Changeux of the Pasteur Institute in Paris, together with Jeffries Wyman of the University of Rome, recognized that the enzymes in the latter class have certain features in common with hemoglobin. They are all made of several subunits, so that each molecule includes several sites with the same catalytic activity, just as hemoglobin includes several hemes that bind oxygen, and they all show similar cooperative effects. Monod and his colleagues knew that deoxyhemoglobin and oxyhemoglobin have different structures, which made them suspect that the enzymes too may exist in two (or at least two) structures. They postulated that these structures should be distinguished by the arrangement of the subunits and by the number and strength of the bonds between them.

If there are only two alternative structures, the one with fewer and weaker bonds between the subunits would be free to develop its full catalytic activity (or oxygen affinity); this structure has therefore been labeled R, for "relaxed." The activity would be damped in the structure with more and stronger bonds between the subunits; this form is called T, for "tense." In either of these structures the catalytic activity (or oxygen affinity) of all the subunits in one molecule should always remain equal. This postulate of symmetry allowed the properties of allosteric enzymes to be described by a neat mathematical theory with only three independent variables: K_R and K_T, which in hemoglobin denote the oxygen equilibrium constants of the R and T structures respectively, and L, which stands for the number of molecules in the T structure divided by the number in the R structure, the ratio being measured in the absence of oxygen. The term allostery (from the Greek roots *allos*, "other," and *stereos*, "solid") was coined because the regulator molecule that switches the activity of the enzyme on or off has a structure different from that of the molecule whose chemical transformation the enzyme catalyzes.

This ingenious theory simplified the interpretation of the cooperative effects enormously. The progressive increase in oxygen affinity illustrated by the parable of the rich and the poor now arises not from any direct interaction between the hemes but from the switchover from the T structure with low affinity to the R structure with high affinity. This transformation should take place either when the second molecule of oxygen is bound or when the third is bound. Chemical agents that do not bind to the hemes might lower the oxygen affinity by biasing the equilibrium between the two structures toward the T form, which would make the transition to the R structure come after, say, three mole-

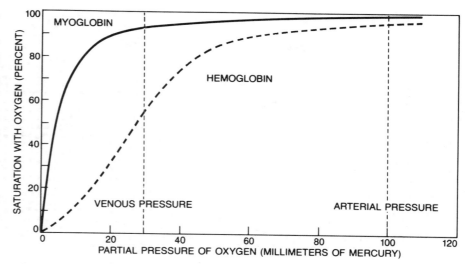

EQUILIBRIUM CURVES measure the affinity for oxygen of hemoglobin and of the simpler myoglobin molecule. Myoglobin, a protein of muscle, has just one heme group and one polypeptide chain and resembles a single subunit of hemoglobin. The vertical axis gives the amount of oxygen bound to one of these proteins, expressed as a percentage of the total amount that can be bound. The horizontal axis measures the partial pressure of oxygen in a mixture of gases with which the solution is allowed to reach equilibrium. For myoglobin (*black*) the equilibrium curve is hyperbolic. Myoglobin absorbs oxygen readily but becomes saturated at a low pressure. The hemoglobin curve (*dashed line*) is sigmoid: initially hemoglobin is reluctant to take up oxygen, but its affinity increases with oxygen uptake. At arterial oxygen pressure both molecules are nearly saturated, but at venous pressure myoglobin would give up only about 10 percent of its oxygen, whereas hemoglobin releases roughly half. At any partial pressure myoglobin has a higher affinity than hemoglobin, which allows oxygen to be transferred from blood to muscle.

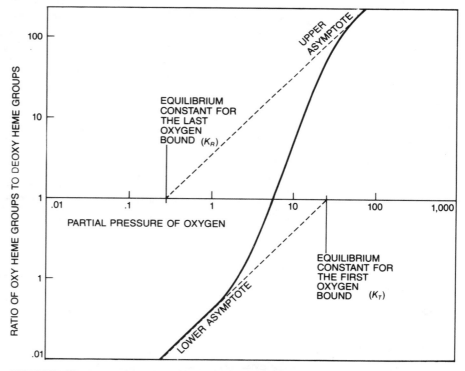

SIGMOID SHAPE of the oxygen equilibrium curve appears more pronounced when the fractional saturation and partial pressure of oxygen are plotted on logarithmic scales. On such a graph the equilibrium curve for myoglobin becomes a straight line at 45 degrees to the axes. The hemoglobin curve begins and ends with straight lines, called asymptotes, at the same angle. Their intercepts with the horizontal line drawn where the concentrations of deoxyhemoglobin and oxyhemoglobin are equal give the equilibrium constants for the first and last oxygen molecules to combine with hemoglobin. In the allosteric interpretation of the curve these are respectively the equilibrium constants of the T structure (K_T) and the R structure (K_R). For the curve shown the two constants are respectively 30 and .3, indicating that the affinity for the last oxygen bound is 100 times the affinity for the first. This ratio determines the free energy of the heme-heme interaction, which is a measure of the influence exerted by the combination of any one of the four hemes with oxygen on the oxygen affinity of the remaining hemes. If the beginning and end of the curve cannot be measured accurately, the maximum slope of the curve, known as Hill's coefficient, indicates the degree of the heme-heme interaction.

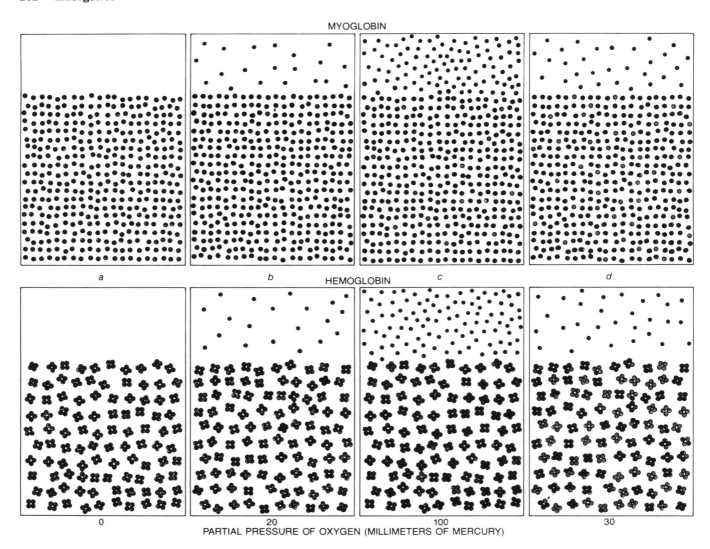

MYOGLOBIN

HEMOGLOBIN

| a | b | c | d |

PARTIAL PRESSURE OF OXYGEN (MILLIMETERS OF MERCURY)

0   20   100   30

HEME-HEME INTERACTION facilitates the release of oxygen by hemoglobin but is not observed in myoglobin. The proteins are shown gray in the deoxy state and black when oxygenated; there are four times as many molecules of myoglobin as hemoglobin but equal numbers of binding sites. In the absence of oxygen (*a*) almost all the binding sites are empty. As the oxygen pressure rises (*b*) myoglobin takes up oxygen more quickly, but when the pressure reaches that of the lungs (*c*), both carriers are approaching saturation. The distinction between the two molecules becomes most apparent when the oxygen pressure falls to a level typical of the capillary system (*d*). The oxygen equilibrium of myoglobin changes little, but hemoglobin sheds about 45 percent of the oxygen it carries. The difference is accounted for by the amplifying effect of the heme-heme interaction. Few hemoglobin molecules bear one or two oxygens: if a hemoglobin molecule takes up one oxygen, it tends to go on and acquire four oxygens, and if a saturated molecule loses one oxygen, two or three more oxygens are usually cast off. The data for hemoglobin are derived from curves calculated by Joyce M. Baldwin of Cambridge.

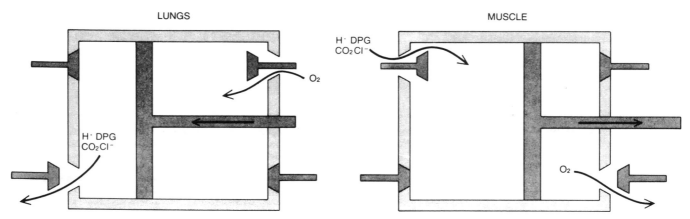

LUNGS

MUSCLE

H· DPG
CO_2 Cl^-

O_2

H· DPG
CO_2 Cl^-

O_2

RECIPROCATING ENGINE serves as a model of the cooperative effects of hemoglobin. The piston is driven to the right by the energy liberated in the reaction of hemoglobin with oxygen (O_2) and to the left by the protons (H^+) and carbon dioxide (CO_2) liberated by respiring tissues. Diphosphoglycerate (DPG) and chloride ions (Cl^-) are passengers riding in company with protons and carbon dioxide.

cules of oxygen have been bound rather than after two molecules have been bound. In terms of allosteric theory such agents would raise L, the fraction of molecules in the T structure, without altering the oxygen equilibrium constants K_T and K_R of the two structures.

Atomic Structures

My own approach to the problem was also influenced by Haurowitz' discovery that oxyhemoglobin and deoxyhemoglobin have different structures. Gradually I came to realize that we would never explain the intricate functions of hemoglobin without solving the structures of both crystal forms at a resolution high enough to reveal atomic detail.

In 1970, 33 years after I had taken my first X-ray-diffraction pictures of hemoglobin, that stage was finally reached. Hilary Muirhead, Joyce M. Baldwin, Gwynne Goaman and I got a good map of the distribution of matter not in oxyhemoglobin but in the closely related methemoglobin of horse, in which the iron is ferric and the place of oxygen is taken by a water molecule. William Bolton and I got a map of horse deoxyhemoglobin, and Muirhead and Jonathan Greer got one of human deoxyhemoglobin. These maps served as guides for the construction of three atomic models, each a jungle of brass spokes and steel connectors supported on brass scaffolding, edifices of labyrinthine complexity nearly four feet in diameter. At first it was hard to see the trees for the forest.

In allosteric terms our methemoglobin model represented the R structure and our two deoxyhemoglobin models the T structure. We scanned them eagerly for clues to the allosteric mechanism but could not see any at first because the general structure of the subunits was similar in all three models. The alpha chains included seven helical segments and the beta chains eight helical segments interrupted by corners and nonhelical segments. Each chain enveloped its heme in a deep pocket, which exposed only the edge where two propionic acid side chains of the porphyrin dip into the surrounding water.

The heme makes contact with 16 amino acid side chains from seven segments of the chain. Most of these side chains are hydrocarbons; the two exceptions are the heme-linked histidines, which lie on each side of the heme and play an important part in the binding of oxygen. The side chain of histidine ends in an imidazole ring made of three carbon atoms, two nitrogen atoms and either four or five hydrogen atoms. One of these histidines, called the proximal histidine, forms a chemical bond with the heme iron [see illustrations on pages 3 and 4]. The other histidine, called the distal one, lies on the opposite side of the heme, in

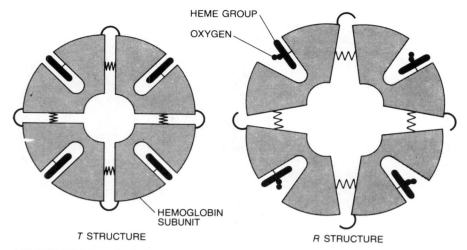

HEME GROUP

OXYGEN

HEMOGLOBIN SUBUNIT

T STRUCTURE   R STRUCTURE

ALLOSTERIC THEORY explains heme-heme interaction without postulating any direct communication between the heme groups. The hemoglobin molecule is assumed to have two alternative structures, designated T for tense and R for relaxed. In the T structure the subunits of the molecule are clamped against the pressure of springs and their narrow pockets impede the entry of oxygen. In the R structure all the clamps have sprung open and the heme pockets are wide enough to admit oxygen easily. Uptake of oxygen by the T structure would strain the clamps until they all burst open in concert and allow the molecule to relax to the R structure. Loss of oxygen would narrow the heme pockets and allow the T structure to re-form.

contact with it and with the bound oxygen but without forming a covalent chemical bond with either. Apart from these histidines, most of the side chains in the interior of the subunits, like those near the hemes, are hydrocarbons. The exterior of the hemoglobin molecule is lined with side chains of all kinds, but electrically charged and dipolar ones predominate. Thus each subunit is waxy inside and soapy outside, which makes it soluble in water but impermeable to it.

The four subunits are arranged at the vertexes of a tetrahedron around a twofold symmetry axis. Since a tetrahedron has six edges, there are six areas of contact between the subunits. The

twofold symmetry leaves four distinct contacts, which cover about a fifth of the surface area of the subunits. Sixty percent of that area is made up of the alpha$_1$-beta$_1$ and alpha$_2$-beta$_2$ contacts, each of which includes about 35 amino acid side chains tightly linked by from 17 to 19 hydrogen bonds. [Hydrogen bonds are made between atoms of nitrogen (N) and oxygen (O) through an intermediate hydrogen atom (H), for instance N–H...N, N–H...O, O–H...O or O–H...N. The hydrogen is bonded strongly to the atom on the left and weakly to the one on the right.]

The numerous hydrogen bonds between the alpha$_1$-beta$_1$ and alpha$_2$-beta$_2$

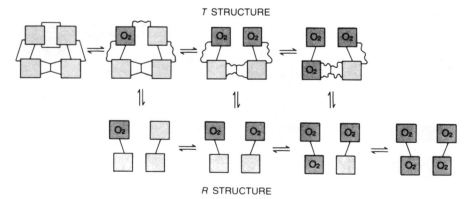

T STRUCTURE

R STRUCTURE

TRANSITION from the T structure to the R structure increases in likelihood as each of the four heme groups is oxygenated. In this more realistic model, salt bridges linking the subunits in the T structure break progressively as oxygen is added, and even those salt bridges that have not yet ruptured are weakened, a process that is represented here by making the lines wavy. The transition from T to R does not take place after a fixed number of oxygen molecules have been bound, but it becomes more probable with each successive oxygen bound. The transition between the two structures is influenced by several factors, including protons, carbon dioxide, chloride and DPG. The higher their concentration is, the more oxygen must be bound to trigger the transition. Fully saturated molecules in the T structure and fully deoxygenated molecules in the R structure are not shown because they are too unstable to exist in significant numbers.

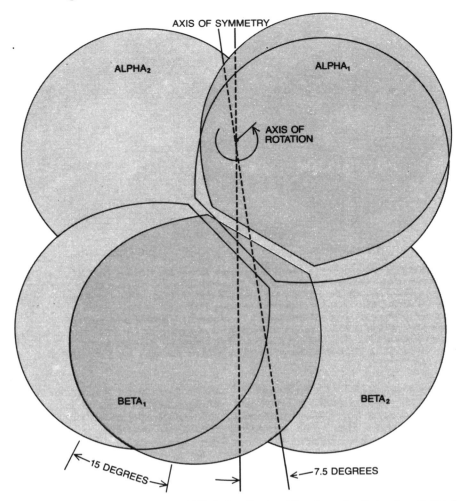

REARRANGEMENT OF SUBUNITS during the transition from the T structure to the R structure consists mainly in a rotation of one pair of subunits with respect to the other pair. Each alpha chain is bonded strongly to one beta chain, and the dimers formed in this way move as rigid bodies. If one dimer is held fixed, the other turns by 15 degrees about an off-center axis and shifts slightly along it. The twofold symmetry of the molecule is preserved, but the axis of symmetry is rotated by 7.5 degrees. The diagram is based on one prepared by Baldwin.

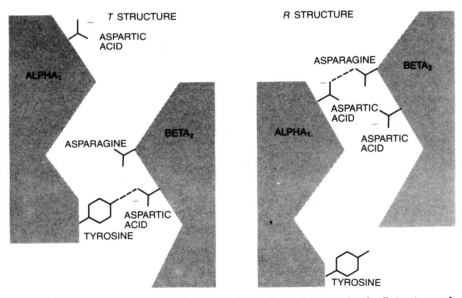

CONTACT between the two dimers has two stable conformations, one for the T structure and the other for the R structure. On transition between the structures the dimers snap from one position to the other. They are stabilized by alternative sets of hydrogen bonds formed between amino acid side chains attached to the opposing faces of the dimers. The two bonds shown here were first discovered by X-ray crystallography. In 1975 Leslie Fung and Chien Ho at the University of Pittsburgh demonstrated the presence of these bonds in solution. This provides evidence that the two structures found in crystals are the same as the structures in red blood cells.

subunits make them cohere so strongly that their contact is hardly altered by the reaction with oxygen, and they move as rigid bodies in the transition between the T and the R structures. On the other hand, the contact alpha$_1$-beta$_2$ in the R structure looked quite different from that in the T structure. This contact includes fewer side chains than alpha$_1$-beta$_1$ and is designed so that it acts as a snap-action switch, with two alternative stable positions, each braced by a different set of hydrogen bonds. We wondered at first whether these bonds were stronger and more numerous in the T structure than they are in the R structure, but that did not seem to be the case.

Where, then, were the extra bonds between the subunits in the T structure that allosteric theory demanded? We spotted them at the ends of the polypeptide chains. In the T structure the last amino acid residue of each chain forms salt bridges with neighboring subunits. (A salt bridge is a bond between a nitrogen atom, carrying a positive charge, and an oxygen atom, carrying a negative charge.) In our maps of the R structure the last two residues of each chain were blurred. At first I suspected this to be due to error, but improved maps made by my colleagues Elizabeth Heidner and Robert Ladner have convinced us that the final residues remain invisible because they are no longer tethered and wave about like reeds in the wind.

Geometrically, the transition between the two structures consists of a rock-and-roll movement of the dimer alpha$_1$-beta$_1$ with respect to the dimer alpha$_2$-beta$_2$. Baldwin has shown that if one dimer is held fixed, the movement of the other one can be represented by a rotation of some 15 degrees about a suitably placed axis together with a small shift along the same axis. The movement is brought about by subtle changes in the internal structure of the subunits that accompany the binding and dissociation of oxygen.

Function of the Salt Bridges

The salt bridges at the ends of the polypeptide chains clearly provide the extra bonds between the subunits in the T structure predicted by Monod, Changeux and Wyman. They also explain the influence on the oxygen equilibrium curve of all the chemical factors that had puzzled us so much. All agents that lower the oxygen affinity do so either by strengthening existing salt bridges in the T structure or by adding new ones. Not all these extra bonds, however, are between the subunits; some are within the subunits and oppose the subtle structural changes the subunits undergo on combination with oxygen.

The salt bridges explain both the lowering of the oxygen affinity by protons and the uptake of protons on release of oxygen. Protons increase the number of nitrogen atoms carrying a positive charge. For example, the imidazole ring of the amino acid histidine can exist in two states, uncharged when only one of its nitrogen atoms carries a proton and positively charged when both do. In neutral solution each histidine has a 50 percent chance of being positively charged. The more acid the solution, or in other words the higher the concentration of protons, the greater the chance of a histidine becoming positively charged and forming a salt bridge with an oxygen atom carrying a negative charge. Conversely, the transition from the R structure to the T structure brings negatively charged oxygen atoms into proximity with an uncharged nitrogen atom and thereby diminishes the work that has to be done to give the nitrogen atom a positive charge. As a result a histidine that has no more than a 50 percent chance of being positively charged in the R structure has a 90 percent chance in the T structure, so that more protons are taken up from the solution by hemoglobin in the T structure.

Hemoglobin includes one other set of groups that behave in this way: they are the amino groups at the start of the polypeptide chains, but their nitrogen atoms take up protons only if the concentration of carbon dioxide is low. If it is high, these nitrogens are liable to lose protons and to combine instead with carbon dioxide to form a carbamino compound. The physiologists F. J. W. Roughton and J. K. W. Ferguson proposed in 1934 that this mechanism plays a part in the transport of carbon dioxide, but their proposal was treated with skepticism until it was confirmed 35 years later by my colleague John Kilmartin, working with Luigi Rossi-Bernardi at the University of Milan. I was pleased that Roughton, who had fathered their experiment, was still alive to see his ideas vindicated. My colleague Arthur R. Arnone, now at the University of Iowa, then showed that in the T structure such carbamino groups, which carry a negative charge, form salt bridges with positively charged groups of the globin and are therefore more stable than they are in the R structure. This finding explains why deoxyhemoglobin has a higher affinity for carbon dioxide than oxyhemoglobin and conversely why carbon dioxide lowers the oxygen affinity of hemoglobin.

The positions in hemoglobin taken up by chloride ions are still uncertain. Arnone has spotted sites in the T structure where other negatively charged ions bind, and these might also be the chloride binding sites. If they are, then chloride ions also brace the T structure by forming additional salt bridges.

The most striking difference between the T and the R structures is the width of the gap between the two beta chains. In the T structure the two chains are widely separated and the opening between

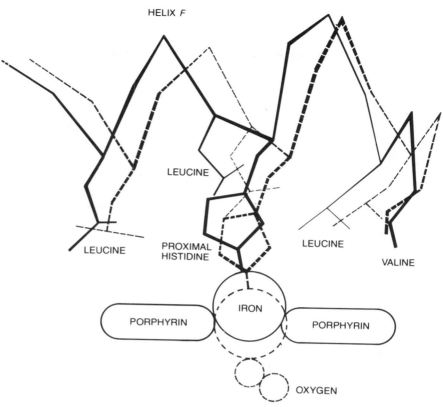

TRIGGERING MECHANISM for the transition from the T to the R structure is a movement of the heme iron into the plane of the porphyrin ring. In the T structure (*black lines*) the center of the iron atom is about .6 angstrom unit above the plane. (One angstrom unit is 10^{-10} meter.) When the molecule switches to the R structure (*dashed lines*), the iron moves into the plane, pulling with it the proximal histidine and helix F. Once the iron has descended into the plane it can readily bind an oxygen molecule. In the reverse transition (from R to T) the iron is pulled out of the plane and the oxygen cannot follow because it bumps against the porphyrin nitrogen atoms. The iron-oxygen bond is thereby weakened and usually breaks. These movements are transmitted to the contacts between the subunits and promote the transitions between the T and R structures. Diagram is based on a drawing by John Cresswell of University College London.

them is lined by amino acid side chains carrying positive charges. This opening is tailor-made to fit the molecule of 2,3-diphosphoglycerate and to compensate its negative charges, so that the binding of DPG adds another set of salt bridges to the T structure. In the R structure the gap narrows, and DPG has to drop out.

The Trigger

How does combination of the heme irons with oxygen make the subunits click from the T structure to the R structure? Compared with the hemoglobin molecule, an oxygen molecule is like the flea that makes the elephant jump. Conversely, how does the T structure impede the uptake of oxygen? What difference between the two structures is there at the heme that could bring about a several-hundred-fold change in oxygen affinity?

In oxyhemoglobin the heme iron is bound to six atoms: four nitrogen atoms of the porphyrin, which neutralize the two positive charges of the ferrous iron; one nitrogen atom of the proximal histidine, which links the heme to one of the helical segments of the polypep-

tide chain (helix F), and one of the two atoms of the oxygen molecule. In deoxyhemoglobin the oxygen position remains empty, so that the iron is bound to only five atoms.

I wondered whether the heme pockets might be narrower in the T structure than in the R structure, so that they had to widen to let the oxygen in. This widening might be geared to break the salt bridges, rather like the childish mechanism shown in the illustration at the top of page 105. When the atomic model of horse deoxyhemoglobin emerged, Bolton and I saw some truth in this idea because in the beta subunits a side chain of the amino acid valine next to the distal histidine blocked the site that oxygen would have to occupy. The alpha subunits, however, showed no such obstruction. Then we noticed the odd positions of the iron atoms. In methemoglobin, which has the R structure, the iron atoms had been displaced very slightly from the porphyrin plane toward the proximal histidine, but in deoxyhemoglobin (with the T structure) the displacement stood out as one of the most striking features of our maps. In each subunit the iron atom had carried the proximal histidine and helix *F* with it, so that they too had moved away from the porphyrin plane. I quickly realized that this might be the long-sought trigger.

Recently Arnone and my colleague Lynn Ten Eyck have obtained an excellent map of human deoxyhemoglobin to which Giulio Fermi fitted an atomic model of heme by computer methods. Fermi's calculations show that each iron atom is displaced by .6 ($\pm$.1) angstrom unit from the mean plane of the porphyrin. (One angstrom unit is 10^{-10} meter.) The nitrogen atom of the proximal histidine, to which the iron atom is bound, lies at a distance of 2.7 ($\pm$.1) angstroms from the same plane. So far we have no direct measure of the corresponding displacements in oxyhemoglobin because

oxyhemoglobin oxidizes to methemoglobin in the X-ray beam. There the iron atoms are displaced from the porphyrin plane by .1 angstrom in the alpha subunits and by .2 angstrom in the beta subunits; the corresponding displacements of the histidine nitrogens are 2.2 and 2.4 angstroms. Judging by the structures of model compounds, the displacement of the histidine nitrogen in oxyhemoglobin should be 2.1 angstroms, which means that the nitrogen would be .6 angstrom closer to the porphyrin plane than it is in deoxyhemoglobin. This shift would trigger the transition from the *T* structure to the *R* structure.

How is this movement transmitted to the contacts between the subunits and to

the salt bridges? One might as well puzzle out how a cat jumps off a wall from one picture of the cat on the wall and another of it on the ground, because our static models of deoxyhemoglobin and methemoglobin do not show what happens in the transition between the *T* and the *R* structures. I tried a bold guess. The second amino acid residue from the end in each chain is a tyrosine whose side chain carries a phenol group, that is, a benzene ring with a hydroxyl group (OH) attached. In the *T* structure the tyrosine in each subunit is wedged into a pocket between helixes *F* and *H* and its hydroxyl group is tethered by a hydrogen bond to an oxygen atom in the *FG* segment of the main polypeptide chain. In carbonmonoxyhemoglobin, which is the nearest relative of oxyhemoglobin and which has the *R* structure, the tyrosines are free. Hence there must be some mechanism that loosens the tyrosines when oxygen is bound.

As I wondered what this mechanism might be I saw that the movement of the proximal histidine toward the porphyrin plane that accompanies oxygen binding pulls helix *F* in a direction that narrows the pocket into which the tyrosine must fit. If the tyrosine were squeezed out of its pocket, it would tear the last amino acid residue of the chain away from its salt-bridged partner. In this way one salt bridge might be broken for each heme that combined with oxygen in the *T* structure. When enough salt bridges have been broken, the *T* structure would become unstable and click to the *R* structure.

If movement of the proximal histidine and the iron toward the porphyrin puts into motion a set of levers that loosens the tyrosines and breaks the salt bridges, then the making of the bridges and the binding of the tyrosines into their pockets must cause the same set of levers to go into reverse and move the histidine and the iron away from the porphyrin. The oxygen molecule on the other side cannot follow because it bumps against the four porphyrin nitrogen atoms, and so the iron-oxygen bond is stretched until it finally snaps.

To be guided by the atomic models toward the molecular mechanism of respiratory transport seemed like a dream. But was it true? Would the mechanism stand the cold scrutiny of experiment? It has been said that scientists do not pursue the truth, it pursues them.

Testing the Salt Bridges

According to allosteric theory, there should be no heme-heme interaction without a transition between the *T* and the *R* structures. This prediction was also tested by Kilmartin. He cleaved the final amino acid residue from the ends of all four polypeptide chains, so that

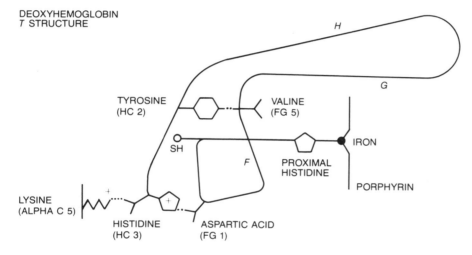

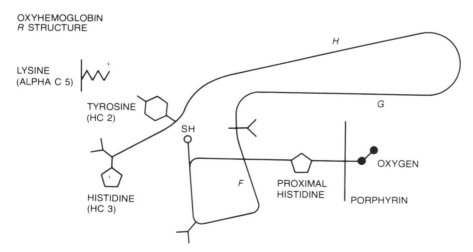

EXTRA BONDS in deoxyhemoglobin are formed by the last two residues of the beta chain. In oxyhemoglobin the iron atom lies in the plane of the porphyrin, the sulfhydryl group (SH) of the amino acid cysteine lies in the pocket between helixes *F* and *H*, and tyrosine *HC* 2, histidine *HC* 3 and lysine *C* 5 are free. In deoxyhemoglobin the iron is displaced from the plane of the porphyrin toward helix *F*, and the tyrosine has displaced the SH group from the pocket between *F* and *H* and forms a hydrogen bond with valine *FG* 5. Finally, the terminal histidine forms salt bridges with aspartic acid *FG* 1 of the same chain and with lysine *C* 5 of the alpha chain. Formation of bridge causes histidine to take up a proton and become positively charged.

there should be no salt bridges to stabilize the T structure. This modified hemoglobin maintained the R structure even in the absence of oxygen and showed a hyperbolic oxygen equilibrium curve with high oxygen affinity. Kilmartin then selected an abnormal human hemoglobin that can be made to maintain the T structure even when it is saturated with oxygen. Again the curve was hyperbolic, but it was shifted to lower oxygen affinity, so that the central thesis of allostery was proved.

The next question concerned the exact role of the salt bridges. At one extreme the strain arising from the combination of any one heme with oxygen might be distributed uniformly throughout the molecule, so that there would be no change in oxygen affinity until all the salt bridges broke in unison, when the structure clicked from T to R. This would fit pure allosteric theory, according to which the salt bridges should do no more than raise L, the fraction of molecules in the T structure. According to my mechanism, on the other hand, a salt bridge should break every time an oxygen combines with the T structure. If that were true, the salt bridges should raise K_T as well as L, that is to say, they should also lower the oxygen affinity of the T structure. To the general reader this may seem like a fine distinction, but to the workers in the field it seemed to cut at the roots of the mechanism and raised passionate controversies that are still going on.

Several experimental results favor my version of the mechanism. In 1965 Eraldo Antonini, Todd M. Schuster, Maurizio Brunori and Jeffries Wyman at the University of Rome, and in 1970 R. D. Gray at Cornell University, showed that binding of oxygen and liberation of Bohr protons go hand in hand right from the start, while hemoglobin is still in the T structure. Kilmartin then showed that most of the Bohr protons come from the rupture of salt bridges. Taken together, the two results prove that salt bridges are broken on binding of oxygen by the T structure, which implies that they lower its oxygen affinity. To test whether they really do we had to make an accurate comparison of the oxygen equilibrium curves of normal hemoglobin and a hemoglobin that lacks one of the salt bridges. Kiyohiro Imai and Hideki Morimoto of the University of Osaka had just developed an ingenious method that allows an oxygen equilibrium curve to be measured precisely and fast with only .1 milliliter of hemoglobin solution. Imai came to Cambridge to build one of his new machines and with Kilmartin measured the equilibrium curves of hemoglobins that lacked specific salt bridges. They found that the absence of any of the bridges left K_R unchanged but low-

ered both L and K_T, in accord with my mechanism.

Paradoxically, another set of observations contradicts these findings. My colleagues Leigh Anderson and Kilmartin, together with Seiji Ogawa of Bell Laboratories, have shown that the salt bridges break only if the hemoglobin molecule is free to click to the R structure but not if that transition is stopped. This happens in certain abnormal human hemoglobins and in fish hemoglobins in acid solution, where the T structure is unusually stable. It seems the T structure must be free to bend and stretch so as to shake off its shackles; if it is laced too tightly, it fails to respond.

I have suggested that the transition from the T to the R structure is triggered

mainly by the movement of the heme iron toward the porphyrin ring. What makes the iron move? There are two reasons, one steric and the other electronic. If the iron is bound to atoms on both sides of the heme, then their attraction by the iron and repulsion by the porphyrin nitrogens tend to balance the iron in the center of the ring. On the other hand, if the iron is bound only to the proximal histidine while the oxygen site is empty, then repulsion between the porphyrin nitrogens and the histidine is not balanced by repulsion between the porphyrin nitrogens and the oxygen, so that the histidine is pushed away from the porphyrin and pulls the iron with it.

The electronic story is more complex. The ferrous iron atom has six outermost

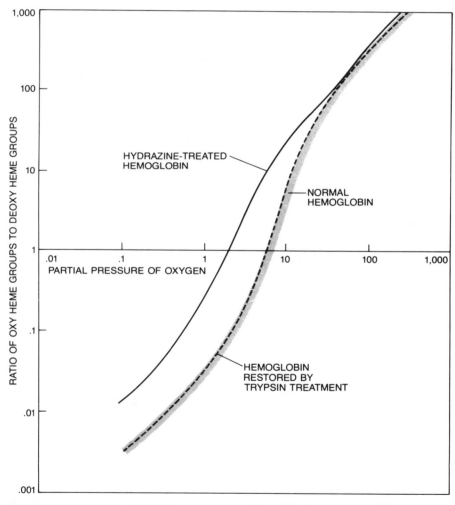

BLOCKING OF SALT BRIDGES in the open position shifts the oxygen equilibrium curve to the left, toward higher oxygen affinity. The initial equilibrium curve for normal hemoglobin is shown in gray. The hemoglobin was then treated with hydrazine (NH_2–NH_2), which prevents the carboxyl terminus of each alpha chain from forming a salt bridge with a lysine on the neighboring alpha chain. When the salt bridge cannot form, the stability of the T structure is diminished. The resulting equilibrium curve is shown in black: affinity for oxygen at low oxygen pressures, when hemoglobin has the T structure, is increased by a factor of about five and the transition from the T to the R structure comes earlier in the reaction. (Both K_T and L are reduced.) There is no effect at high oxygen pressures, where both normal molecules and treated ones are in the R structure. The normal equilibrium was restored (*broken line*) by treating the altered protein with trypsin, which removes hydrazine. The experiment was carried out by **John Kilmartin and Janice Fogg of Cambridge and Arthur R. Arnone of University of Iowa.**

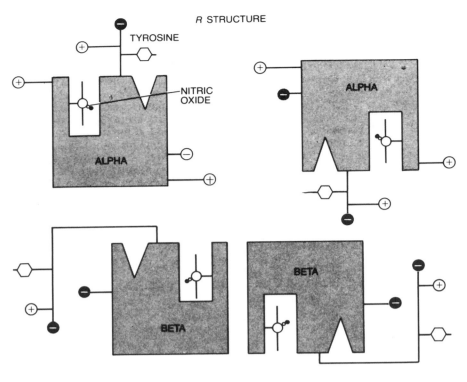

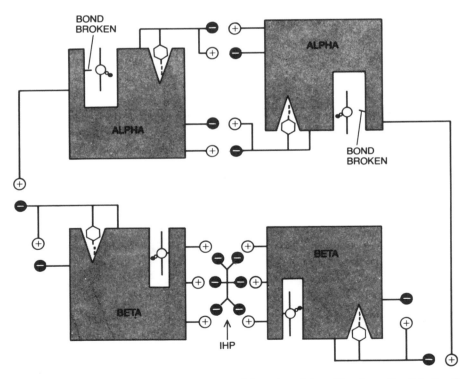

TENSION in the T structure was demonstrated by a forced transition from the R to the T structure. The effect was discovered when the author tested the effect of the transition on all known hemoglobin derivatives. The transition was forced by the addition of inositol hexaphosphate (IHP), which replaces DPG but forms more salt bridges with the beta chains. Oxygen was replaced by nitric oxide (NO), which binds to iron very strongly and also weakens the iron-histidine bond. At the top all the heme irons have bound NO, the molecules are in the R structure and there are no salt bridges between the subunits. At the bottom IHP has converted the molecule to the T structure and the subunits are clamped by salt bridges. The resulting tension has broken the bonds between the iron atoms and the proximal histidines in the alpha subunits, which are much farther away from the IHP binding site than those of the beta subunits, showing that proteins can transmit mechanical effects over large distances.

electrons. In oxyhemoglobin these form three pairs located halfway between the bonds that join the iron to its six surrounding atoms. Repulsion between the electrons of the iron and the electrons of the surrounding atoms is thereby minimized. In deoxyhemoglobin, on the other hand, four of the six electrons are unpaired and two of them lie along bond directions, where they repel the surrounding atoms of the porphyrin ring. This repulsion tends to push the iron farther out of the porphyrin plane than the repulsion between the proximal histidine and the porphyrin nitrogen atoms would do on its own.

Testing the Trigger

Suppose the iron does move in and out of the porphyrin plane every time it combines with or loses a molecule of oxygen. How could we find out if it is really this movement that triggers the allosteric transition between the two structures? I could think of no experiment that would answer the question directly, but I argued that if my proposition were true, then by the laws of action and reaction a forced transition from R to T must put the gears into reverse and pull the iron and the histidine away from the porphyrin ring. In that case the T structure should exercise a tension on the heme, which should be detectable by physical methods. My teacher David Keilin always told me to work with colored proteins because the spectra of the light they absorb can reveal so much. Hemoglobin is doubly blessed because one can feel its pulse both by its absorption spectrum and by the magnetic properties of its iron atoms.

Before we could exploit these properties we had to find a way of switching the structure from R to T other than the usual way of removing the oxygen. Sanford R. Simon of the State University of New York at Stony Brook and I found that this could be done with an analogue of DPG, a substance called inositol hexaphosphate (IHP), which has six phosphate groups in place of the two of DPG and therefore binds to the T structure more strongly.

When IHP was added to oxyhemoglobin, it caused some of the oxygen to be cast off, as was to be expected. I then replaced the oxygen with nitric oxide (NO) because this gas binds to the iron so strongly that the bond, once formed, cannot be broken. When I added IHP to nitric oxide hemoglobin, the structure switched from R to T and the spectrum changed drastically. Analysis of these and other spectral changes told us what had happened: because the strong bond to nitric oxide had held the iron atom tightly to the plane of the porphyrin, the tension exercised by the T structure had snapped the weaker bond between the iron and the proximal histidine instead. Most remarkably, this had happened primarily in the alpha subunits, whose hemes are 35 angstroms away from the phosphate binding site, rather than in the beta subunits, to which the IHP was actually bound. This experiment was done by Kyoshi Nagai, Attila Szabo and me at Cambridge together with John C. Maxwell and Winslow S. Caughey of Colorado State University at Fort Collins. Robert Cassoly of the Institute of Physicochemical Biology in Paris discovered the spectral changes at the same time we did.

Our experiment proved that the tension exists but did not tell us how large it is. To measure it I decided to exploit certain hemoglobin compounds in which the iron atoms are in a state of equilibrium between a weakly and a strongly paramagnetic state. (A paramagnetic substance cannot be permanently magnetized, as metallic iron can, but is drawn into a magnetic field.) At low temperature all the iron atoms are weakly paramagnetic, and the paramagnetism falls as the temperature rises; above a certain temperature the iron atoms begin to oscillate between the two magnetic states, which causes the total paramagnetism to rise as the temperature rises. Today it is known that the bonds between the iron and its surrounding atoms are slightly longer in the strongly paramagnetic state than they are in the weakly paramagnetic one. Therefore if tension in the T structure stretches the bonds to the iron, it should make the proportion of iron atoms in the strongly paramagnetic state larger in the T structure than it is in the R structure and thereby raise the total paramagnetism of the solution.

After several false starts a lucky coincidence finally brought this experiment off. Robert W. Noble appeared at Cambridge from the State University of New York at Buffalo with his pockets full of carp hemoglobin. He showed me how easily the structure of any of the derivatives of this hemoglobin could be switched from R to T by adding a little acid and IHP. Together we set out for Rome, where Massimo Cerdonio and Calogero Messana had just built a highly sensitive superconducting magnetometer at the Snamprogetti Laboratory, but while changing trains at a London Underground station, I left the thermos with our precious samples on the platform and never saw it again. Luckily we had some more carp hemoglobin in our deep freeze at Cambridge, and with it I started off once more for Rome.

The most useful derivative of carp hemoglobin was a ferric form in which the place of oxygen is taken by an azide ion (N_3^-). We measured its paramagnetism in both the R and T structures between -180 and $+30$ degrees Celsius. The results gave us a tremendous thrill: at all temperatures azide methemoglobin of carp was far more strongly paramagnetic in the T structure than in the R structure, which proved that the T structure does favor the state of the heme with the longer iron-nitrogen bonds. The tension at the heme can be gauged from the difference in energy between the two magnetic equilibriums. My colleague Fermi, with my son Robin Perutz of the University of Oxford, worked out that the difference amounts to about 1,000 calories, a third of the free energy of heme-heme interaction. We are not sure where the remaining two-thirds comes from but suspect that the R-to-T transition produces a smaller change of heme structure, and therefore also a smaller change of tension, in azide methemoglobin than uptake and loss of oxygen does.

In the meantime Arieh Warshel of the University of Southern California, Bruce W. Gelin and Martin Karplus of Harvard University, my own colleague Joyce Baldwin and Cyrus Chothia of University College London have tried to disentangle the set of atomic levers that generates the tension in the T structure and relieves it on transition to the R structure. They have demolished some of my early ideas and elaborated others.

All agree that the T structure exerts little or no tension on the deoxygenated heme and that the tension arises only when the iron tries to move toward the porphyrin plane on combination with oxygen, rather as the spring of a screen door is relaxed when it is closed but exerts increasing tension as it is opened. James P. Collman of Stanford University has therefore suggested that one should speak of restraint rather than tension. The restraint may be generated by a lopsided orientation of the proximal histidine with respect to the porphy-

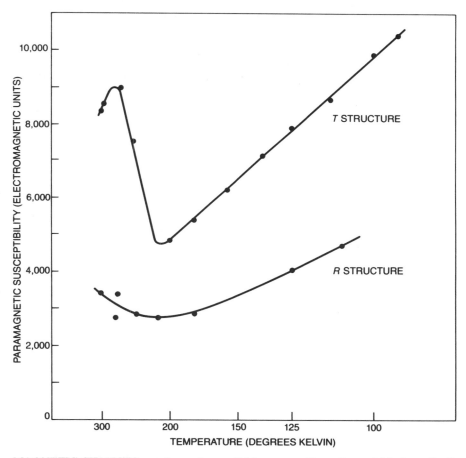

MAGNETIC CHANGES are observed on switching carp azide methemoglobin from the R to the T structure. Paramagnetic susceptibility is plotted on the vertical axis and absolute temperature on the horizontal axis. The paramagnetism of the iron atoms is higher in the T structure than it is in the R structure at all temperatures. In the R structure at low temperatures all the iron atoms are in a weakly paramagnetic state and the susceptibility drops with rising temperature. At about 200 degrees Kelvin the iron atoms begin to oscillate between a strongly and a weakly paramagnetic state, so that the susceptibility rises with rising temperature. In the T structure at low temperature a random mixture of strongly and weakly paramagnetic iron atoms is frozen in. At about 250 degrees K. the fraction of strongly paramagnetic iron atoms begins to rise sharply, only to fall again at higher temperatures for reasons that are not clear. The free-energy equivalent of the tension in the T structure is calculated from the difference in height between the two curves. The magnetic susceptibility measures the force that is exerted on one gram equivalent of iron (55.8 grams) by a magnetic field of one gauss.

rin, which brings one of the histidine carbon atoms close to one of the porphyrin nitrogen atoms. Repulsion between these two atoms would restrain the histidine from moving closer to the porphyrin ring. On transition to the R structure a shift and rotation of the heme in relation to helix F straightens out the histidine, so that it and the iron atom can move toward the porphyrin without restraint.

In the beta subunits a movement of the heme with respect to helix E, which carries the distal valine and distal histidine, may be more important. In the T structure the valine blocks the oxygen combining site, but after the shift to the R structure the site is uncovered. We do not yet know how much this bolting and unbolting contributes to the free energy of the heme-heme interaction.

All these mechanisms are consistent with my early ideas, but my suggestion that the movement of the proximal histidine is transmitted to the salt bridges by squeezing the penultimate tyrosine out of its crevice was too simplistic. Instead the hydrogen bond that holds the tyrosine in place may be stretched, but this loosening may not be enough to break the salt bridges. They may be loosened further by small perturbations of the bonds between the subunits that have so far eluded analysis.

One of the strangest features of both the T and the R structures is the absence of any entrance to the heme pocket wide enough to allow an oxygen molecule to pass. Either the distal histidine or some other group must swing out of the way, but we do not know how this is done because our X-ray analyses portray static structures, which allow us only to guess at the dynamics of the molecule.

John J. Hopfield of Princeton University once said that hemoglobin plays the same role in biochemistry that the hydrogen atom does in physics, because it serves as a touchstone for new theories and experimental techniques. Hemoglobin is the prototype of protein molecules that change their structure in response to chemical stimuli. Scientists will therefore continue to explore its many-faceted behavior. The mechanism I have outlined here will need further refinement before it can explain all their observations, but I am pleased that its main features have stood up to experimental tests and that it accounts reasonably well for the physiological properties of hemoglobin. I have not mentioned here that it also explains the symptoms of patients who have inherited abnormal hemoglobins, because that is another story. I hope that understanding of the structure and mechanism of the hemoglobin molecule will eventually help to alleviate those symptoms and to interpret the behavior of more complex biological systems.

Evidence for Conformational and Diffusional Mean Square Displacements in Frozen Aqueous Solution of Oxymyoglobin

H. Keller and P. G. Debrunner

Department of Physics, University of Illinois at Urbana-Champaign, Urbana, Illinois 61801
(Received 14 March 1980)

The temperature dependence of the Mössbauer spectra of ^{57}Fe in the active center of oxymyoglobin is analyzed. The effectvie mean square displacement $\langle x^2 \rangle$ of the iron is a sum of three parts: a vibrational term $\langle x^2 \rangle_v$, a conformational term $\langle x^2 \rangle_c$, and a diffusional term $\langle x^2 \rangle_d$. For $T > 240$ K diffusional line broadening $\Delta\Gamma \propto \langle x^2 \rangle_d$ is observed, along with a decrease in quadrupole splitting due to random reorientation of the electric field gradient.

Experimental[1-5] and theoretical[6] studies suggest that structural fluctuations are an essential feature of biological macromolecules. Recent x-ray diffraction measurements[1] on the oxygen storage protein myoglobin have shown that even in the crystalline state certain regions of the protein experience large atomic displacements. An understanding of the motions involved is of interest from the molecular dynamics point of view and is needed, eventually, to rationalize the biological activity of these macromolecules. The polypeptide chain of myoglobin forms a pocket for the heme group, a planar molecule with an iron atom at its center. The heme iron is the O_2-binding site and also forms a single covalent bond to the protein. Thus Mössbauer spectroscopy can be used to study the dynamics of the active center in this protein. Specifically, an effective mean square displacement (msd)[7] of the iron can be deduced from the cross section for recoilless absorption, and the line shape is affected by fluctuations of the hyperfine interaction[8] and by diffusion.[9-11]

We report measurements of the msd, linewidth, and quadrupole splitting on frozen aqueous solutions of oxymyoglobin between 4 and 260 K. The data show a clear division into three distinct dynamical regions, which are dominated by vibrational, conformational, and diffusional motion, respectively. Diffusional motion leads to line broadening and a decrease in quadrupole splitting.[8,11] We find empirically that the line broadening $\Delta\Gamma$ is proportional to the diffusional msd.

All Mössbauer measurements were done on ^{57}Fe-enriched sperm whale oxygmyoglobin in frozen aqueous solution.[12] The spectra are approximated by two Lorentzians. The quadrupole splitting $\Delta(T)$ is well reproduced by a polynomial of fourth degree in T up to 240 K. In order to analyze the decrease in Δ above 240 K we use extrapolated values Δ_0 (Table I). The relative

areas $A(T)$ of the absorption lines, corrected for nonresonant background, are taken as a measure of the recoilless fraction $f(T)$. For $f(T)$ we write[7]

$$f(T) = \exp(-k^2 \langle x_{eff}^2 \rangle) \propto A(T), \qquad (1)$$

where $\langle x_{eff}^2 \rangle$ is an effective msd along the wave vector $\vec{k}$ of the γ rays. Figure 1 shows the temperature dependence of the relative msd and the linewidth $\Gamma(T)$. For $T < 230$ K, Γ depends linearly on T.[13] The rapid increase of Γ above 240 K is accompanied by a decrease in Δ as listed in Table I. Figure 2 demonstrates that there are three distinct temperature regions in which $\langle x_{eff}^2 \rangle$ and Γ are linearly related. This result is significant since at high temperatures both $\langle x_{eff}^2 \rangle$ and Γ are nonlinear functions of T.[10,14]

Following Frauenfelder, Petsko, and Tsernaglou,[1] we assume that the total msd of the iron may be approximated by a sum of three statistically independent terms,

$$\langle x_{eff}^2 \rangle = \langle x^2 \rangle_v + \langle x^2 \rangle_c + \langle x^2 \rangle_d, \qquad (2)$$

where the subscripts v, c, and d refer to vibration, conformation, and diffusion, respectively. At low temperatures only $\langle x^2 \rangle_v$ is observed; we assume that it follows, for all T, the classical, linear temperature dependence[7] evident below 160 K. The difference $\langle x_{eff}^2 \rangle - \langle x^2 \rangle_v = \langle x^2 \rangle_{cd}$ [cf. Eq. (2)] is plotted in Fig. 3. The term $\langle x^2 \rangle_c$ is

TABLE I. Comparison of measured quadrupole splittings Δ_M with values Δ_R calculated from Eq. (6). The values Δ_0 are extrapolated from a polynomial of fourth degree valid for $T < 240$ K.

T (K)	Δ_M (mm/s)	Δ_0 (mm/s)	Δ_R (mm/s)
253.3	1.77 ± 0.01	1.82	1.80
256.3	1.75 ± 0.02	1.81	1.76
259.3	1.63 ± 0.05	1.80	1.67

Reprinted from *Physical Review Letters* **45**, 68–71 (1980); © The American Physical Society.

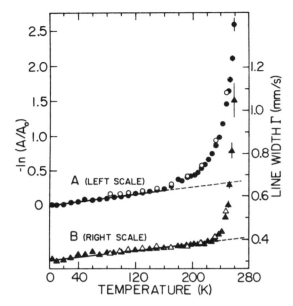

FIG. 1. Temperature dependence of the relative $\langle x_{eff}^2 \rangle$ (curve A) and the linewidth Γ (curve B) [full width at half maximum (FWHM)] (circles, triangles: remeasured points).

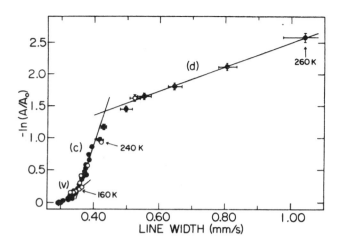

FIG. 2. Correlation of the relative $\langle x_{eff}^2 \rangle$ and the linewidth Γ (FWHM) (circles: remeasured points). (v), (c), and (d) refer to regions dominated by vibrations, conformations, and diffusion, respectively.

postulated to arise from thermally activated transitions between different conformational substates of the molecule.[1] In contrast to conformational fluctuations, diffusive motions lead to a msd $\langle x^2 \rangle_d$ with no upper bound. As a result, the Mössbauer intensity vanishes at 260 K.

It is instructive to compare our msd with that found in crystals. Parak $et\ al.$[4] reported Mössbauer measurements on polycrystalline metmyoglobin. If we treat the msd data of Ref. 4 in the same way as ours by subtracting $\langle x^2 \rangle_v$, the triangles in Fig. 3 are obtained. We note the coincidence of the two data sets up to 240 K and the characteristic increase of the msd in the frozen solution due to diffusion above 240 K.

The temperature dependence of $\langle x^2 \rangle_c$ can be reproduced by the two-state model[15] indicated in Fig. 3. The iron is bound in one of the two potential wells, L and R, and jumps from one to the other at rates k_L and k_R, respectively. If d is the distance between the wells and τ_N the nuclear lifetime, the msd is given by[15]

$$\langle x^2 \rangle_c = d^2 k_L k_R / [(k_L + k_R)(k_L + k_R + \tau_N^{-1})] . \quad (3)$$

If we assume that $k_L \ll k_R$, Eq. (3) reduces to

$$\langle x^2 \rangle_c = d^2 \exp(\Delta S/k_B) \exp(-Q/k_B T) Z(T) , \quad (4)$$

$$Z(T) = k_R / (k_R + \tau_N^{-1}) .$$

Here $\exp(\Delta S/k_B)$ is an entropy term, the transi-

tion rate is $k_R = \nu_0 \exp(-E/k_B T)$, and $\nu_0 \sim 10^{13}$ s^{-1} is a typical vibrational frequency. To test this model we fit both sets of $\langle x^2 \rangle_c$ values in Fig. 3 by Eq. (4). The logarithm of the preexponential factor in k_R was found to be $\log[\nu_0/(\text{s}^{-1})] = 13.1$

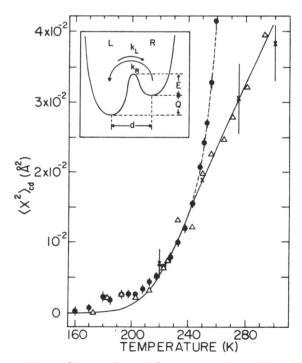

FIG. 3. $\langle x^2 \rangle_{cd} = \langle x^2 \rangle_c + \langle x^2 \rangle_d$ as a function of temperature. Solid circles, MbO$_2$ aqueous solution (this work); triangles, polycrystalline MetMb (see Ref. 4); crosses, MetMb x-ray data (see Ref. 1). See text for inset.

TABLE II. Parameters deduced from Eq. (4) with $\nu_0 = 1.3 \times 10^{13}$ s^{-1}.

Sample	Q (kJ/mole)	E (kJ/mole)	c (Å^2)[b]
MetMb,[a] 200 K $< T <$ 300 K	6.2 ± 1.5	27.8 ± 1	0.5 ± 0.3
MbO$_2$, 200 K $< T <$ 245 K	6 ± 2	27.0 ± 1	0.2 ± 0.1

[a]Data of Ref. 4.   [b]$c = d^2 \exp(\Delta S / k_B)$.

± 0.6. As summarized in Table II, the fitted parameters of the two data sets agree within error and reproduce the measured points quite well (Fig. 3). We note that for x-ray diffraction, which is a fast process, the factor Z in Eq. (4) is unity. Indeed, if we multiply the $\langle x^2 \rangle_c$ values deduced from x-ray studies[1] by Z (crosses in Fig. 3), they agree well with the Mössbauer data. The near identity of $\langle x^2 \rangle_c$ in frozen solutions and crystals in surprising, especially since different chemical states of myoglobin are compared. It suggests that conformational fluctuations occur independently of the matrix and thus are an intrinsic property of the protein.

To estimate the diffusional msd, $\langle x^2 \rangle_d$, which is evident in Fig. 3 for $T > 240$ K, we subtract $\langle x^2 \rangle_c$ of the crystalline metmyoglobin sample[4] from our data using a linear interpolation between the measured points (Fig. 3). We find that both $\langle x^2 \rangle_d$ and the line broadening $\Delta \Gamma$ follow an Arrhenius law with activation energies $E_a = 82 \pm 2$ kJ/mole and $E_a = 79 \pm 4$ kJ/mole, respectively. These values are comparable with $E_a = 66 \pm 8$ kJ/mole found for self-diffusion of ^{3}H in ice.[16] Since the activation energies for $\langle x^2 \rangle_d$ and $\Delta \Gamma$ are the same within error, there is no doubt that the same process is responsible for both $\langle x^2 \rangle_d$ versus $\Delta \Gamma$ in Fig. 4 shows that the two quantities are proportional to each other. Similar empirical correlations have been noted before.[10,14] If $\Delta \Gamma$ of Fig. 4 is expressed as a frequency $\nu = \Delta \Gamma / 2\pi \hbar$, the proportionality constant $\langle x^2 \rangle_d / \nu$ is $(2.5 \pm 0.1) \times 10^{-9}$ Å^2 s. This value is equal to $\tau_N / k^2 = 2.6 \times 10^{-9}$ Å^2 s, where τ_N is the nuclear lifetime, and k is the wave number of the γ ray. Thus we can write the empirical relation[17]

$$\langle x^2(T) \rangle_d \sim (\tau_N / k^2) \nu(T). \tag{5}$$

Evidence for random reorientation of the iron complex as a result of diffusion comes from the decrease in quadrupole splitting Δ observed for $T > 240$ K. The stochastic model of line shape of Dattagupta and Blume[8] predicts a reduced quadrupole splitting Δ_R according to

$$\Delta_R = \Delta_0 [1 - (\Delta \Gamma / \Delta_0)^2]^{1/2}, \tag{6}$$

where $\Delta \Gamma$ is the line broadening and Δ_0 is the quadrupole splitting in the absence of relaxation. The analysis presented in Table I demonstrates that the data are compatible with the fluctuation model.

While all experimental results indicate the presence of diffusive motion, the microscopic mechanism of the process is not clear. The decrease in quadrupole splitting observed above 240 K demonstrates that the surrounding of the iron nucleus is fluctuating and that the motion of the iron complex is not of pure translational character.[8,11] A possible explanation could be ligand fluctuations that may modulate the electric field gradient thermally. These fluctuations are presumably correlated with the melting process of the solvent. Calorimetric measurements show premelting effects starting near 240 K.[18] Most likely the water around the protein starts melting at this temperature, and whole protein or parts of it may perform diffusive motions. The Arrheniuslike temperature dependence of $\langle x^2 \rangle_d$ and $\Delta \Gamma$ suggests a jump diffusion process. Reorientation of H$_2$O molecules or proton jumps may be re-

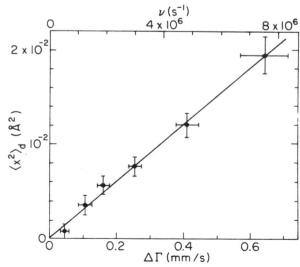

FIG. 4. $\langle x^2 \rangle_d$ vs $\Delta \Gamma$ in the temperature range 240 K $< T <$ 260 K.

sponsible for the diffusion process as suggested by the similarity of the activation energies with that found for ^{3}H in ice.[16]

We wish to thank Dr. L. Eisenstein, Dr. G. Forgacs, Dr. H. Frauenfelder, and Dr. B. Gavish for many helpful discussions. This work was supported by the U.S. Public Health Service under Grant No. GM 16406, and by the National Science Foundation under Grant No. PCM 79-05072.

---

[1]H. Frauenfelder, G. A. Petsko, and D. Tsernoglou, Nature 280, 558 (1979).

[2]K. Wüthrich and G. Wagner, Trends Biochem. Sci. 3, 227 (1978).

[3]P. J. Artymiuk, C. C. F. Blake, D. E. P. Grace, S. J. Oatley, D. C. Phillips, and M. J. E. Sternberg, Nature 280, 563 (1979).

[4]F. Parak, E. N. Frolov, R. L. Mössbauer, and V. I. Goldanskii, to be published.

[5]A. Dwivedi, T. Pederson, and P. G. Debrunner, J. Phys. (Paris) C 2, 531 (1979).

[6]J. A. McCammon, B. R. Gelin, and M. Karplus, Nature 267, 5612 (1977).

[7]R. M. Housley and F. Hess, Phys. Rev. 146, 517 (1966).

[8]S. Dattagupta and M. Blume, Phys. Rev. B 10, 4540 (1974).

[9]K. S. Singwi and A. Sjölander, Phys. Rev. 120, 1093 (1960).

[10]J. H. Jensen, Phys. Kondens. Mater. 13, 273 (1971).

[11]S. Dattagupta, Phys. Rev. B 12, 47 (1975), and B 14, 1329 (1976).

[12]As no phase separation was detectable by calorimetry, the protein is believed to form a solution in the frozen matrix.

[13]D. Bade and F. Parak, Z. Naturforsch. 33C, 488 (1978).

[14]A. Abras and J. G. Mullen, Phys. Rev. A 6, 2343 (1972).

[15]H. Frauenfelder, V. I. Goldanskii, and J. J. Hopfield, private communication.

[16]K. Itagaki, J. Phys. Soc. Jpn. 19, 1081 (1964).

[17]The proportionality between $\langle x^2 \rangle_d$ and $\Delta\Gamma$ may be written in the form of an Einstein relation $\langle x^2 \rangle_d = 2D^*\tau$, where D^* is an effective diffusion constant deduced from $\Delta\Gamma = 2\hbar k^2 D^*$ (see Ref. 9), and $\tau = 22 \pm 1$ ns is a characteristic time for the diffusion process.

[18]H. Keller, unpublished.

Spectral shapes of Mössbauer absorption and incoherent neutron scattering from harmonically bound nuclei in Brownian motion: Applications to macromolecular systems

I. Nowik, E. R. Bauminger, S. G. Cohen,* and S. Ofer*

Racah Institute of Physics, The Hebrew University, 91904 Jerusalem, Israel

(Received 10 October 1984)

The Mössbauer resonance absorption line in solids is of narrow, natural width. In viscous liquids the linewidth is broadened by diffusion. In many biological systems—whole cells, membranes, or proteins—at temperatures above the freezing point of the internal water, a superposition of broad and narrow lines is observed. Here, the Mössbauer spectral shapes expected in this new "phase" of proteinic matter are calculated. The calculation is based on the assumption that the unfrozen conformational degrees of freedom of the macromolecules can be described in terms of damped harmonic oscillators acted upon by random forces. The known classical correlation functions for harmonically bound particles in Brownian motion are utilized to calculate the Mössbauer spectra. The theory predicts the following: a spectrum which can be approximated by a superposition of a narrow and a broad line; and a sharp decrease in the total resonance absorption as a function of temperature and asymmetric quadrupole doublets in ^{57}Fe spectra, even when the Debye-Waller factor is isotropic but the damping frequencies are anisotropic. The presented formulas, with only minor changes, are also applicable to the description of neutron quasielastic scattering from systems in which nuclei diffuse in restricted geometries: bound diffusion on surfaces, lamellar systems, ionic polymers, biopolymers, and membranes. Finally, calculated spectra are compared to recent experimental Mössbauer spectra, and the agreement is outstanding. The spectral shape of particles participating in both bound translational and free rotational diffusion is also calculated.

I. INTRODUCTION

Recent studies of the Mössbauer absorption in biological systems such as proteins and membranes and in polymers[1-9] reveal unusual spectra, different from those observed in ordinary solids or viscous liquids. The spectra can be described as composed of a narrow line, as in solids, and a broad line, as that observed in liquids. Quasielastic incoherent neutron scattering from nuclei in bound geometries display similar spectra.[10,11] In the present paper we apply the theory of damped harmonic oscillators in Brownian motion developed by Uhlenbeck and Ornstein[12] to calculate the expected Mössbauer absorption spectra and neutron scattering spectra in such systems.[13-16] Comparison is made to other formulas developed for Mössbauer and neutron spectroscopy of nuclei performing bound diffusion in solids and liquids.[10,13,15,17-19] The present calculation predicts Mössbauer spectra which can be looked upon, to a good approximation, as composed of a narrow and broad lines with total intensity strongly temperature dependent. When the diffusion is anisotropic it predicts a generalized Karyagin-Goldanskii effect,[20] even when the Debye-Waller factor is isotropic. The calculated Mössbauer spectra are compared to various experimental observations and the agreement is very satisfactory. In the Appendix the spectral shape obtained in spherical particles participating in both bound translational and free rotational diffusion is calculated in detail.

II. THEORY

A. Dynamics of a harmonically bound particle

The general formula for a Mössbauer spectrum in the classical limit is given by[10,21]

$$I(\omega) = \frac{1}{2\pi} \int_{-\infty}^{\infty} dt \, \exp[-i(\omega-\omega_0)t - \tfrac{1}{2}\Gamma\,|\,t\,|\,]$$
$$\times \int_r d\mathbf{r}_0 d\mathbf{r}\, G(\mathbf{r},\mathbf{r}_0,t)$$
$$\times \exp[i\mathbf{k}\cdot(\mathbf{r}-\mathbf{r}_0)]p(\mathbf{r}_0)\,. \qquad (1)$$

Here Γ is the natural width of the Mössbauer line, $\mathbf{k}$ is the γ-ray wave vector, and $G(\mathbf{r},\mathbf{r}_0,t)$ is the probability that at time t the nucleus will be at $\mathbf{r}$ if at time zero it was at $\mathbf{r}_0$. $p(\mathbf{r})$ is the stationary probability of the particle being at $\mathbf{r}$, which equals $G(\mathbf{r},\mathbf{r}',\infty)$ and is of course independent of $\mathbf{r}'$.

A particle of mass m bound to a center by a harmonic force $-mw^2\mathbf{r}$, damped by a frictional force $-m\beta\dot{\mathbf{r}}$, and acted upon by random forces $\mathbf{F}(t)$ will follow the equation of motion

$$m\ddot{\mathbf{r}} + m\beta\dot{\mathbf{r}} + mw^2\mathbf{r} = \mathbf{F}(t)\,. \qquad (2)$$

In the case of one-dimensional motion, Uhlenbeck and Ornstein[12] have derived general formulas for $G(x,x_0,t)$. Their formula for the overdamped case, which seems to be relevant in biological systems, and generalized to three dimensions, yields

$$G(\mathbf{r},\mathbf{r}_0,t) = \left[\frac{\alpha}{2\pi D(1-e^{-2\alpha t})}\right]^{3/2}$$

$$\times \exp\left[-\frac{\alpha(\mathbf{r}-\mathbf{r}_0 e^{-\alpha t})^2}{2D(1-e^{-2\alpha t})}\right]. \qquad (3)$$

Here $\alpha = w^2/\beta$ is the ratio of the harmonic and damping force constants and $D = k_B T/m\beta$ is the diffusion constant. The mean-square displacement after a long period of time is given by $\langle x^2\rangle = k_B T/mw^2 = D/\alpha$. The parameter α is also $1/\tau_c$ where τ_c is the relaxation time of the ensemble average of x,[12] $\langle x\rangle_t = x_0 \exp(-t/\tau_c)$ and $\langle x^2\rangle_t = k_B T/mw^2 + (x_0^2 - k_B T/mw^2)\exp(-2t/\tau_c)$. Thus the two parameters $\langle x^2\rangle$ and τ_c^{-1} characterize completely the statistical averages of the microscopic motion of the particle.

Inserting Eq. (2) into Eq. (1) and performing the Fourier transform of $G(\mathbf{r},\mathbf{r}_0,t)$ and then the integration on $\mathbf{r}_0$ gives the final formula for the spectrum, as a function of angular frequency;[14]

$$I(\omega) = \frac{1}{2\pi}\int_{-\infty}^{\infty} dt \exp\left[-i(\omega-\omega_0)t - \tfrac{1}{2}\Gamma|t| \right.$$

$$\left. -\frac{k^2 D}{\alpha}(1-e^{-\alpha|t|})\right]. \qquad (4)$$

We have used Eq. (4) to compute the absorption spectra as a function of the parameters $\alpha = 1/\tau_c$ and $D = \alpha\langle x^2\rangle$. Examples are shown in Fig. 1 for a fixed value of D and a range of values of α. These spectra are normalized for the sake of display to constant maximum resonance absorption. We observe from Eq. (4) that for a very small value of α (free particle), a Lorentzian spectrum of half-width $\Gamma/2 + k^2 D$, as expected for a freely diffusing particle,[21] is obtained. For large α, $\alpha/\Gamma = w^2/\beta\Gamma \gg 1$, though keeping still within the overdamped limit $w \ll \beta$, $I(\omega)$ becomes a Lorentzian of natural width Γ. At intermediate values of α the spectra have the peculiar shapes shown in Fig. 1. Equation (4) can be presented by an infinite sum of Lorentzian lines which converges very fast. Expanding $\exp(k^2\langle x^2\rangle e^{-\alpha t})$ in a power series[9,15] yields

$$I(\omega) = \exp(-k^2\langle x^2\rangle)$$

$$\times \sum_{n=0}^{\infty} \frac{(k^2\langle x^2\rangle)^n}{n!}\frac{(\tfrac{1}{2}\Gamma+n\alpha)/\pi}{(\tfrac{1}{2}\Gamma+n\alpha)^2+(\omega-\omega_0)^2}. \qquad (5)$$

In Fig. 2 the intensities of the first seven Lorentzian lines in the sum of Eq. (5) are displayed as a function of $k^2\langle x^2\rangle$. In the approximation of a single narrow line

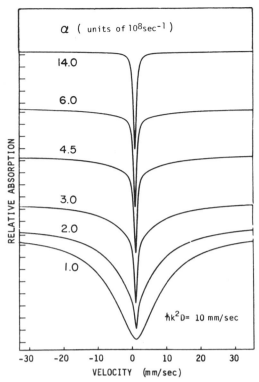

FIG. 1. Mössbauer spectra of harmonically bound particles containing ^{57}Fe in Brownian motion as a function of the parameter $\alpha = w^2/\beta$, for a fixed value of the diffusion constant $D = 1.4\times10^{-9}$ cm^2/sec corresponding to a value of $\hbar k^2 D$ equal to 10 mm/sec in units of the Doppler velocity ($\hbar\Gamma$ was fixed at a value of 0.75 mm/sec).

($n = 0$) and one broad line, the width of the broad line can be calculated by a harmonic average;

$$\frac{2\alpha}{\Gamma_{\rm eff}} = \frac{1}{\exp(k^2\langle x^2\rangle)-1}\sum_{n=1}^{\infty}\frac{(k^2\langle x^2\rangle)^n}{n!}\frac{1}{(\Gamma/2\alpha+n)}. \qquad (6)$$

$\Gamma_{\rm eff}$ was calculated for various $k^2\langle x^2\rangle$ values and is shown in Fig. 3. Thus one can use a two-line approxima-

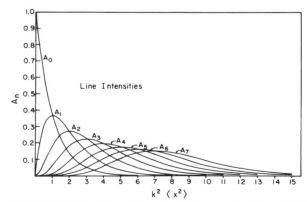

FIG. 2. Relative intensities of the Lorentzian lines of width $\Gamma + 2n\alpha$ appearing in Eq. (5), as a function of $k^2\langle x^2\rangle$.

tion, derive $k^2\langle x^2\rangle$ from the relative intensities of the two lines, and then obtain α from Fig. 3.

We have obtained Eq. (4) for the overdamped case. For the sake of generality, we quote the form of the absorption spectrum for arbitrary w and β:

$$I(\omega)=\frac{1}{2\pi}\int_{-\infty}^{\infty}dt\,\exp[-i(\omega-\omega_0)t]\exp(-\tfrac{1}{2}\Gamma|t|)$$
$$\times\exp\{-k^2\langle x^2\rangle[1-\phi(t)]\}, \qquad (7)$$

where

$$\phi(t)=\begin{cases} e^{-\beta|t|/2}\left[\cos(w_1 t)+\dfrac{\beta}{2w_1}\sin(w_1 t)\right], & w_1^2=w^2-\left[\dfrac{\beta}{2}\right]^2>0 \\[2ex] e^{-\beta|t|/2}\left[1+\dfrac{\beta}{2}t\right], & w=\beta/2 \\[2ex] e^{-\beta|t|/2}\left[\cosh(w't)+\dfrac{\beta}{2w'}\sinh(w't)\right], & (w')^2=\left[\dfrac{\beta}{2}\right]^2-w^2>0. \end{cases}$$

We have seen a closed-form expression for $I(\omega)$ in the overdamped limit, both in an integral and in a sum-of-Lorentzians form. The extreme underdamped case can also be expressed in a convenient form as a sum of Lorentzians. In the extreme underdamped case the expression for $\phi(t)$ can be approximated by

$$\phi(t)=e^{-\beta|t|/2}\cos(wt).$$

A power-series expansion of $\exp[k^2\langle x^2\rangle e^{-\beta|t|/2}\cos(wt)]$ yields

$$I(\omega)=\frac{1}{2\pi}e^{-k^2\langle x^2\rangle}$$
$$\times\sum_{n=0}^{\infty}\frac{(k^2\langle x^2\rangle)^n}{n!}$$
$$\times\left[\frac{(\Gamma+n\beta)/2}{[(\Gamma+n\beta)/2]^2+(\omega-\omega_0-nw)^2}\right.$$
$$\left.+\frac{(\Gamma+n\beta)/2}{[(\Gamma+n\beta)/2]^2+(\omega-\omega_0+nw)^2}\right]. \qquad (8)$$

We observe that in the overdamped case the spectrum is composed of Lorentzian lines of width $\Gamma+2n\alpha$ at ω_0, whereas in the underdamped case one observes sideband lines at $\omega_0\pm nw$ of width $\Gamma+n\beta$. The intensities in both cases are the same.

The model described above can be generalized to treat anisotropic diffusive motions which might be relevant in particular systems, as, e.g., in single-crystal proteins or oriented membranes. For the anisotropic overdamped harmonic oscillator, with independent motions along the three axes, Eq. (4) becomes generalized to the form

$$I(\omega)=\frac{1}{2\pi}\int_{-\infty}^{\infty}dt\,\exp[-i(\omega-\omega_0)t-\tfrac{1}{2}\Gamma|t|$$
$$-k_x^2\langle x^2\rangle(1-e^{-\alpha_x|t|})$$
$$-k_y^2\langle y^2\rangle(1-e^{-\alpha_y|t|})$$
$$-k_z^2\langle z^2\rangle(1-e^{-\alpha_z|t|})], \qquad (9)$$

where $\alpha_x=w_x^2/\beta_x$, and similarly for α_y and α_z.

The line shape, including narrow and wide components, will then depend on the angle of incidence of the γ rays.

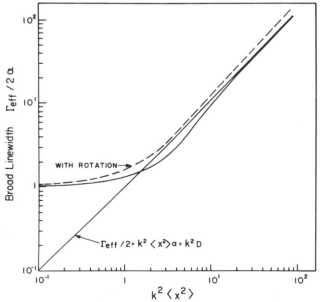

FIG. 3. The effective linewidth of the broad line, in the two-Lorentzian-lines approximation of Eq. (4) or Eq. (5), as a function of $k^2\langle x^2\rangle$.

In polycrystalline proteins or unoriented membranes, in those cases where there are appreciable quadrupole hyperfine interactions, one might expect a generalized Karyagin-Goldanskii[20] effect, the observed line shape being a sum of different lines corresponding to the various transitions between the substates of the hyperfine structure. Thus in a powder sample, the spectral shape of a nuclear transition between quantum numbers M_1 and M_2, with $M_1 - M_2 = \Delta M$ will be given by

$$I(\omega) = \frac{1}{8\pi^2} \int_\phi \int_\theta \int_{-\infty}^{\infty} dt \, \exp\{ -i(\omega - \omega_0)t - \tfrac{1}{2}\Gamma |t| - k^2 \cos^2\theta \langle x^2 \rangle (1 - e^{-\alpha_x |t|})$$

$$- k^2 \sin^2\theta [\cos^2\phi \langle y^2 \rangle (1 - e^{-\alpha_y |t|}) + \sin^2\phi \langle z^2 \rangle (1 - e^{-\alpha_z |t|})]\}$$

$$\times f(\Delta M, \theta) \sin\theta \, d\theta \, d\phi , \tag{10}$$

where $f(\Delta M, \theta)$ for the transition in ^{57}Fe, is given by $f(0, \theta) = \tfrac{3}{2}\sin^2\theta$ and $f(\pm 1, \theta) = \tfrac{3}{4}(1 + \cos^2\theta)$. For axial symmetry ($\langle z^2 \rangle = \langle y^2 \rangle$, $\alpha_y = \alpha_z$) we obtain for the two ^{57}Fe quadrupole lines the spectral shapes ($\eta = \cos\theta$)

$$I_s(\omega) = \frac{1}{2\pi} \int_{-\infty}^{\infty} dt \int_0^1 \exp[-i(\omega - \omega_0)t - \tfrac{1}{2}\Gamma |t| - k^2 \langle x^2 \rangle (1 - e^{-\alpha_x |t|})\eta^2$$

$$- k^2 \langle y^2 \rangle (1 - e^{-\alpha_y |t|})(1 - \eta^2)] f_s(\eta) d\eta , \tag{11}$$

where $f_1(\eta) = \tfrac{3}{4}(1 + \eta^2)$ and $f_2(\eta) = \tfrac{3}{4}(\tfrac{5}{3} - \eta^2)$. Spectra calculated using Eq. (11) for various α_x and α_y values are shown in Fig. 4. The asymmetry in the spectra is observable even when $\langle x^2 \rangle = \langle y^2 \rangle$.

The theoretical treatment given above assumed oscillators of a single frequency (Einstein model). We can extend the treatment to a general distribution of normal frequencies and damping frequencies (Debye model).

If a γ-ray-absorbing nucleus is bound to a macroscopic entity, such as a macromolecule, its motion $\mathbf{r}$ can be expanded in the molecular normal-mode coordinates q_l with $l = 1, 2, \ldots, N$. If the wave vector of the γ ray is $\mathbf{k} = k\boldsymbol{\epsilon}$, we can write

$$(\mathbf{k} \cdot \mathbf{r}) = k(\boldsymbol{\epsilon} \cdot \mathbf{r}) = k \sum_l a_l q_l . \tag{12}$$

If we assume that all normal-mode coordinates follow the harmonic damped motion given by Eq. (2),[13] and if in all cases the overdamped limit is applicable (this requirement can be relaxed, as will be discussed later), then, as all normal modes are statistically uncorrelated, we obtain

$$G(\mathbf{r}, \mathbf{r}_0, t) d\mathbf{r} = \prod_l g(q_l, q_{0l}, t) dq_l , \tag{13}$$

where q_{0l} are normal mode coordinates at time zero,

$$g(q_l, q_{0l}, t)$$
$$= \frac{\exp\{ -(q_l - q_{0l}e^{-\alpha_l t})^2 / [2\langle q_l^2 \rangle (1 - e^{-2\alpha_l t})]\}}{[2\pi \langle q_l^2 \rangle (1 - e^{-2\alpha_l t})]^{1/2}} \tag{14}$$

and

$$\alpha_l = w_l^2 / \beta_l .$$

Performing the double Fourier transform on q_l and q_{0l} we get

$$I(\omega) = \frac{1}{2\pi} \int_{-\infty}^{\infty} \exp\left[-i(\omega - \omega_0)t - \tfrac{1}{2}\Gamma |t| \right.$$
$$\left. - k^2 \sum_l a_l^2 \langle q_l^2 \rangle (1 - e^{-\alpha_l |t|}) \right] dt . \tag{15}$$

If out of the N modes there are m modes for which $\alpha_j \gg \Gamma$ ($j = 1, 2, \ldots, m$), namely, $w_j^2 / \beta_j \gg \Gamma$ or β_j is relatively small yet still $\beta_j \gg \Gamma$, we obtain that within the finite range of experimental observation, $|\omega_{\max} - \omega_0| < 500\Gamma$,

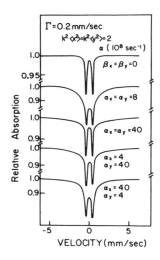

FIG. 4. Mössbauer spectra of harmonically bound particles acted upon by axially symmetric anisotropic damping forces.

$$I_{\text{expt}}(\omega) \approx \left[\exp^{-k^2 \left[\sum_{j=1}^{m} a_j^2 \langle q_j^2 \rangle \right]} \right] \frac{1}{2\pi} \int_{-\infty}^{\infty} \exp \left[-i(\omega - \omega_0)t - \tfrac{1}{2}\Gamma |t| - k^2 \sum_{l\,(\neq j)} a_l^2 \langle q_l^2 \rangle (1 - e^{-\alpha_l |t|}) \right] dt \qquad (16)$$

and

$$f_{\text{tot}} = \int_{\omega_0 - 500\Gamma}^{\omega_0 + 500\Gamma} I_{\text{expt}}(\omega)\,d\omega$$

$$\approx \exp \left[-k^2 \sum_{j=1}^{m} a_j^2 \langle q_j^2 \rangle \right] = \exp(-k^2 \langle x_f^2 \rangle) , \quad (17)$$

where $\langle x_f^2 \rangle$ refers to fast fluctuation rates, $\alpha_j \gg \Gamma$.

Thus again the experimental spectrum consists of a superposition of a narrow and broad line, the narrow line of width Γ of relative intensity

$$f_{\text{nar}} = \exp \left[-k^2 \sum_{l\,(\neq j)} a_l^2 \langle q_l^2 \rangle \right]$$

and the broad line of relative intensity $1 - f_{\text{nar}}$. The broad line is composed of many lines each of which has a half-width $\Gamma/2 + \sum_{l\,(\neq j)} n_l \alpha_l$ and intensity

$$\prod_{l\,(\neq j)} [(k^2 a_l^2 \langle q_l^2 \rangle)^{n_l} / n_l!] f_{\text{nar}} ,$$

where $\{n_l\}$ are series of integers, $\sum_l n_l = n$, $n = 1, 2, 3, 4, \ldots$ The total intensity is given by f_{tot}. Thus the absolute intensity of the narrow line is $f_{\text{tot}} f_{\text{nar}}$ and that of the broad line is $f_{\text{tot}}(1 - f_{\text{nar}})$. In biological systems at the "phase transition" between the solid phase and "viscous liquid" phase, the extra degrees of freedom of the molecular motions melt, become active, and contribute both to the appearance of a broad line and to a sharp decrease in the total resonance absorption f_{tot}.[5]

It is worth mentioning that for very high α values assumed for the m modes, namely large w and small β, the underdamped limit may be reached. However, since $w > \beta \gg \Gamma$ Eq. (8) shows that only the $n = 0$ line will be observable and thus all previous conclusions are still valid.

The theoretical treatment given in the present paper is similar to those given by Knapp et al.[15] and Shaitan and Rubin[16] for Mössbauer spectroscopy and by Rahman et al.[13] for neutron scattering. However, the present treatment is characterized by the following. (a) It presents the final formula in a closed form. (b) It treats the problem in three dimensions and this enables the introduction of anisotropic binding and damping forces in a very simple way. (c) It presents a way to analyze a Mössbauer spectrum in terms of only two lines and yet obtain all the physical information which would have been obtained by applying the exact formula, Eq. (4). (d) It treats the general distribution of restoring and damping frequencies in a scheme which does not require additional parameters in analyzing the experimental spectra, except for the mean-square displacement of the fast fluctuations, $\langle x_f^2 \rangle$.

B. Comparison with dynamics of diffusion in a cage

We have concluded that the relative intensity of the narrow line f_n is given by $\exp(-k^2 \langle x^2 \rangle)$. It is worthwhile to compare this expression to the value given within a model of discrete jumps of the ion among N equivalent sites (diffusion within a cage) which also predicts narrow and broad lines.[17-19] The recoil-free fraction of the absorption line of natural width is given in this model[17,18] as

$$f = f_0 \left| \frac{1}{N} \sum_n e^{i\mathbf{k}\cdot\mathbf{R}_n} \right|^2$$

$$= f_0 \frac{1}{N} \sum_n \frac{1}{N} \sum_m e^{i\mathbf{k}\cdot(\mathbf{R}_n - \mathbf{R}_m)} , \qquad (18)$$

where $\mathbf{R}_n$ are the coordinates of the nth atom.

In a polycrystalline sample the orientation of $\mathbf{k}$ relative to $\mathbf{R}_{nm} = \mathbf{R}_n - \mathbf{R}_m$ is random and can be averaged out to yield

$$f = f_0 \frac{1}{N} \sum_n \left[\frac{1}{N} \sum_m \frac{\sin(kR_{nm})}{kR_{nm}} \right] . \qquad (19)$$

This formula yields, for the case of $N = 2$,

$$f = f_0 \left[\tfrac{1}{2} + \tfrac{1}{2} \frac{\sin(kd)}{kd} \right], \quad R_0^2 = \tfrac{1}{4} d^2 ; \qquad (20a)$$

for $N = 4$ (tetrahedron),

$$f = \tfrac{1}{4} f_0 \left[1 + 3 \frac{\sin(kd)}{kd} \right], \quad R_0^2 = \tfrac{3}{8} d^2 ; \qquad (20b)$$

for $N = 6$ (octahedron),

$$f = \tfrac{1}{6} f_0 \left[1 + \frac{\sin(\sqrt{2}kd)}{\sqrt{2}kd} + 4 \frac{\sin(kd)}{kd} \right],$$

$$R_0^2 = \tfrac{1}{2} d^2 ; \qquad (20c)$$

for $N = 8$ (cube),

$$f = \tfrac{1}{8} f_0 (1 + 3Z_1 + 3Z_2 + Z_3) ,$$

$$Z_l = \frac{\sin(kd\sqrt{l})}{kd\sqrt{l}}, \quad R_0^2 = \tfrac{3}{4} d^2 . \qquad (20d)$$

Here d is the shortest "jump" distance and R_0 is the distance of each site to the center.

If we now allow $N \to \infty$, with all N points equivalent as in the previous cases, and located on a spherical surface of radius R_0, then the sum in Eq. (19) becomes

$$\frac{f}{f_0} = \int_0^{2R_0} \frac{\sin(kR)}{kR} \frac{1}{2R_0^2} R\,dR = \frac{\sin^2(kR_0)}{(kR_0)^2} . \qquad (21)$$

The average value for the square jumping distance is $\langle R^2 \rangle = 2R_0^2$.

According to Eq. (21), R_0 of about 0.4 Å for the 14.4-keV transition of ^{57}Fe makes the narrow absorption line invisible ($< 1\%$). This result is different from that obtained assuming a finite number N. For a finite number

N, even when $\langle R^2 \rangle$ is very large, a finite intensity $1/N$ stays with the absorption line of natural width. In the model based on the overdamped harmonic oscillator, the recoil-free fraction for atoms oscillating around a center within distance R_0 would be given by $\exp(-\frac{1}{3}k^2R_0^2)$ which coincides with Eq. (21) only for low values of kR_0.

Within the above model of jumps among N equivalent sites the Mössbauer spectrum consists of many broad lines which together appear as a single broad line of half-width $\gamma/2 = \Gamma/2 + \Delta/2$ where, as before, Γ is the natural linewidth. The extra width Δ will now be given approximately by the general "diffusion-jump" models,[21,22] $\Delta = 2\lambda[1 - g(\mathbf{k})]$, where λ is the jumping rate and $g(\mathbf{k})$ is the Fourier transform of $g(\mathbf{r})$,[22] the probability to jump a distance $\mathbf{r}$. In our case, if we assume that the probability to jump to any point is the same, the probability to jump a distance between R and $R + dR$ is just the number of points in that region. Thus we obtain for $g(\mathbf{k})$ the same formula as that for f/f_0 and may write

$$\Delta = 2\lambda(1 - f/f_0) . \qquad (22)$$

The diffusion constant is defined by the formula $D = \frac{1}{6}\lambda\langle R^2 \rangle$. For $kR_0 \ll 1$, $\Delta = 2k^2D$, which is consistent with the ordinary diffusion broadening formula. In this case, where $\langle R^2 \rangle = 2R_0^2$ and $R_0 = 3\langle x_0^2 \rangle$, we obtain $D = \lambda\langle x_0^2 \rangle$ and thus λ in the jump model coincides with α of the harmonically bound diffusion model.[16]

C. Application to incoherent neutron scattering

The same formulas presenting the Mössbauer spectra also represent the spectral function for scattered neutrons. The only differences are (a) the natural width Γ can be dropped, (b) ω_0 is zero, and (c) instead of the convolution with the source line in the Mössbauer case, a convolution with the experimental energy resolution function has to be performed. The present formulas can thus be used successfully for representing neutron scattering spectra from diffusing nuclei in bound geometries. It is of interest to make a comparison of the predictions of Eq. (4) with those of Eq. (33) of Ref. 10, which treats the effects of bound diffusion in a sphere of radius R on neutron scattering spectra. Of special interest is the linewidth of the broad line as a function of the momentum transfer $\mathbf{k}$. Within the present model, the half-width of the broad line, Fig. 3, is k independent for low values of k and equals $\alpha = D/\langle x^2 \rangle = 3D/\langle r^2 \rangle$. The model of bound diffusion in a sphere yields at low k a broad line of half-width $4.333D/R^2$ (Fig. 2 in Ref. 10). Considering that $\langle r^2 \rangle$ in our model represents a Gaussian average of the particles' motion amplitude, and R in Ref. 10 represents the largest displacement of the particle, the two formulas are not much different at all. At high k values both models yield a broad line with half-width k^2D.

We conclude that neutron spectroscopy of nuclei in bound diffusion can be treated by Eq. (4) or Eq. (5), with advantages of Eq. (4) for clarity of behavior in extreme cases and in anisotropic cases and of Eq. (5) for computing purposes. The sum converges very fast in comparison to the slowly converging sum of Eq. 33 in Ref. 10.

III. EXPERIMENTAL RESULTS

In Fig. 5 we present typical ^{57}Fe absorption spectra obtained near room temperature in four different biological systems which are

(a) iron storage material in packed cells of chick embryo fibroblasts,[8]

(b) membrane-bound iron storage material from *Mycoplasma capricolum*,[14]

(c) membrane-bound iron storage in packed cells of *Escherichia coli*,[6] and

(d) crystals of deoxy-myoglobin (deoxyMb) highly enriched in ^{57}Fe.[5]

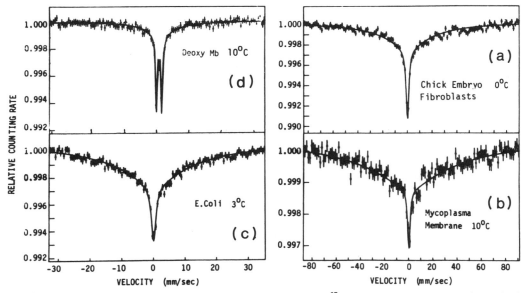

FIG. 5. Mössbauer absorption spectra in various biological systems containing ^{57}Fe. The solid curves are theoretical least-squares fits using Eq. (4). The parameters D (in units of 10^{-10} cm/sec) and α (in units of 10^8 sec^{-1}), are respectively, (a) 40, 10; (b) 87, 13; (c) 26, 5; and (d) 4.3, 2.5.

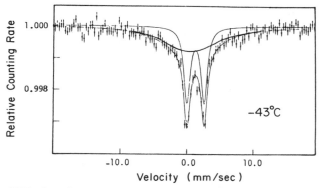

FIG. 6. Mössbauer spectrum of ^{57}Fe in divalent iron Nafion at 230 K. The inner lines show the decomposition of the spectrum into the elastic narrow doublet and the quasielastic broad line. (Nafion is the trade name for perfluorosulfonate membranes developed by DuPond.)

In cases (a), (b), and (c) the iron is in the form of inorganic aggregates of size less than 50 Å bound strongly to the membranes. In (d) each protein molecule contains a single iron atom which is part of the heme group, bound within the protein.

A least-squares fit of Eq. (4) to the experimental data for deoxyMb crystals [Fig. 5(d)] at 283 K yields $D = 4.3 \times 10^{-10}$ cm^2 s^{-1}, $\langle x^2 \rangle = 1.7 \times 10^{-2}$ Å^2, and $\alpha = 2.5 \times 10^8$ sec^{-1} or $\tau_c = 4 \times 10^{-9}$ sec. The diffusive motions of Fe in deoxyMb are large-scale internal motions of structures internal to the protein of which the iron atom is a part.[5] Assuming, therefore, that m is of the order of the mass of the protein (molecular weight 17 000), we obtain $\beta \sim 3 \times 10^{15}$ s^{-1} and $w \sim 9 \times 10^{11}$ s^{-1}. We thus see that indeed, $w \ll \beta$ and $\beta t \gg 1$ (t is of the order of 10^{-7} sec—the lifetime of the ^{57}Fe excited state), as is assumed in deriving Eq. (4). The parameters derived in the three other biological systems displayed in Fig. 5 are given in the figure captions. In all cases consistently better fits were obtained using Eq. (4) compared to fits to the sum of a narrow and a wide Lorentzian line, in accord with simple diffusion-jump models. Another example of the phenomena discussed is observed in Fig. 6 which displays Mössbauer spectra of iron nuclei performing bound diffusion in a polymer.[7]

In Fig. 7 we display the experimental spectra of deoxyMb (Ref. 23) which definitely exhibit an asymmetric shape at temperatures above the freezing point of the internal water, yet none at temperatures below this freezing point, 220 K. Thus one is tempted to fit these spectra with the formula of Eq. (9) with anisotropy in the mean-square deviation. Such least-squares fits are shown in Fig. 7.

IV. CONCLUSIONS

Many aspects of the expected spectra from nuclei performing bound diffusion are not very sensitive to the exact dynamical model for the kind of motion. All theories predict a narrow, elastic line with additional broad quasielastic lines. The relative intensity of the narrow line and the width of the broad line are numerically similar in all theories, though given by different mathematical expressions. The unique properties of the present model are that it considers many modes of motion, as well as anisotropic motion, and yields a closed-form formula which can be easily calculated in extreme limits and easily calculated by computer. Only with very accurate experimental spectra will one be able to distinguish between the relative validity of different dynamical models.

ACKNOWLEDGMENTS

This work was supported in part by the Basic Research Foundation, Israel Academy of Sciences and Humanities.

APPENDIX: SPECTRA FROM NUCLEI IN PARTICLES PERFORMING BOTH BOUND TRANSLATIONAL AND FREE ROTATIONAL DIFFUSION

In cases where the diffusing bound particle is composed of many molecules, as in the case of a rigid macroscopic particle, rotational diffusion may influence the Mössbauer or scattered neutron spectral shapes. The case of bound translational diffusion was treated in great detail in the present paper; the case of free rotational diffusion was treated in detail elsewhere.[24] For a nucleus at a distance r from the center of mass of the particle, the intermediate scattering function, for a γ ray of wave vector **k**, or neutron wave-vector change **k**, is given by

$$F_{rot}(r, \mathbf{k}, t) = \sum_{l=0}^{\infty} (2l+1) j_l^2(kr) e^{-l(l+1)D_{rot}t} \qquad (A1)$$

Here $j_l(x)$ is a spherical Bessel function and D_{rot} is the rotational diffusion constant.

Considering the translational and rotational motion as uncorrelated, we can calculate the spectral shape for the

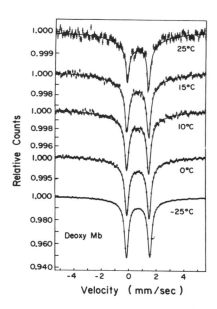

FIG. 7. Mössbauer spectra of Fe in deoxy-myoglobin which display anisotropic harmonically bound motion. The spectra were fitted by Eq. (9) assuming axially symmetric mean-square displacement ($\langle x_{\parallel}^2 \rangle$ and $\langle x_{\perp}^2 \rangle$) and isotropic β.

composed motion. The total intermediate scattering function is given by [see Eq. (4)]

$$F_{\text{tot}}(r,\mathbf{k},t) = F_{\text{rot}}(r,\mathbf{k},t)\exp[-k^2\langle x^2\rangle(1-e^{-\alpha t})] . \quad \text{(A2)}$$

Performing the expansion as in Eq. (5) we obtain that the final spectral function is given by a double infinite sum of Lorentzian lines:

$$I(r,\omega) = \frac{1}{2\pi}\int_{-\infty}^{\infty} e^{-i(\omega-\omega_0)-\Gamma|t|/2}F_{\text{tot}}(r,\mathbf{k},t)\,dt$$

$$= \sum_{l=0}^{\infty}\sum_{n=0}^{\infty}\frac{a_{nl}(r)\Gamma_{nl}/2\pi}{(\Gamma_{nl}/2)^2+(\omega-\omega_0)^2} , \quad \text{(A3)}$$

where

$$\Gamma_{nl}/2 = \Gamma/2 + n\alpha + l(l+1)D_{\text{rot}}$$

and

$$a_{nl} = \exp(-k^2\langle x^2\rangle)\frac{(k^2\langle x^2\rangle)^n}{n!}(2l+1)j_l^2(kr) .$$

In practice we have to average the spectrum over all values of r in which the nuclei are located. For a spherical particle of radius R with a homogeneous density of nuclei, the spectrum of Eq. (A3) will be given by

$$I(\omega = \frac{3}{R^3}\int_0^R I(r,\omega)r^2\,dr .$$

Thus the spectrum is still the same sum of Lorentzians as in Eq. (A3) except that now the intensities are given by

$$A_{nl}(R) = \frac{3}{R^3}\int_0^R a_{nl}(r)r^2\,dr$$

$$= \frac{3}{2}e^{-k^2\langle x^2\rangle}\frac{(k^2\langle x^2\rangle)^n}{n!}(2l+1)$$

$$\times[j_l^2(kR)+j_{l-1}^2(kR)$$

$$-\frac{2l+1}{kR}j_l(kR)j_{l-1}(kR)] . \quad \text{(A4)}$$

In the simple case of a homogeneous spherical particle the spectrum is completely characterized by three parameters; the radius R, the mean-square deviation $\langle x^2\rangle$, and the diffusion constant D (where $\alpha = D/\langle x^2\rangle$ and $D_{\text{rot}} = \frac{3}{4}D/R^2$).

For $k^2\langle x^2\rangle > 5$ the translational diffusion behaves as unbounded diffusion, Fig. 3, and the intermediate spectral function can be taken as $\exp(-k^2Dt)$. Thus the total spectrum will be given by

$$I(r,\omega) = \sum_{l=0}^{\infty}\frac{a_l(r)\Gamma_l/2\pi}{(\Gamma_l/2)^2+(\omega-\omega_0)^2} , \quad \text{(A5)}$$

where

$$a_l(r) = (2l+1)j_l^2(kr)$$

and

$$\Gamma_l/2 = \Gamma/2 + k^2D + l(l+1)D_{\text{rot}} .$$

Again, in the case of a homogeneous spherical particle after averaging over r, one obtains the spectrum of Eq.

(A5) with intensities

$$A_l(R) = \frac{3}{R^3}\int_0^R a_l(r)r^2\,dr$$

$$= \frac{3}{2}(2l+1)[j_l^2(kR)+j_{l-1}^2(kR)$$

$$-\frac{2l+1}{kR}j_l(kR)j_{l-1}(kR)] .$$

Since the narrowest line in Eq. (A5) is of width $\Gamma + 2k^2D$ and since generally $\Gamma \ll k^2D$ and, at least in the Mössbauer case, $D_{\text{rot}} \ll k^2D$ also, we expect that Eq. (A5) can be reproduced by a single Lorentzian line of a width slightly larger than $2k^2D$. In the case of a homogeneous spherical particle, where $D_{\text{rot}} = \frac{3}{4}D/R^2$, one can calculate the effective width of the spectrum given by Eq. (A5) by equating

$$\frac{\Gamma_{\text{eff}}/2\pi}{(\Gamma_{\text{eff}}/2)^2+(\omega-\omega_0)^2} = \sum_{l=0}^{\infty}\frac{A_l(R)\Gamma_l/2\pi}{(\Gamma_l/2)^2+(\omega-\omega_0)^2} . \quad \text{(A6)}$$

At $\omega = \omega_0$, considering $k^2D \gg \Gamma/2$, we obtain

$$\frac{2k^2D}{\Gamma_{\text{eff}}(R)} = \sum_{l=0}^{\infty}\frac{A_l(R)}{1+\frac{3}{4}l(l+1)/k^2R^2} .$$

In Fig. 8, Γ_{eff} (in units of $2k^2D$) is displayed as a function of kR. For $kR \sim 3$, the value of Γ_{eff} is in full saturation, and equals $1.271\times 2k^2D$. Thus for a macroscopic particle, even the size of a small molecule ($R \sim 5$ Å), the Mössbauer spectrum will be affected by rotational diffusion only to the extent of broadening the translational diffusion line by 20–30 %.

In the general case of a sphere performing bound translational and free rotational diffusion, one can represent Eq. (A3) by a superposition of an effective narrow line, the $n = 0$ line broadened by rotation, and a broad line, $n \geq 1$, also broadened by the rotations. The effective

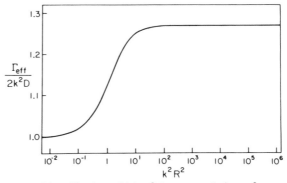

FIG. 8. The effective width of the spectral shape from nuclei in a spherical particle of radius R performing free translational and rotational diffusion.

width of these two lines can be calculated by the harmonic averages

$$\frac{1}{\Gamma_{\text{nar}}} = \sum_{l=0}^{\infty} \frac{A_{0l}}{\Gamma_{0l}}$$

and

$$\frac{1}{\Gamma_{\text{brod}}} = \sum_{n=1}^{\infty} \sum_{l=0}^{\infty} \frac{A_{nl}}{\Gamma_{nl}} \ .$$

These were calculated as functions of $k^2\langle x^2 \rangle$ for $k^2 R^2 > 1000$. In Fig. 9 Γ_{nar} is given in units of Γ as a function of $2k^2 D/\Gamma$. Γ_{brod} is shown in Fig. 3, in units of 2α. We observe that the effect of rotation on the broad line is about the same for all $k^2\langle x^2 \rangle$ values. In a recent paper[25] we have applied the formulas discussed in this appendix to analyze the spectra of iron in magnetic particles located in bacteria.

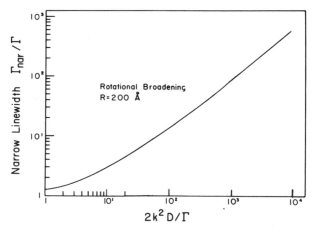

FIG. 9. The line broadening of the $n = 0$ line of Eq. (A3) for spherical particles of radius R ($kR > 10$) as a function of the diffusion constant.

*Deceased.

[1]S. G. Cohen, E. R. Bauminger, I. Nowik, S. Ofer, and J. Yariv, Phys. Rev. Lett. 46, 1244 (1981).

[2]S. G. Cohen, E. R. Bauminger, I. Nowik, S. Ofer, and J. Yariv, in Biomolecular Stereodynamics, edited by R. H. Sarma (Adenine. New York. 1981). Vol. II. p. 299.

[3]E. R. Bauminger, S. G. Cohen, I. Nowik, S. Ofer, and J. Yariv, in Proceedings of the 1981 International Conference on Applications of the Mössbauer Effect, Jaipur, India, edited by V. G. Bhide (Indian National Academy, New Delhi, 1982).

[4]K. H. Mayo, F. Parak, and R. L. Mössbauer, Phys. Lett. 82 A, 468 (1981).

[5]E. R. Bauminger, S. G. Cohen, I. Nowik, S. Ofer, and J. Yariv, Proc. Natl. Acad. Sci. (U.S.) 80, 736 (1983).

[6]E. R. Bauminger, S. G. Cohen, E. Giberman, I. Nowik, S. Ofer, J. Yariv, M. M. Werber, and M. Mevarech, J. Phys. (Paris) Colloq. 37, C6-227 (1976).

[7]E. R. Bauminger, I. Nowik, S. Ofer, and C. Virguin-Heitner, Polymer (to be published).

[8]E. R. Bauminger, S. G. Cohen, S. Ofer, and U. Bachrach, Biochim. Biophys. Acta 720, 133 (1982).

[9]F. Parak, E. W. Knapp, and D. Kucheida, J. Mol. Biol. 161, 177 (1982).

[10]F. Volino and A. J. Dianoux, Mol. Phys. 41, 271 (1980).

[11]F. Volino, M. Pineri, A. J. Dianoux, and A. Degeyer, J. Polym. Sci., Polym. Phys. Ed. 20, 481 (1982).

[12]G. E. Uhlenbeck and L. S. Ornstein, Phys. Rev. 36, 823 (1930).

[13]A. Rahman, K. S. Singwi, and A. Sjölander, Phys. Rev. 126, 997 (1962).

[14]I. Nowik, S. G. Cohen, E. R. Bauminger, and S. Ofer, Phys. Rev. Lett. 50, 1528 (1983).

[15]E. W. Knapp, E. W. Fischer, and F. Parak, J. Chem. Phys. 78, 4701 (1983).

[16]K. V. Shaitan and A. B. Rubin, Biofizika 25, 796 (1980).

[17]A. Bläsius, R. S. Preston, and U. Gonser, Z. Phys. Chem. 115, 187 (1979).

[18]W. Petry, G. Vogl, and W. Mansel, Phys. Rev. Lett. 45, 1862 (1980); W. Petry and G. Vogl, Z. Phys. B 45, 207 (1982).

[19]S. Dattagupta, Solid State Commun. 24, 19 (1977).

[20]V. I. Goldanskii and E. F. Makarov, in Chemical Applications of Mössbauer Spectroscopy, edited by V. I. Goldanskii and R. H. Herber (Academic, New York, 1968), p. 102.

[21]K. S. Singwi and A. Sjölander, Phys. Rev. 120, 1093 (1960).

[22]S. Dattagupta, Phys. Rev. B 14, 1329 (1976); 12, 47 (1975).

[23]E. R. Bauminger, S. G. Cohen, I. Nowik, S. Ofer, and J. Yariv, Hyperfine Interact. 15/16, 881 (1983).

[24]Quasielastic Neutron Scattering for the Investigation of Diffusive Motions in Solids and Liquids, Vol. 64 of Springer Tracts in Modern Physics, edited by T. Springer (Springer, Berlin, 1972), p. 64.

[25]S. Ofer, I. Nowik, E. R. Bauminger, G. C. Papaefthymiou, R. B. Frankel, and R. B. Blakemore, Biophys. J. 46, 57 (1984).

Hemoglobin S Gelation and Sickle Cell Disease

By William A. Eaton and James Hofrichter

THE FUNDAMENTAL cause of sickle cell disease is the decreased deformability of the sickled red cell produced by gelation of hemoglobin S. Partial inhibition of gelation should therefore reduce clinical severity, while complete inhibition should result in a "cure." These basic ideas have stimulated an enormous effort to understand the gelation process in detail and to relate the results of these studies to the pathophysiology of sickle cell disease. Discoveries concerning gelation have also led to new lines of research on a specific therapy. The early finding that fetal hemoglobin inhibits gelation.[1,2] has ultimately led to the development of methods to increase the production of F cells in the bone marrow of sickle cell patients,[3-5] while the discovery of the enormous sensitivity of the rate of gelation to hemoglobin concentration[6] has stimulated studies on the reduction of intracellular hemoglobin concentration as a means of therapy.[6-12] Studies on the structure of the hemoglobin S polymer,[13-18] moreover, have guided the development of agents designed to inhibit gelation by interfering with the formation of intermolecular contacts in the polymer.[19-21]

The purpose of this article is to review recent developments in the relation between hemoglobin S gelation and sickle cell disease. We first present our current understanding of the major features of the gelation process. Since gelation is a physical rather than a chemical process, its description necessarily requires more physical detail than that of most biological processes. From these studies we are able to develop a more rigorous and comprehensive description of the relation between gelation and the pathophysiology than has been possible up to now. By combining the gelation studies with work on the rheology of sickle cells and blood flow in the microvasculature, a clearer picture emerges of the outstanding issues in understanding the mechanism of vaso-occlusion in patients and the resulting cardiovascular response. Finally we discuss the variation in clinical severity and analyze the problem of inhibiting gelation in patients. Throughout this discussion we shall see that the kinetics of gelation is a dominant factor in understanding gelation both in vitro and in vivo, and it will become clear that discussions of the pathophysiology that do not include a kinetic analysis[22,23] are inadequate.

A broader treatment of sickle cell disease, including genetic and clinical aspects, has recently appeared in two excellent books.[24,25] Also the structure; physical chemistry, and rheology of hemoglobin S gelation in solution and in red cells is discussed much more extensively in an article that is being published elsewhere.[26]

GELATION AT EQUILIBRIUM

To understand gelation we first must describe a gel at equilibrium. As shown in Fig 1, a gel can be separated into two phases, a solution phase that contains free hemoglobin molecules and a polymer phase. The structure of the individual polymers has now been determined in considerable detail. It is a fiber made up of 14 intertwined helical strands of hemoglobin S molecules.[14,15,27] The fiber can alternatively be described in terms of seven intertwined double strands of molecules in which the double strands have a structure that, except for a slight helical twist, is nearly identical to the double strands that form the fundamental unit of the deoxy-hemoglobin S single crystal.[16,28,29] In each molecule one of the two $\beta 6$ valines of the $\alpha_2\beta_2$ tetramer is involved in an intermolecular contact with its neighbor in the double strand. The structure of the deoxyhemoglobin S crystal is known to atomic resolution, so that there is a very detailed picture of the intermolecular contacts within the double strand that must be very similar to what occurs in the polymer.[16-18]

A gel at equilibrium behaves very much like a suspension of microscopic protein crystals suspended in a saturated protein solution.[30-32] The concentration of hemoglobin in the solution phase, which is called the solubility, is an accurate measure of the stability of the polymer phase. The solubility is determined experimentally by measuring the hemoglobin concentration in the supernatant obtained after high-speed sedimentation of the polymers.[31-36] Because the concentration of hemoglobin in the polymer phase appears to be relatively constant,[37] the fraction of the total hemoglobin that is polymerized can be calculated from the solubility using a simple mass conservation relation. There have now been systematic investigations of the solubility under a wide variety of solution conditions. These include the dependence on temperature,[31,32,35] pH,[38,39] salts,[40] 2,3-DPG,[39,41,42] carbon monoxide,[36,43,44] oxygen,[45,46] and non-S hemoglobins.[30,37,38,47-63]

The role of non-S hemoglobins in the gelation of hemoglobin mixtures has been the focus of a large number of studies, since early investigations showed that the presence of hemoglobins A, C, and F reduces sickling and is accompanied by decreased clinical severity.[25] For mixtures of hemoglobins S and F (and S + A₂), a detailed analysis of solubility data, including the large contribution of nonideality arising from excluded volume effects,[32,37,54,55] indicates that over the physiologic range of compositions there is little or no copolymerization of either the homotetramers, $\alpha_2\gamma_2$ (and $\alpha_2\delta_2$), or the hybrid tetramers, $\alpha_2\beta^S\gamma$ (and $\alpha_2\beta^S\delta$).[26,37,64] The low probabilities for copolymerization of these molecules can be rationalized as resulting from destabilizing effects on the intermolecular contacts of the double strand that accompany specific amino acid replacements on the molecular surface.[37,65] For S + A and S + C mixtures the analysis indicates that there is little or no copolymerization of the $\alpha_2\beta_2^A$ and $\alpha_2\beta_2^C$ molecules but that the hybrid molecules $\alpha_2\beta^S\beta^A$ and $\alpha_2\beta^S\beta^C$

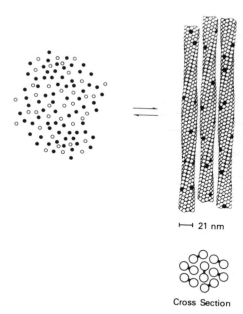

├──┤ 21 nm

Cross Section

Fig 1.   Schematic picture of a gel of hemoglobin S at partial saturation with oxygen. The 64 kD molecule (ie, the tetramer) is represented as a circle. A gel of hemoglobin S contains large polymers, often called fibers, and a concentrated solution of free hemoglobin molecules. The filled circles represent hemoglobin S molecules with one or more oxygen molecules bound. There are relatively fewer filled circles in the polymer because it has a lower oxygen affinity than the solution. The structure was determined by Edelstein and coworkers using electron microscopy and image reconstruction techniques.[14,15] The cross-section shows that the fiber consists of 14 strands and that it can be constructed from seven double strands that are very similar to those found in the deoxyhemoglobin S single crystal.[18,27,29]

copolymerize with a probability that is approximately half that for $\alpha_2\beta_2^S$.[26,37,62] The factor of 2 is consistent with the structural result that a valine residue is required at only one of the two $\beta6$ sites on each molecule for it to be incorporated into the polymer.

Binding of oxygen to hemoglobin S has a dramatic effect on gelation. Experiments performed over 35 years ago demonstrated that fully deoxygenated hemoglobin S gels, while fully oxygenated hemoglobin S does not.[66] To begin to consider the pathophysiology of sickle cell disease, however, it is clear that one needs quantitative data on solutions that are partially saturated with oxygen, particularly over the range of fractional saturations encountered in vivo. Such data have been obtained only relatively recently. The schematic picture of gelation in Fig 1 points out several important questions that must be answered regarding gelation in the presence of oxygen. These include (1) what is the fraction of hemoglobin molecules that are polymerized? and (2) what are the fractional saturations with oxygen of the molecules in the solution and polymer phases? The fraction polymerized can be determined from measurements of the solubility as a

function of solution phase saturation.[46] Also, because there is no aggregation in the solution phase of the gel[67,68] and the $\beta6$ mutation has no effect on the intrinsic affinity of the hemoglobin molecule, the binding curve for the solution phase hemoglobin S molecules is normal.[69,70] The major problem has been to obtain the polymer binding curve, which was accomplished using an optical technique called linear dichroism.[44,46,71-73]

The principal experimental results are shown in Fig 2. The solubility increases slowly at low oxygen saturation, then increases sharply at high saturations (Fig 2 b). The most interesting finding from the binding studies is that the polymer binds oxygen noncooperatively, as evidenced by a slope of unity in a Hill plot.[46] The two-state allosteric model,[74] which has provided an excellent framework for interpreting a wide variety of experiments on hemoglobin, provides a simple molecular interpretation of these results.[46] According to this model a hemoglobin molecule free in solution exists in one of two affinity states at all stages of oxygenation. The low-affinity state, called T, has the quaternary structure of completely deoxygenated hemoglobin, while the high-affinity state, called R, has the quaternary structure of the fully oxygenated molecule. Binding to either quaternary structure is noncooperative. Cooperativeness arises from the continuous conversion of low-affinity T-state molecules to high-affinity R-state molecules as the saturation increases (Fig 2 a). The simplest extension of this model to the gelation of hemoglobin S is to postulate that all T-state molecules polymerize with equal probability independent of the number of oxygen molecules bound and that there is no polymerization of R-state molecules.[46] The model is based on the idea that R-state molecules do not polymerize because their structure is sufficiently different from T-state molecules that they cannot fit into the polymer lattice. Analysis of the structure of the double strand of the deoxyhemoglobin S crystal does indeed show that it is impossible to replace the T-state molecule with an R-state molecule.[18] Since the polymer contains only T-state molecules, the model predicts that it will bind oxygen noncooperatively, exactly as observed. The model is not quantitatively perfect, however, because the affinity of the polymer is slightly lower than that of solution T-state molecules (Fig 2 a). Evidence for this small difference is also found in the solubility data (Fig 2 b), indicating that T-state molecules with oxygen bound are partially discriminated against by polymers. The small conformational changes that are known to take place within the T quaternary structure upon oxygen binding could explain this result.[46] The picture that emerges, then, is that the simplest extension of the two-state allosteric model provides an excellent description of the effect of oxygen on gelation.

The results of these experiments can be used to explain data on polymerization in sickle red cells. The first step is to consider further the oxygen binding curve of a gel. As mentioned above, at a given oxygen pressure the saturation of the gel is a weighted average of the saturations of the solution and polymer phases. Figure 2 c shows gel binding curves under near physiologic conditions that have been constructed from solution and polymer phase binding curves (Fig 2 a) and from the solubility curve (Fig 2 b). As the

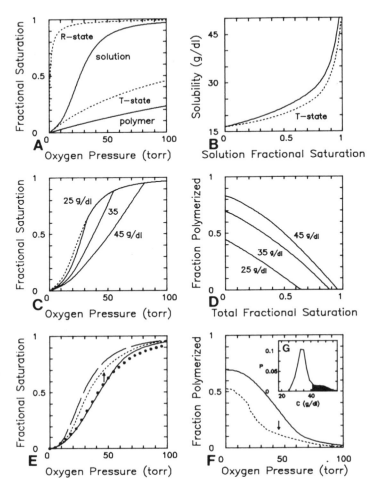

Fig 2.   Effect of oxygen on gelation in solution and in sickle cells. (A) Solution, polymer, and theoretical R- and T-state binding curves. The solution binding curve is the binding curve for normal blood. The polymer binding curve is calculated from the data in phosphate buffer of Sunshine et al.[46] The R- and T-state binding curves are theoretical and were obtained by fitting to the solution binding curve with the two-state allosteric saturation function.[74] (B) Solubility as a function of solution-phase saturation and oxygen pressure from the data of Sunshine et al.[46] The dotted curve shows the theoretical solubility for the hypothetical case in which all T-state molecules polymerize with equal probability, independent of the number of oxygen molecules bound. (C) Gel-binding curves at different total hemoglobin concentrations calculated from the results in (A) and (B).[26,46] The dotted curve is the solution binding curve of A. (D) Fraction polymerized as a function of total fractional saturation calculated from the results in (A) and (B).[26,46] (E) In vivo oxygen binding curve calculated for a population of cells having the distribution of intracellular hemoglobin S concentrations in (G). The binding curve calculated in the absence of polymer (long-dashed curve) and the binding curve calculated when polymer is present at equilibrium (solid curve) are shown for reference. The data points are for SS blood from Winslow.[77] The oxygen unloading curve under in vivo conditions (dotted curve) was calculated by requiring that only the densest 18% of the cell population contain polymer at the average venous Po_2 found in SS patients (46 torr)[78] denoted by the arrow. These conditions were simulated by requiring that each cell be sufficiently supersaturated for polymerization to occur within about 200 ms at each Po_2. (F) Fraction of total hemoglobin S that is polymerized. The fraction is calculated under equilibrium conditions (curve) and under in vivo conditions (dotted curve). (G) The average distribution of intracellular concentrations from a study of 43 patients by Fabry et al[75] used in the calculations in panels (E) and (F). The probability density, P, in dL/g is plotted v the intracellular hemoglobin concentration, C, in g/dL. The blackened area shows the subpopulation of cells that contain polymer at Po_2 of 46 torr. The equations and parameters used in the calculation of all of the above panels are given by Eaton and Hofrichter,[26] which are derived from the work of Sunshine et al.[46]

oxygen pressure increases, not only do the saturations of both phases increase but the solubility also increases, decreasing the contribution of the low affinity polymer phase to the total binding curve. As a result the gel binding curves appear to have lower-than-normal affinity with higher-than-normal cooperativeness.[44] At sufficiently high oxygen pressures the solubility exceeds the total hemoglobin concentration, and the binding curve superimposes on the normal solution

binding curve (Fig 2 c). As the total hemoglobin concentration increases, the affinity of the gel decreases owing to the increased fraction of the low affinity polymer (Fig 2d), and the oxygen pressure at which the polymer disappears also increases.

The oxygen-binding curve of a single red cell should be identical to the binding curve of a gel having the same composition (total hemoglobin S concentration, fraction fetal hemoglobin, pH, 2,3-DPG concentration, etc). Oxygen-binding curves of sickle blood are an average of the gel-binding curves for the individual cells. As in gels, the binding curve for sickle blood is "right shifted" compared to normal blood. Although 2,3-DPG levels are elevated in sickle cell blood, the formation of the low-affinity polymer is the major cause of the right shift in the blood-binding curve.[77,79]

The major difference between the binding curves for cell suspensions (Fig 2 e) and for gels (Fig 2 c) results from the wide distribution of hemoglobin S concentrations, which varies from about 20 g/dL in F cells to almost 50 g/dL.[79-81] This distribution produces a wide range of median affinities within the red cells from a given patient and hence smears the characteristic features of the gel-binding curve (compare Figs 2 c and 2 e). To calculate blood-binding curves it is necessary to utilize the results of recent investigations that have characterized the distribution of total hemoglobin concentrations from density measurements.[62,75,76,82-88] The density distributions for SS cells are broader and more variable than those for normal individuals. Since the distribution of intracellular hemoglobin S and hemoglobin F concentrations was not determined for the cells employed in the oxygen-binding measurements, it is only possible to make qualitative comparisons between the observed binding curves and those calculated from solution data. Figure 2 e compares a whole blood oxygen-binding curve with the binding curve calculated from the solution data using the average concentration distribution from a study of 43 patients.[26,75] The p50s for the curves calculated from concentration distributions for individual patients vary from 37 torr to 46 torr, compared to the 33 torr to 45 torr observed in a study of 14 patients.[77,89] This comparison shows that the patient-to-patient variability in oxygen binding curves can be readily accounted for by the variability in intracellular concentration distributions. A

more direct comparison on a relatively homogeneous cell population obtained by density fractionation, in which the 2,3-DPG and hemoglobin F levels were also measured, gives very good agreement between the cell and solution data.[26,79]

Gelation in isolated solutions and cells has also been compared using nuclear magnetic resonance techniques to measure the average fraction of polymerized hemoglobin as a function of the total saturation of the cells.[76,90-92] The nuclear magnetic resonance measurements take advantage of the fact that the polymerized molecules do not rotate freely, making it possible to selectively measure the spectra of the polymerized and unpolymerized molecules.[93] Measurements on a cell population of known concentration distribution are in good agreement with the curve calculated from the solution data.[76] Agreement is also obtained in a comparison of density fractionated cells, although the experimental uncertainties are much larger.[76]

At this point we should emphasize that the oxygen binding curves and polymer fraction curves that we have discussed are equilibrium or near-equilibrium curves and, as we shall see later, are very different from the in vivo situation in which most cells are very far from equilibrium because of the large kinetic effect of the delay time (Fig 2 e and 2 f).

KINETICS AND MECHANISM OF GEL FORMATION

The most unusual and interesting aspect of the gelation process is the kinetics and mechanism of gel formation. The simplest kinetic experiment takes advantage of the characteristic property that a hemoglobin S solution gels upon heating. A completely deoxygenated solution, having a concentration significantly less than the solubility at 0°C (< 30 g/dL), is heated to some temperature where the concentration exceeds the solubility. Polymer formation can be detected by a variety of techniques, including linear birefrin-

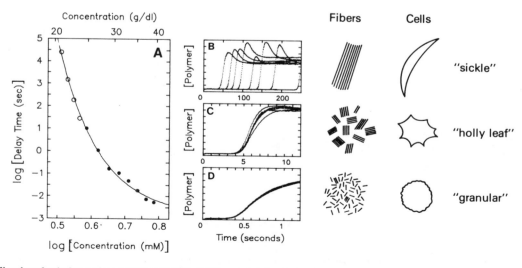

Fig 3.   Kinetics of gelation and morphology of cells. (A) Concentration dependence of the delay time from laser photolysis (filled circles) and temperature-jump measurements (open circles; from Ferrone et al[132]). (B, C, D) Reproducibility of kinetic progress curves for samples having different delay times (from Hofrichter[133]). The schematic at the right shows that in a slowly polymerizing cell the "sickle" morphology is postulated to result from the formation of a single domain of well-aligned fibers; that in a more rapidly polymerizing cell the "holly leaf" morphology results from the formation of a number of smaller domains of shorter, aligned fibers; and that in the fastest polymerizing cells the "granular" morphology results from the formation of a large number of very small domains or randomly oriented short fibers.

gence,[6,43,45,94] turbidity,[36,37,43,44,57,95-112] light scattering (G.W. Christoph and R.W. Briehl, unpublished results),[113-115] viscosity,[59,112,116-122] water proton magnetic resonance linewidths[123] and transverse relaxation times,[124-128] and electron paramagnetic resonance.[129] All of the techniques show the same type of time course. There is an apparent delay period during which there is no evidence for any aggregation, followed by the explosive appearance of polymer.[6,94,95,116,130] Upon lowering the temperature of a preformed gel, depolymerization proceeds much more rapidly and without a delay.[6,94,95] The most striking finding from these studies is that the delay time is enormously sensitive to solution conditions, in particular to the hemoglobin S concentration. The inverse of the delay time is found to be proportional to the 30th to 50th power of the initial hemoglobin concentration.[6,36,37] This is the highest known concentration dependence for a process taking place in solution. The delay time is also found to be directly proportional to the 30th to 45th power of the solubility, independent of the manner in which the solubility is altered.[36,97] For example, in temperature-jump experiments, the delay time for a solution that is 10% saturated with carbon monoxide is increased by a factor of about 10 relative to deoxyhemoglobin S at the same total hemoglobin concentration, and at a saturation of 40% the delay time is increased by a factor of about 10^4.[36,44]

The temperature-jump technique is limited to measuring delay times longer than about 100 seconds. To extend the kinetic measurements to physiologic times and hemoglobin concentrations required the development of a laser photolysis technique that could be used to prepare a completely deoxy-genated hemoglobin S solution in less than a few milliseconds.[131-133] In this technique the carbon monoxide complex of hemoglobin S, which is soluble up to at least 48 g/dL,[134] can be converted to deoxyhemoglobin S by photodissociation under continuous laser illumination. The laser also serves as a source for monitoring gel formation from the change in light scattering. When the laser is turned off, the carbon monoxide recombines, the polymer disassembles to form a solution of monomers (ie, 64 kD hemoglobin S tetramers), and the experiment can be repeated indefinitely. Because the volumes of observation are as small as 10^{-11} cc, the laser photolysis technique can also be used for investigating gelation in single red cells.[135-137]

A combination of the temperature jump and laser photolysis techniques has been used to examine the kinetics of gelation over a wide range of concentrations, temperatures, and times.[132] Figure 3 shows that as the concentration decreases, the delay time increases from about 10 milliseconds at 40 g/dL to about 100,000 seconds at 20 g/dL. An important clue to the mechanism by which gelation occurs comes from a very unusual result, discovered in the course of the laser photolysis experiments, which is described in Fig 3. Highly reproducible delay times are observed for solutions with delay times of a few hundred milliseconds or less. When the delay times become longer than a few seconds, however, the delay times become very *ir*reproducible, despite the fact that the progress curves have very similar shapes once polymerization begins (Fig 3 b to d).[131-133] An important companion observation is that only a single birefringent domain of polymers forms when there are large fluctuations

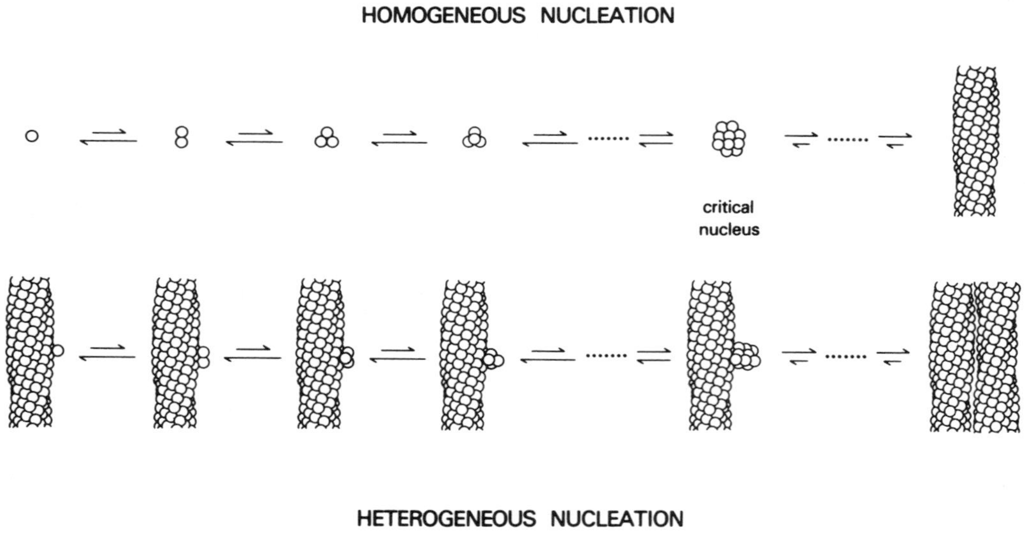

HOMOGENEOUS NUCLEATION

HETEROGENEOUS NUCLEATION

Fig 4. The double nucleation mechanism (Ferrone et al[138]). The two pathways for nucleation of polymers are shown. In the homogeneous pathway nuclei form in the solution, while in the heterogeneous pathway nuclei form on the surface of existing polymers. As more polymers form the increased surface area results in a continuously increasing rate of heterogeneous nucleation. This autocatalytic formation of polymers via the heterogeneous nucleation pathway is responsible for the appearance of a delay period prior to the observation of polymer. For both nucleation pathways there are competing thermodynamic forces. Initially aggregation is unfavorable because entropic forces tend to keep molecules apart. As the nuclei become larger, however, there is an increased number of bonds per monomer, 1/2 for a dimer, 3/3 for a trimer, 6/4 for a tetramer, up to 4.1 in the infinite polymer. As the aggregates grow this increase in the stability from more bonds per monomer finally overcomes the unfavorable entropic forces. The aggregate for which addition of monomer finally becomes favorable is called the critical nucleus.

in the delay time, while reproducible delay times are accompanied by formation of a gel with a large number of domains that are too small to exhibit birefringence (Fig 3).

All of these kinetic observations can be quantitatively explained by the double nucleation mechanism shown schematically in Fig 4.[131,138] According to this mechanism gelation is initiated by the nucleation of a single polymer. This process is called homogenous nucleation because it takes place in the bulk solution, and no surfaces are involved. By nucleation we mean that small aggregates of hemoglobin S molecules are unstable relative to monomers, and addition of a monomer to the aggregate produces a less stable aggregate. Once a certain size, called the critical nucleus, is reached, however, addition of a monomer produces a more stable aggregate, and monomers add endlessly to form a very large polymer. Nucleation results from competition between two thermodynamic forces, an increased freedom of motion (ie, increased entropy), which tends to keep molecules apart, and the favorable free energy of intermolecular, noncovalent bond formation that makes the molecules associate. Initially the entropy dominates, but once a sufficient number of intermolecular bonds *per monomer* are formed, aggregation becomes favorable (Fig 4). Although homogenous nucleation by itself can explain the very high concentration dependence of the rate, it cannot explain the existence of a pronounced delay period.[139] The delay period is produced by the second pathway for nucleation. In this pathway nucleation takes place on the surface of preexisting polymers and is therefore called heterogeneous nucleation.[131,138] As more hemoglobin is polymerized, the surface area on which new polymers can be nucleated continuously increases, resulting in an autocatalytic polymerization for the initial stages of the gelation process.

Mathematical analysis of the double nucleation mechanism shows that the delay period is a manifestation of the autocatalytic formation of polymer via heterogeneous nucleation.[138] For slowly gelling samples the model predicts that throughout the early portion of the measurable progress curve incorporation of monomers into polymers is exponential (region II in Fig 5 b). A consequence of this exponential polymerization is that there is an apparent delay, the length of which depends on the sensitivity of the measurement. No new or different process is occurring during the delay period. The rates of nucleation and growth of polymers are simply less than they are when polymers first become detectable. This has been verified in high-sensitivity–light-scattering measurements (G.W. Christoph and R.W. Briehl, unpublished results).[114]

According to the mechanism, both homogeneous and heterogeneous nucleation rates are proportional to the initial monomer concentration raised to a power that is the size of the critical nucleus. The enormous concentration dependence of the delay time can thus be readily explained as resulting from a large nucleus. As the hemoglobin S concentration increases, aggregation becomes more probable. As a result both homogeneous and heterogeneous nuclei become smaller, and the concentration dependence of the delay time decreases. Because the nuclei are in equilibrium with mon-

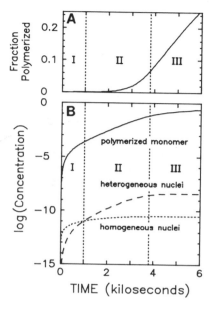

Fig 5.   Theoretical kinetic progress curves for polymer formation calculated from the equations of the double nucleation mechanism (from Ferrone et al[138]). (A) Fraction of polymerized hemoglobin as a function of time. (B) The solid curve is the logarithm of the concentration of polymerized hemoglobin; the dotted curve is the logarithm of the concentration of polymers formed via the homogeneous nucleation pathway; and the dashed curve is the logarithm of the concentration of polymers formed via the heterogeneous nucleation pathway. The concentration of polymers is much less than the concentration of polymerized hemoglobin because the polymer contains a large number of hemoglobin molecules. The average number of hemoglobin molecules per polymer at any time can be obtained by dividing the concentration of polymerized hemoglobin by the sum of the concentrations of the homogeneous and heterogeneous nuclei. The curve for the concentration of polymerized hemoglobin shows that there is a relatively long period during which the formation is exponential (region II, the straight line region) and that the length of the period before polymer is first detected (ie, the delay time) depends on the sensitivity of the measurement. In region I the dominant processes are homogeneous nucleation and growth of polymers, in region II heterogeneous nucleation and growth of polymers, and in region III polymer growth. There is no additional nucleation of polymers in region III.

omers, the large sizes for the critical nuclei also explain the high dependence of the delay time on the solubility.[36] In concentrated phosphate buffer, gels form with a solubility of only 2 g/dL,[102] suggesting stronger intermolecular bonds and smaller nuclei, and the concentration dependence is also much lower.[100-109] Finally, the mechanism explains the irreproducibility of the delay time when single polymer domains are formed in small volumes as resulting from stochastic fluctuations in the time at which single homogeneous nuclei appear.[131,133,138] Under these conditions a single polymer initiates the formation of a domain. The remainder of the polymers, which fill the entire observed volume, are formed by heterogeneous nucleation. This "amplification" of the homogeneous nucleation event allows the stochastic fluctuations to be observed.

We now turn to the important question of whether gelation

inside sickle cells proceeds at the same rates and by the same mechanism as in purified solutions. Several results indicate that the answer is *yes*. First, studies on the addition of red cell membrane components to deoxyhemoglobin S solutions show little or no effect on the delay time.[110,111] Second, the laser photolysis technique has permitted the measurement of the kinetics of gelation in single red cells, yielding results that are in qualitative agreement with those predicted from the solution studies.[135] The shapes of the kinetic progress curves are very similar to those observed in solution,[135-137] and for slowly polymerizing cells there are the expected stochastic fluctuations in the delay times.[135] Figure 6 shows that the distribution of observed delay times, which range from a few milliseconds to over 100 seconds, is almost exactly what is predicted from the solution studies and the known concentration heterogeneity. A more detailed comparison of solution and cell data can be made by calculating the intracellular hemoglobin S concentration distribution from the delay time distribution using the solution delay times. The calculated distribution in Figure 6 is qualitatively the same as the measured distributions (Figure 2g), showing that there are no major differences in the rates of gelation in solution and in cells.

SICKLING AS AN INDICATOR OF INTRACELLULAR GELATION

While it has long been accepted that the deformation of SS red cells upon complete deoxygenation is caused by intracellular polymerization,[66] the detailed relation between gelation and cell deformation has remained somewhat ambiguous. In this section we address two questions. The first is whether or not there is a well-defined relationship between cell shape and intracellular polymerization. The second is whether the wide variety of observed cell shapes can be rationalized in terms of what we now know about the kinetics and thermodynamics of gelation. The answers to these questions are important, since a direct link between cell morphology and polymerization would permit a variety of experiments to be performed by using morphological criteria in place of more complex and difficult physical measure-

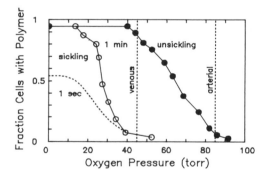

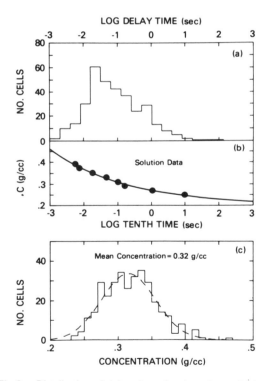

Fig 6.   Distribution of delay times for deoxyhemoglobin gelation in individual cells (from Coletta et al[135]). (A) Distribution of delay times from three homozygous SS patients. (B) Relation between concentration and logarithm of tenth time from solution data (Fig 3). (C) Intracellular hemoglobin S concentration distribution determined by calculating the concentration corresponding to each delay time in (A) using the result in (B). The intracellular hemoglobin S concentration is somewhat underestimated because the delay times in potassium phosphate buffer are shorter than in physiologic buffer (P.L. San Biagio, J. Hofrichter, and W.A. Eaton, unpublished results).

Fig 7.   Fraction of cells containing polymer as a function of calculated oxygen pressure determined by a double laser beam photolysis technique (from Mozzarelli et al[136,137]). In this technique one laser beam illuminating a single red cell is used to prepare hemoglobin S at a steady-state partial saturation with carbon monoxide by continuous photodissociation, while a second more intense laser beam can be switched on at any time to completely photodissociate the remaining carbon monoxide and to measure the kinetics of gelation from the time course of the scattered laser light. If no polymerized hemoglobin is present in the cell at partial saturation, the kinetics of gelation at zero saturation (produced by the second laser beam) are characterized by a delay period, while the presence of polymerized hemoglobin is indicated by the loss of the delay period. Experiments on hemoglobin S solutions show that this technique accurately simulates the gelling behavior of hemoglobin S at partial saturation with oxygen and that the presence of as little as 0.05% polymerized hemoglobin results in a marked shortening of the delay period. The oxygen pressures were calculated from the measured saturations with carbon monoxide using the least squares fit of the two-state allosteric saturation function to the binding curve of normal blood. The filled circles are the equilibrium data obtained in reoxygenation (ie, resaturation with carbon monoxide) experiments, the open circles are data obtained from experiments where deoxygenation (desaturation) is carried out over a period of one minute, and the dashed curve is a theoretical estimate for deoxygenation carried out in one second.[141] The vertical dashed lines indicate the average oxygen pressure found in the arteries and veins of patients with homozygous SS disease.[78]

ments such as light scattering[135-137] or micropipette measurements.[140]

The relationship between cell deformation and intracellular polymerization has recently been studied with a variation on the laser photolysis technique. Gelation in partially saturated single cells was investigated by using the kinetics of gelation after complete photodissociation as a probe for the presence of polymer (Fig 7).[136,137] The delay time provides a very sensitive probe for polymer because even vanishingly small amounts of polymerized hemoglobin (< 0.05%) drastically reduce or eliminate the delay period.[136,137] These experiments demonstrate clearly that sickling accurately reflects the onset of gelation and that unsickling indicates the complete disappearance of polymer.[136,137] As a result, curves that describe the fraction of cells containing polymer as a function of saturation or oxygen pressure may be designated sickling and unsickling curves. Figure 7 shows the fraction of sickled cells, ie, the cells that contain polymerized hemoglobin S, as a function of the oxygen pressure, calculated from the measured saturation with carbon monoxide. There is a very large hysteresis between the sickling and unsickling curves. The oxygen pressure at which polymers are first observed in deoxygenation experiments in an initially polymer-free cell is always much lower than the pressure at which polymers disappear in reoxygenation experiments.The hysteresis occurs because there is a delay period before polymer can be detected upon deoxygenation, but in reoxygenation experiments depolymerization occurs without any delay period. Thus the unsickling curve is very close to an equilibrium curve, while the sickling curve depends on the rate of deoxygenation. The sickling curve in Fig 7 was measured by lowering the saturation over a period of one minute. In the microcirculation, deoxygenation occurs in about one second; sickling curves have not yet been measured on this time scale. However, the theoretical one-second sickling curve shown in Fig 7 is seen to be extremely left shifted,[141] in qualitative accord with the results of kinetic studies that show that about 50% of cells sickle after about one second at zero oxygen pressure.[142,143] Experiments in which cellular deformation occurs much more rapidly have also been performed by deoxygenating cells in a mixer.[144] These data are consistent with the results on intracellular gelation using the laser photolysis technique,[26,135] suggesting that cell deformation is a reliable indicator of intracellular gelation on the second and subsecond time scale as well.

Essentially identical conclusions have been reached from experiments in which rheologic techniques are used to examine individual sickle cells. The most direct data have come from measurements as a function of oxygen pressure using micropipette techniques.[140] Both the static and dynamic rigidities of the cells can be measured. The static rigidity is characterized by the change in the length of the "tongue" aspirated into the pipette with a change in negative pressure, while the dynamic rigidity is characterized by the half-time required to achieve the final tongue length after initiating the pressure change.[145] For oxygenated cells there are only small increases in these quantities, with the largest increases for the irreversibly sickled cells and the densest cells.[146,147] As the oxygen pressure is decreased, cells are observed to undergo a variety of morphological changes, and cells having a given set of morphological characteristics can be examined.[140] Cells that are "spiculated" or have a "granular" surface show markedly altered rheology. In contrast, discocytes that maintain a "smooth surface" show the same static and dynamic rigidities at all oxygen pressures as normal cells, in agreement with the conclusion from the kinetic studies. In cells showing morphologic evidence of gelation, both the static rigidity and the half-time for tongue growth increase with decreasing oxygen pressure. At the lowest oxygen pressures the static rigidity increases by up to a factor of 100, and the half-time for tongue growth increases by a factor of 150 to 1000 relative to normal cells and oxygenated sickle discocytes.[140,148] Since both parameters are much greater for all cells containing polymerized hemoglobin than for polymer-free cells, it appears that the presence or absence of intracellular polymer is much more important in determining cellular rigidity than the extent of intracellular polymerization.[149]

A closely related problem is to understand the enormous variety of cell shapes that are observed in a population of sickled cells. It has been known for over 45 years that slow deoxygenation results in elongated, birefringent cells, while rapid deoxygenation produces a much less distorted cell, originally called a granular form.[150] An unexpected bonus provided by the double nucleation mechanism is that it suggests an explanation for these observations.[138] When cells are rapidly deoxygenated the solubility is suddenly decreased to a low value. The resulting high supersaturation (the ratio of the total concentration to the solubility) causes a high rate of homogeneous nucleation, and the resulting gel contains a very large number of small polymer domains or randomly oriented polymers that could give the cell a granular appearance (Fig 3). In contrast, when deoxygenation is slow, the rate of homogeneous nucleation is reduced to the point that only one homogeneous nucleation event takes place in the cell, and a single polymer domain forms. If it were not for the limited amount of hemoglobin in the cell, this domain would grow to a much larger size than the cell. The cell membrane presumably restricts domain growth to one general direction, resulting in an elongated cell with approximately parallel polymers, the classic "sickle" form (Fig 3). At intermediate rates of deoxygenation, cells may contain a countable number of domains, where, for example, each domain could produce one of the projections of a so-called "holly-leaf" shape (Fig 3).

The appearance of a wide range of morphological forms at a fixed rate of deoxygenation might also be explained as resulting from different rates of polymerization. Remember that the solubility of hemoglobin S decreases very rapidly at high fractional saturations and much more slowly at low fractional saturations (Fig 2 b). Consequently at a fixed deoxygenation rate dense cells become much more supersaturated and consequently polymerize with much shorter delay times than the light cells. As a result dense cells would be expected to contain many more polymer domains and to have a lumpy, granular appearance, as opposed to a classic sickle shape. Cell morphology is therefore expected to be highly correlated with intracellular concentration and cell density. The results of morphological studies on cells polymerized at different rates[151] and on density fractionated cells are consistent with this explanation.[85] The observation of smooth discocytes containing polymer in time-resolved electron microscope studies may represent the initial phase in the formation of a granular form.[152]

GELATION IN VIVO AND VASO-OCCLUSION

We now turn to the question of obstruction of blood flow in the microvasculature resulting from intracellular gelation. Vaso-occlusion is believed to be the cause of pain crises and of the widespread organ damage that contributes substantially to the morbidity and mortality of the disease. Because of the enormous complexity of this problem, the discussion must, of necessity, become much more qualitative and speculative than that which has been presented up to now. We shall see that there are suprisingly little hard data on some of the most basic questions about vaso-occlusion. Nevertheless we believe that a critical examination of this problem is necessary at this point to clarify the important issues and to point to areas where research is most needed. We first discuss gelation and vaso-occulsion, and in the next section we consider the response of the circulatory system to this abnormality.

To gain some perspective on the problem it is instructive to consider the various types of events that have been postulated to occur as a red cell travels through the circulation of an SS patient. In describing these events we shall equate sickling with intracellular gelation. Figure 8 shows a schematic summary. Cells containing no polymerized hemoglobin in the arterial circulation may pass through the microcirculation and return to the lungs without sickling, they may sickle in the veins, or they may sickle in the capillaries. The probability for each of these events will be determined by the delay time for intracellular gelation relative to the appropriate transit time.[7] If it is thermodynamically impossible for gelation to take place (ie, the intracellular concentration is always lower than the solubility so that at equilibrium no polymer can form) or if the delay time at venous oxygen pressures is longer than about 15 seconds, then sickling will not occur. If the delay time is between about one and 15 seconds, then the cell will sickle in the veins, and, if it is less than about one second, the cell will sickle within the capillaries. For cells that sickle within the capillaries a number of possibilities exist, ranging from no effect on its transit time to transient occlusion of the capillary or a more permanent blockage that ultimately results in destruction of the cell. For some cells the intracellular hemoglobin S concentration may be so high that the solubility is exceeded even at arterial oxygen pressures. These cells will still contain polymerized hemoglobin after oxygenation in the lungs. Upon deoxygenation further gelation will occur rapidly and without a delay time because nucleation of polymers is already complete.[7,22,77,136,137] Such cells could become stuck in the arterioles or capillaries or could experience a normal transit time through the microcirculation in spite of the decreased deformability.

Figure 8 points out one fundamental problem in describing the pathophysiology of sickle cell disease is to determine the

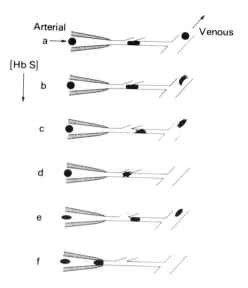

Fig 8.   Possible events in the microcirculation of a patient with homozygous SS disease. A schematic of an arteriole, capillary, and venule is shown. In (a) a cell containing no polymer enters the capillary, deforms to squeeze through, and reaches the venule without polymerization occurring. In (b) the delay time is longer than the capillary transit time, but the cell sickles in the venule. In (c) the delay time is shorter than the capillary transit time, and the cell sickles within the capillary but escapes to the venule, while in (d) intracapillary sickling results in transient or permanent blockage. In (e) and (f), the cell, depicted as an irreversibly sickled cell, already contains polymerized hemoglobin in the arteriole and may pass through the capillary (e) or produce a transient or permanent occlusion (f).

relative probabilities for each of these events. These probabilities will depend on a number of factors, including the total intracellular hemoglobin concentration, the composition of the intracellular hemoglobin, the rate and extent of deoxygenation, and the various transit times involved. For unsickled cells entering the microcirculation, a long capillary transit time will increase the probability of the potentially vaso-occlusive events depicted in Fig 8 in two ways. First, it will permit increased oxygen extraction, which will shorten the delay time. Second, it will increase the probability that a cell with a given delay time will sickle within the capillary. For cells that either enter the microcirculation already sickled or become sickled in the microcirculation, there is a finite probability for occlusion of the small vessels. The duration of an occlusion may be sufficiently long to compromise the oxygen supply to the surrounding tissues and hence may alter the probabilities for sickling and consequent vaso-occlusion in nearby microvessels. This is a somewhat refined version of the "vicious cycle."[153] It is important to recognize that vaso-occlusion is a dynamic process in which the fraction of capillaries that are occluded depends on both the rates of occlusion and the rate of capillary reopening. Thus factors that influence the transit times and the duration of occlusions also play a critical role in the pathophysiology.[7]

With this brief heuristic description as a framework for subsequent discussion, we can now proceed to examine experimental results that help to establish the probabilities for the various events depicted in Fig 8. The most straightfor-ward problem is the determination of the fraction of cells that are sickled in the arteries and the fraction that sickle as a result of deoxygenation in the microcirculation. Morpho-logical examination of cells sampled from arterial blood suggests that the average fraction of sickled cells is about 10%.[78,154-156] This number is, unfortunately, only a rather crude estimate because it is possible that deformed cells such as granular discocytes have not been counted as sickled in some studies; moreover, some irreversibly sickled cells, which may frequently be a major contributor to this count,[156] may contain no polymer.[140] The values for reversibly sickled cells in different patients range from 1% to 16%, while total sickled cell counts range from 9% to 30%.[156] This variation presumably results from differences in the distribution of intracellular hemoglobin composition and concentration, as well as from differences in arterial oxygen saturation.

It would appear, then, that an average of about 90% of cells entering the peripheral circulation contain no polymer and hence would undergo gelation with a delay period if sufficiently deoxygenated. The morphological data suggest that about 20% of cells are sickled in the mixed venous return,[78,154-156] indicating that an additional 10% of cells have sickled as a result of passing through the microcirculation.* Any analysis based on studies of mixed arterial and venous blood is clearly somewhat oversimplified because oxygen extraction in the microcirculation of some tissues, such as in the coronary and hepatic circulations,[155] is considerably greater than average. As a result the cells in these organs will have much shorter delay times leading to a higher number of sickled cells in the veins.

These findings are consistent with existing information on in vitro delay times. Although the ideal in vitro experiment in which gelation and degelation are continuously monitored in individual red cells at physiologic rates of deoxygenation and oxygenation is not yet possible, the laser photolysis experi-ment[136,137] affords an informative preview of the results expected from such experiments. The *unsickling* curves obtained in these experiments (Fig 7) show that *at equilib-rium* only about 5% of the cells contain polymerized hemo-globin S at an oxygen pressure of 85 torr, which is the average arterial value found in homozygous SS patients,[78] while over 90% of cells remain sickled at the average mixed venous pressure of 45 torr. In contrast, the *sickling* curves show that only about 5% of cells (overlapping significantly with the cells that were found to be sickled at 85 torr at equilibrium) are sickled after deoxygenation to venous oxy-gen pressures on physiologic time scales. These results indi-

cate that the delay time is preventing more than 80% of cells from sickling in vivo in this homozygous SS patient. That is, for over 80% of cells gelation would occur if equilibrium were achieved, but the delay times are so long that these cells return to the lungs and are reoxygenated before any signifi-cant amount of polymer has formed. The fact that about 10% of cells already contain polymer in the arterial circulation does not substantially affect the fraction of sickled cells in the microcirculation, since if polymer were not present the delay times for the large majority of these cells would be less than the capillary transit time.† The enormous difference between the unsickling curve and the sickling curve in these experiments (Fig 7) graphically demonstrates the signifi-cance of the delay time for gelation in vivo and simulta-neously shows that equilibrium data or data obtained in slow deoxygenation experiments are not at all representative of the in vivo situation in which the relevant time scale is seconds. Similar large differences are expected for oxygen binding and polymer fraction curves (Fig 2 e, 2 f and discussion below).

This analysis points to the critical need for obtaining much more data relevant to sickling in vivo. An accurate descrip-tion would require direct measurements of the distribution of delay times at physiologic rates and extents of deoxygena-tion. It would also be desirable to have more precise data on the extent of sickling in arterial and venous samples using experimental methods that take into account the kinetics of sickling as well as the recently acquired information on the relation between cellular deformation and intracellular gela-tion.[136,137,140] Most of the data on morphological sickling were obtained before the gelation kinetics were described, and since then very little attention has been given to designing accurate morphological experiments on venous and arterial samples. Such experiments would require rapid fixation of cells and careful examination by scanning electron micros-copy or high resolution optical microscopy. It would, of course, be preferable to develop rapid sampling techniques that could assess the extent of intracellular gelation, or at least the presence or absence of polymer, in individual cells from arterial and venous blood.

We next consider the question of occlusion of the microcir-culation by sickled cells, which is clearly a central problem in understanding the pathogenesis. The results and calculations described above as well as data on red cell survival suggest

*The estimate of 10% additional sickling in the microcirculation is only a very approximate number. Differences in this number are observed for different venous returns of the same patient and for the same venous return in different patients, but the tissue-to-tissue variation is generally smaller than the differences observed between patients.[78,155,156] Since cells are deoxygenated in the microcirculation within one to two seconds, significant variations may result from additional sickling during the time required to sample the cells from the veins and to fix them with glutaraldehyde and possibly some additional unsickling for cells sampled from the arteries.[157,158]

†Shearing forces are known to decrease the delay time by breaking polymers, producing new ends, and thereby increasing the rate of nucleation.[121,139,159,160] In this discussion we have assumed that there is no significant effect of shear on the in vivo delay time. While no direct experimental information on this point exists, two consider-ations suggest that the effect of shear on intracellular polymerization in vivo will be small. First, cells flowing in small tubes concentrate in the low-shear region near the center of the tube, while the high shear regions near the walls are preferentially occupied by plasma.[161,162] The shear field to which cells are exposed is thus very much smaller than the average field in the microvessels. Second, the high internal viscosity of even unpolymerized cells makes coupling of the external shear field to the inside of the cell inefficient, as evidenced by the absence of tank-treading behavior in low viscosity media.[163]

that the densest cells are primarily responsible for occlusion of the microcirculation. These cells are predicted to have a much greater probability for the events in Fig 8 that could lead to vaso-occlusion. Because of the high intracellular hemoglobin S concentration, they have shorter delay times, or no delay times and therefore are more likely to sickle in the capillaries when deoxygenated.[6,7] In addition, if gelation occurs at any given oxygen pressure, the higher intracellular concentrations of polymerized hemoglobin (Fig 2 d) produces stiffer cells (ie, cells with lower static and dynamic deformabilities[140]). The half-life of dense cells (~ two days) is significantly shorter than that of randomly labeled cells from the same patient population (~ five days)[164] or the average half-life of cohort-labeled cells from a wider patient population (~ 17 days).[164-166] Experiments in which the density distribution of a population of labeled reticulocytes was followed over the cell life span also show that the dense cells are the last to appear in the labeled population.[164] These results show clearly that as cells age their density increases and that once they become dense they are removed from the circulation rather quickly.[164] The mechanism by which cells concentrate has been the subject of much recent work.[167] The major contribution appears to result from the loss of cell water associated with potassium loss, but there may also be some contribution from the loss of membrane surface area caused by the sickling-unsickling cycle. If it is assumed that all cells must become dense cells before they are removed from the circulation, then the fraction of dense cells is predicted to be equal to the ratio of the half-life of the dense cells to the mean half-life for new cells. The measured half-life of the dense cell fraction is roughly consistent with this prediction, since the densest 10% to 15% of the cell population has a half-life that is 0.1 to 0.2 times that of labeled reticulocytes.[164] Cells that have an abnormally low probability of sickling, such as low density F cells, are expected to have extremely long delay times even at venous saturations. Since these cells can only be sickled upon stasis or passage through tissues where deoxygenation is extreme, they will presumably concentrate much more slowly than cells having a higher probability of sickling. This protection of F cells results in a longer life span[1] and therefore an increased concentration of F cells in the fractions of intermediate density, which represent the oldest cells in the population, and a reduced concentration in the densest cell fraction.[87,164]

Of the vaso-occlusive events depicted in Fig 8, it is only possible to make even a crude quantitative estimate for the probability of occlusion followed by *destruction* of the cell. Based on a mean cell lifetime of 17 days and a circulation time of 15 seconds, an "average" red cell makes about 100,000 trips through the microcirculation before being removed. If we use the fraction of sickled cells as an estimate of the fraction of dense cells (~20%), the above argument suggests that if dense cells were removed from the circulation only by vaso-occlusive events, they would be trapped and destroyed once in about 20,000 trips through the microcirculation. However, hemolysis data indicate that about 60% to 70% of sickle cells are destroyed in the reticuloendothelial system,[166] so the probability for destructive vaso-occlusion is

reduced to about one in 60,000 trips. While this probability appears, at first glance, to be extremely low, it is large enough to result in the steady-state blockage of a significant fraction of the total number of capillaries, since there are approximately 10^{13} circulating red cells and only about 10^{10} capillaries.[168] The exact fraction blocked is highly dependent on the duration of the destructive blockade, which, by this estimate, occurs in the average capillary about once every hour. If, for example, each blocked vessel remains occluded for one hour, the fraction blocked is 50%, but this value is reduced to 2% if the vessel is blocked for only one minute. This result clearly suggests that factors that influence the duration of a capillary blockade could play a critical role in determining the extent to which tissue oxygen supply is compromised.

It has been very difficult to obtain quantitative information on the frequency of the other events depicted in Fig 8. Recently a series of studies using an excised rat mesocecum preparation has begun to provide some interesting results. This preparation permits control over the tissue oxygen pressure and the perfusing pressure, allowing vascular resistance (defined as the ratio of the arteriovenous pressure difference divided by the venous outflow) changes to be measured quantitatively in both denervated[85,86] and innervated vascular beds.[169] Alternatively, trapping of cells can be measured by first perfusing the innervated bed and subsequently washing out the trapped cells by increasing the perfusion pressure or denervating the prepared bed.[169] These experiments make two important points. The first is that at venous PO_2 the fraction of capillaries blocked at steady state in this preparation can be as large as 80%. The second is that the ability of cells to block the microcirculation is correlated with their density, a finding that is consistent with the description of Fig 8.

These investigations provide the best opportunity to simulate the events occurring in the microcirculation of SS patients. In addition to measuring vascular resistance and cell trapping, cinematographic observations of these preparations permit determination of the sites at which blockage occurs. The limited information obtained so far has not established the relative importance of precapillary and intracapillary sites as the principal sites of occlusion.[170-172] It would be important to extend studies of the microvasculature to use preparations that permit deoxygenation in the tissue[170] to more closely simulate the in vivo situation. It will also be important to use these preparations to determine the factors that affect both the frequency and duration of occlusions, since they are equally important in determining the steady-state fraction of blocked capillaries.

GELATION AND OXYGEN DELIVERY

An important aspect of the pathophysiology of sickle cell disease is to understand how the circulatory system maintains adequate oxygen delivery in the face of anemia and vascular obstruction. In other severe anemias, for example, those arising from blood loss or iron deficiency, there are two primary compensation mechanisms. One is an increased blood flow through the tissues resulting from an expansion of the muscled arterioles.[173,174] The increased blood flow from this decreased peripheral resistance increases filling of the right atrium, thereby increasing the cardiac output.[175] The second is an increase in intracellular 2,3-DPG concentration, which lowers the oxygen affinity and facilitates oxygen unloading in the tissues.[25] As a result there is an abnormally low venous oxygen saturation at or near a normal oxygen pressure.[173,174,176-178] The increased blood flow through the lungs may also result in a decreased oxygen pressure and saturation in arterial blood[173,176,179] from the decreased efficiency of gas exchange,[180] making oxygen delivery less efficient.

In sickle cell disease the abnormalities are somewhat different. The cardiac output is higher than in anemias of comparable severity,[78,173,176,181] the oxygen pressure and saturation of arterial blood are lower,[78,154-156,182,183] and the oxygen pressure and saturation of venous blood are higher.[78,154,156] This decrease in arteriovenous saturation difference, particularly when compared to other anemic states, means that there is significant impairment of oxygen unloading to the tissues.[78] These differences between sickle cell disease and other anemias presumably result from the intracellular gelation and vascular obstruction that are unique to sickle cell disease.

The very low oxygen affinity of polymerized hemoglobin S explains the lower arterial oxygen saturation in sickle cell disease, compared with other anemias, and even the lower arterial oxygen pressure. Infarctive damage to lung tissue of SS patients could also decrease the efficiency of oxygen loading in the lungs, but this does not appear to be a major factor because a comparable degree of arterial unsaturation is observed in children in whom there is no other evidence of impaired lung function.[184,185] We have estimated earlier that about 20% of cells entering the lungs contain substantial amounts of polymerized hemoglobin, which would be expected to decrease both the rate and extent of oxygen binding to the sickled cells in the alveoli. The few in vitro experiments support this contention.[152,157,158,186] Slow depolymerization of the sickled cells after they leave the lungs could also contribute to the lower arterial oxygen tension by scavenging oxygen from the plasma and from the cells that contain no polymer.

In considering oxygen unloading in the tissues, the absence of polymerized hemoglobin in most cells because of the long delay times is an important consideration. The traditional approach has been to ignore this fact and to utilize oxygen dissociation curves measured in vitro over periods of minutes or longer where intracellular gelation is much more extensive than in the in vivo situation. Use of the in vitro curve has led to the conclusion that the large right shift in the equilibrium or quasi-equilibrium dissociation curves substantially compensates for the anemia.[173,187,188] This conclusion is misleading, since oxygen binding to approximately 80% of the cells should be similar to that found for other states of comparable anemia. The calculation in Fig 2 e shows that the in vivo oxygen unloading curve is predicted to be significantly less right shifted than the in vitro equilibrium curve. This smaller right shift is consistent with the observation that the fractional saturation of venous blood from sickle cell patients has near normal values, while the oxygen pressure is higher than in normal blood by about 5 torr.[78]

Perhaps the most puzzling circulatory abnormality is an increase in cardiac output that is larger than is found in anemias of comparable severity. The absence of a significant increase in arterial blood pressure[189] requires that the vascular resistance be *decreased* in direct proportion to the increase in cardiac output.[78] While decreased vascular resistance may appear paradoxical in the presence of capillary blockage,‡ this finding can be rationalized by the fact that the bulk of the peripheral resistance arises from the muscled arterioles and not the capillaries.[192] In severely anemic states the arterioles open to increase blood flow through the tissues, thereby compensating for the low hematocrit.[173,174] A similar response also appears to be the primary mechanism of compensation for the reduced hematocrit in SS disease, but the peripheral resistance is decreased to a significantly greater extent than in other anemias of comparable severity.[78,173] The opening of the muscled arterioles must therefore increase the fraction of capillaries in the tissue bed that are perfused to above normal, even in the presence of blockage, if there is sufficient capillary reserve.

In addition to increasing the number of perfused capillaries, there is evidence that opening of the muscled arterioles increases the pressure drop across the capillaries, thereby increasing the rate at which red cells traverse the capillaries.[192-194] A recent study in which laser Doppler velocimetry was used to measure capillary flow in the forearm skin of SS patients showed that the average rate of *red cell* flow was close to normal.[195] This result implies that the ~40% decrease in hematocrit is almost exactly compensated by a combination of an increase in the number of perfused capillaries and an increase in capillary flow rate. By decreasing the time available for the equilibration of the red cell with the oxygen tension of the capillary wall, an increased capillary flow rate would be expected to decrease oxygen unloading. In spite of this effect a decreased capillary transit time could benefit the patient by decreasing the rate of capillary blockage. The decreased transit time would not only increase the delay time by increasing the final fractional saturation of hemoglobin but it would also decrease the time during which a cell is at risk from sickling within the microcirculation. Such a mechanism provides an attractive explanation for the decreased arteriovenous oxygen saturation difference in SS patients. A frequently invoked explanation for the low arteriovenous saturation difference is shunting through large vessels.[154]

---

‡There is no direct measurement of the fraction of occluded capillaries in any tissue in sickle cell disease. However, a tentative estimate for muscle can be extracted from data on the exercise tolerance of sickle cell patients.[190] In these studies patients were subjected to increasing work loads, and the lactic acid level in the blood was monitored. The work load at which lactate began to increase is defined as the anaerobic threshold. An extension of the Krough model for oxygen delivery to tissues predicts that the work output at this point is nearly directly proportional to the density of perfused capillaries.[191] If it is assumed that the muscle is maximally perfused and that capillary densities in the muscle of SS patients are normal, then the fraction of blocked capillaries can be estimated from the anaerobic threshold to be about 0.4. This fraction decreases to about 0.1 to 0.2 when the fraction of SS cells is decreased to about 50% by exchange transfusion.[190]

While this mechanism might account for the unexpectedly high venous saturation in specific tissues, it cannot explain the fact that similar results are found for vessels such as the femoral vein.[78] The femoral vein primarily drains muscle beds in which there is anatomic evidence that significant shunting is impossible.[196]

Finally, we should point out that there has been only one attempt to quantitatively evaluate the response of the complete circulatory system to the altered properties of SS blood.[197] This study used an established model for microcirculatory control[198] to calculate the changes in peripheral resistance blood flow, and capillary oxygenation.[197] This model incorporates approximate but realistic descriptions of oxygen supply and consumption in the tissues as well as local feedback control of both the arteriolar resistance and capillary density to regulate the tissue oxygen pressure. When anemia is simulated by reducing the hematocrit, the model predicts a compensatory decrease in peripheral resistance and increased blood flow. If, however, the quasi-equilibrium increase in viscosity and reduced equilibrium affinity of SS blood are also introduced, the model predicts a capillary resistance that is about 1.4 times normal and blood flow that is about 80% of normal. The calculated effects are in striking contrast to the observed *decrease* in peripheral resistance and *increase* in blood flow.[78,181] The discrepancy presumably results in part from the incorrect assumption of equilibrium oxygen unloading and viscosity changes made in carrying out these calculations. If the kinetics of intracellular gelation were to be incorporated into this model (Fig 2 e), the predicted effects would be closer to the observed.

VARIATIONS IN CLINICAL SEVERITY

It is now widely recognized that there are large differences in clinical severity among patients with homozygous SS disease, some patients having only the mild symptoms associated with a chronic hemolytic anemia, others suffering from repeated painful episodes and severe organ damage.[24,25,199,200] The reasons for this broad spectrum of clinical manifestations are not at all clear, and it is one of the major areas of current research. To discuss the role of gelation in producing differences in clinical severity among homozygous SS patients, it is useful to briefly summarize the most important results of the preceding discussion.

The picture that emerges is a dynamic one in which a balance between the rate of obstruction and reopening of capillaries results in a steady state in which a certain fraction of capillaries is blocked in each tissue. This balance may be very delicate, with small changes in either the rate of obstruction or reopening capable of significantly altering the fraction of occluded capillaries. Any increase in this fraction, particularly in tissues with inadequate capillary reserve, could result in irreversible hypoxic damage and may be the cause of pain crises. While almost nothing is known about the opening of occluded capillaries, we are beginning to understand the mechanisms that control the rate of capillary obstruction. This rate must depend, at least in part, on the fraction of sickled cells in the microcirculation, which is determined by the times required for intracellular gelation relative to the transit times.[6,7] Factors that favor gelation can increase the steady state number of obstructed capillaries by

decreasing the delay time, thereby increasing the fraction of cells that sickle within the microcirculation and the rate of obstruction. An increased extent of polymerization in a sickled cell could also increase the probability of an occlusion because of the decrease in deformability.[140] Because the delay time is so much more sensitive to changes in physiologic variables than the extent of polymerization (compare, eg, Figs 2 d and 3), it is most probably the dominant factor in determining changes in the rate of capillary obstruction. Simultaneous small changes in a number of physiologic variables could result in a sufficient change in the distribution of delay times to produce the fluctuation in the fraction of blocked capillaries that precipitates a pain crisis. In this way the sensitivity of the delay time could account for the episodic nature of crises.

Another mechanism for increasing the rate of obstruction is to increase the transit time in the microcirculation, which increases the probability of sickling. In this way factors that slow down cells can also affect the rate of obstruction. The only such factor that has been identified so far is the adherence of cells to the vascular endothelium.[201-204] In addition to capillary blockage, other events influence these probabilities by altering the characteristics of the cell population. For example, cells that normally would return to the lungs may sickle in the venous return, particularly in tissues in which the residence times in the veins are long, resulting in an increase in intracellular concentration and therefore a decreased delay time in subsequent trips through the microcirculation.

This picture immediately raises the question of how much of the variation in clinical severity in homozygous SS disease can be explained by variations in intracellular gelation and how much must be attributed to variations in circulatory dynamics. Under the category of intracellular gelation are included the effects of intracellular hemoglobin concentration and composition, arising either from genetic variations or from cell aging. To begin, let us consider the relation between gelation times and clinical severity among the various sickling disorders, where there are easily measurable differences in both gelation and standard hematologic parameters. Figure 9 shows the effect of hemoglobins A, C, and F on the delay time and a comparison of the distribution of delay times at zero saturation for the three most common syndromes: homozygous SS disease, SC disease, and sickle trait. SC disease is generally a much milder sickling disorder than is SS disease, while sickle trait is totally benign.[24,25] The delay times for SC cells are considerably longer than those for SS cells, indicating that many fewer cells sickle in vivo. For sickle trait cells the delay times even at zero saturation are all longer than about one second, indicating that even under totally anoxic conditions cells would escape the microcirculation before polymerization has begun. With the possible exceptions of the hypertonic renal medulla, it would appear that sickle trait cells never sickle in vivo, explaining the lack of any clinical manifestations.

The reasons for the increased sickling of SC cells compared to sickle trait cells are quite interesting. Little or no difference is observed in the gelling properties of hemoglobin S + C mixtures and S + A mixtures. A careful comparison has shown that there are no significant differences in either

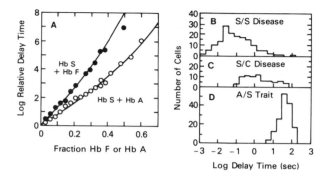

Fig 9.   Effect of non-S hemoglobins on gelation delay times in solutions and cells. (A) Logarithm of the ratio of the delay time of the mixture to the delay time of pure deoxyhemoglobin S at the same total hemoglobin concentration.[57] The effect of hemoglobin C on the delay time is identical to that of hemoglobin A.[62] (B, C, D) Distribution of delay times at zero saturation for cells from a patient with homozygous SS disease (B), hemoglobin SC disease (C), and sickle trait (D) at 37°C. The data in (B) and (C) are taken from Coletta et al,[135] while the data in (D) is from Zarkowsky and Hochmuth[143] after using the temperature dependence of the median delay time to correct the data to 37°C.

the delay times[62] or solubilities.[62,63] The principal reason for the increased sickling of SC cells is that they contain a higher hemoglobin S concentration than sickle trait cells. This increase results from two effects.[62] First, there is a greater fraction of hemoglobin S in SC cells (50/50 S/C) than in sickle trait cells (40/60 to 30/70 S/A). The reduction in the fraction of hemoglobin S in sickle trait cells is caused by a decreased rate of association of α chains to β^S chains relative to β^A chains during the tetramer assembly process.[205,206] When the concentration of α chains is reduced because of coexisting α thalassemia, this competition is enhanced and a disproportionately larger fraction of β^A-containing tetramers are formed. Second, the total intracellular hemoglobin concentration is higher in SC cells.[62,84] Since reticulocytes have nearly the same density distribution as the average cell population, the red cells must emerge from the marrow more concentrated. The reasons for this are not yet completely understood, but it has been suggested that the binding of hemoglobin C to the red cell membrane induces a potassium and water efflux.[207]

Hemoglobin F also has a marked effect on gelation. This is clinically most evident in the uncommon double heterozygous condition of hemoglobin S with pancellular hereditary persistence of fetal hemoglobin, which may be asymptomatic. In this condition hemoglobin F is more evenly distributed, and most cells contain a substantial amount of hemoglobin F (up to 35%).[24] This mixture has gelling properties in vitro that are similar to the 40/60 Hb S + A mixture found in sickle cell trait (Fig 9)[37,57] and would therefore be predicted to have a very mild or asymptomatic clinical course. In homozygous SS disease there is a variable increase in hemoglobin F that results from two factors: an increased production of F reticulocytes and preferential survival of F cells.[81] At hemoglobin F levels above 20%, corresponding to about 60% F cells, there may be some amelioration of the disease, but below 20% there appears to be no significant effect.[201,208]

Because of the tremendous sensitivity of the delay time to

the total hemoglobin concentration, it has been suggested that the clinical severity of homozygous SS disease may be improved by a small dilution of the intracellular hemoglobin.[6,7] The increase in the delay time resulting from a decrease in the intracellular concentration would allow more cells to escape the microcirculation before gelation has begun.[6,7] To estimate the effect of concentration on the delay time we use a 15th power inverse concentration dependence, since this is the concentration dependence found for delay times of about one second (Fig 3). The decrease in MCHC from 32 g/dL to 30 g/dL associated with the coexistence of α thalassemia, in which two of the four α genes are deleted $(-\alpha/-\alpha)$,[209] produces an almost three-fold increase in the delay time for the "average cell." The result is increased red cell survival[87,210] and an indication of fewer episodes of the acute chest syndrome and leg ulceration.[211,212] Also, in SS disease there may be an increased frequency of the α gene deletion with age, suggesting that a decreased total intracellular hemoglobin concentration is associated with a longer life expectancy.[213] In HbS-β^0-thalassemia there is a similar decrease in MCHC, and the clinical course relative to SS disease is "milder in many features."[24]

Thus far we have seen that for genetically different sickling disorders there is a good correlation between intracellular gelation in vitro, for solutions having compositions of the average cell, and disease severity for the average patient.§ To investigate the role of clinical diversity one would ideally want to know at least the distribution of intracellular delay times for patients from a clinically well-characterized population. No such data are yet available. An efficient but limited method of examining distributions of intracellular gelation is to measure density distributions, since the density is proportional to the total intracellular hemoglobin concentrations. Differences in intracellular solvent conditions of pH, 2,3-DPG concentration, etc, are expected to have a much smaller effect on gelation than differences in intracellular hemoglobin concentrations. Consequently the distribution is expected to reflect the distribution of intracellular delay times, except for the effect of F cells.

The only study carried out so far is one in which cell density distributions were compared with the incidence of painful crisis.[214] No correlation was found between the fraction of cells in the highest density range and crisis frequency. This result was interpreted as evidence that the greater probability of intravascular sickling is not the principal cause of increased crisis frequency, but that variations in the anatomy and dynamic properties of the microcirculation are responsible for differences among patients. As pointed out earlier, one factor that could be important in determining transit times in the microcirculation is adherence to the vascular endothelium. A strong correlation has in fact been

§Correlations between gelation and both overall clinical severity and degree of anemia have also been obtained using the in vitro fraction polymerized *at equilibrium* as a measure of gelation in vivo.[209] Although the equilibrium fraction polymerized in vitro is not relevant to the in vivo situation as discussed earlier, these correlations give a very similar result[209] because of the close correlation between the kinetic and equilibrium properties of gelation.[36]

found between overall clinical severity and the tendency of the red cells to adhere to vascular endothelium in in vitro experiments.[201-203] The severity score used in this study included evidence for organ damage resulting from microvascular occlusions as well as the frequency of pain crises. In this same study there was no correlation between severity and hemoglobin F levels or irreversibly sickled cells, which are known to correlate with the fraction of dense cells.[82]

THE PROBLEM OF INHIBITING GELATION IN PATIENTS

The strong correlation between gelation and severity for the "average" patient with the various sickling disorders clearly indicates that inhibition of gelation should result in amelioration of the disease. The data on hemoglobin mixtures shows that it will not be necessary to completely inhibit gelation (ie, increase the solubility such that it equals or exceeds the total intracellular hemoglobin concentrations at all oxygen pressures) but that a therapeutic effect should result from sufficiently increasing the delay time to allow more cells to escape the microcirculation and be reoxygenated in the lungs before gelation has begun.[6,7,57,98] In this way there should be a reduction in the rate of production of dehydrated, rapidly polymerizing cells, which have been generally assumed to be the subpopulation of cells most responsible for initiating vaso-occlusion.[6] To give this concept a quantitative basis we may ask: how much must gelation be inhibited to obtain a specified therapeutic effect in patients? An approximate answer to this question can be obtained from the correlation between in vitro delay times or solubilities in solutions of deoxyhemoglobin mixtures having the compositions found in various sickling disorders and their "average" clinical course.[57]

The data for this comparison are found in Fig 9, and Table 1 shows the increase in delay time and solubility relative to pure deoxyhemoglobin S for solutions having the hemoglobin composition found in sickle-β^+-thalassemia, sickle cell disease with hereditary persistence of fetal hemoglobin, and sickle trait. The results in Table 1 establish a set of criteria for obtaining a specified therapeutic effect. They suggest that the threshold for obtaining a therapeutic effect in SS disease would result from a method that produces an increase in the in vitro delay time of about a factor of 100 (corresponding to a solubility ratio of about 1.2), which is the increase found for solutions having the hemoglobin composition of sickle-β^+-thalassemia; an increase of about 10^3 to 10^5 (solubility ratio of about 1.3) should produce a major therapeutic effect; and a 10^6 to 10^8-fold increase in the in vitro delay time (solubility ratio of about 1.5 to 1.6), found for solutions with the composition of sickle trait cells, is predicted to result in a "cure."

With these estimates we can examine the potential utility of the various strategies that have been proposed to inhibit gelation in patients. Four different approaches have been explored or considered in some detail: (1) blocking intermolecular contact formation in the polymer, (2) raising the oxygen affinity, (3) decreasing the total intracellular hemoglobin concentration, and (4) promoting fetal hemoglobin production. The oldest idea is to develop a competitive or covalent inhibitor that would bind stereospecifically to hemoglobin S and interfere with polymer formation. Two general

Table 1. Clinical Course, Gelation Delay Time and Requirements for Therapy

Disorder	S/β^+-Thalassemia	S/HPHF	A/S Trait
Clinical course relative to S/S disease	Less severe	Much less severe	No disease
Red cell composition*			
% Hb A	20-30	0	60-75
% Hb F	0	20-35	0
% Hb S	80-70	80-65	40-25
Log delay-time ratio†	1.5-2.5	2.5-5.0	6.0-8.0
Solubility ratio‡	1.1-1.2	1.2-1.35	1.45-1.65
Therapy requirements			
Percent saturation of inhibitory site§	20-40	40-55	≥65
Decrease in intracellular concentration (g/dL)‖	3-5	5-9	≥11

*The data are from Serjeant.[24] For S/β^+-thalassemia this is the composition of the non-F cells.

†This is the ratio of the delay time for the mixture to the delay time for pure deoxyhemoglobin S at the same total hemoglobin concentration and is obtained from the data in Fig 9. These ratios are for subphysiologic concentrations using the temperature-jump technique for measuring delay times. For physiologic concentrations where the dependence of the delay time on supersaturation is smaller, these ratios are expected to be smaller, as is found with intact cells (see Fig 9b to d).

‡This is the ratio of the solubility for the mixture to the solubility for pure deoxyhemoglobin S in the limit of no polymerized hemoglobin (from Eaton and Hofrichter[26]).

§This is the fractional saturation of an ideal inhibitory site, ie, one that completely prevents polymerization, required to produce the delay time increase for pure deoxyhemoglobin S.

‖This is the required decrease in intracellular concentration, assuming an intracellular hemoglobin S concentration of 34 g/dL in the cells entering the circulation in S/S disease.

Adapted from Sunshine et al.[57]

types of mechanisms have been considered. In one the "drug" acts directly by binding to an intermolecular contact site in the polymer, thereby competitively inhibiting polymerization. In the other the "drug" inhibits polymerization indirectly by changing the conformation at the intermolecular contact site so that it no longer "fits" into the polymer. The direct approach to inhibiting gelation poses a number of problems. Unlike an enzyme, where a substrate analogue can be a powerful inhibitor of catalysis by binding to the active site, none of the known intermolecular contact sites provide such a target. There are no clefts, grooves, or other obvious structural features that can be used to design molecules with complementary structures that might bind to hemoglobin S with high specificity. Examination of the intermolecular contacts also gives no real clues. This result might have been anticipated because the interactions between molecules in the polymer are weak. One approach would be to determine the structure of hemoglobin-antibody complexes in which polymer contacts are the antigenic determinant. Since hapten-antibody interactions are generally much stronger, the antibody-binding site would be expected to have the structural features of a very effective inhibitor and hence could serve as a model for the ambitious organic chemist attempting to construct molecules that cover the contact sites.

A natural target on the hemoglobin molecule for attack by the indirect mechanism is the pocket between the β subunits, which constitutes a specific, relatively high-affinity binding site for 2,3-DPG. For example, bifunctional aspirin derivatives have been described that crosslink the β subunits by covalently binding to opposite $\beta82$ lysines.[215] Analysis of the three-dimensional structure of the complex by x-ray crystallography shows that this modification causes a shift in residues of the F-helix that are part of the acceptor site for the $\beta6$ contact region, explaining the very large increase in solubility (solubility ratios up 1.5).[19] Although these particular inhibitors may not turn out to be therapeutically useful, this and other recent studies[20,21] demonstrate the power and feasibility of using x-ray crystallography to understand the mechanism of action of inhibitors and to design more effective ones. Most studies of inhibitors of gelation have not taken such a "rational" approach. Nevertheless a number of effective inhibitors have been found, although none has been developed to the point of being a serious candidate for use in patients.[25,216-220]

A second, more speculative strategy for inhibiting gelation is to increase oxygen affinity by shifting the allosteric equilibrium toward the R structure. At any given oxygen pressure there will be a lower concentration of molecules in the T quaternary structure and therefore a decreased tendency to polymerize. Calculations based on the effect of saturation on gelation suggest that therapeutically useful effects might result, although homeostatic responses that maintain oxygen delivery could buffer the inhibitory effect.[57] One interesting way of shifting the allosteric equilibrium toward R and one that would require much lower doses of a drug than directly attacking the hemoglobin molecule would be to inhibit 2,3-DPG synthesis.[57] An additional beneficial effect would result from the fact that 2,3-DPG promotes gelation of T-state molecules.[39,42] It will be important to evaluate the effect of an increase in oxygen affinity in some detail because many inhibitors of deoxyhemoglobin S gelation also increase oxygen affinity.

The third strategy is to decrease the intracellular hemoglobin concentration, an idea directly generated from the kinetic studies.[6,7] This approach takes advantage of the enormous concentration dependence of the delay time. There are two obvious ways that could, in principle, be used to decrease the total intracellular hemoglobin concentration. One is to permanently increase the red cell volume, and the other is to reduce hemoglobin biosynthesis without a decrease in red cell volume, for example by slowly introducing iron deficiency.[7] There are some clinical data to suggest that concomitant iron deficiency is in fact beneficial.[221] The idea of swelling red cells has been tested in a preliminary way. A combination of sodium restriction, high fluid intake, and the use of an antidiuretic reduced the serum sodium to 120 to 125 mg/dL, which resulted in a 2 to 3 g/dL decrease in the MCHC.[9,222] Both the frequency and duration of painful crises appeared to be reduced. Although this study was quite limited, involving only three patients who served as their own controls, it suggests that small reductions in intracellular hemoglobin concentration may indeed have a therapeutic effect, as predicted from the in vitro gelation studies.[7] Another approach to swelling red cells in patients has been to alter the ion transport properties of the red cell membranes so as to affect a net water influx. Several agents have been described.[8,10-12,223,224] Of these the most extensively studied is ceteidil,[8,10,223,225] which may be effective in directly retarding the dehydration that produces the rapidly polymerizing dense cells as a result of sickling-unsickling cycles.[223] In a placebo-controlled, double blind study ceteidil had some effect in reducing the severity and duration of pain crises,[226] but there is yet no information on its effectiveness in decreasing crisis frequency or organ damage.

The fourth strategy for inhibiting gelation in patients is to stimulate the production of γ globin. As discussed earlier, the inhibitory effect results from the inability of the $\alpha_2\gamma_2$ or $\alpha_2\gamma\beta^S$ tetramers to copolymerize with $\alpha_2\beta_2^S$.[37] If γ chains are exchanged for β^S chains in all cells, then some therapeutic effect is expected with hemoglobin F levels of about 10% to 15% (Fig 9 and Table 1). If hemoglobin F is heterogeneously distributed, clinical data from Saudi Arabians, where sickle cell disease is milder, suggest that amelioration would result if the percentage of F reticulocytes exceeds 20%, which results in a steady-state level of about 60% F cells and 20% hemoglobin F.[227] Data on American blacks suggest that at hemoglobin F levels above 10% there is a decreased probability of major organ failure, while the threshold for a decrease in crisis frequency is about 20%.[208] Although the molecular mechanism is not at all well understood, significant stimulation of F reticulocyte production has been achieved in SS patients with two drugs: 5-azacytidine[3,4,228] and hydroxyurea.[5,229,230] With 5-azacytidine hemoglobin F levels of 12% and 20% were achieved in two patients treated for more than 100 days, and there was a concomitant decrease in pain crises.[228]

The preceding analysis indicates that there is cause for optimism, as there are several totally independent and viable approaches to the therapy of sickle cell disease. Too frequently a single approach has been criticized as not being useful because by itself it does not produce a dramatic effect in patients. There is, of course, no reason why a specific treatment for sickle cell disease could not consist of the use of several drugs simultaneously, each inhibiting gelation by a different mechanism and at nontoxic doses that would produce only a small effect if given alone.

ACKNOWLEDGMENT

We thank H. Franklin Bunn for many helpful discussions and criticisms.

REFERENCES

1. Singer K, Fisher B: Studies on abnormal hemoglobins. V. The distribution of type S (sickle cell) hemoglobin and type F (alkali resistant) hemoglobin within the red cell population in sickle cell anemia. Blood 7:1216, 1952

2. Singer K, Singer L: The gelling phenomenon of sickle cell hemoglobin: its biological and diagnostic significance. Blood 8:1008, 1953

3. Ley TJ, DeSimone J, Noguchi CT, Turner PH, Schechter AN, Heller P, Nienhuis AW: 5-azacytidine increases γ-globin synthesis and reduces the proportion of dense cells in patients with sickle cell anemia. Blood 62:370, 1983

4. Charache SG, Dover G, Smith K, Talbot CC, Moyer M, Boyer S: Treatment of sickle cell anemia with 5-azacytidine results in increased fetal hemoglobin production and is associated with non-random hypomethylation of DNA around the γ-δ-β globin gene complex. Proc Natl Acad Sci USA 80:4842, 1983

5. Platt OS, Orkin SA, Dover G, Becordsley GP, Miller B, Nathan DG: Hydroxyurea enhances fetal hemoglobin production in sickle cell anemia. J Clin Invest 74:652, 1984

6. Hofrichter J, Ross PD, Eaton WA: Kinetics and mechanism of deoxyhemoglobin S gelation: A new approach to understanding sickle cell disease. Proc Natl Acad Sci USA 71:4864, 1974

7. Eaton WA, Hofrichter J, Ross PD: Delay time of gelation: A possible determinant of clinical severity in sickle cell disease. Blood 47:621, 1976

8. Asakura T, Ohnishi ST, Adachi K, Ozguc M, Hashimoto K, Singer M, Russell MO, Schwartz E: Effect of ceteidil on erythrocyte sickling: New type of antisickling agent that may affect erythrocyte membranes. Proc Natl Acad Sci USA 77:2955, 1980

9. Rosa RM, Bierer BE, Thomas R, Stoff JS, Kruskall M, Robinson S, Bunn HF, Epstein FH: A study of induced hyponatremia in the prevention and treatment of sickle cell crisis. N Engl J Med 303:1138, 1980

10. Berkowitz LR, Orringer EP: Effect of ceteidil, an in vitro antickling agent, on erythrocyte membrane cation permeability. J Clin Invest 68:1215, 1981

11. Schmidt WF, Asakura T, Schwartz E: Effect of ceteidil on cation and water movements in erythrocytes. J Clin Invest 69:589, 1982

12. Clark MR, Mohandas N, Shohet S: Hydration of sickle cells using the sodium ionophore monensin. J Clin Invest 70:1074, 1982

13. Wishner BC, Ward KB, Lattman EE, Love WE: Crystal structure of sickle-cell deoxyhemoglobin at 5A resolution. J Mol Biol 98:179, 1975

14. Dykes GW, Crepeau RH, Edelstein SJ: Three-dimensional reconstruction of the fibers of sickle cell haemoglobin. Nature 272:506, 1978

15. Dykes GW, Crepeau RH, Edelstein SJ: Three-dimensional recontruction of the 14-filament fibers of hemoglobin S. J Mol Biol 130:451, 1979

16. Edelstein SJ: Molecular topology in crystals and fibers of hemoglobin S. J Mol Biol 150:557, 1981

17. Padlan EA, Love WE: Refined crystal structure of deoxyhemoglobin S: I restrained least-squares refinement at 3.0-A resolution. J Biol Chem 260:8272, 1985

18. Padlan EA, Love WE: Refined crystal structure of deoxyhemoglobin S: II molecular interactions in the crystal. J Biol Chem 260:8280, 1985

19. Chatterjee R, Walder RY, Arnone A, Walder JA: Mechanism for the increase in solubility of deoxyhemoglobin S due to cross-linking the β chains between lysine 82β₁ and lysine 82β₂. Biochemistry 21:5901, 1982

20. Abraham DJ, Perutz MF, Phillips SEV: Physiological and x-ray studies of potential antisickling agents. Proc Natl Acad Sci USA 80:324, 1983

21. Perutz MF, Fermi G, Abraham DJ, Poyart C, and Bursaux S: Hemoglobin as a receptor of drugs and peptides: X-ray studies of the sterochemistry of binding. J Am Chem Soc 108:1064, 1986

22. Noguchi CT, Schechter AN: The intracellular polymeriza-tion of sickle hemoglobin and its relevance to sickle cell disease. Blood 58:1057, 1981

23. Noguchi CT, Schechter AN: Sickle hemoglobin polymerization in solution and in cells. Ann Rev Biophys Biophys Chem 14:239, 1985

24. Serjeant G: Sickle Cell Disease. Oxford, Oxford University Press, 1985

25. Bunn HF, Forget BG: Hemoglobin: Molecular, genetic, and clinical aspects. Philadelphia, Saunders, 1986

26. Eaton WA, Hofrichter, J: Sickle Cell Hemoglobin Gelation. Adv Protein Chem (in press)

27. Rodgers D, Crepeau RH, Edelstein SJ: Fibers of hemoglobin S: Strand pairing and polarity, in Beuzard Y, Charache S (eds): Approaches to the Therapy of Sickle Cell Disease. Paris, INSERM, 1986, p 21

28. Nagel RL, Johnson J, Bookchin RM, Garel MC, Rosa J, Schiliro G, Wajcman H, Labie D, Moo-Penn W, Castro O: β chain contact sites in the haemoglobin S polymer. Nature 283:832, 1980

29. Crepeau R, Edelstein SJ: Polarity of the 14-strand fibers of sickle cell hemoglobin determined by cross-correlation methods. Ultramicroscopy 13:11, 1984

30. Minton AP: A thermodynamic model for gelation of sickle-cell hemoglobin. J Mol Biol 82:483, 1974

31. Ross PD, Hofrichter J, Eaton WA: Calorimetric and optical characterization of sickle cell hemoglobin gelation. J Mol Biol 96:239, 1975

32. Ross PD, Hofrichter J, Eaton WA: Thermodynamics of gelation of sickle cell deoxyhemoglobin. J Mol Biol 115:111, 1977

33. Allison AC: Properties of sickle-cell haemoglobin. Biochem J 65:212, 1957

34. Bertles JF, Rabinowitz R, Doebler J: Hemoglobin interaction: Modification of solid phase composition in the sickling phenomenon. Science 169:375, 1970

35. Magdoff-Fairchild B, Poillon WN, Li T-I, Bertles JF: Thermodynamic studies of polymerization of deoxygenated sickle cell hemoglobin. Proc Natl Acad Sci USA 73:990, 1976

36. Hofrichter J, Ross PD, Eaton WA: Supersaturation in sickle cell hemoglobin solutions. Proc Natl Acad Sci USA 73:3034, 1976

37. Sunshine HR, Hofrichter J, Eaton WA: Gelation of sickle cell hemoglobin in mixtures with normal adult and fetal hemoglobins. J Mol Biol 133:435, 1979

38. Goldberg MA, Husson MA, Bunn HF: The participation of hemoglobins A and F in the polymerization of sickle hemoglobin. J Biol Chem 252:3414, 1977

39. Briehl RW: Gelation of sickle cell hemoglobin IV. Phase transitions in hemoglobin S gels: Separate measures of aggregation and solution-gel equilibrium. J Mol Biol 123:521, 1978

40. Poillon WN, Bertles JF: Deoxygenated sickle hemoglobin. Effects of lyotropic salts on its solubility. J Biol Chem 254:3462, 1979

41. Swerdlow PH, Bryan RA, Bertles JF, Poillon WN, Magdoff-Fairchild B, Milner PF: Effect of 2,3-diphosphoglycerate on the solubility of deoxy sickle hemoglobin. Hemoglobin 1:527, 1977

42. Poillon WN, Kim BC, Walder JA: Deoxygenated sickle hemoglobin: the effects of 2,3-diphosphoglycerate and inositol hexaphosphate on its solubility, in Beuzard Y, Charache S (eds): Approaches to the Therapy of Sickle Cell Anemia: Paris, INSERM, 1986, p 89

43. Hofrichter J, Ross PD, Eaton WA: A physical description of the gelation of deoxyhemoglobin S, in Hercules JI, Cottam GL, Waterman MR, Schechter AN (eds): Molecular and Cellular Aspects of Sickle Cell Disease. Bethesda, DHEW Publication No (NIH) 76-1007, 1976, p 185

44. Hofrichter J: Ligand binding and the gelation of sickle cell hemoglobin. J Mol Biol 128:335, 1979

45. Gill SJ, Spokane R, Benedict RC, Fall L, Wyman J: Ligand-linked phase equilibria of sickle cell hemoglobin. J Mol Biol 140:299, 1980

46. Sunshine HR, Hofrichter J, Ferrone FA, Eaton WA: Oxygen binding by sickle cell hemoglobin polymers. J Mol Biol 158:251, 1982

47. Bookchin RM, Nagel RL, Ranney HM: The effect of $\beta^{73\ Asn}$ on the interaction of sickling hemoglobins. Biochim Biophys Acta 221:373, 1970

48. Bookchin RM, Nagel RL: Ligand-induced conformational dependence of hemoglobin in sickling interactions. J Mol Biol 60:263, 1971

49. Bookchin RM, Nagel RL: Molecular interactions of sickling hemoglobins, in Abramson H, Bertles JF, Wethers DL (eds): Sickle Cell Disease. St Louis, Mosby, 1973, p 140

50. Bookchin RM, Balazs T, Nagel RL, Tellez I: Polymerisation of haemoglobin SA hybrid tetramers. Nature 269:526, 1977

51. Minton AP: Solubility relationships in binary mixtures of hemoglobin variants. Application to the gelation of sickle-cell hemoglobin. Biophys Chem 1:387, 1974

52. Minton AP: Thermodynamic analysis of the chemical inhibition of sickle-cell hemoglobin gelation. J Mol Biol 95:289, 1975

53. Minton AP, Relations between oxygen saturation and aggregation of sickle-cell hemoglobin. J Mol Biol 100:519, 1976

54. Minton AP: Non-ideality and the thermodynamics of sickle-cell hemoglobin gelation. J Mol Biol 110:89, 1977

55. Minton AP: The effect of volume occupancy upon the thermodynamic activity of proteins: Some biochemical consequences. Mol Cell Biochem 55:119, 1983

56. Moffat K: Gelation of sickle cell hemoglobin: effects of hybrid tetramer formation in hemoglobin mixtures. Science 185:274, 1974

57. Sunshine HR, Hofrichter J, Eaton WA: Requirements for therapeutic inhibition of sickle hemoglobin gelation. Nature 275:238, 1978

58. Cheetham RC, Huehns ER, Rosemeyer MA: Participation of haemoglobins A, F, A$_2$, and C in polymerisation of haemoglobin S. J Mol Biol 129:45, 1979

59. Behe MJ, Englander SW: Mixed gelation theory. Kinetics, equilibrium and gel incorporation in sickle hemoglobin mixtures. J Mol Biol 133:137, 1979

60. Benesch RE, Edalji R, Benesch R, Kwong S: Solubilization of hemoglobin S by other hemoglobins. Proc Natl Acad Sci USA 77:5130, 1980

61. Sunshine HR: Effect of other hemoglobins on gelation of sickle cell hemoglobin. Tex Rep Biol Med 40:233, 1980-1981

62. Bunn HF, Noguchi CT, Hofrichter J, Schechter GP, Schechter AN, Eaton WA: The molecular and cellular pathogenesis of hemoglobin SC disease. Proc Natl Acad Sci USA 79:7527, 1982

63. Bookchin RM, Balazs T: Ionic strength dependence of the polymer solubilities of deoxyhemoglobin S + C and S + A mixtures. Blood 67:887, 1986

64. Bunn HF, McDonough M: Asymmetrical hemoglobin hybrids. An approach to the study of subunit interactions. Biochemistry 13:988, 1974

65. Nagel et al (10 authors): Structural bases of the inhibitory effects of hemoglobin F and A$_2$ on the polymerization of hemoglobin S. Proc Natl Acad Sci USA 76:670, 1979

66. Harris JW: Studies on the destruction of red blood cells, VIII. molecular orientation in sickle cell hemoglobin solutions. Proc Soc Exp Biol Med 75:197, 1950

67. Williams RC: Concerted formation of the gel of hemoglobin S. Proc Natl Acad Sci USA 70:1506, 1973

68. Briehl RW, Ewert S: Effects of pH, 2,3-diphosphoglycerate and salts on gelation of sickle cell deoxyhemoglobin. J Mol Biol 80:445, 1973

69. Bunn HF: The interaction of sickle cell hemoglobin with DPG, CO$_2$ and with other hemoglobins: Formation of asymmetrical hybrids, in Brewer GJ (ed): Hemoglobin and Red Cell Structure and Function. New York, Plenum, 1972, p 41

70. Gill SJ, Benedict RC, Fall L, Spokane R, Wyman J: Oxygen binding to sickle cell hemoglobin. J Mol Biol 130:175, 1979

71. Hofrichter J, Hendricker DG, Eaton WA: Structure of hemoglobin S fibers: optical determination of the molecular orientation in sickled red cells. Proc Natl Acad Sci USA 70:3604, 1973

72. Hofrichter J, Eaton WA: Linear dichroism of biological chromophores. Ann Rev Biophys Bioeng 5:511, 1976

73. Eaton WA, Hofrichter J: Polarized absorption and linear dichroism spectroscopy of hemoglobin. Methods Enzymol 76:175, 1981

74. Monod J, Wyman J, Changeux JP: On the nature of allosteric transitions: A plausible model. J Mol Biol 12:88, 1965

75. Fabry ME, Mears JG, Patel P, Schaefer-Rego K, Martinez G, Nagel RL: Dense cells in sickle cell anemia: The effects of gene interaction. Blood 64:1042, 1984

76. Noguchi CT, Torchia DA, Schechter AN: Intracellular polymerization of sickle cell hemoglobin. Effects of cell heterogeneity. J Clin Invest 72:846, 1983

77. Winslow RM: Hemoglobin interactions and whole blood oxygen equilibrium curves in sickling disorders, in Caughey WS (ed): Biochemical and Clinical Aspects of Hemoglobin Abnormalities. New York, Academic, 1978, p 369

78. Lonsdorfer J, Bogui P, Otayeck A, Bursaux E, Poyart C, Cabannes R: Cardiorespiratory adjustments in chronic sickle cell anemia. Bull Eur Physiopathol Respir 19:339, 1983

79. Seakins M, Gibbs WN, Milner PF, Bertles JF: Erythrocyte Hb-S concentration. An important factor in the low oxygen affinity of blood in sickle cell anemia. J Clin Invest 52:422, 1973

80. Chien S, Usami S, Bertles JF: Abnormal rheology of oxygenated blood in sickle cell anemia. J Clin Invest 49:623, 1970

81. Dover GJ, Boyer SH, Charache S, Heintzelman K: N Engl J Med 299:1428, 1978

82. Clark MR, Mohandas N, Embury SH, Lubin BH: A simple laboratory alternative to irreversibly sickled cell (ISC) counts. Blood 60:659, 1982

83. Fabry ME, Nagel RL: Heterogeneity of red cells in the sickler: A characteristic with practical clinical and pathophysiological implications. Blood Cells 8:9, 1982

84. Fabry ME, Kaul DK, Raventos-Suarez C, Chang H, Nagel RL: SC erythrocytes have an abnormally high intracellular hemoglobin concentration. J Clin Invest 70:1315, 1982

85. Kaul DK, Fabry ME, Windisch P, Baez S, Nagel RL: Erythrocytes in sickle cell anemia are heterogeneous in their rheological and hemodynamic characteristics. J Clin Invest 72:22, 1983

86. Kaul DK, Nagel RL, Baez S: Pressure effects on the flow behavior of sickle (HbSS) red cells in isolated (ex-vivo) microvascular system. Microvasc Res 26:170, 1983

87. Noguchi CT, Dover GJ, Rodgers GP, Serjeant GR, Antonarakis SE, Anagnou NP, Higgs DR, Weatherall DJ, Schechter AN. Alpha thalassemia changes erythrocyte heterogeneity in sickle cell disease. J Clin Invest 75:1632, 1985

88. Rodgers GP, Schechter AN, Noguchi CT: Cell heterogeneity in sickle cell disease: Quantitation of the erythrocyte density profile. J Lab Clin Med 106:30, 1985

89. Bookchin RM, Balazs T, Landau LC: Determinants of red cell sickling. Effect of varying pH and of intracellular hemoglobin concentration by osmotic shrinkage. J Lab Clin Med 87:597, 1976

90. Noguchi CT, Torchia DA, Schechter AN: ^{13}C NMR quanti-

tation of polymer in deoxyhemoglobin S gels. Proc Natl Acad Sci USA 76:4936, 1979

91. Noguchi CT, Torchia DA, Schechter AN: Determination of deoxyhemoglobin S polymer in sickle erythrocytes upon deoxygenation. Proc Natl Acad Sci USA 77:5487, 1980

92. Noguchi CT, Torchia DA, Schechter AN: Polymerization of hemoglobin in sickle trait erythrocytes and lysates. J Biol Chem 256:4168, 1981

93. Sutherland JWH, Egan W, Schechter AN, Torchia DA: Carbon-13-proton nuclear magnetic double-resonance study of deoxyhemoglobin S gelation. Biochemistry 18:1797, 1979

94. Hofrichter J, Ross PD, Eaton WA: Kinetic and thermodynamic investigation of deoxyhemoglobin S gelation, in Hercules JI, Schechter AN, Eaton WA, Jackson RE (eds): Proceedings of the National Symposium on Sickle Cell Disease. Bethesda, DHEW Publication No. (NIH) 75-723, 1974, p 43

95. Moffat K, Gibson QH: The rates of polymerization and depolymerization of sickle cell hemoglobin. Biochem Biophys Res Commun 61:237, 1974

96. Pumphrey JG, Steinhardt J: Crystallization of sickle hemoglobin from gently agitated solutions—an alternative to gelation. J Mol Biol 112:359, 1977

97. Noguchi CT, Schechter AN: Effects of amino acids on gelation kinetics and solubility of sickle hemoglobin. Biochem Biophys Res Commun 74:637, 1977

98. Sunshine HR, Ferrone FA, Hofrichter J, Eaton WA: Gelation assays and the evaluation of therapeutic inhibitors, in Rosa J, Beuzard Y, Hercules JI (eds): Development of Therapeutic Agents for Sickle Cell Disease. Amsterdam, INSERM Symposium No 9, Elsevier/North Holland, 1979, p 31

99. Elbaum D, Harrington JP, Bookchin RM, Nagel RL: Kinetics of Hb S gelation. Effect of alkylureas, ionic strength and other hemoglobins. Biochim Biophys Acta 534:228, 1978

100. Adachi K, Asakura T: Demonstration of a delay time during aggregation of diluted solutions of deoxyhemoglobin S and hemoglobin C$_{Harlem}$ in concentrated phosphate buffer. J Biol Chem 253:6641, 1978

101. Adachi K, Asakura T: Gelation of deoxyhemoglobin A in concentrated phosphate buffer. J Biol Chem 254:12273, 1979

102. Adachi K, Asakura T: Nucleation-controlled aggregation of deoxyhemoglobin S. Possible difference in the size of nuclei in different phosphate concentrations. J Biol Chem 254:7765, 1979

103. Adachi K, Asakura T: Polymerization of deoxyhemoglobin C$_{Harlem}$ (β6 Glu → Val, β73 Asp → Asn). The effect of β73 asparagine on the gelation and crystallization of hemoglobin. J Mol Biol 144:467, 1980

104. Adachi K, Asakura T: Kinetics of the polymerization of hemoglobin in high and low phosphate buffers. Blood Cells 8:213, 1982

105. Adachi K, Asakura T: Multiple nature of polymers of deoxyhemoglobin S prepared by different methods. J Biol Chem 258:3045, 1983

106. Adachi K, Asakura T, McConnell ML: Formation of nuclei during delay time prior to aggregation of deoxyhemoglobin S in concentrated phosphate buffer. Biochim Biophys Acta 580:405, 1979

107. Adachi K, Ozguc M, Asakura T: Nucleation-controlled aggregation of deoxyhemoglobin S. Participation of hemoglobin A in the aggregation of deoxyhemoglobin S in concentrated phosphate buffer. J Biol Chem 255:3092, 1980

108. Adachi K, Segal R, Asakura T: Nucleation-controlled aggregation of deoxyhemoglobin S. Participation of hemoglobin F in the aggregation of deoxyhemoglobin S in concentrated phosphate buffer. J Biol Chem 255:7595, 1980

109. Ip YC, Asakura T, Adachi K: Polymerization of carbamylated deoxyhemoglobin S in concentrated phosphate buffer. J Biol Chem 257:12853, 1982

110. Goldberg MA, Lalos AT, Bunn HF: The effect of erythrocyte membrane preparations on the polymerization of sickle hemoglobin. J Biol Chem 256:193, 1981

111. Goldberg MA, Lalos AT, Himmelstein B, Bunn HF: Effects of red cell membrane on the polymerization of sickle hemoglobin. Blood Cells 8:237, 1982

112. Wenger GD, Balcerzak SP: Viscometric and spectrophotometric measurements of hemoglobin S polymerization kinetics. Blood 63:897, 1984

113. Pumphrey JG, Steinhardt J: Formation of needle-like aggregates in stirred solutions of hemoglobin S. Biochem Biophys Res Commun 69:99, 1976

114. Hofrichter J, Gethner JG, Eaton WA: Mechanism of sickle cell hemoglobin gelation. Biophys J 21:20a 1978 (abstr)

115. Madonia F, San Biagio PL, Palma MU, Schiliro G, Musumeci S, Russo G: Photon scattering as a probe of microviscosity and channel size in gels such as sickle haemoglobin. Nature 302:412, 1983

116. Malfa R, Steinhardt J: A temperature-dependent latent period in the aggregation of sickle-cell deoxyhemoglobin. Biochem Biophys Res Commun 59:887, 1974

117. Harris JW, Bensusan HB: The kinetics of the sol-gel transformation of deoxyhemoglobin S by continuous monitoring of viscosity. J Lab Clin Med 86:564, 1975

118. Kowalczykowski S, Steinhardt J: Kinetics of hemoglobin S gelation followed by continuously sensitive low-shear viscosity. J Mol Biol 115:201, 1977

119. Fieschko WM, Measow W, Crandall ED, Litt M, Forster RE: J Lab Clin Med 92:1019, 1978

120. Behe MJ, Englander SW: Sickle hemoglobin gelation. Reaction order and critical nucleus size. Biophys J 23:129, 1978

121. Briehl RW: The effect of shear on the delay time for gelation of hemoglobin S. Blood Cells 8:201, 1982

122. Danish EH, Harris JW: Viscosity studies of deoxyhemoglobin S: Evidence for formation of microaggregates during the lag phase. J Lab Clin Med 101:515, 1983

123. Eaton WA, Hofrichter J, Ross PD, Tschudin RG, Becker ED: Comparison of sickle cell hemoglobin gelation kinetics by NMR and optical methods. Biochem Biophys Res Commun 69:538, 1976

124. Waterman MR, Cottam GL: Kinetics of the polymerization of hemoglobin S: Studies below normal erythrocyte hemoglobin concentration. Biochem Biophys Res Commun 73:639, 1976

125. Cottam GL, Waterman MR, Thompson BC: Kinetics of polymerization of deoxyhemoglobin S and mixtures of hemoglobin A and hemoglobin S at high concentrations. Arch Biochem Biophys 181:61, 1977

126. Cottam GL, Shibata K, Waterman MR: Effectors of the rate of deoxyhemoglobin S polymerization, in Caughey WS (ed): Biochemical and Clinical Aspects of Hemoglobin Abnormalities. New York, Academic, 1978, p 695

127. Shibata K, Waterman MR, Cottam GL: Alteration of the rate of deoxyhemoglobin S polymerization. Effect of pH and percentage of oxygenation. J Biol Chem 252:7468, 1977

128. Waterman MR, Cottam OL, Shibata, K: Inhibitory effect of deoxyhemoglobin A$_2$ on the rate of deoxyhemoglobin S polymerization. J Mol Biol 128:337, 1979

129. Thiyagarajan P, Johnson ME: Saturation transfer electron paramagnetic resonance detection of sickle hemoglobin aggregation during deoxygenation. Biophys J 42:269, 1983

130. Steinhardt J, Malfa R: Anomalous viscosity properties of solutions of deoxy sickle-cell hemoglobin at temperatures below

16°C, in Hercules JI, Schechter AN, Eaton WA, Jackson RE (eds): Proceedings of the National Symposium on Sickle Cell Disease. Bethesda, DHEW Publication No. (NIH) 75-723, 1974, p 135

131. Ferrone FA, Hofrichter J, Sunshine HR, Eaton WA: Kinetic studies on photolysis-induced gelation of sickle cell hemoglobin suggest a new mechanism. Biophys J 32:361, 1980

132. Ferrone FA, Hofrichter J, Eaton WA: Kinetics of sickle hemoglobin polymerization: I. Studies using temperature-jump and laser photolysis techniques. J Mol Biol 183:591, 1985

133. Hofrichter J: Kinetics of sickle hemoglobin polymerization III. Stochastic fluctuations in polymerization progress curves. J Mol Biol 189:553, 1986

134. Briehl RW, Salhany JM: Gelation of sickle hemoglobin III. Nitrosyl hemoglobin. J Mol Biol 96:733, 1975

135. Coletta M, Hofrichter J, Ferrone FA, Eaton WA: Kinetics of sickle haemoglobin polymerization in single red cells. Nature 300:194, 1982

136. Mozzarelli A, Hofrichter J, Eaton WA: Sickling, unsickling, and intracellular polymerization of hemoglobin S, in Beuzard Y, Charache S, Galacteros F (eds): Approaches to the Therapy of Sickle Cell Anemia. Paris, INSERM, 1986, p 39

137. Mozzarelli A, Hofrichter J, Eaton WA: Delay time of hemoglobin S polymerization prevents most cells from sickling in vivo. Science 237:500, 1987

138. Ferrone FA, Hofrichter J, Eaton WA: Kinetics of sickle hemoglobin polymerization: II. A double nucleation mechanism. J Mol Biol 183:611, 1985

139. Eaton WA, Hofrichter J: Successes and failures of a simple nucleation theory for sickle hemoglobin gelation, in Caughey WS (ed): Biochemical and Clinical Aspects of Hemoglobin Abnormalities. New York, Academic, 1978, p 443

140. Nash GB, Johnson CJ, Meiselman HJ: Influence of oxygen tension on the viscoelastic behavior of red blood cells in sickle cell disease. Blood 67:110, 1986

141. Ferrone FA, Cho MR, Bishop MF: Can a successful mechanism for HbS gelation predict sickle cell crises?, in Beuzard Y, Charache S, Galacteros F (eds): Approaches to the Therapy of Sickle Cell Anemia. Paris, INSERM, 1986, p 53

142. Zarkowsky HS, Hochmuth RM: Sickling times of individual erythrocytes at zero PO_2. J Clin Invest 56:1023, 1975

143. Zarkowsky HS, Hochmuth RM: Experimentally-induced alterations in the kinetics of erythrocyte sickling. Blood Cells 3:305, 1977

144. Rampling MW, Sirs JA: The rate of sickling of cells containing sickle cell haemoglobin. Clin Sci Mol Med 45:655, 1973

145. Evans EA, La Celle PL: Intrinsic material properties of the erythrocyte membrane indicated by mechanical analysis of deformation. Blood 45:29, 1975

146. Evans E, Mohandas N, Leung A: Static and dynamic rigidities of normal and sickle erythrocytes. Major influence of cell hemoglobin concentration. J Clin Invest 73:477, 1984

147. Nash GB, Johnson CS, Meiselman HJ: Mechanical properties of oxygenated red blood cells in sickle cell (HbSS) Disease. Blood 63:73, 1984

148. Chien S, Sung K-LP, Skalak R, Usami S, Tozeren A: Theoretical and experimental studies on viscoelastic properties of erythrocyte membrane. Biophys J 24:463, 1978

149. Briehl RW: Solid-like behavior of unsheared sickle haemoglobin gels and the effects of shear. Nature 288:622, 1980

150. Sherman IJ: The sickling phenomenon, with special reference to the differentiation of sickle cell anemia from the sickle cell trait. Bull Johns Hopkins Hosp 67:309, 1940

151. Asakura T, Mayberry J: Relationship between morphologic characteristics of sickle cells and method of deoxygenation. J Lab Clin Med 104:987, 1984

152. Hahn JA, Messer MJ, Bradley TB: Ultrastructure of sickling and unsickling in time-lapse studies. Br J Haematol 34:559, 1976

153. Ham TH, Castle WB: Relation of increased hypertonic fragility and of erythrostasis to the mechanism of hemolysis in certain anemias. Trans Assoc Am Phys 55:127, 1940

154. Sproule BJ, Halden ER, Miller WF: A study of cardiopulmonary alterations in patients with sickle cell disease and its variants. J Clin Invest 37:486, 1958

155. Jensen WN, Rucknagel DL, Taylor WJ: In vivo study of the sickle cell phenomenon. J Lab Clin Med 56:854, 1960

156. Serjeant GR, Petch MC, Serjeant BE: The in vivo sickle phenomenon: A reappraisal. J Lab Clin Med 81:850, 1973

157. Harrington JP, Elbaum D, Bookchin RM, Wittenberg JB, Nagel RL: Ligand kinetics of hemoglobin S containing erythrocytes. Proc Natl Acad Sci USA 74:203, 1977

158. Messer MJ, Hahn JA, Bradley TB: The kinetics of sickling and unsickling of red cells under physiological conditions: Rheological and ultrastructural correlations, in Hercules JI, Cottam GL, Waterman MR, Schechter AN (eds): Proceedings of the Symposium on Molecular and Cellular Aspects of Sickle Cell Disease. Bethesda, DHEW Publication No. (NIH) 76-1007, 1976, p 225

159. Briehl RW: Physical Chemical Properties of Sickle Cell Hemoglobin, in Wallach DFH (ed): The Function of Red Blood Cells: Erythrocyte Pathobiology. New York, Alan R Liss, 1981, p 241

160. Bishop MF, Ferrone FA: Kinetics of nucleation controlled polymerization: a perturbation treatment for use with a secondary pathway. Biophys J 46:631, 1984

161. Cokelet GR: Blood rheology interpreted through the flow properties of the red cell, in Grayson J, Zingg W (eds): Microcirculation, vol 1. New York, Plenum, 1976, p 9

162. Skalak R, Chien S: Capillary flow: History, experiments and theory. Biorheology 18:307, 1981

163. Fischer T, Schmid-Schoenbein H: Tank tread motion of red cell membranes in viscometric flow: Behavior of intracellular and extracellular markers (with film). Blood Cells 3:351, 1977

164. Bertles JF, Milner PFA: Irreversibly sickled erythrocytes: A consequence of the heterogeneous distribution of hemoglobin types in sickle cell anemia. J Clin Invest 47:1731, 1968

165. McCurdy PR, Sherman AS: Irreversibly sickled cells and red cell survival in sickle cell anemia. A study with both $DF^{32}P$ and ^{51}Cr. Am J Med 64:253, 1978

166. Bensinger TA, Gillette PN: Hemolysis in sickle cell disease. Arch Intern Med 133:624, 1974

167. Bookchin RM, Lew VL: Red cell membrane abnormalities in sickle cell anemia. Prog Hematol XIII: p 1, 1983

168. Renkin EM, Michel CC (eds): Handbook of Physiology, Section 2, Cardiovascular System, Volume IV, Microcirculation, Part 1. Bethesda, American Physiological Society, 1984

169. Kaul DK, Fabry ME, Nagel RL: Vaso-occlusion by sickle cells: Evidence for selective trapping of dense red cells. Blood 68:1162, 1986

170. La Celle PL: Pathologic erythrocytes in the capillary microcirculation. Blood Cells 1:269, 1975

171. La Celle PL: Oxygen delivery to muscle cells during capillary occlusion by sickled erythrocytes. Blood Cells 3:273, 1977

172. Klug PP, Lessin LS: Microvascular blood flow of sickled erythrocytes. A dynamic morphologic study. Blood Cells 3:263, 1977

173. Varat MA, Adolph RJ, Fowler NO: Cardiovascular effects of anemia. Am Heart J 83:415, 1972

174. Duke M, Abelman WH: The hemodynamic response to chronic anemia. Circulation 39:503, 1969

175. Little RC: Physiology of the heart and circulation. Chicago, Year Book, 1985

176. Leight L, Snider TH, Clifford GO, Hellems HK: Hemodynamic studies in sickle cell anemia. Circulation 10:653, 1954

177. Bishop JM, Donald KW, Wade OL: Circulatory dynamics at rest and on exercise in the hyperkinetic states. Clin Sci 14:329, 1955

178. Sproule BJ, Halden ER, Miller WF: Cardiopulmonary response to heavy exercise in patients with anemia. J Clin Invest 39:378, 1960

179. Ryan JM, Hickam JB: The alveolar-arterial oxygen pressure gradient in anemia. J Clin Invest 31:188, 1952

180. Roughton FJW: Kinetics of gas transport in the blood. Br Med Bull 19:80, 1963

181. Denenberg BS, Criner G, Jones R, Spann JF: Cardiac function in sickle cell anemia. Am J Cardiol 51:1674, 1983

182. Fowler NO, Smith O, Greenfield JC: Arterial blood oxygenation in sickle cell anemia. Am J Med Sci 234:449, 1957

183. Bromberg PA, Jensen WN: Arterial oxygen unsaturation in sickle cell disease. Am Rev Respir Dis 96:400, 1967

184. Shubin H et al.: Cardiovascular findings in children with sickle cell anemia. Am J Cardiol 6:875, 1960

185. Wall MA, Platt OS, Strieder DJ: Lung function in children with sickle cell anemia. Am Rev Respir Dis 120:210, 1979

186. Rotman HH, Klocke RA, Andersson KK, D'Alecy L, Forster RE: Kinetics of oxygenation and deoxygenation of erythrocytes containing hemoglobin. Respir Physiol 21:9, 1974

187. Milner PF: Oxygen transport in sickle cell anemia. Arch Intern Med 133:565, 1974

188. Nagel RL, Bookchin RM: Oxygen transport and the sickle cell, in Wallach DFH (ed): The Function of Red Blood Cells: Erythrocyte Pathobiology. New York, Alan R. Liss, 1981, p 279

189. Johnson CS, Giorgio AJ: Arterial blood pressure in adults with sickle cell disease. Arch Intern Med 141:891, 1981

190. Miller DM, Winslow RM, Klein HG, Wilson KC, Brown FL, Statham NJ: Improved exercise performance after exchange tranfusion in subjects with sickle cell anemia. Blood 56:1127, 1980

191. Krough A: The number and distribution of capillaries in muscle with calculation of the oxygen pressure head necessary for supplying the tissues. J Physiol (Lond) 52:409, 1919

192. Renkin EM: Control of microcirculation and blood-tissue exchange, in Renkin EM, Michel CC (eds): Handbook of Physiology, Section 2, The Cardiovascular System, Volume IV, Microcirculation, Part 2. Bethesda, American Physiological Society, 1984, p 627

193. Honig CR, Odoroff CL: Calculated dispersion of capillary transit times: significance for oxygen exchange. Am J Physiol 240 (Heart Circ Physiol 9): H199, 1981

194. Honig CR, Odoroff CL, Frierson JL: Capillary recruitment in exercise: rate, extent, uniformity, and relation to blood flow. Am J Physiol 238 (Heart Circ Physiol 7): H31, 1980

195. Rodgers GP, Schechter AN, Noguchi CT, Klein HG, Nienhuis AW, Bonner RF: Periodic microcirculatory flow in patients with sickle-cell disease. N Engl J Med 311:1535, 1984

196. Hammersen F: The terminal vascular bed in skeletal muscle with special regard to the problem of shunts, in Crone C, Lassen NA (eds): Capillary Permeability. Copenhagen, Munksgaard, 1970, p 351

197. Vayo MM, Lipowsky HH, Karp N, Schmalzer E, Chien S: A model of microvascular oxygen transport in sickle cell disease. Microvasc Res 30:195, 1985

198. Granger HJ, Shepherd AP Jr: Intrinsic microvascular control of tissue oxygen delivery. Microvasc Res 5:49, 1973

199. Steinberg MH, Hebbel RP: Clinical diversity of sickle cell anemia: Genetic and cellular modulation of disease severity. Am J Hematol 14:405, 1983

200. Nagel RL, Fabry ME: The many pathophysiologies of sickle cell anemia. Am J Hematol 20:195, 1985

201. Hebbel RP, Boogaerts MAB, Eaton JW, Steinberg MH: Erythrocyte adherence to endothelium in sickle cell anemia. N Engl J Med 302:992, 1980

202. Hebbel RP, Yamada O, Moldow CF, Jacob HS, White JG, Eaton JW: Abnormal adherence of sickle erythrocytes to cultured vascular endothelium. A possible mechanism for microvascular occlusion in sickle cell disease. J Clin Invest 65:154, 1980

203. Hebbel RP, Moldow CF, Steinberg MH: Modulation of erythrocyte-endothelial interactions and the vasocclusive severity of sickling disorders. Blood 58:947, 1981

204. Mohandas N, Evans E, Kukan B: Adherence of sickle erythrocytes to vascular endothelial cells: Requirement for both cell membrane changes and plasma factors. Blood 64:282, 1984

205. Mrabet ND, McDonald MJ, Turci S, Sarkar R, Szabo A, Bunn HF: Electrostatic attraction governs the dimer assembly of human hemoglobin. J Biol Chem 261:5222, 1986

206. Bunn HF: Subunit assembly of hemoglobin: An important determinant of hematologic phenotype. Blood 69:1, 1987

207. Brugnara C, Kopin AK, Bunn HF, Tosteson DC: Regulation of cation content and cell volume in erythrocytes from patients with homozygous C disease. J Clin Invest 75:1608, 1985

208. Powars DR, Weiss JN, Chan LS, Schroeder WA: Is there a threshold level of fetal hemoglobin that ameliorates morbidity in sickle cell disease. Blood 63:921, 1984

209. Brittenham GM, Schechter AN, Noguchi CT: Hemoglobin S polymerization: primary determinant of the hemolytic and clinical severity of the sickling syndromes. Blood 65:183, 1985

210. De Ceulaer K, Higgs DR, Weatherall DJ, Hayes RJ, Serjeant BR, Serjeant GR: α-Thalassemia reduces the hemolytic rate in homozygous sickle-cell disease. N Engl J Med 309:189, 1983

211. Higgs DR, Aldridge BE, Lamb J, Clegg JB, Weatherall DJ, Hayes RJ, Grandison Y, Lowrie Y, Mason KP, Serjeant BE, Serjeant GR: The interaction of alpha-thalassemia and homozygous sickle-cell disease. N Eng J Med 306:1441, 1982

212. Steinberg MH, Rosenstock W, Coleman MB, Adams JG, Platica O, Cedeno M, Rieder RF, Wilson JT, Milner P, West S: Effects of thalassemia and microcytosis on the hematologic and vasoocclusive severity of sickle cell anemia. Blood 63:1353, 1984

213. Mears JG, Lachman HM, Labie D, Nagel RL: Alpha-thalassemia is related to prolonged survival in sickle cell anemia. Blood 62:286, 1983

214. Billett HH, Kim K, Fabry ME, Nagel RL: The percentage of dense red cells does not predict incidence of sickle cell painful crisis. Blood 68:301, 1986

215. Walder JA, Zaugg RH, Iwaoka RS, Watkin WG, Klotz IM: Alternative aspirins as antisickling agents: Acetyl-3,5-dibromosalicylic acid. Proc Natl Acad Sci USA 74:5499, 1977

216. Dean J, Schechter AN: Sickle-cell anemia: Molecular and cellular bases of therapeutic approaches. N Engl J Med 299:752, 804, 863, 1978

217. Klotz IM, Haney DN, King LC: Rational approaches to chemotherapy: Antisickling agents. Science 213:724, 1981

218. Chang H, Ewert SM, Bookchin RM, Nagel RL: Comparative evaluation of fifteen anti-sickling agents. Blood 61:693, 1983

219. Garel MC, Domenget C, Galacteros F, Martin-Caburi J, Beuzard Y: Inhibition of erythrocyte sickling by thiol reagents. Mol Pharmacol 25:559, 1984

220. Edelstein SJ: Sickle Cell Anemia. Ann Rep Med Chem 20:247, 1985

221. Haddy TB, Castro O: Overt iron deficiency in sickle cell disease. Arch Intern Med 142:1621, 1982

222. Charache S, Walker WG: Failure of desmopressin to lower serum sodium or prevent crisis in patients with sickle cell anemia. Blood 58:892, 1981

223. Berkowitz LR, Orringer EP: Effects of ceteidil in monovalent cation permeability in the erythrocyte: An explanation for the efficacy of ceteidil in the treatment of sickle cell anemia. Blood Cells 8:283, 1982

224. Asakura T, Shibutani Y, Reilly MP, DeMeio RH: Antisickling effect of tellurite: A potent membrane-acting agent in vitro. Blood 64:305, 1984

225. Gulley ML, Ross DW, Feo C, Orringer EP: The effect of cell hydration on the deformability of normal and sickle erythrocytes. Am J Hematol 13:283, 1982

226. Benjamin LJ, Peterson CM, Orringer EP, Berkowitz LR, Kreisberg RA, Mankad VN, Prasad AS, Lewkow LM, Chillar RK: Effects of ceteidil in acute sickle cell crisis. Blood 62:53a, 1983(abstr)

227. Dover GJ, Boyer SH, Pembrey ME: F-cell production in sickle cell anemia: Regulation by genes linked to a β-hemoglobin locus. Science 211:1441, 1981

228. Dover GJ, Charache S, Boyer SH, Vogelsang G, Moyer M: 5-Azacytidine increases HbF production and reduces anemia in sickle cell disease: dose-response analysis of subcutaneous and oral dosage regimes. Blood 66:527, 1985

229. Dover GJ, Humphries RK, Moore JG, Ley TJ, Young NS, Charache S, Nienhuis AW: Hydroxyurea induction of hemoglobin F production in sickle cell disease: Relationship between cytotoxicity and F cell production. Blood 67:735, 1986

230. Charache S, Dover GJ, Moyer MA, Moore JW: Hydroxyurea-induced augmentation of fetal hemoglobin production in patients with sickle cell anemia. Blood 69:109, 1987

The Electrocardiogram as an Example of Electrostatics

RUSSELL K. HOBBIE
School of Physics and Astronomy
University of Minnesota
Minneapolis, Minnesota 55455
(Received 7 December 1972)

The relationship of the electrocardiogram (EKG) to the charge distribution in the heart provides an interesting example of electrostatics for a general physics course. The qualitative features of the EKG can be explained by a simple electrostatic model which ignores the electrical conconductivity of the body. This model is developed and used to explain some EKG patterns.

INTRODUCTION

In physics courses which are taken by pre-medical or biology students, it is a simple matter to extend the discussion of electrostatics to provide qualitative understanding of the electrical activity of the heart. If electrical conduction within the body is ignored (see Appendix), Coulomb's law can be used to show that the electrocardiogram (EKG) measures the potential distribution of a set of dipoles within the heart. This approximation is sufficient to explain the qualitative features of both normal and abnormal electrocardiograms. The EKG is seldom discussed in general physics courses because physicists usually are not familiar with it, although laboratory experiments based on it are becoming popular.[1,2] When the EKG is presented in medical school on the other hand, the simplicity of its relation to electrostatics is not considered. This paper will develop the electrostatic model relating the EKG to the electrical activity of the heart. It will also introduce the EKG to physics teachers and show how the model can be used to explain one example of a patho-logical change in the EKG. The arguments will be developed in sufficient detail so that a physics student can use the paper for self-study. Calculus notation will be used, but the discussion can be adapted for a noncalculus course.

DIPOLE LAYERS

Nerve and muscle cells are very similar in their electrical behavior.[3] A dipole layer on the cell wall is an important feature of each. In a muscle cell, changes in this layer precede contraction; in a nerve cell, similar changes occur as the nerve transmits a signal. Such nerve and muscle cells are cylindrical, with diameters ranging from 2×10^{-6} m to 5×10^{-4} m. Lengths range up to 1 m. The wall of each cell is an insulating layer about 5×10^{-9} m thick. In the resting state the cell is polarized, with a charge of $\pm 9 \times 10^{-4}$ C/m^2 distributed on each side of this layer. The potential inside the cell is about -90 mV with respect to the outside. As a nerve cell transmits a signal or a muscle cell prepares to contract, a wave of depolarization sweeps the length of the cell, causing the wall to become temporarily permeable to positive ions. These ions migrate into the cell, making the interior potential zero or somewhat positive.

The simplest model to consider for approximating the electrostatic properties of the cell is two infinite, plane, parallel double layers of charge (see Fig. 1). Each double layer consists of layers of uniform charge density $+\sigma$ and $-\sigma$ separated by a distance a. Application of Gauss' law shows that the electric field in the wall has magnitude σ/ϵ_0 and that the potential in the region between the double layers is therefore[4] $-\sigma a/\epsilon_0$. Similar relationships can be derived for spherical or cylindrical double layers.[5]

Let us now consider the dipole approximation to the potential from a double layer. Consider a dipole of moment $\mathbf{m}$, consisting of charges $\pm q$ separated by a distance a, as shown in Fig. 2(a). The field for $r \gg a$ is

$$V = (1/4\pi\epsilon_0)(\mathbf{m} \cdot \mathbf{r}/r^3) = (1/4\pi\epsilon_0)(m \cos\theta/r^2). \quad (1)$$

Reprinted from *American Journal of Physics* **41**, 824–831 (1973); © American Association of Physics Teachers.

Consider next a dipole consisting of surface charges of density $\pm\sigma$, spread over area dS and again separated by distance a [Fig. 2(b)]. The dipole moment is

$$dm = \sigma a\, dS = \tau\, dS$$

and the potential is

$$dV = (\sigma a/4\pi\epsilon_0)\,(dS\cos\theta/r^2)$$

$$= (\tau/4\pi\epsilon_0)\,(dS\cos\theta/r^2). \qquad (2)$$

Recalling the definition of a solid angle, $d\Omega = dS\cos\theta/r^2$, we can write this as

$$dV = \tau\, d\Omega/4\pi\epsilon_0. \qquad (3)$$

As long as we are a large distance from the double layer compared to its thickness, we can use this equation to calculate the potential by integration. The dipole moment need not be constant, but may vary over the surface. We must remember that there is an algebraic sign associated with $d\Omega$, which is the sign of $\mathbf{m}\cdot\mathbf{r}$.

Consider the application of this equation to the two double layers of Fig. 1. If we are at point A, $\mathbf{m}\cdot\mathbf{r}$ is negative. Since τ is uniform throughout the layer and an infinite sheet subtends a solid

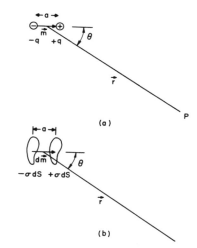

FIG. 2. Dipole of moment $\mathbf{m}$: (a) consisting of two point charges, (b) consisting of double layer of area dS.

angle 2π, the potential due to the left hand layer is

$$V_1 = -2\pi\tau/4\pi\epsilon_0.$$

For the right hand layer $\mathbf{m}\cdot\mathbf{r}$ is also negative and one gets a like contribution. Thus the potential at point A is

$$V = -\tau/\epsilon_0 = -\sigma a/\epsilon_0, \qquad (4)$$

which is in agreement with the earlier calculation. For point B, $\mathbf{m}\cdot\mathbf{r}$ is negative for the left hand double layer and positive for the right hand layer. Thus the potential at B is zero.

Whenever we have a uniform double layer of charge on a closed surface, the exterior potential will be zero. This can be seen from Fig. 3. As long as the surface is closed, for any region dS_1 which contributes to the potential there will be a corresponding region dS_2, subtending the same solid angle, for which $\mathbf{m}\cdot\mathbf{r}$ will have the opposite sign.

Consider now a nerve or muscle cell. We will neglect the fact that it is in a conducting medium and assume that it is in empty space. (The essential features of the results are unchanged by this assumption as discussed in the Appendix.[6]) In the quiescent state the cell may be regarded electrically as a long cylindrical double layer. As a signal passes along the cell, the cell wall becomes permeable to the positive charge on the

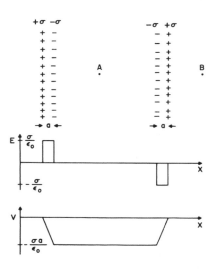

FIG. 1. Electric field and potential for two infinite, parallel, double layers, with charge $\pm\sigma$ per unit area and separation a.

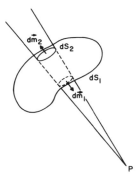

FIG. 3. The potential at an exterior point due to a uniform charged double layer on a closed surface is zero, due to cancellation of contributions for dS_1 and dS_2.

outside. These charges rush in and cause the potential difference to become zero or positive. This wave of depolarization sweeps along the cell at a velocity which is characteristic of the cell. It is about 0.3 m/sec in heart muscle, and 4 m/sec in some nerves in the heart.[7,8] Consider a cell which is depolarized for part of its length, as shown in Fig. 4(a). We wish to find the potential at some exterior point, due to the charges to the right of plane AA. The charge distribution in Fig. 4(a) is a superposition of those shown in Figs. 4(b)–(c). That in Fig. 4(b) will have no exterior potential; thus the desired potential is that due to the double layer "cap" of Fig. 4(c).

ELECTRICAL ACTIVITY OF THE HEART

The heart is an interesting example of a free-running relaxation oscillator.[9] If a heart is removed from an animal and bathed in a nutrient and oxygen supply, it will continue to beat spontaneously. Before a muscle cell contracts, a wave of depolarization sweeps over it. The contraction of the heart is initiated by spontaneous depolarization of some specialized nerve fibers located in the right atrium and called the *sino-auricular node* (SA node). The SA node acts like a free-running relaxation oscillator whose rate is accelerated by the sympathetic nerves to the heart (which release noradrenaline) and slowed by the parasympathetic nerves (which release acetylcholine).[10] Once the SA node has fired, the depolarization sweeps across both atria. There are interconnections between adjacent muscle cells, as well as a

network of faster nerve fibers which allow the depolarization to spread rapidly over the entire atrium.[11] The muscles of the atria are separated from those of the ventricle by fibrous connective tissue which does not transmit the electrochemical impulse. The only connection between the atria and ventricles is some nerve tissue called the *atrioventricular node* (AV node). (There is some controversy as to whether the AV node is stimulated by the adjacent atrial muscle or by direct pathways from the SA node.[10]) After passing through the AV node the depolarization spreads rapidly over the ventricles on bundles of nerve cells on the inner wall of the ventricles. The parts of this network are called the common bundle (or bundle of His), the left and right bundles, and the fine network of Purkinje fibers. The AV node will undergo free running relaxation oscillations at a rate of about 50 beats/min; it usually does not free-run but is triggered by the more rapid beating of the atria. However, in well-trained athletes, the resting pulse rate may be low enough to allow the AV node to fire spontaneously, giving rise occasionally to what are known as nodal escape beats. These are physiologic and no cause for concern.

Let us now consider how the dipole moment of the heart changes with time. Initially all the heart cells are completely polarized and have no net dipole moment. The cells begin to depolarize at the SA node and this wave sweeps across the

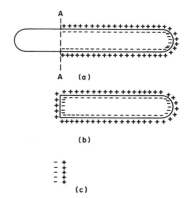

FIG 4. (a) A partially depolarized nerve or muscle cell. The charge distribution on the cell and the potential at any point may be regarded as a superposition of the closed double layer (b) and end cap (c) of opposite polarity.

atria. For each muscle cell, we can imagine a double-layer disc, perpendicular to the cell as in Fig. 4(c), with the dipole moment pointing towards the part of the cell which is still polarized. These discs for all the cells being depolarized constitute an advancing wave front.

We will now assume that we are far from the heart and neglect the fact that different dipoles

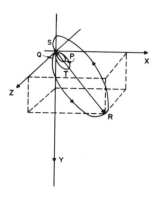

Fig. 6. Locus of the total dipole moment during the cardiac cycle.

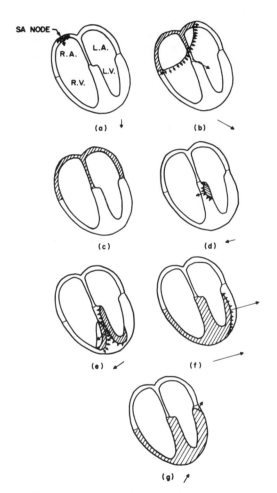

Fig. 5. Wave of depolarization sweeping over the heart. Atrial and ventricular muscle are not connected and contribute separately: (a) Depolarization beginning at SA node. (b) Atria nearly depolarized. (c) Atria completely depolarized. There is no net dipole moment. This state persists while the AV node conducts. (d) Beginning of depolarization of left ventricle. (e), (f) Continuing ventricular depolarization. (g) Ventricular depolarization nearly complete.

are different distances from the point of observation. Then we may speak of the instantaneous total dipole moment as the vector sum of the dipole moments for each cell. This total dipole moment is often referred to in the medical literature as the "electric force vector" or the "activity" of the heart. Figure 5 shows how this net dipole moment changes as the myocardium depolarizes.

The wave of depolarization first travels over the atria as shown in Figs. 5(a) and (b). When the atria are completely depolarized [Fig. 5(c)] they have no net dipole moment. After passing through the AV junction, the depolarization spreads out rapidly on the inner walls of the ventricles [Fig. 5(e)–(f)] and finally travels out through the wall of each ventricle.

Repolarization of the muscle does not spread as an electrochemical wave; rather it is a diffusive process which has not been studied as well as the depolarization process.[12]

The locus of the tip of the net dipole moment **m** in a typical case is shown in the perspective drawing of Fig. 6. The x axis points to the patient's left, the y axis towards his feet, and the z axis from back to front. The small loop labelled P occurs during atrial depolarization, QRS during ventricular depolarization, and T during ventricular repolarization. Atrial repolarization is masked by the ventricular depolarization. A plot of the x, y, z components of **m** as functions of time is shown in Fig. 7. These components are typical; there can be considerable variation in the direction of the various loops of Fig. 6.

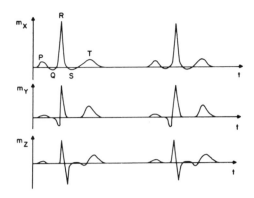

FIG. 7. The three components of the dipole moment as a function of time.

ELECTROCARDIOGRAM

Let us now consider how electrocardiographic measurements are made. We will continue to use a model in which conduction within the torso is neglected, and in which we can speak of a net dipole moment. The potential at position P due to the dipole moment $\mathbf{m}$ is then given by Eq. (1). The potential difference between two points located at $\mathbf{r}_1$ and $\mathbf{r}_2$ each a distance r from the dipole is then

$$V(\mathbf{r}_2 - \mathbf{r}_1) = \mathbf{m} \cdot (\mathbf{r}_2 - \mathbf{r}_1)/4\pi\epsilon_0 r^3$$

$$= \mathbf{m} \cdot \mathbf{R}/4\pi\epsilon_0 r^3.$$

Thus the potential difference between two electrodes separated by a displacement $\mathbf{R}$ and equidistant from the dipole will measure the instantaneous projection of m on R.

The standard EKG involves simultaneous measurement of twelve potential differences using 9 electrodes. This, of course, provides redundant information but does allow fairly simple interpretation. Recent work on vector electrocardiography has been directed towards finding a suitable set of leads to measure three mutually orthogonal components of $\mathbf{m}$ and to display the results in a way which can easily be interpreted.[13]

The first three electrodes are placed one on each wrist and one on the left leg. The limbs in this case serve as conductors, so the three points between which the potential differences are measured are the points where the limbs join the body.

The potential differences between these points correspond to projections of $\mathbf{m}$ on $\mathbf{R}_1$, $\mathbf{R}_2$, and $\mathbf{R}_3$ (assuming the 3 points are equidistant from the heart) as shown in Fig. 8(a). These projections of $\mathbf{m}$ are labelled I, II, and III, respectively, and are called the limb leads. If the three electrodes are assumed to measure the potentials at the vertices of an equilateral triangle, then these three components have been measured along axes which are 120° apart. It is often convenient to know the components along three more axes midway between these. They are linear combinations of the first three. The reference vectors which are traditionally used are shown in Fig. 8(b). The signal which is called aV_L (augmented V_L) is the projection of $\mathbf{m}$ on a vector pointing 30° above the horizontal. (It is called aV_L because the projection is in the direction of a vector from the center of

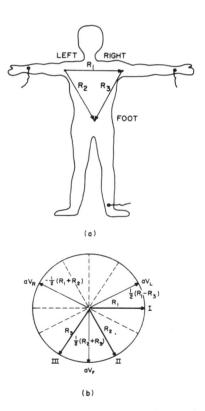

(a)

(b)

FIG. 8. (a) Position of the limb electrodes on the body, showing vectors $\mathbf{R}_1$, $\mathbf{R}_2$, and $\mathbf{R}_3$, along which the dipole moment is projected for tracings I, II and III. (b) Direction of primary projections I, II and III and derived projections aV_L, aV_R, and aV_F.

the equilateral triangle to the left shoulder.) Since I is proportional to $\mathbf{m} \cdot \mathbf{R}_1$, etc., aV_L is $(I-II)/2$. The other two augmented limb leads are aV_R and aV_F. Their directions are shown in Fig. 8(b). All six of these leads measure projections of $\mathbf{m}$ on a plane parallel to the chest and back, called the frontal plane.

The other six leads are placed across the chest in front of the heart and are called precordial leads. The reference potential for these leads is the sum of the potentials of the three limb electrodes, which corresponds to the potential at the center of the equilateral triangle of Fig. 8(a). Hence they measure very nearly horizontal components of $\mathbf{m}$. They are labelled V_1–V_6, as shown in Fig. 9. Since these leads are in some cases rather close to the heart, our assumption of constant r in the integral of Eq. (3) is not satisfied. These leads are influenced most strongly by those portions of the myocardium closest to them. Waveforms for the leads of a typical normal EKG are shown in Fig. 10. The six limb leads measure components of $\mathbf{m}$ every 30° in the xy plane of Fig. 6. In particular, I is the x component and aV_F is the y component. (The graph of aV_F in Fig. 10 is not identical to m in Fig. 7 because they are for different patients.) When $\mathbf{m}$ has its greatest magnitude it is nearly parallel to II. Thus II has the largest spike, and aV_L, which is perpendicular to it, has the smallest. The projection of $\mathbf{m}$ on aV_R is negative.

The precordial leads nominally measure components in a nearly horizontal plane, although the effect of $1/r^2$ is important. The vector is strongest in the direction of V_4 and V_5. V_1 and V_6 are nearly in opposite directions and have opposite signs. In all leads the repolarization spike follows much in the same course as the depolarization, although the amplitude and duration are different.

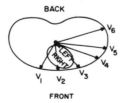

FIG. 9. Location of precordial leads and the direction of components they measure.

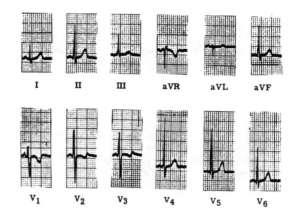

FIG. 10. Normal electrocardiogram. Large divisions are 0.5 mV vertically and 200 msec horizontally.

CHANGES IN THE ELECTROCARDIOGRAM

A detailed discussion of abnormalities in the electrocardiogram and the related pathology is beyond the scope of this paper. However, some changes can be mentioned as illustrative of the general principles involved.

The detailed shape of the QRS spike is dependent on the instantaneous vector sum of the moments in the left and right ventricles. If the right side of the heart has to pump against an abnormally heavy load for a long period of time, this exercise causes enlargement of the myocardium, a condition known as right ventricular hypertrophy. The electrocardiogram for a patient with right ventricular hypertropy is shown in Fig. 11. We see that I now has a negative spike because of the changed direction of m. In this case aV_F shows that there is very little vertical component of m. The precordial leads show the strongest signal in V_1 and V_2, because the right ventricle faces the front-right side of the body. (The extra lead V_4R is often recorded in such patients. It is symmetrical with V_4 but on the right side of the body.)

For further details on the interpretation of EKG tracings, the reader may consult many texts.[14,15] Some general principles may be mentioned. A fault in the conduction bundle, called a bundle branch block, means that the depolarization wave travels through the myocardium rather

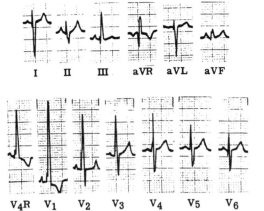

Fig. 11. Electrocardiogram of patient with right ventricular hypertrophy.

than the bundle. Since the velocity in the myocardium is low, the depolarization pulse is wider than usual. In a heart attack or myocardial infarct, a portion of the myocardium is dead and the cells neither depolarize nor repolarize. Therefore one may regard the dipole moment as being the normal one minus the dipole moment of the destroyed area. The exact nature of the changes will depend on the location of the infarct; in fact some very small infarcts are undetectable on the EKG.

SUMMARY

A simplified model of the heart suspended in a nonconducting medium has been used to relate EKG tracings to the charge distribution within the myocardium. This material is suitable for inclusion as an application of Coulomb's law in the general physics course, or as collateral reading for a laboratory experiment on the EKG. It provides a basic physical picture of the process, from which one can go on to include such effects as tissue conductivity. The model is used to correlate one example of an abnormal EKG pattern with the underlying pathological process.

ACKNOWLEDGMENTS

I wish to thank James Moller, MD, for supplying the elctrocardiograms used in Figs. 10 and 11. I am grateful to W. Albert Sullivan, MD, Assistant Dean of the University of Minnesota Medical School, for making the facilities of the Medical School available to me during the past two years.

APPENDIX

It may at first seem surprising to use an electrostatic approximation for the human body, which one usually thinks of as a conductor. An average value for its resistivity ω is 5 ohm-m.[6] However, in a medium for which the current density $\mathbf{J}$ is related to the electric field by $\mathbf{J} = \mathbf{E}/\omega$, the potential in source free regions still satisfies Laplace's equation.[16] Including sources gives

$$\nabla^2 V = \omega \nabla \cdot \mathbf{J}_i$$

where $\mathbf{J}_i$ is the source current. This problem has been discussed in detail.[6] One finds that, if the resistivity inside and outside the cell is uniform,[6]

$$V = (4\pi)^{-1} \int V_T d\Omega,$$

where V_T is the potential difference across the cell membrane ($V_{\text{outside}} - V_{\text{inside}}$). This is identical to the expression in the electrostatic case, which can be obtained by combining Eqs. (3)–(4). The charge sheets of our model are replaced by sources and sinks of current. The formalism for solving the problem in detail, considering the difference in resistivity between the cell and its membrane, is discussed in Ref. 6.

[1] H. Metcalf, Phys. Teach. **10**, 98 (1972).

[2] P. Lafrance, Phys. Teach. **10**, 462 (1972).

[3] B. Katz, *Nerve, Muscle and Synapse* (McGraw-Hill, New York, 1966).

[4] This discussion ignores the dielectric constant, κ, of the cell membrane. If it is included, the potential difference is $-\sigma a/\kappa \epsilon_0$. A typical value of κ is about 7.

[5] For the spherical double layer of radius R and thickness a, the interior potential is $V = -\sigma a R/\epsilon_0 (R-a)$ while for the cylindrical case it is $V = (\sigma a/\epsilon_0) \ln[1 - (a/R)]^{a/R}$.

[6] R. Plonsey, *Bioelectric Phenomena* (McGraw-Hill, New York, 1969), Chap. 5.

[7] *Physiology and Biophysics*, edited by T. C. Ruch and H. D. Patton (Saunders, Philadelphia, 1965), p. 576.

[8] S. Bellet, *Essentials of Cardiac Arrhythmias* (Saunders, Philadelphia, 1972), Chap. 2.

[9] A very early model of some coupled relaxation oscillators which mimicked the EKG was described by B. van der Pol and J. van der Mark, Phil. Mag. Ser. 7 **6**, 763 (1928).

A discussion of the biochemical mechanism of the relaxation oscillation is found in Ref. 8, p. 13.

[10] Katz, Ref. 3, pp. 106–107.

[11] Bellet, Ref. 8, Chap. 1.

[12] Ruch and Patton, Ref. 7, p. 582; Bellet, Ref. 8, p.12.

[13] Plonsey, Ref. 6, Chap. 6.

[14] J. R. Beckwith, *Grant's Clincial Electrocardiography: A Spatial Vector Approach* (McGraw-Hill, New York, 1970), 2nd ed.

[15] T. Winsor, *The Electrocardiogram in Myocardial Infarction* (Ciba Pharmaceutical Co., Summit, N.J., 1968).

[16] W. R. Smythe, *Static and Dynamic Electricity* (McGraw-Hill, New York, 1950), p. 231.

Improved explanation of the electrocardiogram

Russell K. Hobbie

School of Physics and Astronomy, University of Minnesota, Minneapolis, Minnesota 55455

(Received 6 September 1983; accepted for publication 13 December 1983)

A simple model is presented which is used to calculate the potential outside a nerve or muscle fiber embedded in a uniform conducting medium. The resulting integral is evaluated in the dipole approximation and agrees with equations which have been in the literature for a long time. The result can be used to interpret the electrocardiogram.

I. INTRODUCTION

Several years ago[1,2] I described a model which related the electrocardiogram waveform to the electrical changes taking place in the cells of the heart muscle (myocardium). The model assumed that the source of the electrocardiogram was static charges in a body which consists of empty space. Although some discussion was given of the relation of this model to the actual currents, students still have difficulty believing that the body can be approximated by a vacuum. It is not much more difficult to use a more satisfying model which recognizes the fact that the signal is due to superposition of currents which flow in the body as a result of the electrical changes in each myocardial cell.

The model is based on the calculation of the potential outside a cell which is stretched along the x axis and embedded in an infinite, homogeneous conductor. First, the current from a segment of the cell into the infinite conducting medium is related to the potential distribution inside the cell. Then the potential distribution in the medium is calculated by superposing solutions for current injected into a homogeneous conducting medium at different points along the x axis.

The result agrees with calculations in the literature which are obtained by solving Maxwell's equations with inductive and capacitive effects and propagation times ignored. This approach, by ignoring those effects when the model is constructed, makes the calculation accessible to students who have had only one year of calculus.

II. CURRENT FROM A SEGMENT OF A CELL

A resting nerve or muscle cell has an interior potential which is about 80 mV less than outside. The electric field across the cell membrane is generated by a layer of negative charge on the inside and a positive layer of charge on the outside of the membrane. As the nerve cell conducts or the muscle cell prepares to contract, an electrochemical wave of depolarization and subsequent repolarization travels along the cell.[2,3]

Consider a single cell stretched along the x axis. The current density inside can be assumed to be uniform,[4] with total current $i_i(x, t)$ flowing in the x direction. Figure 1 shows a segment of the cell between x and $x + dx$. The current into the segment across surface A is $i_i(x, t)$; current $i_i(x + dx, t)$ crosses surface A'. The difference between these is a current crossing surface AA'. Part of this current flows through the (shaded) membrane. The remainder charges the membrane capacitance:

$$i_i(x, t) - i_i(x + dx, t) = (2\pi a\, dx)\left[c_m\left(\frac{\partial v}{\partial t}\right) + j_m\right]. \quad (1)$$

Here a is the cell radius, c_m the membrane capacitance per unit area, v the potential across the cell membrane, and j_m the membrane current density. Note that surface AA' is just inside the membrane. The two layers of polarization charge are between AA' and BB'. A current di_0 out through surface BB' is made up of the membrane charging (displacement) current and j_m.

$$di_0 = -\left(\frac{\partial i_i(x, t)}{\partial x}\right) dx. \quad (2)$$

The potential across the membrane is $v(x, t) = v_i(x, t) - v_0(x, a, t)$, where $v_0(x, a, t)$ is measured just outside the membrane at radius a. Because the cell contents obey Ohm's law, the interior current is

$$i_i(x, t) = -\pi a^2 \sigma_i\left(\frac{\partial v_i}{\partial x}\right), \quad (3)$$

where σ_i is the conductivity of the material in the cell. The exterior current is

$$di_0 = \pi a^2 \sigma_i\left(\frac{\partial^2 v_i}{\partial x^2}\right) dx. \quad (4)$$

III. POTENTIAL OUTSIDE THE CELL

Consider now the current in the external medium. Imagine that the radius of the cell is so small that the influence of the cell can be replaced by a current distribution $di_0(x)$ along the x axis.

If current di_0 is injected at the origin into an infinite homogeneous conductor of conductivity σ_0, the current in the medium will be directed radially outward and will have spherical symmetry. The current density at distance r will be $j = di_0/4\pi r^2$, the electric field will be j/σ_0, and the potential (with $v = 0$ taken at infinity) will be

$$v(r) = di_0/4\pi\sigma_0\, r. \quad (5)$$

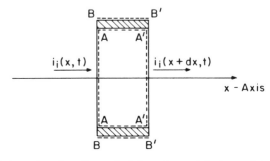

Fig. 1. A portion of a cell stretched along the x axis is shown. It is sliced parallel to the axes of the cylinder, with the cut portions of the membrane shaded. An interior current $i_i(t)$ flows along the axis.

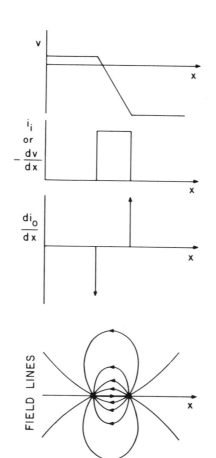

Fig. 2. The potential, current in the cell, current per unit length injected into the external medium, and lines of electric field and current flow in the external medium are shown. For typical cells, the transition takes place in from 1 to 5 mm.

For a distribution of currents $di_0(x)$ along the x axis, the exterior potential at a point far from the cell can be obtained by superposition:

$$v_{ext} = \int dv_{ext} = \int \frac{di_0}{4\pi\sigma_0 \, r}. \tag{6}$$

Using Eq. (4), the potential at a point r_0 is

$$v_{ext}(r_0) = \frac{\pi a^2 \sigma_i}{4\pi\sigma_0} \int \frac{1}{r} \left(\frac{\partial^2 v_i}{\partial x^2} \right) dx. \tag{7}$$

Since r depends on both x and r_0, the coordinate of the observation point, this integral is evaluated by expanding $1/r$ [e.g., Ref. 2, Eq. (7.2)] and keeping only the first-order (dipole) term:

$$\frac{1}{r} \approx \frac{1}{r_0} \left(1 + \frac{x}{r_0} \cos\theta \right). \tag{8}$$

When this is inserted in Eq. (7), the result is

$$v_{ext}(r_0) = \frac{\pi a^2 \sigma_i}{4\pi r_0 \sigma_0} \int_{x_1}^{x_2} \left[\left(\frac{\partial^2 v_i}{\partial x^2} \right) + \frac{x \cos\theta}{r_0} \left(\frac{\partial^2 v_i}{\partial x^2} \right) \right] dx. \tag{9}$$

The first term is proportional to $\partial v_i / \partial x$ evaluated at the end points. The second term is integrated by parts. If x_1 and x_2 are taken to be at points where $\partial v_i / \partial x$ is zero, then

$$v_{ext}(r_0) = \frac{\pi a^2 \cos\theta \, (\sigma_i/\sigma_0)}{4\pi r_0^2} [v(x_1) - v(x_2)]. \tag{10}$$

One can define a vector $\mathbf{p}$ pointing along the cell from x_1 to x_2 with magnitude

$$p = \pi a^2 (\sigma_i/\sigma_0)[v(x_1) - v(x_2)] \tag{11}$$

and write the exterior potential as

$$v_{ext}(\mathbf{r}_0) = \mathbf{p} \cdot \mathbf{r}_0 / 4\pi r_0^3. \tag{12}$$

This has the same form as the potential due to a dipole. Vector $\mathbf{p}$ is called the *electric force vector* by physiologists. "Electrical activity" has the advantage of not suggesting that it is a force. Figure 2 shows the electric field lines outside a cell. One can see that they are the same as those due to a dipole.

The region on the cell where the potential changes often occupies only a small region along the x axis. Therefore we can also define $\Delta\Omega = \pi a^2 \cos\theta / r_0^2$ to be the solid angle of a cross section of the cell where the potential changes, subtended at the point where v_{ext} is measured and write

$$v_{ext} = \left(\frac{\Delta\Omega}{4\pi} \right) \left(\frac{\sigma_i}{\sigma_0} \right) [v(x_1) - v(x_2)]. \tag{13}$$

This is the analog of Eq. (7.4) in Ref. 2 or of Eq. (3) or the second equation in the Appendix in Ref. 1, or of Eq. (5.68) in Ref. 5.

IV. PHYSIOLOGICAL EXAMPLES

In nerve and muscle cells, the initial depolarization is accompanied by a potential change from -80 mV to some small positive value. This occurs over 1–5 mm along the cell. A schematic plot of the potential change, the current in the cell, the current through the membrane, and the electric field lines outside the cell is given in Fig. 2. One sees the dipolelike field associated with the advancing wave front. For a nerve cell and most muscle cells, the cell immediately repolarizes, so that there is another vector $\mathbf{p}$ pointing in the opposite direction, right behind the first one. Thus there is no exterior potential in the dipole approximation. For a myocardial cell, the depolarization lasts for ~100 ms, and the entire cell is completely depolarized for some time. In that case, $v(x_1)$ is 0, $v(x_2)$ is -80 mV, and a relatively large signal exists. In the heart, many cells depolarize at the same time, with the wave of depolarization sweeping through the myocardium, so that $\Delta\Omega$ becomes the solid angle subtended by the entire wave front.

V. CONCLUSION

The currents which give rise to the electrocardiogram have been described in a model which treats the body as an infinite homogeneous conductor. This model is more satisfying to students than the electrostatic model which was proposed earlier. The discussion in Refs. 1 or 2 can be used with Eq. (11) to explain the details of the electrocardiogram signal.

[1] R. K. Hobbie, Am. J. Phys. **41**, 824 (1973).
[2] R. K. Hobbie, *Intermediate Physics for Medicine and Biology* (Wiley, New York, 1978).
[3] R. K. Hobbie, Am. J. Phys. **41**, 1176 (1973).
[4] A. C. Scott, Rev. Mod. Phys. **47**, 487 (1975).
[5] R. Plonsey, *Bioelectric Phenomena* (McGraw-Hill, New York, 1969).

BURSTING, BEATING, AND CHAOS IN AN EXCITABLE MEMBRANE MODEL

TERESA REE CHAY* AND JOHN RINZEL‡

*Department of Biological Sciences, University of Pittsburgh, Pittsburgh, Pennsylvania; and
‡Mathematical Research Branch, National Institute of Arthritis, Diabetes, Digestive and Kidney
Diseases, National Institutes of Health, Bethesda, Maryland 20205

ABSTRACT   We have studied periodic as well as aperiodic behavior in the self-sustained oscillations exhibited by the Hodgkin-Huxley type model of Chay, T. R., and J. Keizer (*Biophys. J.*, 1983, 42:181–190) for the pancreatic β-cell. Numerical solutions reveal a variety of patterns as the glucose-dependent parameter k_{Ca} is varied. These include regimes of periodic beating (continuous spiking) and bursting modes and, in the transition between these modes, aperiodic responses. Such aperiodic behavior for a nonrandom system has been called deterministic chaos and is characterized by distinguishing features found in previous studies of chaos in nonbiophysical systems and here identified for an (endogenously active) excitable membrane model. To parallel the successful analysis of chaos in other physical/chemical contexts we introduce a simplified, but quantitative, one-variable, discrete-time representation of the dynamics. It describes the evolution of intracellular calcium (which activates a potassium conductance) from one spike upstroke to the next and exhibits the various modes of behavior.

INTRODUCTION

A wide variety of oscillation patterns have been observed in membrane potential recordings from β-cells of isolated pancreatic islets. Atwater et al. (1) found that as glucose conentration is increased the response evolves from a steady state of polarization into a bursting pattern with interburst period on the order of seconds. As glucose is increased adequately the bursting gives way to continuous spiking or beating. At very high glucose concentrations one finds a steady depolarization. These features are also found in the theoretical model of Chay and Keizer (2), a model of the Hodgkin-Huxley (3) type in which the depolarizing current is due to calcium rather than sodium ions.

Here, we examine in greater detail the transition between the beating and bursting modes for this theoretical model. We have found that for certain parameter ranges the transition regime is characterized by aperiodic responses—irregular bursting or irregular spiking. We emphasize that the model is deterministic so that this irregularity in the response cannot be explained in terms of any stochastic feature of the biophysical formulation. Rather, it reflects a not uncommon phenomenon in nonlinear dynamical systems often referred to as deterministic chaos.

Aperiodic dynamics and the transition from periodic to aperiodic behavior in other contexts has been studied ambitiously in recent years by chemists, physicists, and mathematicians. Some applications include the transition to turbulence in fluid dynamics (4), sustained oscillations in open chemical systems (5, 6), dynamics of nonlinear

electronic circuits (7), and models of population dynamics in ecology (8). The literature in this area is vast; some recent reviews are (9–12). Experiments and theory have identified some distinguishing routes to chaos, i.e., characteristic sequences of successively more complex patterns as a parameter is adjusted. For example, one may observe a predictable heirarchy of doublet patterns (period-doubling cascade) as in (6, 8, 11). Another possibility is intermittency (13) in which long periods of relative regularity are abruptly and irregularly interrupted by bursts of quite different activity; the mean duration of regular periods grows predictably as a critical parameter value is approached. A useful quantitative tool for studying such chaotic behavior is a discrete time one-variable representation of the observed dynamics (e.g., see reference 5–8, 10–20).

Aperiodic responses have been observed both theoretically and experimentally for excitable cells driven by a sinusoidal stimulating current (14–16, 26). Period-multupling of the cardiac rhythm under drug application has been reported (18). Yet we know of no previous systematic identification of chaos in a model for a nondriven (endogenous) electrically excitable membrane system. Here we report that the Chay-Keizer model exhibits aperiodic responses as the parameter k_{Ca} (the rate of intracellular Ca^{++} uptake) is varied; decreasing this parameter in the model represents decreasing the glucose concentration. We find as k_{Ca} is decreased that loss of the regular beating mode is accompanied by a sequence of period-doublings which is succeeded by chaotic spiking. As k_{Ca} is decreased further we find that bursts of spikes begin to form.

Reprinted from *Biophysical Journal* **47**, 357–366 (1985); © Biophysical Society.

However, these bursting patterns are aperiodic until k_{Ca} decreases adequately. We have supported and complemented these findings by a description of the dynamics in terms of a one-variable, discrete-time model that relates the level of intracellular calcium from one spike upstroke to the next. This map and its solution behavior correlate well with the full five-variable Chay-Keizer model. Although we present quantitative results, our analysis and interpretations are qualitative and we expect that the mechanism for aperiodic behavior which we have found may be operative in other excitable system.

THE MODEL AND ITS DYNAMIC BEHAVIOR

The β-cell model of Chay and Keizer (2) consists of five first-order, simultaneous, nonlinear equations containing the following dynamic variables: (a) n, the degree of activation for the voltage-dependent K$^+$-channel; (b) m, the degree of activation for the voltage-dependent Ca^{++}-channel; (c) h, the inactivation for the same channel; (d) the intracellular calcium concentration, denoted by Ca, whose rate equation consists of the inward calcium current and a first-order disappearance of Ca, written as k_{Ca} [Ca], where k_{Ca} is the rate constant; and (e) the membrane potential, V, whose time derivative is proportional to the sum of ionic currents carried by Ca^{++} and K$^+$ ions.

The differential equations representing these five dynamic variables were solved numerically on a DEC-10 computer (Digital Equipment Corp., Marlboro, MA) with a Gear algorithm, with absolute and relative error tolerances set at 10^{-7}. The parametric values used for our computations were taken from Table 1 of Chay and Keizer (2) with the exception of temperature that is taken here to be 17°C.[1] At this temperature, the plateau membrane potential of the active phase (i.e., spiking phase) becomes very close to the minimum repolarizing membrane potential of the silent phase. Period doublings and chaos shown below occur in the parameter region where the bursting mode becomes the beating mode.

Fig. 1 shows an overall view of the bursting-spiking patterns of the Chay-Keizer model as k_{Ca} is increased from 0.038 to 0.045 ms^{-1}. The left column of plots shows the time course of membrane potential and the right column that of intracellular calcium concentration. The uppermost panels (for k_{Ca} = 0.038 ms^{-1}) exhibit a periodic bursting pattern with the following qualitative features. The spikes appear to ride on a plateau potential and the interspike interval increases dramatically near the end of the burst. With each spike there is a net increase in calcium concentration and thereby further activation of the Ca-dependent potassium conductance. This rising conductance eventually inhibits the spiking mechanism and leads to the K$^+$-dominated silent phase. During the silent phase, calcium concentration decreases (through uptake mecha-

nisms) and a slow depolarization develops. When the membrane potential reaches threshold for spike initiation the active phase is rekindled. The computed burst pattern and this description closely resemble those for experimentally observed β-cell responses in the presence of glucose. A noticeable quantitative difference is that for the temperature chosen here the burst has relatively few pulses (compare to Fig. 1 in reference 2).

The lowermost panels in Fig. 1 correspond to sufficient glucose (k_{Ca} = 0.045 ms^{-1}) to cause continuous spiking or beating. In this case there is zero net increment of calcium from one spike to the next; k_{Ca} is large enough so that calcium uptake by intracellular compartments during the interspike phase just balances the influx from membrane calcium current during the spike's depolarized phase. The two intermediate values of k_{Ca} result in aperiodic responses; even for much longer simulation times than shown in Fig. 1, we find no apparent periodicity. We refer descriptively to these two cases as chaotic bursting (k_{Ca} = 0.04 ms^{-1}) and chaotic spiking (k_{Ca} = 0.0415 ms^{-1}). Such irregular behavior is not due to numerical artifact nor to extraordinary parameter settings.

To obtain additional insight into the model's responses it is useful to view calcium as a dynamic parameter that regulates the faster time scale spiking mechanism and induces switching between the active and silent phases. This is illustrated graphically in the leftmost panels of Fig. 2, where the (five-variable) response trajectory is projected onto the Ca-V plane. Motion here is clockwise. Consider first the upper set of panels; these correspond to the uppermost case of Fig. 1. The periodic burst appears as a closed curve (upper left) whose upper elongated loops are the spikes and whose lower flat portion represents the silent phase. The triple-branched, dashed curve results from a pseudo–steady-state analysis of the model. When Ca is treated as a parameter the remaining variables V, m, h, n form an excitable subsystem that exhibits three different steady state or rest potentials (dashed curves in left panels) when k_{Ca} lies between the approximate values 0.315 and 1.884 μM. With calcium fixed (in this range) the lower steady potential is stable for the subsystem and the intermediate steady state has a saddle point structure with its associated threshold separatrix (21). The upper steady state in this example is not stable, as one might conjecture, but rather unstable. Moreover, for a subinterval of Ca values (0.321 $\geq$ Ca $\geq$ −0.713) it is surrounded (a vague but intuitive notion for the four-variable subsystem) by a stable Ca-dependent oscillation that corresponds to the repetitive spiking of the active phase. The minimum potential of this oscillation is relatively independent of Ca and this corresponds to the plateau potential of the active phase. Note, in this model, the upper steady state does not represent the plateau potential as it does in Plant's (22) treatment of the R-15 bursting pacemaker of *Aplysia*.

Now let us reconsider the burst pattern generated with Ca as a dynamic variable. As Ca increases from one spike

---

[1] Here the temperature 17°C corresponds to $3^{1.07}$.

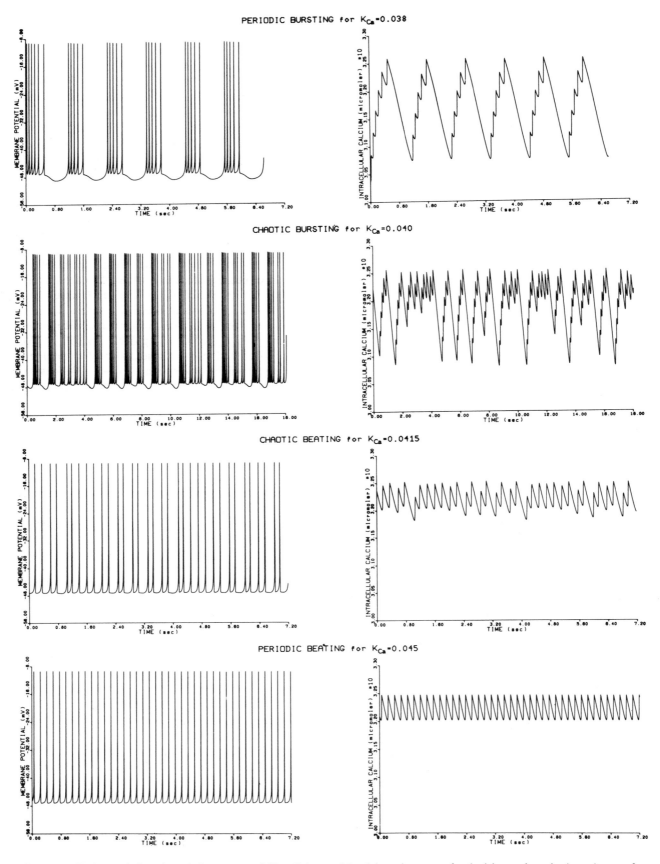

FIGURE 1   Various periodic and aperiodic responses of Chay-Keizer model and dependence upon k_{Ca} the (glucose-dependent) uptake rate of intracellular calcium. From *top* to *bottom* k_{Ca} – 0.038, 0.040, 0.0415, 0.045 ms^{-1}. *Left* column shows membrane potential vs. time; *right* column displays calcium concentration vs. time. Voltage and concentration scales are identical in all panels. Second row of panels has longer time duration (18 s) than others (7.2 s) to illustrate lack of periodicity in burst activity. Note that range of calcium variation in beating cases is much less than in bursting cases.

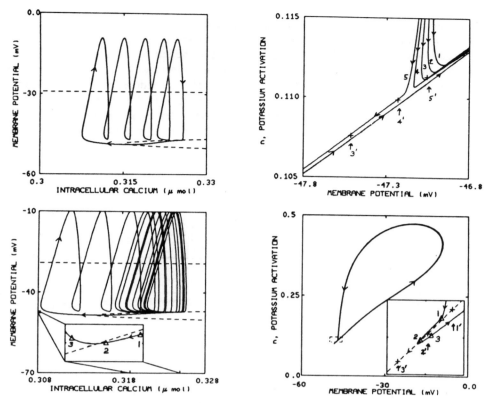

FIGURE 2   Phase space analysis of bursting based on two-variable projections of solutions to five-variable Chay-Keizer model. Two rows of panels here correspond to two top rows of Fig. 1: k_{Ca} = 0.038, 0.040 ms^{-1}. Dashed curves (in *left* panels) represent multiple steady-state potentials of model with calcium as fixed parameter. *Upper left* case shows the V vs. Ca closed orbit of periodic bursting. Silent phase begins when trajectory falls below threshold (*middle* dashed curve); active phase reentered after threshold and lower steady state coalesce and disappear. The corresponding n vs. V projection (*upper right*) illustrates that successive interspike trajectories pass closer to saddle point threshold (crosses, vertical arrows, and primed numbers) until downstroke of fifth spike falls below threshold. *Lower left* case illustrates how trajectory of chaotic bursting pattern reenters active phase prematurely (inset magnifies one example; inset region defined by lines from axes tics). The n vs. V projection (*lower right*) shows mechanism: downward drift of threshold actually overtakes polarizing spike trajectory (see *inset* magnification). Numbered triangles and crosses (*lower* panels) represent trajectory and threshold positions, respectively, at successive times.

to the next during the late active phase the threshold voltage and interspike plateau potential become closer. Finally, the burst terminates when the threshold exceeds the instantaneous plateau potential following the spike downstroke.

These features and the saddle point structure are revealed even more clearly in the upper-right panel of Fig. 2, which shows the spike trajectories of the burst projected onto the V-n plane. Here, we have magnified the interspike phase when voltage is near its plateau value. Successive spikes are numbered one through five. Crosses represent pseudo–steady-state values of V, n for the saddle point corresponding to the calcium level when the spike downstroke enters the picture from above. Because the third and fourth trajectories are above the threshold saddle, they are each followed by another spike. Note how these trajectories (especially the fourth) that pass close to the saddle exhibit the incoming and outgoing directions of that singular point. When the fifth spike downstroke enters the picture it is below threshold so the trajectory sweeps

downward abruptly, the burst terminates, and voltage decreases to its minimum during the silent phase. The active phase is not reentered until after the trajectory passes beyond the dashed curve's knee where the threshold and lower pseudo-steady state coalesce. We remark that, although not for these parameter values, one typically sees the silent phase trajectory more closely track the lower pseudo-steady state of polarization. We further mention that from the above phase space analysis we can interpret more easily the decreased spike frequency near the end of the burst. This occurs because the interspike trajectory slows substantially when it passes close to the threshold saddle point.

Next, we focus on the lower panels of Fig. 2, which correspond to the chaotic bursting case of Fig. 1. A noticeable difference here from the periodic bursting case above is that the active phase is reentered prematurely, i.e., after the trajectory has fallen below threshold it crosses back above threshold before Ca has decreased to below the dashed curve knee. An example of this premature reentry is shown in the inset magnification. To understand this behavior we must examine more closely the dynamics of the Ca-dependent threshold migration and the subsystem trajectory following the spike down stroke. Again we exploit the $V\text{-}n$ projection (lower-right panel) where the many spikes of the aperiodic burst become overlaid. The pointed tail at the lower left of this trajectory corresponds to the interspike phases of near threshold and plateau

potential behavior. The inset magnification isolates one particular threshold crossing event; it corresponds to the inset of the left panel. The triangles represent time marks on the trajectory, whereas the crosses represent the positions of the saddle point determined by the calcium concentration at the corresponding times. At the first two times the trajectory is below threshold. However, the saddle point is migrating leftward faster than the $V\text{-}n$

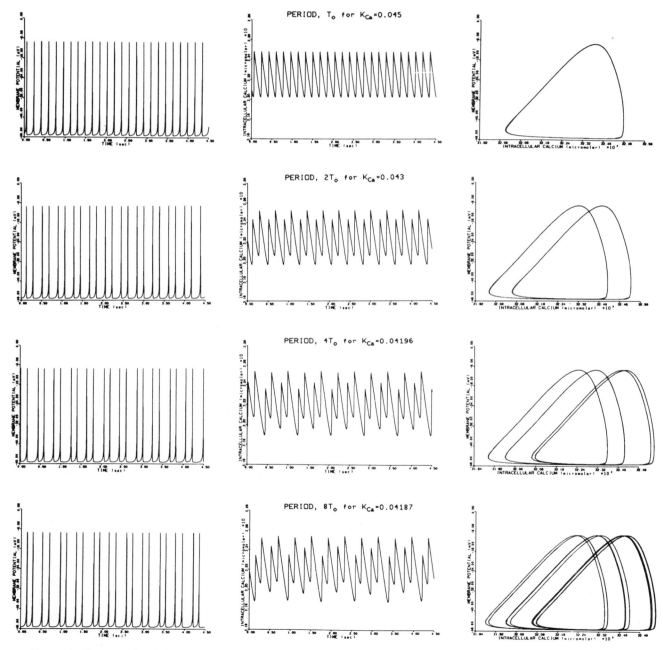

FIGURE 3   Sequence of period-doubling patterns as k_{Ca} is decreased from 0.045 ms^{-1} (the case of periodic beating). *Left* column: membrane potential vs. time; *middle* column: calcium concentration vs. time; *right* column: membrane potential vs. calcium concentration. Voltage time course shows spike doublets, whereas *middle* and *right* columns reveal greater detail of patterns. All panels in a given column have same scale.

trajectory and overtakes it before the third time mark. When this happens the V-n trajectory turns around, a spike upstroke is initiated, and the active phase is reentered prematurely. A major factor for this occurrence is the similar slow time scales of the voltage trajectory and the Ca-dependent saddle point migration. For parameter settings in which the repolarization of the silent phase is very rapid, the threshold has little opportunity to overtake the V-n trajectory and thereby cause premature reentry. Here, the repolarization is slow because the threshold and lower pseudo-steady state are very close together (so $|dV/dt|$ is small when the trajectory passes between them). Also, the larger value of k_{Ca} makes the threshold migration relatively faster, which contributes to the opportunity for reentry.

In Fig. 3 we illustrate another aspect of the model's behavior that complements our observations of chaos in this system. Here the starting point (upper row) is the periodic beating pattern from Fig. 1. As k_{Ca} is then decreased (proceed downward in Fig. 3), we observe a cascade of period doublings; the approximate periods are $T_0, 2T_0, 4T_0, 8T_0$, where T_0 is the interspike interval of the basic beating pattern in the upper panels. Although the voltage time courses show little variation in the doublet patterns from one case to the next, the calcium records (middle column) reveal the fine structure of the successive period doublings. The Ca-V projections (right column) show vividly how a closed orbit bifurcates and is succeeded by a closed orbit with twice as many loops. At bifurcation, the period of the observed response doubles exactly. This bifurcation phenomenon is caused by loss of stability of a periodic solution and we shall discuss it further in the following section. This period-doubling cascade is evidently the route that leads to the chaotic beating pattern of Fig. 1.

APPROXIMATION BY A ONE-VARIABLE DYNAMIC MODEL

To motivate a one-variable dynamic model we again adopt the view of calcium concentration as a regulating dynamic parameter. We observe that throughout the active phase the upstroke-downstroke segments of the successive spikes are nearly identical (e.g., see the V-n trajectories in lower right of Fig. 2). Thus, because V, m, h, n follow essentially the same trajectory during each spike upstroke we may describe approximately the dynamics of this five-variable system by considering the changes in the calcium level from one upstroke to the next. For this we let C denote the calcium concentration at the -45-mV upcrossing of a spike. Then by using the typical values of m, h, n for the -45-mV upcrossing, along with $V = -45$ mV and Ca $= C$, as initial conditions we integrate numerically the five Chay-Keizer equations to obtain the calcium concentration $F(C)$ at the next -45-mV upcrossing. By considering a range of C values we thus generate the graph of $F(C)$. The evolution of C_n the calcium level at the nth spike upstroke to C_{n+1} is then described by the difference equation (or discrete map):

$$C_{n+1} = F(C_n). \tag{1}$$

We remark that although time is discrete in this formulation the implicit time increments are not equal because the interspike interval increases with Ca throughout the active phase. The approach we follow here is similar to that employed in (19, 20) to describe complex oscillation patterns of a chemical system.

In Fig. 4 we present the maps (shown dotted) corresponding to the four period-doubling examples of Fig. 3. They are qualitatively similar: a single-humped curve that crosses the 45° ray or 1:1 line exactly once. Where F is above the 1:1 line, there is a net increase of calcium with each spike. For adequate C_n the curve falls below the 1:1 line and this corresponds to continuous trajectories that fall below threshold after a spike but then cross back above threshold (prematurely) without following a long silent phase. This portion of the right branch of $F(C)$ is steep because the reentry phenomenon occurs for only a narrow range of calcium levels. Where F crosses the 1:1 line, $C_{n+1} = C_n$ and this fixed point corresponds to a discrete periodic solution of Eq 1. This fixed point of the upper left panel approximates the periodic beating response of the continuous model (Fig. 1 bottom and Fig. 3 top). Each of the other panels of Fig. 4 exhibits a similar fixed point, however, in those cases it is unstable and so does not represent stable periodic beating. Note that a fixed point of a discrete dynamical model like Eq. 1 is stable if $|dF/dC| < 1$ (unstable if $|dF/dC| > 1$), where dF/dC is the slope of F at the fixed point. Because this slope becomes steeply negative as k_{Ca} decreases in Fig. 4 the fixed point becomes unstable.

The following features of single-humped maps are now well described in the literature (see reference 8 and the reviews and references in 9–12). When the fixed point becomes unstable a pair of nearby points (one on each side) is born to form a stable 2 cycle (upper-right panel). As the parameter and map are varied further one finds a hierarchy of destabilizations and bifurcations to stable cycles of period 2^n that eventually leads to a regime of chaos.

The stable periodic 2^n-cycles of the map are the discrete analogs of the period-doubled solutions of Fig. 3. We have indicated by crosses on these maps the calcium values at the -45-mV upcrossings for the continuous solutions from Fig. 3 and these points fall on the map. For illustration we have indicated the presumptive discrete trajectories for the $2T_0$ and $4T_0$ cases. Successive iterates C_n alternate to the right and left of the fixed point and are associated with long and shorter succeeding interspike intervals, respectively; this corresponds to the doublet appearance of the continuous voltage records. We have also numerically iterated the map according to Eq. 1 and thereby verified convergence of the discrete solution to the corresponding discrete $2^0, 2^1, 2^2$, and 2^3 cycles.

Next we interpret the patterns of Fig. 1 in terms of the

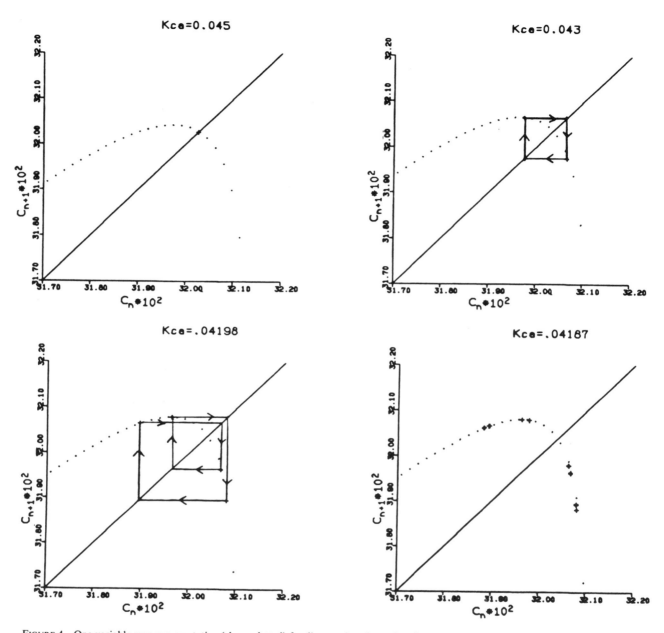

FIGURE 4   One-variable map representation (shown dotted) for discrete time dynamics of calcium concentration from one spike upstroke to the next (see description in text and Eq. 1). Four panels here correspond to period-doubling sequence of Fig. 3. Crosses represent calcium concentration at −45-mV upcrossing of continuous solution of Fig. 3. Fixed point of map (intersection with 45° line) corresponds to stable periodic beating in one case ($k_{Ca} = 0.04$ ms^{-1}); but in other cases fixed point is unstable and corresponds to unstable periodic solution of Chay-Keizer model.

map. For this the graph of F has been computed over a larger range of C values; computed maps are shown in Fig. 5 along with corresponding calcium data (shown as crosses) from the continuous solutions. Again the maps are qualitatively similar but now we clearly see evidence of the silent phase. For large enough C_n there is a large drop in calcium before the next upstroke. This reentry to the active phase occurs in a characteristic way after the voltage threshold and pseudo–steady-state potential of the silent phase have coalesced (see dashed curves of the Ca-V panels

of Fig. 2). Moreover, this reentry point is relatively independent of the initial calcium level when it is high enough. Hence the map is flat for large C_n. As in Fig. 4, the steeply sloping right branch corresponds to premature reentry that occurs for any value of k_{Ca} but only for a narrow range of C_n values.

The effect of decreasing k_{Ca} is to increase the net increment of calcium from one spike to the next and thereby basically to move the map upward. This moves the fixed point onto the steep portion of the right branch; the

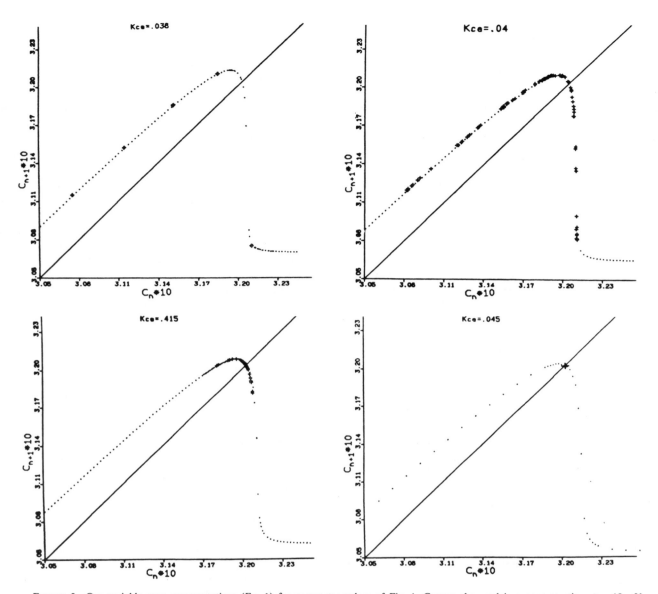

FIGURE 5   One-variable map representations (Eq. 1) for parameter values of Fig. 1. Crosses show calcium concentration at −45-mV upcrossing of continuous solutions of Fig. 1. For example, the case of periodic beating is represented by five points in upper *left* panel. The chaotic patterns nearly fill out subintervals over which the map is defined.

periodic beating pattern becomes unstable. To understand the chaotic patterns we notice carefully what action the map has on points near the relatively flat hump. Suppose such points are mapped onto the destabilizing steep right branch. Intuitively, any perturbation on such a C_n places a large uncertainty on the successor C_{n+1}. A periodic cycle containing such points is likely unstable. In the case of chaotic beating the hump is mapped only part way down the negative slope region and we see only fine-structured, locally contained chaos. For the case of $k_{Ca} = 0.04$ ms^{-1} however, points near the hump are mapped onto nearly the entire length of the destabilizing branch that in turn sends points over a large range of the left branch. Hence we observe premature reentry over a large range of calcium (lower left of Fig. 2) and the aperiodic response has a

bursting appearance. Finally, for $k_{Ca} = 0.038$ ms^{-1}, the map ascends sufficiently for the hump to be mapped onto the lower flat portion and we observe the stable five-spike burst pattern. If k_{Ca} were decreased even further the left branch would move higher corresponding to greater net increment in calcium per spike. Hence fewer spikes would occur during the active phase. For $k_{Ca} = 0.035$ ms^{-1}, we find a periodic four-spike bursting pattern. There may, but not necessarily, be a narrow parameter range of fine-structured period-doubling and/or chaos between the five- and four-spike regimes.

DISCUSSION

We have examined some aspects of a model for excitable membrane electrical oscillations. We have shown that the

Chay-Keizer model for the pancreatic β-cell exhibits aperiodic responses in the parameter range for transition between the bursting and beating modes of activity. This dynamic behavior was illustrated numerically and analyzed via phase space concepts (threshold separatrices, multiple steady states, closed orbits, and reduced subsystems) and in terms of one-variable discrete maps. The success of these approaches rests on the biophysical motivation that calcium plays the influential role of a regulating dynamic parameter. This allows for a pseudo–steady-state analysis and also includes consideration of the subsystem's dynamics. For the discrete model, calcium is the only dynamic variable but the spike action potential and slow calcium uptake mechanisms are both accounted for in this description of the evolution of calcium level from one spike upstroke to the next.

The map representation is computed numerically here although, based on the phase space analysis, its general form could be qualitatively deduced. It allows a compact and simplified way to understand the beating and bursting modes and also the chaotic transition behavior. When k_{Ca} is high there is a smaller net increment in calcium from one spike to the next. This means the map's left branch is close to the 1:1 line and the fixed point is stable (Fig. 5, lower right). When k_{Ca} is low the net increment is greater (because of less uptake between spikes) and the map is higher. Then the fixed point corresponding to periodic beating is unstable and trajectories visit both the left branch and the lower flat right portion to develop a periodic bursting pattern (Fig. 5, upper left). For intermediate parameter values the map sends an interval of points near its maximum to the destabilizing steeply descending right branch and hence aperiodic behavior is observed.

The continuous model and map will maintain their qualitative features over a range of parameter variations so that one may expect aperiodic behavior elsewhere in parameter space. In some regimes, the parameter range for aperiodic activity may be very small or even nonexistent, for example, if transition behavior is dominated by hysteresis in which parameter intervals for different response patterns are overlapping (e.g., see reference 20). Here, we have adjusted the temperature to enhance the k_{Ca} range over which chaos is found. In this case, the spiking plateau potential is close to the silent phase minimum potential. We suggest that this may provide at least one observable signature by which to guide parameter selection for possible systematic experimental investigation of aperiodic phenomena in the transition between bursting and beating. While irregular activity has been observed in the β-cell preparation (e.g., see Figs. 6–7 of reference 23), it has not been studied specifically. In this regard, we note that the conditions described above may be induced in the β-cell system by applying TEA and lowering temperature (I. Atwater, personal communication).

Convincing detection of deterministic chaos is facilitated if parameters can be controlled adequately and appropriate

variables can be measured experimentally to identify (via systematic and slow parameter tuning) features of a known route to chaos such as period doubling. In some situations, the easily accessible variables are sufficiently revealing. In cardiac rhythm studies for example, recordings of membrane potential in the case of in vitro experiments (14) or arterial blood pressure for intact animals (18) have been useful. In our theoretical studies the voltage time courses of quite similar doublet patterns do not reflect visually the details of period-doubling.[2] In such a case recordings of the intracellular calcium level would be quite helpful. However, for some preparations, this experimental possibility awaits further technical development. When such recordings are available then one might even hope to generate an experimental one-variable map analogous to the theoretical ones computed here. We believe that for strongly calcium-dependent systems it is insightful to formulate simplified one-variable models in terms of calcium.

The mechanism for the oscillations explored here may be summarized roughly as follows. There is an underlying subsystem that exhibits some features of excitability and that has multiple steady and/or oscillatory states over a range of values for an identified parameter. This parameter is a dynamic variable (often acting on a slower time scale) for the full system that modulates the subsystem behavior. It exhibits net increase or decrease depending on which pseudo steady or oscillatory state of the subsystem is currently being expressed. When sufficient increases or decreases have accumulated, then activity switches to a different state because the current state of the subsystem has destabilized or disappeared (through coalescence with another state). This general interpretation may be applied in a variety of systems (19, 20, 22).[3] Consequently, such oscillatory systems may exhibit aperiodic behavior especially in parameter regions of transition behavior. Numerical simulations (not described here) of a bursting neuron model (24) also reveal aperiodic solutions. We remark that the models themselves need not be complicated; three variables are sufficient for continuous membrane models (25).

The aperiodic behavior that we report is inherent in the deterministic Chay-Keizer model. An experimental preparation or low-precision numerical calculation may be subject to detectable random noise. Hence, both deterministic and stochastic factors may contribute to observed aperiodicity. (We have verified the deterministic cause for the Chay-Keizer model by comparing our numerical results to

---

[2]To improve identifiability one might use plots (as described in reference 11) of $V(t + T)$ vs. $V(t)$ for T in some reasonable range and/or reliable interspike interval data.

[3]After this paper was submitted for publication the authors learned of a recent modeling study (Hindmarsh, J. L., and R. M. Rose. 1984. *Proc. Roy. Soc. Lond. B.* 221:87–102) of endogenous bursting in which aperiodic behavior was detected. The mathematical model, however, was not based upon a biophysical model as is the Chay-Keizer model.

those obtained with different integration schemes and error tolerances.) Studies of deterministic chaos have revealed parameter regimes with characteristic, and extremely fine-structured, interleaving regions for periodic and aperiodic behavior (e.g. see references 6, 11). The identification of predictable patterns, their parameter regions, and their order of appearance is a typical method by which one presents an argument for deterministic chaos. Dependent on the level of noise, one can expect many of these details to be obscured yet some features of a canonical route to chaos may be discernible just before the first appearance of aperiodicity. Indeed, Guevara et al. (14) identified various complex periodic patterns prior to observing chaos in the heart cell aggregate. The islet preparation however appears to be less tightly coupled (one typically finds near synchrony of burst patterns but not of individual spikes from cell to cell) so that the difference in properties between cells and the presence of noise may preclude distinguishing the two factors for aperiodicity, especially on the time scale of individual spikes. Perhaps a unicellular preparation or tightly-coupled islet might be better for quantitative comparisons between theory and experiment of periodic phenomena in endogenously bursting systems.

One may ask about the implications of aperiodic behavior from deterministic biological systems. In some cases, such as cardiac arythmias or neuronal epileptic activity, physiological pathologies may be direct functional consequences (26). In other cases functional meaning is less clear. If for the normally operating pancreas, islets are not synchronized with each other then periods of chaotic activity of individual islets may have little effect on pancreatic output. Nevertheless, for biophysical interpretation of experimental data on isolated islets or neurons one need not invoke hypotheses about environmental noise to account for observed aperiodic behavior. In many cases, such behavior can be quite consistent with, even supportive of, a simple deterministic model without stochastic elements.

This work was supported by a National Science Foundation grant PCM82 15583 to Dr. T. Chay.

Received for publication 23 April 1984 and in final form 2 October 1984.

REFERENCES

1. Atwater, I., C. M. Dawson, A. Scott, G. Eddlestone, and E. Rojas. 1980. The nature of the oscillatory behavior in electrical activity for pancreatic β-cell. *In* Biochemistry Biophysics of the Pancreatic-β-Cell. Georg Thieme Verlag, New York. 100–107.
2. Chay, T. R., and J. Keizer. 1983. Minimal model for membrane oscillations in the pancreatic β-cell. *Biophys. J.* 42:181–190.
3. Hodgkin, A. L., and A. F. Huxley. 1952. A quantitative description of membrane current and its application to conduction and excitation in nerve. *J. Physiol. (Lond.).* 117:500–544.
4. Fenstermacher, P. R., H. L. Swinney, and J. P. Gollub. 1979. Dynamical instabilities and the transition to chaotic Taylor vortex flow. *J. Fluid Mech.* 94:103–128.
5. Hudson, J. L., and J. C. Mankin. 1981. Chaos in the Belousov-Zhabotinskii reaction. *J. Chem. Phys.* 74:6171–6177.
6. Simoyi, R. H., A. Wolf, and H. L. Swinney. 1982. One-dimensional dynamics in a multicomponent chemical reaction. *Phys. Rev. Lett.* 49:245–248.
7. Testa, J., J. Perez, and C. Jeffries. 1982. Evidence for universal behavior of a driven nonlinear oscillator. *Phys. Rev. Lett.* 48:714–717.
8. May, R. H., and G. F. Oster. 1976. Bifurcations and dynamic complexity in simple ecological models. *Am. Nat.* 110:573–599.
9. Hofstadter, D. 1981. Metamagical themes. Strange attractors: mathematical patterns delicately poised between order and chaos. *Sci. Am.* 245:22–43.
10. Ott, E. 1981. Strange attractors and chaotic motions of dynamical systems. *Rev. Mod. Phys.* 53:655–671.
11. Swinney, H. L. 1983. Observations of order and chaos in nonlinear systems. *Physica.* 7D:3–15.
12. Wolf, A. 1983. Nonlinear dynamics: simplicity and universality in the transition to chaos. *Nature (Lond.).* 305:182–183.
13. Pomeau, Y., and P. Manneville. 1980. Intermittent transition to turbulence in dissipative dynamical systems. *Commun. Math. Phys.* 74:189–197.
14. Guevara, M. R., L. Glass, and A. Shrier. 1981. Phase locking, period-doubling bifurcations, and irregular dynamics in periodically stimulated cardiac cells. *Science (Wash. DC).* 214:1350–1353.
15. Hayashi, H., M. Nakao, and K. Hirakawa. 1982. Chaos in the self-sustained oscillation of an excitable biological membrane under sinusoidal stimulation. *Phys. Lett. A* 88:265–266.
16. Holden, A. V., and M. A. Muhamad. 1984. The identification of deterministic chaos in the activity of single neurons. *J. Electrophysiol. Tech.* 11:135–147.
17. Jansen, J. H., P. L. Christiansen, A. C. Scott, and O. Skovgarrd. 1983. Chaos in nerve. *Proc. IASTED (Intern. Assoc. Sci. Tech. Dev.) Symp., ACI.* 2:15/6–15/9.
18. Ritzenberg, A. L., D. R. Adam, and R. J. Cohen. 1984. Period multupling evidence for nonlinear behaviour of the canine heart. *Nature (Lond.).* 307:159–161.
19. Rinzel, J., and I. B. Schwartz. 1984. One-variable map prediction of Belousov-Zhabotinskii mixed mode oscillations. *J. Chem. Phys.* 80:5610–5615.
20. Rinzel, J., and W. C. Troy. 1983. A one-variable map analysis of bursting in the Belousov-Zhabotinskii reaction. *In* Nonlinear Partial Differential Equations. J. A. Smoller editor. American Mathematical Society, Providence. 411–428.
21. FitzHugh, R. 1960. Thresholds and plateaus in the Hodgkin-Huxley nerve equations. *J. Gen. Physiol.* 43:867–896.
22. Plant, R. E., and M. Kim. 1976. Mathematical description of a bursting pacemaker neuron by a modification of the Hodgkin-Huxley equations. *Biophys. J.* 16:227–244.
23. Meissner, H. P. 1976. Electrical characteristics of the beta-cells in pancreatic islets. *J. Physiol. (Paris).* 72:757–767.
24. Chay, T. R. 1984. Abnormal discharges and chaos in a neuronal model system. *Biol. Cybernetics.* 50:301–311.
25. Chay, T. R. 1985. Chaos in a three-variable model of an excitable cell. *Physica D.* In press.
26. Glass, L., and M. C. Mackey. 1979. Pathological conditions resulting from instabilities in physiological control systems. *Ann. NY Acad. Sci.* 316:214–235.

Section IV
Information Generation and Transfer

Contents

Introduction

Understanding the complex functions of the nervous system is the most challenging problem in the biological physics of neurobiology. Unlike many problems in classical physics that start with simple harmonic oscillators, hydrogen atoms, or the ideal gas law, there is really no simple (in the mathematically tractable sense) system that is easily accessible theoretically and is a good representation of biophysical reality. Nonetheless, the analysis of simple neuronal networks of organisms with low phylogeny has provided valuable information on nerve growth and development and the development of theoretical models. One special case of information generation and transfer, visual photoreception, represents biologically interesting physics from start to finish. The photoreceptors of animals are contained within highly specialized cells that contain the light-sensitive chromophore, retinal. Upon absorption of light the receptor pigment undergoes a photophysical transformation which leads to the initiation of nerve impulses. These nerve impulses are subsequently conducted to the central nervous system. Biological clocks may be defined as physical and/or cellular mechanisms that are used by living organisms to measure time. Although periodicity in biological systems has been known for more than a century, it is only relatively recently biological scientists have recognized the universality of this intriguing phenomenon

in nearly all groups of macroscopic plants, animals, and microorganisms. Many biological rhythms are controlled by internal biological clocks. The periodicity of these clocks closely matches that of environmental periodicities and allows the organism to be coupled to the environmental cycle in a manner beneficial to it.

In Alwyn C. Scott's paper the electrophysics of a nerve fiber, with emphasis on the action potential, is discussed in terms of fundamental concepts of biochemistry and electromagnetic theory. John J. Hopfield's paper on neural networks and physical systems with emergent collective computational abilities discusses computational properties of use to biological organisms as collective properties of systems having a large number of simple equivalent components. Also, the physical meaning of content-addressable memory is described by an appropriate phase space flow of the state of the system. The biological physics of visual photoreception is discussed by Aaron Lewis and Lucian V. Del Priore. Their paper describes how the absorption of a single photon by a pigment molecule, rhodopsin, in a photoreceptor cell in the retina initiates a process of amplification that ends in a neural response. Arthur T. Winfree's paper on resetting biological clocks discusses how the rhythms of plants and animals can be stopped by the proper stimulus delivered at the right time.

The electrophysics of a nerve fiber*

Alwyn C. Scott

Department of Electrical and Computer Engineering, University of Wisconsin, Madison, Wisconsin 53706

The "action potential" is a pulselike voltage wave which carries information along a nerve fiber. Starting with fundamental concepts of biochemistry and electromagnetic theory, the derivation of the nonlinear diffusion equation which governs propagation of the action potential is reviewed. Our current understanding of this equation is discussed, paying particular attention to questions of interest in physics and applied mathematics.

I. INTRODUCTION

In the twenty fourth question added to the second edition of his *Optiks*, Newton (1718) asked

> *Qu.* 24. Is not Animal Motion perform'd by the Vibrations of this Medium, excited in the Brain by the power of the Will, and propagated from thence through the folid, pellucid and uniform Capillamenta of the Nerves into the Mufcles, for contracting and dilating them?

He was fairly close to the mark for, as we shall see, a proper theory for the electrodynamics of the nerve fiber begins with the field equations of Maxwell just as does the science of optics. First, of course, it was necessary to develop the science of electricity, and this was, in turn, profoundly influenced by Galvani's research on animal electricity and Volta's subsequent development of the battery later in the 18th century.

I do not propose to review this early history; the delightful survey by Brazier (1959 [see also Harmon and Lewis (1966)] could not be equaled without an enormous expense of time and effort which would necessarily be subtracted from my main task. I will, however, attempt to place in context those contributions which have directly led to an understanding of the nonlinear wave dynamics associated with propagation of a voltage pulse, or "action potential," along a nerve fiber.

In 1850 Helmholtz used a cleverly designed apparatus (see Fig. 1) to show that the signal velocity on a frog's sciatic nerve is not immeasurably large as was assumed (perhaps due to the continuing influence of Newton's twenty fourth question) but some 27 mps. Details of this

work can be found in Helmholtz (1850), but the basic idea is both simple and elegant. Closure of switch (V) simultaneously breaks the primary (P) initiating a nerve pulse (N), and starts a time measurement on the ballistic galvanometer (G). When the muscle (M) twitches, a mercury contact at k is broken and the measurement terminates. The difference of times measured for inputs at terminals (3–4) and (5–6) divided into the corresponding distance along

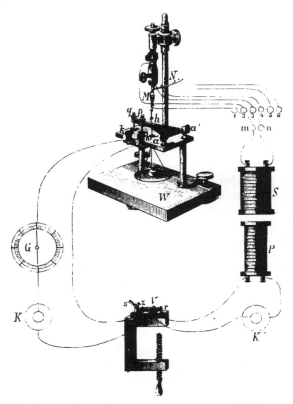

FIG. 1. Apparatus used by Helmholtz to measure the signal velocity on a nerve fiber [Hermann (1879I)].

*This work has been supported by the National Science Foundation under Grant No. GK-37552, by the National Institutes of Health under Grant No. LM-02281, and by the United States Army under Contract No. DA-31-124-ARO-D-462.

Reprinted from *Review of Modern Physics* **47**, 487–533 (1975); © The American Physical Society.

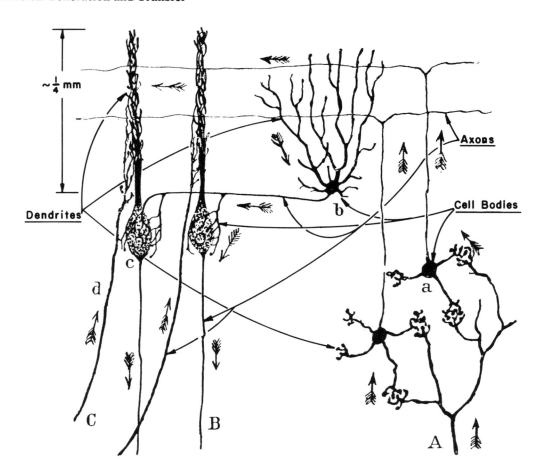

FIG. 2.   A semischematic diagram of neuron structure in the cerebellum by Ramón y Cajal (1908). (A) mossy fiber input, (a) Granule cell (B) Purkinje axon output, (b) basket cell, (c) Purkinje cell body, (C–d) climbing fiber input.

the nerve yields a velocity. Bernstein (1868) described the details of an even more impressive experimental *tour de force*; he measured the *shape* (potential vs time) of the action potential on a frog's nerve and showed that the velocity was equal to the signal velocity measured by Helmholtz. It is a fascinating experience to read over these early papers and appreciate the experimental results which were obtained a full half century before Gasser and Erlanger (1922) introduced the cathode ray oscilloscope into electrophysiological research.

The problem was to understand the physical process involved in the propagation of the action potential. Weber (1873) took an important step with his fundamental study of the flow of electricity in cylinders; indeed, we shall begin our analytic consideration with this calculation in the following section. Hermann (1879 II) seems to have the correct physical ideas in mind. He notes the similarity of nerve propagation to a line of burning powder but rejects a purely chemical explanation since this would seem to require activity throughout the entire cell. He describes circulating currents which excite the neighborhood of a pulse and indicates that these equations would lead to a form of the "heat equation." This line of thought, he wrote in 1879, "*genügt uberhaupt . . . der gestellten Aufgabe nicht.*" Hermann did not appreciate the descriptive power of a *nonlinear* diffusion equation until later and even then he felt such problems would lead to "enormous mathematical difficulties" [Hermann (1905)]. By this time Bernstein (1902), building on studies of charge transport in ionic solutions by Nernst

(1888, 1889), and Planck (1890a,b) had carefully stated in his "membrane hypothesis" that the action potential was the discharge of a (Nernst) diffusion potential caused by an increase in ionic permeability of the membrane.

The concept of a nerve cell or "neuron" as an independently functioning unit was firmly established through the extensive anatomical studies of Ramón y Cajal (1908), and a survey of this work written at the end of his life in 1934 has recently become available in English [Ramón y Cajal (1954)]. Most neurons display an input branching structure of "dendrites" called the dendritic trees, an enlarged cell body, and an output fiber or "axon" which eventually branches into an axonal tree. If appropriate firing conditions are established at the dendritic inputs, the cell body will send a pulse outward on the axon. An idea of the variety of neurons which fall within this basic pattern may be obtained through reference to Fig. 2 which is from the 1906 Nobel lecture of Ramón y Cajal and indicates some of the cerebellar (or motor control) circuitry in the central nervous system of vertebrates. A variety of tree shapes are observed each, presumably, adapted for the function of a particular cell. The size of nerve cells also varies widely. For example the sciatic nerve of a giraffe contains axons which are several meters in length, and the giant axon of the squid can be almost a millimeter in diameter. In this review the term nerve fiber implies both axons and dendrites although most of the available experimental data are for large axons.

The following decades saw: the demonstration of the

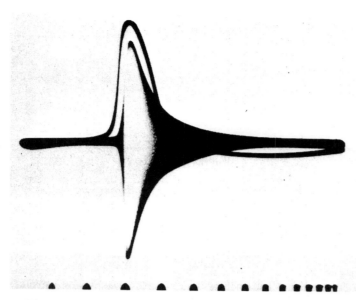

FIG. 3. Direct measurement of the increase in membrane conductance (band) during the action potential (line) on the squid giant axon [Cole and Curtis (1938)]. Time marks are 1 msec.

"all or nothing" nature of nerve fiber response to stimulation [Lucas (1909), Adrian (1914)], confirmation of the existence of the cell membrane, and measurement of its electrical capacitance [Fricke (1923)], discovery of the squid giant axon [Young (1936)], demonstration that the membrane conductance of a squid giant axon increases during the action potential [Cole and Curtis (1938, 1939)] (see Fig. 3), and the observation by Cole (1949) that membrane voltage (rather than current) is the more useful dependent variable for a phenomenological description. The activities of these years are described in detail in the recent book by Cole (1968). Part history, part careful scientific discussion, this book should be studied by everyone who wishes to understand twentieth century electrophysiology. Finally, the pieces of the problem were put together in the brilliant work of Hodgkin and Huxley (1952). They showed how measurements of the conductive parameters of a nerve fiber can be used to directly calculate both the shape and the velocity of an action potential on the squid giant axon.

In retrospect it seems that applied mathematicians forewent an unusual opportunity to make important scientific contributions by ignoring the study of the nonlinear diffusion equation. One exception to this generalization was the work by Kolmogoroff, Petrovsky and Piscounoff (1937) on the equation[1]

$$\phi_{xx} - \phi_t = F(\phi) \qquad (1.1)$$

which was related to the biological problem of genetic diffusion. They showed how steplike initial conditions would evolve into a unique solitary wave solution of the form

$$\phi(x, t) = \phi_T(x - ut) \qquad u \text{ const.,} \qquad (1.2)$$

developed phase plane techniques for determining ϕ_T, and derived explicit formulas for the traveling wave velocity u. This uniquely important contribution was completely over-

---

[1] Equation (1.1) should perhaps be called the KPP equation.

looked by electrophysiologists in the U.S.A.; indeed it is not even noted in the otherwise exhaustive bibliography of the book by Cole (1968). The failure of applied mathematicians to undertake a timely study of (1.1) cannot be ascribed to technical inefficiency in the face of the "enormous mathematical difficulties" envisaged by Hermann (1905). The studies by Boussinesq (1872) and by Korteweg and deVries (1895) of the hydrodynamic solitary waves described by Scott Russell (1844) indicate an ample understanding even before the turn of the century. As Cohen (1971) has suggested, the difficulty may have been the assumption by most mathematicians that the diffusive and nonpropagating behavior of linear diffusion equations would carry over to the nonlinear case.

But one need not turn to Hermann's line of burning powder or the Japanese incense investigated by Kato (1924) for a clear physical representation of nonlinear diffusion; the ordinary candle had been lighting scientific study tables for centuries. Diffusion of heat down the candle releases wax to the flame where it burns to supply the heat. If P is the power (joules/second) necessary to support the flame, and E is the chemical energy stored per unit length of the candle (joules/meter), then the flame (nonlinear wave) will travel at the velocity u for which

$$P = uE. \qquad (1.3)$$

The rate at which energy is eaten (uE) must equal the rate at which it is digested by the flame (P). Equation (1.3) is of more than pedagogical interest; when we turn to the development of formulas for the calculation of nerve pulse propagation velocity we shall use (1.3) to find solutions of (1.1) with the traveling wave character indicated in (1.2).

I take the point of view that nonlinear wave problems can be divided into two main classes: (i) those for which solitary traveling waves imply a balance between rate of energy release by the nonlinearity and its consumption and is indicated by (1.3), and (ii) those for which energy is conserved and therefore obey a conservation law

$$\mathcal{E}_t + \mathcal{P}_x = 0, \qquad (1.4)$$

where $\mathcal{E}$ is energy density, and $\mathcal{P}$ is the power flow. Wave problems of class (ii) include the hydrodynamic waves which were studied by Boussinesq (1972) and by Korteweg and deVries (1895). In this case solitary waves involve a balance between the effects of nonlinearity and dispersion, and the propagation velocity is an adjustable parameter in a family of solutions. Such energy conserving solitary waves sometimes exhibit an infinite number of conservation laws and the nondestructive collisions characteristics of "solitons." Nothing further will be said here about class (ii); the interested reader is referred to Scott, Chu, and McLaughlin (1973) for a review of the current status of this research. Although the present discussion will concentrate upon nonlinear wave problems of class (i), it should not be assumed that conservation laws are unimportant. Indeed we shall find that an approximate conservation law for electric charge can be useful in determining the conditions necessary to stimulate a nerve fiber to the threshold of excitation, and also that a conservation law for pulses

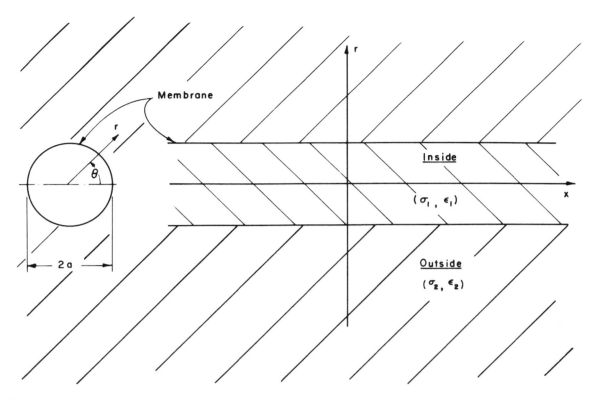

FIG. 4.   Cylindrical geometry for electromagnetic analysis of a nerve fiber.

helps to analyze the evolution of a pulse burst along a fiber.

The Hodgkin and Huxley (1952d) calculation of the action potential shape and velocity from measured parameters of the nerve cell was a scientific achievement of extraordinary significance. They demonstrated that a physical theory for the electrophysiology of a nerve cell could be based on a phenomenological description of the membrane and mooted thereby much of the previous tendency of biomathematicians toward "modeling." The main objective of this review is to present as clearly and simply as possible the elements of such a theory paying particular attention to the contributions of physicists, applied mathematicians, and bioengineers. Thus I begin with an electromagnetic analysis leading to appropriate nonlinear partial differential equations and using ideas familiar to the microwave engineer; and then proceed to a study of ion current flow through a nerve membrane which should be of interest to the solid state physicist. For those who are anxious to get on to an analysis of nonlinear pulse propagation on a nerve fiber, these sections may seem unnecessarily extensive; but the problem of deciding *which* equations to analyze is not at all trivial especially in situations where the traditional geometry of an infinitely long circular cylinder is altered. The phenomenological description of nerve membrane electrodynamics developed by Hodgkin and Huxley is then presented and used as a basis for subsequent mathematical analysis. Emphasis is placed upon those aspects of the mathematical picture where future developments seem likely such as the theory of motor nerves which are "myelinated" to increase pulse velocity and the threshold theory for active fibers. A final section introduces several problems of current research interest involving the interaction of nonlinear pulses on nerve fibers.

II. NONLINEAR PARTIAL DIFFERENTIAL EQUATIONS

Stimulated by attempts of Hermann and Matteucci to understand the manner in which electricity flows through a nerve fiber, Weber (1873a,b) carried out a fundamental study of time independent current density in and near a partially conducting cylinder. The basic coordinate system for this problem is shown in Fig. 4; a cylindrical *membrane* separates an *inside* region with conductivity σ_1 and dielectric constant ϵ_1 from an *outside* region with conductivity σ_2 and dielectric constant ϵ_2. Weber assumed that the electrical potential both inside and outside the membrane satisfies Laplace's equation and applied suitable boundary conditions at the membrane. This approach has been followed by several other investigators up to recent times [Clark and Plonsey (1966, 1968), Geselowitz (1966, 1967), Hellerstein (1968). Lorente de Nó (1947), Plonsey (1964, 1965), Rall (1969), Weinberg (1941, 1942)] and, indeed, is a very good approximation for potentials which vary as slowly as is indicated in Fig. 2. On the other hand it is not more difficult to proceed with the complete Maxwell equations [Pickard (1968, 1969), Rosenfalk (1969), Scott (1972) and this approach allows us to comprehend more precisely the implications of a quasistatic approximation. Thus we write[2]

$$\text{curl } \bar{E}_i = -\mu_0(\partial \bar{H}_i/\partial t) \tag{2.1a}$$

$$\text{curl } \bar{H}_i = \sigma_i \bar{E}_i + \epsilon_i(\partial \bar{E}_i/\partial t) \qquad i = 1, 2. \tag{2.1b}$$

---

[2] The mks system of electromagnetic units will be used throughout this paper. In this section, subscripts denote vector components so partial derivatives will be explicitly indicated.

where $i = 1$ inside the membrane and $i = 2$ outside. These equations are entirely *linear*. The nonlinearity in the problem appears at the membrane boundary where the normal current density, J_{12}, is some nonlinear function of the transverse voltage, v, across the membrane. Thus we can write symbolically

$$J_{12} = N(v), \qquad (2.2)$$

but we must be careful to remember that $N(v)$ can be a rather complex function of v and its time derivatives. In order to appreciate this complexity, the reader might look ahead to the discussion of the Hodgkin–Huxley equations in Sec. IV.

The fact that nonlinear effects occur only on the cylindrical membrane boundary greatly simplifies the study of the electromagnetic problem. For an infinitely long fiber, the regions both inside and outside the membrane are invariant to:

(i) translation in the x direction
(ii) rotation in the θ direction, and
(iii) translation with time (t).

Thus we can compose the fields of elementary functions which vary as $\exp[i(\beta x - \omega t + n\theta)]$ both inside the membrane and outside.

Furthermore we shall begin our analysis by assuming rotational symmetry of the fields as implied by

$$\partial/\partial\theta = 0 \quad \text{or} \quad n = 0. \qquad (2.3)$$

The implications of this assumption will be considered below but for the present it allows us to concentrate our attention upon those TM (transverse magnetic) solutions of (2.1) for which

$$(\bar{H})_r = (\bar{H})_x = 0 \quad \text{and} \quad (\bar{E})_\theta = 0. \qquad (2.4)$$

The TE (transverse electric) modes for which $(\bar{E})_r = (E)_x = 0$ and $(\bar{H})_\theta = 0$ are of little interest since the condition $(\bar{E})_r = 0$ implies zero normal current at the membrane surface. From (2.2) such TE modes would not interact with the nonlinearity of the membrane.

Then we write the θ component of the magnetic intensity vector

$$(\bar{H})_\theta = H_\theta(r) \exp[i(\beta x - \omega t)] \qquad (2.5)$$

and similarly for $(\bar{E})_r$ and $(\bar{E})_x$ (where it should be understood that subscripts 1 or 2 are added for fields inside or outside the membrane) whereupon Maxwell's equations (2.1) reduce to

$$\partial^2 H_\theta/\partial r^2 + r^{-1}(\partial H_\theta/\partial r) - (1/r^2 + k^2)H_\theta = 0, \qquad (2.6a)$$

$$E_r = -(i\beta/\sigma^*)H_\theta, \qquad (2.6b)$$

$$E_x = (1/\sigma^* r)[\partial(r H_\theta)/\partial r]. \qquad (2.6c)$$

In these equations σ^* is the complex conductivity

$$\sigma^* = \sigma + i\omega\varepsilon, \qquad (2.6d)$$

and

$$k^2 = i\omega\mu_0\sigma^* + \beta^2. \qquad (2.6e)$$

Equations (2.6) indicate that H_θ is a rather convenient variable for which to solve. Knowing H_θ, one can determine E_r and E_x through (2.6b) and (2.6c). Equation (2.6a) is Bessel's equation, solutions for which are $I_1(kr)$ and $K_1(kr)$ as defined by Watson (1962). Since K_1 goes to infinity at the origin, I_1 is the appropriate solution inside the membrane; and, since I_1 goes to infinity for large values of r, K_1 is the appropriate solution outside. The magnitude of H_θ at $r = a$ can be easily determined from Ampere's circuital law (which is (2.1b) in integral form) from the total current flowing in the x direction inside the membrane as

$$2\pi a H_\theta = I. \qquad (2.7)$$

Thus a complete solution for H which (i) satisfies Maxwell's equations both inside and outside the membrane, (ii) has no θ variation as required by assumption (2.3), (iii) corresponds to a TM mode with a current component perpendicular to the membrane boundary, (iv) satisfies the appropriate electromagnetic boundary condition at the origin, and (v) goes to zero at large radius, is:

inside

$$H_\theta = \frac{I}{2\pi a} \frac{I_1(k_1 r)}{I_1(k_1 a)}, \qquad (2.8a)$$

outside

$$H_\theta = \frac{I}{2\pi a} \frac{K_1(k_2 r)}{K_1(k_2 a)}, \qquad (2.8b)$$

where $k_1^2 = i\omega\mu_0\sigma_1^* + \beta^2$ inside the membrane and $k_2^2 = i\omega\mu_0\sigma_2^* + \beta^2$ outside the membrane.

At this point in the analysis it is important to recognize that Eqs. (2.8) have been derived without considering the nonlinear aspects of the problem symbolically expressed in (2.2). The appropriate values for ω and β out of which the action potential is to be determined are, as yet, entirely undetermined. We shall now use (2.8) to develop the nonlinear partial differential equations which relate the total longitudinal current flowing inside the membrane, $i(x, t)$, and the (θ independent) voltage across the membrane, $v(x, t)$, as is indicated in Fig. 5a.

To obtain a pde involving the x derivative of v, consider the diagram of the electric field components near the membrane in Fig. 5b where the positive reference directions for the x component of the inside field, $(\bar{E}_1)_x$, and the outside field, $(\bar{E}_2)_x$, are indicated. With these references, the sum of potentials around the path $A\ B\ C\ D$ becomes

$$v(x + dx) + (\bar{E}_1(a, x, t))_x\, dx - v(x)$$
$$+ (\bar{E}_2(a, x, t))_x\, dx = 0$$

for any time. Thus

$$\partial v/\partial x = -(\bar{E}_1(a, x, t))_x - (\bar{E}_2(a, x, t))_x. \qquad (2.9)$$

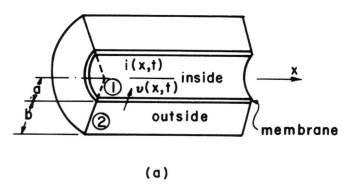

(a)

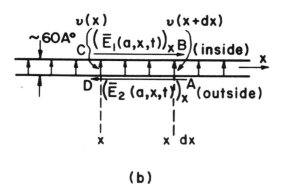

(b)

FIG. 5. (a) Geometry for an idealized nerve fiber, and (b) electric field components near the membrane.

Equation (2.9) is the source of the pde we are after. It can be related to the longitudinal current, $i(x, t)$, in the following way. First consider the expansion of $(\bar{E}_1)_x$ into its spatial and temporal components as

$$(E_1(a, x, t))_x = \iint E_{1x}(a) \exp[i(\beta x - \omega t)] \, d\beta \, d\omega \quad (2.10)$$

and similarly for $(E_2)_x$. Then using Eqs. (2.6c) and (2.7) we can write

$$E_{1x}(a) = z_1 I \quad \text{and} \quad E_{2x}(a) = z_2 I \quad (2.11a, b)$$

where z_1 and z_2 are impedances. Assuming $H_\theta \to 0$ as $r \to \infty$ gives

$$z_1 = \left(\frac{1}{\pi \sigma_1^* a^2}\right) \left[\frac{k_1 a I_0(k_1 a)}{2 I_1(k_1 a)}\right], \quad (2.12a)$$

$$z_2 = \left(\frac{1}{\pi \sigma_2^* a^2}\right) \left[\frac{k_2 a K_0(k_2 a)}{2 K_1(k_2 a)}\right]. \quad (2.12b)$$

Thus (2.9) becomes

$$\partial v / \partial x = -\iint (z_1 + z_2) I(\beta, \omega) \exp[i(\beta x - \omega t)] \, d\beta \, d\omega, \quad (2.13)$$

where $I(\beta, \omega)$ is the spatial and temporal Fourier transform of $i(x, t)$.

Equation (2.13) is not nearly as intractable in practice as it might appear at first glance. First it is important to remember that it is entirely linear; the only nonlinearities appear in connection with current flow through the membrane (2.2) and this effect has not yet been considered. Secondly the temporal frequency components, ω, in a typical action potential are of the order of 10^3 rad/sec (see Fig. 3) and the conductivity, σ, both inside and outside the fiber is approximately that of sea water (4 mho/m). Thus it is a very good approximation to write

$$\sigma^* \approx \sigma \quad \text{(a real constant)}$$

both inside and outside the fiber. Thirdly the radial parameter k which appears in (2.12) is given by (2.6e)

$$k^2 = 4i/\delta^2 + \beta^2 \quad (2.14)$$

where

$$\delta = (2/\sigma \mu_0 \omega)^{1/2} \quad (2.15)$$

is the electromagnetic penetration depth in the conductive medium at frequency ω. For $\sigma \sim 4$ mho/m, $\omega \sim 10^3$ rad/sec and $\mu_0 = 4\pi \times 10^{-7}$ H/m, $\delta \sim 20$/m, which is much greater than the spatial extent of typical action potentials. Thus it is a very good approximation to write (2.14) as

$$k \approx \beta. \quad (2.16)$$

Finally the spatial extent of a typical action potential is typically an order of magnitude or more larger than the fiber radius, a. Small argument approximations are then appropriate for the evaluation of the Bessel functions which appear in (2.12). For example (2.12a) becomes

$$z_1 = (1/\pi a^2 \sigma_1)\{1 + 0[(k_1 a)^2]\}, \quad (2.17)$$

and the most important effect of the $0[(k_1 a)^2]$ terms is to introduce an inductive component into z_1. In a later section the effect of this inductive component will be studied in detail and it will be shown to be entirely negligible. Thus we can write

$$z_1 \approx r_1, \quad (2.18)$$

where

$$r_1 = 1/\pi a^2 \sigma_1 \quad (2.19)$$

is the resistance inside the fiber to longitudinal (x directed) current flow. The ratio of outside to inside longitudinal impedance from (2.12)

$$\frac{z_2}{z_1} = \left(\frac{\sigma_1}{\sigma_2}\right) \left[\frac{K_0(\beta a) I_1(\beta a)}{K_1(\beta a) I_0(\beta a)}\right] \quad (2.20)$$

the square bracket of which is plotted in Fig. 6. Neglecting z_2 with respect to z_1 is seen to introduce an error of no more than a percent.

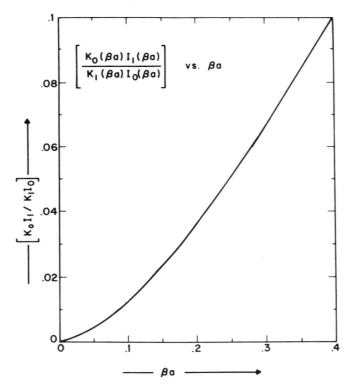

FIG. 6. Plot of the factor $[K_0I_1/K_1I_0]$ as a function of βa. This is approximately equal to the ratio of external to internal series resistance for $\beta b \gg 1$ [Scott (1973)].

Equation (2.13) reduces to

$$\partial v/\partial x = -r_s i, \tag{2.21a}$$

where $r_s = r_1 + r_2 \approx r_1$ as indicated in (2.19). This is one of the two partial differential equations we seek. The other is nonlinear and relates the spatial derivative of i to the membrane voltage v. It is obtained from (2.6b) by noting, first, that $(\bar{H})_\theta$ is proportional to i as indicated in (2.7) and, second, that $(\bar{E})_r$ evaluated at $r = a$ gives the current density normal to the membrane which appears in (2.4). Thus

$$\partial i/\partial x = -2\pi a N(v). \tag{2.21b}$$

It is often useful to combine Eqs. (2.21) to obtain a second-order equation which involves only the membrane voltage

$$\partial^2 v/\partial x^2 = 2\pi a r_s N(v). \tag{2.22}$$

Equation (2.22) is not quite as simple as it looks since $N(v)$ is a rather complex nonlinear function of v. But it is perhaps more simple than one expects for the geometry indicated in Fig. 5(a). Thus it may be useful at this point to recapitulate the assumptions which were involved in the derivation of (2.22).

A. Rotation symmetry of fields

A basic assumption connected with (2.22) is that the membrane voltage is a function only of x and t and is not a function of the angle of rotation around the cylinder axis, θ. In the course of the analysis, this restriction allowed us to exclude non-TM modes from consideration. Rall (1969) has studied the question of angle dependence in detail for a cylinder of fixed length. He has shown that the time constant, τ_n, for angle variation [as $\exp(in\theta)$] to disappear is related to the basic time constant of the membrane, τ, by

$$\frac{\tau_n}{\tau} \approx \frac{a}{n}\left(\frac{1}{\sigma_1} + \frac{1}{\sigma_2}\right) G, \tag{2.23}$$

where $G \equiv \partial N/\partial v$ is a conductance per unit area of the membrane. For typical values of the parameters, the right-hand side of (2.23) is something like $10^{-4}/n$. Thus, for uniform cylindrical geometry, we can expect angularly dependent fields to relax to the angularly independent case in a time which is very short compared with the time scale for solutions of (2.22).

B. Uniform fiber cross section

A real nerve fiber is often not shaped as the uniform circular cylinder indicated in Fig. 5(a); angular bends, local distention, tapering, and collapse into a ribbon shaped cross section are some of the deviations easily observed. Judgment is required to determine the degree of confidence which one can place in (2.22) in such cases. First, of course, the πa^2 which appears in (2.12) and (2.19) should be replaced by the cross sectional area of the fiber, and the $2\pi a$ in (2.21b) and (2.22) should be replaced by the fiber circumference. [Some calculations for flat cells are presented by Minor and Maksimov (1969)]. A more serious difficulty arises from the scattering of TM fields [described by (2.22)] into non-TM modes; this effect is not represented at all. Furthermore if the nonuniformities vary with x on a scale short with respect to β^{-1} (the length of the action potential), the easy transition from (2.13) to (2.21a) will no longer be valid. On the other hand, some progress has been made with the solution of the nonlinear problem with a gradual exponential taper [Lindgren and Buratti (1969)]. Another important case is the so called "myelinated axon" for which $N(v)$ is approximately zero except at periodically spaced active nodes. This situation is also considered in detail in Sec. VIII.

C. Infinite external medium

In the development of the expression for outside impedance (2.12b) we assumed that the dimension "b" in Fig. 5(a) is large enough to insure that $K_1(k_2 b)$ is zero in (2.8b). From (2.16) a more precise statement of this requirement is

$$b \gg \beta^{-1}, \tag{2.24}$$

where, as was noted before, β^{-1} is of the order of the length of the action potential. Although this condition is easily satisfied in experiments on isolated fibers, it also easily violated. Furthermore cells and fibers are often closely packed in functioning neural systems; thus the situation when (2.24) is not satisfied deserves careful attention.

If the external current is constrained to flow in a region $b \ll \beta^{-1}$ (i.e., very close to the membrane surface), the outside resistance will increase from approximately zero to

$$r_2 \approx 1/A_0\sigma_2, \tag{2.25}$$

where A_0 is the cross-sectional area outside the membrane. However if A_0 does not exhibit rotational symmetry, the TM fields will again be scattered into non-TM modes in a manner which is not described by (2.22). Furthermore if the changes in A_0 take place on a distance scale short compared with β^{-1} the easy transition from (2.13) to (2.21a) will again no longer be valid. Qualitative effects of various experimental restrictions in the external geometry have been reviewed by Taylor (1963). Often nerve fibers are not isolated but arranged in bundles surrounded by a sheath of connective tissue. The sciatic nerve of vertebrates (see Fig. 35) is constructed this way to permit the transmission of a multicomponent message from the spinal cord to the muscle. This situation has been carefully investigated by Clark and Plonsey (1968) who present several numerical calculations which help to determine the effect of fiber geometry upon r_2.

D. Resistive approximation for the longitudinal impedances

Equation (2.22) specifically assumes that the sum of the inside and outside longitudinal impedances can be approximated by a single real number

$$r_s \equiv r_1 + r_2 \approx z_1 + z_2. \qquad (2.26)$$

This approximation ignores terms of order $(ka)^2$ in evaluating the small argument expressions for z_1 and z_2 in (2.12). Physically this implies neglect of the effect of time dependent magnetic field on the electric field, or inductive effects. In a later section, after the nature of the nonlinear propagation process has been clarified, we shall see that the only sensible effect of this inductive correction is to preclude a pulse velocity greater than the velocity of light.

A transmission line equivalent circuit can easily be constructed which corresponds to Eqs. (2.13) and (2.21b). For example in the differential ladder network of Fig. 7(a), the change in series current over a differential distance, dx, is found from Kirchoff's current law (or conservation of electric charge) to be

$$i(x) - i(x + dx) = 2\pi a N(v) dx \qquad (2.27)$$

which implies (2.21b). In a similar way the change in shunt voltage over a differential distance, dx, is obtained from Kirchoff's voltage law (or conservation of energy) to be

$$v(x) - v(x + dx) = dx \mathfrak{I}^{-1}\{(z_1 + z_2)\mathfrak{I}[i]\} \qquad (2.28)$$

where $\mathfrak{I}$ and $\mathfrak{I}^{-1}$ respectively represent the Fourier transform on both x and t and its inverse. Equation (2.28) implies (2.13).

Transmission line equivalent circuits (TLEC) of this sort have found wide application in electronics since the development of the electric telegraph [Kelvin (1855)] and in electrophysiology since the turn of the century [Hoorweg (1898), Hermann (1905)]. For rather complete reviews see Taylor (1963) and Cole (1968). Various attitudes may be taken toward the TLEC, two of which are as follows:

(1) The TLEC can be considered simply a mnemonic device through which the partial differential equations

under consideration, (2.13) and (2.21b), are represented pictorially. It is often useful to suggest reasonable higher approximations for further study [Scott (1970)].

(2) The TLEC can be taken as the starting point for analysis. Equations (2.27) and (2.28) are then considered fundamental equations from which (2.21b) and (2.13) are derived. This attitude has characterized much of past research in electrophysiology [Cole (1968)].

It is my opinion that the problems which arise in studying the electrophysics of the nerve cell are sufficiently difficult that neither attitude should dominate. For a nerve fiber which approximates the idealized geometry of Fig. 5(a), it is clearly more satisfying (for the physicist, at least) to begin the analysis with Maxwell's equations. Various approximations can be itemized and explicit analytic expressions can be obtained for z_1 and z_2. This analysis, on the other hand, can eventually lead to the nonlinear pde (2.22) which is also obtained directly from (2.27) and (2.28). In situations with more complex geometry, where the electromagnetic analysis may not be tractable, one can begin with a TLEC and appeal to the results for simpler geometry as a justification. Rall (1962, 1964) has demonstrated the power of this approach through his application of "compartmental analysis" to study the rather complex geometrical effects which arise in dendritic fibers.

The general TLEC to be considered in this review is shown in Fig. 7(b) for which suitable expressions to determine r_1 and r_2 are given in (2.19) and (2.25). With the series inductances, l_1 and l_2, equal to zero, this TLEC was studied by Offner as early as 1937 and serves as the basis for the calculation of conduction velocity for an action potential by Offner, Weinberg, and Young (1940). We will continue to assume these inductances equal to zero for the initial development of the nonlinear analysis. In a later section explicit expressions and values will be calculated, and a nonlinear propagation problem will be solved in order to demonstrate that it is a valid assumption to take these inductances equal to zero. Notice that the shunt element in Fig. 7(b) is represented differently than that in Fig. 7(a). The reason for this change is that in Fig. 7(b) it is explicitly recognized that membrane current consists of two distinct components: displacement current and ion current. Equating the shunt currents in the two figures yields

$$2\pi a N = c(\partial v / \partial t) + j_i, \qquad (2.29)$$

where j_i is the ion current, and $c(\partial v / \partial t)$ the displacement current passing through the membrane both per unit length in the x direction. The decomposition indicated in (2.29) is especially interesting because there is substantial experimental evidence [Cole (1968)] to show that c is a constant throughout the course of the action potential [see, however, FitzHugh and Cole (1973)]. Substituting (2.29) into (2.22) yields a new form for the basic equation of nerve propagation

$$\partial^2 v / \partial x^2 - r_s c(\partial v / \partial t) = r_s j_i. \qquad (2.30)$$

Notice that (2.30) has the form of the nonlinear diffusion equation discussed briefly in the introduction. In the following sections we will consider the chemical physics of the nerve membrane and the development of phenomenological theories to describe the nonlinear dependence of j_i upon v.

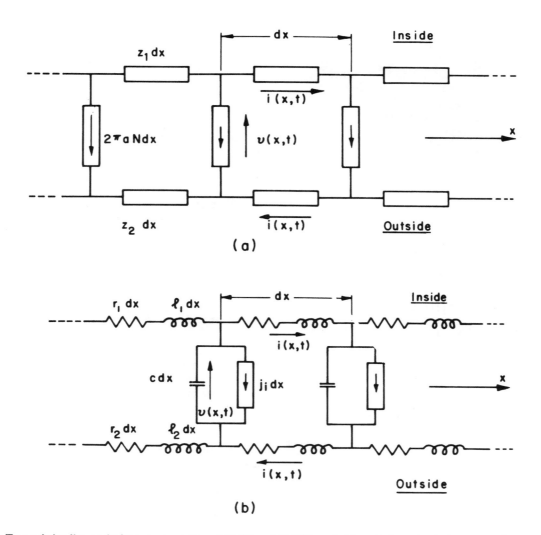

FIG. 7.   (a) Transmission line equivalent representation of (2.13) and (2.21b), and (b) equivalent circuit for a nerve fiber to be considered in this review.

III. PHYSICS OF A CELL MEMBRANE

Our next task is to become acquainted with the physical character of the cell membrane which is indicated merely as a surface in Fig. 4, and as a homogeneous region in Fig. 5. The existence of a membrane for red blood cells was confirmed by the measurements of Fricke (1923, 1925a,b) on the conductivity vs frequency of cell suspensions. He measured a membrane capacitance of 0.81 $\mu F/cm^2$ which, for an assumed relative dielectric constant of 3, implied a membrane thickness of 33Å. At about the same time Gorter and Grendel (1925) demonstrated that these cells "are covered by a layer of fatty substances that is two molecules thick." It is well to devote a moment to the measurement technique of Gorter and Grendel because it exemplifies nicely the energetics of membrane structures. The general structure of a lipid (fatty) molecule is "cigar shaped" with a charged head group localized at one end of a hydrocarbon tail. [See Chap. 10 of Lehninger (1970) for many chemical details.] Building upon a previous demonstrated by Lord Rayleigh (1899) that oil films on a water surface become monomolecular, Langmuir (1917) [see also Adam (1921, 1922)] showed that the structure of the monolayer is with the charged head groups oriented

toward the water surface where the electric field energy can be reduced by the high dielectric constant of water (ca. 80 ϵ_0), and the hydrocarbon tails maintained in a closely packed, vertical structure by transverse van der Waals attraction. Gorter and Grendel distilled the lipid material from a known quantity of blood cells and found that the area of the monolayer which could be obtained with this lipid at an air–water interface was about twice the area of the cell surfaces. Thus the red blood cell membrane appeared to be largely the lipid bilayer shown in Fig. 8. This same structure was proposed [Danielli and Davson (1935), Danielli (1936)] from an energetic comparison of various lipid organizations, as the basic structure of biological cell membranes. Membrane distillates always contain a substantial fraction (>50%) of protein [Bretscher (1973), Kilkson (1969)]; and if these are located within the lipid phase they are called *intrinsic* [Green (1971)] or *integral* [Singer and Nicolson (1972)]. Proteins attached weakly to the surface of the lipid bilayer, called *extrinsic* or *peripheral* are considered to be of less importance for membrane function. Green, Ji, and Brucker (1972) have emphasized the importance of protein domains through which long-range ordering of (perhaps octal) protein subunits is established [Vanderkooi and Green (1970)] as indicated in

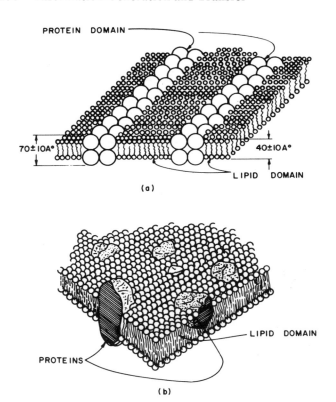

FIG. 8.   Membrane models. (a) The "structure-function unitization model" redrawn from Green, Ji, and Brucker (1972). Domain geometry is assumed to be highly variable from membrane to membrane. (b) The "fluid mosaic model" redrawn from Singer and Nicolson (1972).

Fig. 8(a). Singer and Nicolson, on the other hand, have suggested that the proteins may be considered to float in the two-dimensional lipid liquid [Fig. 8(b)]. Good general surveys of biological membranes are given in the recent books by Cereijido and Rotunno (1970), Jain (1972), and Nystrom (1973), and many of the historically important papers have been collected by Branton and Park (1968). The direct synthesis of a biological membrane was attempted by Bungenberg de Jong and Bonner (1935), Devaux (1936), Dannielli (1936), Teorell (1936), Langmuir and Waugh (1938), and Dean (1939) who produced bulayer films with a capacitance of about 1 μF/cm^2 [Dean, Curtis and Cole (1940)]. This work lay dormant for more than two decades until the ease with which lipid bilayers can be formed was demonstrated by Mueller, Rudin, Tien, and Wescott (1962). The key idea was an observation in Newton's *Optiks* on the color patterns of soap bubbles. He had observed that: "after all the Colours were emerged at the top, there grew in the center of the Rings a small round black Spot . . . which continually dilated itself till it became sometimes more than 1/2 or 3/4 of an inch in breadth." Newton was observing that it is energetically favorable for a soap film to thin into a lipid bilayer. In this case, the charged head groups are oriented inward toward a remnant layer of water. Such a soap film appears "black" (i.e., almost reflectionless) because its thickness ($\sim$100Å) is very much less than the wave length of light. [I can only avoid the temptation to say more about this subject by directing the reader to the delightful descriptions prepared by Lawrence (1929) and by Mysels, Shinoda and Frankel (1959).] Mueller, Rudin *et al.* (1962) showed that the same result could be obtained for lipid films between aqueous phases.

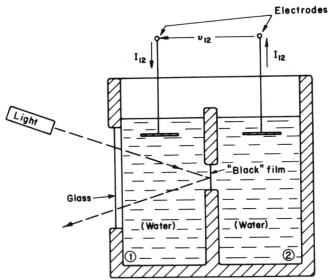

FIG. 9.   A basic experimental arrangement for physical measurements on artificial lipid bilayers.

A diagram of the basic arrangement for measurements on artificial lipid bilayers is given in Fig. 9. A camel's hair brush is dipped in the lipid solution and then stroked across a small ($\sim$1 mm) hole in a two chamber vessel. The resulting thick lipid film thins in about 10 minutes, as Newton described, to a lipid bilayer of black film. Optical measurements of film thickness and electrical measurements of capacity and conductivity can then proceed. The experimentalist who wishes to begin such an investigation is referred to the review by Goldup, Ohki, and Danielli (1970), the careful discussion of experimental details by Howard and Burton (1968), and the recent book by Jain (1972).

The processes by which ions flow across the membranes of living cells are often divided into *passive* and *active* mechanisms. Passive transport is considered to be in response to a gradient of the electrochemical potential. Active transport involves the flow of ions against the electrochemical potential; a good discussion of such processes can be found in Chapter 27 of Lehninger (1970). During the propagation of an action potential along a nerve fiber (Fig. 3) only passive transport is involved; active processes merely recharge the energy sources. My objective here is to present a simple phenomenological description of passive transport from which the ionic current components in (2.2) can be constructed.

It should be understood from the start that intrinsic membrane proteins completely dominate ion flow in a living membrane. To appreciate the truth of this assertion, it is instructive to begin with an investigation of passive transport of only sodium ions across an ideal lipid bilayer as is indicated in Fig. 10. The steady state current density from chamber ① to ②, J_{12}, will be proportional to the ion density [Na$^+$] and to the gradient of the electrochemical potential, ψ. Thus we can write the Nernst–Planck equation [Nernst (1888, 1889), Planck (1890a, b), Smith (1961)]

$$J_{12} = \mu q[\text{Na}^+](d\psi/dr), \qquad (3.1)$$

where q is the electronic charge and μ is the ionic mobility.

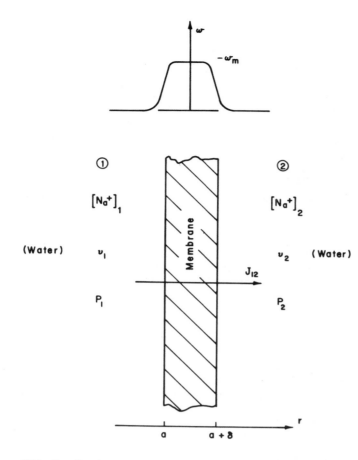

FIG. 10.   Simple geometry for passive transport of a single ion through a uniform membrane.

The expression for sodium ion concentration which satisfies these boundary conditions and maintains J_{12} constant has been determined by Neumke and Läuger (1969) as [see also Boltaks, Vodyanoi and Fedorovich (1971) and Markin, Grigor'ev and Yermishkin (1971)]

$$[Na^+] = \exp[-(v+w)q/kT] \Big[[Na^+]_1 \exp(v_1 q/kT)$$

$$+ \{ [Na^+]_2 \exp(v_2 q/kT) - [Na^+]_1 \exp(v_1 q/kT) \}$$

$$\times \left(\int_a^{a+r} \exp[(v+w)q/kT] \, dr \right) \Big/$$

$$\left(\int_a^{a+\delta} \exp[(v+w)q/kT] \, dr \right) \Big] \qquad (3.5)$$

which upon substitution into (3.4) yields

$$J_{12} = \left(\mu kT \Big/ \int_a^{a+\delta} \exp[(v+w)q/kT] \, dr \right)$$

$$\times \{ [Na^+]_1 \exp(v_1 q/kT) - [Na^+]_2 \exp(v_2 q/kT) \}. $$

$$(3.6)$$

For a detailed discussion of the effect of barrier shape, $w(r)$, on volt-ampere characteristics see Hall, Mead and Szabo (1973). Under experimental conditions for which the membrane structure remains independent of the applied voltage, they have demonstrated that $w(r)$ can be computed from volt-ampere measurements. The measured barrier height is close to the difference in electrostatic energy of the ion in lipid and in water; the shape is trapezoidal as indicated in Fig. 10.

Introducing the notational definitions

$$v_{12} \equiv v_1 - v_2, \qquad (3.7)$$

and

$$V = (kT/q) \log([Na^+]_2/[Na^+]_1), \qquad (3.8)$$

we obtain from (3.6)

$$J_{12} = (kT/q)G[\exp[(v_{12} - V)q/kT] - 1] \qquad (3.9)$$

$$\approx G(v_{12} - V) \quad \text{for} \quad |v_{12} - V| < kT/q, \qquad (3.10)$$

where

$$G \equiv \mu q [Na^+]_2 \exp(v_2 q/kT) \Big/ \int_a^{a+\delta} \exp[(v+w)q/kT] \, dr.$$

$$(3.11)$$

From (3.2) it is clear that $(v_{12} - V)$ is the change in electrochemical potential from chamber ② to chamber ①. For a small enough difference in electrochemical potential, (3.10) indicates that the relation between voltage and ion current density should be linear. If the concentration gradient is zero, $[Na^+]_1 = [Na^+]_2$, this linear relation

Assuming that the pressure gradient can be neglected, the electrochemical potential is

$$\psi = (kT/q) \log[Na^+] + v + w, \qquad (3.2)$$

where v is the externally applied electrical potential, and w is the contribution to the electrochemical potential arising from the presence of the membrane. Since the dielectric constant of water ($\sim 80\,\epsilon_0$) is much greater than that of the lipid, a major contribution to w will be the image force at the water-membrane boundary. [In direct physical terms, the electrostatic field energy associated with the ion is much lower in the water phase where the ionic charge can be neutralized by rotating water molecules.] The factor (kT/q) in (3.2) [where k is the Boltzmann constant and T is absolute temperature] appears from the Einstein (1905) relation between diffusion constant, and mobility.

Substituting (3.2) into (3.1) gives

$$J_{12} = -\mu kT \left\{ \frac{d[Na^+]}{dr} + \frac{[Na^+]q}{kT} \frac{d}{dr}(v+w) \right\}, \qquad (3.3)$$

which, in steady state, must be independent of r. The boundary conditions to be satisfied at the edges of the membrane are

$$r = a: \quad [Na^+] = [Na^+]_1, \quad v = v_1, \quad w = 0 \qquad (3.4a)$$

$$r = a + \delta: \quad [Na^+] = [Na^+]_2, \quad v = v_2, \quad w = 0. \qquad (3.4b)$$

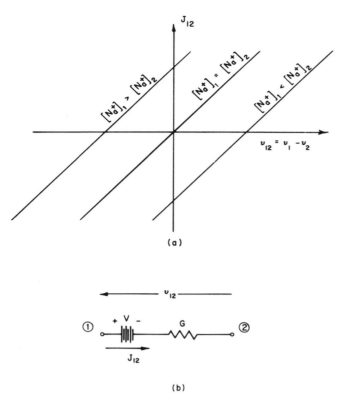

(a)

(b)

FIG. 11. (a) Sodium ion current density at a small difference of electrochemical potential. (b) An equivalent circuit for the current density carried by sodium ions.

should go through the origin as indicated in Fig. 11(a). For positive ions if $[Na^+]_1 < [Na^+]_2$ and the current density is zero, chamber ① will have a positive potential with respect to ②. If $[Na^+]_1 > [Na^+]_2$, the polarity of the zero current voltage difference will be reversed. Thus each ionic species appears in the membrane as [see Fig. 11(b)] a *battery* of voltage given by (3.8) with its positive terminal directed toward decreasing (increasing) ion concentration for positive (negative) ions. In general there will be several species of ions present which makes the analysis considerably more difficult. In 1943 Goldman derived a generalization of (3.6) under the assumption of a constant electric field (electroneutrality), and Offner (1971) has recently discussed numerical techniques which do not require this assumption. See Rosenberg (1969) for a comparison of resting potential formulas.

Let us now consider an experiment in which a pure lipid bilayer is carefully prepared in the apparatus of Fig. 9 [Howard and Burton (1968)] with equal concentrations for all ions so the ionic batteries are zero. The initial slope of the current density–voltage curve can be as low as [Goldup, Ohki and Danielli (1970)]

$$G \sim 10^{-9} \text{ mho/cm}^2$$

which, for a membrane thickness of about 100Å ($\sim 10^{-6}$ cm), implies a membrane resistivity

$$\rho \sim 10^{15} \text{ ohm-cm.}$$

Thus a clean lipid bilayer should be classified as a *very* good insulator, and the importance of the protein complex in facilitating ionic conduction through biological membranes cannot be overemphasized.

Mueller, Rudin, Tien and Wescott (1962) showed that the addition of small amounts of properly chosen and refined proteinaceous material (called EIM for "excitability inducing material") will increase the membrane conductivity by many orders of magnitude, and can introduce the nonlinearity essential for generation of an action potential. At low protein concentrations the conductance has been observed to increase in quantum units of about 4×10^{-10} mhos [Goldup *et al.* (1970)]. When alamethicin (a circular polypeptide with molecular weight ~ 1800) is added to the aqueous phase of a clean experiment, the membrane conductance is found to increase with the sixth power of concentration [Mueller and Rudin (1968b)]. These observations suggest that the alamethicin molecules may be coördinated in groups of six to permit ionic conduction through the membrane. Hille (1970) has surveyed a wide variety of kinetic, electrochemical, and pharmacological data for biological nerve membranes and concluded that the conductance changes observed during the action potential (see Fig. 3) are caused by the opening and closing of localized conductance channels. The term "pore" is often used in a generic sense to indicate a localized region of high conductivity on the membrane. For such a porous membrane (3.6) is no longer useful. The barrier potential, $w(r)$, and the ionic mobility, μ, depend strongly upon the position on the membrane surface and also upon the membrane voltage.

In this situation it is helpful to return to (3.1) and write it in the form

$$J_{12} = G(v_{12} - V) \qquad (3.12)$$

where, as was previously noted, $(v_{12} - V)$ is the negative of the change in electrochemical potential from chamber ① to ②. The conductivity is not a constant but a nonlinear function of the experimental variables. The form of (3.12) merely makes explicitly evident the zero in ion current which appears when the electrochemical difference for that ion is zero. Often, the conductance per unit area, G, appears as a function only of the transmembrane volrage, v_{12}. An exceptionally clear example of this has recently been published by Eisenberg, Hall, and Mead (1973) in connection with their careful study of the effect of alamethicin on artificial lipid bilayer membranes. The volt–ampere curve in Fig. 12(a) exhibits a distinct region of negative differential conductance; but the conductance [see in Fig. 12(b)] shows a simple exponential rise throughout this region. The experimental rise is the same as that observed without an ion imbalance. Thus it is clear that in this case we can write (3.12) in the form

$$J_{12} = G(v_{12})(v_{12} - V). \qquad (3.13)$$

As has been pointed out by Cole (1968, p. 289) and by Mueller and Rudin (1968a, b), the condition for negative differential conductance can then be expressed by differentiating (3.13) with respect to v_{12}

$$dJ_{12}/dv_{12} = G'(v_{12} - V) + G$$

so

$$dJ_{12}/dv_{12} < 0 \Rightarrow G'(V - v_{12}) > G. \qquad (3.14)$$

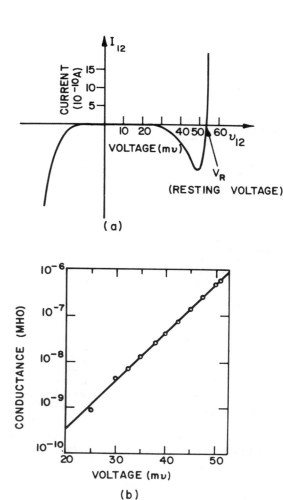

(a)

(b)

FIG. 12. Measurements on an artificial lipid bilayer membrane in a 100:1 KCl gradient. Chamber ①: 0.5 m KCl and 6×10^{-7} g/ml alamethicin. Chamber ②: 0.005 m KCl and 9×10^{-6} g/ml alamethicin. From Eisenberg, Hall and Mead (1973). (a) Current vs voltage. (b) Conductance vs voltage.

This condition for negative differential conductance was first demonstrated for an alamethicin-doped artificial lipid bilayer membrane by Mueller and Rudin (1968b). Whenever membrane current (J_{12}) is related to membrane voltage (v_{12}) as in (3.13), the condition can be expressed in the following simple physical terms: *negative differential conductance will appear when G is rising rapidly enough below the resting voltage.* Since the resting voltage depends upon ion concentrations, negative differential conductance of a membrane can be made to appear or disappear simply by changing the composition of the external solutions! Thus, as Agin (1969) has emphasized, the mere appearance of a negative conductance need not depend upon exotic effects such as interaction of divalent ions, conformational changes of macromolecules, micelle transformations of lipid systems, enzyme reactions, ion specific carriers, redistributions of pores, chemical gates, etc.

Cole (1968, pp. 287–290) points out that the functional form in (3.12) is especially useful for description of a squid axon membrane since G remains constant for times up to the order of 100 μsec. The current flow in response to more rapid changes in voltage is simply ohmic.

It should be noted that the current indicated in Fig.

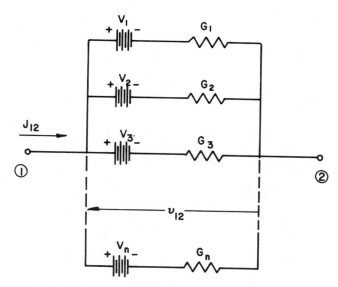

FIG. 13.   Membrane equivalent circuit for n ionic species.

12(a) is due to *both* potassium and chlorine ions. In general a membrane which separates n ionic species can be represented as in Fig. 13 whereupon current is related to transmembrane potential by [Cole (1968, pp. 193–7)]

$$J_{12} = \left[\sum_{i=1}^{n} G_i\right]v_{12} - \sum_{i=1}^{n} (G_i V_i) \qquad (3.15)$$

which has the same form as (3.12). From the discussion related to Fig. 11 it should be clear that for positive ions of concentrations $[C^+]_1$ and $[C^+]_2$

$$V_i = (kT/q) \log([C^+]_2/[C^+]_1). \qquad (3.16a)$$

as in (3.8). For negative ions of concentrations $[C^-]_1$ and $[C^-]_2$

$$V_i = (kT/q) \log[C^-]_1/[C^-]_2). \qquad (3.16b)$$

The resting potential (i.e., the value of v_{12} for $J_{12} = 0$) is

$$V_R = \sum (G_i V_i)/\sum G_i; \qquad (3.17)$$

thus if the conductance, G, for a particular ion becomes large, the resting potential will approach the battery voltage for that ion. To see how these equations can be used, consider the data of Fig. 12(a). The resing potential, $V_R = 53$ mV and, from the ion concentration ratios and (3.16), $V_K = +115$ mV and $V_{Cl} = -115$ mV. Thus from (3.17) we find at the resting potential that $G_K/G_{Cl} = 2.7$ so about 73% of the ion current flowing in the vicinity of the resting potential should be carried by potassium ions.

Depending upon one's point of view, (3.15) can be considered as (i) a flexible and useful description of multicomponent ion flow, or (ii) a phenomenological representation without physical meaning. The second attitude has been presented in detail by Tasaki (1968). He points out that if no restrictions are placed upon the functional dependence of the G's, then (3.15) says nothing more than (3.12). Furthermore (3.17) is of no value for calculation of a resting potential unless other information about the

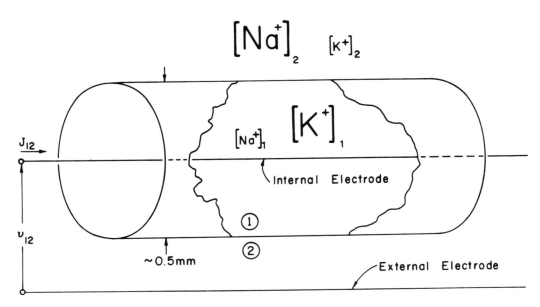

FIG. 14.   Geometry for a space clamped measurement on a squid nerve membrane.

membrane permeability to various ions is available. Tasaki (1968) carefully considers the calculation of resting potentials from physical considerations under a variety of simplifying assumptions. A complementary discussion is presented in the recent book by Khodorov (1974).

As an example of the kind of equation which can be derived for the resting voltage, Hodgkin and Katz (1949) assumed that each ion obeys the Nernst–Planck equation (3.3) and that the ion concentration just inside the membrane is a partition coefficient, γ, times the corresponding concentration outside the membrane. Then for univalent ions

$$V_R = (kT/q) \log \frac{\sum^+ \mu_i \gamma_i [C^+]_2 + \sum^- \mu_i \gamma_i [C^-]_1}{\sum^+ \mu_i \gamma_i [C^+]_1 + \sum^- \mu_i \gamma_i [C^-]_2}. \quad (3.18)$$

where $\sum^+$ ($\sum^-$) indicates summation over the positive (negative) ions.

So far we have been considering only passive (i.e., nonmetabolic) mechanisms for ion transport across a cell membrane. Active ion transport is extremely important in the operation of a living cell; and, although the details of such processes are not yet well understood, the broad outlines are emerging [(Lehninger (1970)]. The inside of a nerve cell, for example, is usually some 60–70 mV negative with respect to the outside. Using the convention of Figs. 5 and 10

$$J_{12} = 0 \quad \text{for} \quad v_{12} = V_R \approx -65 \text{ mV}.$$

For the squid giant axon [Hodgkin and Huxley (1952)]

$$[Na^+]_2/[Na^+]_1 \approx 7.5 \Rightarrow V_{Na} = +50 \text{ mV}$$

and

$$[K^+]_1/[K^+]_2 \approx 30 \Rightarrow V_K = -77 \text{ mV}.$$

Thus metabolic energy must be expended to pump sodium ions outward and potassium ions inward against the resting potential. We shall see that the electric field energy associated with the resting potential is expended in the propagation of an action potential [Hodgkin (1964)]. Current knowledge of the processes for outward pumping of Na^+ and inward pumping of K^+ has recently been reviewed by Thomas (1972). There are indications that three sodium ions are removed for each two potassium ions which enter. The energy for this process is supplied by the conversion of ATP (adenosine*tri*phosphate) to ADP (adenosine*di*phosphate). The ATP, in turn, is reconstituted in the membranes of subcellular units known as mitochondria.

IV. ELECTRODYNAMICS OF AN ACTIVE NERVE MEMBRANE

The most extensive nerve membrane measurements have been made on the giant axon of the squid [see Cole (1968) for a thorough discussion of the literature and a beautiful color photograph of the animal]. This fiber is between 0.5 and 1 mm in diameter, and several centimeters in length. It is easily removed from the squid and continues to function for at least several hours and often as long as a day.[3]

A typical experimental arrangement for measuring the electrodynamic properties of a membrane is indicated in Fig. 14 [Hodgkin, Huxley, and Katz (1952)]. This is called a "space-clamped" measurement because the electrode arrangement eliminates the possibility of longitudinal variation of voltage and the associated wave propagation effects; it is also called a "voltage clamped" measurement if a negative feedback amplifier is introduced to reduce the source impedance and permit v_{12} to be independently specified. We are interested in interpreting the relationship between J_{12} and v_{12} to extract the nonlinear character of the membrane indicated simply by $J_{12} = N(v_{12})$ in (2.2). As was previously mentioned in connection with (2.29), J_{12} is composed of a displacement current component through

---

[3] An introduction to the surgical procedures for removal of a nerve fiber is provided by the two part film loop *Nerve Impulse* available from Ealing Corp., 2225 Massachusetts Ave., Cambridge, Mass. 02140.

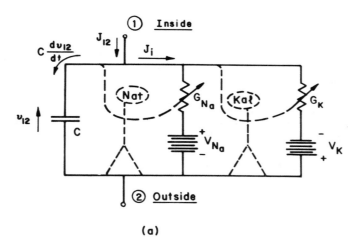

(a)

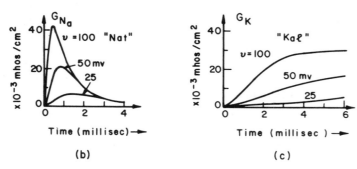

(b)   **(c)**

FIG. 15. (a) A simplified equivalent circuit for a unit area of squid membrane. (b) The reaction of Nat to displacement of membrane voltage from the resting value. (c) Ditto for Kal [Cole (1968), p. 272].

the membrane capacity, and an ion current through the membrane. Thus

$$J_{12} = C(dv_{12}/dt) + J_i, \qquad (4.1)$$

where C is the capacitance per unit area of the membrane (about 1 μF/cm²), and J_i is the sum of all the individual ionic currents through the membrane. If v_{12} is independent of time, the displacement current is zero, and the ion current should be a sum of terms as in (3.15). In measuring the ionic currents, it is therefore convenient to hold the membrane voltage fixed. It was this voltage-clamp measurement [Cole (1949), Marmot (1949)] which led Hodgkin and Huxley (1952) to a representation of J_i which could be used to solve (2.30) for a propagating action potential.

The sodium and potassium ion currents are most interesting because they respond nonlinearly to changes of voltage across the membrane. The behavior of these nonlinear currents has been described in a simple and appealing way by Katz (1966) and by Cole (1968) using the equivalent circuit shown in Fig. 15(a). This representation includes a sodium battery of about 50 mV directed inward and a potassium battery of about 77 mV directed outward. As was noted in the previous section, the ion batteries account for the tendency of sodium ions to diffuse inward and for potassium ions to diffuse outward. These batteries are in series with a sodium conductance per unit area, G_{Na}, and a potassium conductance per unit area, G_K, respectively as is indicated in Fig. 13. A small boy (named "Nat") senses the voltage across the membrane and adjusts G_{Na} according

to some rules of his own, and another small boy (named "Kal") does the same for G_K. What Nat and Kal do is conveniently described in terms of the change of potential inside the membrane with respect to its resting value. Thus we define

$$v \equiv v_{12} - V_R. \qquad (4.2)$$

If the voltage inside the membrane is made more negative (*hyperpolarized*), the membrane conductances remain small with little change in value. If the voltage inside the membrane is made less negative (*depolarized*), the reactions of Nat and Kal are indicated in Fig. 15(b) and (c). The individual ion current components can be measured by assuming the validity of (3.15) and adjusting the external salt solution to make V_{Na} or V_K equal to zero.

The curves in Fig. 15 indicate the way G_{Na} and G_K change with time for a *fixed* change in voltage. If the circuit is not voltage clamped, however, it will "switch." The reason for this is that a small depolarizing voltage ($v > 25$ mV) increases the conductance of the membrane to sodium ions. Thus sodium ions flow *into* the membrane which *increases* the depolarizing voltage causing the sodium ion conductance to increase even more. It is a positive feedback effect; once initiated the membrane will rapidly approach the sodium ion battery voltage $v_{12} = V_{Na}$ or $v = V_{Na} - V_R \simeq$ 115 mV due to the inrush of sodium ions. Then G_{Na} will fall back toward zero [Fig. 15(b)] and G_K will rise [Fig. 15(c)] allowing an outflow of potassium ions. This outward potassium ion current will bring the membrane potential back to its resting value. Increasing the potential inside the membrane by 25 mV or more is something like pulling the chain on the hopper; once the process starts it goes through the complete cycle. In large fibers the total ionic flow during one switching cycle is a very small fraction of the total ion concentration; many hundreds of thousands of firings can occur in a squid giant axon before the ionic batteries become discharged. In smaller fibers, such as those shown in Fig. 2, the ionic flow per impulse can be a substantial fraction of the total ion concentration.

This is a description of *what* happens. *Why* it happens is not yet understood, but some interesting clues can be gleaned from an investigation of the total ion current which flows in response to a fixed voltage (so $J_{12} = J_i$). From Fig. 16 it can be seen that if the voltage v_{12} is held at a value less than V_{Na}, the current J_{12} is first negative (inward), then positive (outward). From these curves it is possible to define an initial peak, J_p, and a final steady state value, J_{ss} as is indicated for the curve at $v_{12} = -20$ mV in Fig. 16. Both J_p and J_{ss} can then be plotted against the corresponding value of the voltage step as is indicated in Fig. 17(a). The early, J_p, branch of the curve is primarily sodium ion current; while the steady state, J_{ss}, branch is primarily potassium ion current. The membrane appears to be in a high conductance state for $v_{12} > -40$ mV and a low conductance state for $v_{12} < -50$ mV. Returning to the inequality condition for differential negative conductance expressed in (3.14), we see that in the range -50 mV $< v_{12} < -40$ mV the conductance is rising "rapidly enough."

Similar data for other electrically active biological membranes are plotted in Figs. 17(b)–(f). In each case there is an early current density (J_p) or current (I_p) which exhibits negative differential conductivity and eventually relaxes

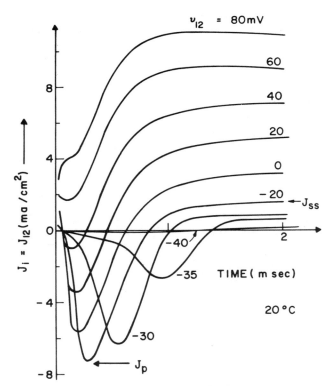

FIG. 16. Typical response of squid membrane current density to fixed steps of voltage [Cole (1968), p. 326].

into a steady state current density (J_{ss}) or current (I_{ss}) with only positive differential conductivity. In Figs. 17(a)–(d) the early current is carried primarily by sodium ions and the later current is carried primarily by potassium ions. In the measurement on *Aplysia californica* shown in Fig. 17(e), Geduldig and Gruener (1970) find clear evidence for a calcium ion contribution to the early current [see also Kryshtal', Magura and Parkhomenko (1969) and Chap. 5 of Khodorov (1974)]. The data in Fig. 17(f) are from a plant cell, the fresh water alga *Nitella*. This plant, which produces giant internodal cells with about the same dimension as the squid giant axon, has been described in detail by Scott (1962). For *Nitella* it appears that the early current is carried by an outward flux of chloride ions, while the later current is primarily outward potassium. The time required to relax from the J_p branch to the J_{ss} branch is the order of seconds for *Nitella* in contrast with a time of the order of milliseconds for the animal fibers in Figs. 17(a)–(c).

No universally acceptable theory has yet been proposed to explain the relation between membrane electrodynamics (Fig. 17) and membrane biochemistry (Fig. 8). An important recent contribution to this quest, however, is the review of various proposed mechanisms in Chapter 9 of the book by Khodorov (1974). These mechanisms include (i) mobile carriers with affinities for particular ions, (ii) special pores with ionic selectivity and the ability to open and close, (iii) conformational changes in membrane micromolecules, and (iv) special mechanisms for artificial membranes. Khodorov's discussion is particularly valuable because it brings the work of Russian scientists into focus.

As Tasaki (1968, 1974b) and Changeux (1969) have demonstrated, there is a considerably body of experimental evidence to suggest that the basic process of excitation in natural membranes involves a transition between two conformational states of the membrane. Figure 17 certainly suggests the ubiquitous nature of two conductivity states, and more detailed data includes: (i) direct observation of two conductivity states when Ca^{++} is used as the external cation [Inoue, *et al.* (1973)], (ii) observation of switching between these states by variation of the temperature, (iii) changes in extrinsic fluorescence during the time course of an action potential [Tasaki 1974a], (iv) electron micrographs of configurational transitions involving collapse and extension of "headpiece stalks" in mitochondrial membranes [Hatase *et al.* (1972)], and of lattice structure on electrically excitable membranes of insect photoreceptors [Gemmel (1969)], (v) nonaxoplasmic birefringence changes during the action potential [Cohen *et al.* (1970), Watanabe *et al.* (1973), Sato *et al.* (1973)], (vi) protein binding of a nontoxic dye during the action potential [Levin *et al.* (1968)], and (vii) direct observation of spatial nonuniformity during switching of a squid axon [Inoue *et al.* (1974)]. The absence of birefringence change in pure lipid bilayers reported by Berestovskii *et al.* (1970) reinforces the attitude of Green *et al.* (1972) that protein complexes play the key role in membrane function. The physical ideas recently suggested by Frölich (1970) may clarify the understanding of protein conformational states.

It should be emphasized that the concept of a conformational change during activity of a natural membrane does not conflict with the idea that ions flow through channels or "pores" in the membrane which was discussed in detail by Hille (1970). The two points of view can be considered as complementary aspects of a more complex reality. On the other hand, one should not conclude that the switching observed on the leading edge in Fig. 3 is direct evidence of membrane macromolecular dynamics [Changeux *et al.* (1967); Lehninger (1968); Nachmansohn and Neuman (1974)]. The basic positive feedback mechanism which drives an action potential is that discussed above and diagrammed in Fig. 18 [Hodgkin (1951, 1964)]. Several scientists have indicated how one might proceed from an essentially conformational membrane model to the ionic current data of Fig. 16 which, in turn, implies the feedback mechanism of Fig. 18 [Goldman (1964), Jain *et al.* (1970), Chizmadzhev *et al.* (1972, 1973)].

In 1952 Hodgkin and Huxley introduced a phenomenological expression for the ion current density through a squid membrane with the form

$$J_i = \bar{G}_K n^4 (v_{12} - V_K) + \bar{G}_{Na} m^3 h (v_{12} - V_{Na})$$
$$+ G_L (v_{12} - V_L), \tag{4.3}$$

where $\bar{G}_K$ and $\bar{G}_{Na}$ are, respectively, the maximum potassium and sodium conductances per unit area, and G_L is a constant leakage conductance. The phenomenological variables n, m, and h lie between zero and unity; the potassium conductance is "turned on" by n, and the sodium conductance is "turned on" and "turned off" by m and h, respectively. It is assumed that n, m, and h are independently relaxing toward equilibrium values n_0, m_0, and h_0, with characteristic times τ_n, τ_m, and τ_h. Thus

$$dn/dt = -(n - n_0)/\tau_n; \quad dm/dt = -(m - m_0)/\tau_m;$$
$$dh/dt = -(h - h_0)/\tau_h. \tag{4.4a,b,c}$$

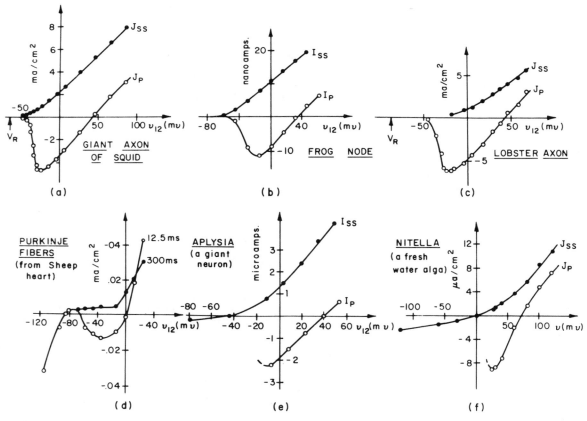

FIG. 17.   Voltage clamp data from various active biological membranes.   (a), (b), and (c) redrawn from Cole (1968), (d) redrawn from Deck and Trautwein (1964), (e) redrawn from Geduldig and Gruener (1970), and (f) redrawn from Kishimoto (1965). J_p and J_{ss} are defined in Fig. 16.

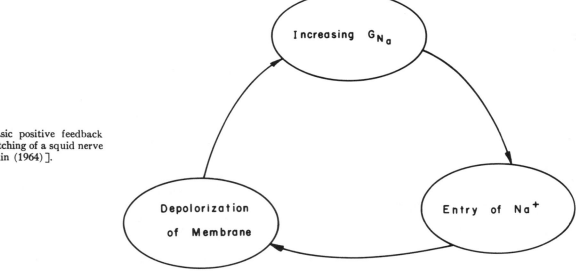

FIG. 18.   The basic positive feedback mechanism for switching of a squid nerve membrane [Hodgkin (1964)].

The relaxation parameters (n_0, m_0, h_0, τ_n, τ_m, and τ_h) can be determined as functions of voltage such that (4.3) will reproduce voltage clamp data as in Fig. 16. The nature of this functional dependence is shown in Fig. 19(a) where the constant values given by Cole (1968) are also indicated.

When the corresponding variables are determined for the active node of a frog myelinated axon (area $\sim 20\ \mu^2$), the results are strikingly similar as shown in Fig. 19(b). This

result might be anticipated from a comparison of Figs. 17(a) and 17(b).

Hodgkin and Huxley obtained analytic expressions for the parameters in (4.4) of the form

$$dn/dt = \alpha_n(1 - n) - \beta_n n, \qquad (4.4'a)$$

$$dm/dt = \alpha_m(1 - m) - \beta_m m, \qquad (4.4'b)$$

$$dh/dt = \alpha_h(1 - h) - \beta_h h. \qquad (4.4'c)$$

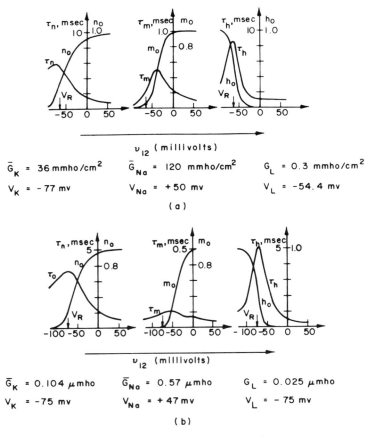

FIG. 19. The functional dependence of Hodgkin–Huxley phenomenological parameters on membrane voltage. (a) For a typical squid axon and (b) for a frog mode of area $\approx 20\ \mu^2$. Redrawn from Cole (1968) p. 283, 479.

Then, as functions of the voltage $v = v_{12} - V_R$ defined in (4.2) and measured in millivolts,

$$\alpha_n = \frac{0.01(10 - v)}{[\exp(10 - v)/10 - 1]}, \tag{4.5a}$$

$$\beta_n = 0.125 \exp(-v/80), \tag{4.5b}$$

$$\alpha_m = \frac{0.1(25 - v)}{[\exp(25 - v)/10 - 1]}, \tag{4.5c}$$

$$\beta_m = 4 \exp(-v/18), \tag{4.5d}$$

$$\alpha_h = 0.07 \exp(-v/20), \tag{4.5e}$$

$$\beta_h = \frac{1}{[\exp(30 - v)/10 + 1]}, \tag{4.5f}$$

where the units are $msec^{-1}$. Equations (4.5) give the rate constants measured at a temperature of 6.3°C. For other temperatures they should be multiplied by the factor κ where

$$\kappa = \exp[(T - 6.3)/10]. \tag{4.6}$$

Clearly (4.3) and (4.4) provide wide flexibility for fitting voltage clamp data similar to that displayed in Figs. 15 and 16. Although the ion battery potentials (V_K, V_{Na} and V_L) are fixed by the respective concentration ratios, the maximum conductivities ($\bar{G}_K$, $\bar{G}_{Na}$ and G_L) can be adjusted in addition to the six functions of v required to specify (4.4). Furthermore the choice of powers appearing in (4.3) is somewhat arbitrary. The fourth power of n was chosen to yield the "sigmoidicity" in the initial rise of potassium conductance evident from Fig. 15(c); and, as Hodgkin and Huxley note, "better agreement might have been obtained

with a fifth or sixth power, but the improvement was not considered to be worth the additional complication." A later study by Cole and Moore (1960) suggested that the twenty-fifth power of n is more appropriate in order to reproduce the time delay which appears when the membrane is switched on from the hyperpolarized state.[4] Similar considerations apply to the m^3h factor in (4.3). The task is to represent a sodium conductance which first rises then falls as is indicated in Fig. 15(b). Such an experimental result can be described by dependence upon a single variable which obeys a second-order differential equation or upon two variables each of which obeys a first-order differential equation. Hodgkin and Huxley note, "the second alternative was chosen since it was simpler to apply the experimental results."

The Hodgkin–Huxley expression for ion current density (4.3) is well defined, and useful for a variety of numerical and intuitive checks on experimental results. It has stimulated an everwidening analytical study which extends far beyond the professional boundaries of neurophysiology. Thus there is an inevitable (and regrettable) tendency to consider (4.3–5) as "graven on a stone tablet." The applied mathematician should be more concerned with the qualitative features of (4.3) than the algebraic details. The biochemist, on the other hand, should concentrate upon the development of a fundamental theory of membrane dynamics which can reproduce the voltage clamped data as displayed in Fig. 16. Useful reviews of the Hodgkin–Huxley equations include Noble (1966) and Moore (1968) in addition to the books by Cole (1968) and Khodorov (1974).

Various other suggestions for analytical representations

---

[4] A suggestion which "wasn't recognized for tongue-in-cheek" (Cole, 1975).

of the potassium conductance include the work of:

(i) Tille (1965) who takes

$$I_K = \bar{G}_K N \tag{4.7}$$

where

$$dN/dt = a_1 N + a_2 N^2 + a_3 N^3 + a_4 N^4 + a_5 N^5 \tag{4.8}$$

and the a_i are appropriately chosen functions of membrane voltage,

(ii) FitzHugh (1965) who obtains (4.7) with $N = \exp(-\mu)$ and

$$d\mu/dt = \alpha - \beta\mu \tag{4.9}$$

where α and β are functions of the membrane voltage, and

(iii) Hoyt (1963) who uses (4.4a) from Hodgkin and Huxley (1952) then empirically determines the functional form for $G_K(n)$. She finds deviation from a power law at larger values of n.

In a discussion following the presentation of FitzHugh (1965), Cole points out the wide range of functional expressions which can represent the sigmoid nature of the potassium conductance rise with roughly equal accuracy. During this discussion Cole, Hoyt and FitzHugh are in agreement that there is no uniquely superior analytical form. For the Purkinje fibers in the mammalian heart, however, Noble (1962) has described a modified representation for the potassium which accounts for the slow recovery indicated in Fig. 17(d). This slow recovery is necessary for the generation of heartbeats.

Analytical study of the rise and fall of sodium conductance [Fig. 15(b)] has been of more fundamental importance. Frankenhaeuser and Huxley (1964) have shown for myelinated axons of the toad (*Xenopus laevis*) that an m^2h dependence is more appropriate. Hoyt (1963, 1968) and Hoyt and Adelman (1970) have demonstrated that for a squid giant axon the sodium conductance is somewhat better represented by dependence upon a single variable which satisfies a second-order differential equation or, equivalently, two variables which satisfy coupled first order equations. Hoyt and Adelman state: "These conclusions imply that the mechanism responsible for the increase in sodium conductance is more likely to be dependent upon the production of an intermediate state than on the competition of two antagonistic but independent processes..."; but see also Jakobsson (1963). Molecular theories leading to coupled equations include the work of Mullins (1959), Goldman (1964), Fishman et al. (1972), and Chizmadzhev et al. (1972, 1973). Other models for membrane dynamics with varying degrees of phenomenology and membrane biochemistry include the work of Jain, Marks, and Cordes (1970), Offner (1970, 1972, 1974), Moore and Jakobsson (1971), and Jakobsson and Scudiero (1975). Hoyt and Strieb (1971) and Landdowne (1972) have independently suggested that the time course of the sodium conductance [Fig. 15(b)] may be explained by assuming the current to be carried primarily by ions stored *within* the membrane. This implies a temperature dependence of ion flux which is much weaker than is indicated by (4.6), and initial

experiments seem to confirm this prediction [Landowne (1973), Cohen and Landowne (1974)].

V. THE HODGKIN–HUXLEY AXON

We are now in a position to discuss the nonlinear dynamics of the nerve fiber shown in Fig. 5(a). The first order partial differential equations are (2.21) together with (4.4). Combining (2.21b) with (2.29) we can write these as

$$\partial v/\partial x = -r_s i, \tag{5.1a}$$

$$\partial i/\partial x + c(\partial v/\partial t) = -j_i(v, n, m, h), \tag{5.1b}$$

$$\partial n/\partial t = -[n - n_0(v)]/\tau_n(v), \tag{5.1c}$$

$$\partial m/\partial t = -[m - m_0(v)]/\tau_m(v), \tag{5.1d}$$

$$\partial h/\partial t = -[h - h_0(v)/\tau_h(v)], \tag{5.1e}$$

where j_i in (5.1b) is the membrane ion current per unit length. From here on it is typographically convenient to use the voltage variable $v = v_{12} - V_R$ defined in (4.2); evidently this makes no difference on the left-hand sides of (5.1a) and (5.1b). From (4.3)

$$j_i = \bar{g}_K n^4(v - V_R - V_K) + \bar{g}_{Na} m^3 h(v - V_R - V_{Na})$$
$$+ g_L(v - V_R - V_L), \tag{5.2}$$

where $\bar{g}_K = 2\pi a \bar{G}_K$, $\bar{g}_{Na} = 2\pi a \bar{G}_{Na}$ and $g_L = 2\pi a G_L$.

The "average axon" chosen for numerical study by Hodgkin and Huxley (1952) had the following parameters in addition to those specified in the previous section.

Resting potential:	$V_R = -65$ mV.
Axoplasm conductivity:	$\sigma = 2.9$ mho/m.
Axon radius:	$a = .238$ mm.
Membrane capacitance:	$C = 1$ μF/cm².

One approach to the analysis of these equations is to seek traveling wave solutions where all dependent variables (v, i, h, m, h) are functions only of a moving spatial variable

$$\xi = x - ut. \tag{5.3}$$

This can be considered as a special case of the more general independent variable transformation

$$x \to \xi = x - ut \qquad \partial/\partial x \to \partial/\partial \xi,$$

so

$$t \to \tau = t \qquad \partial/\partial t \to \partial/\partial \tau - u(\partial/\partial \xi). \tag{5.4}$$

Assuming independence with respect to τ in the $(\xi-\tau)$ system, we can replace $\partial/\partial x$ by $d/d\xi$, and $\partial/\partial t$ by $-ud/d\xi$, whereupon Eqs. (5.1) become the ordinary differential

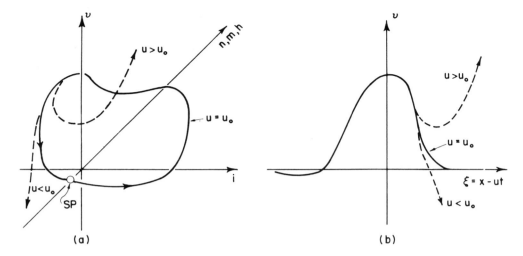

(a)   (b)

FIG. 20.   (a) Phase space trajectory corresponding to (b) an action potential. The phase space actually has five dimensions, but n, m, and h are indicated along a single axis.

equations

$$dv/d\xi = -r_s i,$$

$$di/d\xi = -r_s c u i - j_i,$$

$$dn/d\xi = (n - n_0)/u\tau_n,$$

$$dm/d\xi = (m - m_0)/u\tau_m,$$

$$dh/d\xi = (h - h_0)/u\tau_h. \tag{5.5}$$

This is an *autonomous* set of equations [Hurewicz (1958), Lefschetz (1962)] since the derivatives are uniquely defined as functions of the dependent variables. Thus phase space techniques can be helpful in understanding the structure of solutions [Kolmogoroff *et al.* (1937)]. It is important to note, however, that u (the velocity of the moving spatial coordinate in (5.4)) appears as an adjustable parameter in (5.5). In general one can expect the topological character of the phase space trajectories to depend upon the value chosen for the velocity u. Only those trajectories for which the dependent variables are bounded will be of physical interest. In particular a trajectory corresponding to the action potential shown in Fig. 3 should have the qualitative character indicated in Fig. 20(a). The values $v = 0$, $i = 0$, and $(n, m, h) = (0.35, 0.06, 0.6)$ are a solution of (5.1) so the corresponding point in the phase space of (5.5) is a *singular point* (SP) at which all the ξ derivatives are equal to zero. The task of finding a pulselike traveling wave solution for (5.1) involves determining the proper value of the velocity u at which a trajectory which emanates from this singular point (at $\xi = -\infty$) eventually returns to it (as $\xi \rightarrow +\infty$). Such a trajectory is sometimes called *homoclinic*, while a *heteroclinic* trajectory would pass between two different singular points.

A homoclinic trajectory was determined by Hodgkin and Huxley (using a hand calculator) in 1952. Voltage and membrane conductance are plotted as a function of time from this calculation in Fig. 21 for the proper value of 18.8 mps. This value is in satisfactory agreement with the measured value of 21.2 mps; and, as a comparison of Figs. 3 and 21 will show, so also are the waveforms $v(t)$ and $G(t)$.

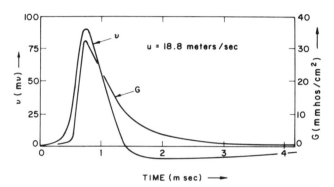

TIME (m sec) ——

FIG. 21.   Waveforms of the action potential and membrane conductance calculated from (5.5) at 18.5°C. Redrawn from Hodgkin and Huxley (1952d).

From a theoretical point of view, the discovery of a pulselike traveling wave solution for (5.1) from an investigation of the phase space topology associated with (5.5) does not mean that the pulse is stable to perturbations of its shape. Such *waveform instability* involves dependence upon τ, and (5.5) was derived with the specific assumption of independence with respect to τ. We will study this question in detail below. Another form of instability which appears in these calculations is *numerical instability* during the integration of (5.5). This arises because the assumed pulse velocity, u, is an adjustable parameter in the analysis. Choosing u slightly too small or too large may cause the computed waveform to diverge as is indicated in Fig. 20. Such numerical instability of a solution to (5.5) seems to be a necessary condition to avoid a waveform instability in the corresponding solution of (5.1) [Scott (1970)].

Machine computations for the space clamped membrane were first reported by Cole, Antosiewicz, and Rabinowitz in 1955, and for the propagating axon by FitzHugh and Antosiewicz and by Huxley in 1959. Huxley demonstrated the existence of a second pulse solution (shown in Fig. 22) which propagates with only 30% of the velocity of the full action potential. This pulse has an unstable waveform; it will either decay to zero or rise to the full action potential

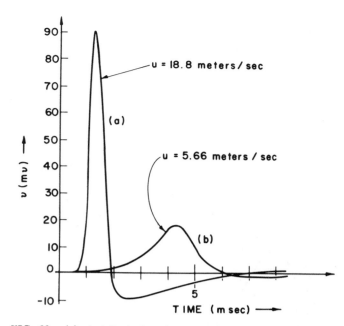

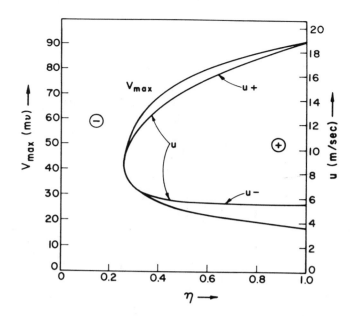

FIG. 22. (a) A full sized action potential and (b) an unstable threshold pulse for the Hodgkin–Huxley axon at 18.5°C. Redrawn from Huxley (1959).

FIG. 23. Amplitude and velocity for a traveling wave pulse on a Hodgkin–Huxley axon vs a "narcotization factor," η, which reduces the sodium and potassium conductances. Redrawn from Cooley and Dodge (1966).

and thus represents a boundary or threshold state of the fiber. Huxley (1959) also indicated the possibility of a subthreshold wave train which would correspond to a closed cycle in the phase space sketched in Fig. 20. The observation of a threshold pulse was confirmed by Cooley and Dodge (1966) through direct integration of (5.1). They extended the result by assuming that the effect of a narcotic agent would be to lower $\bar{g}_{Na}$ and $\bar{g}_K$ by a factor η. The results are plotted in Fig. 23 where it can be seen that *no* attenuation-less propagation or threshold effect obtains for $\eta < \eta_0 = .261$. At smaller values of this "narcotization factor" a "decremental" pulse [Lorente de N6 and Condouris (1959)] propagates with slowly diminishing amplitude as shown in Fig. 24. Since this pulse is not a function only of the argument $x - ut$, it is not represented by solutions of (5.5) and requires the complete set (5.1) for its description. Such decremental or "graded" pulses have also been extensively discussed by Leibovic (1972). Decremental pulses have been of great interest to physiologists in decades past and some of the flavor of these discussions is captured in the accounts by Kato (1924, 1970). Numerical analysis of the Hodgkin–Huxley axon not only indicates the possibility of decremental conduction, but the experimental conditions under which it should be observed. Recently Kashev and Bellman (1974) have introduced a new method of "differential quadrature" for more rapid integration of (9.1).

It is interesting to relate the results of these numerical studies to the notion of power balance which was introduced in (1.3). The $u - \eta$ locus in Fig. 23 indicates where pulse solutions can be found which satisfy (1.3). Since the lower branch is unstable, the inside $\oplus$ region is where pulse solutions for (5.1) can be found with $uE > P$. In the outside $\in$ region, $uE < P$ for all pulse solutions of (5.1). For η slightly less than η_0, we expect that uE for a pulse with appropriate shape and velocity will be almost equal to P. In this case an approximate calculation using only data from the traveling wave analysis may be useful. To see

this, note that the data in Fig. 23 are fairly well represented by

$$[u - u_0(\eta)]^2 = k(\eta - \eta_0), \qquad (5.6)$$

where $u_0 \equiv \frac{1}{2}(u_+ + u_-)$ and $k = 74 \ \mathrm{m^2/sec^2}$. When $\eta = \eta_0$, there will be a traveling wave pulse with the Fourier transform, $F_0(\beta)$.

$$v_0[x - u_0(\eta_0)t] = \int_{-\infty}^{\infty} F_0(\beta) \exp[i\beta(x - u_0 t)] \, d\beta. \quad (5.7)$$

When $\eta < \eta_0$, (5.6) indicates a *complex* value for the traveling wave velocity

$$u = u_0 \pm iu_i \qquad (5.8)$$

where $u_i \equiv [k_0(\eta_0 - \eta)]^{1/2}$. The primary effect of the imaginary component of velocity is to modify the magnitude of the Fourier transform. Thus an approximate expression for the evolution of a decremental pulse is

$$v(x, t) \approx \int_{-\infty}^{\infty} F_0(\beta) \exp(- |\beta| u_i t) \exp[i\beta(x - u_0 t)] \, d\beta.$$

$$(5.9)$$

The reason for taking the absolute value of β is to keep the Fourier transform symmetric so $v(x, t)$ remains real; the justification is that the roots in (5.8) may be interchanged without introducing a physical discontinuity when $\beta = 0$. Equation (5.9) may of course be written as the convolution of $v_0(x - u_0 t)$ with the Lorentzian pulse

$$\left\{ (\pi u_i t) \left[1 + \left(\frac{x - u_0 t}{u_i t} \right)^2 \right] \right\}^{-1}$$

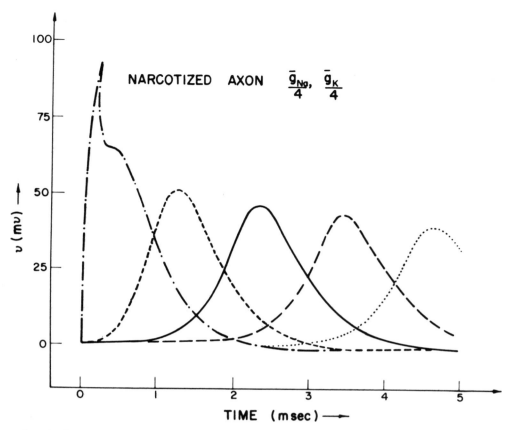

FIG. 24.   Propagation of a decremental pulse on a Hodgkin–Huxley axon narcotized by a factor of 0.25. Curves are voltage waveforms at 1 cm intervals. Redrawn from Cooley and Dodge (1966).

which has unit area and a half-width of $2u_i t$. For large t, (5.9) implies decay as t^{-1} rather than exponentially which is clearly incorrect. A numerical evaluation of (5.9) is currently being made for intermediate values of time.

Impedance bridge measurements by Cole and Baker (1941) indicated that the membrane appears to have an inductive current component at small ac amplitudes between 30 cps and 200 kc. For the membrane equivalent circuit representing 1 cm² of membrane shown in Fig. 25(a), they found $C = 1\ \mu F$, $R = 400\ \Omega$ and $L = 0.2\ H$. Hodgkin and Huxley (1952) investigated the dynamical relation between small changes in voltage and current in (4.3) and directly calculated in values $R = 820$ ohms and $L = 0.39\ H$ with a threefold increase in L for a 10° fall in temperature. Such an inductance is much too large to have any connection with magnetic fields; thus a physical interpretation is illustrated in Fig. 25(b) which depends upon the experimental fact that membrane conductance [G in Eq. (3.12)] remains constant for times of the order of 100 μsec or less [Mauro (1961)]. If the current is concave in the direction of depolarization, a sudden change of current from J_1 to J_2 must be associated with a change of voltage from v_1 to v_2'. The voltage will then slowly relax toward a smaller difference v_2. These conditions are met by the n and h dependencies in (4.3) both of which contribute to the inductance indicated in Fig. 25(a). Extensive studies of this effect include those by Chandler et al. (1962) and Mauro et al. (1970). Offner (1969) has related membrane reactance to the dynamics of internal ions.

The phenomenological inductance also influences the propagation of alternating subthreshold waves on the axon; this is evident from the "overshoot" in the return to rest of the action potential in Fig. 21. Subthreshold oscillatory propagation has been studied in detail by Sabah and Leibovic (1960) [Leibovic (1972)] using Laplace transform techniques and by Mauro, Freeman, Cooley and Cass (1972). Mauro et al. use both numerical analysis of (5.1) and experimental observations on squid axons to show that phase velocity of an oscillatory subthreshold wave is rather closely related to the pulse velocity of an action potential as indicated in Table I. [See also Optowski (1950) in connection with this relation.] In electronic jargon the squid axon looks like a low Q, bandpass filter tuned to about 100 cps when it is stimulated by a subthreshold, oscillatory current.

Cooley and Dodge (1966) also computed the response of a Hodgkin–Huxley axon to a steady stimulation by longitudinal current [$i(0, t) = $ const in Fig. 5(a)]. For a steady current around 3.4 μA a periodic train of spikes was generated with a frequency rather insensitive to the stimulation. This result is in contrast to the real axon which generates a burst of only a few spikes. FitzHugh (1969) has suggested that the real axon exhibits an "adaptation" effect which tends to decrease excitability with a time constant of the order of a second. Such an effect, which is not represented by the Hodgkin–Huxley equations, may be connected with slow changes in ion concentration or in temperature.

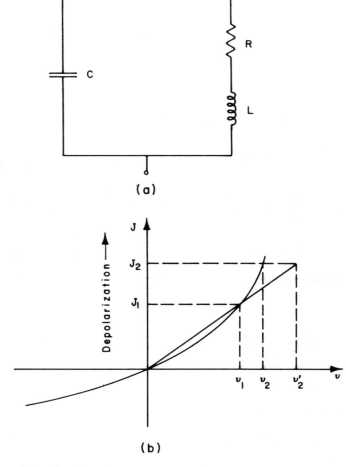

(a)

(b)

FIG. 25. (a) Membrane small signal equivalent circuit measured by Cole and Baker (1941). (b) Physical explanation of the phenomenological inductance.

VI. PROPAGATION OF THE LEADING EDGE

Comparison of numerical results reported in the previous section with corresponding experimental data indicates that the Hodgkin–Huxley equations (5.1) are of considerable value in describing the facts of electrophysiology, but it is also of interest to consider approximate forms of (5.1) which can be analytically investigated. Physical motivation for one such approximation stems from the observations (see Fig. 21) that (i) the most rapid dynamical change occurs on the leading edge of an action potential, (ii) this leading edge transition carries the membrane potential from its resting potential to approximately the sodium diffusion potential, V_{Na}, and (iii) the velocity of the leading edge determines the velocity of the entire action potential. For the squid giant axon the functions n_0, m_0, h_0, τ_n, τ_m, and τ_h are sketched in Fig. 19(a) from which it is evident that the relaxation time, τ_m, for sodium turn on is about an order of magnitude less than τ_n and τ_h for potassium turn on and sodium turn off respectively. Thus it is interesting to consider the approximation [FitzHugh (1969)]

$$\tau_m = 0, \qquad \tau_n = \tau_h = \infty,$$

whereupon the ion current through the membrane (5.2)

TABLE I. Velocities of the action potential and a subthreshold oscillation vs temperature for the H–H axon. [Mauro et al. (1972)].

Temperature (°C)	Pulse velocity of action potential (mps)	Phase velocity of subthreshold oscillation (mps)
18.5	18.8	16.1
12.5	16.1	14.6
6.3	12.7	13.3

becomes simply a function of voltage $j_i \approx j(v)$, where

$$j(v) = \bar{g}_K n_0^4(V_R)(v - V_R - V_K)$$
$$+ \bar{g}_{Na} m_0^3(v) h_0(V_R)(v - V_R - V_{Na})$$
$$+ g_L(v - V_R - V_L). \tag{6.1}$$

This approximation is valid only for dynamical processes which occur in times long compared with τ_m and short compared with τ_n and τ_h, but, as reference to Fig. 21 indicates, the leading edge transition comes close to fulfilling these requirements. Equation (2.30) then takes the form[5]

$$v_{xx} - r_s c v_t = r_s j(v) \tag{6.2}$$

which is the equation for nonlinear diffusion discussed in the Introduction. Together with (6.2) it is convenient to write (2.21) in the form

$$v_x = -r_s i \tag{6.3a}$$

$$i_x + c v_t = -j(v) \tag{6.3b}$$

as an equivalent set of first order pde's.

Equation (6.1) does not have a particularly convenient analytic form, but we expect it to go through zero at the origin (the resting potential), at a higher voltage $V_2 = V_{Na} - V_R$, and at a voltage, V_1, somewhere between. With this in mind, let us apply the transformation (5.4) discussed in the previous section to (6.3) with the assumption that $\partial/\partial\tau = 0$. Then the set of ordinary equations which are equivalent to (5.5) becomes

$$dv/d\xi = -r_s i \tag{6.4a}$$

$$di/d\xi = -r_s c u i - j(v). \tag{6.4b}$$

Singular points for this set occur where $i = 0$ and $j(v) = 0$, i.e., at $v = 0$, V_1 and V_2. If we define

$$g(v) \equiv dj/dv, \tag{6.5}$$

then $g(0)$ and $g(V_2)$ will be positive, and $g(V_1)$ will be negative as is indicated in Fig. 26(a). From this one can show [Scott (1970), McKean (1970)] that the singular points at $(i, v) = (0, 0)$ and $(0, V_2)$ are saddle points, while the intermediate singular point at $(0, V_1)$ is an inward (outward) node or focus for $u > 0$ (<0). Kunov (1967) used "Bendixon's negative criterion" [Andronov et al. (1966)] to show that (6.4) has a homoclinic trajectory, corresponding to a "pulselike" solution of (6.3), only for zero velocity. Thus the basic solutions with nonzero velocity

[5] From here on the conventional subscript notation for partial differentiation will be used wherever it is typographically convenient.

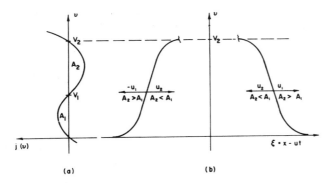

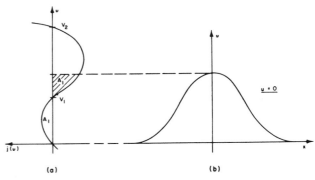

FIG. 26.   (a) A representation of $j(v)$ as in (6.1). (b) Propagating waves which change the voltage level.

FIG. 27.   (a) $j(v)$ with $A_2 > A_1$. (b) A stationary pulse solution.

are the "level change" waves shown in Fig. 26(b). From the phase space point of view, the velocity of such a transition is fixed by the condition that an isolated trajectory leaving one saddle point (at $\xi = -\infty$) must become an isolated trajectory approaching the other saddle point (as $\xi \to \infty$). Yoshizawa (1971) has demonstrated that these waves can either charge the membrane capacitance when area A_2 is greater than area A_1, or discharge the capacitance for $A_1 > A_2$. In either case, the power balance condition (1.3) must be satisfied.

If $A_1 = A_2$ these velocities are equal to zero which is a special case of the zero velocity pulse indicated in Fig. 27 for the case $A_2 > A_1$. From (6.4) with $u = 0$, it is easily seen that a pulse like solution is obtained by substituting into (6.4a) the homoclinic trajectory

$$i = \pm \left[\frac{2}{r_s} \int_0^v j(v') \, dv' \right]^{1/2}. \tag{6.6}$$

Although, as we shall see below, this solution is unstable, it is of interest because it specifies the condition for threshold stimulation of a fiber. Lindgren and Buratti (1969) have shown the pulse velocity to be nonzero for a tapered fiber.

A family of analytic solutions for the wave forms and velocities indicated in Fig. 26 can be obtained by writing [Scott (1974)]

$$dv/d\xi = T(v), \tag{6.7}$$

whereupon (6.4) requires that T must satisfy

$$T' = j(v)/T - r_s cu. \tag{6.8}$$

For $u = 0$, the pulselike trajectory of (6.6) is recovered. Now suppose $u \neq 0$ and $j(v)$ is a polynomial of order n, and $T(v)$ is a polynomial of order m, then T' is of order $(m-1)$, and from (6.8)

$$n = 2m - 1. \tag{6.9}$$

The case $m = 2$ implies $n = 3$ so $j(v)$ must be approximated by a cubic polynomial [Nagumo, Arimoto and Yoshizawa (1965), FitzHugh (1969)]

$$j(v) = Bv(v - V_1)(v - V_2), \tag{6.10}$$

where B is a constant (with units of mho/V²) chosen to make $j(v)$ approximate $2\pi a J_p$ from Fig. 17(a) as closely as possible. Since $m = 2$, a suitable quadratic trajectory is

$$i = Kv(v - V_2)$$

which, upon differentiation, gives

$$di/dv = 2Kv - KV_2.$$

But di/dv can also be evaluated by dividing left- and right-hand sides of (6.4) to obtain

$$di/dv = cu + (B/r_s K)(v - V_1).$$

Thus $K = -(B/2r_s)^{1/2}$ so

$$u = (B/2r_s c^2)^{1/2}(V_2 - 2V_1) \tag{6.11}$$

and (6.4a) can be integrated to

$$v = \tfrac{1}{2} V_2 \{ 1 + \tanh[\tfrac{1}{2} V_2 (\tfrac{1}{2} Br_s)^{1/2}(x - ut)] \}. \tag{6.12}$$

Note that the velocity given by (6.11) changes sign as V_1 becomes greater than $V_2/2$. This corresponds to the area condition indicated on Fig. 26(b). Similar results have been obtained for other nonlinear wave systems simulating the nerve axon by Il'inova and Khokhlov (1963) and by Parmentier (1969).

Another approximation for $j(v)$ which permits an analytic solution for (6.4) corresponds to the case $m = 1$; so from (6.9) $n = 1$ and we have a piecewise linear curve indicated in Fig. 28. Below a voltage V_1 the membrane is assumed to remain in a *resting* state with low conductance; above V_1 it is assumed to switch into an *active* state of much higher conductance. Such an approximation is certainly suggested by several of the curves for J_p vs v_{12} in Fig. 17. Using the notation of Tasaki (1968), we write [Scott (1962), McKean (1970)]

$$j(v) = g_r v \qquad \text{for } v < V_1$$
$$= g_a(v - V_2) \qquad \text{for } v > V_1. \tag{6.13}$$

The discontinuity at V_1 is acceptable because Eqs. (6.2) and (6.3) do not involve derivatives of $j(v)$. With $j(v)$ approximated as in (6.13), Eq. (6.2) is *linear* both above and below V_1. Thus the nonlinearity in the problem mani-

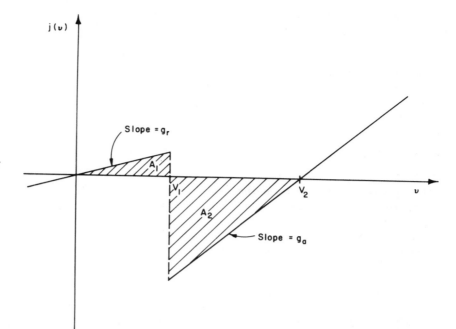

FIG. 28. Piecewise linear approximation for $j(v)$ [Tasaki (1968)].

fests itself only where $v = V_1$. To simplify the discussion we will begin by assuming that $g_r = 0$. Equation (10.4) can be written

$$d^2v/d\xi^2 + r_s cu(dv/d\xi) - r_s j(v) = 0$$

which becomes

$$d^2v/d\xi^2 + r_s cu(dv/d\xi) = 0 \qquad \text{for } v < V_1, \qquad (6.14a)$$

and

$$d^2v/d\xi^2 + r_s cu(dv/d\xi) - r_s g_a(v - V_2) = 0 \qquad \text{for } v > V_1. \qquad (6.14b)$$

If, for convenience, we choose $\xi = 0$ to be where $v = V_1$, a leading edge which makes a transition between zero and V_2 [see Fig. 26(b)] and satisfies (6.14) is easily constructed. Thus

$$v = V_1 \exp(-\gamma_1 \xi) \qquad \text{for } v < V_1, \qquad (6.15a)$$

and

$$v = V_2 - (V_2 - V_1) \exp(\gamma_2 \xi) \qquad \text{for } v > V_1, \qquad (6.15b)$$

where $\gamma_1 = r_s cu$ and $\gamma_2 = (r_s cu/2)[-1 + (1 + 4g_a/r_s c^2 u^2)^{1/2}]$. The velocity of propagation is not yet determined in (6.15) but it may be computed in either of two ways [Scott (1962)]:

(i) Equate the total power being produced by $j(v)$ and absorbed by r_s over the waveform to $\frac{1}{2}cV_2^2 u$, the power being absorbed by the membrane capacitance at velocity u; or

(ii) Demand continuity in the longitudinal current, i, at $\xi = 0$.

Approach (i) is employment of the power balance idea behind (1.3) which was discussed in the Introduction. The

leading edge must absorb energy (electrical energy in the membrane capacitance) at the same rate it is being produced for a steady traveling wave to exist. Approach (ii) is equivalent to (i) and somewhat more convenient. From (6.15) and (6.4a), $i(\xi)$ is easily calculated for the ranges $\xi > 0$ and $\xi < 0$, and current continuity at $\xi = 0$ implies $\gamma_2(V_2 - V_1) = \gamma_1 V_1$ which is readily solved for the velocity as

$$u = [g_a/r_s c^2]^{1/2}[(V_2 - V_1)/(V_2 V_1)^{1/2}]. \qquad (6.16)$$

The case $g_r \neq 0$ has been studied in detail by Kunov (1966) and by Vorontsov, Kozhevnikova and Polyakov (1967) who use a similar technique to find

$$u = \frac{\{[(V_2 - V_1)/V_1]^2 g_a - g_r\}}{(r_s c^2 [V_2(V_2 - V_1)/V_1^2]\{[(V_2 - V_1)/V_1]g_a + g_r\})^{1/2}} \qquad (6.17)$$

[see also Kompaneyets (1971)]. It can be seen that a necessary condition for a steady wave of transition from $v = 0$ to V_2 is $(V_2 - V_1)^2 g_a > V_1^2 g_r$. This again implies again that the areas A_2 and A_1 in Fig. 28 must satisfy the inequality

$$A_2 > A_1. \qquad (6.18)$$

The effect of "narcotization," discussed in the previous section in connection with Figs. 23 and 24, is to reduce g_a. Eventually the inequality (6.18) is violated and only decremental conductance can take place.

The value of (6.16) can be assessed by using it to calculate the velocity of the action potential for the Hodgkin–Huxley axon shown in Fig. 21. From the Hodgkin–Huxley axon parameters given in the previous section, the factor $(g_a/r_s c^2)^{1/2}$ is equal to 33.2 m/sec. Taking $V_2 = V_{Na} - V_R = 115$ mV and (from Fig. 17) $V_1 = 40$ mV gives $u = 36.6$ m/sec which is almost a factor of 2 higher than that calculated by Hodgkin and Huxley. The source of this error seems to be the assumption that $\tau_m = 0$ which was made at the beginning of this section. This assumption implies that sodium current will begin to flow fully as soon as the membrane voltage changes by 40 mV. But inspection of Fig. 19 or Fig. 21 indicates that this is not so. The time delay associated with sodium turn on requires the membrane voltage to change by about 60 mV before the membrane conductance rises to half of its full active value. Taking $V_1 = 60$ mV gives $u = 22$ m/sec which is quite satisfactory considering the nature of the approximations which have been made.

The importance of time delay in the conductance rise was emphasized by Offner, Weinberg, and Young (1940) who developed a velocity formula similar to (6.16) shortly after Cole and Curtis (1939) recorded the waveforms displayed in Fig. 3. This delay is also of theoretical importance since (6.16) and (6.17) imply

$$u \to [(g_a/r_s c^2)(V_2/V_1)]^{1/2} \to \infty \qquad \text{as } V_1 \to 0; \qquad (6.19)$$

but, with $\tau_m \neq 0$, the effective value of V_1 cannot reach zero. Thus an infinite propagation velocity is prevented by the nonzero value of τ_m.

Early attempts to calculate the propagation velocity of an action potential have been reviewed by Offner, Weinberg, and Young (1940). Since that time, additional approaches have been developed by Rosenblueth, Wiener, Pitts, and Garcia Ramos (1948), Huxley (1959) Kompaneyets and Gurovich (1965), Balakhovskii (1968), Namerow and Kappl (1969), Smolyaninov (1969), Pickard (1966), and Markin and Chizmadzhev (1967), of which the last two references relate propagation velocity to the rate of rise on the leading edge of the action potential. Such a relation is easily obtained from (6.15a) since

$$\partial v/\partial t \mid_{\max} = -u \, dv/d\xi \mid_{\xi=0} = \gamma_1 u V_1.$$

Thus

$$u = [v_{t,\max}/r_s c V_1]^{1/2}, \qquad (6.20)$$

as is readily verified for the waveform in Fig. 21. This is the formula used by Zeeman (1972).

We now turn out attention briefly to the effect of magnetic fields, which are associated with the longitudinal currents and represented as the inductors $l_i + l_0 = l_s$ in Fig. 7(b), upon the propagation velocity. This question arises because it has been suggested [Lieberstein (1967a, b, 1973), Brady (1970), Isaacs (1970), Lieberstein and Mahrous (1970)] that (2.30) should be augmented to the form

$$\partial^2 v/\partial x^2 - l_s c(\partial^2 v/\partial t^2) = r_s(c(\partial v/\partial t) + j_i) + l_s(\partial j_i/\partial t). \qquad (6.21)$$

The numerical instability discussed in connection with Fig. 20 can then be avoided if *both* sides are individually set to zero at a velocity

$$u = [l_s c]^{-1/2}. \qquad (6.22)$$

Van Der Pol (1957) has proposed a similar model for propagation on a nerve fiber. To examine this question [Scott (1971)] we will again ignore turn on delay and assume $\tau_m = 0$, $\tau_n = \tau_n = \infty$. The first order partial differential equations corresponding to Fig. 7(b) and (6.21) become

$$v_x = -l_s i - r_s i,$$

$$i_x = -cv_t - j(v). \qquad (6.23)$$

Taking $j(v)$ as in Fig. 28 with $g_r = 0$ and assuming a steady wave of propagation, $v(x - ut) = v(\xi)$, then yields [Scott (1963, 1970)]

$$u = \left\{ \left(\frac{g_a}{r_s c^2} \right) \frac{(V_2 - V_1)^2}{V_2 V_1} \left[1 - \frac{g_a l_s}{r_s c} \frac{(V_2 - V_1)}{V_2} \right] \right\}^{1/2}. \qquad (6.24)$$

This implies that series inductance will have a negligible effect upon velocity if it satisfies the inequality

$$l_s \ll \left(\frac{r_s c}{g_a} \right) \left(\frac{V_2}{V_2 - V_1} \right). \qquad (6.25)$$

The left-hand side of (6.25) can be evaluated from (2.12) and (2.14) using small argument approximations for the Bessel functions as

$$z_i + z_0 \approx (1/\pi \sigma_1^* a^2) + i\omega(\mu_0/4\pi)[1 - 2\log(\beta a)], \qquad (6.26)$$

the second term of which gives the series reactance from magnetic fields both inside and outside the fiber. Thus

$$l_s = (\mu_0/4\pi)[1 - 2\log(\beta a)] \qquad (6.27)$$

where $\mu_0 = 4\pi \times 10^{-7}$ H/m is the mks magnetic permeability of nonmagnetic materials. Taking $\beta a \sim 10^{-2}$ implies $l_s \sim 10^{-6}$ H/m. The right-hand side of (6.25) is greater than 100 H/m, thus the inequality is satisfied by eight orders of magnitude, and magnetic energy storage will have no measurable effect upon the normal propagation of an action potential. This conclusion is further supported by the numerical studies of Kaplan and Trujillo (1970). Solutions of (6.21) at the velocity given in (6.22) for which both sides of the equation go to zero would correspond to a decoupling of high frequency electromagnetic waves from the membrane. While this may have been what Newton (1718) had in mind when he posed his "twenty-fourth question," it does not correspond to normal nerve activity.

VII. THE FITZHUGH–NAGUMO EQUATION

The previous two sections have bracketed (in the sense of an artilleryman) the representation of a propagating nerve fiber. The Hodgkin–Huxley equations, (5.1) and (5.2), give a fairly accurate description of spike propagation but are somewhat difficult to analyze without the aid of an automatic computer. The nonlinear diffusion equation (6.2) is simple enough for analytical investigation and yields some useful results [e.g., Eq. (6.16) for the conduction velocity], but it fails to reproduce the qualitatively important feature of pulse recovery which is necessary for repeated firing of the fiber. In this situation FitzHugh

(1961) and Nagumo, Arimoto and Yoshizawa (1962) proposed a modification of the nonlinear diffusion equation which would retain its simplicity but allow the action potential to return to a resting level. In properly chosen units of space, time and voltage, (6.2) can be written $V_{xx} - V_t = F(V)$, where $F(V)$ is a function with the character indicated in Figs. 26(a) or 28. Augmenting this equation with a new "recovery" variable R to [FitzHugh (1969)]

$$V_{xx} - V_t = F(V) + R,$$

where

$$R_t = \epsilon(V + a - bR) \qquad \text{(7.1a, b)}$$

yields the desired recovery. To see this note that R in (7.1a) acts as an outward ion current which tends to decrease the area A_2 in Figs. 26(a) or 28. With reference to the Hodgkin–Huxley equations (5.1) and (4.4a), there is a correspondence between

$$R \sim n,$$

$$\epsilon b \sim \kappa \tau_n^{-1},$$

$$\epsilon V \sim \kappa n_0 \tau_n^{-1},$$

where κ is the "temperature factor" indicated in (4.6). The constant a in (7.1b) can be absorbed into the definition of R and F so there is no loss of generality in setting it to zero. The constant b is often arbitrarily assumed equal to zero. Since ϵ is proportional to κ, it can be considered as a parameter which increases with temperature.

Equation (7.1) is beginning to assume the role with respect to nerve fiber propagation that the equation of Van Der Pol (1926, 1934) has played with respect to oscillator theory. "Van Der Pol's equation" displays the qualitative features of many oscillators (spontaneous excitation, limit cycle, continuous transition between sinusoidal and blocking behavior, etc.) without necessarily being an exact representation of any particular dynamical system. As recent studies [Cohen (1971), Hastings (1972), Greenberg (1973)] indicate, such a model is very stimulating and useful for the applied mathematician. Equation (7.1) is often called "Nagumo's equation" [McKean (1970), Greenberg (1973)] although FitzHugh (1968, 1969) refers to it as the "BVP equation" in recognition of the introduction by Bonhoeffer (1948) of phase plane analysis into the study of the passive iron nerve model, and of Van Der Pol. The reference to Van Der Pol, however, is somewhat unfortunate for in 1957 he introduced his own modification for application to nerve problems which failed to consider the diffusive character of the nerve fiber. Thus the name "FitzHugh-Nagumo equation" used by Cohen (1971), Rinzel and Keller (1973), and Hastings (1975a) seems most appropriate.

The general utility of (7.1) can be appreciated by considering the design of a neuristor or electronic analog of the active nerve fiber proposed by Crane (1962). Equations (7.1) describe the most natural technique for achieving pulse return in an electronic neuristor [Nagumo *et al.* (1962), Crane (1962), Scott (1962, 1964), Berestovskiy (1963), Noguchi, Kumagai and Oizumi (1963), Yoshizawa

and Nagumo (1964), Sato and Miyamoto (1967)] and are closely related to the dynamical equations for active superconducting transmission lines which employ tunneling of either normal electrons (Giaever-type) or superconducting electrons (Josephson-type) [Scott (1964, 1970), Parmentier (1969, 1970), Johnson (1968), Nakajima, Yamashita and Onodera (1974), Nakajima, Onodera, Nakamura and Sato (1974)]. Considered as a model for the nerve axon, (11.1) neglects (i) turn-on delay for the sodium current, (ii) the fourth power dependence of potassium current upon n, and (iii) the dependence of τ_n upon v. More exact second-order systems have recently been considered by Krinskii and Kokoz (1973). A good general survey of these problems is given in the thesis by Kunov (1966).

The analysis of (11.1) was begun by Nagumo, Arimoto and Yoshizawa (1962) who considered the ordinary differential equations for traveling wave solutions of the form $V = V(x - ut) = V(\xi)$ and $R = R(x - ut) = R(\xi)$ as indicated in (5.3). Then V and R must satisfy

$$dV/d\xi = W, \qquad \text{(7.2a)}$$

$$dW/d\xi = F(V) + R - uW, \qquad \text{(7.2b)}$$

$$dR/d\xi = \frac{\epsilon}{u}(bR - V - a). \qquad \text{(7.2c)}$$

They assumed $F(V)$ to be cubic, took $b = 0$, and obtained numerical evidence for the existence of two homoclinic trajectories for sufficiently small values of ϵ. At a critical value, ϵ_c, these solutions merged and for $\epsilon > \epsilon_c$ no homoclinic orbits were found, just as in Fig. 23. Such results suggest the existence of two pulse like traveling wave solutions to (7.1), as in Fig. 22, and experiments on an electronic analog indicated that only the pulse with higher velocity is stable. These results were confirmed by FitzHugh (1968, 1969) through numerical studies of (7.1) and (7.2) with $b \neq 0$ and

$$F(V) = \tfrac{1}{3}V^3 - V. \qquad \text{(7.3)}$$

Velocities of the two branches vs. the "temperature parameter" ϵ are shown in Fig. 29. FitzHugh (1968) also made a motion picture entitled "Impulse propagation in a nerve fiber"[6] which is based upon numerical integration of (7.1). Some selected frames from this film are reproduced in Fig. 30 which show the propagation of two pulses away from a point of stimulation. In the fully developed pulses [Figs. 30(f), (g) and (h)], the recovery variable, R, follows behind the voltage, V. These pulses correspond to the upper velocity (A') at $\epsilon = 0.08$ in Fig. 29. The lower velocity pulse (B') is unstable. Again the locus of allowed traveling wave velocities in the $u - \epsilon$ plane indicates where the power balance condition (1.3) is satisfied. For $\epsilon > \epsilon_c$, only decremental conduction is possible.

Arima and Hasegawa (1963) have considered a generalized form of (7.1) with $R_t = G(V)$. With suitable restrictions on F, G, and the smoothness of the initial data, they show that a unique solution exists in the half-space $|x| > \infty$

---

[6] Available on loan from the National Medical Audiovisual Center (Annex) Station K, Atlanta, Georgia 30333.

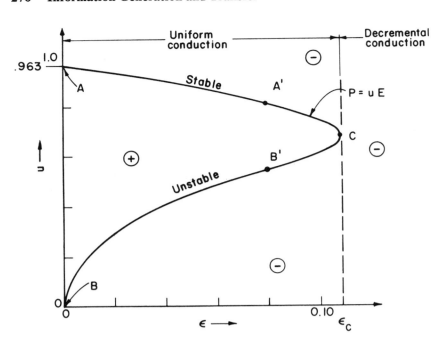

FIG. 29. Propagation velocity for traveling wave pulse solutions of the FitzHugh-Nagumo equations (7.1) vs the temperature parameter ϵ for $a = 0.7$ and $b = 0.8$. Redrawn from FitzHugh (1969).

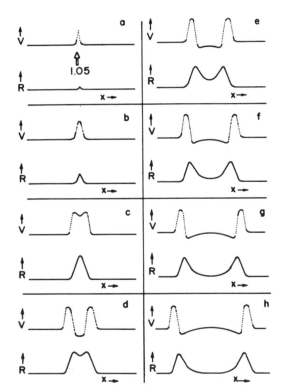

FIG. 30. Frames from computer movie of FitzHugh (1968) showing results of a local stimulation of (7.1) 5% above threshold with $\epsilon = 0.08$, $a = 0.7$, and $b = 0.8$.

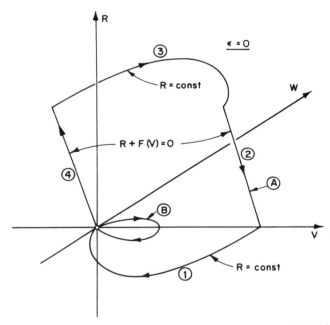

FIG. 31. Phase space sketch of homoclinic trajectories for (7.2) with $\epsilon = 0$.

and $t > 0$. Building on this result Yamaguti (1963) showed that solutions of (7.1) with $a = 0$, $b = 0$ and $VF \geq CV^2$ tend uniformly to zero. A related result was obtained by Yoshizawa and Kitada (1969) who considered (7.1) with $b = 0$ but $F(V)$ a cubic polynomial. They confirm the existence of a threshold by showing that every solution in some neighborhood of zero converges to zero with increasing time. Green and Sleeman (1974) have established upper and lower bounds for the velocity of a traveling wave.

The existence of homoclinic trajectories for Eqs. (7.2) has been studied in detail by Carpenter (1974) and Hastings (1975b); and important analytical results have been obtained by Casten, Cohen and Lagerstrom (1975) using singular perturbation methods. To see this consider Fig. 31 which shows two homoclinic orbits in the limit $\epsilon = 0$ which implies (7.2c) $R = $ const. Orbit B corresponds to point B in Fig. 29 and is just the trajectory given in (6.6) for the zero velocity "threshold pulse" shown in Fig. 27. Orbit A which corresponds to point A in Fig. 29 is somewhat more complex. It is the singular orbit approached as $\epsilon \to 0$ of a family of homoclinic orbits which correspond to the pulse shown in Fig. 32. Going backward in ξ or forward in time, this pulse can be described as follows: ① The "leading edge"

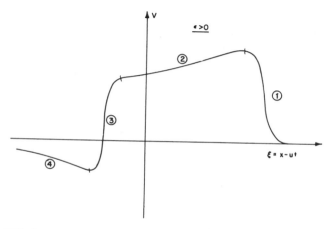

FIG. 32. Voltage pulse corresponding to orbit A in Fig. 31 with $\epsilon > 0$.

involves a rapid transition between the outer zeros of $F(V)$ as was discussed in detail in the previous section. ② A "slow relaxation" from $R = 0$ to a new value R_1 determined by (7.2c) with the condition $F(V) + R \approx 0$. ③ A rapid downward voltage transition between the two outer zeros of $F(V) + R_1$. The value of R_1 must be such that this trailing edge will have the same velocity as the leading edge (see Fig. 26). ④ Finally there is a slow relaxation from $R = R_1$ back to zero.

The velocity, u_0, of the singular orbit A is just that velocity discussed in the previous section. Assuming $a = 0$, $b = 0$, and $\epsilon > 0$, we can write

$$u = u_0 + \epsilon u_1^A + \epsilon^2 u_2^A + \cdots, \tag{7.4a}$$

$$V = V_0 + \epsilon V_1 + \epsilon^2 V_2 + \cdots, \tag{7.4b}$$

$$R = R_0 + \epsilon R_1 + \epsilon^2 R_2 + \cdots. \tag{7.4c}$$

We can then substitute into (7.2), and equate powers of ϵ to obtain

$$d^2V_0/d\xi^2 + u_0(dV_0/d\xi) - [F(V_0) + R_0] = 0, \tag{7.5}$$

$$R_0 = \text{const}, \tag{7.6}$$

$$d^2V_1/d\xi^2 + u_0(dV_1/d\xi) - V_1 F'(V_0)$$
$$= R_1 - u_1^A(dV_0/d\xi), \tag{7.7}$$

$$dR_1/d\xi = -V_0/u_0. \tag{7.8}$$

Using (7.5) and some integration by parts, it is not difficult to show that the left-hand side of (7.7) is orthogonal to

$$(dV_0/d\xi)\exp(u_0\xi).$$

Thus, from (7.7) and (7.8), u_1^A is determined as

$$u_1^A = -\frac{1}{u_0} \int_{-\infty}^{\infty} \left(\int_{-\infty}^{\xi} V_0(\xi')\,d\xi' \right) \frac{dV_0}{d\xi} \exp(u_0\xi)\,d\xi \Big/$$

$$\int_{-\infty}^{\infty} \left(\frac{dV_0}{d\xi} \right)^2 \exp(u_0\xi)\,d\xi. \tag{7.9}$$

From Fig. 29 it is clear that the approximation

$$u \approx u_0 + u_1^A \epsilon \tag{7.10}$$

is useful over a substantial portion of the upper (stable) branch; and, it is important to notice, the determination of u_1^A in (7.9) requires only knowledge of the singular pulse $V_0(\xi)$.

For the orbit B in Fig. 31, $u_0 = 0$ and (7.9) cannot be used. In this case Casten et al. write

$$V = V_0 + \epsilon^{1/2}V_1 + \cdots, \tag{7.11a}$$

$$R = \epsilon^{1/2}R_1 + \cdots, \tag{7.11b}$$

$$u = \epsilon^{1/2}u_1^B + \cdots, \tag{7.11c}$$

to obtain

$$d^2V_0/d\xi^2 - F(V_0) = 0, \tag{7.12}$$

$$d^2V_1/d\xi^2 - F'(V_0) = R_1 - u_1^B(dV_0/d\xi), \tag{7.13}$$

$$dR_1/d\xi = -V_0/u_1^B. \tag{7.14}$$

The orthogonality condition still holds for the left-hand side of (7.13) so

$$u_1^B = +\left(\int_{-\infty}^{\infty} V_0^2\,d\xi \Big/ \int_{-\infty}^{\infty} \left(\frac{dV_0}{d\xi} \right)^2 d\xi \right)^{1/2}. \tag{7.15}$$

Again we see that the approximation

$$u \approx \epsilon^{1/2}u_1^B \tag{7.16}$$

is useful over much of the lower branch in Fig. 29.

Closed trajectories satisfying (7.2) correspond to the periodic wave solutions which were originally suggested by Huxley (1959) for the Hodgkin–Huxley equations. The existence of such closed orbits has been studied by Hastings (1974a) and by Carpenter (1974) using the concept of "isolating blocks" [Conley (1973)] around a singular orbit. Rinzel and Keller (1973) have studied solutions of (7.2) with $a = 0$, $b = 0$, and

$$F(V) = V \text{ for } V < V_1$$
$$= V - 1 \text{ for } V > V_1. \tag{7.17}$$

This is the function of Fig. 28 with $g_a = g_r$ so the phase space equations are linear except along the plane $V = V_1$. For a periodicity defined by

$$V(\xi) = V(\xi + \lambda) \tag{7.18}$$

some numerical values for velocity, u, and amplitude, A, are shown as functions of λ in Fig. 33. Again there are two waves for each period, the slower wave being unstable.

Currently it is of great interest to extend such exact results to the full Hodgkin–Huxley equations (5.1) or to the corresponding ordinary differential equations for traveling wave solutions (5.5). Evans and Shenk (1970) have

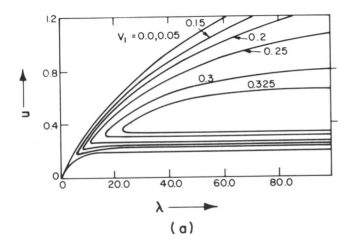

(a)

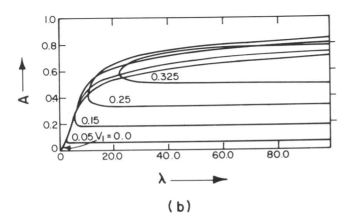

(b)

FIG. 33. (a) Velocity, and (b) amplitude vs pulse spacing, λ, for periodic solutions for the FitzHugh–Nagumo equation with $\epsilon = 0.05$, $a = 0$, $b = 0$, and $F(V)$ as in (7.17). Redrawn from Rinzel and Keller (1974).

shown that (5.1) has a unique solution for arbitrary bounded initial conditions with continuous dependence on the initial conditions. Quite recently Carpenter (1974) has extended the concept of isolating blocks to the higher dimensional phase space associated with (5.5) and indicated in Fig. 20. To do this she takes m to be a fast variable so v, i, and m vary along branches ① and ③, and only n and h vary along branches ② and ④. The small parameters are then τ_m, τ_n^{-1}, and τ_h^{-1}, and both homoclinic and periodic orbits are established. The analysis by Hastings (1974b) which does not assume τ_m small is probably closer to physiological reality as we saw in the previous section; however periodic orbits were not obtained.

VIII. THE MYELINATED AXON

Examination of (6.16) reveals a major design difficulty of the smooth nerve fiber. Since $g_a = 2\pi a G_a$, $c = 2\pi a C$ and $r_s = (\pi a^2 \sigma_1)^{-1}$, where a is the radius of the fiber, the conduction velocity is proportional to

$$u \propto a^{1/2} \tag{8.1}$$

or to the fourth root of the cross-sectional area. [See Fitz-

Hugh (1973) for a careful application of dimensional analysis to nerve problems.] In order to double the velocity, the area (and therefore the volume) of the fiber must increase by a factor of sixteen; to triple the velocity requires a factor of eighty-one. Since the giant axons of the squid transmit "escape signals" (generated in forward nerve cell complexes) to the appropriate muscles (located aft) there is evolutionary pressure to increase the speed, and this probably explains the unusually large size of the fiber. But clearly the fibers can't get much faster without using an unacceptable fraction of the squid's cross section, and only a single bit of information ("leave" or "stay") is being transmitted at any instant of time. Equation (6.16) also indicates a way out of this dilemma. If the fiber is partially covered by an insulating material so only a fraction f of the active membrane remains exposed, g_a/c and r_s would remain the same. But c would be proportional to f, so the conduction velocity should depend upon the exposed fraction roughly as

$$u \propto f^{-1/2}. \tag{8.2}$$

Thus velocity can be increased without changing the cross-sectional area by making f small.

Something like this takes place in the design of the motor axons of vertebrates. The structure of the fiber appears as in Fig. 34 where the fiber is almost everywhere covered by a relatively thick insulating coat of *myelin* consisting of a couple of hundred layers of cell membrane [Hodgkin (1951, 1964)]. Only at small active nodes (nodes of Ranvier) can the membrane function in the normal way and these are spaced apart by a distance $D \sim 1$ mm. In this manner the diameter of the fiber can remain as small as 10 μ, while the conduction velocity is as large as that on the squid fiber. The frog nerve studied by Helmholtz (1850) and shown in Fig. 1 is actually a bundle of many axons myelinated as in Fig. 34. Young (1951) has prepared a graphic comparison of the squid giant axon and the sciatic nerve of a rabbit which is reproduced in Fig. 35. The conduction velocity is about the same in both cases, so the myelinated nerve bundle can carry at least two orders of magnitude more bits of information per unit time. This high information rate permits the fine muscular control which is one of the striking features of higher animals.

The role of isolated active nodes in increasing conduction speed was first recognized by Lillie (1925, 1936) in connection with his experiments on the passive iron wire analog for the nerve fiber. He showed that the conduction velocity on this model was greatly increased when the wire was enclosed in a glass tube broken into segments; and he noted that the excitation seemed to "jump" quickly from one opening in the glass tubing to the next, an effect called "saltatory" conduction by physiologists.[7] Shortly thereafter Osterhout and Hill (1930) demonstrated that conduction in *Nitella* which had been blocked in fresh water by chloroform could be restored by introducing a salt bridge around the block, and Kato (1934) isolated in the conductable state a single fiber from the sciatic nerve of the Japanese toad. Building on these classic results, Tasaki (1939) demonstrated that conduction jumped from node to node in a single Japanese toad fiber. For general surveys of experi-

---

[7] *Saltare* is the Latin and modern Italian verb "to jump."

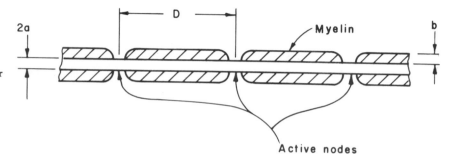

FIG. 34.   Structure of a myelinated nerve fiber (not to scale).

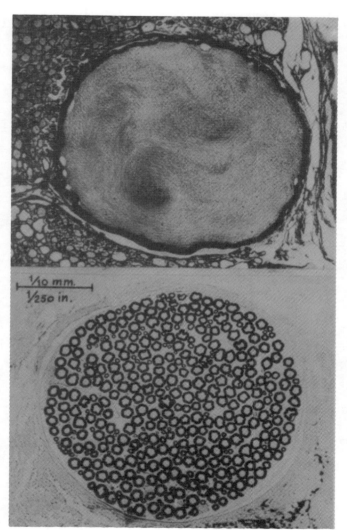

FIG. 35.   Comparison of cross sections for the squid giant axon (above) and the sciatic nerve bundle controlling the calf muscle of a rabbit (below). There are about 400 myelinated fibers in the rabbit nerve each conducting pulses at about 80 meters per second [Young (1951)].

TABLE II.   Data on frog myelinated fiber.

Fiber radius (a)	7μ
Myelin thickness (b)	2μ
Distance between active nodes (D)	2 mm
Area of active node	2.2×10^{-7} cm^2
Internal resistance per unit length	140 MΩ/cm
Capacity of myelin per unit length	10–16 pF/cm
Conductance of myelin per unit length	2.5–4×10^{-8} mho/cm
Capacity of active node	0.6–1.5 pF
Resting resistance of node	40–80 MΩ
Conduction velocity	23 m/sec

The square root of the ratio of squid fiber radius (238μ) to that in Table II is 5.83. The inverse square root of the fraction of exposed area multiplied by the ratio of total capacitance to node capacitance is 10.6. Thus (8.1) and (8.2) imply that the squid axon velocity should be 0.55 times that of the frog, axon whereas in fact they are about equal. This simple estimate ignores: (i) the effect of concentrating the active membrane at isolated points, (ii) the differences between frog and squid membrane dynamics indicated in Fig. 19, and (iii) differences in conductivity of the axoplasm. The first of these corrections can be brought into focus by noting, from the considerations of Sec. II, that the myelinated fiber is closely approximated by a linear diffusion equation which is periodically loaded by the active nodes [Pickard (1966), Markin and Chizmadzhev (1967)]. This picture can be further simplified by lumping the internode capacitance of the myelin together with the nodal capacitance. This leads to the equivalent circuit indicated in Fig. 36, where

$$R = 28 \ M\Omega,$$

$$C = 2.6 - 4.7 \ pF,$$

and $I(i)$ is the ion current calculated at the ith node from Eqs. (4.3) and (4.4) using the data in Fig. 19(b). Equations (5.1a, b) are then replaced by the *difference differential equations*

$$v_i - v_{i-1} = -i_i R, \tag{8.3a}$$

$$i_{i+1} - i_i + C(dv_i/dt) = -I(i). \tag{8.3b}$$

To determine a conduction velocity the traveling wave assumption, displayed in (5.3) and (5.4), must be replaced by a search for solutions which satisfy the condition

$$v_{i-1}(t) = v_i(t - T), \tag{8.4a}$$

$$i_{i-1}(t) = i_i(t - T), \tag{8.4b}$$

ments on myelinated fibers the reviews by Tasaki (1959) and Hodgkin (1951, 1964) are suggested in addition to the discussion by Cole (1968); here we list some representative data on the frog myelinated fiber collected by Hodgkin (1964).

It is interesting to note how close this average conduction velocity is to the value of 27 m/sec measured by Helmholtz in 1850.

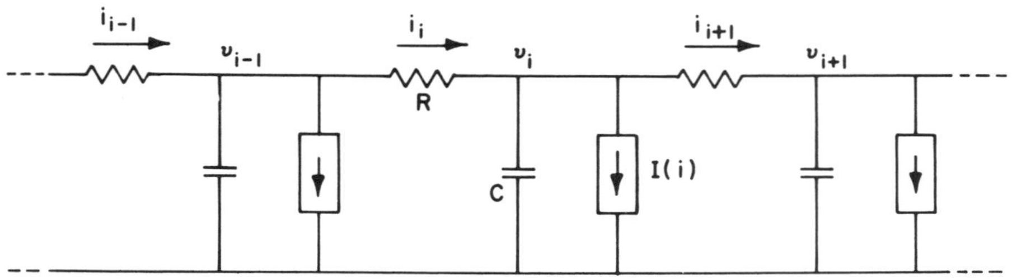

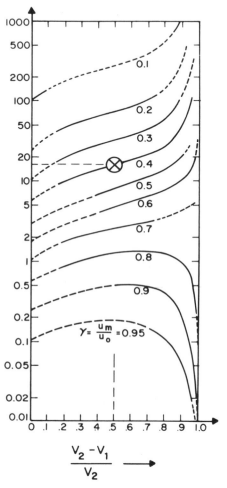

FIG. 36.   A difference differential representation of the myelinated nerve fiber.

where T is a *section delay*. If T can be found, the conduction velocity for the myelinated fiber is evidently

$$u_m = D/T. \tag{8.5}$$

In solving for the section delay, it is interesting to begin by assuming $I(i) = I(v_i)$, where

$$
\begin{aligned}
I(v_i) &= 0 && \text{for } v_i < V_1, \\
&= G(v_i - V_2) && \text{for } v_i > V_1, \tag{8.6}
\end{aligned}
$$

as we did in (6.3) for the smooth axon. Then, for R and G sufficiently small, the differences in (8.3) can be approximated as x derivatives and (6.16) gives

$$T \approx [RC^2/G]^{1/2}[(V_2 V_1)^{1/2}/(V_2 - V_1)]. \tag{8.7}$$

The problem is to determine T as a function of R, C, G, V_1, and V_2 when the approximation of (8.7) is *not* valid. This problem was carefully studied by Kunov and Richer at the Electronics Laboratory of the Technical University of Denmark during 1964–65. A detailed description of this work is included in the thesis by Kunov (1966) from which some of the salient points have been published [Kunov (1965), Richer (1965, 1966)]. Kunov's thesis describes a variety of analytical studied including: (i) numerical integration of (8.3) for a finite number of sections, (ii) an iterative computation to find solutions with the form (8.4), (iii) a Laplace transform solution, and (iv) measurements on an electronic analog [Kunov (1965)]. These studies indicate that the ratio of conduction velocity on the myelinated axon, u_m, to that calculated from (6.16) for the smooth axon, u_0, is a function of the parameters RG, and $(V_2 - V_1)/V_2$. Thus

$$u_m/u_0 = \gamma[RG, (V_2 - V_1)/V_2], \tag{8.8}$$

and curve specifying γ are reproduced in Fig. 37. For the frog axon, the data in Fig. 19(b) give $G = 0.57\ \mu$mho so

$$RG = 16,$$

and in Sec. VI the value of V_1 which seemed to account for delay in sodium turn on was about 60 mV. Thus

$$(V_2 - V_1)/V_2 = 0.5.$$

From Fig. 37 these two values indicate a reduction in velocity of the myelinated fiber over that of a smooth axon by the factor

$$\gamma = 0.4,$$

FIG. 37.   Ratio of myelinated conduction velocity (u_m) to that of the corresponding smooth fiber (u_0) given by (6.16). Dashed lines indicate extrapolated or interpolated values.

whereas our rough estimate obtained above by comparison of squid and frog fibers was

$$\gamma = 0.55.$$

This is rather close agreement considering the uncertainty in the capacitance C and the indication in Fig. 19(b) that the frog membrane responds somewhat more quickly than that of the squid. Furthermore the appropriate value for G may not be as large as $0.57\ \mu$mho since potassium and leakage currents flow in the opposite direction and, in addition, leakage current through the myelin and the resting conductance may have a noticeable effect as indicated in (6.17) [Kompaneyets (1971)].

Richer (1966) has made an important contribution to

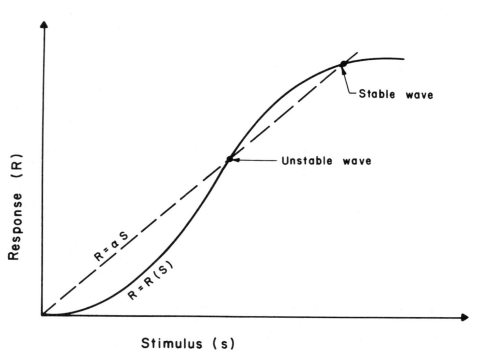

FIG. 38. Nasonov diagram for a mye-
inated nerve fiber.

this problem by finding an *exact* solution for the case $G = \infty$ which he calls "switch-line." This solution gives an implicit relation between normalized section delay, T/RC, and $(V_2 - V_1)/V_2$ as

$$(V_2 - V_1)/V_2 = \exp\left(-\int_0^{\pi/2} F(\alpha, T/RC)\,ctn\alpha\,d\alpha\right),$$

$$(8.9)$$

where

$$F(\alpha, T/RC) \equiv (2/\pi)\,\tan^{-1}[(ctn\alpha)\,\tanh(2T/RC\,\sin^{-2}\alpha)].$$

$$(8.10)$$

Equation (8.9) appears a bit unwieldy, but fortunately it can be closely approximated by the much simpler expression

$$(V_2 - V_1)/V_2 \approx (1 + T/RC)^{-1} \qquad (8.11)$$

which is found to be asymptotically correct for both large and small values of T, and overestimates T by about 10% at T/RC equal to unity. A simple algebraic relation which interpolates between (8.11) and (8.7) is

$$\frac{T}{RC} = \left(\frac{V_2}{V_2 - V_1}\right)\left[\left(\frac{V_1}{V_2}\right)\left(\frac{1}{RG} + \frac{V_1}{V_2}\right)\right]^{1/2}. \qquad (8.12)$$

This equation agrees well with digital computer solutions for a long but finite system and also with the results of analog simulation [Kunov (1966)]. Richer (1965) has also considered the addition of resting conductance as in Fig. 28 and has shown that only a positive or negative level change can propagate (not a pulse) just as in the smooth axon. It is interesting to note that he finds an intermediate range for which neither wave can propagate.

Kunov (1966) considers recovery models or discrete FitzHugh–Nagumo systems, and Markin and Chizmadzhev (1967) discuss propagation when the internodes are described by the linear diffusion equation. FitzHugh (1962) computed the initiation and conduction of pulses on a linear diffusion equation periodically loaded with Hodgkin–Huxley nodes, and improved computations have recently been reported by Goldman and Albus (1968). The high velocity (stable) and low velocity (unstable) pulses which appear in Figs. 23 and 29 for the Hodgkin–Huxley and FitzHugh–Nagumo equations can be appreciated on the myelinated fiber by considering the "Nasonov diagram" [Averbach and Nasonov (1950), FitzHugh (1969)] in Fig. 38. If it is assumed that: (i) each node has a "sigmoid" stimulus-response curve, and (ii) a fraction, $1/\alpha$, of the response for each node is presented as a stimulus to the next, then *stationary* levels of activity occur where the sigmoid curve intersects the line $R = \alpha S$. The lower amplitude intersection is unstable since a small increase in S will lead to a larger increase in R, etc. The upper intersection, on the other hand, appears to be stable. As the parameter α is increased, these two intersections eventually merge; and above this critical value of α only decremental conduction obtains.

IX. WAVEFORM STABILITY

In Sec. V–VIII we have considered the problem of finding traveling wave solutions for the partial or difference differential equations describing nerve fibers. For pde's the analytic technique was to *assume* that dependent variables are functions of x and t only through the argument $\xi = x - ut$ as indicated in (5.3). This is equivalent to introducing the independent variable transformation (5.4) and then assuming no dependence upon τ ($\partial/\partial\tau = 0$). Having found such traveling wave solutions, it is interesting to know whether or not they are *stable* with respect to perturbations which might reasonably be expected to arise in an experimental situation. To study the time evolution of such perturbations, it is necessary to consider the τ dependence.

To introduce the basic ideas of waveform stability analysis we will investigate the KPP form of the nonlinear diffusion equation (1.1),

$$V_{xx} - V_t = F(V), \tag{9.1}$$

traveling wave solutions for which were considered in detail in Sec. VI assuming a cubic form for the function F(V). Equation (9.1) is simple enough for exposition and the results to be obtained serve as a basis for stability investigation of the FitzHugh–Nagumo and Hodgkin–Huxley traveling waves.

Under the transformation (5.4), (9.1) becomes

$$V_{\xi\xi} + uV_\xi - V_\tau = F(V), \tag{9.2}$$

where V is now considered a function of ξ (space in a coordinate system moving with velocity u) and τ (the same time scale as t). The traveling wave solution $V_T(\xi)$ must satisfy

$$V_{T,\xi\xi} + uV_{T,\xi} = F(V_T) \tag{9.3}$$

and a general solution of (9.2) can be considered as the sum of a traveling wave solution and a perturbation $V_P(\xi, \tau)$. Thus

$$V(\xi, \tau) = V_T(\xi) + V_P(\xi, \tau). \tag{9.4}$$

Substituting (9.4) into (9.2) gives

$$V_{P,\xi\xi} + uV_{P,\xi} - V_{P,\tau} = F(V_P + V_T) - F(V_T) \tag{9.5}$$

as a nonlinear and ξ-dependent pde for the evolution of the perturbation. It is important to recognize that *no* approximations have been made in going from (9.1) to (9.5).

Investigation of (9.5) for the evolution of $V_P(\xi, \tau)$ subject to prescribed initial and boundary conditions constitutes the "waveform stability problem" for a traveling wave solution to (9.1) with velocity u. This equation has been studied in connection with the propagation of: (i) flames [Zeldovich and Barenblatt (1959), Kanel' (1962)], (ii) "Gunn effect" domains in bulk semiconductors, [Knight and Peterson (1967), Eleonskii (1968)], and (iii) traveling waves on "neuristors" and electronic analogs for the nerve fiber [Parmentier (1967, 1968, 1969, 1970), Buratti and Lindgren (1968), Lindgren and Buratti (1969), Maginu (1971)].

One approach to the study of (9.5) is to assume the perturbation small enough so the right-hand side can be approximated by

$$F(V_P + V_T) - F(V_T) \approx dF/dV |_{V=V_T} \times V_P$$
$$\equiv G(V_T)V_P, \tag{9.6}$$

whereupon (9.5) is "linearized" to

$$V_{P,\xi\xi} + uV_{P,\xi} - V_{P,\tau} = G[V_T(\xi)]V_P. \tag{9.7}$$

Elementary solutions to (9.7) will either decay exponentially with time, grow exponentially with time, or remain constant. Thus, *with respect to the linearized equation*, we can say the system is: (i) *asymptotically stable* if *all* elementary solutions decay, (ii) *unstable* if *any* elementary solution grows, and (iii) *stable* if (i) and (ii) are not satisfied.

This is a neat scheme but we must be wary of drawing conclusions from (9.7) which are not relevant to the application of (9.5) in a real situation. While we might conclude asymptotic stability with respect to (9.7), for example, it may not be reasonable to assume perturbations small enough for (9.7) to apply. As Eckhaus (1965) puts it "infinitesimal disturbances are certainly unavoidable, but not all unavoidable disturbances may be considered infinitesimal."[8] On the other hand if (9.7) indicates elementary solutions which grow, these will eventually be bounded by the nonlinear character of (9.5). Such a bound may be so close to the original solution that the system is, in effect, stable. With these caveats in mind, let us proceed to the analysis of (9.7).

If V_P is constructed from elementary product solutions of the form

$$V_P \sim \phi(\xi) \exp(-\lambda\tau), \tag{9.8}$$

then ϕ must satisfy the eigenvalue equation

$$\phi_{\xi\xi} + u\phi_\xi + \{\lambda - G[V_T(\xi)]\}\phi = 0. \tag{9.9}$$

The condition for asymptotic stability is that all the eigenvalues, λ, which are allowed for solutions of (9.9) must have positive real parts. This would require that the magnitude of the corresponding elementary solution (9.8) will decay exponentially with time. In a certain sense asymptotic stability is never possible. To see this, differentiate (9.3) for the traveling wave solution with respect to ξ to obtain

$$(V_{T,\xi})_{\xi\xi} + u(V_{T,\xi})_\xi - G(V_T)V_{T,\xi} = 0, \tag{9.10}$$

and note that this is the same equation obeyed by ϕ when $\lambda = 0$. Thus the eigenfunction of (9.9) with zero eigenvalue is

$$\phi = V_{T,\xi} \quad \text{for } \lambda = 0. \tag{9.11}$$

The physical meaning of this result is seen by considering an infinitesimal translation, α, of V_T along the ξ axis. Since

$$V_T(\xi + \alpha) = V_T(\xi) + \alpha V_{T,\xi}, \tag{9.12}$$

this is equivalent to adding an infinitesimal amount of the $\lambda = 0$ eigenfunction. But we expect a translational perturbation neither to grow nor decay. The observation that *the perturbation eigenfunction corresponding to zero eigenvalue is the derivative of the traveling wave* is quite general and not at all restricted to solutions of (9.1). Many investigators avoid this situation by defining stability with respect to a metric which permits arbitrary translations with ξ [Zeldovich and Barenblatt (1959), Kanel' (1962), Maginu (1971), Evans (1972), Brooke Benjamin (1972), Sattinger (1975)].

---

[8] Those who experiment with real nerve fibers will probably agree.

Next it is of interest to determine whether or not $\lambda = 0$ is the lowest eigenvalue; if it is not, (9.8) indicates instability. We shall make this determination with respect to the boundary conditions

$$\phi \to 0 \qquad \text{as } |\xi| \to \infty, \tag{9.13}$$

which imply perturbations of finite energy. If the change of dependent variable

$$\phi = \exp[-(u/2)\xi]\psi \tag{9.14}$$

is introduced into (9.9) [Parmentier (1967)], ψ must satisfy the Schrödinger equation

$$\psi_{\xi\xi} + \{\lambda - \tfrac{1}{4}u^2 - G[V_T(\xi)]\}\psi = 0 \tag{9.15}$$

for which the eigenvalues are real and bounded from below [Morse and Feshbach (1953) pp. 766–8]. If $\lambda = 0$ and $G \to G_1 > 0$ as $\xi \to +\infty$, ψ must also satisfy the boundary condition (9.13). Then $\lambda = 0$ is the lowest eigenvalue if the corresponding eigenfunction $dV_T/d\xi$ has no zero crossings. This condition is satisfied for the "level change" waves in Fig. 26 but not for the pulse wave in Fig. 27. Thus the smooth level change waves are *stable* with respect to the linearized equation, but any solution for which V_T is not monotone increasing with ξ will have eigenvalues $\lambda < 0$ and, from (9.8) will be *unstable*. This conclusion is independent of the form of the function $F(V)$ in (9.1).

This result can be extended to perturbations which are not infinitesimal by expressing the right-hand side of (9.5) by the Taylor series

$$\begin{aligned} F(V_T + V_P) &- F(V_T) \\ &= F'(V_T)V_P + \tfrac{1}{2}F''(V_T)V_P^2 + \cdots \end{aligned} \tag{9.16}$$

for V_P within the appropriate range of convergence. Lindgren and Buratti (1969) have constructed a Lyapunov functional which implies nonlinear stability from linear stability if V_P is small enough compared with the first positive eigenvalue in (9.15). Maginu (1971) has obtained a stronger result. He writes

$$V_P = V_P^{(1)} + V_P^{(2)} + \cdots, \tag{9.17}$$

where $V_P^{(1)} + V_P^{(2)} + \cdots + V_P^{(n)}$ satisfies (9.5) to nth order with the right-hand side approximated up to the nth derivative in the Taylor series (13.16). Then he shows that as $\tau \to \infty$, $V_P^{(1)} \to \alpha V_{T,\xi}$, $V_P^{(2)} \to \tfrac{1}{2}\alpha^2 V_{T,\xi\xi}$, ..., $V_P^{(n)} \to (\alpha^n/n!)V_{T,\xi}n$. Thus

$$V_P \to V_T(\xi + \alpha) - V_T(\xi) \qquad \text{as } \tau \to \infty. \tag{9.18}$$

This is nonlinear stability with respect to a metric which permits translations in the ξ direction. The only restriction on V_P is that it must lie within the range of convergence in (9.16).

To see how this proof goes, note first that we have already demonstrated, through analysis of (9.7), that $V_P^{(1)} \to$

$\alpha V_{T,\xi}$ as $\tau \to \infty$. To second order $V_P^{(2)}$ must satisfy

$$\begin{aligned} V_{P,\xi\xi}^{(2)} + u V_{P,\xi}^{(2)} &- V_{P,\tau}^{(2)} \\ &= F'(V_T)V_P^{(2)} + \tfrac{1}{2}F''(V_T)(V_P^{(1)})^2. \end{aligned} \tag{9.19}$$

Differentiating (9.3) twice with respect to ξ gives

$$V_{T,\xi\xi\xi\xi} + u V_{T,\xi\xi\xi} = F'(V_T)V_{T,\xi\xi} + F''(V_T)V_{T,\xi}^2. \tag{9.20}$$

The variable

$$w \equiv V_P^{(2)} - \tfrac{1}{2}\alpha^2 V_{T,\xi\xi} \tag{9.21}$$

obeys the equation $[(9.19)-\tfrac{1}{2}\alpha^2 \; (9.20)]$ or

$$\begin{aligned} w_{\xi\xi} + u w_\xi - w_\tau &= F'(V_T)w \\ &+ \tfrac{1}{2}F''(V_T)[(V_P^{(1)})^2 - \alpha^2 V_{T,\xi}^2]. \end{aligned} \tag{9.22}$$

But, as $\tau \to \infty$, this approaches

$$w_{\xi\xi} + u w_\xi - w_\tau = F'(V_T)w \tag{9.23}$$

which is identical to (9.7), so $w \to \alpha_1 V_{T,\xi}$. Then from (9.21)

$$V_P^{(1)} + V_P^{(2)} \to (\alpha + \alpha_1)V_{T,\xi} + \tfrac{1}{2}\alpha^2 V_{T,\xi\xi} \tag{9.24}$$

as $\tau \to \infty$. The addition of α_1 to α in the first term constitutes a second-order correction to the translation caused by the initial perturbation; it can be absorbed simply by redefining α in (9.21) and (9.22). Higher order estimates are treated in a similar manner.

Consider finally the nonlinear bounds on those traveling waves, $V_T(\xi)$, which are not monotone increasing and therefore unstable with respect to the linearized equation (9.7). These will grow no further than the stable, monotone increasing transition wave and they will decay no further than zero. It seems reasonable to speculate that these are the bounds of interest.

It should be emphasized that these conclusions do not apply to transition waves between 0 and V_1 in Fig. 26. Since the singular point at V_1 corresponds to negative differential conductance of the membrane, it is unstable even under space clamped conditions. The stability of such waves is studied in connection with a problem of genetic diffusion where the dependent variable must be less than or equal to its value at the singular point [Fisher (1936), Kolmogoroff *et al.* (1937), Canosa (1973), Rosen (1974)]. Aronson and Weinberger (1975) have carefully compared the asymptotic behavior of (9.1) for $F(V)$ equal to $V(1 - V)$ with $V(1 - V)(V - V_1)$.

A corresponding stability investigation for a traveling wave solution of the FitzHugh–Nagumo equation (7.1) is considerably more difficult because the linearized problem is third order. Thus the eigenvalue problem, corresponding to (9.9), cannot be made self-adjoint and the eigenvalues are in general complex. The eigenfunction for $\lambda = 0$ is still $V_{T,\xi}$, but there is no simple relation between the number of zero crossings of the eigenfunctions and the order of the real parts of the corresponding eigenvalues. However we have already shown branches ① and ③ of the singular orbit Ⓐ in Fig. 31 to be stable, which is consistent with the numerical results of FitzHugh (1969) and Rinzel and

Keller (1973) indicating stability along the high velocity branch for particular functions $F(V)$.

In a series of papers, Evans (1972) has investigated a generalization, of the Hodgkin–Huxley equations with the form suggested by FitzHugh (1969)

$$V_{xx} - V_t = F_0(V, w_1, \ldots w_n),$$
$$w_{i,t} = F_i(V, w_1, \ldots w_n) \qquad i = 1, \ldots n, \qquad (9.25)$$

where the F's are twice continuously differentiable. This set reduces to (i) the KPP equation for $n = 0$, (ii) the FitzHugh–Nagumo equations with $n = 1$, and (iii) the Hodgkin–Huxley equations with $n = 3$. Writing $W \equiv col(V, w_1, \ldots w_n)$ and assuming a traveling wave solution of the form $W(x, t) = W_T(x - ut) = W_T(\xi)$, a general solution can be written $W(\xi, \tau) = W_T(\xi) + W_P(\xi, \tau)$. The linearized equation for W_P is then [as in (9.7)]

$$\begin{bmatrix} V_{P,\xi\xi} \\ 0 \\ 0 \\ \cdot \\ \cdot \\ \cdot \\ 0 \end{bmatrix} + uW_{P,\xi} - W_{P,\tau} = AW_P, \qquad (9.26)$$

where A is an $(n + 1) \times (n + 1)$ matrix with elements obtained by differentiating the F's with their arguments and evaluating at W_T. Evans shows:

(i) The solution for (9.25) decays exponentially to $W_T(\xi + \alpha)$ (from a suitably small initial perturbation) if and only if the solution for (9.26) decays exponentially to $W_{T,\xi}$.
(ii) The solution for (9.26) decays exponentially to $W_{T,\xi}$ if and only if the associated eigenvalue equation

$$\begin{bmatrix} \phi_{0,\xi\xi} \\ 0 \\ 0 \\ \cdot \\ \cdot \\ \cdot \\ 0 \end{bmatrix} + u\Phi_\xi + (\lambda - A)\Phi = 0, \qquad (9.27)$$

where

$$\Phi \equiv col(\phi_0, \phi_1, \ldots, \phi_n)$$

has no eigenvalues with negative real parts, and $\Phi = W_{T,\xi}$ is the *only* eigenfunction for $\lambda = 0$.

A similar result has quite recently been obtained by Sattinger (1975) for a more general system which allows the F's in (9.25) to depend upon the $w_{i,x}$. The zero eigenvalue of the linear operator must be isolated at the origin of the complex plane, and the remaining eigenvalues must lie within a certain parabola in the right half-plane. Evans (1974) has extended his work to show that there must be an unstable pulse as well as a stable pulse.

The stability investigation of waveforms on myelinated fibers is yet to begin. Beyond the speculations associated with Nasonov diagrams (see Fig. 38), there is only the work of Predonzani and Roveri (1968) which treats equilibrium stability of a lossless transmission line that is periodically loaded with active bipoles. Thus much remains to be done before the study of waveform stability is complete. This work should not be dismissed by the experimentalist as merely of mathematical interest. The point of Sec. IV is that a fundamentally correct theory of ion currents has not yet been established. Stability theory is necessary to decide what a given description of the membrane will predict to occur in the laboratory.

X. THRESHOLD FOR AN ACTIVE FIBER

The classic experimental procedure for determining threshold conditions of a piece of nerve membrane is the "strength-duration" measurement. A current of strength (I) is applied for a time duration (τ) which just causes the membrane to fire [see Fig. 39(a)]. Then both I and τ are adjusted to find their functional relation under this condition. For the space clamped membrane shown in Fig. 14 the relation between I and τ is easily understood. When $\tau \ll \tau_n$ and τ_h, the current pulse must supply the charge, Q_θ, necessary to change the potential across the membrane to a value at which the ion current flows inward. From the curves of Fig. 17, this is about 20 mV. Thus

$$I\tau = Q_\theta. \qquad (10.1)$$

As I is reduced, the duration necessary for threshold excitation increases; and, eventually, I reaches a level below which steady application will never cause inward ion current. This level is traditionally called the "rheobase." If the stimulating current is turned on slowly (with respect to τ_n and τ_h), the outward potassium current begins to flow which offsets the inward sodium current and increases the rheobase. These effects have been phenomenologically described by the "two factor" theory of Rashevsky (1960) and Hill (1936). See Katz (1939) for an excellent survey of the early studies; recent work has recently been carefully reviewed by FitzHugh (1969).

Here our attention will be focused on similar calculations for the nerve fiber. The experimental situation is as indicated in Fig. 39(b) where the longitudinal stimulating current, $i(t)$, is conveniently chosen to have the character indicated in Fig. 39(a). In this case also the strength-duration curve is given by (10.1) for small values of τ, and reaches a rheobase for large τ. Computations by Cooley and Dodge (1966) for the Hodgkin–Huxley axon are presented in Fig. 40 which agrees well with experimental results [Noble and Stein (1966), Cole (1968)]. As Noble (1966) has emphasized, the threshold condition for a fiber cannot be calculated from the condition that the voltage at the end of the fiber change by a fixed amount. Indeed, attempts to derive a strength-duration curve from this condition invariably lead to a relation $I(t)^{1/2} = $ constant for small τ which is manifestly incorrect [Kunov (1966), Scott (1973)].

One simple and fundamental way to evaluate Q_θ for a propagating axon is to notice that for small τ and large I, $|c \, \partial v/\partial t| \gg |j_i|$ in (5.1b). Thus (5.1b) can be written as

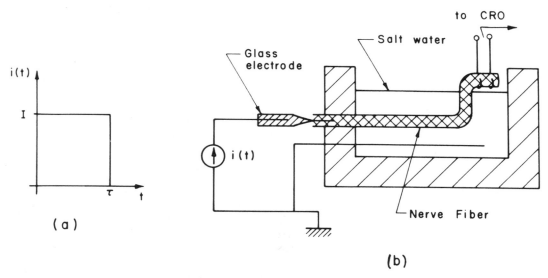

FIG. 39.   (a) Strength (I) and duration (τ) for a threshold measurement. (b) An experiment to measure strength duration curves for a nerve fiber.

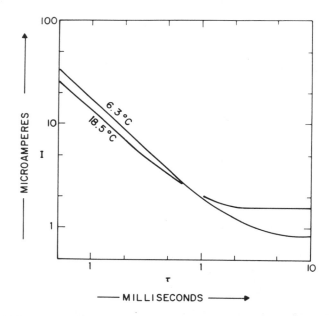

FIG. 40.   Calculated strength duration curves for the Hodgkin–Huxley axon. Redrawn from Cooley and Dodge (1966).

the *approximate conservation law* [Scott (1973a)]

$$\partial i/\partial x + \partial(cv)/\partial t \approx 0. \qquad (10.2)$$

Longitudinal current, i, is the *flow*, and (cv) is the *density* of the approximately conserved quantity which is, therefore, a quantity of charge. Equation (10.2) is approximately satisfied on the leading edge of an action potential since the displacement current is greatest (i.e., $\partial v/\partial t$ is maximum) and the turn on of sodium current is delayed by τ_m [Hodgkin–Huxley (1952)]. Thus the amount of approximately conserved charge carried by the leading edge of a pulse can be evaluated as

$$Q_0 = \int_{\text{[leading edge]}} i \, dt. \qquad (10.3)$$

From (5.1a), $i = -v_x/r_s$; and, from (5.3), $v_x = -v_t/u$ so

TABLE III.   Comparison of Q_0 for fully developed action potention until Q_0 threshold charge.

Temperature (°C)	Q_0 (C)	Q_θ (C)	Q_θ/Q_0
18.5	2.52×10^{-9}	1.33×10^{-9}	0.53
6.3	4.23×10^{-9}	1.71×10^{-9}	0.41

(10.3) is readily evaluated as

$$Q_0 = V_{\max}/ur_s, \qquad (10.4)$$

where $V_{\max}$ is the height of the action potential. Estimates of Q_0 for the fully developed action potential on the Hodgkin–Huxley axon are compared with the corresponding values of threshold charge, Q_θ, (from Fig. 40) in Table III.

The fact that conserved charge carried in the leading edge is about twice as large as the threshold charge should not be surprising. This "safety factor" is necessary in order to insure reliable propagation of the pulse in the presence of inhomogeneities of the fiber [Smolyaninov (1968), Markin and Patushenko (1969), Patushenko and Markin (1969), Khodorov, *et al.* (1969–1971), Berkinblit, *et al.* (1970), Aronov and Kheifets (1971), Polyakov (1973)]. Table III implies the relation

$$Q_\theta = \alpha Q_0, \qquad (10.5)$$

where α is a constant approximately equal to $1/2$. In general it can be estimated as the ratio of leading edge charge for a threshold pulse to that of a stable action potential. On this basis the curves in Fig. 22 indicate $\alpha = 0.63$. The discrepancy is probably connected with the fact that the approximate conservation law (10.2) is not so well satisfied for the leading edge of the threshold pulse. More precise measures of threshold pulses using an appropriately defined Lyapunov functional [Parmentier (1970), Elias and Ghausi (1972)] may be useful in improving these estimates. In myelinated fibers the threshold conditions are somewhat more complex [Tasaki (1959), BeMent and Ranck (1969)] but a recent study by Bean (1974) indicates that a threshold condition

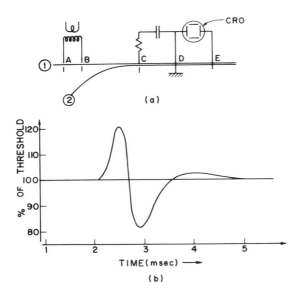

FIG. 41.   (a) Experiment of Katz and Schmitt (1939) to measure interaction in parallel fibers. (b) Change in threshold on ② caused by presence of a pulse on ①.

of fixed voltage change, corresponding to a nodal charge, is appropriate.

XI. PULSE INTERACTIONS

A. Single fiber interactions

The well established experimental fact that two oppositely directed nerve pulses will annihilate each other upon collision is readily understood from our previous development of leading edge dynamics. Consider the interaction of two oppositely directed leading edge transitions shown in Fig. 26. If the approximate conservation law (10.2) is assumed, then together with (5.1a) the leading edge interaction is governed by the linear diffusion equation which can be written

$$\partial^2/\partial x^2 (v - V_2) \approx r_s c (\partial/\partial t)(v - V_2). \qquad (11.1)$$

Thus we expect a relaxation toward $v = V_2$ for $j(v)$ as indicated in Fig. 26(a) if (11.1) remains valid until the voltage rises above V_1. As soon as $(v - V_2)$ lies within the range of convergence for the Taylor series expansion for $j(v)$ about V_2, v must decay to V_2. In terms of (10.4), we can say that the *net* approximately conserved charge for the leading edges is zero. Referring back to Fig. 32 for the action potential of the FitzHugh–Nagumo equation, we expect next a slow relaxation with a time constant τ_n. The third stage is the interaction of the trailing edges which, according to the same argument employed for the leading edges, should bring the voltage to a negative value followed by a slow relaxation toward zero.

B. Parallel fiber interactions

No more than a glance at the lower photograph in Fig. 35 should be necessary to justify an interest in the interaction of pulses which are traveling on parallel fibers. The study of this effect was initiated in an elegant series of experiments by Katz and Schmitt (1939, 1940, 1942). Working on a pair of naturally adjacent fibers from the limb nerve of a crab, their basic apparatus was as shown in Fig. 41. A

reference pulse was initiated at AB on fiber ① at a fixed time; and at an adjusted later time the threshold for pulse excitation on fiber ② at CD was measured. The result is recorded in Fig. 41(b) and is interpreted as a stimulation of fiber ② which is roughly proportional to the derivative of the voltage (or from (5.1a) the membrane current) in fiber ①. They also observed the effects of mutual pulse interaction between impulses simultaneously initiated on the two fibers which produced various combinations of speeding or slowing depending upon the phase relation. In particular, *synchronization* of the pulses could be observed if their independent velocities did not differ by more than about 10%. All interaction effects could be increased by reducing the conductivity of the interstitial fluid. Similar effects have been observed by Crane (1964) on neuristors and by Kunov (1966) on electronic analogs for nerve fibers.

Recently Markin (1970a, b) has developed a nonlinear theory for parallel fiber interactions. Starting from a TLEC representing two fibers which share the external medium, he derived a pair of coupled nonlinear diffusion equations with the form

$$(1/\gamma)[(r_2 + r_3)v_{1,xx} - r_3 v_{2,xx}] - c_1 v_{1,t} = j_1, \qquad (11.2a)$$

$$(1/\gamma)[(r_1 + r_3)v_{2,xx} - r_3 v_{1,xx}] - c_2 v_{2,t} = j_2, \qquad (11.2b)$$

where r_1, c_1, j_1, and v_1 are the series resistance/length, shunt capacitance/length, membrane ion current/length, and transmembrane voltage for fiber ① and similarly for fiber ②. The interstitial resistance/length is r_3 and $\gamma \equiv r_1 r_2 + r_1 r_3 + r_2 r_3$; so as $r_3 \to 0$, (11.2a, b) become two uncoupled equations with the form (2.30). Nonlinear pulse interaction was studied by representing $j_1(t)$ as

$$
\begin{aligned}
j_1(t) &= 0 &&\text{for } t < 0, \\
&= -J_1 &&\text{for } 0 < t < \tau_1, \\
&= +J_2 &&\text{for } \tau_1 < t < \tau_1 + \tau_2, \\
&= 0 &&\text{for } \tau_1 + \tau_2 < t, \qquad (11.3)
\end{aligned}
$$

where the condition $J_1\tau_1 = J_2\tau_2$ is imposed for zero net charge transfer and pulse return. A similar form was assumed for $j_2(t)$ but with an adjustable time delay. This simple description of the nerve pulse was previously shown [Markin and Chizmadzhev (1967), Undrovinas *et al.* (1972)] to give both the stable (upper) and unstable (lower) velocities which arise in the FitzHugh–Nagumo description. They found that two pulses on adjacent fibers can have three stable bound (collective) states if the uncoupled velocities are sufficiently close together. More recently Markin (1973a, b) has extended this approach to the study of interactions in fiber bundles. The derivation and solution of coupled nonlinear diffusion equations should be of considerable interest to physicists and mathematicians during the next few years.

C. Interactions at branching points of axons and dendrites

As we saw in Fig. 2, the action potential propagates away from the cell body along an axon or outgoing fiber which may or may not be myelinated. This outgoing pulse travels up to the axonal tree and eventually delivers inputs to many other cells through chemical contacts called *synapses*. On the input end, the dendrites and cell body receive many synaptic inputs which somehow contribute to a firing decision by the cell body or the main axon. Extensive branching occurs both in the axonal (output) tree and dendritic (input) trees. The behavior of pulses near these branching points does not yet appear to be well understood.

On the axonal side it is often assumed that the "parent" fiber excites all "daughters" at each branching point so the signal travels without interruption to every distal (distant) twig, but experiments by Barron and Matthews (1935), Krnjevic and Miledi (1959), Chung, Raymond, and Lettvin (1970), Parnas (1972), and Grossman, Spira and Parnas (1973) cast doubt on this simple picture. In these studies, the branch points of some axons emerge as regions of low safety factor where high frequency blockage and alternate firing can take place. Some understanding of this situation may be obtained considering the concept of "threshold charge" expressed in (10.5). From (10.4) it can be seen that Q_0 is proportional to $d^{3/2}$ (where d is the fiber diameter) so we can write [Scott (1973b)]

$$Q_0 = kd^{3/2}. \tag{11.4}$$

For conduction from a parent of diameter d_2 to two daughters each of diameter d_1, the leading edge charge carried by the parent must equal the sum of the threshold charges required by the daughters. This requirement implies

$$d_2/d_1 \geq (2\alpha)^{2/3}, \tag{11.5}$$

where the equality indicates marginal transmission. From our approximate estimate $\alpha \approx 1/2$ (see Table III), marginal conduction should occur when the parent and daughters are of roughly equal size. Conduction through the branch point under marginal conditions might be influenced by small changes in local geometry and electric coupling from pulses on neighboring fibers as well as fatigue from repetition. Thus axonal branch points might provide a location for modification of neural transmission or learning.

On the dendritic side of the nerve cell, the situation is even less clear. Much of the confusion is connected with the implications of the "all or nothing law" of propagation [Lucas (1909), Adrian (1914)] on an active fiber which has dominated the thinking of electrophysiologists for over half a century [Lorente de Nó and Condouris (1959)]. If this "law" is interpreted as implying that an action potential will fire all active fibers to which it is connected, then the integrative function of the dendritic trees cannot be understood unless they are assumed to be passive or at least decremental. But the situation is not so simple. In the first place, as FitzHugh (1955, 1969) has pointed out, the continuity properties for the Hodgkin–Huxley equations [Lefschetz (1962)] do not permit a discontinuous jump from "off" to "on" as the initial conditions are changed. Either the latent period before firing goes to infinity or the latent period is bounded and the derivative of response with respect to stimulus is also bounded. Of course discontinuous response could be invoked by assuming fast regeneration in the phase change of the membrane which was discussed in Sec. IV; and, on the other hand, a continuous rise of response can be so steep that it is indistinguishable from a discontinuous jump in the presence of unavoidable laboratory noise. In sum, therefore, it seems that threshold problems should be approached through careful study rather than imprecise generalities.

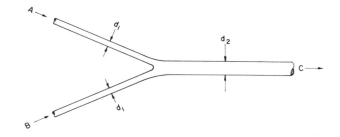

FIG. 42.   A simple dendritic bifurcation.

There are experimental results which indicate that information proceeds through the dendritic trees of some neurons by purely passive means [Purpura and Grundfest (1956), von Euler, Green and Ricci (1956), Grundfest (1958)], and a corresponding mathematical theory of passive dendrites has been developed [Rall (1959, 1962a, b, 1964, 1967), Pokrovskii (1970), Pickard (1947)] which essentially involves solving a linearized version of (2.30) with the coefficients taken as functions of x. Rall (1959, 1962a) paid particular attention to impedance matching conditions and pointed out that the characteristic admittance Y_0 (defined as the square root of the ratio of shunt admittance/length to r_s) is proportional to the $3/2$ power of the fiber diameter. Thus

$$Y_0 \propto d^{3/2} \tag{11.6}$$

which was used to define an "equivalent dendritic cylinder [Rall (1962b)], satisfying the condition $\Sigma d_i^{3/2} = \text{const}$ at each successive branching, in order to simplify dendritic computations.

But experiments indicating passive dendritic conduction are open to various interpretations [Bishop (1958), Eccles (1960), Rall and Shepherd, (1968), Rall (1970), Bogdanov and Golovchinskii (1970)], and there have been several studies which imply that action potentials can propagate at least on the larger branches of some dendritic trees [Lorente de Nó (1947), Cragg and Hamlyn (1955), Eyzaquirre and Kuffler (1955), Fatt (1957), Hild and Tasaki (1962), Anderson, Holmquist and Voorhoeve (1966), Llinás, Nicholson, Freeman and Hillman (1968), Luk'yanov (1970), Nicholson and Llinás (1971), Llinás and Nicholson (1971)]. Lorente de Nó (1960), Arshavskii et al. (1965), Gutman (1971), Waxman (1972), Scott (1973b), Llinás et al. (1969) and Gutman and Shimolinuas (1973) have pointed out that the dendrites should be able to perform elementary logical operations at branching points if they can propagate action potentials or even decremental pulses. However the elementary application of the "all or nothing law" must be replaced by a consideration of threshold conditions at each branching point.

A simple argument to indicate the nature of active dendritic logic can be presented in connection with the bifurcation shown in Fig. 42 [Scott (1973b)]. The "OR" condition obtains if an incoming pulse on either branch A or branch B can provide the charge necessary to stimulate an active pulse on branch C. From (11.4) the leading edge charge coming in on a daugher branch is $kd_1^{3/2}$. This charge will divide between the parent and the other daughter in a ratio which is fixed by their respective characteristic ad-

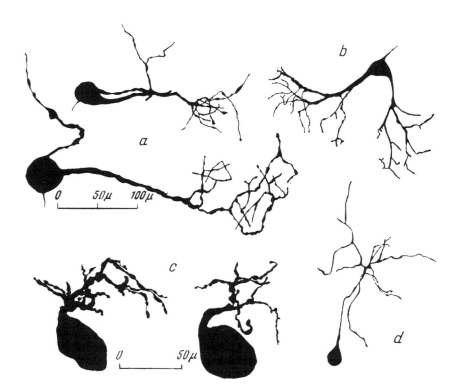

FIG. 43. Cochlear neurons of (a) monkey, (b) hedgehog, (c) owl and (d) bat from Bogoslovskaya *et al.* (1973).

mittances below threshold as given in (11.6). Then a fraction $d_2^{3/2}/(d_1^{3/2} + d_2^{3/2})$ of the incoming charge will reach the parent and this must exceed $\alpha k d_2^{3/2}$ in order to fire the parent. This condition is equivalent to

$$d_1/d_2 > [\alpha/(1 - \alpha)]^{2/3} \qquad (11.7)$$

as the requirement on the diameter ratios for "OR" logic at the branch point. If the inequality in (11.7) is not satisfied, input pulses on both daughters A "AND" B are necessary in order to fire the parent. Corresponding expressions are easily obtained when the daughter fibers are unequal, or for the threshold number of daughter fibers which must be excited on the "tufted" branching points which Ramón–Molinar (1962) describes as being typical for dendrites of sensory neurons. Pastushenko, Markin, and Chizmadzhev (1969a, b) have conducted a much more detailed analysis of this problem using (2.30) to describe the nerve, but with j_i as in (11.3). Their basic boundary condition was Kirchhoff's current law and their threshold requirement was that the voltage rise should reach a preassigned level at $t = 0$ in (11.3). They derive relations corresponding to (11.7), and they account for nonsynchronous effects in "AND" junctions. Berkinblit *et al.* (1971) have studied the problem numerically using the Hodgkin–Huxley equations (5.1) to represent the three fibers. In addition to confirming previous results, they were able to demonstrate *inhibition* by a subthreshold pulse on one daughter of a properly delayed pulse on the other daughter. Some time ago, Tauc and Hughes (1963) demonstrated similar effects during experiments with axons of nerve cells in a mollusk.

The possibility of dendritic logic opens intriguing lines for speculation and future research. As an example, consider the dendrites of the cerebellar Purkinje cell shown in Fig. 2. These trees lie in a plane about $\frac{1}{4} \times \frac{1}{4}$ mm² and about 6 μ thick for man and receive some 80 000 synaptic inputs from perpendicular parallel fiber axons [Eccles (1973), Szentágothai (1968)]. The output axon provides inhibitory signals for muscle control; and, as has been suggested by Marr (1969) and Albus (1971), the cell may function as a "Perceptron" [Block (1962), Block, Knight and Rosenblatt (1962)] which merely calculates a weighted sum of the inputs and decides whether or not it is above a threshold for firing the cell body and/or the axon. But if each of the branching points can function as a logic gate, the computing power would be much greater than that of a Perceptron. Rall (1962) has suggested that careful dendritic studies may also be relevant to the problem of learning and memory. Rose *et al.* (1960), for example, have suggested that the regrowth of cortical dendrites observed after radiation damage may be due to a "normal, continuous growth of central neurons." Thus the logical character of an existing branch point might be modified by changes in its geometry or in the geometry of neighboring cells. It might be possible to observe such effects in tissue culture experiments similar to those conducted by Hild and Tasaki (1962).

Efforts to understand the nature of propagation on nonuniform fibers [Smolyaninov (1968), Khodorov *et al.* (1969), Pastushenko and Markin (1969), Khodorov *et al.* (1970), Khodorov and Timin (1970, 1971), Pastushenko and Markin (1973), Parnas *et al.* (1973), Goldstein and Rall (1974), Khodorov (1974)] should be viewed in relation to the question of axonal and dendritic logic. A widening of the fiber leads to a propagation delay [Markin and Pastushenko (1969), Berkinblit *et al.* (1970), Khodorov *et al.* (1971), Goldstein and Rall (1974)] which appears to be caused by charging of the extra membrane capacitance to a threshold level. Bogoslovskaya *et al.* (1973) suggest that varicose regions in the dendrites of cochlear neurons (see Fig. 43) may be related to information processing functions.

D. Pulse burst dynamics

Whitham (1974) has developed a technique for finding solutions to nonlinear wave problems that are locally periodic, but for which the frequency, wave length and amplitude are *slowly varying* functions of space and time. Such periodic solutions are not sinudoidal (often they are elliptic functions) and the corresponding dispersion equation is of the form $\omega = \Omega(\beta, A)$, where

$$\omega = 2\pi/T \quad \text{and} \quad \beta = 2\pi/\lambda \qquad (11.8a, b)$$

and T, λ, A are, respectively, the wave time and space periods, and the amplitude. Two quasilinear equations for the slow evolution of ω, β, and A are obtained from variation of a Lagrangian density which has been averaged over a cycle of the periodic wave. Such a Lagrangian density can be obtained from an energy conservation law (1.4). A third quasilinear equation is *conservation of wave crests*

$$\partial \beta / \partial t + \partial \omega / \partial x = 0. \qquad (11.9)$$

For nerve fiber problems, we do not have conservation of energy; propagation is governed instead by the power balance condition (1.3). Furthermore, as Fig. 33 indicates, the frequency, propagation constant and amplitude for a stable periodic wave are fixed by the local propagation velocity

$$u = \omega/\beta. \qquad (11.10)$$

Thus $\omega = \omega(u)$, $\beta = \beta(u)$, and $A = A(u)$, so only (11.9) is needed to describe the slow evolution of ω, β, and A. Conservation of wave crests becomes

$$\partial u / \partial t + U(u)(\partial u / \partial x) = 0, \qquad (11.11)$$

where

$$U(u) \equiv d\omega/d\beta \qquad (11.12)$$

is a *nonlinear group velocity*. For the periodic wave described in Fig. 33(a), a typical $\omega - \beta$ diagram is sketched in Fig. 44(a). Along the stable (high velocity) branch it is clear that

$$U(u) < u \qquad (11.13)$$

as was noted by Rinzel and Keller (1973). Thus, as is indicated in Fig. 44(b), the compressed region of a pulse burst should drift to the rear. The question of "rear end collisions" [Crane (1964)] may be important for a nerve fiber just as it is in a corresponding study of automobile traffic dynamics [Whitham (1974)].

XII. CONCLUSION

It may be appropriate to end this review by presenting some reasons for which I believe a physicist should be interested in studying the nerve fiber. First, as we saw in Secs. III and IV, a fundamental connection has not yet been established between a dynamical description of the nerve membrane and the underlying biochemistry. The special knowledge of solid state physics may be helpful to

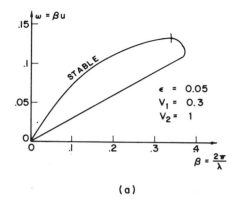

(a)

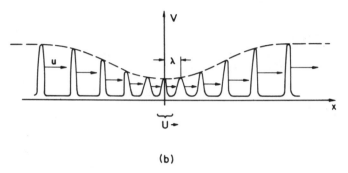

(b)

FIG. 44.   (a) ω–β diagram for a curve from Fig. 33(a). (b) A slowly varying train of pulses.

biochemists who are attempting to solve this intriguing riddle. Second, nerve fiber studies present a number of well defined problems (e.g., pulse properties, pulse interactions, threshold effects, decremental conduction, stability, electromagnetic consideration of nonuniform fibers, etc.) which should be challenging for many physicists and applied mathematicians to consider. Finally there is the program, outlined by Caianiello (1961), which begins with an appropriate description of neural elements and proceeds toward an understanding of brains. In addition to providing a sound "atomic theory" for this program, study of the nerve fiber provides an excellent example of the "hierarchy of boundary conditions" which Polanyi (1962, 1965, 1968) finds characteristic of life. The organization of protein bearing lipid membranes into branching tubes with different internal and external salt solutions clearly introduces "higher principles" associated with special forms of the nonlinear diffusion equation. These principles must necessarily be understood in order to describe the dynamics of nerve pulses and they exist not in conflict with the principles of physics and chemistry but in addition to them. Problems of perceiving and understanding such higher principles become acute as one considers more complex dynamic systems, but the danger for those who miss the point has been emphasized by Goethe[9]

> Wer will was Lebendig's erkennen und beschreiben,
> Sucht erst den Geist heraus zu treiben,
> Dan hat er die Teile in seiner Hand,
> Fehlt leider nur das geistige Band.

[9] Quoted by Franz Boas in his introduction to Ruth Benedict's *Patterns of Culture.*

ACKNOWLEDGMENT

While accepting full responsibility for any errors, I express my appreciation to the following scientists for detailed and valuable comments: G. A. Carpenter, K. S. Cole, R. FitzHugh, D. E. Green, S. Hastings, A. F. Huxley, M. K. Jain, J. B. Keller, B. Lavenda, F. F. Offner, W. F. Pickard, J, W. Rall, Rinzel, I. Tasaki, and S. Yoshizawa.

REFERENCES

Abbott, B. C., A. V. Hill and J. V. Howarth, 1958, "The positive and negative heat production associated with a nerve impulse," Proc. Roy. Soc. (London) B **148**, 149.

Adam, N. K., 1921, 1922, "The properties and molecular structure of thin films," Proc. Roy. Soc. (London) Ser. A. **99**, 336; **101**, 452; and 516.

Agin, D., 1969, "An approach to the physical basis of negative conductance in the squid axon," Biophys. J., **9**, 209.

Albus, J. S., 1971, "A theory of cerebellar function," Math. Biosci **10**, 25.

Anderson, P., B. Holmquist and P. E. Voorhoeve, 1966, "Entorhinal activation of dentate granule cells," Acta. Physiol. Scand. **66**, 448.

Andronov, A. A., A. A. Vitt and S. E. Khaikin, 1966, *Theory of Oscillators*, (Addison Wesley, Reading, Mass.)

Arima, R., and Y. Hasegawa, 1963, "On global solutions for mixed problem of a semi-linear differential equation," Proc. Japan Acad., **39**, 721.

Aronov, I. Z. and L. M. Kheifets, 1971, "Spread of an impulse in a model of the nerve fibre," Biophysics, **16**, 766.

Aronson, D. G., and H. F. Weinberger, 1975, "Nonlinear diffusion in population genetics, combustion, and nerve propagation," Proc. Tulane Program in PDE.

Arshavskii, Yu. I., M. B. Berkinblit, S. A. Kovalev, V. V. Smolyaninov, and L. M. Chailakhyan, 1965, "The role of dendrites in the functioning of nerve cells," Doklady Akademii Nauk SSSR **163**, 994.

Averbach, M. S., and D. N. Nasonov, 1950, "The law of self regulation of spreading stimulation ("all or nothing")" Fiziologicheskii Zhurnal SSSR **36**, 46 (in Russian).

Balakhovskii, I. S., 1968, "Constant rate of spread of excitation in an ideal excitable tissue," Biophysics **13**, 864.

Bean, C. P., 1974, "A theory of micro-stimulation of myelinated fibers," (to be published).

BeMent, S. L., and J. B. Ranck, Jr., 1969, "A model for electrical stimulation of central myelinated fibers with monopolar electrodes," Exp. Neurol. **24**, 171.

Berestovskiy, G. N., 1963, "Study of single electric model of neuristor," Rad. Eng. & Elec. Phys. **18**, 1744.

Berestovskii, Liberman, Lunevskii, and Frank, 1970, "Optical investigations of change in the structure of the neural membrane on passage of a nerve impulse," Biophysics **15**, 60.

Berkinblit, M. B., I. Dudzyavichus and L. M. Chailakhyan, 1971, "Dependence of the rate of spread of an impulse in a nerve fiber on the capacitance of its membrane," Biophysics **16**, 594.

Berkinblit, M. B., N. D. Vvedenskaya, L. S. Gnedenko, S. A. Kovalev, A. V. Kholopov, S. V. Fornin and L. M. Chailakhyan, 1970, "Computer investigation of the features of conduction of a nerve impulse along fibres with different degrees of widening," Biophysics, **15**, 1121.

Berkinblit, M. B., N. D. Vvedenskaya, L. S. Gnedenko, S. A. Kovalev, A. V. Kholopov, S. V. Formin, and L. M. Chailakhyan, 1971, "Interaction of the nerve impulses in a node of branching (investigation on the Hodgkin-Huxley model)" Biophysics, **16**, 105.

Bernstein, J., 1868, "Ueber den zeitlichen Verlauf der negativen Schwankung des Nervenstroms," Arch. ges. Physiol., **1**, 173.

Bernstein, J., 1902, "Untersuchungen zur Thermodynamik der bioelektrischen Ströme," Arch. ges. Physiol. **92**, 521.

Bishop, G. H., 1958, "The dendrite: receptive pole of the neurone," Electroenceph. Clin. Neurophysiol. Suppl. **10**, 12.

Block, H. D., 1962, "The Perceptron: A model for brain functioning," Rev. Mod. Phys. **34**, 123.

Block, H. D., B. W. Knight, Jr., and F. Rosenblatt, 1962, "Analysis of a four-layer series-coupled Perceptron," Rev. Mod. Phys. **34**, 135.

Bogdanov, K. Yu., and V. B. Golovchinskii, 1970, "Extracellularly recorded action potential and the possibility of spread of excitation over the dendrites," Biophysics **15**, 672.

Bogoslovskaya, L. S., I. A. Lyubinskii, N. V. Pozin, Ye. V. Putsillo, L. A. Shmelev and T. M. Shura-Bura, 1973, "Spread of excitation along a fiber with local inhomogeneities (results of modelling)" Biophysics **18**, 944.

Bonhoeffer, K. F., 1948, "Activation of passive iron as a model for the excitation of nerve," J. Gen. Physiol., **32**, 69.

Boltaks, B. I., V. Ya. Vodyanoi, and N. A. Fedorovich, 1971, "Mechanisms of conductivity of phospholipid membranes," Biophysics **16**, 856.

Boussinesq, M. J., 1872, "Théorie des oudes qui se propagent le long d'un canal rectangulaire horizontal, en communiquant au liquide contenu dans ce canal des vitesses sensiblement pareilles de la surface au fond," J. Math. Pures Appl. **17**, 55.

Brady, S. W., 1970, "Boundedness theorems and other mathematical studies of differential equations," Math. Biosci. **6**, 209.

Branton, D., and R. B. Park, 1968, *Papers on Biological Membrane Structure*, (Little Brown Co., Boston).

Brazier, M. A. B. 1959 , "The historical development of neurophysiology," in *Handbook of Physiology*, Am. Physiol. Soc., Sec. 1, Vol. I, p. 1.

Bretscher, M. S., 1973, "Membrane structure: some general principles," Science **176**, 622.

Brooke Benjamin, T., 1972, "The stability of solitary waves," Proc. R. Soc. Lond. A. **328**, 153.

Bungenberg de Jong, H. G. and J. Bonner, 1935, "Phosphatide autocomplex coacervates as ionic systems and their relation to the protoplasmic membrane," Protoplasma **15**, 198.

Buratti, R. J., and A. G. Lindgren, 1968, "Neuristor wave forms and stability by the linear approximation," Proc. IEEE **56**, 1392.

Caianiello, E. R., 1961, "Outline of a theory of thought-processes and thinking machines," J. Theor. Biol. **2**, 204.

Canosa, J., 1973, "On a nonlinear diffusion equation describing population growth," IBM J. Res. Dev. **17**, 307.

Carpenter, G., 1974, "Traveling wave solutions of nerve impulse equations," Ph.D. Thesis, U. of Wisconsin.

Casten, R. G., H. Cohen and P. A. Lagerstrom, 1975 "Perturbation analysis of an approximation to the Hodgkin-Huxley theory," Quart. Appl. Math **32**, 365.

Cereijido, M., and C. A. Rotunno, 1970, *Introduction to the Study of Biological Membranes*, (Gordon and Breach, New York).

Chandler, W. K., R. Fitzhugh and K. S. Cole, 1962, "Theoretical stability properties of a space clamped axon," Biophys. J. **2**, 105.

Changeux, J. P., Thiéry, J., Tung, Y, and Kittel, C., 1967, "On the cooperativity of biological membranes," Proc. Natl. Acad. Sci. **57**, 335.

Chizmadzhev, Yu. A., Markin, V. S., and Muler, A. L., 1973, "Conformational model of excitable cell membrances—II. Basic equations," Biophysics **18**, 70.

Chizmadzhev, Yu. A., Muler, A. L., and Markin, D. S., 1972, "Conformational model of excitable cell membranes—I. Ionic permeability," Biophysics **17**, 1061.

Chung, S. H., S. A. Raymond and J. Y. Lettvin, 1970, "Multiple meaning in single visual units," Brain Behav. Evol. **3**, 72.

Clark, J., and R. Plonsey, 1966, "A mathematical evaluation of the core conductor model," Biophs. J. **6**, 95.

Clark, J., and R. Plonsey, 1968, "The extracellular potential field of the single active nerve fiber in a volume conductor," Biophs. J. **8**, 842.

Cohen, H., 1971, "Nonlinear diffusion problems," in *Studies in Applied Mathematics* (A. H. Taub, ed.) Math. Assoc. Am., Prentice-Hall, 27.

Cohen, L. B., R. D. Keynes, and B. Hille, 1968, "Light scattering and birefringence changes during nerve activity," Nature, 218, 438.

Cohen, L. B., B. Hille, and R. D. Keynes, 1970, "Changes in axon birefringence during the action potential," J. Physiol. **211**, 495.

Cohen, L. B. and D. Landowne, 1974, "The temperature dependence of the movement of sodium ions associated with nerve impulses," J. Physiol. **236**, 95.

Cole, K. S., 1949, "Dynamic electrical characteristics of the squid axon membrane," Arch. Sci. Physiol. **3**, 253.

Cole, K. S., 1968, *Membranes, ions and impulses*, (University of California Press, Berkeley and L. A.).

Cole, K. S., 1975, "Neuromembranes: paths of ions," *Neurosciences: Paths of Discovery*, edited by G. Adelman (Rockefeller U. P., New York).

Cole, K. S., H. A. Antosiewicz and P. Rabinowitz, 1955, "Automatic computation of nerve excitation," SIAM Jour. **3**, 153.

Cole, K. S., and R. F. Baker, 1941, "Longitudinal impedance of the squid giant axon," J. Gen. Physiol. **24**, 771.

Cole, K. S. and H. J. Curtis, 1938, "Electrical impedance of nerve during activity," Nature 142, 209.

Cole, K. S. and Curtis, H. J., 1939, "Electric impedance of the squid giant axon during activity," J. Gen. Physiol. 22, 649.

Cole, K. S., and Moore, J. W., 1960, "Potassium ion current in the squid giant axon: Dynamic characteristic," Biophys. J. 1, 1.

Conley, C., 1973, "On the existence of bounded progressive wave solutions of the Nagumo equation," (unpublished notes).

Cooley, J., F. Dodge, and H. Cohen, 1965, "Digital computer solutions for excitable membrane models," J. Cell. Comp. Physiol. 66 (Supp.), 99.

Cooley, J. W. and F. A. Dodge, Jr., 1966, "Digital computer solutions for excitable and propagation at the nerve impulse," Biophys. J. 6, 583.

Cragg, B. G., and L. H. Hamlyn, 1955, "Action potential of the pyramidal neurons in the hippocampus of the rabbit," J. Physiol. London, 129, 608.

Crane, H. D., 1962, "Neuristor—a novel device and system concept," Proc. IRE, 50, 2048.

Crane, H. D., 1964, "Possibilities for signal processing in axon system," in Neural Theory and Modeling, edited by R. F. Reiss, (Stanford U. P., Stanford).

Danielli, J. F., 1936, "Some properties of lipoid films in relation to the structure of the plasma membrane," J. Cellular Comp. Physiol. 7, 393.

Danielli, J. F. and H. Davson, 1935, "A contribution to the theory of permeability of thin films," J. Cellular Comp. Physiol. 5, 495.

Dean, R. B., 1939, "Potentials at oil water interfaces," Nature 144, 32.

Dean, R. B., H. J. Curtis and K. S. Cole, 1940, "Impedance of bimolecular films," Science 91, 50.

Deck, K. A., and W. Trautwein, 1964, "Ionic currents in cardiac excitation," Arch. ges. Physiol. 280, 63.

Devaux, H., 1936, "Détermination de l'épaisseur de la membrane d'albumine formée entre l'eau et la benzine et propriétés de cette membrane," C. R. Acad. Sci., 202, 1957.

Eccles, J. C., 1960, "The properties of the dendrites," in Structure and Function of the Cerebral Cortex, (American Elsevier, New York).

Eccles, J. C., 1973, The Understanding of the Brain, (McGraw-Hill, New York).

Eckhaus, W., 1965, Studies in Non-Linear Stability Theory, (Springer-Verlag, New York).

Einstein, A., 1905, "Über die von der molekular-kinetischen Theorie der Wärme geforderte Bewegung von in ruhenden Flüssigkeiten suspendierten Teilchen," Ann. der Physik 17, 549.

Eisenberg, M., Hall, J. E., and C. A. Mead, 1973, "The nature of the voltage-dependent conductance induced by alamethicin in black lipid membranes," J. Membrane Biol. 11, 1–34.

Eleonskii, V. M., 1968, "Stability of simple stationary waves related to the nonlinear diffusion equation," Sov. Phys. JETP 26, 382.

Elias, N. J., and M. S. Ghausi, 1972, "An analysis of excitation thresholds in nonlinear distributed neuristor circuits," J. Franklin Inst. 293, 421.

von Euler, C., J. D. Green, and G. Ricci, 1956, "The role of hippocampal dendrites in evoked responses and after discharges," Acta. Physiol. Scand. 42, 87.

Evans, J., and N. Shenk, 1970, "Solutions to axon equations," Biophys. J. 10, 1090.

Evans, J. W., 1972, "Nerve axon equations: I. Linear approximations, II. Stability at rest, III. Stability of the nerve impulse, Ind. U. Math. J., 21, 877; 22, 75 and 577.

Evans, J. W., 1974, "Nerve axon equations: IV. The stable and the unstable impulse," Ind. U. Math. J. (to be published).

Eyzaguirre, C., and S. W. Kuffler, 1955, "Further study of some, dendrite and axon excitation in single neurons," J. Gen. Physiol. 39, 121.

Falk, G. and P. Fatt, 1964, "Linear electrical properties of striated muscle fibres observed with intracellular electrodes," Proc. Roy. Soc. (London) Ser. B., 160, 69.

Fatt, P., 1957, "Electrical potentials occurring around a neurone during its antidromic activation," J. Neurophysiol. 20, 27.

Fisher, R. A., 1937, "The wave of advance of advantageous genes," Ann. Eugenics (now Ann. Human Genetics) 7, 355.

Fishman, S. N., B. I. Khodorov and M. V. Volkenshtein, 1972, "Molecular mechanisms of change in the ionic permeability of an electro-excitable membrane—II. Model of the process of activation," Biophysics 17, 637.

FitzHugh, R., 1955, "Mathematical models of threshold phenomena in the nerve membrane," Bull. Math. Biophys. 17, 257.

FitzHugh, R., 1961, "Impulses and physiological states in theoretical models of nerve membrane," Biophys. J., 1, 445.

FitzHugh, R., 1962, "Computation of impulse initiation and saltatory conduction in a myelinated nerve fiber," Biophys. J., 2, 11.

FitzHugh, R., 1965, "A kinetic model of the conductance changes in nerve membrane," J. Cell. and Comp. Physiol. 66, 111.

FitzHugh, R., 1969, "Mathematical models of excitation and propagation in nerve," in Biological Engineering, edited by H. P. Schwan (McGraw Hill, New York), 1–85.

FitzHugh, R., 1973, "Dimensional analysis of nerve models," J. Theor. Biol. 40, 517.

FitzHugh, R. and H. A. Antosiewicz, 1959, "Automatic computation of nerve excitation—detailed corrections and additions," SIAM Jour. 7, 447.

FitzHugh, R. and K. S. Cole, 1973, "Voltage and current clamp transients with membrane dielectric loss," Biophys. J. 13, 1125.

Frankenhaeuser and A. F. Huxley, 1964, "The action potential in the myelinated nerve fibre of Xenopus Laevis as computed on the basis of voltage clamp data," J. Physiol. 171, 302.

Fricke, H., 1923, "The electric capacity of cell suspensions," Phys. Rev. 21, 708.

Fricke, H., 1925a, 1926, "A mathematical treatment of the electric conductivity and capacity of disperse systems," Phys. Rev. 24, 575; 26, 678.

Fricke, H., 1925b, "The electric capacity of suspensions with special reference to blood," J. Gen. Physiol. 9, 137.

Frölich, H., 1970, "Long range coherence and the action of enzymes," Nature 228, 1093.

Gasser, H. S. and J. Erlanger, 1922, "A study of the action currents of nerve with a cathode ray oscillograph," Am. J. Physiol. 62, 496.

Geduldig, D. and R. Gruener, 1970, "Voltage clamp of the Aplysia giant neurone: early sodium and calcium currents," J. Physiol. 211, 217.

Geselowitz, D. B., 1966, "Comment on the core conductor model," Biophys. 6, 691.

Geselowitz, D. B., 1967, "On bioelectric potentials in an inhomogeneous volume conductor," Biophys. J. 7, 1.

Goldman, D. E., 1943, "Potential, impedance and rectification in membranes," J. Gen. Physiol. 27, 37.

Goldman, D. E., 1964, "A molecular structural basis for the excitation properties of axons," Biophys. J. 4, 167.

Goldman, L. and J. S. Albus, 1968, "Computation of impulse conduction in myelinated fibers; theoretical basis of the velocity diameter relation," Biophys. J., 8, 596.

Goldstein, S. S. and W. Rall, 1974, "Changes of action potential shape and velocity for changing core conductor geometry," Biophys. J. 14, 731.

Goldup, A., S. Ohki and J. F. Danielli, 1970, "Black lipid films," in Recent Progress in Surface Science, edited by J. F. Danielli, A. C. Riddiford and M. D. Rosenberg, New York, p. 193.

Gorter, E. and F. Grendel, 1925, "On biomolecular layers of lipoids on the chromocytes of the blood," J. Exptl. Med. 41, 439.

Green, D. E., 1971, "Membrane structure," Science 174, 863.

Green, D. E., S. Ji and R. F. Brucker, 1972, "Structure-function unitization model of biological membranes," Bioenergetics 4, 527.

Greenberg, J. M., 1973, "A note on the Nagumo equation," Quart. J. Math. 24, 307.

Grossman, Y., M. E. Spira and I. Parnas, 1973, "Differential flow of information into branches of a single axon," Brain Research 64, 379.

Grundfest, H., 1958, "Electrophysiology and pharmacology of dendrites," Electroenceph. Clin. Neurophysiol. Suppl., 10, 22.

Gutman, A. M., 1971, "Further remarks on the effectiveness of the dendrite synapses," Biophysics 16, 131.

Gutman, A., and A. Shimoliunas, 1973, "Finite dendrite with an N-shaped current-voltage characteristic for the membrane," Biophysics 18, 1013.

Hall, J. E., C. A. Mead, and G. Szabo, 1973, "A barrier model for current flow in lipid bilayer membranes," J. Membrane Biol. 11, 75.

Harmon, L. D. and E. R. Lewis, 1966, "Neural modeling," Physiol. Rev. 46, p. 513.

Hastings, S. P., 1972, "On a third order differential equation from biology," Quart. J. Math. 23, 435.

Hastings, S. P., 1974a, "The existence of periodic solutions to Nagumo's equation," Quart J. Math. (to be published).

Hastings, S. P., 1974b, "On travelling wave solutions of the Hodgkin-Huxley equations," Arch. Rat. Mech. Anal. (to be published).

Hastings, S. P., 1975a, "Some mathematical problems from neurobiology," Am. Math. Month. (to be published).

Hastings, S. P., 1975b, "The existence of homoclinic orbits for Nagumo's equations" (to be published).

Hatase, O., T. Wakabayashi, H. Hayashi and D. E. Green, 1972, "Collapse and extension of the headpiece-stalk projections in mitochondrial electron transport partiles," Bioenergetics, **3**, 509.

Hearon, J. Z., 1964, "Application of results from linear kinetics to the Hodgkin-Huxley equations," Biophys. J., **4**, 69.

Hellerstein, D., 1968, "A generalization of the theory of electrotonus," Biophys. J. **8**, 358.

Helmholtz, H., 1850, "Messungen über den zeitlichen Verlauf der Zuchung animalischer Muskeln und die Fortpflanzungsgeschwindigkeit der Reizung in den Nerven," Arch. Anta. Physiol. 276.

Hermann, L., 1879, *Handbuch der Physiologie* I. Bewegungsapparate, II. Nervensystems. (Leipzig).

Hermann, L., 1905, "Beiträge zur Physiologie und Physik des Nerven," Arch. ges. Physiol. **109**, 95.

Hild, W., and I. Tasaki, 1962, "Morphological and physiological properties of neurons and glial cells in tissue culture," J. Neurophysiol. **25**, 277.

Hill, A. V., 1936, "Excitation and accommodation in nerve," Proc. Roy. Soc. (London), B **119**, 305.

Hille, B., 1970, "Ionic channels in nerve membranes," Prog. Biophys. Mol. Biol. **21**, 3.

Hodgkin, A. L., 1951, "The ionic basis of electrical activity in nerve and muscle," Biol. Rev. **26**, 339.

Hodgkin, A. L., 1964, *The Conduction of the Nervous Impulse*, (Liverpool U. Press. Liverpool).

Hodgkin, A. L. and A. F. Huxley, 1952a, "Currents carried by sodium and potassium ions through the membrane of the giant axon of *Loligo*," J. Physiol., **116**, 449.

Hodgkin, A. L. and A. F. Huxley, 1952b, "The components of membrane conductance in the giant axon of *Loligo*," J. Physiol., **116**, 473.

Hodgkin, A. L. and A. F. Huxley, 1952c, "The dual effect of membrane potential on sodium conductance in the giant axon of *Loligo*," J. Physiol. **116**, 497.

Hodgkin, A. L. and A. F. Huxley, 1952d, "A quantitative description of membrane current and its application to conduction and excitation in nerve," J. Physiol., **117**, 500.

Hodgkin, A. L., A. F. Huxley and B. Katz, 1952, "Measurement of current-voltage relations in the membrane of the giant axon of *Loligo*," J. Physiol., **116**, 424.

Hodgkin, A. L., and B. Katz, 1949, "The effect of sodium ions on the electrical activity of the giant axon of the squid," J. Physiol. **108**, 37.

Hoorweg, J. L., 1898, "Ueber die elektrischen Eigenschaften der Nerven," Arch. ges. Physiol. **71**, 128.

Howard, R. E. and R. M. Burton, 1968, "Thin lipid membranes with aqueous interfaces: apparatus design and methods of study," J. Am. Oil Chem. Soc. **45**, 202.

Hoyt, R. C., 1963, "The squid giant axon; mathematical models," Biophys. J. **3**, 399.

Hoyt, R. C. 1968, "Sodium inactivation in nerve fibers," Biophys. J. **8**, 1074.

Hoyt, R. C., and W. J. Adelman, Jr., 1970, "Sodium inactivation; experimental test of two models," Biophys. J. **10**, 610.

Hoyt, R. C. and J. D. Strieb, 1971, "A stored charge model for the sodium channel," Biophys. J. **11**, 868.

Hurewicz, W., 1958, *Lectures on Ordinary Differential Equations*, (Wiley, New York).

Huxley, A. F., 1959a, "Ion movements during nerve activity," Ann. N.Y. Acad. Sci., **81**, 221.

Huxley, A. F., 1959b, "Can a nerve propagate a subthreshold disturbance?" J. Physiol. **148**, 80P.

Il'inova, T. M. and R. V. Khokhlov, 1963, "Wave processes in lines with nonlinear shunt resistance," Rad. Eng. Elec. Phys., **8**, 1864.

Inoue, I., Y. Kobatake and I. Tasaki, 1973, "Excitability, instability and phase transitions in squid axon membrane under internal perfusion with dilute salt solutions," Biochemica et Biophysica Acta **307**, 471.

Inoue, I., I. Tasaki and Y. Kobatake, 1974, "A study of the effects of externally applied sodium-ions and detection of spatial nonuniformity of the squid axon membrane under internal perfusion," Biophysical Chemistry **2**, 116.

Isaacs, C. D., 1970, "Analog-digital-hybrid studies of the reformulated equations of Hodgkin-Huxley," Math. Biosci. **7**, 305.

Jain, M. K., 1972, *The Bimolecular Lipid Membrane*, (Van Nostrand Reinhold, New York).

Jain, M. K., R. H. L. Marks and E. H. Cordes, 1970, "Kinetic model of conductance changes across excitable membrances," Proc. Nat. Acad. Sci. **67**, 799.

Jakobsson, E., 1973, "The physical interpretation of mathematical models for sodium permeability changes in excitable membranes," Biophys. J. **13**, 1200.

Jakobsson, E. and C. Scudiero, 1975, "A transient excited state model for sodium permeability changes in excitable membranes," (to be published).

Johnson, W. J., 1968, "Nonlinear wave propagation on superconducting tunneling junctions," Ph.D. Thesis, U. of Wisconsin.

Kanel' Ya. I., 1962, "On the stability of the Cauchy problem for equations occurring in the theory of flames," Matematicheskii Sbornik 59 (**101**), 246.

Kaplan, S. and D. Trujillo, 1970, "Numerical studies of the partial differential equations governing nerve impulse conduction: The effect of Lieberstein's inductance term," Math. Biosci. **7**, 379.

Kashef, B. and R. Bellman, 1974, "Solution of the partial differential equation of the Hodgkin-Huxley model using differential quadrature," Math. Biosci. **19**, 1–8.

Kato, G., 1924, *The Theory of Decrementless Conduction in Narcotized Region of Nerve*, (Nankōdō, Tokyo, Japan).

Kato, G., 1934, *Microphysiology of Nerve*, (Maruzen, Tokyo, Japan).

Kato, G., 1970, "The road a scientist followed," Ann. Rev. Physiol. **32**, 1.

Katz, B., 1939, *Electric Excitation of Nerve*, (Oxford Univ. Press, London).

Katz, B., 1966, *Nerve, Muscle and Synapse*, (McGraw-Hill, New York).

Katz, B. and O. H. Schmitt, 1939, "Excitability changes in a nerve fibre during the passage of an impulse in an adjacent fiber," **96**, 9p.

Katz, B., and O. H. Schmitt, 1940, "Electric interaction between two adjacent nerve fibers," J. Physiol. **97**, 471.

Katz, B. and O. H. Schmitt, 1942, "A note on the interaction between nerve fibers," J. Physiol., **100**, 369.

Kelvin, Lord (William Thompson), 1855, "On the theory of the electric telegraph," Proc. Roy. Soc. (London) **7**, 382.

Khodorov, B. I., 1974, *The Probability of Excitability*, (Plenum Press, New York).

Khodorov, B. I., Ye. N. Timin, S. Ya. Vilenkin, and F. B. Gul'ko, 1969, "Theoretical analysis of the mechanisms of conduction of a nerve pulse over an inhomogeneous axon. I. Conduction through a portion with increased diameter," Biophysics, **14**, 323.

Khodorov, B. I., Ye. N. Timin, S. Ya. Vilenkin and F. B. Gul'ko, 1970, "II. Conduction of a single impulse across a region of the fiber with modified functional properties," Biophysics **15**, 145.

Khodorov, B. I. and Ye. N. Timin, 1970, "III. Transformation of rhythms in the cooled part of the fibre," Biophysics, **15**, 526.

Khodorov, B. I., Ye. N. Timin, N. V. Pozin, and L. A. Shmelev, 1971, "IV. Conduction of a series of impulses through a portion of the fibre with increased diameter," Biophysics, **16**, 96.

Khodorov, B. I. and Ye. N. Timin, 1971, "V. Phenomena of Vvedenskii in portions of the fiber with reduced sodium and potassium conductivity of the membrane," Biophysics, **16**, 513.

Kilkson, 1969, R., "Symmetry and function of biological systems at the macromolecular level," in *Nobel Symposium* 11, edited by A. Engström and B. Strandberg (Wiley-Interscience, New York), 257.

Kishimoto, U., 1965, "Voltage clamp and internal perfusion studies on *Nitella* internodes," J. Cell. Comp. Physiol. 66 *Supp.* 2, 43.

Knight, B. W. and G. A. Peterson, 1967, "Theory of the Gunn effect," Phys. Rev. **155**, 393.

Kokoz, Yu. M. and V. I. Krinskii, 1973, "Analysis of equations of excitable membranes—II. Methods of analysing the electrophysiological characteristics of the Hodgkin-Huxley membrane," Biophys. **18**, 937.

Kolmogoroff, A., I. Petrovsky, and N. Piscounoff, 1937, "Étude de l'équation de la diffusion avec croissance de la quantité de matière et son application á un problème biologique," Bull. Univ. Moscou, Série Int., A **1**, 1.

Kompaneyets, A. S., 1971, "Influence of ohmic resistance of the nerve fiber membrane on certain effects of nervous excitation," Biophysics **16**, 926.

Kompaneyets, A. S. and V. Ts. Gurovich, 1965, "Propagation of an impulse in a nerve fiber," Biophysics **11**, 1049.

Korteweg, D. J. and G. de Vries, 1895, "On the change of form of long waves advancing in a rectangular canal, and on a new type of long stationary wave," Phil. Mag. **39**, 422.

Krinskii, V. I., and Yu. M. Kokoz, 1973, "Analysis of equations of excitable membranes—I. Reduction of the Hodgkin-Huxley equations to a second order system," Biophysics, **18**, 533.

Kryshtal', O. A., I. S. Magura and N. T. Parkhomenko, 1969, "Ionic currents across the membrane of the soma of giant neurones of the edible snail with clamped shifts of the membrane potential," Biophysics, **14**, 987.

Kunov, H., 1965, "Controllable piecewise-linear lumped neuristor realization," Electronics Lett. **1**, 134.

Kunov, H., 1966, "Nonlinear transmission lines simulating nerve axon," Thesis, Electronics Laboratory, Technical University of Denmark.

Kunov, H., 1967, "On recovery in a certain class of neuristors," Proc. IEEE, **55**, 428.

Landowne, D., 1972, "A new explanation of the ionic currents which flow during the nerve impulse," J. Physiol. **222**, 46P.

Landowne, D., 1973, "Movement of sodium ions associated with the nerve impulse," Nature **242**, 457.

Langmuir, I., 1917, "The constitution and fundamental properties of solids and liquids," **39**, 1848.

Langmuir, I., and D. F. Waugh, 1938, "The adsorption of proteins at oil-water interfaces and artificial protein-lipoid membranes," J. Gen. Physiol. **21**, 745.

Lawrence, A. S. C., 1929, *Soap Films*, (Bell, London).

Lefschetz, S., 1962, *Differential Equations: Geometric Theory*, (Wiley-Interscience, New York).

Leibovic, K. N., 1972, *Nervous System Theory*, (Academic Press, New York).

Leibovic, K. N., and N. H. Sabah, 1969, "On synaptic transmission, neural signals and psychophysiological phenomena," in *Information Processing in the Nervous System*, edited by K. N. Leibovic, (Springer-Verlag, New York).

Lehninger, A. L., 1968, "The neuronal membrane," Proc. Nat. Acad. Sci. **60**, 1069.

Lehninger, A. L., 1970, *Biochemistry*, (Worth Pub. Co., New York).

Levin, S. V., D. L. Rozenfal' and Ya. Yu. Komissarchik, 1968, "Structural changes in the axon membrane on excitation," Biophysics **13**, 214.

Lieberstein, H. M. 1967a, "On the Hodgkin-Huxley partial differential equation," Math. Biosci. **1**, 45.

Lieberstein, H. M., 1967b, "Numerical studies of the steady-state equations for a Hodgkins-Huxley model," Math. Biosci. **1**, 181.

Lieberstein, H. M., 1973, *Mathematical Physiology: Blood Flow and Electrically Active Cells*, (Am. Elsevier, New York).

Lieberstein, H. M. and M. A. Mahrous, 1970, "A source of large inductance and concentrated moving magnetic fields on axons," Math. Biosci. **7**, 41.

Lillie, R. S., 1925, "Factors affecting transmission and recovery in the passive iron nerve model," J. Gen. Physiol. **7**, 473.

Lillie, R. S., 1936, "The passive iron wire model of protoplasmic and nervous transmission and its physiological analogues," Biol. Rev. **11**, 181.

Lindgren, A. G. and R. J. Buratti, 1969, "Stability of wave forms on active nonlinear transmission lines," Trans IEEE on Circuit Theory, CT-**16**, 274.

Llinás, R., C. Nicholson, J. A. Freeman, and D. E. Hillman, 1968, "Dendritic spikes and their inhibition in alligator Purkinje cells," Science **160**, 1132.

Llinás, R., C. Nicholson and W. Precht, 1969, "Preferred centripetal conduction of dendritic spikes in alligator Purkinje cells," Science **163**, 184.

Llinás, R., and C. Nicholson, 1971, "Electrophysiological properties of dendrites and somata in alligator Purkinje cells," J. Neurophysiol. **34**, 532.

Lorente de Nó, 1947a, "A study of nerve physiology," Studies from the Rockefeller Institute, **132**, 384.

Lorente de Nó, R., 1947b, "Action potential of the motoneurons of the hypoglossus nucleus," J. Cell. Comp. Physiol. **29**, 207.

Lorente de Nó, R., 1960, "Decremental conduction and summation of stimuli delivered to neurons at distant synapses," in *Structure and Function of the Cerebral Cortex*, (American Elsevier, New York).

Lorente de Nó, R., and G. A. Condouris, 1959, "Decremental conduction in peripheral nerve. Integration of stimuli in the neuron," Proc. N.A.S. **45**, 592.

Lucas, K., 1909, "The 'all-or-none' confraction of amphibian skeletal muscle," J. Physiol. **38**, 113.

Luk'yanov, A. S., 1970, "Generation of action potentials by dendrites in the frog optic tectum," Fiziologicheskii Zhurnal SSSR **56**, 1130–1135; translated in Neurosciences Translations, **16**, 74 (1970–71).

Maginu, K., 1971, "On asymptotic stability of wave forms on a bistable transmission line," Professional Group on Nonlinear Problems NLP **70-24** (in Japanese).

Markin, V. S. and Yu. A. Chizmadzhev, 1967, "On the propagation of an excitation for one model of a nerve fiber," Biophysics **12**, 1032.

Markin, V. S., P. A. Grigor'ev, and L. N. Yermishkin, 1971, "Forward passage of ions across lipid membranes. I. Mathematical model," Biophysics **16**, 1050.

Markin, V. S. and V. F. Pastushenko, 1969, "Spread of excitation in a model of an inhomogeneous nerve fibre. I. Slight change in dimensions of fibre," Biophysics **14**, 335.

Markin, V. S., 1970a, "Electrical interaction of parallel nonmyelinated nerve fibres. I. Change in excitability of the adjacent fibre," Biophysics, **15**, 122.

Markin, V. S., 1970b, "II. Collective conduction of impulses," Biophysics, **15**, 713.

Markin, V. S., 1973a, "III. Interaction in bundles," Biophysics, **18**, 324.

Markin, V. S., 1973b, "IV. Role of anatomical inhomogeneities of nerve trunks," Biophysics **18**, 539.

Marmont, G., 1949, "Studies on the axon membrane. I. A new method," J. Cellular Comp. Physiol. **34**, 351.

Marr, D., 1969, "A theory of cerebellar cortex," J. Physiol. **202**, 437.

Mauro, A., 1961, "Anomalous impedance, a phenomenological property of time-variant resistance," Biophys. J., **1**, 353.

Mauro, A., F. Conti, F. Dodge, and R. Schor, 1970, "Subthreshold behavior and phenomenological impedance of the squid giant axon," J. Gen. Physiol., **55**, 497.

Mauro, A., A. R. Freeman, J. W. Cooley and A. Cass, 1972, "Propagated subthreshold oscillatory response and classical electrotonic response of squid giant axon," Biophysik, **8**, 118.

McKean, H. P., Jr., 1970, "Nagumo's equation," Adv. Math., **4**, 209.

Minor, A. V. and V. V. Maksimov, 1969, "Passive electrical properties of the model of a flat cell," Biophysics, **14**, 349.

Moore, J. W., 1968, "Specifications for nerve membrane models," Proc. IEEE **56**, 895.

Moore, L. E. and E. Jakobsson, 1971, "Interpretation of the sodium permeability changes of myelineated nerve in terms of linear relaxation theory," J. Theoret. Biol., **33**, 77.

Morse, P. M. and H. Feshbach, 1953, *Methods of Theoretical Physics*, (McGraw-Hill, New York).

Mueller, P. and D. O. Rudin, 1968a, "Resting and action potentials in experimental bimolecular lipid membranes," J. Theoret. Biol. **18**, 222.

Mueller, P. and D. O. Rudin, 1968b, "Action potentials induced in bimolecular lipid membranes," Nature, **217**, 713.

Mueller, P., D. O. Rudin, H. T. Tien and W. C. Wescott, 1962, "Reconstitution of cell membrane structure *in vitro* and its transformation into an excitable system," Nature, **194**, 979.

Mullins, L. J., 1959, "An analysis of conductance changes in squid axon," J. Gen. Physiol. **42**, 1013.

Mysels, K. J., K. Shinoda and S. Frankel, 1959, *Soap Films, Studies of Their Thinning and a Bibliography*, (Pergamon Press, New York).

Nachmansohn, D. and E. Neuman, 1974, "Properties and function of proteins in excitable membranes: An integral model of nerve excitability," Ann. N.Y. Acad. Sci., **227**, 275.

Nagumo, J., S. Arimoto and S. Yoshizawa, 1962, "An active pulse transmission line simulating nerve axon," Proc. IRE, **50**, 2061.

Nagumo, J., S. Yoshizawa and S. Arimoto, 1965, "Bistable transmission lines," Trans. IEEE on Circuit Theory, CT-**12**, 400.

Nakajima, K., Y. Onodera, T. Nakamura and R. Sato, 1974, "Numerical analysis of vortex motions on Josephson structures," J. Appl. Phys. **45**, 4095.

Nakajima, K., T. Yamashita and Y. Onodera, 1974, "Mechanical analogue of active Josephson transmission line," J. Appl. Phys. **45**, 3141.

Namerow, N. S., and J. J. Kappl, 1969, "Conduction in demyelinated axons—a simplified model," Bull. Math. Bioph. **31**, 9.

Nernst, W., 1888, "Zur Kinetik der in Lösung befindlichen Köper: Theorie der Diffusion," Zeit. Physik Chem. **2**, 613.

Nernst, W., 1889, "Die electromotorische Wirksamkeit der Ionen," Zeit physik chem., **4**, 129.

Neumke, B., and P. Läuger, 1969, "Nonlinear electrical effects in lipid bilayer membranes, II. Integration of the generalized Nernst-Planck equations," Biophys. J. **9**, 1160.

Newton, I., 1718, Optiks; or, a treatise of the reflections, refractions, inflections and colours of light, 2nd ed. with additions, p. 328.

Newton, I., 1952, *Optiks* (based on the fourth edition, London, 1730) (Dover Publications, Inc., New York).

Nicholson, C., and R. Llinás, 1971, "Field potentials in the alligator cerebellum and theory of their relationship to Purkinje cell dendritic spikes," J. Neurophysiol. **34**, 509.

Noble, D., 1962, "A modification of the Hodgkin-Huxley equations applicable to Purkinje fibre action and pacemaker potentials," J. Physiol. **160**, 317.

Noble, D., 1966, "Applications of Hodgkin-Huxley equations to excitable tissues," Physiol. Rev. **40**, 1.

Noble, D. and R. B. Stein, 1966, "The threshold conditions for initiation of action potentials by excitable cells," J. Physiol. **187**, 129.

Noguchi, S., Y. Kumagai and J. Oizumi, 1963, "General considerations on the neuristor circuits," Sci. Rept. Res. Inst. Tohoku U. **14**, 155.

Nystrom, R. A., 1973, *Membrane Physiology*, (Prentice-Hall, New Jersey).

Offner, F. F., 1969, "Ionic forces and membrane phenomena," Bull. Math. Biophys. **31**, 359.

Offner, F. F., 1970, "Kinetics of excitable membranes. Voltage amplification in a diffusion regime," J. Gen. Physiol. **56**, 272.

Offner, F. F., 1971, "Nernst-Planck-Poisson diffusion equation: numerical solution of the boundary value problem," J. Theor. Biol. **31**, 215.

Offner, F. F., 1972, "The excitable membrane. A physiochemical model," Biophys. J. **12**, 1583, 1629.

Offner, F. F., 1974, "Solution of the time-dependent ionic diffusion equation," J. Theoret. Biol. **45**, 81.

Offner, F., A. Weinberg, and G. Young, 1940, "Nerve conduction theory: some mathematical consequences of Bernstein's model," Bull. Math. Biophys. **2**, 89.

Opatowski, I., 1950, "The velocity of conduction in nerve fiber and its electric characteristics," Bull. Math. Biophys. **12**, 277.

Osterhout, W. J. V., and S. E. Hill, 1930, "Salt bridges and negative variations", J. Gen. Physiol. **13**, 547.

Parmentier, R. D., 1967, "Stability analysis of neuristor waveforms," Proc. IEEE **55**, 1498.

Parmentier, R. D., 1968, "Neuristor waveform stability analysis by Lyapunov's second method," Proc. IEEE **56**, 1607.

Parmentier, R. D., 1969, "Recoverable neuristor propagation on superconductive tunnel junction strip lines," Solid-State Electron., **12**, 287.

Parmentier, R. D., 1970a, "Neuristor analysis techniques for nonlinear distributed electronic systems," Proc. IEEE **58**, 1829.

Parmentier, R. D., 1970b, "Estimating neuristor excitation thresholds," Proc. IEEE **58**, 605.

Parnas, I., 1972, "Differential block at high frequency of branches of a single axon innervating two muscles," J. Neurophysiol. **35**, 903.

Pastushenko, V. F. and V. S. Markin, 1973, "Spread of excitation in a nerve fiber with septa—II. Blocking of the impulse by the septum," Biophysics **18**, 740.

Pastushenko, V. F. and V. S. Markin, 1969, "Propagation of excitation in a model of an inhomogeneous nerve fibre. II. Attenuation of pulse in the inhomogeneity," Biophysics **14**, 548.

Pastushenko, V. F., V. S. Markin and Yu. A. Chizmadzhev, 1969a, "Propagation of excitation in a model of the inhomogeneous nerve fibre. III. Interaction of pulses in the region of the branching mode of a nerve fibre," Biophysics **14**, 929.

Patushenko, V. F., V. S. Marking and Yu. A. Chizmadzhev, 1969b, "IV. Branching as a summator of nerve pulses," Biophysics **14**, 1130.

Pickard, W. F., 1966, "On the propagation of the nervous impulse down medullated and unmedullated fibers," J. Theoret. Biol. **11**, 30.

Pickard, W. F., 1968, "A contribution to the electromagnetic theory of the unmyelinated axon," Math. Biosci. **2**, 111.

Pickard, W. F., 1969a, "The electromagnetic theory of electrotonus along an unmyelinated axon," Math. Biosci. **5**, 471.

Pickard, W. F., 1969b, "Estimating the velocity of propagation along myelinated and unmyelinated fibers," Math. Biosci. **5**, 305.

Pickard, W. F., 1974, "Electrotonus on a nonlinear dendrite," Math Biosci. **20**, 75.

Planck, M., 1890a, "Ueber die erregung von Elektricität und Wärme in Elektrolyten," Ann. Physik. Chem. **39**, 161.

Planck, M., 1890b, Ueber die Potential differenz zwischen zwei verdünuten Lösungen binärer Elektrolyte," Ann. Physik. Chem. **40**, 561.

Plonsey, R., 1964, "Volume conductor fields of action currents," Biophys. J., **4**, 317.

Plonsey, R., 1965, "An extension of the solid angle potential formulation for an active cell," Biophys. J., **5**, 663.

Pokrovskii, A. N., 1970, "Mechanism of origin of electric potentials in nerve tissue," Biophysics **15**, 914.

Van Der Pol, B., 1926, "On relaxation oscillations," Phil. Mag. Ser 7, **2**, 978.

Van Der Pol, B., 1934, "The nonlinear theory of electric oscillations," Proc. IRE **22**, 1051.

Van Der Pol B., 1957, "On a generalisation of the non-linear differential equation $U_{tt} - \epsilon(1 - u^2)U_t + u = 0$," Proc. Acad. Sci. Amsterdam, **A60**, 477.

Polanyi, M., 1961, "Knowing and being," Mind **70**, 458.

Polanyi, M., 1962, "Tacit knowing: its bearing on some problems of philosophy," Rev. Mod. Phys. **34**, 601.

Polanyi, M., 1965, "The structure of consciousness," Brain **88**, 799.

Polyakov, I. V., 1973, "Rising, quasi-stationary voltage waves in nonhomogeneous lines with nonlinear resistance," Rad. Eng. Elec. Phys. **18**, 722.

Predonzani, G., and Roveri, A., 1968, "Criteri di stabilita per una linea contenente bipoli attive concentrati non-lineari," Note, Receusioni e Notizie, **17**, 1433.

Purpura, D. P. and H. Grundfest, 1956, "Nature of dendritic potentials and synaptic mechanisms in cerebral cortex of cat," J. Neurophysiol. **19**, 573.

Rall, W., 1959, "Branching dendritic trees and montoneuron membrane resistivity," Exp. Neurol. **1**, 491.

Rall, W., 1962a, "Theory of physiological properties of dendrites," Ann. N.Y. Acad. Sci., **96**, 1071.

Rall, W., 1962b, "Electrophysiology of a dendritic neuron model," Biophys. J., **2**, 145.

Rall, W., 1964, "Theoretical significance of dendritic trees for neuronal input-output relations," *Neural Theory and Modeling*, edited by R. F. Reiss, (Stanford U. Stanford) p. 73.

Rall, ·W., 1967, "Distinguishing theoretical synaptic potentials computed for different soma-dendritic distributions of synaptic input," J. Neurophysiol. **30**, 1138.

Rall, W., 1969, "Distributions of potential in cylindrical coordinates and time constants for a membrane cylinder," Bioph. J. **9**, 1509.

Rall, W., 1970, "Dendritic neuron theory and dendrodendritic synapses in a simple cortical system," *The Neurosciences: Second Study Program*, edited by F. O. Schmitt, (The Rockefeller U. P., New York).

Rall, W. and G. M. Shepherd, 1968, "Theoretical reconstruction of field potentials and dendrodendritic synaptic interactions in olfactory bulb," J. Neurophysiol. **31**, 884.

Ramón y Cajal, 1908, "Structure et connexious des neurons," Anchivio di Fisiologia **5**, 1.

Ramón y Cajal, S., 1954, *Neuron Theory or Reticular Theory?* Cous. Sup. de Invest. Cientificas, Madrid.

Ramon-Moliner, E., 1962, "An attempt at classifying nerve cells on the basis of their dendritic patterns," J. Comp. Neurol. **119**, 211.

Rashevsky, N., 1960, *Mathematics Biophysics* (Dover, New York).

Richer, I., 1965, "Pulse transmission along certain lumped nonlinear transmission lines," Electron. Lett. **1**, 135.

Richer, I., 1966, "The switch-line: a simple lumped transmission line that can support unattenuated propagation," Trans. IEEE on Circuit Theory, **CT-13**, 388.

Rinzel, J. and J. B. Keller, 1973, "Traveling wave solutions of a nerve conduction equation," Biophys. J. **13**, 1313.

Rinzel, J. and W. Rall, 1974, "Transient response in a dendritic neuron model for current injected at one branch," Biophys. J. **14**, 759.

Rose, J. E., L. I. Malis, L. Kruger and C. P. Baker, 1960, "Effects of heavy, ionizing, monoenergetic particles on the cerebral cortex. II. Histological appearance of laminar lesions and growth of nerve fibers after laminar destructions," J. Comp. Neurol. **115**, 243.

Rosen, G., 1974, "Approximate solution to the generic initial value problem for nonlinear reaction-diffusion equations," SIAM J. Appl. Math. **26**, 221.

Rosenberg, S. A., 1969, "A computer evaluation of equations for predicting the potentials across biological membranes," Biophys. J. **9**, 500.

Rosenblueth, A., N. Wiener, W. Pitts and J. Garcia Ramos, 1948, "An account of the spike potential of axons," J. Cell. Comp. Physiol. **32**, 275.

Rosenfalk, P., 1969, *Intra- and extracellular potential fields of active nerve and muscle fibres*, (Akademisk Forlag, Copenhagen).

Sabah, N. H. and K. N. Leibovic, 1969, "Subthreshold oscillatory responses of the Hodgkin-Huxley cable model for the squid giant axon," Biophys. J. **9**, 1206.

Sato, R. and H. Miyamoto, 1967, "Active transmission lines," Elec. and Comm. in Japan, **50**, 131.

Sato, H., I. Tasaki, E. Carbone and M. Hallett, 1973, "Changes in axon birefringence associated with excitation: implications for the structure of the axon membrane," J. Mechanochem. Cell Motility **2**, 209.

Sattinger, D. H., 1975, "On the stability of waves of nonlinear parabolic systems," (to be published).

Scott, A. C., 1962, "Analysis of nonlinear distributed systems," Trans. IRE, *CT-9*, 192.

Scott, A. C., 1963, "Neuristor propagation on a tunnel diode loaded transmission line," Proc. IEEE 51, 240.

Scott, A. C., 1964ª, "Steady propagation on nonlinear transmission lines," Trans. IEEE on Circuit Theory, *CT-11*, 146.

Scott, A. C., 1964ᵇ, "Distributed device applications of the superconducting tunnel junction," Solid State Elec. **7**, 137.

Scott, A. C., 1970, *Active and Nonlinear Wave Propagation*, (Wiley, New York).

Scott, A. C., 1971, "Effect of the series inductance of a nerve axon upon its conduction velocity," Math. Biosci. **11**, 277.

Scott, A. C., 1972, "Transmission line equivalent for an unmyelinated nerve axon," Math. Biosci. **13**, 47.

Scott, A. C., 1973a, "Strength duration curves for threshold excitation of nerves," Math. Biosci. **18**, 137.

Scott, A. C., 1973b, "Information processing in dendritic trees," Math. Biosci. **18**, 153.

Scott, A. C., 1974, "The application of Bäcklund transforms to physical problems," Proc. of NSF Workshop on Contact Transformations, Vanderbilt U., Sept. 1974 (to be published).

Scott, A. C., F. Y. T. Chu, D. W. McLaughlin, 1973, "The soliton: a new concept in applied science," Proc. IEEE 61, 1443.

Scott, B. I. H., 1962, "Electricity in plants," Sci. Am. 207, No. 4, 107.

Scott Russell, J., 1844, "Report on waves," Proc. Roy. Soc., Edinburgh, 319.

Singer, S. J. and G. L. Nicolson, 1972, "The fluid mosaic model of the structure of cell membranes," Science 175, 720.

Smith, R. A., 1961, *Semiconductors*, (Cambridge Univ. Press, Cambridge, England) p. 234.

Smolyaninov, V. V., 1968, "Omission of pulses in an elementary fiber model," Biophysics, **13**, 587.

Smolyaninov, V. V., 1969, "Speed of conduction of excitation along a fibre and a syncytium," Biophysics **14**, 357.

Szentágothai, J., 1968, "Structuro-functional considerations of the cerebellar neuron network," Proc. IEEE **56**, 960.

Tasaki, I., 1939, "The electro-saltatory transmission of the nerve impulse and the effect of narcosis upon the nerve fiber," Am. J. Physiol. **127**, 211.

Tasaki, I., 1959, "Conduction of the nerve impulse," Handbook of Physiology, Section 1, 75.

Tasaki, I., 1968, *Nerve Excitation, A Macro-molecular Approach*, (C. C. Thomas, Springfield, Illinois).

Tasaki, I., 1974a, "Energy transduction in the nerve membrane and studies of excitation processes with extrinsic fluorescence probes," Ann. N.Y. Acad. Sci., **227**, 247.

Tasaki, I., 1974b, "Nerve excitation. New experimental evidence for the macromolecular hypothesis," in *Actualités Neurophysiologiques*, (Masson et Cie, Paris), 79.

Tauc, L. and G. M. Hughes, 1963, "Modes of initiation and propagation of spikes in the branching axons of molluscan central neurons," J. Gen. Physiol. **46**, 533.

Taylor, R. E., 1963, "Cable theory," in *Physical Techniques in Biological Research*, edited by W. L. Nastuk, Vol. 6, ch. 4, (Academic Press, New York), 219.

Teorell, T., 1936, "Electrical changes in interfacial films," Nature **137**, 994–995.

Thomas, R. C., 1972, "Electrogenic sodium pump in nerve and muscle cells," Physiol. Rev. **52**, 563.

Tille, J., 1965, "A new interpretation of the dynamic changes of the potassium conductance in the squid giant axon," Biophys. J. **5**, 163.

Undrovinas, A. I., V. F. Pastushenko and V. S. Markin, 1972, "Calculation of the shape and speed of nerve impulses," Doklady Biophysics 204, 47.

Vanderkooi, G. and D. E. Green, 1970, "Biological membrane structure, I. The protein crystal model for membranes," Proc. Nat. Acad. Sci. **66**, 615.

Vorontsov, Yu. I., M. I. Kozhevnickova, and I. V. Polyakov, 1967, "Wave processes in active RC-lines," Rad. Eng. Elec. Phys., **11**, 1449.

Watanabe, A., S. Terakawa and M. Nagano, 1973, "Axoplasmic origin of the birefringence change associated with excitation of a crab nerve," Proc. Japan Acad. **49**, 470.

Watson, G. N., 1962, *Theory of Bessel fnctions*, (Cambridge U. Press, Cambridge, England) 2nd ed.

Waxman, S. G., 1972, "Regional differentiation of the axon: a review with special reference to the concept of the multiplex neuron," Brain Research 47, 269.

Weber, H., 1873a, "Ueber die Besselschen Functionen und ihre Anwendung auf die theorie der elektrischen Strome," J. Reine Angew. Math. 75, 75.

Weber, H., 1873b, "Ueber die stationären Strömungen der Elektricität in Cylindern," J.F.D. reine und angew. Mathematik, **76**, 1.

Weinberg, A. M., 1941, "Weber's theory of the kernleiter," Bull. Math. Bioph. 3, 39.

Weinberg, A. M., 1942, "Green's functions in biological potential problems," Bull. Math. Biophy. 4, 107.

Whitham, G. B., 1974, *Linear and Nonlinear Waves*, (Wiley-Interscience, New York).

Yamaguti, M., 1963, "The asymptotic behavior of the solution of a semi-linear partial differential equation related to an active pulse transmission line," Proc. Japan Acad. **39**, 726.

Yoshizawa, S., 1971, "Asymptotic behavior of a bistable transmission line," IECE Professional Group on Nonlinear Problems, NLP 71-1 (in Japanese).

Yoshizawa, S. and Y. Kitada, 1969, "Some properties of a simplified nerve equation," Math. Biosci. 5, 385.

Yoshizawa, S. and J. Nagumo, 1964, "A bistable distributed line," Proc. IEEE, **52**.

Young, J. Z., 1936, "Structure of nerve fibers and synapses in some invertebrates," Cold Spring Harbor Symp. Quart. Biol. 4, 1.

Young, J. Z., 1951, *Doubt and Certainty in Science*, (Oxford Clarendon Press).

Zeeman, E. C., 1972, "Differential equations for the heartbeat and the nervous impulse," *Toward a Theoretical Biology*, edited by C. H. Waddington, (Aldine Publ. Co., Chicago, Ill.)

Zeldovich, Y. B. and G. I. Barenblatt, 1959, "Theory of flame propagation," Combust. Flame, **3**, 61.

ADDITIONAL REFERENCES

Barron, D. H. and B. H. C. Matthews, 1935, "Intermittent conduction in the spinal cord," J. Physiol. **85**, 73.

Changeux, J. P., 1969, "Remarks on the symmetry and cooperative properties of biological membranes," in *Nobel Symposium* 11, edited by A. Engström and B. Strandberg (Wiley-Interscience, New York), 235.

Gemme, G., 1969, "Axon membrane crystallites in insect photoreceptors," in *Nobel Symposium* 11, edited by A. Engström and B. Strandberg (Wiley-Interscience, New York), 305.

Green, M. W. and B. D. Sleeman, 1974, "On FitzHugh's nerve axon equations," J. Math. Biol. 1, 153.

Krnjević, K. and R. Miledi, 1959, "Presynaptic failure of neuromuscular propagation in rats," J. Physiol. **149**, 1.

Parnas, I., S. Hochstein, H. Parnas and M. Spira, 1973, "Numerical solution to the Hodgkin–Huxley equations applied to spike train transmission across an axon inhomogeneity," Israel J. Med. Sci. **9**, 681.

Neural networks and physical systems with emergent collective computational abilities

(associative memory/parallel processing/categorization/content-addressable memory/fail-soft devices)

J. J. HOPFIELD

Division of Chemistry and Biology, California Institute of Technology, Pasadena, California 91125; and Bell Laboratories, Murray Hill, New Jersey 07974

Contributed by John J. Hopfield, January 15, 1982

ABSTRACT   Computational properties of use to biological organisms or to the construction of computers can emerge as collective properties of systems having a large number of simple equivalent components (or neurons). The physical meaning of content-addressable memory is described by an appropriate phase space flow of the state of a system. A model of such a system is given, based on aspects of neurobiology but readily adapted to integrated circuits. The collective properties of this model produce a content-addressable memory which correctly yields an entire memory from any subpart of sufficient size. The algorithm for the time evolution of the state of the system is based on asynchronous parallel processing. Additional emergent collective properties include some capacity for generalization, familiarity recognition, categorization, error correction, and time sequence retention. The collective properties are only weakly sensitive to details of the modeling or the failure of individual devices.

Given the dynamical electrochemical properties of neurons and their interconnections (synapses), we readily understand schemes that use a few neurons to obtain elementary useful biological behavior (1–3). Our understanding of such simple circuits in electronics allows us to plan larger and more complex circuits which are essential to large computers. Because evolution has no such plan, it becomes relevant to ask whether the ability of large collections of neurons to perform "computational" tasks may in part be a spontaneous collective consequence of having a large number of interacting simple neurons.

In physical systems made from a large number of simple elements, interactions among large numbers of elementary components yield collective phenomena such as the stable magnetic orientations and domains in a magnetic system or the vortex patterns in fluid flow. Do analogous collective phenomena in a system of simple interacting neurons have useful "computational" correlates? For example, are the stability of memories, the construction of categories of generalization, or time-sequential memory also emergent properties and collective in origin? This paper examines a new modeling of this old and fundamental question (4–8) and shows that important computational properties spontaneously arise.

All modeling is based on details, and the details of neuroanatomy and neural function are both myriad and incompletely known (9). In many physical systems, the nature of the emergent collective properties is insensitive to the details inserted in the model (e.g., collisions are essential to generate sound waves, but any reasonable interatomic force law will yield appropriate collisions). In the same spirit, I will seek collective properties that are robust against change in the model details.

The model could be readily implemented by integrated circuit hardware. The conclusions suggest the design of a delo-calized content-addressable memory or categorizer using extensive asynchronous parallel processing.

The general content-addressable memory of a physical system

Suppose that an item stored in memory is "H. A. Kramers & G. H. Wannier *Phys. Rev.* **60**, 252 (1941)." A general content-addressable memory would be capable of retrieving this entire memory item on the basis of sufficient partial information. The input "& Wannier, (1941)" might suffice. An ideal memory could deal with errors and retrieve this reference even from the input "Vannier, (1941)". In computers, only relatively simple forms of content-addressable memory have been made in hardware (10, 11). Sophisticated ideas like error correction in accessing information are usually introduced as software (10).

There are classes of physical systems whose spontaneous behavior can be used as a form of general (and error-correcting) content-addressable memory. Consider the time evolution of a physical system that can be described by a set of general coordinates. A point in state space then represents the instantaneous condition of the system. This state space may be either continuous or discrete (as in the case of N Ising spins).

The equations of motion of the system describe a flow in state space. Various classes of flow patterns are possible, but the systems of use for memory particularly include those that flow toward locally stable points from anywhere within regions around those points. A particle with frictional damping moving in a potential well with two minima exemplifies such a dynamics.

If the flow is not completely deterministic, the description is more complicated. In the two-well problems above, if the frictional force is characterized by a temperature, it must also produce a random driving force. The limit points become small limiting regions, and the stability becomes not absolute. But as long as the stochastic effects are small, the essence of local stable points remains.

Consider a physical system described by many coordinates $X_1 \cdots X_N$, the components of a state vector X. Let the system have locally stable limit points $X_a, X_b, \cdots$. Then, if the system is started sufficiently near any X_a, as at $X = X_a + \Delta$, it will proceed in time until $X \approx X_a$. We can regard the information stored in the system as the vectors $X_a, X_b, \cdots$. The starting point $X = X_a + \Delta$ represents a partial knowledge of the item X_a, and the system then generates the total information X_a.

Any physical system whose dynamics in phase space is dominated by a substantial number of locally stable states to which it is attracted can therefore be regarded as a general content-addressable memory. The physical system will be a potentially useful memory if, in addition, any prescribed set of states can readily be made the stable states of the system.

The model system

The processing devices will be called neurons. Each neuron i has two states like those of McCullough and Pitts (12): $V_i = 0$

("not firing") and $V_i = 1$ ("firing at maximum rate"). When neuron i has a connection made to it from neuron j, the strength of connection is defined as T_{ij}. (Nonconnected neurons have $T_{ij} \equiv 0$.) The instantaneous state of the system is specified by listing the N values of V_i, so it is represented by a binary word of N bits.

The state changes in time according to the following algorithm. For each neuron i there is a fixed threshold U_i. Each neuron i readjusts its state randomly in time but with a mean attempt rate W, setting

$$\begin{matrix} V_i \to 1 \\ V_i \to 0 \end{matrix} \quad \text{if} \quad \sum_{j \neq i} T_{ij} V_j \quad \begin{matrix} > U_i \\ < U_i \end{matrix} \qquad [1]$$

Thus, each neuron randomly and asynchronously evaluates whether it is above or below threshold and readjusts accordingly. (Unless otherwise stated, we choose $U_i = 0$.)

Although this model has superficial similarities to the Perceptron (13, 14) the essential differences are responsible for the new results. First, Perceptrons were modeled chiefly with neural connections in a "forward" direction $A \to B \to C \to D$. The analysis of networks with strong backward coupling $A \rightleftarrows B \rightleftarrows C$ proved intractable. All our interesting results arise as consequences of the strong back-coupling. Second, Perceptron studies usually made a random net of neurons deal directly with a real physical world and did not ask the questions essential to finding the more abstract emergent computational properties. Finally, Perceptron modeling required synchronous neurons like a conventional digital computer. There is no evidence for such global synchrony and, given the delays of nerve signal propagation, there would be no way to use global synchrony effectively. Chiefly computational properties which can exist in spite of asynchrony have interesting implications in biology.

The information storage algorithm

Suppose we wish to store the set of states V^s, $s = 1 \cdots n$. We use the storage prescription (15, 16)

$$T_{ij} = \sum_s (2V_i^s - 1)(2V_j^s - 1) \qquad [2]$$

but with $T_{ii} = 0$. From this definition

$$\sum_j T_{ij} V_j^{s'} = \sum_s (2V_i^s - 1) \left[\sum_j V_j^{s'} (2V_j^s - 1) \right] \equiv H_i^{s'} \qquad [3]$$

The mean value of the bracketed term in Eq. 3 is 0 unless $s = s'$, for which the mean is $N/2$. This pseudoorthogonality yields

$$\sum_j T_{ij} V_j^{s'} \equiv \langle H_i^{s'} \rangle \approx (2V_i^{s'} - 1) N/2 \qquad [4]$$

and is positive if $V_i^{s'} = 1$ and negative if $V_i^{s'} = 0$. Except for the noise coming from the $s \neq s'$ terms, the stored state would always be stable under our processing algorithm.

Such matrices T_{ij} have been used in theories of linear associative nets (15–19) to produce an output pattern from a paired input stimulus, $S_1 \to O_1$. A second association $S_2 \to O_2$ can be simultaneously stored in the same network. But the confusing simulus $0.6 S_1 + 0.4 S_2$ will produce a generally meaningless mixed output $0.6 O_1 + 0.4 O_2$. Our model, in contrast, will use its strong nonlinearity to make choices, produce categories, and regenerate information and, with high probability, will generate the output O_1 from such a confusing mixed stimulus.

A linear associative net must be connected in a complex way with an external nonlinear logic processor in order to yield true computation (20, 21). Complex circuitry is easy to plan but more difficult to discuss in evolutionary terms. In contrast, our model obtains its emergent computational properties from simple properties of many cells rather than circuitry.

The biological interpretation of the model

Most neurons are capable of generating a train of action potentials—propagating pulses of electrochemical activity—when the average potential across their membrane is held well above its normal resting value. The mean rate at which action potentials are generated is a smooth function of the mean membrane potential, having the general form shown in Fig. 1.

The biological information sent to other neurons often lies in a short-time average of the firing rate (22). When this is so, one can neglect the details of individual action potentials and regard Fig. 1 as a smooth input–output relationship. [Parallel pathways carrying the same information would enhance the ability of the system to extract a short-term average firing rate (23, 24).]

A study of emergent collective effects and spontaneous computation must necessarily focus on the nonlinearity of the input–output relationship. The essence of computation is nonlinear logical operations. The particle interactions that produce true collective effects in particle dynamics come from a nonlinear dependence of forces on positions of the particles. Whereas linear associative networks have emphasized the linear central region (14–19) of Fig. 1, we will replace the input–output relationship by the dot-dash step. Those neurons whose operation is dominantly linear merely provide a pathway of communication between nonlinear neurons. Thus, we consider a network of "on or off" neurons, granting that some of the interconnections may be by way of neurons operating in the linear regime.

Delays in synaptic transmission (of partially stochastic character) and in the transmission of impulses along axons and dendrites produce a delay between the input of a neuron and the generation of an effective output. All such delays have been modeled by a single parameter, the stochastic mean processing time $1/W$.

The input to a particular neuron arises from the current leaks of the synapses to that neuron, which influence the cell mean potential. The synapses are activated by arriving action potentials. The input signal to a cell i can be taken to be

$$\sum_j T_{ij} V_j \qquad [5]$$

where T_{ij} represents the effectiveness of a synapse. Fig. 1 thus

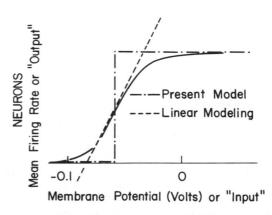

FIG. 1. Firing rate versus membrane voltage for a typical neuron (solid line), dropping to 0 for large negative potentials and saturating for positive potentials. The broken lines show approximations used in modeling.

becomes an input–output relationship for a neuron.

Little, Shaw, and Roney (8, 25, 26) have developed ideas on the collective functioning of neural nets based on "on/off" neurons and synchronous processing. However, in their model the relative timing of action potential spikes was central and resulted in reverberating action potential trains. Our model and theirs have limited formal similarity, although there may be connections at a deeper level.

Most modeling of neural learning networks has been based on synapses of a general type described by Hebb (27) and Eccles (28). The essential ingredient is the modification of T_{ij} by correlations like

$$\Delta T_{ij} = [V_i(t)V_j(t)]_{\text{average}} \qquad [6]$$

where the average is some appropriate calculation over past history. Decay in time and effects of $[V_i(t)]_{\text{avg}}$ or $[V_j(t)]_{\text{avg}}$ are also allowed. Model networks with such synapses (16, 20, 21) can construct the associative T_{ij} of Eq. **2**. We will therefore initially assume that such a T_{ij} has been produced by previous experience (or inheritance). The Hebbian property need not reside in single synapses; small groups of cells which produce such a net effect would suffice.

The network of cells we describe performs an abstract calculation and, for applications, the inputs should be appropriately coded. In visual processing, for example, feature extraction should previously have been done. The present modeling might then be related to how an entity or *Gestalt* is remembered or categorized on the basis of inputs representing a collection of its features.

Studies of the collective behaviors of the model

The model has stable limit points. Consider the special case $T_{ij} = T_{ji}$, and define

$$E = -\frac{1}{2} \sum_{i \neq j} \sum T_{ij} V_i V_j \;. \qquad [7]$$

ΔE due to ΔV_i is given by

$$\Delta E = -\Delta V_i \sum_{j \neq i'} T_{ij} V_j \;. \qquad [8]$$

Thus, the algorithm for altering V_i causes E to be a monotonically decreasing function. State changes will continue until a least (local) E is reached. This case is isomorphic with an Ising model. T_{ij} provides the role of the exchange coupling, and there is also an external local field at each site. When T_{ij} is symmetric but has a random character (the spin glass) there are known to be many (locally) stable states (29).

Monte Carlo calculations were made on systems of $N = 30$ and $N = 100$, to examine the effect of removing the $T_{ij} = T_{ji}$ restriction. Each element of T_{ij} was chosen as a random number between -1 and 1. The neural architecture of typical cortical regions (30, 31) and also of simple ganglia of invertebrates (32) suggests the importance of 100–10,000 cells with intense mutual interconnections in elementary processing, so our scale of N is slightly small.

The dynamics algorithm was initiated from randomly chosen initial starting configurations. For $N = 30$ the system never displayed an ergodic wandering through state space. Within a time of about $4/W$ it settled into limiting behaviors, the commonest being a stable state. When 50 trials were examined for a particular such random matrix, all would result in one of two or three end states. A few stable states thus collect the flow from most of the initial state space. A simple cycle also occurred occasionally—for example, $\cdots A \rightarrow B \rightarrow A \rightarrow B \cdots$.

The third behavior seen was chaotic wandering in a small region of state space. The Hamming distance between two binary states A and B is defined as the number of places in which the digits are different. The chaotic wandering occurred within a short Hamming distance of one particular state. Statistics were done on the probability p_i of the occurrence of a state in a time of wandering around this minimum, and an entropic measure of the available states M was taken

$$\ln M = -\sum p_i \ln p_i \;. \qquad [9]$$

A value of $M = 25$ was found for $N = 30$. *The flow in phase space produced by this model algorithm has the properties necessary for a physical content-addressable memory* whether or not T_{ij} is symmetric.

Simulations with $N = 100$ were much slower and not quantitatively pursued. They showed qualitative similarity to $N = 30$.

Why should stable limit points or regions persist when $T_{ij} \neq T_{ji}$? If the algorithm at some time changes V_i from 0 to 1 or vice versa, the change of the energy defined in Eq. 7 can be split into two terms, one of which is always negative. The second is identical if T_{ij} is symmetric and is "stochastic" with mean 0 if T_{ij} and T_{ji} are randomly chosen. The algorithm for $T_{ij} \neq T_{ji}$ therefore changes E in a fashion similar to the way E would change in time for a symmetric T_{ij} but with an algorithm corresponding to a finite temperature.

About 0.15 N states can be simultaneously remembered before error in recall is severe. Computer modeling of memory storage according to Eq. **2** was carried out for $N = 30$ and $N = 100$. n random memory states were chosen and the corresponding T_{ij} was generated. If a nervous system preprocessed signals for efficient storage, the preprocessed information would appear random (e.g., the coding sequences of DNA have a random character). The random memory vectors thus simulate efficiently encoded real information, as well as representing our ignorance. The system was started at each assigned nominal memory state, and the state was allowed to evolve until stationary.

Typical results are shown in Fig. 2. The statistics are averages over both the states in a given matrix and different matrices. With $n = 5$, the assigned memory states are almost always stable (and exactly recallable). For $n = 15$, about half of the nominally remembered states evolved to stable states with less than 5 errors, but the rest evolved to states quite different from the starting points.

These results can be understood from an analysis of the effect of the noise terms. In Eq. 3, $H_i^{s'}$ is the "effective field" on neuron i when the state of the system is s', one of the nominal memory states. The expectation value of this sum, Eq. 4, is $\pm N/2$ as appropriate. The $s \neq s'$ summation in Eq. 2 contributes no mean, but has a rms noise of $[(n-1)N/2]^{1/2} \equiv \sigma$. For nN large, this noise is approximately Gaussian and the probability of an error in a single particular bit of a particular memory will be

$$P = \frac{1}{\sqrt{2\pi\sigma^2}} \int_{N/2}^{\infty} e^{-x^2/2\sigma^2}\, dx \;. \qquad [10]$$

For the case $n = 10$, $N = 100$, $P = 0.0091$, the probability that a state had no errors in its 100 bits should be about $e^{-0.91} \approx 0.40$. In the simulation of Fig. 2, the experimental number was 0.6.

The theoretical scaling of n with N at fixed P was demonstrated in the simulations going between $N = 30$ and $N = 100$. The experimental results of half the memories being well retained at $n = 0.15\,N$ and the rest badly retained is expected to

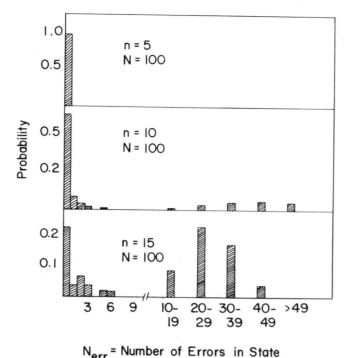

FIG. 2. The probability distribution of the occurrence of errors in the location of the stable states obtained from nominally assigned memories.

be true for all large N. The information storage at a given level of accuracy can be increased by a factor of 2 by a judicious choice of individual neuron thresholds. This choice is equivalent to using variables $\mu_i = \pm 1$, $T_{ij} = \Sigma_s \mu_i^s \mu_j^s$, and a threshold level of 0.

Given some arbitrary starting state, what is the resulting final state (or statistically, states)? To study this, evolutions from randomly chosen initial states were tabulated for $N = 30$ and $n = 5$. From the (inessential) symmetry of the algorithm, if $(101110\cdots)$ is an assigned stable state, $(010001\cdots)$ is also stable. Therefore, the matrices had 10 nominal stable states. Approximately 85% of the trials ended in assigned memories, and 10% ended in stable states of no obvious meaning. An ambiguous 5% landed in stable states very near assigned memories. There was a range of a factor of 20 of the likelihood of finding these 10 states.

The algorithm leads to memories near the starting state. For $N = 30$, $n = 5$, partially random starting states were generated by random modification of known memories. The probability that the final state was that closest to the initial state was studied as a function of the distance between the initial state and the nearest memory state. For distance ≤ 5, the nearest state was reached more than 90% of the time. Beyond that distance, the probability fell off smoothly, dropping to a level of 0.2 (2 times random chance) for a distance of 12.

The phase space flow is apparently dominated by attractors which are the nominally assigned memories, each of which dominates a substantial region around it. The flow is not entirely deterministic, and *the system responds to an ambiguous starting state by a statistical choice* between the memory states it most resembles.

Were it desired to use such a system in an Si-based content-addressable memory, the algorithm should be used and modified to hold the known bits of information while letting the others adjust.

The model was studied by using a "clipped" T_{ij}, replacing T_{ij} in Eq. 3 by ± 1, the algebraic sign of T_{ij}. The purposes were to examine the necessity of a linear synapse supposition (by making a highly nonlinear one) and to examine the efficiency of storage. Only $N(N/2)$ bits of information can possibly be stored in this symmetric matrix. Experimentally, for $N = 100$, $n = 9$, the level of errors was similar to that for the ordinary algorithm at $n = 12$. The signal-to-noise ratio can be evaluated analytically for this clipped algorithm and is reduced by a factor of $(2/\pi)^{1/2}$ compared with the unclipped case. For a fixed error probability, the number of memories must be reduced by $2/\pi$.

With the μ algorithm and the clipped T_{ij}, both analysis and modeling showed that the maximal information stored for $N = 100$ occurred at about $n = 13$. Some errors were present, and the Shannon information stored corresponded to about $N(N/8)$ bits.

New memories can be continually added to T_{ij}. The addition of new memories beyond the capacity overloads the system and makes all memory states irretrievable unless there is a provision for forgetting old memories (16, 27, 28).

The saturation of the possible size of T_{ij} will itself cause forgetting. Let the possible values of T_{ij} be 0, ± 1, ± 2, ± 3, and T_{ij} be freely incremented within this range. If $T_{ij} = 3$, a next increment of $+1$ would be ignored and a next increment of -1 would reduce T_{ij} to 2. When T_{ij} is so constructed, only the recent memory states are retained, with a slightly increased noise level. Memories from the distant past are no longer stable. How far into the past are states remembered depends on the digitizing depth of T_{ij}, and 0, $\cdots$, ± 3 is an appropriate level for $N = 100$. Other schemes can be used to keep too many memories from being simultaneously written, but this particular one is attractive because it requires no delicate balances and is a consequence of natural hardware.

Real neurons need not make synapses both of $i \rightarrow j$ and $j \rightarrow i$. Particular synapses are restricted to one sign of output. We therefore asked whether $T_{ij} = T_{ji}$ is important. Simulations were carried out with only one ij connection: if $T_{ij} \neq 0$, $T_{ji} = 0$. The probability of making errors increased, but the algorithm continued to generate stable minima. A Gaussian noise description of the error rate shows that the signal-to-noise ratio for given n and N should be decreased by the factor $1/\sqrt{2}$, and the simulations were consistent with such a factor. This same analysis shows that the system generally fails in a "soft" fashion, with signal-to-noise ratio and error rate increasing slowly as more synapses fail.

Memories too close to each other are confused and tend to merge. For $N = 100$, a pair of random memories should be separated by 50 ± 5 Hamming units. The case $N = 100$, $n = 8$, was studied with seven random memories and the eighth made up a Hamming distance of only 30, 20, or 10 from one of the other seven memories. At a distance of 30, both similar memories were usually stable. At a distance of 20, the minima were usually distinct but displaced. At a distance of 10, the minima were often fused.

The algorithm categorizes initial states according to the similarity to memory states. With a threshold of 0, the system behaves as a forced categorizer.

The state $00000 \cdots$ is always stable. For a threshold of 0, this stable state is much higher in energy than the stored memory states and very seldom occurs. Adding a uniform threshold in the algorithm is equivalent to raising the effective energy of the stored memories compared to the 0000 state, and 0000 also becomes a likely stable state. The 0000 state is then generated by any initial state that does not resemble adequately closely one of the assigned memories and represents positive recognition that the starting state is not familiar.

Familiarity can be recognized by other means when the memory is drastically overloaded. We examined the case $N = 100$, $n = 500$, in which there is a memory overload of a factor of 25. None of the memory states assigned were stable. The initial rate of processing of a starting state is defined as the number of neuron state readjustments that occur in a time $1/2W$. Familiar and unfamiliar states were distinguishable most of the time at this level of overload on the basis of the initial processing rate, which was faster for unfamiliar states. This kind of familiarity can only be read out of the system by a class of neurons or devices abstracting average properties of the processing group.

For the cases so far considered, the expectation value of T_{ij} was 0 for $i \neq j$. A set of memories can be stored with average correlations, and $\overline{T}_{ij} = C_{ij} \neq 0$ because there is a consistent internal correlation in the memories. If now a partial new state X is stored

$$\Delta T_{ij} = (2X_i - 1)(2X_j - 1) \quad i,j \leq k < N \quad [11]$$

using only k of the neurons rather than N, an attempt to reconstruct it will generate a stable point for all N neurons. The values of $X_{k+1} \cdots X_N$ that result will be determined primarily from the sign of

$$\sum_{j=1}^{k} c_{ij} x_j \quad [12]$$

and X is completed according to the mean correlations of the other memories. The most effective implementation of this capacity stores a large number of correlated matrices weakly followed by a normal storage of X.

A nonsymmetric T_{ij} can lead to the possibility that a minimum will be only metastable and will be replaced in time by another minimum. Additional nonsymmetric terms which could be easily generated by a minor modification of Hebb synapses

$$\Delta T_{ij} = A \sum_{s} (2V_i^{s+1} - 1)(2V_j^s - 1) \quad [13]$$

were added to T_{ij}. When A was judiciously adjusted, the system would spend a while near V_s and then leave and go to a point near V_{s+1}. But sequences longer than four states proved impossible to generate, and even these were not faithfully followed.

Discussion

In the model network each "neuron" has elementary properties, and the network has little structure. Nonetheless, collective computational properties spontaneously arose. Memories are retained as stable entities or *Gestalts* and can be correctly recalled from any reasonably sized subpart. Ambiguities are resolved on a statistical basis. Some capacity for generalization is present, and time ordering of memories can also be encoded. These properties follow from the nature of the flow in phase space produced by the processing algorithm, which does not appear to be strongly dependent on precise details of the modeling. This robustness suggests that similar effects will obtain even when more neurobiological details are added.

Much of the architecture of regions of the brains of higher animals must be made from a proliferation of simple local circuits with well-defined functions. The bridge between simple circuits and the complex computational properties of higher nervous systems may be the spontaneous emergence of new computational capabilities from the collective behavior of large numbers of simple processing elements.

Implementation of a similar model by using integrated circuits would lead to chips which are much less sensitive to element failure and soft-failure than are normal circuits. Such chips would be wasteful of gates but could be made many times larger than standard designs at a given yield. Their asynchronous parallel processing capability would provide rapid solutions to some special classes of computational problems.

The work at California Institute of Technology was supported in part by National Science Foundation Grant DMR-8107494. This is contribution no. 6580 from the Division of Chemistry and Chemical Engineering.

1. Willows, A. O. D., Dorsett, D. A. & Hoyle, G. (1973) *J. Neurobiol.* 4, 207–237, 255–285.
2. Kristan, W. B. (1980) in *Information Processing in the Nervous System*, eds. Pinsker, H. M. & Willis, W. D. (Raven, New York), 241–261.
3. Knight, B. W. (1975) *Lect. Math. Life Sci.* 5, 111–144.
4. Smith, D. R. & Davidson, C. H. (1962) *J. Assoc. Comput. Mach.* 9, 268–279.
5. Harmon, L. D. (1964) in *Neural Theory and Modeling*, ed. Reiss, R. F. (Stanford Univ. Press, Stanford, CA), pp. 23–24.
6. Amari, S.-I. (1977) *Biol. Cybern.* 26, 175–185.
7. Amari, S.-I. & Akikazu, T. (1978) *Biol. Cybern.* 29, 127–136.
8. Little, W. A. (1974) *Math. Biosci.* 19, 101–120.
9. Marr, J. (1969) *J. Physiol.* 202, 437–470.
10. Kohonen, T. (1980) *Content Addressable Memories* (Springer, New York).
11. Palm, G. (1980) *Biol. Cybern.* 36, 19–31.
12. McCulloch, W. S. & Pitts, W. (1943) *Bull. Math Biophys.* 5, 115–133.
13. Minsky, M. & Papert, S. (1969) *Perceptrons: An Introduction to Computational Geometry* (MIT Press, Cambridge, MA).
14. Rosenblatt, F. (1962) *Principles of Perceptrons* (Spartan, Washington, DC).
15. Cooper, L. N. (1973) in *Proceedings of the Nobel Symposium on Collective Properties of Physical Systems*, eds. Lundqvist, B. & Lundqvist, S. (Academic, New York), 252–264.
16. Cooper, L. N., Liberman, F. & Oja, E. (1979) *Biol. Cybern.* 33, 9–28.
17. Longuet-Higgins, J. C. (1968) *Proc. Roy. Soc. London Ser. B* 171, 327–334.
18. Longuet-Higgins, J. C. (1968) *Nature (London)* 217, 104–105.
19. Kohonen, T. (1977) *Associative Memory—A System-Theoretic Approach* (Springer, New York).
20. Willwacher, G. (1976) *Biol. Cybern.* 24, 181–198.
21. Anderson, J. A. (1977) *Psych. Rev.* 84, 413–451.
22. Perkel, D. H. & Bullock, T. H. (1969) *Neurosci. Res. Symp. Summ.* 3, 405–527.
23. John, E. R. (1972) *Science* 177, 850–864.
24. Roney, K. J., Scheibel, A. B. & Shaw, G. L. (1979) *Brain Res. Rev.* 1, 225–271.
25. Little, W. A. & Shaw, G. L. (1978) *Math. Biosci.* 39, 281–289.
26. Shaw, G. L. & Roney, K. J. (1979) *Phys. Rev. Lett.* 74, 146–150.
27. Hebb, D. O. (1949) *The Organization of Behavior* (Wiley, New York).
28. Eccles, J. G. (1953) *The Neurophysiological Basis of Mind* (Clarendon, Oxford).
29. Kirkpatrick, S. & Sherrington, D. (1978) *Phys. Rev.* 17, 4384–4403.
30. Mountcastle, V. B. (1978) in *The Mindful Brain*, eds. Edelman, G. M. & Mountcastle, V. B. (MIT Press, Cambridge, MA), pp. 36–41.
31. Goldman, P. S. & Nauta, W. J. H. (1977) *Brain Res.* 122, 393–413.
32. Kandel, E. R. (1979) *Sci. Am.* 241, 61–70.

THE BIOPHYSICS OF VISUAL PHOTORECEPTION

Aaron Lewis and Lucian V. Del Priore

Vision is awe inspiring. The wondrous nature of this sensory process becomes clear when we consider just a few of its features. In the terminology of today's technology, the human visual system has a set of stereoscopic foreoptics that are instantly and automatically focusable to a few centimeters and are fully corrected for geometric aberrations, a servo-controlled two-axis scanning mechanism, a millisecond framing rate, sensitivity to brightnesses varying by a factor of 100 billion, the ability to detect a single photon, nearly 100% quantum efficiency and a spatial as well as temporal image processor that could not be matched by the fastest supercomputer.

The retina is where the excitation of visual sensation begins, where single photons are amplified, where the eye adapts to light levels and generates a partially processed signal to the brain. Except for the partial processing of the image, these functions of the retina occur largely in a cell called the photoreceptor (figure 1), which absorbs and amplifies the photon and transduces it to a neural response.

Physics provides the key to understanding this process of visual transduction. In this article we will focus on the photoreceptor cell and attempt to integrate the results of intensive recent physical and biochemical research. We also will touch on sensitivity control in photoreceptors and discuss possible ways in which these cells attain such control. We will follow a quantum of light from an object to the retina and into a photoreceptor cell, where it is absorbed by a pigment molecule called rhodopsin, producing a neural response. We will look in detail at the chemical amplification process that leads to this response, which consists of reduction in a critical sodium conductance through the plasma membrane, or outer membrane, of this cell. There is little doubt that we are now at the threshold of understanding for the first time how this exquisitely sensitive biological system detects single photons.

This is the first time that many of the steps in a cell's path from excitation to response have been elucidated, and striking parallels are becoming obvious between the electrical response of a photoreceptor to a single photon and a variety of other cellular responses, such as the triggering of signal transmission in nerves by single acetylcholine molecules and the triggering of cells by

Reprinted from *Physics Today* **41**, 38–46 (January 1988); © American Institute of Physics.

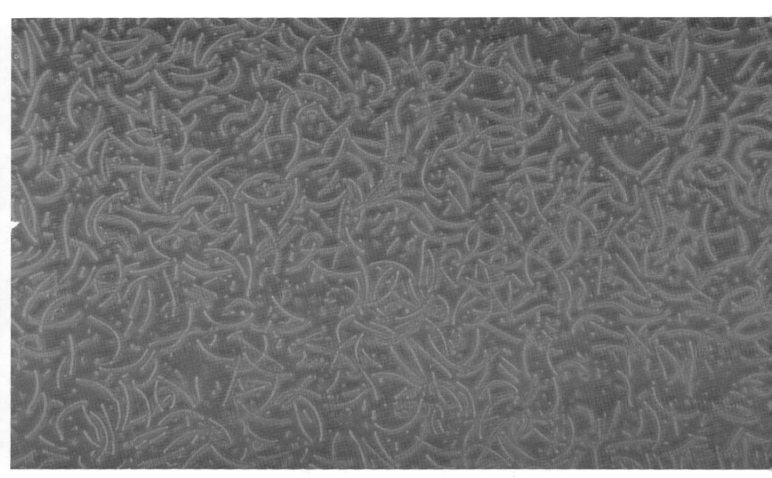

Photoreceptor cells from the eye of a toad. Thousands of such cells detach when a retina is peeled from the back of an eye and jiggled in an aqueous solution. The cells are about 6 microns in diameter and 60 microns long.   **Figure 1**

single hormone molecules. All of these cellular processes involve an amplified response to a weak signal. Understanding these processes is clearly a monumental achievement, one that has been attained through a synthesis of elegant biochemistry and forefront applied physics.

From object to photoreceptor

As figure 2a indicates, the first part of the eye that light hits is the cornea, which has an index of refraction of 1.376. The cornea accounts for a large part of the focusing of the light. The light then goes through the lens, which provides additional focusing by changing shape according to the distance between the eye and the object. The light next traverses a clear, jelly-like substance called the vitreous humor, arriving at a thin strip of tissue called the retina, which lines the back of the eye. In the retina, the light is absorbed by photoreceptors, but first it must pass through the neurons that integrate the signal from the photoreceptors: The light-sensitive material is located in the last cellular layer of the retina, farthest from the front of the eye and pointing toward the back of the eye, away from the light source, as figure 2b shows.

Vertebrates possess two types of photoreceptors, cone and rod cells, shown schematically in figure 2c. The cone cells exist principally in the fovea, which is an indentation in the primate retina located within a half-degree of the optical axis of the eye. These cells, which are packed with a density of 150 000 cones/mm^2, are responsible for color vision and vision in bright light, that is, for high visual acuity. The 10^8 photoreceptor cells distributed through-

out the rest of the retina are mainly rods. These cells are responsible for vision in dim light. Their electrical response, which is much slower than that of cone cells, saturates at a light level incident on the cornea of 500 photons/μm^2/sec. This light level corresponds to about 7000 single-photon absorptions/sec/rod, which closely matches the lowest intensities at which cone cells can be excited.

Structure of cells

Rod and cone cells differ in size and cytology, or cellular structure. These two cells are compared in the drawing in figure 2c and in the scanning electron micrograph in figure 3, which shows the photoreceptor side of the retina, the side farthest from the light. A major cytological difference lies in the region of the cell called the outer segment. In rod outer segments, light-absorbing pigment molecules called rhodopsins are embedded in discs, which are membranous structures resembling Mediterranean pita bread. The discs form continuously from the external, or plasma, membrane at the base of the outer segment, where the outer and inner segments join. Discs are digested away at the top of the outer segment by a layer of tissue called the pigment epithelium (figure 2b), which is also involved in the renewal of rhodopsin. The constant renewal process of digestion and formation occurs at a rate of one disc every 30 minutes in cold-blooded animals. Throughout this extremely rapid renewal process the size of the photoreceptor remains constant. In other words, discs are scavenged from the tip of the rod at the same rate that they are infolded from the plasma membrane. In rods the discs separate from the plasma membrane from which they form, whereas the disc-like

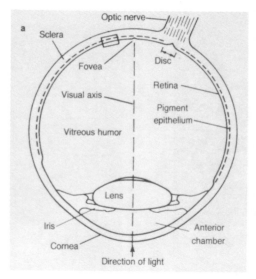

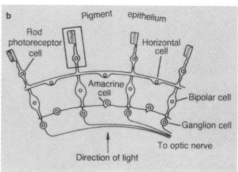

Eye, retina and photoreceptors. **a:** Schematic diagram of an eye. **b:** Expanded view of the region of the retina in the box in **a**. Light has to pass through the integrating neurons of the retina before it reaches the light-absorbing photoreceptor cells, which point backward. The tops of these cells adjoin a layer of tissue called the pigment epithelium, which helps in the regeneration of the light-absorbing pigment rhodopsin. **c:** Three-dimensional comparison of rod and cone photoreceptor cells, adapted from a drawing by Alan Fein of Woods Hole Oceanographic Institution and Boston University. The outer segment, which is the center of visual transduction, is cut away. The inner segment contains the nucleus, which includes the genetic material for renewal of the outer segment and the mitochondria for the final breakdown of the food that provides energy for the cell. The cells end in a terminal called a synapse, which chemically transfers the light-induced response to the neuronal network in the retina. The calycal processes on the outside of the cell emanate from the inner segment and contain the contractile protein actin. Their physiological function is unknown. **Figure 2**

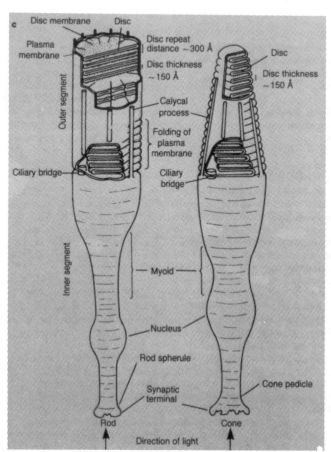

infoldings in cones, which are called sacs, never separate from the plasma membrane. Rods and cones both obtain a high absorption cross section by having the light traverse many layers of stacked membranes containing pigment.

Because rods far outnumber cones in all regions of the retina except the fovea, much of the available information on visual transduction comes from experiments on these cells. The unique cytological features of rods and their ability to amplify single photons have patterned our understanding of the molecular basis of visual transduction for the past decade. Thus we now focus our discussion on rod cells.

It has become apparent in the last several years that cells contain structural proteins that form a cellular framework known as the cytoskeleton (see PHYSICS TODAY, April 1985, page 68), but the role of such structural proteins in the transduction of light by photoreceptors has generally been ignored. In fact, the most widely expounded hypothesis of how rod cells transduce light energy is the internal transmitter or second messenger hypothesis. This hypothesis, which was formulated over two decades ago, was based on early evidence that the rod discs, containing rhodopsin, were physically and electrically isolated from the plasma membrane across which light eventually causes a crucial electrical response.[1] Thus a molecule or ion was required to transmit to the plasma membrane an amplified message that a rhodopsin molecule had absorbed a photon.

The electrical isolation of the plasma membrane from the disc membrane is now firmly established. However, within the past several years it has become apparent that rods, like all cells, contain a detailed cytoskeleton formed of protein filaments. In rods this cellular skeleton

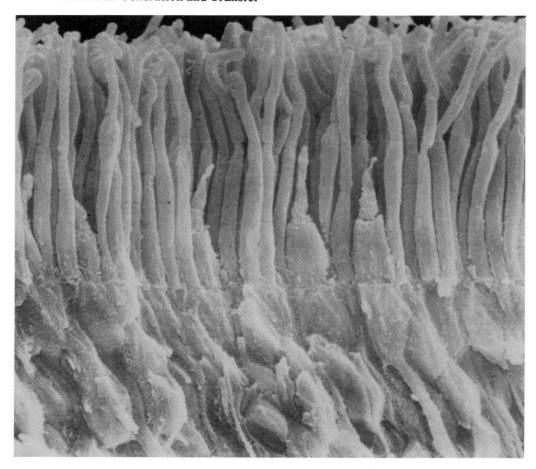

Photoreceptors in place in the retina. Most of the cells in this scanning electron micrograph are rods. Three small cells near the center of the photo can be identified as cones by their characteristic shape. (Photograph by William Miller, Yale University.)   **Figure 3**

connects discs to one another and to the plasma membrane. On the periphery of each disc in the outer segment of a toad rod there are about 8000 filaments that connect one disc to another, and additional filaments connect each disc to the plasma membrane.[2] Experiments in our laboratory at Cornell have elucidated the three-dimensional organization of one component of the rod cytoskelton, using a fluorescent mushroom toxin that binds to the cytoskeleton protein F-actin.[3] Our results (see the photograph on the cover of this issue) are only a first step, and work to identify fully all the proteins that make up the photoreceptor cytoskeleton continues in several laboratories. Whether these cytoskeletal interconnections of the discs and the plasma membrane have any physiological role is still to be determined. However, as we note below, light-induced alterations in photoreceptors involve changes that in other cells are known to affect cellular physiology.

The pigment. Within the last few years there has been an explosion in our knowledge of the structure and molecular cytology of the visual pigments, that is, the rhodopsins. These pigments consist of a light-absorbing molecule, called retinal, chemically linked to the protein opsin (see figure 4). Converging lines of evidence indicate that the rhodopsin molecule has a prolate ellipsoid structure (the dotted line in figure 4) that spans the disc membrane in which it is imbedded.[4] Conventional biochemical techniques have given us the sequence of amino acid residues—the monomers—that form the protein polymer in cow rhodopsin. And in a scientific *tour de force* Stanford graduate student Jeremy Nathans used the techniques of genetic engineering to isolate from his blood the genes for the visual pigments in his own rod and cone cells.[5]

From these studies and other physical measurements we have begun to understand in detail how the same retinal molecule can have a different absorption properties in the red-, green- and blue-sensitive cones and in the rods, which are not sensitive to color. It appears that differences in the opsin polymer chains alter the dipolar or charge interactions between retinal and opsin, and that these interactions in turn perturb the absorption properties of retinal.

Light and rhodopsin

How does light change rhodopsin? This is the central question of visual excitation that has fascinated a whole generation of physicists, chemists and biologists. Rhodopsin undergoes a series of color changes after it absorbs a single photon. The first color change, which represents the direct product of the photochemistry, was recently measured using femtosecond lasers. This measurement was made in a bacterial retinal–protein complex called bacteriorhodopsin (related to rhodopsin) within a few hundred femtoseconds after excitation.[6] Most researchers believe that visual pigments will exhibit changes as fast as those seen in bacteriorhodopsin when similar experiments are performed on them.

Rhodopsin accomplishes this superfast photochemistry with a quantum yield of about 70%. The action of light is over in a fraction of a picosecond, and calorimetric evidence shows that as a result of this subpicosecond, super-efficient event, 60% of the photon energy is stored as chemical energy to be used in the later steps of the visual response.[7] Thus the problem of visual excitation not only has implications for vision but also is a fascinating question in structural chemistry, one that will require physicists to develop new techniques of investigation.

How does rhodopsin store that much energy in such a short time? There is still no unequivocal answer, in spite of years of work throughout the world since George Wald's discovery of rhodopsin in 1936. The earliest hypothesis, which still has the strongest support,[8] was suggested by Wald. He showed that eventually, several minutes after light absorption, the retinal breaks off from the protein and assumes a structural, or isomeric, form with a linear (trans) chain rather than the bent (11-cis) chain, shown in figure 4, that he extracted from the pigment in the "dark" state. Based on this evidence and the known photochemical behavior of free retinal in solution, Wald suggested that the only action of light in vision is to cause a conformational or structural change: the 11-cis to trans isomerization of the retinal molecule. In his hypothesis, the light-induced change in the retinal structure causes structural changes in the protein polymer: The retinal is altered and kicks the protein into action.

One important aspect of the photoexcitation of rhodopsin is difficult to fully account for by this simple isomerization mechanism—the large energy storage that is a critical component of visual pigment photochemistry. To account for this energy storage, Lewis[9] proposed over a decade ago a mechanism based on what was a generally accepted mechanism of color regulation in visual pigments. All visual pigments have the same chromophore, or color-producing chemical group—namely the retinal molecule—but these pigments can exhibit absorptions that extend from the ultraviolet in bees to the deep red in snakes. The controlling factor in this color regulation is the protein that surrounds the chromophore. In one proposed mechanism of color regulation, excitation causes a large alteration in the dipole moment of the chromophore, which is then stabilized or destabilized by charges on the protein.[10] Lewis extended this notion and noted that large changes in the chromophore's dipole moment, in addition to its isomerization, could affect the protein structure; the changes in the retinal dipole moment could separate charged groups on the protein, and thereby store large amounts of energy. Thus the crucial step in this mechanism is a large alteration in the electron distribution in retinal upon excitation.

In the years since Lewis proposed this mechanism, we have not been able to prove or disprove it unequivocally. It has driven us to search for new laboratory approaches, from resonance Raman spectroscopy to femtosecond spectroscopy, but until recently none of the methods available allowed us to probe whether the chromophore in the protein indeed underwent as large a change in dipole moment on excitation as had been observed for free retinal in solution.[10]

Recently we have turned to nonlinear laser spectroscopy to develop a technique that has the potential to reveal the nature of the chromophore dipole alteration on excitation in both free retinal and retinal–protein complexes. Our method is based on the elegant experiments Yuen-Ron Shen and his colleagues at Berkeley did to measure second-harmonic generation by monolayers of molecules. In a similar vein we have formed monolayers of retinal and retinal–protein complexes in Langmuir troughs.[11] Cornell graduate student Jung Huang[12] has obtained good correlations between the second-harmonic intensity and the dipole change on light-induced excitation of various retinals as measured by detecting electric field shifts in the optical absorption.[10]

When we, in collaboration with Theo Rasing and Shen at Berkeley, extended these experiments to bacteriorhodopsin-containing membranes, we found that such membranes (known as purple membranes because of the color of the bacteriorhodopsin pigment molecules) exhibit a signal that points to the similarity of the chromophore dipole change in and out of the membrane. This result depends critically upon the dielectric constant of the protein medium surrounding the retinal. Nonetheless, choosing a value for the dielectric constant consistent with the protein absorption indicates that the chromophore in the protein does indeed exhibit a large change in dipole upon excitation. How and whether the proton responds to such an instantaneous (less than a femtosecond) and large alteration in dipole is still an enigma, especially in view of our recent discovery that the intramolecular dephasing time of the chromophore (38 fsec in and 17 fsec out of the protein) is essentially independent of the protein. Our studies clearly indicate that a combination of nonlinear and linear spectroscopic methods,[13] together with protein alteration by gene engineering and conventional chemical modification, offer the best hope of understanding in detail this fascinating problem of photon energy storage by rhodopsin.

Single-photon amplification

After a single photon induces a photochemical transformation in rhodopsin, which takes less than 1 picosecond, as discussed above, several thermally generated intermediate conformations are produced with varying absorption spectra. Eventually, about 1 millisecond after light absorption, a blue-absorbing, yellow state called metarhodopsin II is produced. The purple bacterial pigment mentioned above, which is the only protein found in the crystalline purple membrane generated by the salt-loving

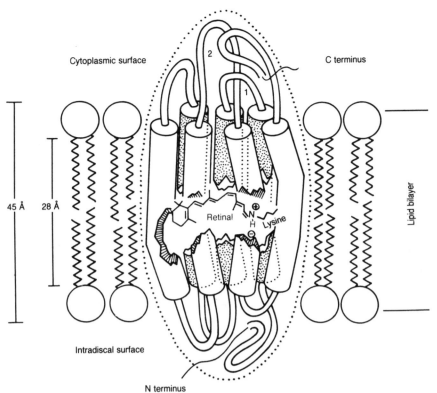

Cytoplasmic surface

C terminus

2

1

45 Å   28 Å

Retical   Retinal N⊕ H Lysine

Lipid bilayer

Intradiscal surface

N terminus

Rhodopsin molecule in the lipid bilayer of a disc membrane. In this proposed structure and positioning of the molecule, cylinders represent connected helical transmembrane protein segments. The two termini of the protein opsin, labeled N and C, are at opposite ends of the disc membrane. Retinal, a structure related chemically to vitamin A, is responsible for the light absorption. It is chemically linked to lysine, an amino acid residue in the opsin chain. Absorption of light by retinal causes a crucial protein in the photoreceptor cytoplasm, or intracellular space, to bind to the loop regions labeled 1 and 2, initiating the chemical changes in the cell involved in light amplification. This drawing, based on similar representations by Paul Hargrave, Edward Dratz and their coworkers,[4] is highly schematic, but there is growing evidence for its overall validity. **Figure 4**

bacterium *Halobacterium halobium*, is a unique and rugged system that allows us to illustrate dramatically this light-induced color change, which occurs in all visual pigments and in bacteriorhodopsin. For this demonstration we use the purple membrane films that we have developed for erasable, extremely high-density optical memories and for experimental implementations of rapidly reprogrammable neural networks.[14] In figure 5 the purple image of the letters "HU" is formed by bacteriorhodopsin molecules that have not absorbed light, and the yellow background arises from those molecules that have been altered by light into the yellow intermediate. This stable, room temperature image, which is erased when light is absorbed by the purple or yellow molecules, was made by Zhongping Chen, a Cornell graduate student working with Lewis. These technological applications of our research on retinal pigments are the focus of a joint effort with Isaiah Nebenzahl of The Hebrew University of Jerusalem.

The generation in visual pigments of the yellow metarhodopsin II intermediate, which is stable for more than 10 seconds at physiological temperatures, sets the stage for the rapid amplification of the chemical effects that light produces in the single rhodopsin molecule that has absorbed the photon. The final effect of this chemical amplification is the generation of an electrical response by the photoreceptor cell. The outline below of the steps in this cellular amplification of a single photon is summarized in figure 6.

Herman Kuhn in Jülich, West Germany, was the first to wonder if photon-induced alteration in a single rhodopsin could change the strength of the binding of the disc to other proteins in the cytoplasm, or intracellular space. He was able to show that the conformational alteration in a single rhodopsin molecule (labeled "metarhodopsin II" in

figure 6) causes a protein known as the G protein to complex to rhodopsin in the disc.[15] Proteins of the G type occur universally in cells. In photoreceptors the G protein is often called transducin, and this accounts for our use of the symbols T_α, T_β and T_γ to identify subunits of the photoreceptor G protein.

Growing evidence suggests that the regions labeled 1 and 2 in the diagrammatic representation of rhodopsin in figure 4 are involved in complexing the G protein. Lubert Stryer at Stanford, among others, has shown that this complexing catalyzes the exchange of a relatively small molecule called guanosine diphosphate, or GDP, which is normally bound in the dark to the G protein, with an energy-rich form called guanosine triphosphate, or GTP. After this exchange, the metarhodopsin II–G–GTP complex dissociates to produce T_α–GTP, which is a component, or subunit, of the G protein. The single metarhodopsin can catalyze about 500 such exchanges before it is turned off by the addition of phosphates to a portion of rhodopsin's C-terminus tail. This phosphorylation allows a protein called arrestin (in a ratio of 10 rhodopsins to 1 arrestin, not shown in figure 6) to compete with G protein for rhodopsin.

Each of the 500 released T_α–GTP molecular complexes now binds to a protein called phosphodiesterase, or PDE, which is a complex of three subunits, α, β and γ. The interaction of T_α–GTP with PDE splits off the PDE_γ subunit, and T_α–GTP·$PDE_{\alpha\beta}$ can then catalyze the breakup of a molecule named cyclic guanosine monophosphate, or cGMP, which is related to a monomer of the polymeric genetic material RNA. This breakup reaction of cGMP produces one noncyclic GMP molecule plus one H^+ ion for each cGMP broken up, or hydrolyzed.

Eight hundred cGMP molecules are hydrolyzed before the T_α component of the T_α–GTP·$PDE_{\alpha\beta}$ complex under-

Image on an optically switchable film of bacteriorhodopsin molecules. The purple image is formed by molecules that have not been exposed to light; the yellow background consists of molecules that have been exposed to light. The image is stable at room temperature over time, and can be erased either by blue light, which switches the yellow background to purple, or by yellow light, which switches the purple letters to yellow. **Figure 5**

goes a self-timed deactivation. This deactivation occurs through the breaking up of GTP into GDP, resulting in the release of $PDE_{\alpha\beta}$, which then reassociates with the PDE_{γ} subunit. The T_{α}, which now once again has GDP bound to it, reassociates with the T_{β} and T_{γ} subunits to complete the cycle initiated by the rhodopsin molecule's absorption of a photon and storage of energy. This two-stage amplification cycle, powered by the energy released by the T_{α}-induced breakup of GTP to GDP, was worked out in a series of experiments by Mark Bitensky, Paul Liebman, Derik Bownds, William Miller, Kuhn, Stryer and others. The final result of this finely tuned mechanism is the hydrolysis of 400 000 cGMP molecules within one second of the absorption of a single photon.

Physics of photoreceptor response

Photoreceptors show a variety of physical phenomena. These include osmotic changes and other mechanical alterations such as rapid, light-induced, longitudinal contractions along the outer segment. For vision, however, the most important of these physical phenomena is the electrical reponse.

The essential question in the biophysics of photoreception is how a single photon can stimulate the electrical response of the photoreceptor cell. Although investigators since the 1940s have thought that rod cells can respond to single photons, it was not until the late 1970s that Dennis Baylor, Trevor Lamb and King-Wai Yau at Stanford were able to measure the photocurrent of a single rod cell by sucking the cell into a pipette, as shown in figure 7a. From these experiments they determined that a single photon induces a transient, 1-picoamp reduction in the current of sodium ions that continuously flows through the plasma membrane of a photoreceptor cell even in the dark.

Such experiments are also capable of measuring the dark noise of the photoreceptor—the fluctuations in the sodium current that occur in the absence of light. There are two types of dark noise. First, there is a component of dark noise whose source, as indicated by the component's amplitude and spectral composition, is a fluctuation in one of the chemical amplification steps in the transduction process. This fluctuation causes the coordinated closure of the channels through which the sodium current flows. In bright light, when most of the plasma membrane channels are closed, this noise is suppressed.

Second, there is a noise component that is composed of individual discrete events, which look very much like the single-photon response. In toad photoreceptor cells these events occur once every 50 seconds at 20 °C, whereas in monkey rods at 37 °C the events occur once every two and a half minutes. If these discrete events actually correspond to thermal instability in the rhodopsin molecule, this indicates that a monkey rhodopsin molecule spontaneously turns on once every 420 years and the only reason we see such events in an isolated photoreceptor is because 10^8 molecules of rhodopsin are present in every rod outer segment. We should emphasize, however, that the half-life of rhodopsin molecules based on this measurement of photoreceptor noise is only a lower bound because other steps in the amplification scheme could also be contributing to the observed spontaneous dark events that resemble the photoreceptor's response to the absorption of a photon.

Measuring a membrane patch. Although the suction electrode technique provided new perspectives on photoreceptor electrophysiology, it remained for a method called patch clamping to make a final connection between the conductance of the plasma membrane and the biochemistry of the cell. The suction electrode technique was actually an outcome of the more generally applicable patch clamp method developed in Germany by Erwin Neher and Bert Sackmann. In patch clamping one brings a pipette in contact with the plasma membrane of the cell, making a tight seal that allows one to measure picoamp conductance changes in the "patch," or area of contact. The experiments on photoreceptors completed by Evgeniy Fesenko and his group at the Institute of Biological Physics in Moscow used an "inside-out patch." In this method, after making the tight seal between the plasma membrane and the pipette, the experimenter pulls the pipette away, as illustrated in figure 7b. The membrane patched to the pipette tip breaks off from the cell, exposing the inside of the plasma membrane to the solution in which the cells are suspended.

This system allowed Fesenko to expose the inside of the plasma membrane to solutions containing a variety of the prime players in photoreceptor biochemistry. Of these substances, which included calcium, hydrogen ions, cGMP, GMP and so on, one and only one substance caused the membrane patch to conduct sodium ions, and this was cGMP. These and subsequent experiments showed that

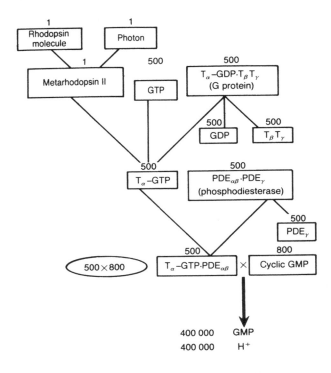

Amplification. The diagram shows the sequence of chemical steps involved in amplifying the structural change that a single photon induces in rhodopsin. The molecules of this scheme are found in the photoreceptor cell in a ratio of 1000 rhodopsins to 100 G proteins to 14 phosphodiesterases. **Figure 6**

the channel was opened cooperatively by three or more cGMP molecules. The presence of divalent cations has an effect on the conductance induced by cGMP. Under physiological conditions, about 4 femtoamps of current flow through an open channel. Because a channel remains open for about 1 millisecond on the average, approximately 25 sodium ions may enter. In view of these results, the 1-picoamp reduction in the current corresponding to the peak of the response to a single photon indicates that 250 sodium channels must lose their cGMP and close in response to the absorption of a single photon. As Stryer has noted, this permits the rod outer segment to transmit amplitude information without significant time and amplitude jitter in response to small numbers of photons.[15]

The obvious connection between the electrical response of photoreceptor cells and their chemistry is cGMP. We now know that photon absorption causes a progressively amplified decrease in the concentration of free cGMP in the cell, and that the plasma membrane contains an exquisitely sensitive switch that is turned on and off by cGMP. This gives us a simple mechanism for single-photon amplification and response in photoreceptors: The amplified decrease in the free cGMP concentration caused by the absorption of a single photon involves a kinetic release of cGMP from about 250 sodium channels, and this results in the 1-picoamp reduction in the sodium current that is the neural response transmitted to the brain. There is a tremendous feeling of exhilaration among the biologists, chemists and physicists working on visual transduction, reflecting the fact that after many years of sustained effort by groups all over the world we have

finally made the ultimate connection between excitation by light and the amplified response of the photoreceptor.

A word of caution

In all this excitement some caution is certainly in order. First, there are unresolved problems; for example, certain experiments under physiological conditions do not detect a reduction in cGMP. Second, cellular changes involving all sorts of other molecules take place in response to light. These include alterations in the concentration of calcium and magnesium, the activation by light of an enzyme that actually synthesizes cGMP, and the activation by light-activated G protein of an enzyme called phospholipase A_2, which affects molecules important as messengers in other cells. Many of these changes could have independent or interrelated effects on single-photon amplification. Third, photoreceptors undergo a variety of physical and chemical changes connected with the mechanics of the cells. These include the light-induced contraction of the outer segment and even the light-induced growth of filaments. Except for a couple of suggestions in the literature,[16] researchers have given very little attention to the effects of mechanical coupling in the photoreceptor cell. Many of the cellular cytoskeleton proteins that are involved with cellular mechanics are affected by substances that are known to undergo changes in photoreceptors exposed to light. Although such mechanical effects are unlikely to be involved in the rising phase of the electrical response, they may very well be involved in the falling phase of a single-photon response and in the photoreceptor cell's reduced sensitivity to subsequent photons. This reduced sensitivity is partially responsible for the lack of sensitivity that one perceives on entering a dark room.

Caution notwithstanding, there is cause for great joy in the community of photoreceptor lovers. We have come tantalizingly close to solving major problems without putting ourselves out of business. We have even begun to see that the behavior of cones is similar to that of rods in certain respects, such as in the existence of a cGMP-controlled conductance. But here too, numerous questions remain, such as why the cones respond more quickly than the more sensitive rods.

Parallels in biology

Plasma membrane conductances controlled by cGMP now seem to be appearing everywhere. There are reports of their existence in the olfactory and other systems. It has recently become obvious that proteins controlling functions as diverse as cellular regulation by hormones and even nerve transmission behave similarly to rhodopsin in

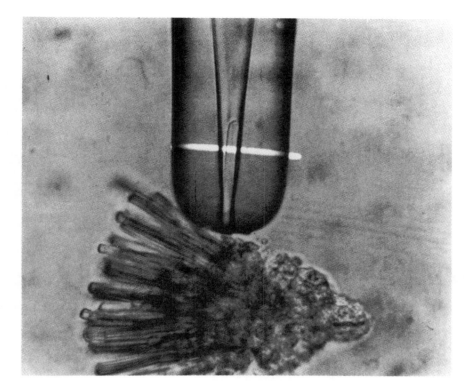
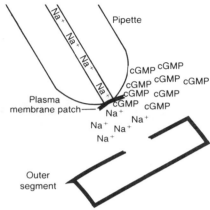

Electrical measurement. **a:** Suction electrode method for measuring currents through a photoreceptor membrane. This photograph shows a single photoreceptor outer segment that has been sucked into the end of a tightly fitting glass pipette. An excellent seal is effected between the glass and the photoreceptor membrane, allowing measurement of picoamp currents. The beam of light perpendicular to the cell changes the current. (Photograph courtesy of Dennis Baylor, Stanford University.) **b:** Technique for measuring currents in an "inside-out" patch of membrane.   **Figure 7**

activating G protein and causing an amplified response. These proteins exhibit striking structural and functional homologies to rhodopsin, including:
▷ the same seven excursions of the protein through the membrane indicated in figure 4
▷ conservation of the amino acids in the region of the protein where the retinal, the hormone or the neurotransmitter (signaler) molecules bind
▷ conservation of regions of the loops labeled 1 and 2 in figure 4, where the G protein may bind
▷ phosphorylation as a turn-off mechanism
▷ conservation of amino acids in the N-terminus loops of the protein, probably for structural reasons
▷ possible binding of calcium ions
▷ attachment of sugar groups in the N-terminus region shown in figure 4.
These similarities indicate that the controlling proteins, which have diverse functions, come from a single ancestral gene. Thus the visual photoreceptor has become a model for a class of transduction mechanisms, and research in this field is leading the way in our understanding of how natural systems detect, amplify and respond to signals.

* * *

We acknowledge the support of the United States Army (contract numbers DAMD 17-79C-9041 and DAMD 17-85C-5136) and especially Edwin Beatrice for his constant encouragement and support. Some of our nonlinear laser experiments are also supported by the US Air Force, under contract number AFOSR-84-0314. Our work using retinal pigments as neural networks is supported by the Office of Naval Reseach, under grant number N00014-87-G-0236. Part of the research discussed in this article was done by Lewis in the vision research laboratory of the recently established Hadassah Hospital Laser Center, in Jerusalem, Israel. Lewis thanks the School of Applied and Engineering Physics of Cornell University, which for the last 15 years has provided a most fertile environment, where his appreciation for the beauty of visual photoreception could grow and expand.

References

1. For an excellent review, see E. N. Pugh Jr, W. H. Cobbs, Vision Res. **26**, 1613 (1986).
2. J. Usukura, E. Yamada, Biomed. Res. **2**, 177 (1981). D. J. Roof, J. E. Heuser, J. Cell Biol. **95**, 487 (1982).
3. L. V. Del Priore, A. Lewis, S. Tan, W. W. Carley, W. W. Webb, Invest. Ophthalmol. **28**, 633 (1987).
4. E. A. Dratz, P. A. Hargrave, Trends Biochem. Sci. **8**, 128 (1983).
5. J. Nathans, Annu. Rev. Neurosci. **10**, 163 (1987).
6. M. Downer, M. Islam, C. V. Shank, A. Harootunian, A. Lewis, in *Ultrafast Phenomena IV*, D. H. Austin, K. B. Eisenthal, eds., Springer-Verlag, Berlin (1984), p. 500. H. J. Polland, M. A. Franz, W. Zinth, W. Kaiser, E. Kolling, D. Oesterhelt, Biophys. J. **49**, 651 (1986).
7. A. Cooper, Nature **282**, 531 (1979).
8. B. Honig, T. Ebrey, R. H. Callender, V. Dinur, M. Ottolenghi, Proc. Natl. Acad. Sci. USA **76**, 2503 (1979).
9. A. Lewis, Proc. Natl. Acad. Sci. USA **75**, 549 (1978).
10. R. Mathies, L. Stryer, Proc. Natl. Acad. Sci. USA **73**, 2169 (1976).
11. J. Schildkraut, A. Lewis, Thin Solid Films **134**, 13 (1985).
12. J. Huang, A. Lewis, Th. Rasing, J. Phys. Chem. **92** (March 1988), in press.
13. Methods in Enzymology **81** and **88**; these two issues are devoted to the techniques used to study rhodopsin and bacteriorhodopsin.
14. C. Mobary, A. Lewis, Proc. SPIE **700**, 304 (1986).
15. L. Stryer, Annu. Rev. Neurosci. **9**, 87 (1986).
16. L. V. Del Priore, A. Lewis, Biophys. Soc. Abstr. **47**, 103a (1985). R. J. Bert, B. Oakley, Assoc. Res. Vision Ophthalmol. Abstr. **26**, 248 (1985).

∎

Resetting biological clocks

Arthur T. Winfree

A pendulum can be stopped by a single impulse of the right magnitude, delivered at the proper time; started again, its phase will have changed. A biological oscillation, although it is a vastly more complicated phenomenon, can likewise be arrested by a single stimulus of a definite strength delivered at the proper time. This remarkable fact emerges from experiments I have conducted on two biological clocks, in organisms in the plant and animal kingdoms. The most interesting potential applications are, of course, those that involve the internal clocks of Man.

Biologists have long known that the rhythms that plants and animals exhibit continue at about the same rate when the periodic external stimuli, such as diurnal light and temperature variations, are experimentally removed. These "endogenous" rhythms must therefore depend on internal mechanisms, the biological clocks.

In one of the examples used below a rhythm whose period is approximately 24 hours, the so-called "circadian" rhythm, terminates the metamorphosis of fruit flies. Under conditions of constant darkness, these flies emerge from their pupae in pulses about 24 hours apart. In my experiments I have shifted the phase of these periodic emergences by stimulating the pupae with a single pulse of blue light. Figure 1 shows about 200 of these pupae on a balsa-wood plate, ready to be inserted into the "time machine" that provides a timed, measured light pulse and later collects the emerging flies for counting.

However, before we proceed to the fruit-fly experiments, we will examine phase resetting in a simpler system, the metabolic cycle by which yeast cells convert sugar to alcohol. In this case the absence of oxygen induces an oscillation of half-minute period which, as we shall see below, can be reset by a single pulse of oxygen.

The pendulum analogy

The rephasing patterns of these two systems may at first appear surprising and unfamiliar, but in certain respects they are very simple. Their mathematics can be illustrated by a simple pendulum, although that analogy is misleading in some respects.

A pendulum swings with unit period. Giving the bob a small shove only alters the momentum of the pendulum, so that it continues to oscillate with the same period but now with a new amplitude and a reset phase. Figure 2 shows a phase diagram for the pendulum; the units are chosen so that the ellipses in the position–momentum plane become circles. An impulse moves the phase point from one circle to another along a vertical line.

A shove of exactly the right magnitude applied at exactly the right time can even stop the pendulum—or, one might say, reduce its amplitude to zero and make its phase ambiguous. A slight error in administering that critical shove, however, leaves the pendulum in motion with a small but nonzero amplitude and a phase that is impossible to predict. On a plot of the new phase against the old phase and the impulse, the critical stimulus would be a phase singularity.

As can be seen from figure 2, the relation between the phases is

$$\tan (\text{new phase}) = \frac{\sin (\text{old phase}) - \text{impulse}}{\cos (\text{old phase})}$$

in suitable dimensionless coordinates. The three-dimensional representation of this equation is a *helicoid*, a surface shaped like a spiral staircase. It climbs around a pole, corresponding to the critical stimulus, that is located at *old phase* $= \pi/2$, *impulse* $= 1$. Because the phases are periodic, each of these surfaces occupies a unit cell of a periodic *lattice of helicoids*.

Oscillations in yeast

A diversity of living organisms responds similarly to a perturbation of

Reprinted from *Physics Today* **28**, 34–39 (March 1975); © American Institute of Physics.

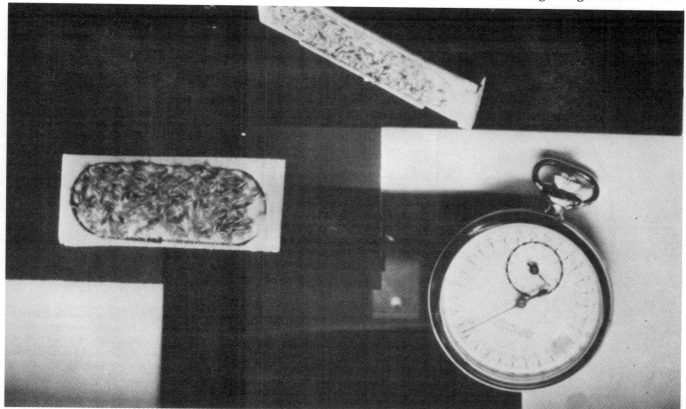

Fruit fly pupae are mounted on balsawood plates to study the way in which the daily time pattern of emergence of the adult flies can be shifted by suitably timed and measured flashes of blue light. The pupae, looking like small grains, are about 2 mm long. Also shown are (under the stopwatch) the blue rephasing filter and (under the plate on the left) the red safelight filter.   Figure 1

their biological clocks, apparently because of topological peculiarities that many oscillating systems share with the pendulum. Detailed measurements of this effect have been carried out with an oscillation of glycolysis, the metabolic splitting of sugar molecules, in yeast. When glucose is injected into a yeast suspension, it enters the cells, where it undergoes a complex series of biochemical reactions. In the first reaction, the glucose molecule attaches a phosphate radical at the expense of the cell's adenosine triphosphate (ATP); now it cannot get back out through the plasma membrane. More important for purposes of observing the reaction is the fact that one reaction reduces the enzyme cofactor NAD (nicotinamide-adenine dinucleotide) to NADH (reduced NAD), while later the process is reversed, NADH being oxidized to NAD, as the final product, ethyl alcohol, is formed. In this way ethanol is built up steadily, as the successful operation of breweries attests.

What happens if the supply of oxygen should be shut off? When the dissolved oxygen is used up, the NADH concentration in a 1-ml sample rises dramatically and then oscillates with a period of about 30 seconds. This oscillation damps out in about 15 minutes to an anaerobic steady-state level with an amplitude on the order of 0.1 micromole per gram of wet cells.

As Britton Chance showed, this process can be studied by monitoring the green fluorescence of NADH with a recording photometer while the cells are irradiated with ultraviolet light. Figure 3 shows a graph of an experiment of this type, with NADH fluorescence, a measure of its concentration, plotted against time. With a glucose concentration of 100 mM, oxygen is shut off; when it is used up, the NADH oscillations begin.

Now that the biological oscillator is on, let us give it a shove. This is done by injecting oxygen dissolved in buffer. The cells turn out more ATP and oxidize NADH back to NAD. This quickly upsets the otherwise regular oscillations of all reaction rates, kicking cell metabolism off its limit cycle and inflicting a phase shift.

Figure 3 depicts the NADH fluorescence record of such a sample experiment: The cells are vigorously oxygenated, glucose is injected, NADH is made and recovered, and then soars up as the free oxygen vanishes. Eight seconds after the second NADH peak in this experiment, I added one-third of a micromole of oxygen per gram of cells. The NADH concentration plummets as it becomes oxidized and then recovers. The oscillation continues with a *permanent phase adjustment* relative to an unperturbed control, but the rhythm damps out at the same rate as in the un-

perturbed controls. This is a chemical perturbation that selectively affects *timing* without otherwise altering the cell's cycle.

Now let us examine the results of a hundred such experiments in which the time T and dosage D of oxygen were systematically varied, and the times θ of subsequent NADH peaks were recorded. Again, let us be very clear about the symbols and coordinate system for plotting these results; the oxygen is given at time T, measured in seconds after the second NADH peak, and θ is the time at which NADH peaks up again, measured in seconds after the oxygen injection. Note that we have a whole series of θ values, about 30 seconds apart as the oscillation continues. The disturbance dosage D is measured in micromoles of oxygen per gram of wet cells.

Now we plot θ against T and D, forming a three-dimensional graph in which reset phase is a surface arching over the time–dose plane.

A spiral staircase of data

I injected from zero to about one-half micromole of oxygen per gram of cells, throughout the 30-second cycle after the second NADH peak. Above each of these hundred time and dose points the resulting peak times were plotted, in seconds measured from the injection. The overall pattern of these points

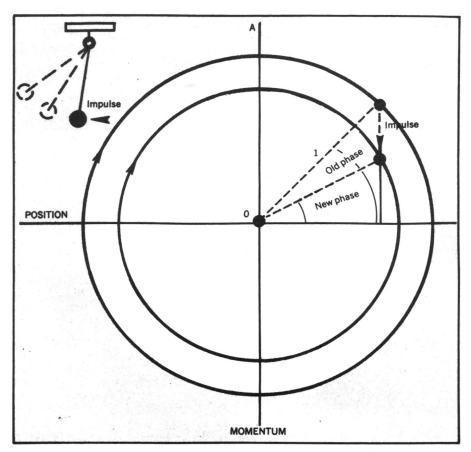

To stop a swinging pendulum with a single blow (upper left) requires an impulse, delivered at its equilibrium position, equal to the momentum the pendulum has there. On the phase diagram, this corresponds to an impulse at point A with strength AO; this, in the units used, is equal to 1. What if timing and impulse are off by a random error? Then the new phase can be obtained from the geometry of the figure; the equation for this relation is given in the text. When the pendulum is stopped, however, the new phase is indeterminate.   Figure 2

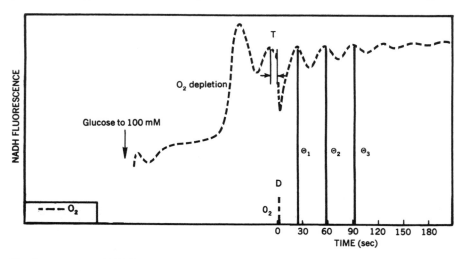

The fluorescence of NADH, the reduced form of an enzyme cofactor, in yeast under ultraviolet irradiation. The fluorescence (here in arbitrary units) is proportional to the concentration of NADH in the sample. The exhaustion of dissolved oxygen starts a damped oscillation of approximately 30-sec period. A pulse of oxygen of dosage $D = \frac{1}{3}$ μM/g is injected at time $T = 8$ sec after the second peak, resetting the phase, θ.   Figure 3

forms the screw-shaped surface shown in figure 4.

Let us examine this figure, a sort of unit cell of a biological time crystal, in some detail. We will start by tracing the border of the surface of this unit cell. Consider experiments at the NADH peak, at $T = 0$. It turns out that oxygen causes a great *transient* disturbance, but when the oscillation recovers, regardless of the dose it shows no phase shift with respect to the con-

trols: θ is thus the same for all doses at $T = 0$, and so the $T = 0$ edge of the cube is horizontal. At doses of 0.5 μM/g and above, regardless of T, the oscillation is restarted at the NADH maximum: θ is the same at all T's, so the edge of the cube for $D = 0.5$ μM is also horizontal. At $T = 30$ seconds, we find the third NADH maximum, where there is no phase response to oxygen, exactly as at the previous maximum $T = 0$.

In contrast, at zero dosage, θ does depend on T; specifically, $T + \theta$ is constant, since the time from NADH peak to injection (T) plus the time from injection to next peak (θ) is always 30 seconds when no oxygen disturbs it. These unperturbed controls therefore lie neatly along the diagonal $\theta + T = 30$ sec. Now we're back to the starting place, but one unit cell lower along the θ-axis.

That describes the spiral boundary of the cloud of data points on this three-dimensional graph; now let us look at experiments with time and dose combinations within this boundary, starting with cross sections of the unit cell consisting of experiments at fixed initial phase, T. At any time up to 25 seconds after an NADH peak, oxygen retards the following peak, that is, increases θ (downward in figure 4), in a way that depends on the initial phase T but never exceeds a full cycle's delay. In the figure this is labeled "Type D" resetting, to remind us that more oxygen produces a greater delay. In contrast, oxygen given a few seconds before the NADH peak (25 sec $< T <$ 30 sec), further *advances* the following peaks, diminishing θ (upward) in a way that depends on initial phase T but never exceeds one cycle. On the drawing this is called "Type A" resetting to remind us that more oxygen produces a greater advance.

The lattice of helicoids

Now let us turn 90 degrees and look at cross sections consisting of experiments at fixed dosage D. To the right of the screw axis, at doses less than 0.2 μM/g, every phase can be reached by giving the chosen oxygen dose at a properly selected phase in the cycle: this is labeled "Type 1" resetting, to point out that, as initial phase T is varied through one full cycle, so does the resulting phase θ vary through exactly one full cycle, and in the same direction. In contrast, to the left of the screw axis, at doses exceeding 0.2 μM/g, θ rises and falls as T is scanned through the cycle: only certain phases can be reached, but each can be reached at two distinct T's with the same dose. Since θ varies through zero net cycles per cycle of T, this part of the rephasing surface is labeled "Type 0."

A critical time $T^*(= 25$ sec) and a

critical dose $D*(= 0.2\ \mu M/g)$ divide the phase response pattern into the four zones: D in front, A behind, 0 to the left, and 1 to the right. The critical stimulus is labeled "$T*, D*$." This is the *screw axis*, where θ tilts up infinitely steeply and becomes indeterminate; it is the singularity I was looking for. What happens when the singular combination $(T*, D*)$ is administered? In those experiments in which (T, D) was near the singularity and only in them, was the NADH rhythm's amplitude after perturbation different from the control at the same time: it was strikingly less. Figure 5 shows one such experiment (black line) and its control (colored line): the oscillation is almost switched off—the phase is intermediate.

A second oxygen injection will restart the NADH rhythm, but until it is administered, there is no rhythm.

This singularity is a necessary topological consequence of having a periodic surface with a spiral boundary. The surface within the spiral boundary is a helicoid and the vertical line $(T*, D*)$ is its symmetry axis. Remember that both T and θ are periodic coordinates—since these experiments could have been done in any cycle of the NADH rhythm and because, after each experiment, NADH peaks up every 30 seconds. Thus this helicoid-shaped unit cube is repeated along both these axes. To summarize, *the overall pattern of phase control of metabolic energy production by an oxygen pulse is a lattice of helicoids whose symmetry axes are at the isolated critical stimuli, (T*, D*)*.

Now consider a radically different rhythmic system, about which virtually nothing is known at the levels of physiology and biochemistry: the circadian clocks that regulate physiological and behavioral rhythms of periods of about 24 hours in the greatest imaginable diversity of one-celled organisms and higher plants and animals, including Man.

Metamorphosis

One of the best-studied circadian clocks keeps time in the fruit fly *Drosophila*, a photograph of which is shown on the cover of this issue of PHYSICS TODAY. You may remember the fly's life history: it starts as a worm in garbage, then pupates, forming a little cocoon. The pupa then dissolves all its wormy organs and reassembles itself in the pattern of an adult fly. When done with this metamorphosis, it emerges, inflates its wings and flies off in search of a mate.

If metamorphosing pupae are transferred from growth conditions under constant light to constant darkness at the same temperature, adult flies are found to emerge only in discrete pulses

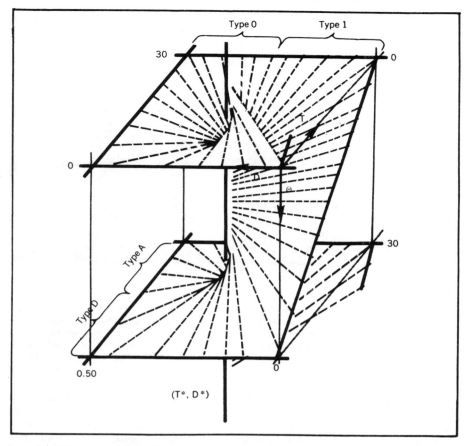

An idealized sketch of the helicoidal surface that gives the phase shift θ as a function of the time of disturbance T and the oxygen-pulse dosage D. The units of θ and T are seconds, while D is given in micromoles per gram of wet cells. At points on the symmetry axis $(T*,D*)$ the oscillation is arrested, so that the new phase is indeterminate. The projections of this axis divides the dosage into "type-0" and "type-1" behavior, and the time T into "type-A" and "type-D" behavior, which are further discussed in the text. Because of the periodicity of θ and T, the figure shown is a unit cell of a lattice of helicoids.   Figure 4

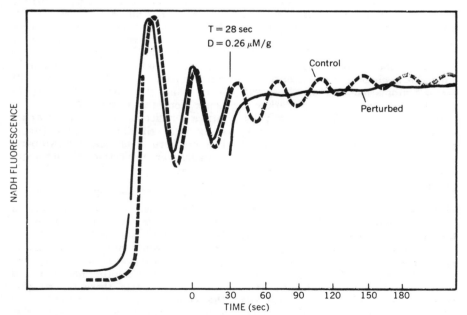

The critical combination of time and dose $(T*,D*)$ is closely approximated in this experiment. The perturbed curve (black) shows that, when 0.26 micromoles of oxygen per gram of cells is administered 28 sec after the second peak, the oscillation virtually stops. The control experiment (dashed lines) shows the oscillations of the unpulsed system. Once stopped, the oscillations can be restarted with arbitrary phase. Will phase resetting be of help in overcoming problems associated with human cycles, such as menstrual timing and the "jet lag"?   Figure 5

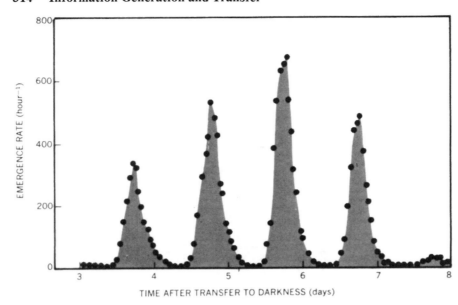

The number of fruit flies that emerge per hour from the pupal stage is plotted in days after transferring the pupae from a constant-light environment to one of constant darkness. The data, taken from 39 independent and indistinguishable experiments, show that emergence peaks are 6–8 hours wide and occur every 24 hours until all of the pupae have matured and opened. The emerging flies have been caught in a "flytrap" for counting.   Figure 6

24 hours apart, each about 6 hours wide, as shown in figure 6. The phasing of these daily emergence bursts can be reset at will by a brief pulse of blue light. The situation is analogous to the oxygen dose given to yeast cells in otherwise oxygenless conditions, some minutes after the oxygen suddenly ran out: Here we have a light pulse in otherwise dark conditions some days after the light was turned off.

In a way that is formally identical to the glycolysis experiments discussed above, we can measure how the rephasing of this rhythm depends on the time T when the pupae are exposed and the number of photons of light admitted D. I did this in about 300 significantly different combinations of T and D, spread over the first three cycles of darkness, with doses of 0 to 10 000 ergs/cm^2; this is the range in which the resulting emergence time θ proved to depend sensitively on time and dose.

Figure 7 shows the distribution of experiments sampling time and dosage combinations in the first 24 hours after darkness begins. The emergence times are shown as functions of both the timing and the strength of the stimulus. The normally narrow peaks of daily emergence activity are replaced by continuous emergence following exposure to exactly the right dose D^* at a critical time T^*, indicated by the vertical axis of the spiralling cloud of data. Consequently no peak time and data can be shown near this axis. T^* repeats every 24 hours.

The critical dose D^* turns out to be remarkably small by the standards we had been accustomed to: 6000 ergs/

cm^2 is accumulated under normal sunlight in less than a second. It must be administered within one half hour of T^*. It is not hard to understand how such a small effect was overlooked until theoretical considerations predicted the existence of these critical combinations of time and dose and prescribed a recipe for finding them.

At the singularities, the phase is indeterminate, but everywhere else on the time–dose plane, a light pulse merely rephases a rhythm that is otherwise indistinguishable from undisturbed controls. Plotting emergence times θ, in hours past the light pulse, above the time–dose plane we again obtain a screw surface that repeats every 24 hours along the θ axis, and every 24 hours along the T axis. It is another lattice of helicoids, with their three axes at the three critical annihilating stimuli. It is not hard to see that the two parts of figure 7 are projections of such a spiral surface.

The language of topology

Is this rephasing pattern peculiar to the systems we have been considering? Hardly; reports are available in the literature in which circadian rhythms of the most varied organisms are rephased by light pulses of appropriate colors and energies. None of these reports give the complete resetting pattern, but many give one-dimensional sections at fixed times or doses. All of these sections obtained at various laboratories from diverse animals and plants belong to the rather restrictive class of shapes obtained by sectioning a lattice of helicoids.

In experiments at fixed dose, for example, θ is recorded as T is varied. Both θ and T are periodic, which is the same as saying they are phases, defined on the unit circle. So the experimental curves obtained must be mappings of the circle onto itself. They therefore fall into the discrete types of such mappings, such as the type-1 and type-0 curves found in yeast. In fact only those two types are found among all the circadian-rhythm resetting curves: Type 1 at small doses, then a discontinuity at some critical dose and then type 0 at large dose, just as in yeast.

The reasons could scarcely be the same in all these cases. I believe the reason is very fundamental, and is better expressed in the language of topology than in the language of chemical mechanisms.

In fact, recent experiments by Wolfgang Engelmann in West Germany and Anders Johnsson in Sweden gives the complete resetting pattern for the circadian rhythm of flower opening and closing in the plant *Kalanchoe:* It is again a spiral staircase, repeated along both T and θ axes, organized around a series of phase singularities.

Sleepless flies

Returning to flies, we find that not only phase, but also "amplitude" or "intensity" or "vigor" of the clock oscillation is reset: it is lastingly reduced in proportion to the nearness of (T, D) to the singularity. The continuous emergence from populations given a near-singular pulse is partly due to dispersion of phase within the population, but it also reflects severe attenuation of the circadian time-keeping process within each pupa. This process is, in fact, switched off at the singularity—until reinitiated by a second pulse, as in yeast glycolysis.

Adult flies also have a circadian activity rhythm. Emergence may be only its first manifestation: They sleep every 24 hours, even in constant darkness. How would the singularity show up in this physiological rhythm? If the clock is reduced to a neutral, motionless state, we might expect the fly to suffer from chronic insomnia, or at least to have very irregular sleeping habits. This sounds positively occult when taken together with the observation that for flies (and apparently also for other organisms) T^* is close to subjective midnight, and D^* is equivalent to a few minutes' full moonlight!

That ends what I want to say about experiments. What properties must a system have so that we can expect its resetting pattern to resemble a lattice of helicoids containing critical stimuli? Unlike Urbain Leverrier's anticipation that an unknown planet would appear at a specific place and time in the sky, these phenomena were not predicted

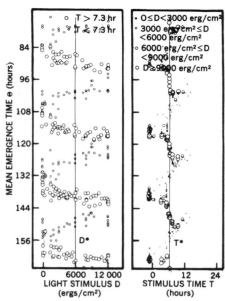

Projections of fruit fly emergence data onto two planes, corresponding to two sides of the unit cell of figure 4. Depth is roughly indicated by the symbol size, as shown in the key; $T^* = 7.3$ hr, $D^* = 6000$ erg/cm². For clarity, data points for 15 hr $< T < 24$ hr are omitted. **Figure 7**

from a detailed understanding of the dynamics involved. The kinetics of metabolic control are terribly complicated, the mechanism of circadian chronometry is altogether unknown. But purely topological arguments indicate that in a stably oscillating dynamical system satisfying the sort of continuity conditions typical of chemical reactions, there must exist sets of initial conditions from which the system cannot return spontaneously to the prior mode of oscillation. This set of points, analogous to the equilibrium point of the pendulum that we considered at the beginning of this article, is the *phaseless set*. However, this set, of dimensionality one or two less than the state space,

must "thread" the cycle regardless of the dynamical complexity of the oscillator.

Now consider a disturbance that, if indefinitely prolonged, forces the oscillation to a halt at some equilibrium state—as prolonged oxygenation does for yeast metabolism, and as prolonged light does for *Drosophila*'s circadian clock. The set of states accessible by application of that disturbance for various durations or at various doses is a two-parameter set bounded only by the cycle. Thus it intersects the phaseless set along a curve or at a point we call the singularity. Following a stay at this critical (T^*, D^*), the oscillation reappears only after an unpredictable lapse and at an unpredictable phase, if at all.

Applications

Because the helicoidal lattice and its singularities depend for their existence only on such abstract and general properties of the system, phase singularities of the type we have been discussing should play a central role in many other systems. One of these is the cell cycle of proliferating tissues. Stuart Kauffman has undertaken such experiments, and appears to be encountering this kind of singularity.

One suggested application to human circadian rhythms is that, when a "pill" is discovered that rephases the circadian clock in travelers to avoid the discomfort and inconvenience of "jet lag," the helicoid may well provide a fundamental principle for dosage scheduling.

It also appears reasonable to enquire whether the month-long female cycle shares the general features noted above as sufficient for helicoidal resetting. Prolonged estrogen therapy does in fact bring the menstrual rhythm to a halt. Recent experiments illustrate that rephasing experiments are feasible and that a modest dose of estrogen can

evoke a large phase shift, depending on when it is given. But, as far as I know, such experiments have never been carried out in a way sufficiently systematic and comprehensive to establish whether the resetting pattern actually does have the topology of a spiral staircase and, if so, whether the implicit neutral phase condition is sufficiently stable to have a practical utility. If there is a critical time and dose of suitable hormones, and if it resets the female endocrine oscillator to near-zero amplitude stably, a single, well-timed "pill" would suffice to prevent ovulation.

The phenomena discussed in this article thus may have some important applications; but whether these become realities or not, I believe that these explorations of singularities in biological time capture something that is both simple and fundamental about biological clocks.

Bibliography

- A. T. Winfree: "The Temporal Morphology of a Biological Clock," in *Lectures on Mathematics in the Life Sciences* (M. Gerstenhaber, ed.) Am. Math. Soc., Providence, R.I. (1970).
- A. T. Winfree: "Corkscrews and Singularities in Fruitflies," in *Biochemistry* (M. Menaker, ed.), Nat. Acad. Sci., Washington, D.C. (1971).
- A. T. Winfree, Arch. Biochem. Biophys. **149**, 388 (1972).
- A. T. Winfree: "Time and Timelessness in Biological Clocks," in *Temporal Aspects of Therapeutics*, (J. Urquardt, F. E. Yates, eds.) Plenum, N.Y. (1973).
- W. Engelmann, H. G. Karlsson, A. Johnsson, Int. J. Chronobiol. **1**, 147 (1973).
- E. Bunning, *The Physiological Clock*, (third edition), Springer, New York (1973).
- S. Kauffman, J. Wille, J. Theor. Biol. (in press, 1975).
- A. T. Winfree, Nature **253**, 315 (1975). □

Section V
Experimental Techniques

Contents

Introduction

The eight papers in this section typify the range of techniques used by physicists to study living systems. The measurement of the magnetic field associated with the evoked potential in humans using a superconducting interference device (SQUID) enables a mapping of cortical activity stimulated by auditory detection. By measuring magnetic fields of the order of femtotesla (10^{-15} tesla, i.e., 10^{-11} gauss) a three-dimensional localization of the source of the evoked signal within the brain is possible. The foundation of proton NMR imaging which has developed into an important noninvasive diagnostic tool now known as MRI and used in all large hospitals is described in the paper by Lauterbur. The extension of this technique to follow metabolism of Phosphorus 31 in muscle tissue demonstrates that MRI can be used to assay chemical activity during a dynamical process. Laser spectroscopy which has enabled the study of fast transfer of energy within molecules is represented also by two papers, the first studying photosynthesis, the second studying the energy barriers controlling the release of O_2 and CO_2 from myoglobin. The use of ultraviolet fluorescence to characterize amino acids in solution is the basis of fluorescence polarization studies which are now routinely used to study protein dynamics. The next paper on spectral broadening in biological molecules is an illustration of the use of optical phenomena to observe conformational motion, in this case, librational and torsional, and helps unify concepts of conformational motion and reaction kinetics. One of the major successes in research on biological systems is the deciphering of the quantum events associated with energy transfer in photosynthesis. The large number of photosynthetic papers in this volume reflect this success. The last paper in this section describes a technique which applies our understanding of photosynthesis to produce a simple model photosynthetic system.

Characterization of the Human Auditory Cortex by the Neuromagnetic Method*

G.L. Romani[1], S.J. Williamson, L. Kaufman, and D. Brenner[2]

Neuromagnetism Laboratory, Depts. of Physics and Psychology, New York University, 4 Washington Place, New York, NY 10003, USA

Summary. Neuromagnetic studies show that the location of cortical activity evoked by modulated tones and by click stimuli in the steady state paradigm can be determined non-invasively with a precision of a few millimeters. The progression of locations for tones of increasing frequency establish an orderly tonotopic map in which the distance along the cortex varies as the logarithm of the frequency. The active region responding to clicks lies at a position that is consistent with this map if the stimulus is characterized by the frequency of the peak of its power spectrum. A latency of about 50 ms observed for the response to clicks is in close correspondance with a strong component of the transient response to an isolated click reported in the literature. Monaural stimulation of the ear contralateral to the hemisphere being monitored produces a latency which is about 8 ms shorter than stimulation of the ipsilateral ear, in agreement with previous studies of transient responses. The amplitudes of the responses for binaurally presented clicks for sleeping subjects is substantially diminished for repetition rates above 20 Hz but is enhanced for lower rates.

Key words: Magnetoencephalography – Evoked fields – Evoked potentials – Steady state responses – Audition

Introduction

Since Marshall and Talbot (1942) first proposed a one-to-one mapping of the retina onto the visual cortex, the general notion that peripheral signals are mapped onto primary projection areas of the cerebral cortex has won wide acceptance. It is now recognized from studies of evoked potentials detected on the cortex that there are several different visual areas (Van Essen 1979), and each of these is an isomorphic representation of the retina. The pioneering work of Penfield and Rasmussen (1950) with electrical stimuli applied to various areas of the body demonstrated an orderly projection of the various portions of the body on the cortex (the somatosensory "homunculus"). Recent studies of auditory cortex of monkey (Merzenich and Brugge 1973), cat (Merzenich et al. 1975) and squirrel (Merzenich et al. 1976) reveal a similarly ordered projection area in which tones of various frequencies activate different sets of neurons (Woolsey and Walzl 1942). Part of the interest in such a systematic ordering of the projection areas lies in the possibility that it may have significance for information processing at the cortical level (Schwartz 1980). Apart from that possibility, our general knowledge of the organization of the human nervous system would be enhanced if it were possible to describe the way in which auditory signals are mapped in the human auditory cortex.

To our knowledge the method of scalp-detected evoked potentials has not yet been successfully used to discern such a mapping. One study in which electrical recordings were made from the exposed pial surface of the brain of 19 patients during surgery could find no differences in responses to tones of 600 and 1,000 Hz (Celesia 1976). Since the neuromagnetic technique has been successfully used to localize small and functionally discrete regions of the brain,

---

* Supported in part by Office of Naval Research Contract N00014-76-C-0568

1 Supported by Consiglio Nazionale delle Ricerche and by Progetto Finalizzato Superconduttivitá – C.N.R.; Permanent address: Istituto di Elettronica dello Stato Solido – C.N.R., Via Cineto Romano 42, I–00156 Roma, Italy

2 Present address: Rockwell International, 3370 Mira Loma Avenue, Anaheim, CA, USA

Offprint requests to: S.J. Williamson, ScD (address see above)

Reprinted from *Experimental Brain Research* **47**, 381–393 (1982); © Springer-Verlag, Inc.

we applied the method and mapped the auditory cortex in two subjects (Romani et al. 1982a).

Previous neuromagnetic studies identified specific areas of the somatosensory cortex representing different regions of the body (Brenner et al. 1978; Kaufman et al. 1981; Okada et al. 1981a) and the positions of activity in visual cortex responding to patterned visual stimuli (Brenner et al. 1981). After discovery of a magnetic field near the auditory cortex by Reite et al. (1978), Farrell et al. (1980) interpreted a 50 ms component evoked by click stimuli as due to a current dipole lying within a few centimeters of the auditory cortex. A 100 ms component following the onset of a 1,000 Hz tone and a sustained component of the field occurring during its presentation have both been interpreted as arising within the Sylvian fissure (Hari et al. 1980; Elberling et al. 1980).

In the first part of the present paper we give a detailed account of the method used to determine the position and depth of cortical activity that varies in response to repeated clicks or to a tone when its amplitude is modulated at a low rate. We then describe the measurements and data analysis which determine this tonotopic map and the total strength and orientation of the individual cortical field sources for individual tones. This is followed by a study of responses to clicks presented at various rates which provides a measure of the latency for the cortical activity. In the course of these measurements we observed a marked difference in response amplitudes between waking and sleeping states of three subjects which varied with repetition rate.

Theory of Determining Source Location

It is well known that the problem of determining the location of a source of potentials on the scalp or the magnetic field outside the head does not have a unique solution, as many different sources could provide exactly the same distributions. However, a particularly suitable model for activity evoked in cerebral cortex is that of a current dipole immersed in a conducting medium, for instance a sphere with homogeneous conductivity or with a radially dependent conductivity. Using this model it is possible to compute the distribution of potentials and fields at the surface of the medium. A comparison between these patterns and the observed ones can give information on the source and on its location inside the brain.

There is a fundamental difference between the potential and field patterns generated by a current dipole in a sphere: the distribution of potentials at the surface is associated with volume currents, whose pattern is influenced by the distribution in conductivity. In addition, the orientation of the dipole affects the symmetry of the pattern of isopotentials across the surface (Williamson and Kaufman 1981a). On the other hand, the component of the magnetic field normal to the scalp is unaffected by volume currents (Cuffin and Cohen 1977) and is due solely to the current dipole itself. In simple models for neural activity, the dipole represents the relatively high-density intracellular currents flowing in the active neural tissue (Plonsey 1981). Furthermore, to generate a magnetic field outside the sphere it is necessary that the current dipole lie tangential to the surface. Indeed, for a radially oriented dipole, the boundary of the sphere perturbs the pattern of the volume current in such a way as to create a magnetic field outside that exactly cancels the field from the dipole itself (Grynszpan and Geselowitz 1973). Thus for a dipole oriented in an arbitrary direction it is only the tangential component that generates the field outside. As a consequence, the shape and the position of the pattern will remain the same if the dipole is tilted from a tangential direction to a radial one, the only difference being that the field strength will monotonically decrease to zero. The actual pattern of the field normal to the surface is characterized by two identical regions symmetrically arranged on either side of the dipole, one representing the field directed outward from the sphere and the other the field directed inward. A unique relationship between the location of the maxima for the outward and inward field and the depth of the dipole in the sphere has been deduced (Williamson and Kaufman 1981a). It is clear from the above considerations that a similar simple relation cannot be inferred in the case of a potential pattern because its shape may be asymmetrical. For the analysis of magnetic field data to determine the depth and strength of a dipole the effect of the geometry of the magnetic field sensor must be taken into account.

It must be remarked that the sphere model represents only a first approximation to the head. Indeed, as indicated in topographic atlases, actual heads vary in the ratio of their length to width (Delmas and Pertuiset 1959). Nevertheless, the portion of the scalp posterior to the frontal lobe, including the region near the auditory cortex of concern here, is well approximated by a sphere. We have assumed that the "best fit" is achieved with a diameter matching the width of the head at the ear canal, and with the center of the sphere located so as to touch the top and the back of the head. It must be emphasized that even if this assumption of the spherical model seems to be oversimplified, the values of source depth which are obtained are quite reasonable, as will be shown later. The agreement with anatomical features is *a posteriori* confirmation that the method is appropriate.

Methods

The techniques for biomagnetic measurements were fully described in a recent review (Williamson and Kaufman 1981b). Our magnetic field sensor consisted of a second-derivative gradiometer with 2.4 cm diameter and 3.2 cm baseline between adjacent coils coupled to a Superconducting Quantum Interference Device, or SQUID (S.H.E. Corporation, San Diego, CA). The use of such a sensor provides both the sensitivity required to measure evoked fields and a satisfactory reduction of the environmental noise without the aid of any magnetic shielding. All the superconducting circuitry as well as the other cryogenic facilities were contained in a superinsulated fiberglass dewar which permits positioning the pickup coil of the gradiometer as close as 8 mm to the scalp. The dewar was oriented so that the magnetic field component perpendicular to the scalp was monitored by the gradiometer. The output voltage from the SQUID electronics, which is simply proportional to the net field sensed by the gradiometer, was applied to a bandpass filter tuned at the stimulus repetition rate, with 48 db/octave rolloff on high and low frequency sides. Consequently in this "steady-state" paradigm the output representing the response was a sine wave at the stimulus repetition rate whose amplitude and phase relative to the stimulus cycle were to be determined.

Auditory stimuli were presented binaurally to the subjects by means of standard airline plastic earphones, the transducer of which had been tested previously to confirm that it produces no magnetic artifacts. Two different kinds of acoustic stimuli were used: clicks and amplitude-modulated pure tones. Clicks were produced by feeding the transducer a train of short (100 μs) electric pulses. The repetition rate could be suitably varied in a relatively wide range of 4–55 Hz. The power spectrum of the clicks was measured with a 2 cm^3 earphone coupler and a General Radio 1900A wave analyzer. It had a broad maximum centered at ≈ 900 Hz with most of the power between 400 and 4,000 Hz. The amplitude was typically set at a total power level of about 80 db SPL. To obtain a satisfactory signal-to-noise ratio for the evoked field a digital signal averager was used to average responses at the stimulus repetition rate. Thus the averaged amplitude and phase of the component of the response at the repetition rate were determined. A preliminary investigation indicated that strongest responses were provided for repetition rates between 25 and 40 Hz; for this reason a repetition rate of 32 Hz was chosen for mapping the field about the scalp. At least 500 responses were averaged for each trial, and at least two trials were obtained at each position over the scalp.

To present modulated tones the electrical signal provided to the transducer consisted of an audio frequency sine wave whose amplitude was sinusoidally modulated at slightly less than 100% at a frequency of 32 Hz. Our choice of this stimulus was motivated by microelectrode studies of animals (e.g., Merzenich et al. 1976) which show populations of neurons that respond continuously to the presence of a steady tone, provided that the frequency is sufficiently high. The response is not modulated at the tone frequency, unlike the behavior seen in other populations which respond to slowly repeated clicks (de Ribaupierre et al. 1972). The observed response of the latter population diminishes as the repetition rate is increased into the domain of 100 to 200 Hz and then disappears. Thus to study the high frequency population we chose to turn the tone "on" and "off" in a slow, sinusoidal fashion

to monitor neural activity that follows the envelope of the stimulus. The modulation frequency was much lower than the carrier frequency, so that the resulting Fourier spectrum of the sound was confined to a very narrow bandwidth. The spectrum consisted of a carrier at the frequency of the tone and side bands at higher and lower frequencies shifted from the carrier by an amount equal to the modulation frequency. Four different carrier frequencies of 200, 600, 2,000 and 5,000 Hz were used for one subject (SW). For the second subject (CP) only a weak response to the 5,000 Hz stimulus was observed while a relatively strong response was measured at 100 Hz, which was therefore chosen as the fourth frequency. The amplitude for the 2,000 Hz tone was set at a power level of about 80 db SPL, and the amplitudes of the other stimuli were set so that they were subjectively equal in loudness to it. Responses were averaged for 1,000 modulation cycles to achieve a satisfactory signal-to-noise ratio. The amplitude and phase of the sinusoidal response at 32 Hz were measured for different positions of the gradiometer over the scalp, for each carrier frequency. At least two measurements were made at each carrier frequency at each position over the scalp.

To obtain a satisfactory field mapping over the appropriate region of the scalp it was necessary to measure amplitude and phase of the evoked signal at about 40 positions. The typical distance between adjacent positions was 1 cm and a region of about 10 × 10 cm was covered by the mapping. For this purpose a spherical coordinate system was chosen to define a convenient measuring grid. The origin of the system was located at the ear canal, the horizontal position was referenced to an "equator" connecting the origin to the corner of the eye; the vertical position was measured along the appropriate "meridian" pointing from the equator to the vertex.

Results

Tonotopic Projection

A map displaying isofield contours for the observed amplitude of the magnetic response of one subject to click stimuli is shown in Fig. 1. The contours were calculated by computer using an interpolation fit of the data by the Laplacian method. Field strength is expressed in femtotesla (fT), where 1 fT in SI units is equivalent to 10^{-11} gauss in the emu system of units. Two separated extrema of opposite field direction (i.e. π radians phase difference) are evident, but the assignment of which represents field emerging from the head is not defined by the steady state method. We have arbitrarily chosen one to represent the outward field and indicated it by positive field values. Similar patterns were obtained for three other subjects. The shape of the pattern approximates that which would be produced by a current dipole lying beneath the scalp midway between the two maxima. The orientation of the dipole corresponding to our choice of field polarity is indicated by the arrow in the figure. The model described in the theory section was used to evaluate the depth of the equivalent dipole beneath the scalp. The value R = 7.5 cm was determined as the average radius of the head of this

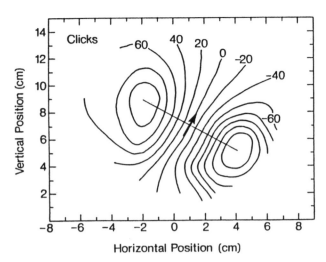

Fig. 1. Isofield contours for the field evoked by binaurally presented clicks repeated at 32 Hz. Field values are indicated in femtotesla for the amplitude (half of the peak-to-peak value) of the component normal to the scalp over the right hemisphere of subject SW. The origin of the coordinate system is the ear canal with the horizontal position measured toward the corner of the eye (at position +9) and vertical position measured from this line toward the vertex

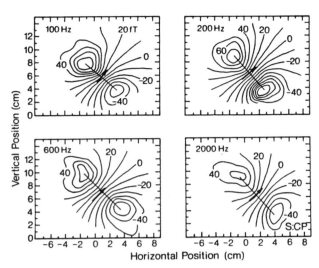

Fig. 2. Isofield contours for responses evoked by modulated tones at the indicated frequencies for subject CP. The coordinate system in each panel is the same as in Fig. 1

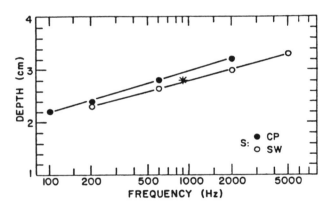

Fig. 3. Depth beneath the scalp of the current dipole representing cortical activity evoked by modulated tones versus the logarithm of the frequency, for two subjects. The star represents the depth of the response of subject SW to clicks

subject (SW) on the basis of the measured local curvature of that portion of the head. This value was found to be well in agreement with half of the ear to ear distance. The total value of an 8.5 cm radius was used for the calculation, to include the distance between the pickup coil and the scalp. Once the computer fit had established the positions of the extrema, the actual distance between the extrema was measured directly on the head and the relative angles were computed. The effect of the finite baseline of the gradiometer was taken into account in calculating the depth of the current dipole (Romani et al. 1982b). The effect of the finite radius of the pickup coil was found to be negligible. A depth of 2.8 cm beneath the scalp for the current dipole was computed from the data in Fig. 1.

Figure 2 shows isofield contour maps obtained from subject CP after stimulation with modulated tones of 100, 200, 600 and 2,000 Hz, respectively. As for the preceding study, the pattern in each of these maps approximates that of a current dipole lying beneath the scalp midway between the maxima, as indicated by the arrow. An analogous set of maps was obtained from subject SW for stimulation with tones of 200, 600, 2,000 and 5,000 Hz (Romani et al. 1982a). Following the procedure illustrated above and using the appropriate values for the average radii of the head (R = 7.5 cm for both subjects) it was possible to evaluate the depth of the evoked dipole for each condition. Figure 3 shows these depths as

they vary with frequency for the two subjects. The progression of depths for both subjects is adequately described by a logarithmic dependence on frequency.

In the same figure the depth corresponding to the dipole evoked by click stimuli for subject SW is also shown. The value of 900 Hz on the horizontal axis was chosen to represent the click since that frequency corresponds to the peak in its power spectrum. The indicated depth of this dipole is then found to be in excellent agreement with the sequence of depths determined by the modulated sine wave stimuli. The principal source of inaccuracy is believed to be a possible systematic bias that would arise from choosing an inappropriate radius for the head. That our deduced depths for the dipole sources are in reason-

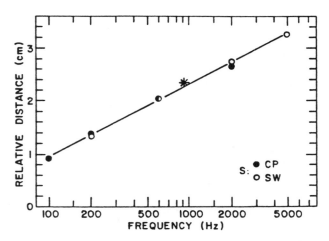

Fig. 4. Cumulative straight-line distance between positions of the current dipoles representing cortical activity evoked by modulated tones of various frequencies. The star represents the position of the response of subject SW to clicks

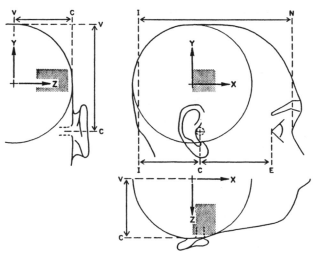

Fig. 5. Three views along axes of an orthogonal coordinate system whose origin is the center of the sphere best approximating the posterior portion of the head. Landmarks include: C – ear canal, E – corner of the eye, I – inion, N – nasion, V – vertex

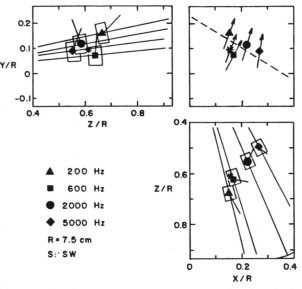

Fig. 6. Expanded rectangular sections corresponding to the stippled areas in Fig. 6. Positions within the frontal (Z-Y), sagittal (X-Y), and transverse (X-Z) planes are indicated in units of the radius R of the fitted sphere (R = 7.5 cm). The projections of the position of the current dipole representing cortical activity evoked by each modulated tone are shown for subject SW, together with the estimated uncertainty in their determinations. Stars represent the projections of the computed position for the source evoked by a click stimulus. Arrows denote the projections of each dipole vector

able agreement with the range of depths of the primary auditory cortex (Barr 1979) is an *a posteriori* justification for our procedure of determining this radius.

Although the radius of curvature of the heads of the two subjects was the same, both the lateral translation and the rate of increase in depth of the dipole sources were slightly different. Differences in cortical anatomy could well account for this finding. To take better account of such topological differences, we computed the three-dimensional straight-line distance between adjacent sources as a better estimate of the actual distances along the cortex. These "relative distances" added together sequentially represent a physiologically more meaningful parameter for denoting positions on a tonotopic map. The vertical axis of Fig. 4 shows the cumulative relative distance as it varies with the logarithm of the frequency. For purposes of comparison the data for each subject were shifted vertically to assign the same origin when the trend of each set of data is extrapolated to a tone of 20 Hz. As above, the relative distance corresponding to the dipole evoked by click stimuli for subject SW has been inserted at 900 Hz for comparison.

From the three-dimensional localization of the dipoles provided by these results, it is possible to delineate the sequence of source positions for comparison with typical anatomical features of the cortex. For this purpose three perpendicular views of a head are schematically illustrated in Fig. 5, together with the sphere fitted into the posterior portion of the head. The three views refer to the frontal (Z-Y) plane, sagittal (X-Y) plane, and transverse (X-Z) plane. The center of the sphere also serves as the

origin of the cartesian coordinate system shown in the figure. Enlarged versions of the stippled rectangular regions lying in the three sections are shown in Fig. 6 together with the cartesian coordinates whose distances are expressed in the units of the radius of

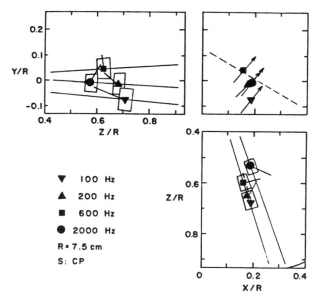

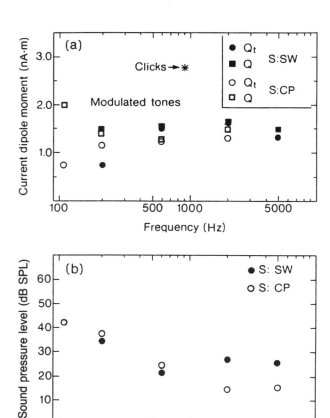

Fig. 7. Projections onto three orthogonal planes of the positions of the dipole source representing the location of activity evoked by modulated tones of various frequencies, as in Fig. 6, but for subject CP

the sphere. The deduced positions of the evoked current dipoles are shown in these enlarged sections for subject SW. The indicated uncertainties are based on the scatter in the data. Arrows indicate the directions of the dipoles projected onto the sagittal plane. An estimate for the location of the Sylvian fissure based on the description in a standard anatomy textbook (Gray 1977) is also shown as a dashed line in the X-Y section.

The same procedure repeated for subjects CP yielded the projections shown in Fig. 7. It is evident in the X-Y sections of Figs. 6 and 7 that the dipoles for both subjects are oriented approximately normal to the Sylvian fissure, strongly suggesting that the evoked field arises from current flowing perpendicular to the surface of the cortex forming the floor of the fissure.

The magnitude Q_t of the component of each dipole lying tangential to the scalp was deduced from the maximum field amplitude at one of the extrema once the position of the dipole was established. If the strengths of the two observed extrema did not exactly match, the average of the two values was used. Corrections were made to take into account the effect of the finite baseline of the gradiometer (Romani et al. 1982b). The values of Q_t measured for subjects SW and CP as they vary with frequency are reported in Fig. 8a. For subject SW the value of Q_t evoked by click stimulation is reported as well. In Fig. 8b the audiograms of the same subjects for tones

Fig. 8. a Strength of the component Q_t of the dipole lying tangential to the scalp for modulated tones of various frequencies, for each of two subjects. The deduced strength of the total dipole Q at various frequencies is also shown. The corresponding value of Q_t for a click stimulus is also shown for subject SW. **b** Audiograms for subjects CP and SW with the vertical scale given as db SPL

presented by high-quality earphones are provided for comparison.

As has been emphasized in the theory section, measurements of the field from a single current dipole give direct information only on the tangential component Q_t of the dipole. However, in the present case where data on an orderly sequence of dipoles are available, by means of a simple and reasonable hypothesis we can deduce for each stimulus condition a value for the total dipole, including its radial component. We accept the hypothesis that the current dipole at each position is normal to the cortical surface in three dimensional space. Thus it is normal to the line on the surface tracing the locus of successive dipoles. With this condition and the known magnitude and orientation of Q_t, the radial component Q_r of the dipole is uniquely determined. As an approximation to the direction of this line at a given dipole we take the direction of the vector connecting the position of one of its neighbors to the position of the neighbor on its other side. For a dipole at the end of the sequence, we approximate the direction of the line by the vector to its sole neighbor. It is then straightforward to calculate Q_r for a given dipole from the mathematical expression that the scalar product of the total dipole vector and the direction of the line at its position be zero, subject to the constraint that the transverse component Q_t of the dipole vector is known. With the value of Q_r determined by this method and the known value for Q_t, the strength Q of the total dipole was calculated.

The value of Q as it varies with the frequency of the tone is shown in Fig. 8a. For both subjects the values of Q are essentially independent of frequency and have a common value of about 1.4 nanoampere-meter. The one exception is the value at the lowest frequency (100 Hz) for subject CP, which lies 30% higher. However, the uncertainty for this value is substantially greater than for the others: this dipole has the greatest radial component of the group, and because it lies at the end of the sequence there is a comparatively large uncertainty in its value owing to the imprecision in deducing the direction of the normal to the cortex at its position. Figures 6 and 7 show the components of each dipole projected onto three orthogonal planes.

Temporal Features

Studies with binaural click stimuli were carried out on four subjects to determine the relationship between the phase lag of the evoked field and the stimulus repetition rate. Fields were measured by positioning the pickup coil over one location on the right hemisphere where the field strength was an extremum. The stimulus repetition rate was varied from 4 Hz to typically 50 Hz, and a minimum of 500 responses were averaged for each trial with at least two trials being performed at each frequency. The SPL was maintained at 80 db as for the previous mapping studies, to keep it well above the ambient noise. The amplitude of the evoked field was found to be insensitive to the SPL in the range 70–80 db.

Figure 9 shows a typical phase trend for one subject. The increase in phase lag is proportional to the repetition rate for rates above 18 Hz. Below this rate a more complicated relation seems to be present. This feature was observed for all subjects, the threshold rate for the onset of linearity ranging from 16 to 20 Hz. From the slope of the "high" frequency data it was possible to evaluate a characteristic latency for each subject (Regan 1972): 45 ms for CS;

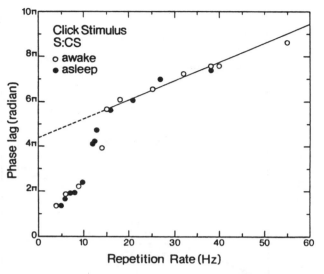

Fig. 9. Variation in the phase lag of the steady state response with repetition rates of a click stimulus. The field was measured over the right hemisphere at the position (4.5, 5). A phase lag which is a multiple of 2π indicates that the maximum field emerging from the head occurs simultaneously with a click. The modulus of 2π chosen for the origin insures that the phase lag extrapolated to zero repetition rate lies between zero and 2π. Open circles are for the subject awake, and closed circles for the subject asleep

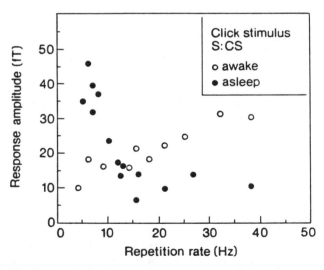

Fig. 10. Amplitude of the steady state responses (half of the peak-to-peak value) described in Fig. 9

56 ms for SW; 42 ms for RS; and 52 ms for RR, with a standard deviation of 2 ms for each value. The overall average is 49 ms for the group. A strong dependence of the amplitude of the magnetic response on the stimulus repetition rate was observed for all subjects, as illustrated in Fig. 10 for one of them. In particular the amplitude was found to be maximum for rates between 28 and 40 Hz; below and

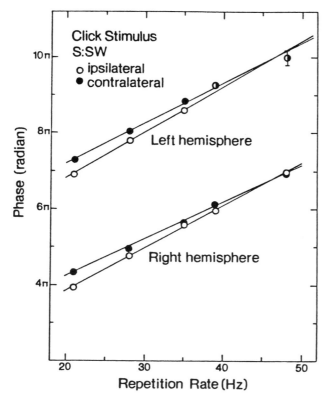

Fig. 11. Phase lag measured over the right and left hemispheres of subject SW near position (4, 5.5) for monaural stimulation to the ear ipsilateral and contralateral to the hemisphere being monitored. For clarity the phases for right hemisphere responses were decreased by 2π

above this range the amplitude decreases gradually to undetectable levels.

Monaural Stimulation

Previous studies of transient potentials (Butler et al. 1969; Majkowski et al. 1971) and fields (Elberling et al. 1981) evoked by a short monaural tone burst revealed that the first strong component, with a nominal latency of 100 ms, has a slightly shorter latency when the stimulus was presented to the ear contralateral to the hemisphere from which the recordings were made. The difference between contralateral and ipsilateral responses is about 8–9 ms. This leads to the question of whether or not the latency of the response as deduced from steady state measurements reflects this same type of difference for the nominal 50 ms latency, since if it is present it would support the validity of this method for a determination of latency.

Figure 11 shows the data for one subject (SW) obtained during ipsilateral and contralateral stimulation by clicks presented at rates between 20 and 48

Hz. A clear difference in slope is evident, indicating that responses from the left hemisphere had a latency that is 51 ± 3 ms for contralateral presentation and 58 ± 4 ms for ipsilateral presentation. Similarly, responses over the right hemisphere had a latency of 48 ± 1 ms for contralateral and 56 ± 1 ms for ipsilateral presentation. For subject CP the left hemisphere latencies were 42 ± 3 ms for contralateral and 49 ± 1 ms for ipsilateral presentation, and the right hemisphere latencies were 42 ± 1 ms for contralateral and 50 ± 1 ms for ipsilateral presentation. Thus the response latencies are on the average 7.5 ms shorter for these subjects when the stimulus is contralateral. These results confirm that the latencies computed from steady state responses differ by about the same amount as the latencies of the 100 ms component in the transient response for monaural presentation.

Effect of Sleep

The effect of sleep on the neuromagnetic response was studied in three subjects. No attempt was made to monitor the stage of sleep in these initial studies other than to insure that the subject did not respond to verbal questions posed by the experimenter. However, great care was taken to avoid inadvertent movements of the subject that could modify the position of the earphones with respect to the ear canals. Indeed, once the subject was awakened the changes which had been noticed during sleep disappeared with no apparent movement of the subject having occurred in the interval.

No influence of the sleep state was found on the phase trend over the full range of repetition rates from 4 to 50 Hz, as shown in Fig. 9. On the other hand, a dramatic effect on the amplitude was observed in all three subjects. The relation between the amplitude and repetition rate measured for subject CS when awake and asleep is shown in Fig. 10. The amplitude of the evoked signal for rates above 15 Hz is strongly reduced and, by contrast, below 15 Hz it is markedly increased. A similar effect was shown by the second subject (RS), but with the "crossing" point at 20 Hz. Subject SW showed an analogously strong reduction in amplitude above 20 Hz but only a modest increase at lower rates. For subjects SW and CS the effect was reproduced on two separate days. It is perhaps significant that for all three subjects the "crossing" point is also where the linear phase trend begins.

Discussion

There has been a longstanding interest in developing a non-invasive technique for pinpointing the locations of neural activity in sensory areas of the human brain. Such information is desired in part to determine whether such maps will provide insight into the functional mechanisms by which sensory information is organized and processed. Additional motivation is the possibility for clinical applications in the diagnosis of sensory defects.

Tonotopic Map

The present study employed the measurement and analysis of evoked magnetic field patterns to establish the tonotopic map of the human auditory cortex for modulated sine wave stimuli. By representing the evoked source for a given frequency as a current dipole and modeling the head as a sphere, the position of each dipole was determined with a precision of a few millimeters. The deduced positions of these sources is unaffected by exact variation of the conductivity throughout the sphere so long as the conductivity depends only on the distance from its center (Grynszpan and Geselowitz 1973). As a verification of the more general applicability of this tonotopic map, the location of the response to click stimuli of high repetition rate was also determined. The evoked source is found to lie at the position on the map corresponding to the frequency of the peak in the power spectrum of the stimulus.

Figure 3 provides evidence from two subjects that different tones evoke activity in specific areas of the cortex, with the depth of responses increasing as the frequency increases. Taking account of the displacement of one evoked source from the next in three dimensions, we show in Fig. 4 that the total distance from one response to the next along the cortex increases approximately as the logarithm of the frequency, within the frequency range studied; of course these results do not preclude the possibility that neurons within this area of the cortex may respond selectively to other features of an auditory stimulus (e.g., to an abrupt pressure increase), nor do we claim that this length of cortex embraces the entire primary projection area. The position of the response evoked by a click stimulus repeated at 32 Hz is in agreement with the tonotopic map if the stimulus is characterized by the frequency where the peak in its power spectrum lies. This implies that the response observed in the present measurements is primarily determined by a place-coding and not by a time-coding synchronized with the repetition rate (de Ribaupierre et al. 1972).

There is evidence from microelectrode studies made directly on the brain of the monkey (Merzenich and Brugge 1973), cat (Merzenich et al. 1975), and squirrel (Merzenich et al. 1976) for a similar tonotopic map. In these animals, as well, the projection sequence is essentially logarithmic over the middle decade of the spectrum. Previous efforts to detect different active regions in humans apparently have not been successful, even with direct measurement on the cortex (Celesia 1976).

The logarithmic tonotopic projection presented in Fig. 4 is a physiological counterpart to known logarithmic measures of perception. For example, the just noticeable frequency difference within the bandwidth 500–5,000 Hz is nearly a fixed percentage of the frequency, implying that on a logarithmic scale, the intervals between just discriminable frequencies are constant (Shower and Biddulph 1931). However, in the extremities of this frequency range and beyond the intervals increase. It would thus be interesting to extend the present measurements to much higher frequencies to see whether there is a corresponding departure from the logarithmic sequence of projections. However, over the range of frequencies we studied, if we assume the active cortex to have a uniform width and density of neurons, we conclude that the same number of neurons is dedicated to each octave in frequency span. It is interesting to notice that in the cochlea the point of maximum sensitivity indicated by electrical measurements shows a similar displacement with the logarithm of the frequency (Honrubia and Ward 1968). This suggests there may be a direct mapping of the cochlea on the cortex.

Relationship Between Field and Potential

We are unaware of any previous field or potential studies of steady state responses for humans that were analyzed to deduce source positions within the head. Therefore we are limited to comparing our results with studies of transient responses. Various stimuli have been used in prior work, and the latencies of the evoked components are affected by the type of stimulus. Vaughan and Ritter (1970) investigated the 100 ms component of the response to a 1 kHz tone burst at select positions over the scalp and interpreted their results in terms of a source lying near the Sylvian fissure, because the polarity of several components inverted on crossing that region. However, a debate has ensued (Kooi et al. 1971; Picton et al. 1974; Streletz et al. 1977) over the question as to whether this inversion denotes the source or merely an active equipotential contour of the reference electrode which was placed on the nose. Indeed, several authors have concluded that the lemniscal activity of primary auditory cortex cannot be resolved in scalp recordings or even by a depth electrode passing close to this area (Goff et al. 1977). More recent studies of responses to click stimuli by Wood and Wolpaw (1981) with a multiple electrode array yielded patterns of equipotentials suggesting to them that the polarity reversal is better characterized as the summation of potentials from several sources than as a reversal due to a single dipole source. The biomagnetic technique registers the magnetic field within a single pickup coil and as such requires no reference.

The first report of an auditory evoked magnetic field by Reite et al. (1978) did not reveal the exact position of the source, because only one field extremum was observed and not two. Subsequently Farrell et al. (1980) reported the observation by magnetic measurements of a 50 ms component for a click stimulus and concluded that its source can be modeled by a vertically oriented current dipole near or within the auditory area of each hemisphere. This model is consistent with the polarity of the evoked potential reported by Picton et al. (1974) which has a positive amplitude at the vertex. By assuming an inverse cube law for the variation in field amplitude with distance from the source, Farrell et al. (1980) estimated that the dipole lies about 3 cm beneath the scalp. In recent studies of the magnetic response to a click stimulus Zimmerman et al. (1981) deduced that the orientation of an equivalent dipole source is tipped away from the vertical by an amount that is consistent with the dipole being oriented perpendicular to the Sylvian fissure.

Additional studies of magnetic responses, but in this case to a 1 kHz tone burst of 800 ms duration, have been reported by Hari et al. (1980). The first strong component is not at 50 ms but at 100 ms. Measurements of the field normal to the scalp at positions along a track running nearly parallel to the Sylvian fissure display a reversal of direction near the primary auditory cortex for this component, as well as a component at 180 ms and the sustained field which continues during the presentation of the tone. These results and complementary data for the distribution of scalp potentials led Hari et al. (1980) to conclude that the sources of at least the 100 ms component and the sustained field could be modeled by current dipoles located in or near the primary auditory cortex. A study of the 100 ms component by Elberling et al. (1980) is in general agreement with these results, but the source was inferred to be aligned vertically (along the Y-axis of Fig. 6) at a depth of 1 cm beneath the scalp, close to the T4 location (Bak et al. 1981). More recent measurements and analyses show hemispheric differences in the source locations (Reite et al. 1981) and place the

current dipole at a depth of 2 cm (Elberling et al. 1981).

Physiological Origin of Evoked Field

Our results depicted in the X-Y sections of Figs. 6 and 7 are evidence that the evoked current dipole in steady state measurements is not vertical but is tipped forward at an angle that places it essentially perpendicular to the Sylvian fissure. An analogous situation was found for steady state somatically evoked responses (Brenner et al. 1978), where the equivalent current dipole source for finger stimulation lies perpendicular to the Rolandic fissure, and for steady state visually evoked responses to stimulation with a half-field contrast reversing grating, where the evoked dipole lies perpendicular to the longitudinal fissure (Brenner et al. 1981). Thus there is a common relationship between the direction of the current dipole and the orientation of fissures for these three sensory modalities. This implies that the current source in the active neural tissue flows perpendicular to the cortical surface.

Furthermore the relative polarities of evoked field and potentials for responses evoked by auditory (Farrell et al. 1980; Hari et al. 1980), somatic (Kaufman et al. 1981), and visual (Okada et al. 1981b) stimuli shows that the current giving rise to the field flows in a direction that is opposite to the current associated with the scalp potential. Thus it is likely that the evoked field arises from *intracellular* current in the active neurons, whereas the skin potential is associated with the *extracellular,* or volume current, flowing in the surrounding conducting tissue in the reverse direction. The detailed evidence presented here that the component of the current dipole in the sagittal plane lies perpendicular to the Sylvian fissure suggests that neurons preferentially aligned in that direction play the dominant role. The most likely population consists of the pyramidal cells whose apical dendrites and basal axons preferentially have that direction. The volume currents arising from activity in this population have been implicated as a source of some early components of evoked potentials (Goff et al. 1978).

For a spherical head, measurements of the component of the field normal to the surface can determine only the component of the dipole lying tangential to the scalp (Cuffin and Cohen 1978). We have shown in this study how our data for the transverse components and positions of dipoles lying in an orderly sequence, together with the hypothesis that the direction of each dipole is normal to the line smoothly joining these positions permits the three orthogonal vector components of each dipole to be determined. The projections of each dipole lying in three orthogonal planes are shown in Figs. 6 and 7. While there is a preferred orientation in the direction normal to the Sylvian fissure, the polarity of the Z-components of the dipoles varies systematically in correspondence with the presumed contours of the cortical gyri.

Source Strength

The strength of the current dipole representing each source evoked by a modulated sine wave was also determined. The variation with frequency of the tangential component Q_t and the total strength Q is summarized in Fig. 8a for stimuli of nearly equal perceived loudness. For all but one case (100 Hz for subject CP) the dipoles are predominantly tangential to the scalp. Over a factor of 20 span in frequency, the strength Q for all responses in the two subjects varies by less than 10% from the value 1.4 nA-m. Thus the strength Q in these suprathreshold measurements does not follow the variation in the sensitivity of the auditory system, as defined by the reciprocal of the sound level threshold in Fig. 8b (compare Figs. 8a and 8b). The observed constancy of response strength suggests that an equal number of cortical neurons is activated by each modulated tone, regardless of frequency.

Compared with the strength of cortical responses evoked by stimuli of other sensory modalities, the moment Q of the current dipole found in the present studies is comparatively weak. Earlier studies of the steady state somatic response to electrical stimulation of the little finger (Brenner et al. 1978) and the transient auditory response to a sustained tone (Bak et al. 1981) detected dipole moments of about 10 nA-m, almost one order of magnitude greater.

Influence of Sleep

We turn now to consider functional aspects of the observed responses. Our data for the increase in phase lag with repetition frequency for click stimuli suggest two domains where different mechanisms determine the response. For rates below about 20 Hz the phase trend is non-linear (Fig. 9), but above 20 Hz a linear trend is evident, permitting a latency to be defined for that entire range. It may be no accident that the ≈50 ms latency computed from the steady state response is the same as the period of the 20 Hz stimulus dividing the linear from the non-linear domains. This threshold rate corresponds to the

traditional lower end of the auditory frequency range where sensitivity decreases dramatically to zero.

Studies of the effect of sleep on the amplitude (Fig. 10) provide additional evidence for domains of high and low repetition rate characterizing different sensory functions: when a subject is asleep the response amplitude decreased markedly in the high-rate domain; whereas for two of our three subjects the amplitude increased in the low-rate domain.

The increase in response amplitude at low rates, although observed here in steady state measurements, must characterize the transient magnetic response as well. The reason is that Farrell et al. (1980) reported no change in the transient waveform for repetition rates extending as high as 4 Hz where we have conducted steady state measurements. Thus the enhancement that we observe can be identified with a similar increase by a factor of 2–3 in the 50 ms component of the transient scalp potential previously reported by Williams et al. (1962). Our results extend this earlier finding by showing that the enhancement at low repetition rates during sleep is superseded by a suppression when the rate is sufficiently high.

It may be of some interest to note that this finding is consistent with the common observation that one adjusts readily to continuous sounds, e.g., air conditioners, and can sleep, whereas infrequently recurring sounds (low-frequency clicks) may have an arousing effect on the sleeper. The sleep effect reported here suggests that the "gain" of the auditory system is greater for stimuli with low repetition rate during sleep than those with high rates while it is relatively higher for stimuli with high repetition rates during wakefulness. The potential adaptive value of such a variation in gain is obvious.

Steady State Latency

The latencies of responses in the high-rate domain were studied to assess their relationship to latencies of components of the transient response. The latency for binaural stimulation varies across subjects, ranging from 42 to 58 ms, with an average of 49 ms for our group of five subjects. These values are comparable to the latency of ≈50 ms reported for the first strong component of the transient responses to click stimuli (e.g., Picton et al. 1974; Farrell et al. 1980). The present results show that the steady state latency for a monaural stimulus is approximately 7.5 ms shorter for contralateral presentation than for ipsilateral. We are unaware of comparable published studies of the 50 ms component for transient responses to a click with which this may be compared, but our value is close to the difference reported for the 100 ms component in transient responses to a tone burst (Elberling et al. 1981). Thus the magnitude of the steady state latency and its difference for monaural contralateral and ipsilateral presentation correspond to characteristics of transient responses; however, a direct connection between the functional significance of the steady state and transient latencies is yet to be established.

In summary we have shown how the neuromagnetic method provides a non-invasive means for determining a tonotopic map of human auditory cortex. Separate areas of the cortex responding to specific tones can be identified. This establishes a basis for more sophisticated studies with superimposed tones, sounds with different temporal features, and elements of language to determine the locations and relationship of responding areas. Measurements of the latencies for steady state responses established a correspondence with characteristics of transient responses, but the functional significance of these latencies is yet to be determined. The dramatic effect of sleep on the amplitudes of steady state responses reveals that this paradigm may serve as a means for studying the effect of attention on the level of evoked cortical activity.

Acknowledgements. The authors thank Y. Okada for many helpful discussions and his skilled guidance, J. A. Movshon for useful suggestions on data processing, M. Perkins for the audiogram recording, M. Pavel for earphone characterization and C. Paulsen for assistance with the experiments.

References

Bak C, Kofoed B, Lebech J, Saermark K, Elberling C (1981) Auditory evoked magnetic fields from the human brain. Source localization in a single-dipole approximation. Phys Lett 82A: 57–60

Barr ML (1979) The human nervous system: An anatomic viewpoint. 3rd edn. Harper and Row, Hagerstown

Brenner D, Lipton J, Kaufman L, Williamson SJ (1978) Somatically evoked magnetic fields of the human brain. Science 199: 81–83

Brenner D, Okada Y, Maclin E, Williamson SJ, Kaufman L (1981) Evoked fields reveal different visual areas in human cortex. In: Erné SN, Hahlbohm HD, Lübbig H (eds) Biomagnetism. de Gruyter, Berlin, pp 431–444

Butler RA, Keidel WD, Spreng M (1969) An investigation of the human cortical evoked potential under conditions of monaural and binaural stimulation. Acta Oto-Laryngol 68: 317

Celesia GG (1976) Organization of auditory cortical areas in man. Brain 99: 403–414

Cuffin BN, Cohen D (1977) Magnetic fields of a dipole in special volume conductor shapes. IEEE Trans Biomed Eng BME-24: 372–381

Delmas A, Pertuiset B (1959) Cranio-cerebral topometry in man. In: Masson C (ed) C.C. Thomas, Springfield

de Ribaupierre F, Goldstein Jr MH, Yeni-Komshiam G (1972) Cortical coding of repetitive acoustic pulses. Brain Res 48: 205–225

Elberling C, Bak C, Kofoed B, Lebech J, Saermark K (1980) Magnetic auditory responses from the human brain. A preliminary report. Scand Audiol 9: 185–190

Elberling C, Bak C, Kofoed B, Lebech J, Saermark K (1981) Auditory magnetic fields from the human cortex: Influence of stimulus intensity. Scand Audiol 10: 203

Farrell DE, Tripp JH, Norgren R, Teyler TJ (1980) A study of the auditory evoked field of the human brain. Electroenceph Clin Neurophysiol 49: 31–37

Goff WR, Allison T, Lyons W, Fisher TC, Conte R (1977) Origins of short latency auditory evoked potentials in man. Prog Clin Neurophysiol 2: 30–44

Goff WR, Allison T, Vaughan Jr HE (1978) The functional neuroanatomy of event related potentials. In: Callaway E, Tueting P, Koslow S (eds) Event-related brain potentials in man. Academic Press, New York, pp 1–79

Gray H (1977) Anatomy, descriptive and surgical. Bounty Books, New York

Grynszpan F, Geselowitz DB (1973) Model studies of the magnetocardiogram. Biophys J 13: 911–925

Hari R, Aittoniemi K, Järvinen ML, Katila T, Varpula T (1980) Auditory evoked transient and sustained magnetic fields of the human brain. Exp Brain Res 40: 237–240

Honrubia V, Ward PH (1968) Longitudinal distribution of the cochlear microphonics inside the cochlear duct (guinea pig). J Acoust Soc Am 44: 951–958

Kaufman L, Okada Y, Brenner D, Williamson SJ (1981) On the relation between somatic evoked potentials and fields. Int J Neuroscience 15: 223–239

Kooi KA, Tipton AC, Marshall RE (1971) Polarities and field configurations of the vertex components of the human auditory evoked response: A reinterpretation. Electroenceph Clin Neurophysiol 31: 166–169

Majkowski J, Bochenek Z, Bochenek W, Knapik-Fijalkowska D, Kopec J (1971) Latency of averaged potentials to contralateral and ipsilateral auditory stimulation in normal subjects. Brain Res 25: 416–419

Marshall WH, Talbot SA (1942) Recent evidence for neural mechanisms in vision leading to a general theory of sensory acuity. In: Klüver H (ed) Visual mechanisms. Biological symposium, vol 7. Cattell Press, Lancaster, PA, pp 117–164

Merzenich MM, Brugge JF (1973) Representation of the cochlear partition on the superior temporal plane of the macaque monkey. Brain Res 50: 275–296

Merzenich MM, Knight PL, Roth GL (1975) Representation of cochlea within primary auditory cortex in cat. J Neurophysiol 38: 231–249

Merzenich MM, Kaas JH, Roth GL (1976) Comparison of tonotopic maps in animals. J Comp Neurol 166: 387–402

Okada Y, Kaufman L, Brenner D, Williamson SJ (1981a) Application of a SQUID to measurements of somatically evoked fields: transient responses to electrical stimulation of the median nerve. In: Erné SN, Hahlbohm HD, Lübbig H (eds) Biomagnetism. de Gruyter, Berlin, pp 445–456

Okada Y, Kaufman L, Brenner D, Williamson SJ (1981b) Modulation transfer functions of the human visual system revealed by magnetic field measurements. Vision Res 22: 319–333

Penfield W, Rasmussen T (1950) The cerebral cortex of man. MacMillan, New York

Peronnet F, Michel F, Eschallier JF, Girod J (1974) Coronal topography of human auditory evoked responses. Electroenceph Clin Neurophysiol 37: 225–230

Picton TW, Hillyard SA, Krausz HI, Galambos R (1974) Human

auditory evoked potentials. I. Evaluation of components. Electroenceph Clin Neurophysiol 36: 179–190

Plonsey R (1981) Magnetic field resulting from action currents on cylindrical fibers. Med Biol Eng Comput 19: 311–315

Regan D (1971) Evoked potentials in psychology, sensory physiology, and clinical medicine. Chapman and Hall, London

Reite M, Edrich J, Zimmerman JT, Zimmerman JE (1978) Human magnetic auditory evoked fields. Electroenceph Clin Neurophysiol 45: 114–117

Reite M, Zimmerman JT, Zimmerman JE (1981) Magnetic auditory evoked fields: Interhemispheric asymmetry. Electroenceph Clin Neurophysiol 51: 388–392

Romani GL, Williamson SJ, Kaufman L (1982a) Tonotopic organization of the human auditory cortex. Science 216: 1339–1340

Romani GL, Williamson SJ, Kaufman L (1982b) Biomagnetic instrumentation. Rev Sci Instrum (in press)

Schwartz EL (1980) A quantitative model of the functional architecture of human striate cortex with application to visual illusion and cortical texture analysis. Biol Cybernetics 37: 63–76

Shower EG, Biddulph R (1931) Differential pitch sensitivity of the ear. J Acoust Soc Am 3: 275–287

Streletz LJ, Katz L, Hohenberger M, Cracco RQ (1977) Scalp recorded evoked potentials and sonomotor responses: An evaluation of components and recording techniques. Electroenceph Clin Neurophysiol 43: 142–206

Van Essen DC (1979) Visual areas of the mammalian cerebral cortex. Ann Rev Neurosci 2: 227–263

Vaughan Jr HG, Ritter W (1970) The sources of auditory evoked responses recorded from the human scalp. Electroenceph Clin Neurophysiol 28: 360–367

Williams HL, Tepas DI, Morlock Jr HC (1962) Evoked responses to clicks and electroencephalographic stages of sleep in man. Science 139: 685–686

Williamson SJ, Kaufman L (1981a) Evoked cortical magnetic fields. In: Erné SN, Hahlbohm HD, Lübbig H (eds) Biomagnetism. de Gruyter, Berlin, pp 353–402

Williamson SJ, Kaufman L (1981b) Biomagnetism. J Magn Mat 22: 129–202

Wood CC, Wolpaw JR (1980) Scalp distribution of human auditory evoked potentials: A reassessment. Proc Soc Neurosci Abstr 203: 595

Woolsey CN, Walzl EM (1942) Topical projection of nerve fibers from local regions of the cochlea to the cerebral cortex of the cat. Bull Johns Hopkins Hosp 71: 315–344

Zimmerman JT, Reite M, Zimmerman JE (1981) Magnetic evoked fields: Dipole orientation. Electroenceph Clin Neurophysiol 52: 151–156

Received November 10, 1981

Image Formation by Induced Local Interactions: Examples Employing Nuclear Magnetic Resonance

An image of an object may be defined as a graphical representation of the spatial distribution of one or more of its properties. Image formation usually requires that the object interact with a matter or radiation field characterized by a wavelength comparable to or smaller than the smallest features to be distinguished, so that the region of interaction may be restricted and a resolved image generated.

This limitation on the wavelength of the field may be removed, and a new class of image generated, by taking advantage of induced local interactions. In the presence of a second field that restricts the interaction of the object with the first field to a limited region, the resolution becomes independent of wavelength, and is instead a function of the ratio of the normal width of the interaction to the shift produced by a gradient in the second field. Because the interaction may be regarded as a coupling of the two fields by the object, I propose that image formation by this technique be known as zeugmatography, from the Greek ζευγμα, "that which is used for joining".

The nature of the technique may be clarified by describing two simple examples. Nuclear magnetic resonance (NMR) zeugmatography was performed with 60 MHz (5 m) radiation and a static magnetic field gradient corresponding, for proton resonance, to about 700 Hz cm^{-1}. The test object consisted of two 1 mm inside diameter thin-walled glass capillaries of H_2O attached to the inside wall of a 4.2 mm inside diameter glass tube of D_2O. In the first experiment, both capillaries contained pure water. The proton resonance line width, in the absence of the transverse field gradient, was about 5 Hz. Assuming uniform signal strength across the region within the transmitter–receiver coil, the signal in the presence of a field gradient represents a one-dimensional projection of the H_2O content of the object, integrated over planes perpendicular to the gradient direction, as a function of the gradient coordinate (Fig. 1). One method of constructing a two-dimensional projected image of the object, as represented by its H_2O content, is to combine several projections, obtained by rotating the object about an axis perpendicular to the gradient direction (or, as in Fig. 1, rotating the gradient about the object), using one of the available methods for reconstruction of objects from their projections[1-5]. Fig. 2 was generated by an algorithm, similar to that of Gordon and Herman[4], applied to four projections, spaced as in Fig. 1, so as to construct a 20 × 20 image matrix. The representation shown was produced by shading within contours interpolated between the matrix points, and clearly reveals the locations and dimensions of the two columns of H_2O. In the second experiment, one capillary contained pure H_2O, and the other contained a 0.19 mM solution of $MnSO_4$ in H_2O. At low radio-frequency power (about 0.2 mgauss) the two capillaries gave nearly identical images in the zeugmatogram (Fig. 3a). At a higher power level (about 1.6 mgauss), the pure water sample gave much more saturated signals than the sample whose spin-lattice relaxation time T_1 had been shortened by the addition of the paramagnetic Mn^{2+} ions, and its zeugmatographic image vanished at the contour level used in Fig. 3b. The sample region with long T_1 may be selectively emphasized (Fig. 3c) by constructing a difference zeugmatogram from those taken at different radio-frequency powers.

Applications of this technique to the study of various inhomogeneous objects, not necessarily restricted in size to those commonly studied by magnetic resonance spectroscopy, may be anticipated. The experiments outlined above demonstrate the ability of the technique to generate pictures of the distributions of stable isotopes, such as H and D, within an object. In the second experiment, relative intensities in an

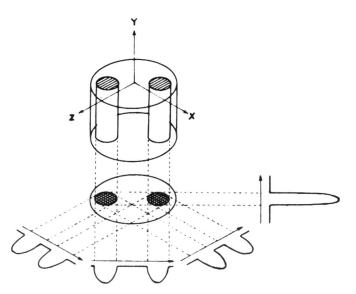

Fig. 1 Relationship between a three-dimensional object, its two-dimensional projection along the Y-axis, and four one-dimensional projections at 45° intervals in the XZ-plane. The arrows indicate the gradient directions.

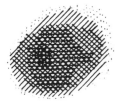

Fig. 2 Proton nuclear magnetic resonance zeugmatogram of the object described in the text, using four relative orientations of object and gradients as diagrammed in Fig. 1.

image were made to depend upon relative nuclear relaxation times. The variations in water contents and proton relaxation times among biological tissues should permit the generation, with field gradients large compared to internal magnetic inhomogeneities, of useful zeugmatographic images from the rather sharp water resonances of organisms, selectively picturing the various soft structures and tissues. A possible application of considerable interest at this time would be to the *in vivo* study of malignant tumours, which have been shown to give proton nuclear magnetic resonance signals with much longer water spin-lattice relaxation times than those in the corresponding normal tissues[6].

The basic zeugmatographic principle may be employed in many different ways, using a scanning technique, as described above, or transient methods. Variations on the experiment, to be described later, permit the generation of two- or three-dimensional images displaying chemical compositions, diffusion coefficients and other properties of objects measurable by spectroscopic techniques. Although applications employing

Reprinted from *Nature* 242, 190–191 (1973); © Macmillan Magazines, Ltd.

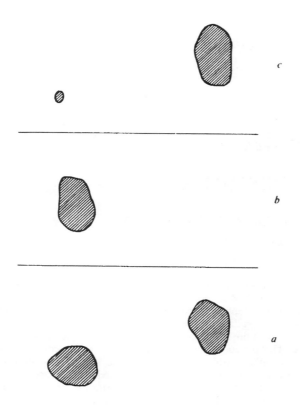

nuclear magnetic resonance in liquid or liquid-like systems are simple and attractive because of the ease with which field gradients large enough to shift the narrow resonances by many line widths may be generated, NMR zeugmatography of solids, electron spin resonance zeugmatography, and analogous experiments in other regions of the spectrum should also be possible. Zeugmatographic techniques should find many useful applications in studies of the internal structures, states, and compositions of microscopic objects.

P. C. LAUTERBUR

Department of Chemistry,
State University of New York at Stony Brook,
Stony Brook, New York 11790

Received October 30, 1972; revised January 8, 1973.

[1] Bracewell, R. N., and Riddle, A. C., *Astrophys. J.*, **150**, 427 (1967).
[2] Vainshtein, B. K., *Soviet Physics–Crystallography*, **15**, 781 (1971).
[3] Ramachandran, G. N., and Lakshminarayan, A. V., *Proc. US Nat. Acad. Sci.*, **68**, 2236 (1971).
[4] Gordon, R., and Herman, G. T., *Comm. Assoc. Comput. Mach.*, **14**, 759 (1971).
[5] Klug, A., and Crowther, R. A., *Nature*, **238**, 435 (1972).
[6] Weisman, I. D., Bennett, L. H., Maxwell, Sr., L. R., Woods, M. W., and Burk, D., *Science*, **178**, 1288 (1972).

Fig. 3 Proton nuclear magnetic resonance zeugmatograms of an object containing regions with different relaxation times. *a*, Low power; *b*, high power; *c*, difference between *a* and *b*.

Author's Note:

The last sentence in my manuscript read "Zeugmatographic techniques should find many useful applications in studies of the internal structures, states and compositions of microscopic and macroscopic objects." When I received proof, I found that "and macroscopic" had been removed. I reinserted it, only to find that it had vanished again from the published paper.

Observation of tissue metabolites using ^{31}P nuclear magnetic resonance

D. I. Hoult, S. J. W. Busby, D. G. Gadian, G. K. Radda, R. E. Richards & P. J. Seeley

Department of Biochemistry, University of Oxford, South Parks Road, Oxford, OX1 3QU, UK

^{31}P NMR spectra of intact biological tissues can now be observed. The use of the spectra to study the course of reactions within the tissues is illustrated by experiments on muscle and its glycogen particle fraction.

ALTHOUGH phosphorus NMR has only $^1/_{15}$ the sensitivity of proton NMR, it is attractive because studies may be done in aqueous solution and the spectra are relatively simple because of the small number of different chemical environments in which phosphorus atoms are found. In addition, the chemical shift range is much larger for phosphorus than for protons.

As a result of considerable instrumental improvements achieved recently in both the magnitude of the magnetic field and the sensitivity of the detection[1] we are now in a position to study ^{31}P resonances in a large variety of systems at concentrations found naturally in biology, using Fourier transform and impulse response techniques[2]. Useful spectra can now be obtained of systems varying in complexity from solutions of purified enzymes to intact tissues.

Phosphate resonances

At the high magnetic field strength employed (7.5 tesla) the ^{31}P resonances of a large number of biologically important phosphate-containing compounds can be resolved. The chemical shifts of the phosphate groups are in the range of about 30 p.p.m., and many sugar phosphates and glycolytic intermediates can be resolved. Furthermore, the state of ionisation of the phosphates and their interaction with metal ions, such as Mg^{2+}, affect the positions of the resonances. Figure 1 shows ^{31}P spectra of a mixture of compounds recorded at various pH values, the individual resonances having been assigned by observing the spectra of the individual components.

The resolution of the resonances from various phosphorus-containing small molecules allows rapid assay of the components of mixtures of these molecules. Measurements are carried out without destruction or dilution of the sample, and provide a method of monitoring turnover and interconversions of these molecules in organelles.

In the glycogen particulate fraction isolated from rabbit muscle, covalent enzyme regulation has been exhaustively studied[3,4] and transient phosphorylation of phosphorylase b has been shown to be the principal trigger for glycogen breakdown. The enzymes concerned are also regulated by small ligands. Therefore, knowledge of concentrations of these regulators at any point in the transient covalent activation is important for full understanding of the control mechanism[5]. Figure 2 shows the turnover of the phosphorus containing

ligands during a typical transient activation, obtained from a series of ^{31}P NMR spectra, recorded on a single sample. The active form of phosphorylase immediately catalyses glycogen breakdown, leading to the production of glucose-1-phosphate and so (by phosphoglucomutase activity) to glucose-6-phosphate. The production of glucose-6-phosphate is concomitant with phosphate utilisation. During the phosphorylation of phosphorylase b, ATP is converted to ADP. Because of the presence of adenylate kinase, which catalyses the reaction $2ADP \rightleftharpoons ATP + AMP$, and of AMP-deaminase, which catalyses the reaction $AMP \rightarrow IMP + NH_3$, the ADP initially formed is depleted and IMP—not distinguishable from AMP by ^{31}P NMR—is produced. The instability of the nucleotide levels indicates that the system is not a very good model for extrapolation to the *in vivo* situation.

The assay technique outlined above has also been used to estimate the separate activities in a crude extract of homogenised rabbit muscle by following the sequential production of glycolytic intermediates. The mass action ratios for some of the

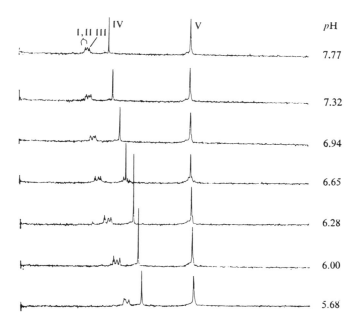

Fig. 1 ^{31}P NMR spectra of a mixture of fructose-1, 6-diphosphate (I, II), IMP (III), inorganic phosphate (IV), creatine phosphate (V) at various pH values, recorded at 129 MHz. No buffer present; total phosphate concentration 60 mM. Sweep width 5 kHz, pulse interval 2 s, 200 scans. Spectrum recorded without proton irradiation.

Reprinted from *Nature* **252**, 285–287 (1974); © Macmillan Magazines, Ltd.

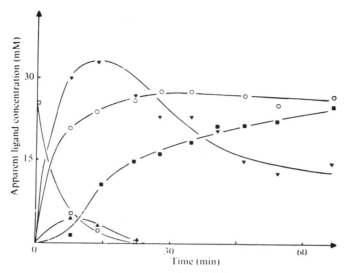

Fig. 2 Variations in concentrations of □, ATP; ▲, ADP; ○, AMP and IMP; ■, glucose-6-phosphate; and ▼, inorganic phosphate during flash activation of phosphorylase in glycogen particles. (27 mM ATP, 25 mM $MgCl_2$, 1 mM $CaCl_2$, 50 mM triethanolamine, 100 mM potassium chloride, 1 mM EDTA pH 6.9). ATP was added at time zero. Sweep width 5 kHz, Pulse interval 3 s, 50 scans. Spectrum recorded without proton irradiation. Differential saturation of the resonances can lead to small errors in the relative concentrations of ligands as measured from the areas under the peaks. The measured concentration of any ligand, however, is always a fixed proportion of its true concentration.

individual reactions have been deduced from the results obtained. The technique has also rapidly revealed the presence of contaminant activities in supposedly pure enzymes.

Metabolites in whole muscle

Our attempts to follow enzyme activities in intact tissues using phosphorus NMR have, to date, been principally concerned with observations on muscle from the hind leg of the rat. Figure 3 shows the ^{31}P spectrum of an intact, relaxed muscle, freshly excised from a rat killed by etheration. Assignments are made on the basis of the pH titrations mentioned in section 1. Part of the 'sugar phosphate' peak is attributed to

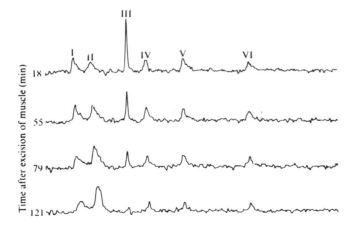

Fig. 3 ^{31}P NMR spectrum of an intact muscle from the hind leg of the rat recorded at 129 MHz, without proton irradiation. Temperature 20° C and pulse interval 16 s. Peak assignments: I, sugar phosphate and phospholid; II, inorganic phosphate; III, creatine phosphate; IV, γ ATP; V, α ATP; VI, β ATP. The times are the midpoints of the 50 scan spectral accumulations (referred to excision time as zero). The muscle was bathed in a minimum volume of calcium-free Locke ringer.

phospholipid as there is a broad peak of identical chemical shift in the phosphorus spectrum of debris from muscle extracted with aqueous buffer. Assays on this aqueous extract record a concentration of about 1 mM for glucose-6-phosphate. The frequency of the inorganic phosphate resonance defines the apparent pH of its environment (7.1 in this spectrum) and the frequencies of the ATP peaks correspond to those for the Mg^{2+}-ATP complex at the phosphate pH. (The frequency of the creatine phosphate resonance is independent of pH around 7.) *In vitro* studies of Mg^{2+} binding to ATP show that there are large shifts (2–3 p.p.m.) in the β- and γ-phosphate resonances on complex formation. These shifts show that the ATP observed in muscle is almost entirely complexed to magnesium ion. Changes in ionic strength also produce measurable spectral shifts, but these are much too small to alter our assignments for the intact muscle spectrum.

The signal peaks of the muscle phosphates are broader than those for the corresponding compounds free in solution, and the inorganic phosphate peak is markedly broader than the ATP or creatine phosphate resonances. The explanation for this phosphate broadening cannot lie in a disturbance of magnetic field homogeneity caused by the nature of the sample as the width of the creatine phosphate signal gives an upper limit for

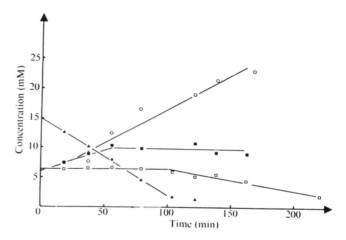

Fig. 4 Variation of phosphorus metabolite levels in an intact rat leg muscle with time after excision. The integrals of spectra shown in Fig. 3 are plotted in this graph. ○, Inorganic phosphate; ▲, creatine phosphate; ■, sugar phosphate and phospholipid; □, ATP. An absolute concentration scale was established by running a standard sample of 10 mM phosphate in the same conditions as the muscle.

the field inhomogeneity. Inorganic phosphate has a pK at 6.9 whereas the nearest pK for creatine phosphate is at 4.6. The frequency of the former resonance is therefore strongly pH dependent around pH 7 whilst that of the latter is pH independent in this region. It seems likely that the large width of the phosphate line may be due to a distribution of pH within the muscle, that is to some partial compartmentation of inorganic phosphate. The pH range that would account for the linewidth is about 0.5 pH units.

We have measured the approximate concentrations ($\pm 10\%$) of phosphorus metabolites in the intact muscle by integration of the absorption spectrum, using a sample of known concentration as a standard after taking care to avoid saturation of the resonances. There is some variation from muscle to muscle, and the creatine phosphate : phosphate ratio proves to be a sensitive index of the degree of stimulation of the rat muscle before the animal's death, that is a high creatine phosphate : phosphate ratio in rats killed without excessive stimulation. We have measured the concentrations of 'sugar phosphate', inorganic phosphate, creatine phosphate and ATP at intervals between 16 and 170 min from excision of the muscle (Fig. 4). Extrapolation along the time axis shows the metabolite levels at the time the muscle was excised to be: sugar phosphate+ phospholipid, 6 mM; inorganic phosphate, 6 mM; creatine phosphate, 15 mM; and ATP, 6.5 mM.

Creatine kinase maintains the ATP level constant at the expense of creatine phosphate until all the latter substrate has been used up, demonstrating the ability of the kinase to buffer the muscle ATP concentration[6]. Ageing of the muscle is accompanied by a fall in pH which is monitored by changes in the frequencies of the phosphate and ATP resonances. The pH at the start of data accumulation is 7.1, and after 160 min it has fallen to 6.2. Acid accumulation seems to accelerate at the time when the creatine phosphate concentration falls to zero and glycolysis presumably begins.

These results can only be obtained with intact muscle. Samples which are even slightly lacerated during handling have only an inorganic phosphate peak in the phosphorus spectrum. Breakdown of organic phosphates by phosphatases is assumed to have occurred in the damaged muscle.

Scope

On the basis of these experiments and other data we have obtained from a variety of biological systems, we believe that ^{31}P NMR can yield general information about metabolite levels, turnover, interactions and compartmentation. As it is also possible to distinguish between signals from extracellular and intercellular materials, in favourable cases the study of certain transport processes is now also possible. The delineation of unusual pathways, the detection of possible new intermediates and the presence of bound nucleotides also seems within the scope of the method. To what extent the kind of structural information that is beginning to emerge from solution studies can be extended to the study of intercellular components remains to be determined.

This work is supported by grants from the Science Research Council (Oxford Enzyme Group) and from the Medical Research Council. The spectrometer was built in Oxford with a grant from the Paul Instrument Fund of the Royal Society.

Received July 8; revised August 28, 1974.

[1] Hoult, D. I., *thesis*, Oxford University, (1973).
[2] Ernst, R. R., and Anderson, W. A., *Rev. Sci. Inst.*, **37**, 93–102 (1966).
[3] Meyer, F., Heilmeyer, L. M. G., Haschke, R. H., and Fischer, E. H., *J. biol. Chem.*, **245**, 6642–6648 (1970).
[4] Heilmeyer, L. M. G., Meyer, F. Haschke, R. H., and Fischer, E. H., *J. biol. Chem.*, **245**, 6649–6656 (1970).
[5] Fischer, E. H., Heilmeyer, L. M. G., and Haschke, R. H., *Curr. Top. cell. Reg.*, **4**, 211–251 (1971).
[6] Ennor, A. H., and Morrison, J. F., *Physiol. Rev.*, **38**, 631–674 (1958).

Femtosecond spectroscopy of excitation energy transfer and initial charge separation in the reaction center of the photosynthetic bacterium *Rhodopseudomonas viridis*

(primary donor photooxidation/bacteriochlorophyll/electron transfer/stimulated emission)

J. BRETON[†], J.-L. MARTIN[‡], A. MIGUS[‡], A. ANTONETTI[‡], AND A. ORSZAG[‡]

[†]Service de Biophysique, Département de Biologie, Centre d'Etudes Nucléaires de Saclay, 91191 Gif-sur-Yvette Cedex, France; and [‡]Laboratoire d'Optique Appliquée, Institut National de la Santé et de la Recherche Médicale U275, Ecole Polytechnique, Ecole Nationale Supérieure des Techniques Avancées, 91128 Palaiseau Cedex, France

Communicated by Pierre Joliot, April 2, 1986

ABSTRACT   Reaction centers from the photosynthetic bacterium *Rhodopseudomonas viridis* have been excited within the near-infrared absorption bands of the dimeric primary donor (P), of the "accessory" bacteriochlorophylls (B), and of the bacteriopheophytins (H) by using laser pulses of 150-fsec duration. The transfer of excitation energy between H, B, and P occurs in slightly less than 100 fsec and leads to the ultrafast formation of an excited state of P. This state is characterized by a broad absorption spectrum and exhibits stimulated emission. It decays in 2.8 ± 0.2 psec with the simultaneous oxidation of the primary donor and reduction of the bacteriopheophytin acceptor, which have been monitored at 545, 675, 815, 830, and 1310 nm. Although a transient bleaching relaxing in 400 ± 100 fsec is specifically observed upon excitation and observation in the 830-nm absorption band, we have found no indication that an accessory bacteriochlorophyll is involved as a resolvable intermediary acceptor in the primary electron transfer process.

The initial separation and stabilization of electric charges, which constitute the key processes of photosynthesis, occur in a transmembrane chlorophyll–protein complex named the reaction center. Recently the organization of the prosthetic groups within the protein scaffold of the reaction center isolated from the photosynthetic bacterium *Rhodopseudomonas viridis (Rps. viridis)* has been solved to 3-Å resolution (1). The pigments exhibit an approximate *C-2* symmetry, with two closely interacting bacteriochlorophyll molecules forming a special pair (P). Two bacteriopheophytins (H_A and H_B) are located on either side of P, while two other bacteriochlorophylls (B_A and B_B) are arranged approximately in between H_A or H_B and P. The primary quinone electron acceptor (Q_A) lies at a greater distance on the "branch" occupied by the B_A and H_A molecules. This structural organization of the pigments suggests that the charge separation initially occurs between P and B_A and is followed by migration of the electron to H_A and then to Q_A. Indeed, using pulse-probe experiments on *Rps. viridis* reaction centers with 150-fsec pulses at 620 nm, a wavelength at which all four bacteriochlorophyll molecules absorb, Zinth *et al.* (2) have observed a transient absorption change, which they have attributed to the $P^+B_A^-$ state. Such a state has been proposed previously on the basis of picosecond spectroscopy on *Rhodopseudomonas sphaeroides* reaction centers (3). However, doubts have also been raised regarding the existence of the $P^+B_A^-$ state (4, 5).

In a recent femtosecond spectroscopy study of the initial charge separation in *Rps. sphaeroides* reaction centers with direct excitation of P at 850 nm (6), we have observed the generation in less than 100 fsec of an excited state of P (P*) that decays directly to $P^+H_A^-$ in 2.8 ± 0.2 psec. We found no experimental evidence for a transient state $P^+B_A^-$. Investigating the same system but using 0.8-psec pulses at 610 nm for excitation, Woodbury *et al.* (7) have also proposed the same reaction scheme but reported a 4.1 ± 0.2-psec time constant for the initial charge separation. Furthermore, they observed a transient bleaching in the absorption band of B, which could in principle represent the state $P^+B_A^-$. However, they favored an alternative interpretation in which the transient bleaching is assigned to the initially excited B molecules before energy transfer to P occurs (within about 1.5 psec). Such an energy transfer step might explain, at least in part, the longer time constant for charge separation observed in ref. 7 as compared to ref. 6, because in the latter study P was directly excited by the 150-fsec pulse. A comparison of the kinetics of P* formation upon excitation in B or in P should resolve this question, while observation at longer times should also allow the role of B in the electron transfer sequence to be better characterized.

Previous picosecond absorption spectroscopy studies on the reaction center of *Rps. viridis* (8, 9) have demonstrated that P^+ and H_A^- appear in less than 10 psec and less than 20 psec, respectively, while the subsequent electron transfer step to Q_A occurs in $\approx$200 psec. Although there are considerable analogies between the reaction centers of *Rps. sphaeroides* and *Rps. viridis* regarding the polypeptide primary sequence (10), the spectroscopy and organization of the pigments (11), and the electron transfer processes (12), one cannot exclude that subtle modifications in the arrangement of the chromophores lead to larger changes in the most primary energy transfer and electron transfer reactions. It is thus important to assess whether the differences in the initial electron transfer mechanisms and kinetics reported for *Rps. viridis* (2) and for *Rps. sphaeroides* (6, 7) can be rationalized in terms of differences in the experimental conditions used in these femtosecond spectroscopy studies or if these discrepancies come from a genuine difference between the two types of reaction centers.

In the present work, reaction centers of *Rps. viridis* have been excited at a variety of wavelengths (between 800 and 930 nm), and the photoinduced absorbance changes have been monitored in the 545- to 1310-nm spectral range with 100-fsec time resolution in order to investigate both the excitation energy transfer from the H and B molecules to P and the initial steps and kinetics of the electron transfer.

Abbreviations: B, accessory reaction center bacteriochlorophyll (B_A or B_B); H, reaction center bacteriopheophytin (H_A or H_B); P, primary electron donor; Q_A, first quinone acceptor; Q_X, $S_2 \leftarrow S_0$ electronic transition of chlorophyllous pigments; Q_Y, $S_1 \leftarrow S_0$, electronic transition of chlorophyllous pigments.

Reprinted from *Proceedings of The National Academy of Sciences* 83, 5121–5125 (1986); © The National Academy of Sciences.

MATERIALS AND METHODS

The femtosecond laser pump–probe setup has been described previously (6). For continuum amplification at 900 and 930 nm the dye LDS 867 (Exciton, Dayton, OH) has been used. At 930 nm the energy per pulse was 0.5 mJ·cm^{-2}. The full width at half-maximum of the pump pulses was 6 nm. At all the excitation and probe wavelengths the duration of the pulse was close to 150 fsec. Handling of the data and the fitting procedure of the kinetics of absorbance changes were carried out essentially as described in ref. 13.

Reaction centers from *Rps. viridis*, prepared according to ref. 11, were suspended in Tris·HCl buffer (20 mM, pH 8.0) containing 0.5% sodium cholate.

RESULTS

In the near infrared region the absorption spectrum of *Rps. viridis* reaction centers (Fig. 1a) shows the main $S_1 \leftarrow S_0 (Q_Y)$ absorption band of P around 960 nm and of the B molecules at ≈ 830 nm. The shoulder around 800 nm is assigned to the H molecules, whose $S_2 \leftarrow S_0 (Q_X)$ transitions absorb around 540 nm. The 600-nm band is due to the Q_X transitions of both the B and the P bacteriochlorophylls. In Fig. 1b the rise of the bleaching at 960 nm representing the disappearance of the ground state of P is shown after excitation of P (930 nm), of B (827 nm), or predominantly of H (803 nm). On a 2-psec time scale these three kinetics all present an instrument-limited risetime and are well fitted with a 150-fsec pulse duration, indicating an "instantaneous" bleaching—i.e., occurring in less than 100 fsec. Such an instantaneous rise has also been observed upon an excitation at 900, 854, 837, and 797 nm (data not shown).

When observed on extended time scales, the initial bleaching at 960 nm shows a small relaxation phase that becomes more pronounced toward the long-wavelength side of the 960-nm band (data not shown). At 1050 nm the asymptotic value of this apparent bleaching is zero (Fig. 2a) and a monoexponential time constant of 2.8 psec gives the best fit to the relaxation kinetics. Because this apparent bleaching is significantly larger than the absorbance at this wavelength, it has to be assigned to stimulated emission. The instantaneous rise followed by the 2.8-psec decay of the stimulated emission has been observed both for relatively high energy pulses at

803 nm causing the photooxidation of about 40% of the reaction centers (Fig. 2a) and for lower energy pulses (causing less than 10% photooxidation) at either 854 nm or 930 nm (data not shown).

Fig. 2b shows the induced absorption increase at 1310 nm after excitation at 803 nm. The 1310-nm absorption band has been assigned at equilibrium to P$^+$ (8, 14). The kinetics is biphasic with an instantaneous contribution and a 2.8-psec component. More precisely, the best fit corresponds to the sum of two species with different absorption coefficients; the first species appears instantaneously and populates, in 2.8 psec, the second one, whose lifetime is much larger than the time domain investigated. Fig. 2c depicts the induced absorption at 675 nm, which has been attributed to the formation of H$_A^-$ (14). Within our experimental uncertainty this kinetics is identical to the one measured at 1310 nm. The formation of H$_A^-$ in the states P$^+$H$_A^-$ and PH$_A^-$ is accompanied by a bleaching of the Q_X band of H$_A$ at 545 nm. Fig. 2d shows the measurement at 545 nm after excitation at 854 nm. It reveals a biphasic response with an initial instantaneous absorption increase followed by a recovery phase that leads to an induced bleaching with a formation time of 2.8 psec. After an instantaneous absorbance rise, the blue shift of the 830-nm band occurring upon P$^+$H$_A^-$ formation (measured at 815 and 830 nm following excitation at 930 nm) appears with a 2.8-psec time constant (Fig. 2 e and f). There is no evidence for an appreciable transient bleaching in the 100-fsec to 3-psec time domain.

When both the excitation and observation wavelengths are within the 830-nm band, rather complex kinetics are observed (Fig. 3). In Fig. 3 a–c, which shows the signals observed at 850 nm upon excitation at 827 nm with pulses of increasing energy, the three kinetic traces have been fitted with two exponentials: an instantaneous bleaching recovering with a 400-fsec time constant and a slower bleaching that develops with a 2.8-psec time constant. The amplitude of the fast transient bleaching exhibits an approximately linear dependence versus the energy of the pump pulse when the latter is decreased (Fig. 3a) or increased (Fig. 3c) by a factor of 4 from the normal intensity used (Fig. 3b), with which about 20% of the reaction centers are photooxidized. In contrast, the amplitude of the 2.8-psec phase of bleaching appears less linear, especially at the highest intensity used, with which a saturation effect is observed. Such saturation of photochemistry has been previously reported for *Rps. sphaeroides* reaction centers excited with pulses of 10- or 25-psec duration (15, 16). At least for the high excitation energies, the kinetics of relaxation of the transient bleaching cannot be fitted with a 100-fsec time constant, as demonstrated in Fig. 3d. Upon detection at 832 nm, the kinetics are very dependent upon the excitation wavelength: when the excitation is at 854 nm, no fast transient bleaching is observed, while a kinetic trace identical to that shown in Fig. 3d is found upon excitation at 803 nm. On the other hand, when both the excitation (803 nm) and the detection (810 nm) are located predominantly in H, a fast transient bleaching is observed, the decay of which can be fitted with a 100-fsec time constant (data not shown).

DISCUSSION

Excitation Energy Transfer to the Special Pair. The identical kinetics of the bleaching of the band of P at 960 nm observed upon excitation either directly in P at 930 nm or at several wavelengths within the B or H absorption bands (Fig. 1b and Results) demonstrate that the transfer of excitation energy from the B and H molecules to the special pair occurs in less than 100 fsec. This ultrafast process of energy transfer, which to our knowledge constitutes the fastest direct measurement reported so far in a biological system, implies a very close proximity of the chromophores, as is indeed observed in the

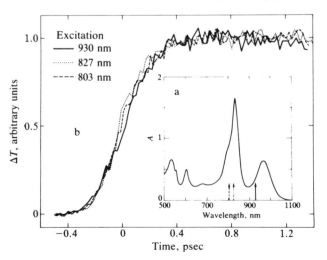

FIG. 1. (a) Room temperature absorption spectrum of the reaction centers from *Rps. viridis* used for the kinetic measurements (light path, 0.1 cm). (b) Kinetics of induced transmission at 960 nm upon excitation of the reaction centers at 930, 827, and 803 nm. The curves are normalized in amplitude and the respective zero-delay positions have been placed at the inflection point of the rise.

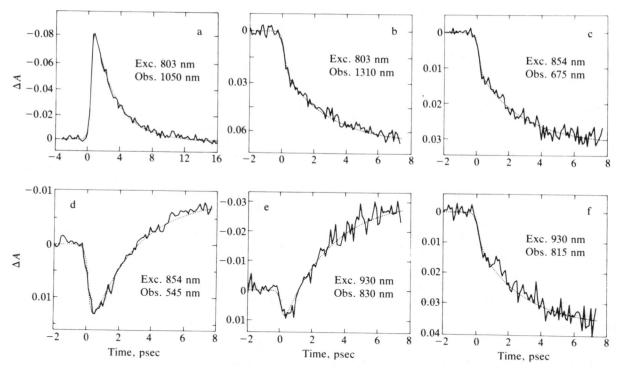

FIG. 2. (a) Kinetics of stimulated emission at 1050 nm. The relaxation phase is best fitted (·····) with a single exponential decay of 2.8 psec. Exc., excitation; Obs, observation. (b–f) Different kinetics at the indicated wavelengths demonstrating an instantaneous induced absorption relaxing in 2.8 psec superimposed upon an absorption increase (b, c, f) or decrease (d, e) developing with a 2.8-psec time constant. The asymptotic values for the kinetics b–f have been determined on a full scale of 20 psec.

molecular model of the reaction center of *Rps. viridis* (1). Although in this system the distances and relative orientations between the various pigments are now well characterized, the precise assignment of the various transitions underlying the 830-nm band and the degree of electronic coupling of the chromophores is still debated (11, 17, 18). The resolution of four distinct absorption bands at very low temperature in the 830-nm spectral range (11) should permit a more precise characterization of the kinetics of energy transfer amongst the various spectral components. However, to conclude which mechanism of energy transfer is dominant in this system, Förster or electron exchange, it would be necessary to know other parameters such as the overlap between the emission spectrum of the donor and the absorption spectrum of the acceptor as well as the precise orbital overlaps.

Several picosecond measurements on *Rps. sphaeroides* reaction centers have been previously interpreted in terms of finite energy transfer steps among the various pigments. For example, Moskowitz and Malley (15) have estimated a time

constant of ≈10 psec for the energy transfer from H to P, while Akhmanov *et al.* (16) have proposed values that are one order of magnitude shorter. More recently, Woodbury *et al.* (7) have also proposed an energy transfer step occurring within about 1.5 psec between the B molecules and P in order to explain a transient bleaching of the 800-nm band of B observed upon excitation at 610 nm. The <100-fsec transfer between H or B and P in *Rps. viridis* determined in the present study is thus significantly shorter than the values or upper limits previously proposed. These faster kinetics observed here cannot be due to a difference in the bacterial species used, as we have also observed ultrafast energy transfer to P in *Rps. sphaeroides* upon excitation in the H and B molecules at 760 nm and at 800 nm, respectively (19).

The Initial $P^+H_A^-$ Charge Separation. The instantaneous absorbance increase observed at a variety of wavelengths (Fig. 2 b–f) is assigned to an excited state of P, called P^* for simplification (6, 7). This state could correspond to a quantum mechanical mixture of the pure electronically excited state and of an internal charge transfer state $P^\pm$ within the

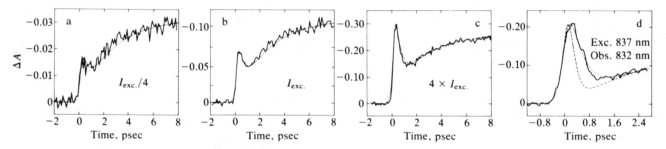

FIG. 3. (a–c) Kinetics of induced bleaching observed at 850 nm upon excitation at 827 nm. The intensity of the excitation, $I_{exc.}$, has been varied over a factor of 16, demonstrating the change of shape due to the saturation of the formation of the state $P^+H_A^-$ (2.8-psec component). The fast transient bleaching, whose amplitude is linear with excitation, is best fitted (·····) with a 400-fsec relaxation time. (d) Kinetics of induced bleaching observed at 832 nm upon excitation at 837 nm. In addition to the 2.8-psec phase of bleaching the fits include a fast transient bleaching that decays with a 400-fsec (·····) or a 100-fsec (- - -) time constant. Exc., excitation; Obs, observation.

dimer making up P. Alternatively, P* might correspond to a charge transfer state P± generated from the electronically excited state. These interpretations are consistent with recent reports from photochemical hole-burning experiments (20, 21), which are interpreted as indicating an ultrafast (15- to 25-fsec) relaxation of the initial singlet excited state of P. The stimulated emission observed within the long-wavelength band of P is assigned to the state P* (6, 7) and its kinetics of decay can be best measured at 1050 nm, which represents an isosbestic point for the P − P+ transition (Fig. 2a). An instantaneous rise followed by a monoexponential 2.8-psec decay has been observed at 1050 nm upon excitation with pulses of various wavelengths and energies (see *Results*). Thus, provided the fast transient bleachings discussed in the following section are taken into account, the decay of P* and the generation of the ionized species can be monitored upon excitation either directly in P or within the 830-nm band.

The similarity of the kinetics of the absorbance changes at 1310 nm (Fig. 2b) and at 675 nm (Fig. 2c) demonstrates that P is oxidized simultaneously to the reduction of H_A with a time constant of 2.8 psec, which also corresponds to the decay of P*. This is confirmed by the 2.8-psec phase of the kinetics of the absorbance decrease at 545 nm (Fig. 2d) and the same kinetics for the blue shift of the 830-nm band (Fig. 2 e and f). The involvement of a transient state such as $P^+B_A^-$ in the electron transfer pathway between P* and $P^+H_A^-$ should manifest itself as a transient bleaching in the absorption band of B around 830 nm. Thus, the absorbance increase observed in the 830-nm band upon excitation at 930 nm (Fig. 2 e and f) excludes the participation of B_A as a distinct intermediary electron acceptor operating in the 100-fsec to 2.8-psec time range—i.e., between the appearance of P* and the reduction of H_A.

Fast Transient Bleaching Associated with Excitation and Detection Within the 830-nm Band. Fast transient bleachings can be expected before the excited species H* or B* transfer their excitation energy to P. However, while we know from Fig. 1b that this transfer should not take more than 100 fsec, it is also clear (Fig. 3 b–d) that some of the transient bleaching of B* decays more slowly than this. If we assume that only one relaxation is involved, the best fit gives a 400-fsec decay time, which cannot correspond to the decay of the main precursor of the P* state (Fig. 1b). Thus, we have to conclude that at least two parallel processes are occurring. These observations suggest that most of the states B*P decay to BP* in a time $\tau_1 < 100$ fsec, while a small fraction of B* enters a competitive channel and relaxes back in one or more steps to the ground state of B in a time $\tau_2 \approx 500$ fsec. Although the signal-to-noise ratio in Fig. 1b does not allow us to exclude that about 10% of P* is generated in ≈500 fsec, we note that a quantum yield of P+ formation $r = 0.93$ has been reported for *Rps. sphaeroides* reaction centers excited at 800 nm in the band of B, compared to a yield of essentially 1.0 when P is excited directly (22). Using 800-nm excitation, we have observed the same transient bleaching in *Rps. sphaeroides* (19) as reported here for *Rps. viridis* reaction centers. This decreased yield compared to that observed upon direct excitation of P could correspond to the small loss of B* described in our scheme provided $\tau_1/\tau_2 = 1 - r$. The fast transient bleaching would then include the contributions (i) of a small fraction of the B* population relaxing to the B ground state in about 500 fsec and (ii) of most of the B* states transferring to P in ≈50 fsec.§ Due to the fact that the kinetics are measured with pulses longer than this characteristic time, the maximum amplitude of the ≈50-fsec contribution is

§The limits for the latter time constant are estimated from a model involving the branching ratio described above, which gives a lower limit of 35 fsec, and from our experimental time resolution, which gives an upper limit of 100 fsec.

attenuated by roughly a factor of 6, while the 500-fsec component is almost not affected. With 5–10% of B* participating in the 500-fsec kinetics, this would lead to an apparent bleaching of 20–25% of the absorption of B. However, taking into account the absorption cross-sections at 850 and 960 nm for the states P (Fig. 1a) and P+ (9, 23), the amplitude of the fast transient bleaching at 850 nm (Fig. 3 a–c) can be estimated at ≈30% of the absorption of the fraction of excited reaction centers at this wavelength. Thus, even without considering a probable component of absorption from P at 850 nm (11, 23), it seems very likely that the transient signal also includes a contribution of stimulated emission from B*. This effect also would explain the absence of a fast transient bleaching upon excitation at 854 nm and observation at 830 nm (see *Results*), while a large transient bleaching is observed upon the almost symmetrical condition of excitation at 827 nm and observation at 850 nm (Fig. 3 a–c).

An alternative to the competitive channel scheme discussed above would be a situation in which a small fraction (≈10%) of photooxidized reaction centers is present. However, this interpretation can be ruled out in view of the observation of this transient relaxing in ≈400 fsec even when the redox state of the reaction centers is strongly modified by chemical oxidation ($P^+H_AQ_A$), chemical reduction ($PH_AQ_A^-$), and photochemical reduction ($PH_A^-Q_A^-$). This striking observation (unpublished data) can be rationalized with our model, which primarily involves excited states of B and not of P or H. Furthermore the ultrafast energy transfer B*P → BP* offers a plausible explanation of the different intensity dependence of the two phases seen in Fig. 3 a–c if the decreasing yield of photochemistry observed at high intensity is due to the absorption of 830-nm light by reaction centers that have already entered the state BP* during the pulse.

As already mentioned, the fast transient bleaching of B* relaxing in ≈400 fsec has been detected in both *Rps. viridis* and *Rps. sphaeroides* reaction centers (19). It is thus justified to ascribe the transient bleaching around 800 nm reported upon excitation of *Rps. sphaeroides* reaction centers at 610 nm with 0.7- to 0.8-psec pulses (7, 24) to the same effect as discussed here. In both types of reaction centers the 600-nm region corresponds to the Q_X transitions of the four bacteriochlorophylls. A bleaching of the Q_Y transitions of the B molecules should thus be accompanied by a corresponding bleaching of their Q_X transitions. This constitutes in our view the simplest interpretation of the fast transient bleaching at 620 nm reported by Zinth *et al.* (2) for *Rps. viridis*.

CONCLUSIONS

Using a pulse–probe technique featuring a 100-fsec resolution and tunability of the excitation in the near-infrared, we have investigated the excitation energy transfer and the initial charge separation occurring in the reaction center of the photosynthetic bacterium *Rps. viridis*. Energy transfer from H to B and then to P, occurring on an ultrafast time scale, leads to the formation of the state P* in slightly less than 100 fsec. This state, characterized by a broad absorption spectrum, is capable of stimulated emission and decays in 2.8 ± 0.2 psec with formation of the radical pair $P^+H_A^-$. Upon direct excitation of P at 930 nm no fast transient bleaching is observed around 830 nm, thus excluding any significant contribution of the state $P^+B_A^-$ as a spectrally or kinetically resolvable intermediate. This conclusion contradicts the reaction scheme recently proposed by Zinth *et al.* (2).

Upon excitation in the H or B absorption bands the initial charge separation also proceeds from the state P* and its kinetics is unaffected when the excitation energy is varied by more than one order of magnitude. However, under these conditions, a large transient bleaching recovering in 400 ± 100 fsec is detected in the spectral range 830–850 nm and could be mistakenly interpreted as a transient state such as $P^+B_A^-$. The magnitude and spectral characteristics of this transient bleaching, as well as the observation that it is essentially unaffected when the redox state of P, Q_A, and H_A is altered, lead us to propose that it is due to the combination of a fast (≈ 50 fsec) component of energy transfer $B^*P \rightarrow BP^*$ that affects most of the B molecules and of a slower (≈ 500-fsec) relaxation to the ground state that affects a much smaller fraction of B^*, both components probably being enhanced by stimulated emission. Additional information on a possible difference in the involvement of B_A and B_B in this process as well as on the kinetics of energy transfer from H_A or H_B to B and P might be obtained by kinetic measurements at low temperature.

An analogous fast transient bleaching has also been observed around 800 nm in the reaction center from *Rps. sphaeroides*. More generally, the excitation energy transfer among H, B, and P as well as the characteristics of the initial charge separation (kinetics, nature of the ionized species involved) appear identical in the reaction centers of *Rps. sphaeroides* (6, 19) and of *Rps. viridis*. These conclusions thus further strengthen the proposal (11) that the geometrical organization of the chromophores is essentially the same in the reaction center of these two organisms.

We acknowledge stimulating discussions with W. W. Parson and thank S. Andrianambinintsoa for preparing the reaction centers. This work was supported by the Institut National de la Santé et de la Recherche Médicale, the Ecole Nationale Supérieure des Techniques Avancées, the Ecole Polytechnique, and the Agence Française pour la Maîtrise de l'Energie.

1. Deisenhofer, J., Epp, O., Miki, K., Huber, R. & Michel, H. (1984) *J. Mol. Biol.* **180,** 385–398.
2. Zinth, W., Nuss, M. C., Franz, M. A., Kaiser, W. & Michel, H. (1985) in *Antennas and Reaction Centers of Photosynthetic Bacteria,* ed. Michel-Beyerle, M. E. (Springer, Berlin), pp. 286–291.
3. Shuvalov, V. A. & Klevanik, V. A. (1983) *FEBS Lett.* **160,** 51–55.
4. Borisov, A. Y., Danielus, R. V., Kudzmauskas, S. P., Piskarskas, A. S., Razjivin, A. P., Sirutkaitis, V. A. & Valkunas, L. L. (1983) *Photobiochem. Photobiophys.* **6,** 33–38.
5. Kirmaier, C., Holten, D. & Parson, W. W. (1985) *FEBS Lett.* **185,** 76–82.
6. Martin, J.-L., Breton, J., Hoff, A. J., Migus, A. & Antonetti, A. (1986) *Proc. Natl. Acad. Sci. USA* **83,** 957–961.
7. Woodbury, N. W., Becker, M., Middendorf, D. & Parson, W. W. (1985) *Biochemistry* **24,** 7516–7521.
8. Netzel, T. L., Rentzepis, P. M., Tiede, D. M., Prince, R. C. & Dutton, P. L. (1977) *Biochim. Biophys. Acta* **460,** 467–479.
9. Holten, D., Windsor, M. W., Parson, W. W. & Thornber, J. P. (1978) *Biochim. Biophys. Acta* **501,** 112–126.
10. Deisenhofer, J., Epp, O., Miki, K., Huber, R. & Michel, H. (1985) *Nature (London)* **318,** 618–624.
11. Breton, J. (1985) *Biochim. Biophys. Acta* **810,** 235–245.
12. Parson, W. W. (1982) *Annu. Rev. Biophys. Bioeng.* **11,** 57–80.
13. Martin, J.-L., Migus, A., Poyart, C., Lecarpentier, Y., Astier, R. & Antonetti, A. (1983) *Proc. Natl. Acad. Sci. USA* **80,** 173–177.
14. Davis, M. S., Forman, A., Hanson, L. K., Thornber, J. P. & Fajer, J. (1970) *J. Phys. Chem.* **83,** 3325–3332.
15. Moskowitz, E. & Malley, M. M. (1978) *Photochem. Photobiol.* **27,** 55–59.
16. Akhmanov, S. A., Borisov, A. Y., Danielus, R. V., Kozlovskij, V. S., Piskarskas, A. S. & Ravzjivin, A. P. (1978) in *Picosecond Phenomena,* eds. Shank, C. V., Ippen, E. P. & Shapiro, S. L. (Springer, Berlin), pp. 134–139.
17. Parson, W. W., Scherz, A. & Warshel, A. (1985) in *Antennas and Reaction Centers of Photosynthetic Bacteria,* ed. Michel-Beyerle, M. E. (Springer, Berlin), pp. 122–130.
18. Zinth, W., Knapp, E. W., Fischer, S. F., Kaiser, W., Deisenhofer, J. & Michel, H. (1985) *Chem. Phys. Lett.* **119,** 1–4.
19. Breton, J., Martin, J.-L., Migus, A., Antonetti, A. & Orszag, A. (1986) in *Ultrafast Phenomena V,* eds. Fleming, G. R. & Siegman, A. E. (Springer, Berlin), in press.
20. Meech, S. R., Hoff, A. J. & Wiersma, D. A. (1985) *Chem. Phys. Lett.* **121,** 287–292.
21. Boxer, S. G., Lockhart, D. J. & Middendorf, T. R. (1986) *Chem. Phys. Lett.* **123,** 476–482.
22. Wraight, C. A. & Clayton, R. K. (1973) *Biochim. Biophys. Acta* **333,** 246–260.
23. Paillotin, G., Vermeglio, A. & Breton, J. (1979) *Biochim. Biophys. Acta* **545,** 249–264.
24. Holten, D., Hoganson, C., Windsor, M. W., Schenck, C. C., Parson, W. W., Migus, A., Fork, R. L. & Shank, C. V. (1980) *Biochim. Biophys. Acta* **592,** 461–477.

Dynamics of Ligand Binding to Myoglobin[†]

R. H. Austin, K. W. Beeson, L. Eisenstein, H. Frauenfelder,[*] and I. C. Gunsalus

ABSTRACT: Myoglobin rebinding of carbon monoxide and dioxygen after photodissociation has been observed in the temperature range between 40 and 350 K. A system was constructed that records the change in optical absorption at 436 nm smoothly and without break between 2 μsec and 1 ksec. Four different rebinding processes have been found. Between 40 and 160 K, a single process is observed. It is not exponential in time, but approximately given by $N(t) = (1 + t/t_0)^{-n}$, where t_0 and n are temperature-dependent, ligand-concentration independent, parameters. At about 170 K, a second and at 200 K, a third concentration-independent process emerge. At 210 K, a concentration-dependent process sets in. If myoglobin is embedded in a solid, only the first three can be seen, and they are all nonexponential. In a liquid glycerol–water solvent, rebinding is exponential. To interpret the data, a model is proposed in which the ligand molecule, on its way from the solvent to the binding site at the ferrous heme iron, encounters four barriers in succession. The barriers are tentatively identified with known features of myoglobin. By computer-solving the differential equation for the motion of a ligand molecule over four barriers, the rates for all important steps are obtained. The temperature dependences of the rates yield enthalpy, entropy, and free-energy changes at all barriers. The free-energy barriers at 310 K indicate how myoglobin achieves specificity and order. For carbon monoxide, the heights of these barriers increase toward the inside; carbon monoxide consequently is partially rejected at each of the four barriers. Dioxygen, in contrast, sees barriers of about equal height and moves smoothly toward the binding site. The entropy increases over the first two barriers, indicating a rupturing of bonds or displacement of residues, and then smoothly decreases, reaching a minimum at the binding site. The magnitude of the decrease over the innermost barrier implies participation of heme and/or protein. The nonexponential rebinding observed at low temperatures and in solid samples implies that the innermost barrier has a spectrum of activation energies. The shape of the spectrum has been determined; its existence can be explained by assuming the presence of many conformational states for myoglobin. In a liquid at temperatures above about 230 K, relaxation among conformational states occurs and rebinding becomes exponential.

1. Myoglobin and Its Ligands

Myoglobin (Mb),[1] a globular protein of about 17200 molecular weight and 153 amino acids containing one protoheme, plays an important role in the mammalian cell where it stores' (Theorell, 1934) and transports (Wittenberg, 1970) oxygen and possibly also carries energy (Hills, 1973). An understanding of the reactions of ferrous Mb with ligands, particularly dioxygen and carbon monoxide, is desirable because, as the simplest protein capable of reversible oxygenation, it can serve as a prototype for more complex systems. Dynamic studies are particularly meaningful be-cause the primary and tertiary structures have been determined (Kendrew et al., 1958) and many properties of the active center are known (Weissbluth, 1974).

The reactions of various ligands with Mb have been investigated extensively with stopped-flow, flash-photolysis, and T-jump techniques. The pioneering work has been performed by Gibson (Gibson, 1956); his and later experiments are reviewed and referenced in the monograph by Antonini and Brunori (Antonini and Brunori, 1971). Our study of the binding of O_2 and CO to sperm whale Mb by flash photolysis extends earlier work in three directions, temperature, time, and dynamic range. Phenomena change so rapidly with temperature that measurements are needed at 10-K intervals between 40 and 350 K. Since processes can encompass more than nine orders of magnitude in time, we constructed a system capable of recording over this range in one sweep. Our equipment records data over more than three orders of magnitude in optical density and we thus can see even processes with relative intensities of less than 1%.

The experimental data are rich and complex but they can be unraveled to give a coherent description of the dynamics of ligand binding to Mb. The essential *experimental* fact is the discovery of four distinct processes (Austin et al., 1973). The central *interpretative* assumption is that ligand binding is governed by successive barriers (Frauenfelder, 1973). The four processes depend differently on temperature and concentration; thus all essential parameters of our model can be determined and all observed features can be understood.

[†] From the Department of Physics and the Department of Biochemistry, University of Illinois at Urbana—Champaign, Urbana, Illinois 61801. *Received August 7, 1974.* This work was supported in part by the U.S. Department of Health, Education, and Welfare under Grants No. GM 18051 and No. AM 00562, and the National Science Foundation under Grant No. GB 41629X.

[*] To whom correspondence should be addressed at the Department of Physics.

[1] Abbreviations, symbols, and units: Mb, ferrous sperm whale myoglobin; L, ligand molecule; PVA, poly(vinyl alcohol). Processes I–IV are defined in section 4, barriers I–IV and wells A–E in Figure 11. $N(t)$ denotes the fraction of Mb molecules that have not rebound a ligand at the time t after photodissociation. $N_a(t)$, for instance, gives the probability of finding a ligand molecule in well A at time t, k_{ab} is the first-order rate for transitions from well A to B. Second-order rates are denoted by primes, for instance k_{ed}' (eq 6). E_{ab} and A_{ab}, for instance, denote the activation energy and the frequency factor for the transition from well A to B. Energies are given in kcal/mol; 1 kcal/mol = 0.043 eV = 4.18 kJ/mol. Entropies are given in terms of the dimensionless ratio S/R, where $R = 1.99$ cal mol^{-1} K^{-1} is the gas constant.

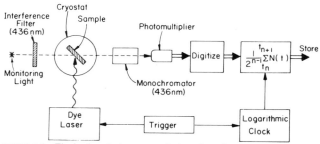

FIGURE 1: Flash photolysis system designed to observe processes that extend over many decades in time.

Preliminary data indicate that the processes discussed here for Mb occur also in other heme proteins, namely hemoglobin, the separated hemoglobin monomers, cytochrome P450, and carboxymethylated cytochrome c. Thus our model may well apply to many biomolecules. We will show that the rate constants for ligand binding at physiological temperatures are determined by multiple barriers whose characteristics can be separated only if photodissociation is studied in the entire temperature range from about 40 to 320 K. Low-temperature experiments are thus necessary to understand the physiological behavior of heme proteins.

Section 2 describes the experimental approach, section 3 the results. In section 4, we introduce a model capable of explaining all observed features. The data evaluation follows in sections 5–8; activation energies and entropies are collected in section 9. Section 10 provides a summary.

2. Experimental Approach

The idea underlying studies of heme protein dynamics by flash photolysis is simple. Consider Mb with bound ligand L, denoted by MbL. Irradiation with light absorbed by MbL breaks the bond; L dissociates from Mb and later rebinds. Photodissociation and rebinding can be followed optically. The Soret absorption band is at 434 nm for Mb, 423 nm for MbCO, and 418 nm for MbO_2; the absorbancy near 434 nm thus indicates the fraction of Mb molecules free of L. The sample is placed in a cryostat with optical windows and the transmission at 436 nm monitored with a photomultiplier. Photodissociation is initiated with a 590-nm flash of 2 μsec duration from a 0.5-J rhodamine 6G–methanol dye laser. Sample heating by the laser flash is less than 0.5 K at 300 K. A long-pass filter, 540–3000 nm, between laser and sample prevents the laser from disturbing the monitoring at 436 nm.

Customarily the photomultiplier output is fed into a storage oscilloscope and the data are taken from the scope tracing. Since the time bases of oscilloscopes are linear, only a limited range in time is observed after a single flash and data from several flashes must be pieced together for complete coverage. Such an approach is time consuming and introduces errors. We therefore have developed a system for smooth observation of ligand rebinding over nine decades in time (R. H. Austin et al., to be published). The approach and components are explained with Figure 1. A trigger simultaneously fires the laser and starts a "logarithmic clock". The flash illuminates the sample in the cryostat and photodissociates MbL. Rebinding is followed by detecting the transmitted 436-nm beam with a photomultiplier. The light source, a tungsten–iodide lamp, is stabilized by monitoring the intensity with a temperature-controlled photodiode. After triggering, the crystal-controlled logarithmic

clock emits signals in exponentially increasing intervals. The first m intervals have a length Δ, the second m intervals 2Δ, the third $2^2\Delta$, and the nth set of m intervals $2^{n-1}\Delta$. Here m is an integer adjustable from 1 to 10. The photomultiplier output is integrated over time Δ, digitized, and summed over a given interval of length $2^{n-1}\Delta$. The sum is divided by 2^{n-1} and the result stored. Thus, even though the interval length increases exponentially with time, a constant input signal results in a constant output. In the present system, the maximum $n = 24$ and the minimum $\Delta = 2$ μsec. If the first interval is 2 μsec, the longest interval is $2^{24-1} \times 2$ μsec = 16.8 sec, and the entire measurement extends over $m(2^n - 1)\Delta$ or 336 sec when $m = 10$. Kinetics at longer times can be observed by increasing the length Δ of the basic interval. From the observed intensity as a function of time, the optical density is computed. The result is expressed in terms of $N(t)$, the fraction of Mb molecules that have not rebound a ligand molecule at time t after the flash.

Our system is capable of observing $N(t)$ over an intensity range of more than three orders of magnitude. To avoid the inability of most photomultipliers to reproduce rapid and large changes in intensity well, an anti-hysteresis photomultiplier, RCA Type 4837, is used. The transient response to a fast increase in absorbance was tested by switching off a green LED (light emitting diode) with a mercury relay in the presence of a steady light. The overshoot or undershoot after the LED shut-off was less than 0.01%.

Samples were prepared from Sigma Type 2 sperm whale Mb, dissolved in 0.1 M phosphate (pH 7.6), and filtered through 0.2-μm filters. (Further purification by ion-exchange chromatography did not affect results.) Three types of ferrous Mb samples were used: buffered aqueous solution, glycerol–water solution, and Mb embedded in poly(vinyl alcohol) (PVA). Reduction was accomplished by adding freshly prepared anaerobic solutions of sodium dithionite to anaerobic samples. Samples were rendered anaerobic by stirring in an argon atmosphere. The dithionite concentration was five times the Mb concentration for CO samples. For O_2, a twofold excess of dithionite was added for initial reduction of Mb, then removed by stirring in the desired O_2 atmosphere. To check for dithionite effects in the oxygen runs an enzymatic reduction system (NADPH, spinach ferredoxin–NADP oxidoreductase, and ferredoxin) was also employed, with identical results.

CO or O_2 was introduced by stirring the samples in a thermostated cell exposed to the desired gaseous atmosphere. For CO equilibration, at 20°C, 10 min of stirring was sufficient for aqueous solutions; the viscous glycerol samples required 1 hr for saturation. The O_2 samples were prepared at 5°C. The oxygenated Mb was stabilized by adding 10 mM EDTA. The aqueous O_2 solutions were equilibrated for 0.5 hr while for glycerol runs a concentrated aqueous Mb sample was oxygenated and then added to the previously equilibrated glycerol, thus keeping equilibration time to a minimum.

The aqueous samples contained 100 mM phosphate buffer (pH 7.6) with a freezing point of 270 K. The glycerol samples contained 3 ml of reagent grade glycerol and 1 ml of 100 mM phosphate buffer (pH 7.6). The PVA samples contained 10% by weight of poly(vinyl alcohol), dissolved in 1 mM phosphate buffer by boiling, and cooled to room temperature. The Mb was then added, reduced, and equilibrated with the desired gas, and the sample was placed in a desiccator under the desired gaseous atmosphere and allowed to dry to a hard film. The process took several days.

Several concentrations of myoglobin were used in the experiments. For temperatures above 200 K, 10 μM Mb was placed in a 1-cm long cell. For measurements extending below 200 K, a 0.5-mm thick cell containing 200 μM Mb minimized problems with cracking of the sample. With such a short cell, light transmission after cracking was reduced by only about a factor 5. Thus a monitoring light level could be chosen that had no measurable effect on the kinetics.

3. Results

Flash photolysis experiments by Chance et al. (Chance et al., 1965) had already shown in 1965 that rebinding of CO to heme proteins after a photoflash can occur at a temperature of 77 K. Our preliminary work (Austin et al., 1973, 1974) revealed three different processes in the temperature range between 40 and 300 K. The present measurements imply four different processes, denoted by I–IV and distinguished by their temperature and concentration dependences.[2] Below about 180 K, only process I appears. At intermediate temperatures, between about 180 and 280 K, up to four different processes can be seen. Above about 280 K, only one process (III or IV, depending on conditions) remains directly observable. Processes I–III are independent of the ligand concentration and IV is proportional to it.

In each experiment, we measure the intensity of the transmitted beam as a function of time and then compute $N(t)$, the fraction of Mb molecules that have not rebound a ligand molecule at the time t after a flash. Usually, log $N(t)$ is plotted vs. t. However, our rebinding data extend over many orders of magnitude in time, and a plot of log $N(t)$ vs. t either compresses the fast or eliminates the slow part. Consequently, nearly all data are presented by plotting log $N(t)$ vs. log t. In a log–log plot a straight line corresponds to a power law; $N(t) \propto t^{-n}$ leads to log $N(t) \propto -n$ log t. An exponential, $N(t) = \exp(-kt)$, leads to log $N(t) = -0.434 \exp[2.30(\log k + \log t)]$ so that the shape of log $N(t)$ vs. log t does not depend on the rate k; k only determines the position along the log t axis.[3] After these preliminary remarks, we present the experimental results. The curves in Figures 2–9 form a small selection from all the data that were taken; additional information, such as the concentration dependence of the various processes, is given in the text.

Low-Temperature Region. Rebinding data for MbCO and MbO$_2$ below 200 K are given in Figures 2a, 2b, and 4a. In taking these data, two precautions are observed. (1) Between any two runs, the sample is warmed up to at least 120 K in order to allow all ligand molecules to rebind before the next photodissociation.[4] (2) The intensity of the monitoring light is adjusted so that ligand molecules that have rebound are not driven off again.

---

[2] The classification introduced here differs from our earlier one (Austin et al., 1973) and is based on more data.

[3] We use the independence of the shape on k to construct a simple tool for rapid preliminary data evaluation. By cutting a cardboard exponential $-0.434 \exp(2.30 \log t)$ and shifting it along the log t axis, we check if a part of the experimental curve can be approximated by an exponential; if yes, the position of the sample exponential gives a preliminary value of k.

[4] If this precaution is neglected, the long-time components do not rebind. The next flash then can drive off only the ligand molecules that have returned quickly; an apparent temperature dependence of the quantum yield and a short recombination time are observed. This effect may explain the discrepancy between our data and those of Iizuka et al. (Iizuka et al., 1974).

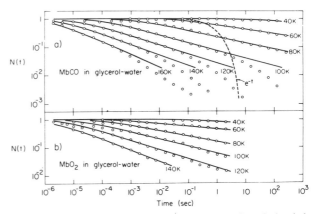

FIGURE 2: (a) Rebinding of CO to Mb after photodissociation below 160 K. The solvent is glycerol-water (3:1, v/v). We denote this low-temperature process by I; the solid lines are least-squares fit to the data with $N(t) = (1 + t/t_0)^{-n}$. At each temperature, the two parameters t_0 and n are determined. The dashed line is an exponential with mean return time 1 sec. Exponentials with other mean times are obtained by shifting the dashed curve along the t axis. (b) Rebinding of O$_2$ to Mb after photodissociation.

Figures 2a, 2b, and 4a show only one rebinding process below about 180 K; we denote it by I. It is independent of the ligand molecule concentration in the solvent and depends only weakly on solvent nature. The shape of $N(t)$ is remarkable. A sample exponential is shown in Figure 2; the experimental curves deviate markedly from exponentials and are much closer to power laws. Indeed, for $N(t) > 0.05$, all experimental curves below about 180 K can be approximated by expressions of the form

$$N_0^I(t) = (1 + t/t_0)^{-n} \qquad (1)$$

where parameters t_0 and n change smoothly with temperature. The parameter n gives the slope of $N(t)$ in the "straight" part of the log–log plot shown in Figure 2a; t_0 very approximately gives the time where $N(t)$ breaks away from the horizontal. Computer fits of eq 1 to the observed points are shown in Figure 2 as solid lines; these fits yield values of t_0 and n for each temperature. Comparison of Figure 2a and b indicates that O$_2$ and CO rebind similarly; both ligands display the same nonexponential behavior, but O$_2$ is somewhat faster. At a first glance, the curves in Figures 2a and 4a appear identical. A closer look, however, shows that the solvent affects the shape of $N(t)$ somewhat: the curves in PVA break away more gradually from the horizontal and drop off less rapidly at long times.

In Figure 3, the result of a multiple flash experiment is shown. The sample is kept at 70 K and flashing is repeated every 18 sec, before all CO molecules have rebound.

Intermediate and High Temperature Regions. Figure 4 shows the rebinding of CO to Mb after photodissociation for MbCO embedded in PVA. The sample is solid up to at least 370 K. Below 180 K, only process I is present. Above 180 K, process II emerges and it constitutes the major fraction of $N(t)$ at 230 K. At 240 K, process III appears. As the temperature is further increased, III becomes dominant. All three processes are nonexponential and independent of CO concentration.

Figure 5 concerns MbCO in water. At low temperatures, only partially shown, processes I–III appear as in PVA, nonexponential and CO-concentration independent. Above the melting point, $N(t)$ becomes slower, exponential, and with a return rate that is proportional to the CO concentra-

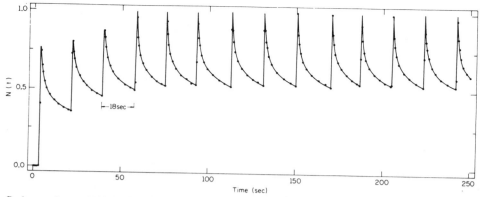

FIGURE 3: Multiple flash experiment: $N(t)$ is plotted vs. time. Sample: MbCO in glycerol–water (3:1, v/v). $T = 70$ K. Flashing is repeated approximately every 18 sec.

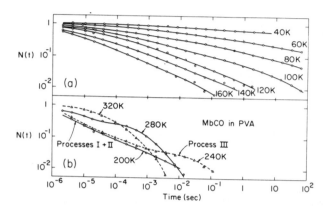

FIGURE 4: Rebinding of CO to Mb after photodissociation. MbCO is embedded in poly(vinyl alcohol); the solvent is solid up to at least 350 K. (a) Low-temperature region, where only process I is visible. (b) At temperatures above 200 K, three processes can be seen; all are nonexponential. Lines are drawn to guide the eye.

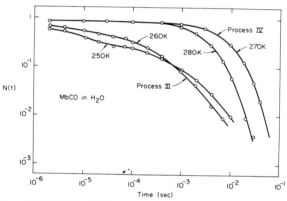

FIGURE 5: Rebinding of CO to Mb in water (ice). Above the melting point, 270 K, the ligand concentration-proportional and exponential process IV is seen. Below the freezing point, diffusion into the solvent is inhibited and the faster, concentration-independent and nonexponential, process III appears. At lower temperatures, I and II, not shown here, are similar to the corresponding processes in Figures 2 and 4. [CO] = 10^{-4} M.

tion in the solvent after the flash. We identify this process with IV.

Figure 5 suggests that process IV, which implies ligand diffusion into the solvent, takes place if the surrounding of Mb is liquid. More information on this point is obtained by observing a PVA sample during solidification (Figure 6). Immediately after preparation, while the sample is still liquid, the concentration-dependent exponential process IV is present. After about 2 hr, IV vanishes and is replaced by the faster and concentration-independent process III. The sample at this point is still liquid and III is exponential. If the sample is prevented from further drying, III remains exponential. If the PVA sample is hardened by drying, III becomes nonexponential as indicated in Figure 6 and also in Figure 4b.

To study process IV in more detail, we use solvents with low melting (glass) points, for instance glycerol–water (3:1 by volume) as in Figure 7. Here, only I exists up to about 180 K. Between 180 and 200 K, I and II are present simultaneously. At about 200 K, III sets in, and at about 210 K, IV appears. I–III are independent, IV is proportional to CO concentration. The decomposition of $N(t)$ at 215 K into four components, done by computer, is shown in Figure 8. The curves I–IV do not depend on the glass transition temperature of the glycerol–water solvent (Bohon and Conway, 1972).

So far, we have concentrated on CO. Figure 9 gives rebinding data for O_2 to Mb in glycerol–water. Only processes I, II, and IV can be seen clearly; I and II are independent of, IV is proportional to, the O_2 concentration.

4. A Model

The experimental curves in the previous section demonstrate that ligand binding to Mb is complex. To understand and evaluate the data, a model is needed. We have constructed one that reproduces all experimental data and can be correlated with structural features of Mb. The salient features are multiple energy barriers, interplay of entropy and enthalpy, presence of an energy spectrum, and occurrence of conformational relaxation. It is possible that equivalent models exist, but we believe that they will contain the same central features.

The data in section 3 show four processes in the rebinding of ligands to Mb after photodissociation.[5] Processes I–III are independent of ligand concentration. IV is proportional to the ligand concentration and slower than a diffusion-limited process. The rates of all four processes are thus most likely governed by barriers. Four barriers are needed to produce four distinct processes. They can be in series, parallel,

---

[5] The number four is not universal. In some other proteins, we find fewer processes; in MbCO, there is evidence for a fifth one. It is likely that further improvements in equipment will lead to the discovery of additional processes.

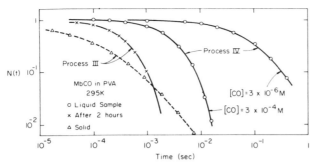

FIGURE 6: Rebinding of CO to Mb in PVA. The curves denoted by IV are obtained in a freshly prepared sample that is still liquid. After about 2-hr drying time, IV disappears and the exponential, but CO independent and faster, process III appears. On complete drying, when PVA becomes solid, the nonexponential curve III is observed.

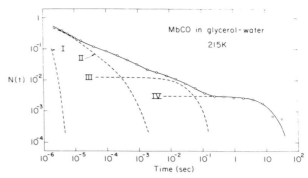

FIGURE 8: Separation of $N(t)$ for $T = 215$ K into the components I–IV. MbCO in glycerol-water, 3:1, v/v. [CO] $= 3 \times 10^{-5}$ M. The figure also shows some tentative evidence for a fifth process between II and III.

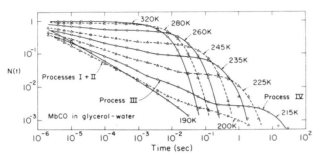

FIGURE 7: Rebinding of CO to Mb after photodissociation. Solvent: glycerol-water, 3:1, v/v. [CO] $= 3 \times 10^{-5}$ M. All four processes discussed in the text are recognizable.

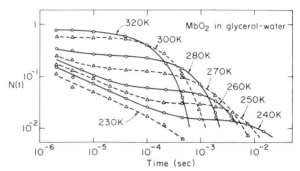

FIGURE 9: Rebinding of O_2 to Mb. Solvent: glycerol-water, 3:1, v/v. [O_2] $= 6 \times 10^{-5}$ M. Processes I, II, and IV are indicated; III is not very prominent and is neglected.

or mixed; they can be caused by different parts of the system biomolecule–solvent, or by conformational transformations. We postulate four barriers in sequence.

Four barriers in sequence can be discussed without referring to the actual physical system, but we find it constructive to think in terms of a concrete model, as sketched in Figure 10. The structure of Mb shows the prosthetic group, protoheme, in a pocket of the protein with the iron approximately 1 nm from the protein surface (Watson, 1968). The pocket is lined with nonpolar groups; the polar residue of the proximal histidine His-64 is close to the heme iron. The protein is surrounded by the hydration shell, a layer of water molecules about 0.4 nm thick with properties different from bulk water (Cooke and Kuntz, 1974, Kuntz and Kauzmann, 1974). The Mb-solvent system thus can be pictured as in Figure 10. This figure makes it plausible that a ligand molecule on its way to the binding site from the solvent can encounter multiple barriers. We will discuss the properties of the four barriers and a possible identification in section 10; here we assume that the potential (enthalpy) seen by a ligand molecule on its approach to the iron looks as in Figure 11. The abscissa represents the reaction coordinate. In each well, the ligand thermalizes before making the next move.

The processes that a ligand molecule can undergo can be described with Figures 10 and 11. Entering from the solvent corresponds to *association*. In *dissociation* a ligand molecule initially bound to the iron atom in well A moves to the outside by thermally overcoming all barriers. In *photodissociation*, L initially also occupies well A. An incident photon excites the iron atom into an antibonding state, changing the attractive well A into a repulsive one (Zerner et al., 1966), and L moves into well B. Processes I–IV in section 3

can also be interpreted with the help of Figures 10 and 11. Immediately after photodissociation, L is in well B and the following processes can take place: I, L rebinds directly from B to A; II, L jumps over barrier II to well C; from there it rebinds by first returning to B and then to A; III, L jumps to D via C and then rebinds, D → C → B → A. IV, L jumps to E via C and D and diffuses into the solvent. All ligand molecules in the solvent then compete for the vacant binding site via the chain E → D → C → B → A. In this interpretation, processes I–III are independent of ligand concentration, IV is proportional to it. One feature observed in section 3 is thus reproduced. A second striking feature, the appearance of only one process below about 180 K, is intuitively plausible. At very low temperatures, the Mb molecule is "frozen shut" and only intramolecular processes occur. At high temperatures, the ligand can leave the Mb molecule. To understand this feature in more detail, consider the two-barrier situation obtained in Figure 11 if barrier III is very high. A ligand in well B then can either rebind directly (process I), or first move to C and then rebind via B (process II). We assume that all rates obey Arrhenius equations

$$k(T) = A e^{-E/RT} \qquad (2)$$

A is a frequency factor, E an activation energy, and $R = 1.99$ cal mol^{-1} K^{-1} the gas constant. A and E are taken to be temperature independent. The ratio of transition rates from B to A and C is now determined by

$$\frac{k_{ba}}{k_{bc}} = \frac{A_{ba}}{A_{bc}} e^{(E_{bc} - E_{ba})/RT} \qquad (3)$$

Here, k_{ba}, for instance, denotes the transition rate from B

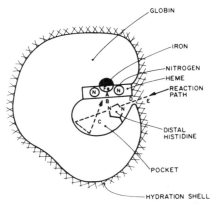

FIGURE 10: Binding of a ligand molecule to myoglobin. A possible reaction path is indicated by the dashed line. A tentative identification of the barriers and wells with known structural features of Mb is given in subsection 10.2.

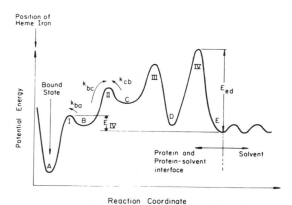

FIGURE 11: Potential encountered by a ligand molecule approaching the heme iron from the solvent. The barriers are numbered I–IV, the wells labeled A–E. The energy E_{ed}, for instance, measures the activation energy for the step E → D. The energy E_{IV} is defined by eq 48.

to A, E_{ba} is the activation energy, and A_{ba} the frequency factor for the same transition. The data evaluation in sections 5–8 indicates that

$$A_{ba}/A_{bc} \ll 1, \quad E_{bc} > E_{ba} \quad (4)$$

Consequently, k_{ba}/k_{bc} becomes large at low and tends to zero at high temperatures. At low temperatures, ligand molecules in pocket B rebind directly; at higher temperatures, they make the round trip via well C before rebinding. This argument thus can explain why only one process is present below about 180 K. To describe the occurrence of all four processes, the treatment must be extended to four barriers. A detailed comparison between model and experiment then requires the solution of the four coupled differential equations that describe the motion of a particle over the four barriers shown in Figure 11:

$$dN_a/dt = -k_{ab}N_a + k_{ba}N_b$$

$$dN_b/dt = k_{ab}N_a - k_{ba}N_b \cdots \cdots {}_bN_c$$

348 Experimental Techniques

$$\cdot N_d \quad (5)$$

$${}_lN_e$$

$$dN_e/dt = k_{de}N_d - k_{ed}N_e$$

Here, N_a denotes the probability of finding a ligand molecule in well A at time t and k_{ab} is the rate for transitions from well A to well B. $N(t)$, the quantity defined in section 3 as the fraction of Mb molecules that have not rebound a ligand, is given by $N(t) = 1 - N_a$. The equations are valid if the ligand concentration [L] in the solvent is sufficiently large so that process IV can be treated as a pseudo-first-order reaction and k_{ed} is a pseudo-first-order rate. Otherwise, k_{ed} must be replaced:

$$k_{ed} \rightarrow k_{ed}'[L] \quad (6)$$

where k_{ed}' is a second-order rate. We will restrict our discussion to the limiting case of high ligand concentrations.

Photodissociation moves the ligand molecules from well A to B so that

$$N_b(0) = 1, \quad N_a(0) = N_c(0) = N_d(0) = N_e(0) = 0 \quad (7)$$

In general, the solution of eq 5 must be done numerically. Dissociation is slow at all temperatures and satisfies $k_{ab} \ll k_{ba}$ so that we always set $k_{ab} = 0$ in eq 5. In the following sections we use eq 5 to deduce the various rates from the ex-

perimental data. The temperature dependence of the rates yields the activation energies and frequency factors of the four barriers. The approach rests on the crucial assumption that the Mb molecule does not undergo a major conformational transformation in the temperature range between 40 and 300 K, except for relaxation effects to be described in section 7. The fact that we can fit all data with a consistent set of energies and entropies supports the assumption, but further experiments will be needed to prove it unambiguously.

5. Direct Rebinding and Activation Energy Spectrum

At temperatures below about 180 K, only one process, identified with the direct return B → A (Figure 11), is observed. Direct rebinding implies $k_{bc} \ll k_{ba}$ and eq 5 reduces to a one-barrier problem:

$$dN_a/dt = k_{ba}N_b = k_{ba}(1 - N_a) \quad (8)$$

where we have assumed $k_{ab} = 0$. The solution of eq 8 with the initial condition eq 7 is an exponential:

$$N(k_{ba},t) = 1 - N_a = e^{-k_{ba}t} \quad (9)$$

The curves in Figures 2 and 4, however, are not exponentials, but closer to power laws, as expressed by eq 1. The observed shape and temperature dependence of these curves are explained (Austin et al., 1974) if the energy of barrier I is not sharp, but given by a distribution $g(E_{ba})$, where $g(E_{ba})$ denotes the probability of finding a Mb molecule with activation energy between E_{ba} and $E_{ba} + dE_{ba}$. For a distributed activation energy, eq 9 is generalized to read[6]

$$N(t) = \int_0^\infty dE_{ba}g(E_{ba})N(k_{ba},t) =$$

$$\int_0^\infty dE_{ba}g(E_{ba})e^{-k_{ba}t} \quad (10)$$

where k_{ba} is related to the activation energy E_{ba} and the

---

[6] Nonexponential curves have been observed in many fields and their explanation in terms of integrals as in eq 10 dates back to at least 1913 (Wagner, 1913). Detailed theoretical treatments can be found in papers by Macdonald (1962, 1963, 1964) and Primak (1955).

frequency factor A_{ba} by eq 2.[7] We assume here that A_{ba} is independent of E_{ba}. Equation 2 gives $dE_{ba} = -RTdk_{ba}/k_{ba}$ and $N(t)$ becomes

$$N(t) = RT \int_0^{A_{ba}} dk_{ba}[g(k_{ba})/k_{ba}]e^{-k_{ba}t} \quad (11)$$

By eq 2, the energy E_{ba} is related to k_{ba} by $E_{ba} = RT \ln (A_{ba}/k_{ba})$. For a given frequency factor A_{ba}, the distribution function $g(E_{ba})$ thus can be considered a function of k_{ba}. We will write it alternately as $g(E_{ba})$ or $g(k_{ba})$, depending on whether we express it in terms of E_{ba} or k_{ba}. In either case, g is different from the function $f(k_{ba})$ described in footnote 7.

For the times involved in our measurements, A_{ba} satisfies the relation $A_{ba}t \gg 1$. The upper limit of the integral can then be taken to be infinite and $N(t)$ becomes the Laplace transform of $g(k_{ba})/k_{ba}$:

$$N(t) = RT \int_0^\infty dk_{ba} \frac{g(k_{ba})}{k_{ba}} e^{-k_{ba}t} = $$
$$RT\mathcal{L}\{g(k_{ba})/k_{ba}\} \quad (12)$$

From the measured $N(t)$, the distribution function can be determined in terms of k_{ba} by the inverse Laplace transform

$$g(k_{ba}) = (k_{ba}/RT)\mathcal{L}^{-1}\{N(t)\} \quad (13)$$

To find the distribution as a function of the activation energy E_{ba}, A_{ba} must be known. A_{ba} is determined by measuring $N(t)$ at a number of temperatures, as will be shown below.

The inversion eq 13 can be performed analytically if the observed data can be fitted by a function $N(t)$ that can be Laplace inverted. In general, the determination of $g(k)$ or $g(E)$ must be done by computer. An approximate form for $g(k)$ is found by noting that $N_0^l(t)$ in eq 1 represents the observed curves reasonably well over a fair range in time. Inserting $N_0^l(t)$, eq 1, into eq 13 yields

$$g(k_{ba}) = \frac{(t_0 k_{ba})^n e^{-t_0 k_{ba}}}{RT\Gamma(n)} \quad (14)$$

where $\Gamma(n)$ is the gamma function. The peak of this distribution, given by $dg/dk_{ba} = 0$, occurs at

$$k_{ba}^{peak} = n/t_0 \quad (15)$$

By eq 2, the peak energy is related to $k_{ba}^{peak} = n/t_0$ by

$$RT \ln (n/t_0) = RT \ln A_{ba} - E_{ba}^{peak} \quad (16)$$

The parameters n and t_0 are determined at each temperature by fitting eq 1 to the measured curves in Figures 2 and 4. A plot of $RT \ln (n/t_0)$ vs. T is shown in Figure 12 for Mb in a glycerol-water solvent. The slope of the straight line determines A_{ba}, the intercept E_{ba}^{peak}. From Figure 12, and from analogous data for MbCO in PVA, the following values are obtained:

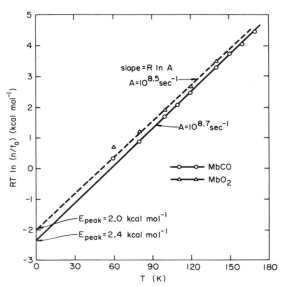

FIGURE 12: Plot of $RT \ln (n/t_0)$ vs. T determines A_{ba} and E_{peak}. MbCO and MbO$_2$ in glycerol-water (3:1, v/v).

$$\text{MbCO(glyc) } E_{ba}^{peak} = 2.4 \text{ kcal/mol}, \quad A_{ba} = 10^{8.7} \text{ sec}^{-1}$$
$$\text{MbO}_2\text{(glyc) } E_{ba}^{peak} = 2.0 \text{ kcal/mol}, \quad A_{ba} = 10^{8.5} \text{ sec}^{-1}$$
$$\text{MbCO(PVA) } E_{ba}^{peak} = 1.4 \text{ kcal/mol}, \quad A_{ba} = 10^{7.7} \text{ sec}^{-1}$$
$$\quad (17)$$

The statistical errors in these numbers are small; variations from sample to sample lead to an error of about ±0.2 kcal/mol in E_{ba}^{peak} and ±0.5 in log A_{ba}. With A_{ba} known, k_{ba} in eq 14 can be expressed in terms of E_{ba} and the desired $g(E_{ba})$ thus is found. This procedure can be performed at all temperatures where t_0 and n have been determined. If eq 1 were a very good approximation at all temperatures, all of these $g(E_{ba})$ should be identical. Figure 2a shows, however, that the fit is not equally good at all temperatures. We therefore use the following approach to find the "best" $g(E_{ba})$. We first determine by visual inspection the temperature at which eq 1 fits the data points best, and then compute $g(E_{ba})$ from eq 14 at this fitting temperature T_f. For all further computer calculations, we use this $g(E_{ba})$. The fitting temperature is 120 K for MbCO in glycerol-water, and 300 K for MbCO in PVA and MbO$_2$ in glycerol-water.

Figure 2 shows that the energy spectra thus determined must still have a shortcoming: eq 1, on which the spectra are based, fits the experimental data only to about $N(t) = 0.05$; for smaller values of $N(t)$, the experimental points fall off more rapidly than eq 1. In fact, below $N \approx 0.05$, $N(t)$ is closer to an exponential than to a power law. An exponential corresponds to one single activation energy and can be approached if the spectral function $g(E_{ba})$ cuts off at a maximum energy E_{ba}^{max}.[8] The cut-off is introduced by setting the upper limit of the integral in eq 10 equal to E_{ba}^{max}. E_{ba}^{max} is then varied till best agreement is obtained with the experimental data at all temperatures. For MbCO in glycerol-water, the cut-off energy becomes $E_{ba}^{max} = 5.4 \pm 0.2$ kcal/mol. Three energy spectra are shown in Figure 13. With these spectra and eq 10 and 17, $N(t)$ can be computed numerically at any desired temperature. As an example,

---

[7] Instead of eq 10, it is also possible to write $N(t) = \int_0^\infty dk_{ba}f(k_{ba})e^{-k_{ba}t}$; $g(E)_{ba}$ is related to $f(k_{ba})$ by $g(E_{ba}) = -f(k_{ba})dk_{ba}/dE_{ba} = k_{ba}f(k_{ba})/RT$. At *each* temperature, $f(k_{ba})$ can be found as the Laplace inverse of $N(t)$, $f(k_{ba}) = \mathcal{L}^{-1}\{N(t)\}$. Such an inversion makes no assumption about the nature of the phenomenon giving rise to the nonexponential rebinding curves; it simply replaces one function, $N(t)$, by another one, $f(k)$. Our approach is more stringent. We must fit all data over a wide temperature range (in PVA from 40 to 350 K) with one temperature-independent energy spectrum.

[8] Of course, an abrupt cut-off as shown in Figure 13 is unnatural; a real cut-off will be smoother. However, the essential features are reproduced by the approach given here.

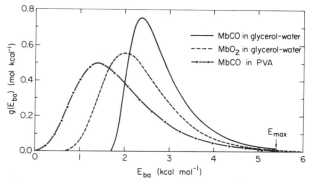

FIGURE 13: Activation energy spectra for MbCO and MbO$_2$.

$N(t)$ curves for MbCO in glycerol–water are shown in Figure 14. Theoretical curves and experimental points agree very well in the entire temperature range where process I can be observed, from 40 to 160 K. MbCO in aqueous solution yields an energy spectrum that is close to MbCO in glycerol–water. The energy spectrum for MbO$_2$ in glycerol–water displays the same general shape. The rebinding curve $N(t)$ for MbCO in PVA in Figure 4a is generally similar to that of MbCO in glycerol–water, but breaks away earlier from the horizontal and curves downward later. A good fit is obtained with the spectrum shown in Figure 14; it has a lower peak energy and higher fitting temperature than MbCO in glycerol–water.

Figure 14 demonstrates that a temperature-independent energy spectrum can explain the nonexponential rebinding curves in the temperature range from 40 to 160 K. The temperature independence of $g(E_{ba})$ can be partially checked in a different way. The rebinding at times $t \gg t_0$ is characterized by rates $k_{ba} \ll 1/t_0$ and the corresponding distribution function $g(k_{ba})$, eq 14, can be approximated by

$$g(k_{ba}) = (t_0 k_{ba})^n / RT\Gamma(n) \quad (18)$$

The spectral function $g(E_{ba})$ that governs rebinding for times $t \gg t_0$ thus becomes

$$g(E_{ba}) = \frac{(A_{ba} t_0)^n}{RT\Gamma(n)} e^{-n E_{ba}/RT}, \quad E_{ba} < E_{ba}{}^{max}$$
$$g(E_{ba}) = 0, \qquad E_{ba} \geq E_{ba}{}^{max} \quad (19)$$

If the energy spectrum is temperature independent, the approximation eq 19 should be temperature independent also and assume, for $E_{ba} < E_{ba}{}^{max}$, the form

$$g(E_{ba}) = \beta e^{-\alpha E_{ba}} \quad (20)$$

Here α and β are two constants defined through eq 20. Comparison of eq 19 and 20 shows that $g(E_{ba})$ can be temperature independent only if n satisfies the relation

$$n = \alpha RT \quad (21)$$

In Figure 15, n is plotted vs. T. Within errors, n is proportional to T and eq 21 is satisfied. The values of α are:

MbCO in: glycerol–water $\alpha = 1.4$ mol/kcal

water $\alpha = 1.4$ mol/kcal

PVA $\alpha = 1.1$ mol/kcal

MbO$_2$ in glycerol–water $\alpha = 1.4$ mol/kcal (22)

The temperature independence of the activation energy spectrum speaks against a frequency-factor spectrum. In such an explanation, E_{ba} would have a unique value, A_{ba} would be described by a spectrum, and eq 10 would be replaced by $N(t) = \int dA_{ba} g(A_{ba}) \exp(-k_{ba} t)$. In order to fit the observed curves, A_{ba} would have to span a range from about 10^5 to 10^{20} sec^{-1} and $g(A_{ba})$ would change with temperature. Consequently we prefer a temperature-independent energy spectrum and keep A_{ba} sharp.[9]

What causes the energy spectrum? Two possibilities come to mind. (a) Each biomolecule possesses a number of sites for L in well B and the rebinding rate depends on the site occupied after the flash-off. (b) Myoglobin does not exist only in one conformation; a given primary sequence gives rise to different conformational states, with different activation energies (Klotz, 1966; Weber, 1972). To decide between these possibilities, we have performed multiple-flash experiments where successive flashes are triggered before all CO molecules have rebound. In case (a), where all Mb molecules are assumed to be identical, each successive flash would pump more ligand molecules into states with long return times, fewer ligand molecules would return fast to the binding site, and the signal would become progressively smaller. In case (b), where each Mb molecule is assumed to have one rate k_{ba}, the molecules with small k_{ba} would be removed from the game by the first flash and only the ones with large k_{ba} and hence short return times would continue to flash off and rebind. The result of a multiple-flash experiment is shown in Figure 3. During the first four flashes, the flash-off increases, indicating that the light intensity is not large enough to remove all CO at once. After the fourth flash, the behavior is repetitive and indicates that no pumping into long-lived states occurs. We consequently favor explanation (b) and assume that we deal with an assembly of Mb conformers, each with a well-defined activation energy.[10] The data do not delineate how many different states exist but the smoothness of the curves in Figures 2 and 4 indicates that there must be more than, say, ten; we assume for simplicity that the number is so large that $g(E_{ba})$ can be treated as continuous.

We finally return to the temperature dependence of the energy spectrum. Figure 14 shows that a temperature-independent $g(E_{ba})$ fits the experimental data very well. An equivalent fit is obtained by assuming the shape of $g(E_{ba})$ to be temperature independent, but that the spectrum shifts with temperature so that $E_{ba}(T) = E_{ba}(0) - S_0 T$ (A. P. Minton, private communication). We consider this possibility unlikely for the following reason. We will show in sections 7 and 8 that transitions from one conformational state to another occur with a rate given by eq 52. Below 160 K, this rate is smaller than 10^{-10} sec^{-1} so that each Mb molecule remains frozen in a given conformational state for a time long compared to our experiments and no shift of the energy spectrum occurs. The distribution of Mb molecules over all possible conformational states is *not* an equilibrium distribution but depends on the thermal history of the sample. We will return to this point in subsection 10.6.

---

[9] We cannot rule out the possibility that both E_{ba} and A_{ba} are continuous. The theory can be extended to this case (Primak, 1955), but we will not use this more general approach, since we can explain all data with our expressions.

[10] Actually, the result of the multiple-flash experiment is more general than stated in the text. It indicates that the activation energy spectrum is caused by a heterogeneity in the ensemble of Mb molecules. This heterogeneity can be due to conformational states, but other explanations are not excluded.

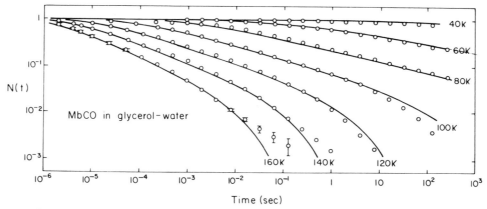

FIGURE 14: Comparison of experimental points and theoretical curves for MbCO in glycerol-water. The experimental points are the same as in Figure 2a. The solid lines are calculated from eq 10 assuming the energy spectrum given in Figure 13. Note the fundamental difference between the solid lines in Figures 2a and 14. In Figure 2a, they represent separate fits at each temperature; in Figure 14, two parameters ($g(E_{ba})$ and A_{ba}) determine all curves at all temperatures.

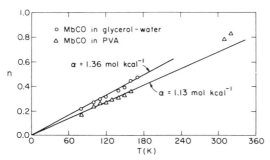

FIGURE 15: Plot of the exponent n as a function of temperature for MbCO in PVA and in glycerol-water. The fact that n follows eq 21 up to at least 200 K implies that the activation energy spectrum is unchanged to at least this temperature.

6. Binding without Diffusion and Relaxation

Between 40 and 180 K, only process I is present and the innermost barrier can thus be studied in detail and without interference from other processes. Investigation of the other barriers is complicated by two facts. All four processes may be present simultaneously, and transitions among the conformational states of Mb may affect the shape of $N(t)$. To unravel the phenomena above 180 K, we first discuss the experiments shown in Figure 4 in which MbCO is embedded in a solid matrix, PVA. Ligand diffusion into the solvent (process IV) and transitions among conformational states (relaxation) are absent. Below 180 K, the properties of barrier I can be found. Above 180 K, processes I–III are present. Since the crucial features of barrier I, $g(E_{ba})$ and A_{ba}, are known, the rates over the barriers II and III can be extracted numerically at each temperature, and the corresponding activation energies and frequency factors can be found. We describe the procedure in subsection 6.1. At temperatures above about 300 K, calculations become simpler and we treat this high-temperature limit in subsection 6.2. In subsection 6.3, we give the main conclusions.

(6.1) Activation Energies and Frequency Factors. Between about 180 and 220 K, it seems as if process I had become approximately independent of temperature, as can be seen in Figure 4. What happens is the subtle emergence of process II. The quantitative evaluation below indicates that process I continues to become faster as the temperature increases, but process II appears and simulates a slowing down of $N(t)$. At about 230 K, process III appears and by

280 K, dominates. The concentration-dependent process IV is absent in the solid matrix so that $k_{de} = 0$. We are thus left with a three-barrier problem, with the innermost barrier characterized by the activation energy spectrum $g(E_{ba})$ and the frequency factor A_{ba}.

The extraction of the rates k_{bc}, k_{cb}, k_{cd}, and k_{dc} at a given temperature from the rebinding curve $N(t)$ is performed by computer. In matrix form and with $k_{ab} = 0$, eq 5 for three barriers ($k_{de} = 0$) is

$$dN/dt = MN \qquad (23)$$

where

$$\mathbf{N} = \begin{pmatrix} N_a \\ N_b \\ N_c \\ N_d \end{pmatrix}$$

$$\mathbf{M} = \begin{pmatrix} 0 & k_{ba} & 0 & 0 \\ 0 & -k_{ba} - k_{bc} & k_{cb} & 0 \\ 0 & k_{bc} & -k_{cb} - k_{cd} & k_{dc} \\ 0 & 0 & k_{cd} & -k_{dc} \end{pmatrix} \qquad (24)$$

The general solution of eq 23 is found by determining the eigenvalues λ_i and eigenvectors $\mathbf{v}_i$ of the eigenvalue equation

$$(\mathbf{M} - \lambda_i \mathbf{I})\mathbf{v}_i = 0 \qquad (25)$$

and expanding

$$\mathbf{N}(t) = \sum_{i=1}^{4} c_i \mathbf{v}_i e^{\lambda_i t} \qquad (26)$$

$\mathbf{I}$ in eq 25 is the unit matrix. The coefficients c_1–c_4 are determined by the initial condition eq 7. The rebinding function is obtained from $\mathbf{N}(t)$ by integrating $(1 - N_a(t))$ with the proper weight $g(E_{ba})$

$$N_{calcd}(t) = \int_0^{E_{ba}^{max}} dE_{ba} g(E_{ba}) [1 - \bar{N}_a(t)] \qquad (27)$$

The bar over $N_a(t)$ implies that an average over the experimental time interval has been taken. For fixed values of the rates k_{bc}, k_{cb}, k_{cd}, and k_{dc}, eq 27 yields $N_{calcd}(t)$. To obtain the actual values of the rates, the computer is instructed to fit $N_{calcd}(t)$ to the experimental points $N(t_i)$ by minimizing

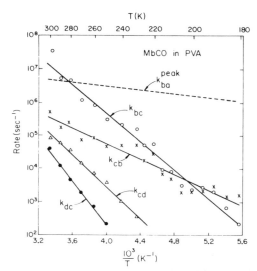

FIGURE 16: Rates for processes I–III for MbCO embedded in a solid PVA matrix. Activation energies and frequency factors deduced from these curves are given in Table II.

$$\chi^2 = \sum_{\text{all } i} \left[\frac{N(t_i) - N_{\text{calcd}}(t_i)}{\Delta N(t_i)} \right]^2. \tag{28}$$

Here $N(t_i)$ is the value of $N(t)$ measured at time t_i, with error $\Delta N(t_i)$. This procedure yields the four unknown rates at all temperatures where processes I–III are involved. The rates are plotted vs. $1/T$ in Figure 16. Slopes and intercepts give the activation energies and frequency factors for MbCO in PVA listed in Table II in section 10. With all relevant parameters of the three barriers known, the rebinding curve $N_{\text{calcd}}(t)$ at any temperature can be calculated from eq 26 and 27. A few such curves are shown in Figure 17, together with experimental points. The theoretical curves agree nearly everywhere within error with the experimental points. At a few temperatures, the experimental behavior is smoother than the theoretical one; a possible explanation will be mentioned in subsection 6.3. The curves in Figure 17 are not separate fits at each temperature; the data at *all* temperatures, from 40 to over 330 K, are faithfully reproduced with one set of activation energies and frequency factors.

(6.2) High-Temperature Limit. The numerical solution of the three-well problem provides us with rates, but not with a deeper understanding of the kinetics of the binding process and the effect of the activation energy spectrum. Fortunately, there exists a limit in which the solution of eq 5 can be found easily, namely if the outbound rates are much larger than the inbound ones:

$$k_{\text{bc}} \gg k_{\text{ba}}; \quad k_{\text{cd}} \gg k_{\text{cb}}, k_{\text{dc}} \tag{29}$$

If these conditions are satisfied, then essentially all ligand molecules placed into well B after a flash move first to well D before ultimately rebinding in well A. Figure 16 shows that eq 29 is approached above about 330 K. We call the situation where essentially all ligand molecules move to the outside before rebinding the *high-temperature limit*.

In the high-temperature limit, a quasi-equilibrium among wells B, C, and D is established rapidly after a flash, with most ligand molecules occupying well D. Rebinding to well A occurs slowly by leakage from B. The situation can be treated by the steady-state approximation where $dN_b/dt = dN_c/dt \approx 0$ and $N_a + N_d = 1$. The approximate solution to eq 5 then becomes with $k_{\text{ab}} = k_{\text{de}} = 0$

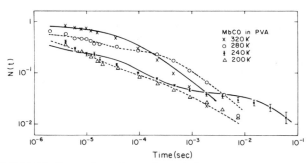

FIGURE 17: Comparison of some theoretical curves with experimental data. The experimental points are from Figure 4, the theoretical curves are computer calculations based on eq 5, the energy spectrum for PVA given in Figure 13, and the activation energies and frequency factors extracted from Figure 16. The experimental points are smoother than the theoretical curves, suggesting that also barrier II could possess an activation energy spectrum.

$$N(k_{\text{ba}}, t) = 1 - N_a(t) = e^{-\lambda_{\text{III}} t} \tag{30}$$

where

$$\lambda_{\text{III}} = k_{\text{ba}} \frac{k_{\text{cb}} k_{\text{dc}}}{k_{\text{bc}} k_{\text{cd}}} \tag{31}$$

For a distributed energy E_{ba}, eq 30 is generalized to

$$N^{\text{III}}(t) = \int_0^\infty dE_{\text{ba}} g(E_{\text{ba}}) e^{-\lambda_{\text{III}} t} \tag{32}$$

The factor $k_{\text{cb}} k_{\text{dc}} / k_{\text{bc}} k_{\text{cd}}$ in λ_{III} is independent of the energy E_{ba} of barrier I so that for fixed barriers II and III

$$\frac{d\lambda_{\text{III}}}{\lambda_{\text{III}}} = \frac{dk_{\text{ba}}}{k_{\text{ba}}} = -\frac{dE_{\text{ba}}}{RT} \tag{33}$$

Inserting $dE_{\text{ba}} = -RT \, d\lambda_{\text{III}}/\lambda_{\text{III}}$ into eq 32 gives as in eq 12

$$N^{\text{III}}(t) = RT \mathcal{L}\{g(\lambda_{\text{III}})/\lambda_{\text{III}}\} \tag{34}$$

In the absence of relaxation and in the high-temperature limit process III should have the same *shape* as process I. Computer calculations bear out this prediction, but it can also be verified approximately in a simpler way. The form of $g(\lambda_{\text{III}})$, without cut-off, is given by eq 14 with k_{ba} replaced by λ_{III}. The Laplace transform then gives

$$N^{\text{III}}(t) \approx (1 + t/t_0^*)^{-n} \tag{35}$$

where the parameter n is the same as for process I. A relation between n, t_0^*, and λ_{III} is obtained as in eq 15 by setting $dg/d\lambda_{\text{III}} = 0$; this condition leads to

$$\frac{n}{t_0^*} = \lambda_{\text{III}}^{\text{peak}} = \frac{k_{\text{ba}}^{\text{peak}} k_{\text{cb}} k_{\text{dc}}}{k_{\text{bc}} k_{\text{cd}}} \tag{36}$$

The parameters n and t_0^* are obtained by fitting eq 35 to the data.

(6.3) MbCO in a Solid Matrix. The results obtained with MbCO in PVA lead to the following conclusions. (i) In a solid matrix, the activation energy spectrum at the innermost barrier determines even the shape of process III, which corresponds to rebinding from the outer well D. (ii) The energy spectrum $g(E_{\text{ba}})$ is essentially unchanged up to at least 320 K. This fact can be seen from the agreement of the experimental curves with the predictions of eq 34. Moreover, the derivation leading to eq 35 implies that the exponent n should be given by eq 21 as $n(T) = \alpha RT$. Values of n for MbCO in PVA, taken from Figure 4 and

shown in Figure 15, indeed lie close to the straight line drawn through the process I data. (iii) Process III, as stated in (i), reflects the activation energy spectrum at the inner barrier. This feature supports our model in which it is assumed that the various barriers are in sequence. If process III led to rebinding by bypassing barrier I, it would be unlikely that it would display the same energy spectrum as process I. (iv) The experimental curves in Figures 4 and 17 are smooth and show little structure. The theoretical curves in Figure 17, in contrast, show small bumps that correspond to processes II and III. A smoothing of the bumps would be expected if not only barrier I, but also one or both of the other barriers were distributed in energy. It will be difficult to explore this aspect in detail. Process I occurs alone over about a factor 4 in temperature and consequently can be studied easily. Processes II and III can only be observed in a much smaller range and they never occur alone; they are also always influenced by the innermost barrier.

7. Conformational Relaxation

We have explored the properties of the barriers I–III in sections 5 and 6 and we should now turn to the outermost one. Before being able to do so, we must solve another problem. Figures 4–7 show that the rebinding of a ligand to Mb after a photoflash is exponential in a liquid sample, but close to a power law in a solid one. The difference can be understood by assuming that in solid samples each Mb molecule remains in a given conformational state whereas in a liquid one a given Mb molecule changes rapidly from one conformational state to another. We call this process conformational relaxation.

Relaxation can be characterized by a correlation time, τ_r, or rate, $k_r = 1/\tau_r$. Loosely defined, τ_r is the time required for a substantial change in the activation energy E_{ba} of a given molecule, for instance by a rearrangement in the tertiary structure. If no relaxation occurs, the rebinding process reflects the activation energy spectrum of the inner barrier and the rebinding function $N(t)$ is nonexponential. If relaxation is complete, $N(t)$ is exponential.

The influence of relaxation is easiest to discuss if process I is alone. The maximum time spent by a ligand molecule in well B before rebinding then is given by $\tau_{max} = 1/k_{ba}^{min}$, where k_{ba}^{min} is the smallest rate corresponding to the energy E_{ba}^{max} in Figure 13. In terms of rates, we have *no relaxation in process I* if

$$k_r \ll k_{ba}^{min} \qquad (37)$$

In contrast, with fast relaxation, all Mb molecules undergo many conformational changes before their ligands rebind. The condition for *complete relaxation in process I* is

$$k_r \gg k_{ba}^{max} \qquad (38)$$

where k_{ba}^{max} is the maximum return rate. At any time, the ensemble of Mb molecules will be in dynamic equilibrium, and the energy spectrum given by $g(E_{ba})$.[11] The number of Mb molecules with activation energies between E_{ba} and $E_{ba} + dE_{ba}$ at time t will consequently be $dN(t) = N(t) \cdot (E_{ba})dE_{ba}$. The number of these that rebind during the time interval between t and $t + dt$ is $d(dN(t)) = -k_{ba}dN(t)dt$

or $d^2N(t) = -k_{ba}N(t)g(E_{ba})dE_{ba}dt$. Integration with the initial condition $N(0) = 1$ yields

$$N(t) = e^{-\int dE_{ba}g(E_{ba})k_{ba}t} \equiv e^{-k_{ba}^{mean}t} \qquad (39)$$

Regardless of the form of $g(E_{ba})$, rebinding in the limit $k_r \gg k_{ba}^{max}$ will be exponential. The "effective energy spectrum" thus has narrowed to a delta function and the process is similar to motional line narrowing in nuclear magnetic resonance (Bloembergen et al., 1948; Pines and Slichter, 1955). With $k_{ba}dE_{ba} = -RTdk_{ba}$, and with $g(k_{ba})$ given by eq 14, the integral can be done; the result is with eq 15 in a good approximation

$$N(t) = e^{-k_{ba}^{peak}t} \qquad (40)$$

If condition (38) is satisfied, the ligand molecules will rebind exponentially with a rate corresponding to the activation energy E_{ba}^{peak}.

Another case that is easy to solve is the high-temperature limit, where again only one process dominates. As an example, we shall use the case discussed in section 6 where $k_{de} = 0$, process IV is absent, and $N(t) \simeq N^{III}(t)$. *No relaxation of process III* implies

$$k_r \ll \lambda_{III}^{min} = \frac{k_{ba}^{min}k_{cb}k_{dc}}{k_{bc}k_{cd}} \qquad (41)$$

The rebinding function $N(t)$ is nonexponential and given by eq 34. *Complete relaxation in process III* implies

$$k_r \gg \lambda_{III}^{max} = \frac{k_{ba}^{max}k_{cb}k_{dc}}{k_{bc}k_{cd}} \qquad (42)$$

Again, the energy spectrum of the Mb molecules will be in dynamic equilibrium and $d^2N(t) = -k_{ba}(k_{cb}k_{dc}/k_{bc}k_{cd})N(t)g(E_{ba})dE_{ba}dt$ which leads to

$$N(t) = \exp\left[-\frac{k_{cb}k_{dc}}{k_{bc}k_{cd}}\int dE_{ba}g(E_{ba})k_{ba}t\right] \qquad (43)$$

Solving the integral in eq 43 gives with eq 36

$$N(t) = e^{-\lambda_{III}^{peak}t} \qquad (44)$$

and rebinding will be exponential.

Equations 40 and 44 imply that problems in which relaxation is complete are solved by the replacement

$$\int dE_{ba}g(E_{ba})k_{ba} \to k_{ba}^{peak} \qquad (45)$$

The prescription should hold even if more than one process is present.

The general solution of the intermediate case in which relaxation is present but not complete during a significant part of the rebinding process requires a theoretical model for relaxation at times near τ_r. However, for times $t \ll \tau_r$ and $t \gg \tau_r$, the solution should be model independent. We can therefore get an approximate solution for $N(t)$ and an approximate relaxation time at a given temperature by making τ_r one of the parameters in the computer fitting routine and setting up the following problem. (a) Assume no relaxation of $N(t)$ for times $t \le \tau_r$. (b) Assume complete relaxation for times $t > \tau_r$. (c) Use the final conditions of the unrelaxed solution at $t = \tau_r$ as the initial conditions for the completely relaxed solution. This procedure will be used in section 8 for the analysis of the data in glycerol–water solvent.

8. Binding with Diffusion and Relaxation

In a liquid solvent, two phenomena can take place in ad-

---

[11] In contrast, if no relaxation occurs, the ensemble is not in dynamic equilibrium. At the start of the reaction, the Mb molecules that have small activation energies react first and are depleted from the ensemble represented by $g(E_{ba})$. At long times, only the high energy tail of $g(E_{ba})$ contributes to the reaction.

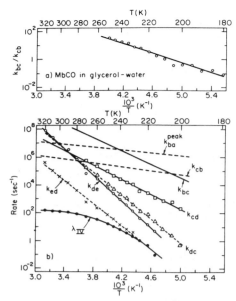

FIGURE 18: Rates for processes I–IV for MbCO in glycerol-water, 3:1, v/v. (a) Ratio of rates over barrier II, k_{bc}/k_{cb}. (b) Arrhenius plots for all rates. Note that λ_{IV}, describing the observed rate for process IV, does not describe an elementary step, as is discussed in subsection 8.1. Activation energies and frequency factors deduced from these Arrhenius plots are given in Table I. [CO] = 3×10^{-5} M.

dition to those discussed in section 6. (1) Some or all of the ligand molecules released by a photoflash can diffuse into the solvent. All ligand molecules in the solvent then compete for the vacant binding site. The resulting process IV thus is second order, proportional to the ligand concentration, and becomes pseudo-first-order at high ligand concentrations. (2) Conformational relaxation can set in and the rebinding curves can become exponential. The evaluation of rebinding after a photoflash must in general be performed numerically; the relevant ideas have been sketched at the beginning of section 6 and the end of section 7. Details will be given separately (Beeson, 1975). In the high-temperature limit, an explicit formula for the rebinding rate can be given. In this section, we will first treat the high-temperature limit and then present the results for MbCO and MbO$_2$.

(8.1) The High-Temperature Limit. We define the high-temperature limit as in eq 29 through the condition that the outbound rates are much bigger than the inbound ones:

$$k_{bc} \gg k_{ba}{}^{peak}; \quad k_{cd} \gg k_{cb}; \quad k_{de} \gg k_{dc}, k_{ed} \quad (46)$$

The rate λ_{IV} for process IV then follows from eq 5 as

$$\lambda_{IV} = k_{ba}{}^{peak} \frac{k_{cb}}{k_{bc}} \frac{k_{dc}}{k_{cd}} \frac{k_{ed}}{k_{de}} \quad (47)$$

so that

$$E_{IV} = E_{ba}{}^{peak} - E_{bc} + E_{cb} - E_{cd} + E_{dc} - E_{de} + E_{ed}$$

$$A_{IV} = A_{ba}A_{cb}A_{dc}A_{ed}/A_{bc}A_{cd}A_{de} \quad (48)$$

E_{IV} is the energy difference between the bottom of well E and the top of the innermost barrier, as indicated in Figure 11.

Equations 47 and 48 permit us to understand the behavior of the rate λ_{IV}. Figure 7 shows that process IV for MbCO appears at about 210 K, becomes larger as the temperature increases, and is the sole directly observable process above 270 K. In the entire temperature range IV is ex-

ponential: $N^{IV}(t) = N^{IV}(0) \exp(-\lambda_{IV}t)$. The rate λ_{IV} can be determined from Figure 7, for instance with the tool described in footnote 3, and is plotted in Figure 18b vs. $10^3/T$. The rate λ_{IV} does *not* obey an Arrhenius relation, but shows a pronounced bend centered at about 270 K. At low temperatures, the slope yields

$$E_{IV}' = 18.9 \text{ kcal/mol}, \quad A_{IV}' = 10^{18.3} \text{ sec}^{-1} \quad (49)$$

Above 300 K, the result is

$$E_{IV} = 2.9 \text{ kcal/mol}, \quad A_{IV} = 10^{4.1} \text{ sec}^{-1} \quad (50)$$

This change in apparent activation energy and frequency factor can be understood as follows. At low temperatures, around 220 K, where very few CO molecules diffuse into the solvent, their return is governed by the outermost barrier. Above 300 K, Figure 7 shows that all CO molecules leave Mb after photodissociation; the conditions for the high-temperature limit are approached or satisfied. The activation energy measured above 300 K therefore does not describe one barrier, but measures the energy difference between the bottom of well E and the top of barrier I.

(8.2) MbCO in Glycerol-Water. Rates for rebinding of CO to Mb after photodissociation, computed from the data in Figure 7, are plotted in Figure 18b vs. $10^3/T$. The elementary rates follow Arrhenius relations very well, in some cases over as many as seven decades in rate. The activation energies and frequency factors deduced from the slopes and intercepts of the various $k(T)$ are collected in Tables I and II. Conclusions will be given in section 10, but a number of remarks are in order here.

(i) In MbCO, the rates for barriers I, III, and IV can be determined unambiguously. The rates for barrier II are hard to obtain and the values for E_{bc}, E_{cb}, A_{bc}, and A_{cb} in Table I consequently have large errors. The ratio k_{bc}/k_{cb}, however, can be extracted well; it is shown in Figure 18a and yields

$$E_{bc} - E_{cb} = 7.7 \text{ kcal/mol}$$

$$A_{bc}/A_{cb} = 10^{8.2} \quad (51)$$

The difficulty of finding k_{bc} and k_{cb} separately is caused by the time limitation of our equipment and the nature of barrier II. Between 180 and 195 K, where process II can be observed well, k_{cb} is considerably larger than k_{bc} as is evident from Figure 18a. A CO molecule that jumps from well B to C returns immediately to B again. Process II therefore "hugs" process I closely and does not stand out as for instance process IV. Extension of observations to times shorter than 2 μsec would permit a much better determination of the individual rates. The inability of obtaining the separate activation energies and frequency factors for barrier II does not affect our conclusions in section 10.

(ii) Figure 7 provides evidence for conformational relaxation. Process IV is exponential at all temperatures where it can be observed well, but process II is nonexponential up to at least 230 K. The relaxation rate k_r consequently must lie between the corresponding rates. If we assume that the relaxation rate follows an Arrhenius relation, a computer fit at all temperatures is obtained by assuming

$$E_r \geq 24 \text{ kcal/mol}, \quad A_r \geq 10^{23} \text{ sec}^{-1} \quad (52)$$

(iii) The rates in Figure 18 have been obtained by assuming that conformational relaxation takes place with a single rate satisfying eq 52. All elementary rates in Figure 18 follow Arrhenius plots from 200 to 320 K. For k_{ed} and k_{de}, the lines are straight over about seven decades. These features suggest that no major conformational change occurs in Mb in this temperature range, apart from the conformational relaxation already discussed in (ii).

(8.3) *MbO₂ in Glycerol–Water.* The evaluation of the O_2 data in Figures 2b and 9 proceeds as for MbCO and leads to the rates shown in Figure 19, but some special problems are involved. (i) The preparation of a ferrous MbO_2 sample is more difficult than that of MbCO and small variations from sample to sample lead to somewhat different high-temperature rebinding curves. (ii) The quantum yield for photodissociation of O_2 is substantially smaller than one and the energy of the pulsed laser is not sufficient to photodissociate all MbO_2. (iii) We have studied the optical spectrum of MbO_2 at temperatures below 10 K. The spectrum after complete photodissociation differs from that of deoxymyoglobin. The change in optical density at the monitoring wavelength, 436 nm, in going from oxymyoglobin to the photodissociated state is only 60% of the corresponding change from oxy- to deoxymyoglobin. Thus, even if all bound O_2 molecules are photodissociated, the signal observed in the photomultiplier at low temperatures is 40% smaller than at higher temperatures. We assume that the optical spectrum observed at low temperatures after flashoff is associated with well B and treat the data accordingly. (iv) Photodissociation of MbO_2 is initiated with a 590-nm (rhodamine 6G) or 540-nm (coumarin 6) dye laser. Both have disadvantages. At low temperatures, the α band in the optical spectrum of MbO_2 sharpens and the overlap with the 590-nm light decreases. The 540-nm laser gives good overlap with the β band, but the laser energy is low. Combination of the problems (ii)–(iv) results in a change in the optical density for MbO_2 that is about five times smaller than for MbCO. (v) As Figure 9 shows, process III can barely be seen. We have therefore evaluated the MbO_2 data assuming only three barriers, I, II, and IV. The identification of the barely visible process with III is based on its position in the time sequence and on the similarities of the other three with processes I, II, and IV in MbCO. Neglecting barrier III implies that the activation energy determined from the rate k_{ce} describes the transition from the bottom of well C to the top of barrier IV. (vi) As in MbCO, barrier II is not as well determined as barriers I and IV. As a result of the difficulties listed here, the barriers for MbO_2 are not as well fixed as for MbCO.

9. Activation Energies and Entropies

We present in this section the relevant equations governing rate constants and summarize activation energies and entropies of the ligand reaction with Mb. Transition-state theory (Glasstone et al., 1941) gives for the rate of a transition B → A

$$k_{ba} = \nu e^{-G_{ba}/RT} \tag{53}$$

where G_{ba} is the free-energy change between the initial state B and the transition state (intermediate state or activated complex) and ν is an approximately constant factor.[12] Expressing the free-energy change in terms of enthalpy and

---

[12] The popular relation $\nu = kT/h$, where k is Boltzmann's and h Planck's constant, predicts ν to be proportional to T. Modern treatments indicate, however, that the temperature dependence of the frequency factor is strongly model dependent (Menzinger and Wolfgang, 1969; Lin and Eyring, 1972); assuming temperature independence appears to be a good compromise.

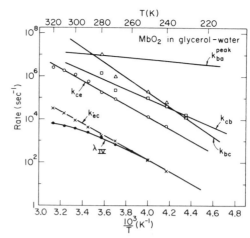

FIGURE 19: Rates for processes I, II, and IV for MbO_2 in glycerol-water, 3:1, v/v. Activation energies and frequency factors deduced from these curves are given in Table I. $[O_2] = 6 \times 10^{-5} \, M$.

entropy changes, $G_{ba} = H_{ba} - TS_{ba}$, leads to

$$k_{ba} = \nu e^{S_{ba}/R} e^{-H_{ba}/RT} \tag{54}$$

We take ν, S, and H to be temperature independent and assume for a first-order transition

$$\nu = 10^{13} \, \text{sec}^{-1} \tag{55}$$

Comparison of eq 2 and 54 gives

$$H_{ba} = E_{ba}, \quad S_{ba}/R = \ln (A_{ba}/\nu) \tag{56}$$

The entropy change from B to the transition state can be obtained from the measured frequency factor by assuming $\nu = 10^{13} \, \text{sec}^{-1}$. This approach is modified for process IV. The rate $k_{ed} = k_{ed}'[L]$ (eq 6) refers to a second-order reaction in a liquid (or glass). Reaction theory for this case is not well-developed (Weston and Schwarz, 1972) and the approximation we are forced to make is even cruder than eq 55. In a gas, eq 55 is replaced by

$$\nu' = 10^{14} \, \text{cm}^3/(\text{mol sec}) = 10^{11} \, (M \, \text{sec})^{-1} \tag{57}$$

where ν' refers to a second-order rate. We will use eq 57 to obtain crude values for the entropy change in the transition E → D, even though the solvent is liquid.

Activation energies, frequency factors, and entropy changes for the reaction of Mb with CO and O_2 are collected in Tables I and II. Entropy changes, obtained from eq 55–57, are listed in terms of the dimensionless quantity S/R. Most of the numbers in Tables I and II come from the experiments discussed in sections 4–8. Errors in the activation energies and frequency factors for barriers III and IV are approximately given by

$$\Delta E/E \approx \pm 0.10, \quad \Delta \log A/\log A \approx 0.10 \tag{58}$$

The corresponding relative error in (S/R) is also given by eq 58; however, the calculation of S/R rests on eq 55–57 and hence is theory dependent. Errors for barrier I are given after eq 17. Errors in the values for barrier II are considerably larger than eq 58, but the differences $E_{bc} - E_{cb}$ and $S_{bc} - S_{cb}$ have relative errors of only about 10%. Energy and entropy differences between wells B and C are thus much better known than the corresponding values for barrier II.

Data for the overall transition E → A are taken from published experiments (Keyes et al., 1971; Rudolph et al., 1972). Since these were performed in solvents different

Table I: Activation Energies (E), Frequency Factors (A), and Entropy Changes (S/R) for the Binding of CO and O$_2$ to Myoglobin in Glycerol–Water, 3:1, v/v.[h]

Transition[a]	Carbon Monoxide[b]				Dioxygen[b]			
	E (kcal/mol)	Log A (sec^{-1})	Log A' (M sec)$^{-1}$	S/R	E (kcal/mol)	Log A (sec^{-1})	Log A' (M sec)$^{-1}$	S/R
A → B	24[c]	16[d]		+7.1	15	13[d]		−0.3
B → A	2.4	8.7		−9.9	2.0	8.5		−10.4
B → C	11.0	16.5		+8.1	15.6	18.9		+13.6
C → B	3.3	8.3		−10.8	10.2	13.8		+1.8
C → D	12.1	15.5		+5.8				
D → C	19.3	20.9		+18.2				
D → E[f]	21.7	22.7		+22.3	12.5	15.0		+4.6
E → D[f]	18.9	18.3[e]	22.8	+27.2	12.7	13.2[e]	17.4	+14.8
E → A[g]	−21			−17	−18			−16

[a] Wells are defined in Figure 11. Unless otherwise noted, values are measured to the top of the barriers. [b] Values for CO are taken from eq 17 and Figure 18, for O$_2$ from eq 17 and Figure 19. [c] E_{ab} is obtained by subtraction: eq 59 with eq 48 and 50 becomes $E_{ab} = -E_{ea}^0 + E_{IV}$. [d] Obtained from eq 60. For O$_2$, $k_{ab} \approx 5k_{dis}$. [e] Pseudo-first-order rates, concentrations: [CO] = $3 \times 10^{-5}\,M$, [O$_2$] = $6 \times 10^{-5}\,M$. [f] For O$_2$, values in these rows actually refer to the transitions C → E and E → C. [g] Data for the transition E → A have been taken from Keyes et al. (1971) and Rudolph et al. (1972). The values represent differences between initial and final, not transition, states. [h] Relative errors are given by eq 58 as about ±0.1 except for the transitions A → B, B → C, C → B, and E → A, where they are hard to estimate.

from ours, the values for E_{ea}^0 and S_{ea}^0 in Table I should be considered approximate. The activation energy E_{ab} is obtained by subtraction:

$$E_{ab} = -E_{ea}^0 + E_{ed} - E_{de} + E_{dc} - E_{cd} + E_{cb} - E_{bc} + E_{ba} \quad (59)$$

as is evident from Figure 11. Note that the energy change E_{ea}^0 and the entropy change S_{ea}^0 are *not* differences between initial and transition, but between initial and final states. The frequency factor A_{ba} is extracted from the dissociation rate. In dissociation, a ligand molecule initially in well A moves to the outside by thermally overcoming all barriers. In the high-temperature limit, a ligand that has jumped from A to B will then nearly always move to the outside; the dissociation rate k_{dis} is therefore given by the rate-limiting step A → B:

$$k_{dis} \approx k_{ab} \quad (60)$$

In general, however, k_{dis} will be smaller than k_{ab}, and k_{ab} must be found from k_{dis} and the other rates by numerically solving eq 5. Once k_{ab} is known, an approximate value for A_{ab} is determined by setting $k_{ab} = A_{ab} \exp(-E_{ab}/RT)$. The experimental values for k_{dis} at 293 K are (Antonini and Brunori, 1971) k_{dis}(MbCO) = 0.02 sec^{-1}, k_{dis}(MbO$_2$) = 10 sec^{-1}. The resulting values of A_{ab} and S_{ab}/R are given in Table I for MbCO and MbO$_2$ in glycerol–water. Since k_{dis} has not been measured in PVA, the corresponding values in Table II are missing.

10. Summary and Interpretation

In the present section, we summarize and interpret the main results of our work. The section is largely self-contained, but we refer to earlier sections for amplification and proofs. In each of the eight subsections, a specific facet is described.

(10.1) Multiple Barriers. We have observed the rebinding of carbon monoxide and dioxygen to Mb after photodissociation over a wide range in temperature and time. The rebinding functions $N(t)$, displayed in Figures 4, 7, and 9, show four different processes, I–IV, characterized in section 4. With one potential barrier alone, four distinct processes cannot be explained. In section 4, we therefore have proposed a model in which four barriers are arranged in se-

Table II: Activation Energies (E), Frequency Factors (A), and Entropy Changes (S/R) for the Binding of Carbon Monoxide to Myoglobin Embedded in a Solid PVA Matrix.[a]

Transition	E (kcal/mol)	Log A (sec^{-1})	S/R
B → A	1.4	7.7	−12.2
B → C	9.7	14.1	+2.5
C → B	5.3	9.4	−8.3
C → D	10.9	13	0
D → C	15	15.4	+5.5

[a] Values are taken from eq 17 and Figure 16.

quence, as sketched in Figure 11. In sections 5–8, we have determined all relevant barrier parameters; they are listed in Table I for MbCO and MbO$_2$ in glycerol–water and Table II for MbCO in a solid PVA matrix. In Figure 20a, the barriers for MbCO are drawn by giving the enthalpies H, the entropies S in units of R, and the free energies $G = H - TS$ at 310 K. The values of H, S/R, and G are only measured in the wells and at the transition states (top of the barriers); the curves are drawn to guide the eye. Figure 20b shows the corresponding wells and barriers for MbO$_2$.

The free energy G_{ed} and the entropy S_{ed} require one remark. The values of S_{ed}/R given in Table I are based on eq 6 and 57 and consequently imply a standard state of unit mole fraction for the ligand. G_{ed} thus also refers to the same standard state. The wells E in Figure 20 are drawn for two situations; the dashed lines refer to unit mole fraction, the solid ones to concentrations [CO] = $3 \times 10^{-5}\,M$ and [O$_2$] = $6 \times 10^{-5}\,M$.

In Figure 21, the *free energy* for binding of CO to Mb is shown at two temperatures. At 50 K, the outer barriers dwarf the inner one; at 310 K, the inner is slightly larger than the outer ones. The change makes it clear why CO molecules in B rebind directly at low, but move preferentially to the outside at high temperatures.

In our discussions we have assumed that all four barriers are in sequence. In subsection 6.3, we have already partially justified this model by showing that the three barriers in solid PVA are likely to be in series. Figures 5 and 6 suggest that barriers III and IV are also in series: when process IV is blocked, process III is seen. Further support comes from

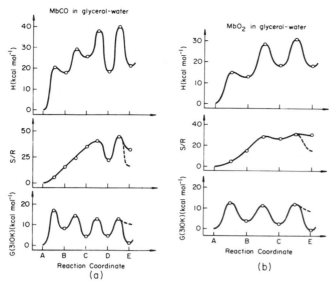

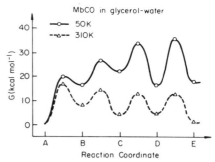

FIGURE 21: The free energy for the binding of CO to Mb as a function of reaction coordinate. At 50 K, a ligand in well B will move only to A; at 310 K, it will predominantly go to well E.

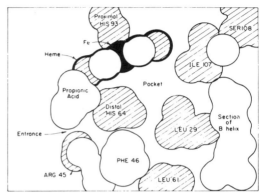

FIGURE 20: Enthalpy, entropy, and free energy (calculated at $T = 310$ K) as a function of reaction coordinate for (a) CO and (b) O_2 binding to Mb. The solid lines in well E for S/R and G are calculated assuming a pseudo-first-order reaction ($[CO] = 3 \times 10^{-5}$ M, $[O_2] = 6 \times 10^{-5}$ M), the dashed ones describe a second-order reaction calculated for unit mole fraction.

FIGURE 22: Cross section through the region near the active center of Mb. A pocket on the distal side of the heme can be entered from the outside through a narrow hole formed between His-64 and the propionic side chain. Shaded groups are in, unshaded ones above, the plane containing the iron.

consideration of all enthalpies and entropies. Equation 50 shows that the energy difference between the bottom of well E and the top of barrier I (Figure 11) can be obtained in two different ways, either from the slope of λ_{IV} in the high-temperature limit or by adding the proper energies from Table I. The first approach gives $E_{IV} = 2.9$ kcal/mol (eq 50), the second -0.9 kcal/mol. The two values agree within the limits of error implied by eq 58. Similarly the frequency factor $A_{IV} = 10^{4.1}$ sec^{-1} agrees well with the combination $A_{ba}A_{cb}A_{dc}A_{ed}/A_{bc}A_{cd}A_{de} = 10^{1.5}$ sec^{-1}. An additional test comes from the entropy difference between wells E and A, where the value $(S/R) = -17$ (Keyes et al., 1971, Rudolph et al., 1972) agrees within errors with the algebraic sum of the kinetically determined numbers from Table I, $(S/R)_{kin} = -18.6$. Finally we note that it would be unlikely for the Arrhenius plots in Figure 18 to be straight over as many as seven orders of magnitude in rate unless the model had some validity.

Why does Mb possess multiple barriers? Efficient oxygen storage probably demands a highly ordered state. The transition from the free O_2 (or CO) in the solvent to the bound state then requires a large negative change in S. For a given k_{ba}, enthalpy changes to the transition state are connected by eq 54 as

$$\frac{H_{ba}}{RT} = \ln\left(\frac{\nu}{k_{ba}}\right) + \frac{S_{ba}}{R} \qquad (61)$$

In order to achieve a fixed reaction rate, H_{ba} must decrease with increasing order, i.e., increasing *negative* S. This consideration explains the small activation energy at the inner barrier. If no outer barriers were present, the small inner one would probably be overwhelmed by all kinds of molecules in the solvent. If this speculation is correct, then we would expect to find multiple barriers in many enzymes. Indeed, we have seen such multiple barriers in hemoglobin, the separated hemoglobin chains, carboxymethylated cytochrome c, and cytochrome P450 with and without camphor substrate.

(10.2) Barriers and Mb Structure. So far we have interpreted the experimental results of section 3 by postulating the existence of four barriers, but without identifying these with specific parts of Mb. X-Ray and neutron diffraction data (Kendrew et al., 1958; Watson, 1968; Schoenborn, B. P., Norwell, J., and Nunes, A. C., personal communication) provide evidence for features that could produce these wells and barriers. Figure 22 shows a cross section through part of Mb, approximately at right angles to the heme plane. The heme extends to the Mb surface, but the iron in its center is well hidden. On the distal side of the iron is a prominent pocket with linear dimensions of at least 0.5 nm lined with hydrophobic residues. The distal histidine His-64 inside the pocket is present in all mammalian Mb; its nitrogen $N_{\epsilon 2}$ is about 0.4 nm away from the heme center. The entrance to the pocket is narrow and partially blocked by the propionic side chain of the heme. The entire molecule is surrounded by a hydration shell. A plausible, but by no means unique, scenario for the binding of a ligand to Mb is as follows. Moving from the solvent (well E) toward the entrance, L first overcomes barrier IV, formed by the hydration shell. Well D thus could be between the hydration shell and the entrance to the pocket. Support for the identification of barrier IV with the hydration shell comes from experiments with heme (Alberding et al., to be published) which show two barriers, with properties similar to the Mb barriers I and IV. We therefore assign I to the heme and IV to the solvent. Barrier III could be formed by the narrow entrance to the pocket, well C by a weak bond to the surface of the cavity. Breaking this bond and thus overcoming

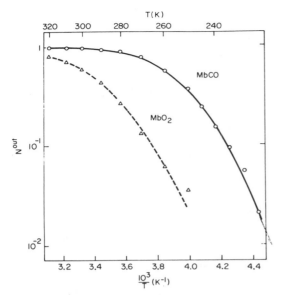

FIGURE 23: Plot of N^{out} vs. $10^3/T$. N^{out} is the total fraction of ligand molecules that do not rebind directly, but first move into the solvent.

barrier II, L moves to well B close to the heme iron and due possibly to His-64. Evidence for such an assignment comes from optical (Iizuka et al., 1974) and Mössbauer (Lang, private communication) experiments which indicate that photodissociation of CO at low temperatures does not lead to the deoxygenated form. Finally in well A, L is bound to the heme iron, and possibly also to His-64. The innermost barrier, I, could be caused by the active center since there is a connection between spin state and position of the heme iron, planarity of the heme ring, and occupation of the active site (Huber et al., 1970; Perutz, 1970). Of the six ligand positions of the iron atom, four are bound to four nitrogen atoms in the heme plane. Position 5 connects the iron to the polypeptide chain at the proximal histidine. The molecules O_2 or CO bind at position 6. X-Ray, neutron diffraction, and Mössbauer data indicate that the states of the iron and the heme depend on whether position 6 is free or occupied. The iron atom with free position 6 is displaced out of the heme plane, in a high-spin state, $S = 2$, and the heme plane is puckered. With CO or O_2 bound, the iron is more nearly in the plane, in an $S{=}0$ low-spin state, and the heme plane is flat. As the ligand molecule approaches the heme center with position 6 not yet occupied, the iron must move into the plane, change spin state, and the heme must alter shape. Possibly the activation energy required for these changes accounts for the inner barrier. If so, the height of the barrier could be influenced by the protein structure via position 5 and the heme nitrogen atoms.

(10.3) Specificity and Order. Two essential attributes of enzymes are specificity and order. The native substrate is preferentially selected and guided to a state of high order. Even though Mb is not an enzyme, it illustrates how specificity and order are produced.

Specificity is studied here by comparing the binding of CO and O_2. The free-energy profiles in Figure 20 tell the following story. A CO molecule entering from the solvent encounters monotonically increasing free energy barriers. In each intermediate well the outbound rates are larger than the inbound ones: $k_{de} > k_{dc}$, $k_{cd} > k_{cb}$, and $k_{bc} > k_{ba}$. CO therefore proceeds to the binding site A by a random walk. After the first step, E → D, CO has a larger probabil-

ity to return to E than to proceed to C and it takes a few tries before CO makes it to C. Once there, it is again more likely to return to D and from there to E than to continue to B. CO consequently shuttles many times among the various wells before coming to rest at the heme iron. At 310 K, the rate λ_{IV} is about 600 times smaller than k_{ed}; CO moves a few hundred times over barrier IV before binding. Binding of O_2 proceeds in a much smoother way. Barriers I and IV dominate and they have essentially equal free-energy heights, barrier II is smaller. At 310 K, λ_{IV} is only about four times smaller than k_{ed}; an O_2 molecule that moves for instance from the solvent over barrier IV has nearly equal probability of returning to the solvent or reaching the binding site A and requires only a few tries to get from E to A. Specificity in Mb thus appears to be achieved through a collaboration of active center (barrier I), globin structure (barriers II and III), and hydration shell (barrier IV). The design is optimal for rejecting the intruder CO and accepting the natural ligand O_2: CO is discriminated against at each barrier. O_2, on the other hand, moves just as easily in as out so that Mb is an efficient oxygen acceptor *and* donor. The difference between CO and O_2 can also be seen in Figure 23 which gives $N^{out}(T)$, the fraction of ligand molecules that move into the solvent after photodissociation. All CO molecules are ejected from Mb after a flash at temperatures above about 270 K. A major fraction of all O_2 molecules, however, do not leave Mb, but return directly to the binding site after each flash.

Figure 20 also shows how *order* is established. Consider a CO molecule coming from the solvent. The entropy S/R of the system Mb + CO increases over the first two barriers (IV and III) and then steadily decreases till it reaches a minimum when CO is at the binding site A. The increase in S/R at barrier IV can be due to breaking of hydrogen bonds when CO passes through the hydration shell. The increase at barrier III can be caused by breaking of bonds or displacement of residues. S/R is larger in well C than in D or E; this fact can find a natural interpretation in the size and construction of the pocket, as implied by Figure 22. The number of states, W, for the CO molecule in the pocket can be large enough so that the Boltzmann relation, $S/R = \ln W$, can explain all or part of the observed value. A contribution can also come from the interaction of the CO with the protein. A problem arises in the step B → A: it appears impossible to obtain the large entropy change, $S/R = -17$, by considering only the states of the small ligand molecule. We suggest the following explanation. In state B, the ligand molecule is bound to the partially charged nitrogen $N_{\epsilon 2}$ of the distal histidine His-64. In well A, the ligand is additionally bound to the heme iron and thus links the heme group and the distal histidine as has been suggested by Pauling (Pauling, 1964). The state B would be the precursor to the final state A; with binding enthalpy CO–$N_{\epsilon 2}$ given by $E_{bc} \simeq 11$ kcal/mol. A precursor could facilitate the final binding step, B → A, which involves a spin change of two units and thus requires presumably times of the order of nanosecond or longer. A number of observations support this model. (a) The bond Fe–CO–$N_{\epsilon 2}$ can stabilize a major part of the protein; the drop in entropy then is caused not only by restricting the states of CO but also of a number of residues, and the observed value, $S/R \approx -17$, can be explained. (b) CO fits well between the heme and His-64. Using the known positions of Fe and $N_{\epsilon 2}$ (Antonini and Brunori, 1971), a bond length of about 0.14 nm for 0–$N_{\epsilon 2}$ results and this value appears reasonable. (c) It is known that CO and

O_2 stabilize heme proteins against denaturation (Cassatt and Steinhardt, 1971). A $Fe-N_{\epsilon 2}$ link could produce stabilization. (d) The stretching frequency of CO bound to Mb (1944 cm^{-1}) is considerably smaller than for CO bound to free heme (1974 cm^{-1}) or heme proteins lacking the distal histidine (1970 cm^{-1}) (Alben and Caughey, 1968; Caughey et al., 1969). Bonding with charge transfer from $N_{\epsilon 2}$ of His-64 could explain the frequency shift (Franceschetti, D. R. and Yip, K. L., to be published). (e) We have also studied low-temperature photodissociation of CO in carboxymethylated cytochrome c and in heme (Alberding et al., to be published). In both cases, no distal histidine is present. The frequency factor A_{ba} is of the order of 10^{12} sec^{-1}, corresponding to an entropy change $S_{ba}/R \approx 2$, considerably smaller than for Mb.

MbO_2 displays features similar to MbCO, but the entropy changes are generally smaller. Here also, the ligand O_2 could form a bridge between the heme and the distal histidine. Evidence for such a bridge has been obtained by Yonetani and coworkers in a related system, cobalt myoglobin (Yonetani et al., 1974).

(10.4) Activation Energy Spectra. At temperatures below about 180 K, only process I, the direct rebinding from B, takes place after photodissociation. Figure 21 explains that the dominance of I is due to a shrinking of the innermost and a growth of the outer free-energy barriers with decreasing temperature. Against expectation, process I is not exponential, as can be seen dramatically in Figures 2 and 4. Our explanation is to ascribe the observed time dependence $N(t)$ to the existence of an energy spectrum at the inner barrier. From $N(t)$, the spectral distribution $g(E_{ba})$ can be found and it is given for CO and O_2 in Figure 13. The result of a multiple flash experiment, Figure 3, demonstrates that the activation energy spectrum must be due to a heterogeneity in the ensemble of Mb molecules.

We have already made some comments concerning the activation energy spectrum in section 5, and add some further remarks here. (a) Nonexponential rebinding at low temperatures occurs not only in Mb, but in all heme proteins that we have studied so far (N. Alberding et al., to be published). The energy spectra depend strongly on the protein and differ, for instance, considerably for the separated α and β chains of hemoglobin. Substrate also affects the spectrum and cyt P450 with and without the camphor substrate display very different spectra. Low-temperature spectra thus may become useful for fingerprinting proteins. (b) The innermost barrier can be studied extremely well because it can be seen alone over an enormous temperature range (in Mb from 40 to 160 K) and because $N(t)$ extends at most temperatures over more than six orders of magnitude in time. The other barriers cannot be investigated in similar detail and it is not possible to assert unambiguously that they are described by only one energy. In fact, some tentative evidence in section 6 suggests that at least barrier II is also distributed. (c) The solid lines in Figure 14 are based on the energy spectrum for MbCO in glycerol-water shown in Figure 13. The curves are *not* separate fits at each temperature. One measurement of $N(t)$ at one temperature yields $g(k_{ba})$ through eq 11; a second temperature fixes A_{ba}. The energy spectrum $g(E_{ba})$ is then known and all curves at any desired temperature are unambiguously predictable. In this sense the fit in Figure 14 is given by two parameters, the shape of $g(E_{ba})$ and A_{ba}; the curves represent the experimental data extremely well over an enormous range in $N(t)$, t, and T. (d) We based the determination of

$g(E_{ba})$ on eq 1. This approximation works well for Mb, but is less satisfactory for some other proteins and should not be considered a universal solution. For some proteins, it may be better to guess a $g(E)$ and then let a computer generate $N(t)$ curves through eq 10. The function $g(E_{ba})$ is then varied till a good fit is obtained at all temperatures. (e) In some proteins, return after photodissociation can be observed below 2 K. The observed $N(t)$ curves can only be explained by eq 10 down to about 20 K. At lower temperatures, marked deviations occur which we ascribe to tunneling of either CO or Fe (N. Alberding et al., to be published).

(10.5) Conformational Relaxation. An attractive explanation for the energy spectrum is the existence of many different conformational states of Mb. At low temperatures or in a solid matrix, transitions from one conformation to another are extremely slow and each Mb molecule remains in a particular state, with a corresponding value of the activation energy. At temperatures above about 230 K in glycerol-water, relaxation sets in and each molecule changes rapidly from one state to another. Absence or presence of relaxation can be determined from the shape of the function $N(t)$; complete relaxation implies exponential $N(t)$ as eq 44 proves. The relaxation rate for MbCO in glycerol-water is characterized by the values $E_r \gtrsim 24$ kcal/mol, $A_r \gtrsim 10^{23}$ sec^{-1}, as given in eq 52. The denaturation enthalpy of Mb has been determined approximately as $H_{den} = 40$ kcal/mol (Hermans and Acampora, 1967). It is therefore tempting to speculate that conformational relaxation as observed in our experiments is connected to denaturation.

One question is raised by the suggestion in subsection 10.3 that ligands stabilize the protein. If stabilization indeed occurs, it will probably also affect conformational relaxation, and the relaxation process may be different depending on whether or not a ligand is in well A. Investigations of relaxation rates with independent tools could clarify this point.

The connection between relaxation and diffusion also deserves more attention. The extremes are clear. In a solid surrounding, both diffusion and relaxation are absent, as is seen in section 6. At high temperatures and in proper liquid solvents, diffusion is present (process IV can be seen) and relaxation is rapid (all observed rebinding curves are exponentials). In liquid PVA, however, diffusion can be absent while relaxation still persists (Figure 6).

(10.6) Conformational Energy. The total free energy of an Mb molecule in a given conformational state j can be denoted by F_j. The equilibrium number of Mb molecules in state j at temperature T will be proportional to the Boltzmann factor $\exp(-F_j/RT)$. To find F_j, we assume that each conformational state j gives rise to a unique activation energy E_{ba}; we can then label the states F_j with the corresponding activation energy E_{ba} and replace F_j by $F(E_{ba})$. To get the explicit form of $F(E_{ba})$, we consider the energy spectrum for MbCO in Figure 13. According to our assumption, the spectrum represents a Boltzmann distribution in a potential well described by $F(E_{ba})$. The temperature at which the distribution is established follows from eq 52. In a typical experiment with a glycerol-water solvent, about 10^3 sec are required to cool the sample to below about 200 K. Equation 52 implies that $k_r < 10^{-3}$ sec^{-1} at about 200 K. The distribution will therefore freeze at about $T_0 = 200$ K, and $g(E_{ba})$, normalized to $F(E_{ba}^{peak}) = 0$, will be given by

$$g(E_{ba}, T_0) = g(E_{ba}^{peak}) \exp[-F(E_{ba})/RT_0] \quad (62)$$

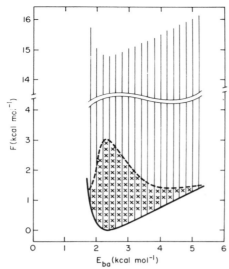

FIGURE 24: Conformational energy F_j as a function of the activation energy E_{ba}. The spikes represent the energy barriers in relaxation. The crosses correspond to a Boltzmann equilibrium distribution at 200 K.

Table III: Comparison of the Rates for Binding of CO to Mb in PVA (Solid) and Glycerol–Water (3:1, v/v) at 310 K.

Transition Barrier	Rate	Log Rate (sec⁻¹)		Ratio of Rates in Glycerol–Water to PVA
		PVA	Glycerol–Water	
I	k_{ba}^{peak}	6.7	7.0	2
II	k_{bc}	7.3	(8.8)	(30)[a]
	k_{cb}	5.6	(6.0)	(2.5)[a]
	k_{bc}/k_{cb}	1.7	2.7	10
III	k_{cd}	5.3	6.9	40
	k_{dc}	4.8	7.3	320
IV	k_{de}	0	7.4	∞

[a] Rates in parentheses have larger errors than the other ones.

With $g(E_{ba},T_0)$ from Figure 13, $F(E_{ba})$ can be calculated numerically; the result is displayed in Figure 24. The solid line represents $F(E_{ba})$, and the spikes are the barriers (activation energies) that have to be overcome in conformational relaxation. The crosses show a Boltzmann equilibrium distribution at 200 K. As the temperature is lowered below 200 K, the Mb molecules remain frozen in their states and this distribution remains unchanged. In contrast, as the temperature rises above 200 K relaxation sets in, Mb molecules jump over the barriers, and the distribution assumes the equilibrium shape corresponding to the ambient temperature T.

(10.7) Influence of Solvent and Relaxation. The dynamical action of Mb cannot be treated without considering the solvent an integral part of the system. The following examples bear out this assertion. (a) The outermost barrier is intimately connected to the properties of the solvent and it may, in fact, be produced by the hydration shell (subsection 10.2). Changes in the solvent thus will affect the rebinding directly. (b) The surrounding can prevent diffusion into the solvent and thus block the interchange between the interior of the biomolecule and the outside. An example is shown in Figure 6: PVA, before it hardens, permits relaxation, but process IV is absent. In solid surroundings, both relaxation and diffusion are stopped, as is shown in Figures 5 and 6. (c) Activation energies and frequency factors for the various barriers are considerably different in solid PVA and in liquid glycerol–water, as is illustrated by Figures 16 and 18. (d) The inner parts of the protein are better shielded than the outer ones. To demonstrate this trend, we give in Table III the rates at 310 K for MbCO in solid PVA and in liquid glycerol–water. The ratio of rates in the two solvents shows that the core, near the active center, is only slightly affected while the outer barriers are strongly changed. (e) As Table III shows, the rates in the liquid surrounding are larger than in the solid one; relaxation makes access to the active center easier. Nevertheless, Mb does not behave like a sieve and the path to the binding site is well controlled as is demonstrated by Figures 18 and 19.

(10.8) Low-Temperature Biochemistry. Our work implies that it is not possible to understand the dynamics of ligand binding to Mb if measurements are taken only in a

narrow temperature interval, say between 278 K ($10^3/T = 3.6$ K⁻¹) and 320 K ($10^3/T = 3.13$ K⁻¹). Some examples will prove this assertion. (a) Between 278 and 320 K, only one process is observed after a photoflash; its rate, denoted by λ_{IV}, is shown in Figure 18 for CO and Figure 19 for O₂. If only points between 278 and 320 K are considered, the curvature in λ_{IV} is overlooked and activation energies of 3 and 6 kcal/mol are obtained for CO and O₂, respectively. They would be interpreted as barrier heights in a single-step process. Our investigation shows, however, that the binding process involves at least four barriers and that the activation energies observed at high temperatures must be broken down into individual components. In the high-temperature limit, defined by eq 46, the observed activation energies and frequency factors are given by eq 48. (b) Figure 18 shows that nearly all rates are approximately equal near physiological temperatures (compensation effect). The one rate that is much smaller, k_{ed}, is a second-order rate and depends on the ligand concentration. It can therefore also be shifted toward $k \approx 10^7$ sec⁻¹ if desired. The fact that all relevant rates are approximately equal implies that there is not one rate-limiting step, but that all barriers must be taken into account in any discussion of CO binding. The rebinding curves in Figure 7 show only one process near 310 K; the complex interplay among barriers is hidden. Only if a wide temperature range is studied can the various processes be separated. (c) Figure 23 displays the fraction of ligand molecules that do not rebind directly after a flash, but move outside Mb. For CO, this fraction is very close to 1 above 278 K and the existence of the fast process I would be difficult to detect. For O₂ process I still contributes a considerable fraction at high temperatures and probably has been seen in fast laser-flash work (Alpert et al., 1974). Without low-temperature data, however, the interpretation is difficult. These three examples suggest that the investigation of biochemical reactions in many systems will require experiments extending over a wide range in temperature even though the actual physiological range is small.

Acknowledgments

Many of our friends have helped us during the course of this work by guiding us through unfamiliar terrain, scrutinizing our ideas, pointing out related research, or generously loaning us equipment. We should like to thank all of them very much, in particular we express our appreciation to J. O. Alben, N. Alberding, S. S. Chan, P. G. Debrunner, G. DePasquali, I. Dezsi, P. Douzou, H. G. Drickamer, M. Eigen, C. P. Flynn, D. R. Franceschetti, V. I. Goldanskii, M. Gouterman, J. S. Koehler, D. Lazarus, V. P. Marshall,

R. A. Marcus, A. P. Minton, T. M. Nordlund, J. Norvell, T. Pederson, D. G. Ravenhall, J. R. Schrieffer, C. P. Slichter, and L. Sorensen.

References

Alben, J. O., and Caughey, W. S. (1968), *Biochemistry 7*, 175.

Alpert, B., Banerjee, R., and Lindqvist, L. (1974), *Proc. Natl. Acad. Sci. U.S.A. 71*, 558.

Antonini, E., and Brunori, M. (1971), Hemoglobin and Myoglobin in Their Reactions with Ligands, Amsterdam, North-Holland Publishing Co.

Austin, R. H., Beeson, K., Eisenstein, L., Frauenfelder, H., Gunsalus, I. C., and Marshall, V. P. (1973), *Science 181*, 541.

Austin, R. H., Beeson, K., Eisenstein, L., Frauenfelder, H., Gunsalus, I. C., and Marshall, V. P. (1974), *Phys. Rev. Lett. 32*, 403.

Beeson, K. W. (1975), Unpublished Thesis; available as Technical Report from the Department of Physics, University of Illinois at Urbana—Champaign, Urbana, Ill.

Bloembergen, N., Purcell, E. M., and Pound, R. V. (1948), *Phys. Rev. 73*, 679.

Bohon, R. L., and Conway, W. T. (1972), *Thermochim. Acta 4*, 321.

Cassatt, J. C., and Steinhardt, J. (1971), *Biochemistry 10*, 264.

Caughey, W. S., Alben, J. O., McCoy, S., Boyer, S. H., Charache, S., and Hathaway, P. (1969), *Biochemistry 8*, 59.

Chance, B., Schoener, B., and Yonetani, T. (1965), in Oxidase and Related Redox Systems, King, T. E. Mason, H. S., and Morrison, M., Ed., New York, N.Y., Wiley, p 609.

Cooke, R., and Kuntz, I. D. (1974), *Annu. Rev. Biophys. Bioeng. 3*, 95.

Frauenfelder, H. (1973), *Moessbauer Eff. Proc. Int. Conf. 5th*, 401.

Gibson, Q. H. (1956), *J. Physiol. 134*, 112.

Glasstone, S., Laidler, K. J., and Eyring, H. (1941), The Theory of Rate Processes, New York, N.Y., McGraw-Hill.

Hermans, J., and Acampora, G. (1967), *J. Am. Chem. Soc 89*, 1547.

Hills, B. A. (1973), *Science 182*, 823.

Huber, R., Epp, O., and Formanek, H. (1970), *J. Mol. Biol. 52*, 349.

Iizuka, T., Yamamoto, H., Kotani, M., and Yonetani, T. (1974), *Biochim. Biophys. Acta 371*, 126.

Kendrew, J. C., Bodo, G., Dintzis, H. M., Parrish, R. G., and Wyckoff, H. (1958), *Nature (London) 181*, 662.

Keyes, M. H., Falley, M., and Lumry, R. (1971), *J. Am. Chem. Soc. 93*, 2035.

Klotz, I. M. (1966), *Arch. Biochem. Biophys. 116*, 92.

Kuntz, I. D., and Kauzmann, W. (1974), *Adv. Protein Chem. 28*, 239.

Lin, S. H., and Eyring, H. (1972), *Proc. Natl. Acad. Sci. U.S.A. 69*, 3192.

Macdonald, J. R. (1962), *J. Chem. Phys. 36*, 345.

Macdonald, J. R. (1963), *J. Appl. Phys. 34*, 538.

Macdonald, J. R. (1964), *J. Chem. Phys. 40*, 1792.

Menzinger, M., and Wolfgang, R. (1969), *Angew. Chem., Int. Ed. Engl. 8*, 438.

Pauling, L. (1964), *Nature (London) 203*, 182.

Perutz, M. F. (1970), *Nature (London) 228*, 726.

Pines, D., and Slichter, C. P. (1955), *Phys. Rev. 100*, 1014.

Primak, W. (1955), *Phys. Rev. 100*, 1677.

Rudolph, S. A., Boyle, S. O., Dresden, C. F., and Gill, S. J. (1972), *Biochemistry 11*, 1098.

Theorell, H. (1934), *Biochem. Z. 268*, 73.

Wagner, K. W. (1913), *Ann. Physik. 40*, 817.

Watson, H. C. (1968), *Prog. Stereochem. 4*, 299.

Weber, G. (1972), *Biochemistry 11*, 864.

Weissbluth, M. (1974), Hemoglobin, New York, N.Y., Springer-Verlag.

Weston, R. E., and Schwarz, H. A. (1972), Chemical Kinetics, Englewood Cliffs, N.J., Prentice-Hall.

Wittenberg, J. B. (1970), *Physiol. Rev. 50*, 559.

Yonetani, T., Yamamoto, H., and Iizuka, T. (1974), *J. Biol. Chem. 249*, 2168.

Zerner, M., Gouterman, M., and Kobayashi, H. (1966), *Theor. Chim. Acta 6*, 363.

Ultraviolet Fluorescence of the Aromatic Amino Acids

By F. W. J. TEALE AND G. WEBER

Department of Biochemistry, The University, Sheffield 10

(*Received* 25 *June* 1956)

Compounds resembling the aromatic amino acids are known to show appreciable fluorescence in the near ultraviolet [Ley & Englehardt (1910), Kowalski (1911), Marsh (1924)], but the fluorescence of the aromatic amino acids themselves has not so far been unequivocally characterized. Debye & Edwards (1952) have made observations of the phosphorescence of the aromatic amino acids. The position of the phosphorescence bands and the decay times found by these authors indicate the probable existence of fluorescence bands in the near ultraviolet with normal decay times (McClure, 1949).

It is shown in this paper that the three aromatic amino acids exhibit characteristic ultraviolet fluorescence. The fluorescence-excitation spectra, fluorescence spectra and quantum yields of aqueous solutions have been studied. Further papers will deal with aspects of the ultraviolet fluorescence of peptides and proteins.

Characterization of a substance as a fluorescent entity

It is often necessary to determine whether the fluorescence shown by a solution is due to a given substance present in it. To ascribe the observed fluorescence to a given component of the system the following criteria are proposed:

Fluorescence-excitation spectrum. The quantum yield of the fluorescence of a substance in *solution*, defined as the ratio of the number of quanta emitted to the number of quanta absorbed, is known to be independent of the exciting wavelength, at least for excitation with light in the air ultraviolet ($\lambda > 2000$Å) and visible regions of the spectrum (Wavilov, 1927; Neporent, 1947; Weber & Teale, in preparation). Therefore,

$$F(\lambda) = kq A(\lambda), \qquad (1)$$

in which $F(\lambda)$ is the fluorescence intensity set up by excitation with light of wavelength λ, $A(\lambda)$ the number of photons absorbed in the solution, q the quantum yield and k a constant depending on the general geometry of the system and the distribution of the intensity along the exciting beam. A solution of optical density $E(\lambda)$ illuminated with light of this wavelength of intensity $I(\lambda)$ photons absorbs

$$A(\lambda) = I(\lambda)\,(1 - 10^{-E(\lambda)}). \qquad (2)$$

From (1) and (2)

$$F(\lambda) = qkI(\lambda)\,(1 - 10^{-E(\lambda)}). \qquad (3)$$

For an arbitrary wavelength $\bar{\lambda}$

$$qk = F(\bar{\lambda})/I(\bar{\lambda})\,(1 - 10^{-E(\bar{\lambda})}). \qquad (4)$$

If (4) is used to eliminate qk from (3)

$$F(\lambda) = \frac{F(\bar{\lambda})\,I(\lambda)}{I(\bar{\lambda})}\left[\frac{1 - 10^{-E(\lambda)}}{1 - 10^{-E(\bar{\lambda})}}\right].$$

$\log \Omega$ may now be defined as

$$\log \Omega \equiv \log 1 \Big/\!\left\{1 - \frac{F(\lambda)/F(\bar{\lambda})}{I(\lambda)/I(\bar{\lambda})}\,(1 - 10^{-E(\bar{\lambda})})\right\} = E(\lambda). \quad (5)$$

$\bar{\lambda}$ may be conveniently chosen to be the wavelength of the maximum of the absorption band of least frequency. Then $F(\lambda)\,I(\bar{\lambda})/F(\bar{\lambda})\,I(\lambda)$ is simply the fluorescence relative to that set up by excitation with the wavelength of the absorption maximum referred to equal photon excitation. Equation (5) shows that a plot of $\log \Omega$ against λ should reproduce the absorption spectrum of the fluorescence entity which may be identified in this way. It must be stressed that k is very sensitive to a change in the distribution of the fluorescent intensity along the exciting beam, and therefore equation (5) is only applicable when $E(\lambda)$ is small. In practice $E(\lambda)$ must be less than 0·1, preferably less than 0·05. When $E(\bar{\lambda})$ may be reduced to 0·05 or less equation (5) may be reduced to the simpler form

$$E(\bar{\lambda})\,\frac{F(\lambda)\,I(\bar{\lambda})}{F(\bar{\lambda})\,I(\lambda)} = E(\lambda). \qquad (6)$$

Fluorescence spectrum. It is known that the fluorescence spectrum continues the absorption spectrum towards longer wavelengths. Although the difference between the maximum of absorption λ_A and the maximum of emission λ_F varies considerably from one substance to another and often in the same substance depending on the solvent, always $\lambda_F > \lambda_A$. This provides a useful criterion to exclude certain components of a solution as responsible for the observed fluorescence.

Quantum yield. The possibility that the fluorescence of a solution may be due to an impurity can often be resolved by a knowledge of the overall quantum yield. In a system consisting of several components the overall quantum yield due to excitation by light of wavelength λ is

$$\overline{q(\lambda)} = q_a\,\frac{E_a}{E(\lambda)} + q_b\,\frac{E_b}{E(\lambda)} + \ldots + q_n\,\frac{E_n}{E(\lambda)},$$

where q_a, q_b, ..., are the quantum yields and $E_a/E(\lambda)$, $E_b/E(\lambda)$, ..., the fractional absorptions of the respective components. Let the principal component of the system be a. Since all q's $\leqslant 1$ and

$$E(\lambda) = E_a + E_b + \ldots + E_n,$$

$$\overline{q(\lambda)} \leqslant q_a\,\frac{E_a}{E(\lambda)} + \frac{E(\lambda) - E_a}{E(\lambda)};$$

but also $E_a/E\lambda \leqslant 1$. Therefore

$$\left.\begin{aligned}\overline{q(\lambda)} &\leqslant q_a + \left[1 - \frac{E_a}{E(\lambda)}\right]\\[6pt]\text{and}\qquad q_a &\geqslant \overline{q(\lambda)} - \left[1 - \frac{E_a}{E(\lambda)}\right].\end{aligned}\right\} \qquad (7)$$

Thus in stable substances purified by crystallization a quantum yield of a few per cent indicates that the fluorescence is due to the main component of the system.

EXPERIMENTAL

The fluorescence-excitation spectra were determined by means of the apparatus the block diagram of which is shown in Fig. 1. Light from a high-pressure Nester-type hydrogen arc H enters the monochromator after passing through the collimating system K. The monochromatic beam emerges from the exit slit X into the light-tight box B containing the cell C filled with the fluorescent solution. The cell axis makes an angle of 15° with the line of view of the detector so that reflected exciting light is thrown on to the blackened wall of the box opposite to the detector. This is a 27M3 Mazda photomultiplier. The envelope of this type of photomultiplier is transparent to wavelengths longer than 2200 Å. A stabilized power supply was used to energize the photomultiplier (Fellgett, 1954) and the photoelectric currents were read directly on a Scalamp galvanometer. The stability of the system permitted the determination of photoelectric currents to within $\pm 2 \cdot 10^{-3}\,\mu\text{A}$. In solutions of amino acids the scattering of the exciting light was negligible and no filter was required to separate the fluorescent from the exciting light. Ordinary fused-quartz cells such as those used in ordinary spectrophotometric work were found to give a blue fluorescence when irradiated with light of wavelength 230–260 mμ. The response of the photomultiplier was thus a response to the fluorescence of both cell and solution. Although a correction for the cell fluorescence is simply done by determining the response to the cell and solvent on one hand and to the cell and solution on the other, it is important to keep this correction as small as possible. The fluorescence of cells made of Ultrasil Haereus

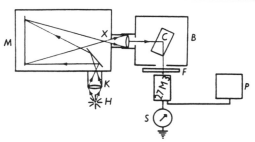

Fig. 1. Experimental arrangement for determination of excitation spectra of fluorescence. H, Nester-type hydrogen arc; K, collimating system; M, grating monochromator; X, exit slit; B, light-tight box containing the silica cell C with the fluorescence solution; F, filter to eliminate exciting light; 27M3, Mazda photomultiplier with P, power pack and S, Scalamp galvanometer.

(Quartz Glass Ltd., Barkingside, Essex) was found to be six to eight times weaker than that of cells of ordinary fused quartz. Such cells were used throughout this work. The monochromator was a Bausch & Lomb instrument with an aluminium grating blazed for the first order in the ultraviolet, giving a dispersion of 3·3 mμ./mm. at the exit slit. To determine excitation spectra the relative distribution of energy in the spectrum of the source was obtained by two methods. The first was the blackening of a photographic plate (e.g. Weissberger, 1946), the validity of the reciprocity law being assumed. The second method consisted in the use of a concentrated solution of a fluorescent substance as a proportional photon counter (Bowen, 1936; Bowen & Sawtell, 1937). As the fluorescent yield and spectrum are independent of the exciting wavelength, if the concentration of the solution is such that total absorption of the beam takes place in a very thin layer at the front of the cell, the dimensions of the fluorescent volume, and therefore the geometry of the system, remain approximately independent of wavelength and the response of the detector will be proportional to the number of photons absorbed by the solution. At the concentrations required for complete absorption of the light in a very thin layer ($>10^{-2}$M) many fluorescent substances form an appreciable proportion of non-fluorescent dimers. As the absorption spectrum of the dimers differs from that of the monomers the fluorescent yield will be found to vary with wavelength. The selection of a suitable fluorescent substance for this purpose was therefore a matter of trial and error. Substances in which the fluorescent yield of the concentrated solution increases markedly with temperature or in which the absorption spectrum of the concentrated solutions shows absorption bands absent in the dilute solutions are therefore inadequate for this purpose (Förster, 1951). If water solutions were used, a relatively small molecular ion offered the best chances of remaining unaggregated at high concentrations. Solutions (10^{-2}M) of sodium 1-dimethylaminonaphthalene-5-sulphonate, 1-dimethylaminonaphthalene-7-sulphonate and 1-aminonaphthalene-3:6:8-trisulphonate in water were used. These substances have the further advantages that the overlap of the absorption and emission bands is negligible and that the quantum yields of the first two are very high (0·75 for 1-dimethylaminonaphthalene-7-sulphonate, 0·53 for the 1:5 isomer) (Weber & Teale, 1956).

Fig. 2 shows that the results obtained with these salts agreed very satisfactorily with the photographic method. Finally, if the direct response of the photomultiplier to the source was corrected, with the spectral-response curve provided by the manufacturers, good agreement with the other methods was obtained for the wavelengths to which the photomultiplier envelope was transparent ($\lambda > 2200$Å).

Fluorescence spectra

These were determined with the monochromator as shown in Fig. 3. The cell C containing the fluorescent solution was mounted in front of the collimator K of the monochromator. The fluorescence was excited by light from the low-pressure mercury arc (Westinghouse Sterilamp 794) filtered through a 3 cm. layer of 0·01 M p-nitrophenol and M-NiSO$_4$ solution in water. This combination effectively isolated the 2537Å line, which carried over 90% of the unfiltered energy output of the source from the other ultraviolet mercury lines. When very weak ultraviolet fluorescences were studied with this excitation filter the weak mercury lines at 2980, 3130 and

3350Å were recorded and appeared superimposed on the fluorescent spectrum. Fortunately with the aromatic amino acids the fluorescence was sufficiently strong to make unnecessary any correction for the breakthrough of these lines. As shown in Fig. 3 the fluorescent light leaves the cell after traversing a variable thickness of solution where some of the fluorescent light may be reabsorbed to excite further fluorescence. The spectrum of the fluorescent light entering the monochromator is the technical spectrum (Birks, 1954) which differs from the molecular spectrum in two respects: the emission at wavelengths at which the absorption and fluorescence spectra overlap appreciably appears attenuated, and since the fluorescence spectrum is independent of the exciting wavelength, the re-emission of part of the absorbed light increases the emission at those wavelengths where overlap is negligible. Thus the corrections to be applied to obtain the molecular from the technical spectrum are an attenuation correction and a re-emission correction.

Attenuation correction. If $\alpha(\lambda)$ is the transmission coefficient of the solution for wavelength λ, $\bar{l}$ the average thickness of solution traversed by the rays entering the

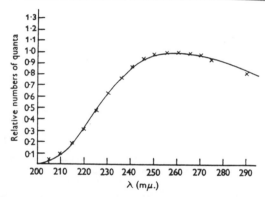

Fig. 2. Spectral photon distribution of the hydrogen arc, as emerging from grating monochromator. The continuous line is the plot of the distribution obtained by the use of fluorescent solutions of naphthylaminesulphonic acids as photon counters. The crosses are the values from the Ilford ultraviolet Q3 plates, using the reciprocity law.

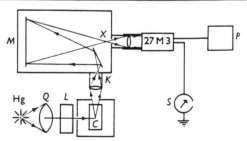

Fig. 3. Experimental arrangement for determination of fluorescence spectra. Hg, Low-pressure mercury arc (Sterilamp 974); Q, quartz lens; L, liquid filter (3 cm. layer of 0·01 M p-nitrophenol and M-NiSO$_4$); C, cell containing fluorescent solution; K, M, P, S and X as in Fig. 1.

monochromator and $F(\lambda)$ the intensity at the exit slit corrected for the change of response of the detector with wavelength and for losses in the grating, then $F(\lambda)/\alpha(\lambda)^l$ is the intensity corrected for attenuation.

Re-emission correction. Although it is difficult to derive an exact expression for the re-emission correction, the order of this correction may be estimated as follows. The ratio of the areas under the emission bands with and without correction for attenuation is

$$\beta = \int F(\lambda)/\alpha(\lambda)^l d\lambda \Big/ \int F(\lambda)\, d\lambda. \qquad (8)$$

The fractional contribution of the secondary fluorescence to the total fluorescence entering the monochromator is $(\beta - 1)\, q$, and this will be the fractional distortion of the band when the correction is ignored. If a sufficiently dilute solution is used $\beta - 1$ can be reduced to a few per cent. In such cases if q itself is small the re-emission correction may be altogether ignored as it falls within the experimental error of the determinations.

Correction for attenuation by the grating. The relative attenuation by the grating was determined as follows: light emerging at the exit slit of the grating monochromator was allowed to fall upon the entrance slit of a D. 246 Hilger prism monochromator, the detector being placed at the exit slit of the latter instrument. The band width of the light emerging from the grating instrument was kept large so that the effective band width of the light reaching the detector was always determined by the prism instrument. The source was then placed directly in front of the entrance slit of the prism instrument and the response of the detector to the range of wavelengths again determined. The ratio of the response to light reaching the detector through grating and prism to the response through prism alone is proportional to the grating transmission for this wavelength. The slit widths of both instruments were kept constant throughout the calibration. The relative attenuation by the grating was found to change slowly with wavelength, less than $0.5\%/\text{m}\mu$. in the region of most rapid change. Thus the distortion introduced by ignoring the grating correction is small for the relatively narrow fluorescent bands but becomes important in the determination of relative quantum yields, when the areas under the bands emitted at very different wavelengths have to be compared.

Quantum yields of fluorescence

The method of Weber & Teale (1956) was used to determine the quantum yield of the fluorescence of the amino acids. In this method the intensity of the fluorescence emitted at right angles to the direction of excitation is compared with the intensity of the light scattered in the same direction by a glycogen solution. If the exciting radiation is monochromatic and the apparent optical densities of the scattering and fluorescent solutions are the same, the glycogen solution acts as a standard of quantum yield 1. Then the ratio of the fluorescent to the scattered intensity is proportional to the quantum yield, after a correction for the unequal spatial distribution of the radiation has been applied. This correction may be computed from measurements of the polarization of the radiation, and a detailed analysis leads to the equation:

$$q = \frac{S_F(3 + p_F)\, f(\lambda_0)}{S_S(3 + p_s)\, \overline{f(\Delta\lambda)}}. \qquad (9)$$

In this equation S_S and S_F are the slopes in the plot of the signal from the scattering and fluorescent solutions respectively against the optical density E when E tends to zero. p_s and p_F are the linear polarizations of the right-angle scattering and fluorescence respectively; $f(\lambda_0)$ and $\overline{f(\Delta\lambda)}$ are factors characterizing the response of the detector to a photon of scattered exciting light of wavelength λ_0 and the average response to a photon of fluorescent light respectively. Thus

$$\overline{f(\Delta\lambda)} = \int_{\Delta\lambda} f(\lambda)\, I(\lambda)\, d\lambda \Big/ \int\int_{\Delta\lambda} I(\lambda)\, d\lambda. \qquad (10$$

In the last equation $I(\lambda)$ is the intensity of the fluorescence emitted at wavelength λ. From a knowledge of the fluorescence spectrum and the spectral response of the photomultiplier, $\overline{f(\Delta\lambda)}$ may be easily computed by graphical integration. On the other hand if a solution of a fluorescent substance is used as a proportional photon counter, $f(\lambda_0)/\overline{f(\Delta\lambda)} = 1$. Both methods have been used in the present case. A 10^{-2}M solution of 1-dimethylaminonaphthalene-5-sulphonate was used as a proportional photon counter as already described. Glycogen solutions were used to scatter the $2537\,\text{Å}$ radiation. The polarizations of the scattering and fluorescence were measured by means of a polarizer of fourteen quartz coverslips mounted at the Brewster angle. The experimental values were corrected for transmission of the unwanted component which was approximately 4% (Conn & Eaton, 1954). Relative quantum yields may be determined from the areas under the emission bands. If $E(\lambda)$ is the energy received by the detector from an interval of wavelength $\Delta\lambda$, $E(\lambda)/\Delta\lambda$ is the energy emitted in unit wavelength interval about λ. If $n(\lambda)$ is the number of photons in the same interval,

$$\frac{E(\lambda)}{\Delta\lambda} = n(\lambda)\,\frac{hc}{\lambda}, \qquad (11)$$

with $F(\lambda)$ denoting the signal from the detector corrected for change in response with wavelength and for losses in the monochromator, the quantity $\lambda F(\lambda)$ is proportional to $n(\lambda)$ if $\Delta\lambda$ is kept constant throughout. The areas under the plot of $\lambda F(\lambda)$ against λ are proportional to the number of photons emitted and the ratio of two such areas obtained under *identical conditions of excitation* is the ratio of the quantum yields. The condition of constant $\Delta\lambda$ is simply fulfilled with a grating monochromator by keeping the slit width constant, and this is the reason for the choice of λ rather than ν in the present instance. The optical densities were measured by means of a Uvispek spectrophotometer. Of the substances used the amino acids were commercial samples several times recrystallized. The dimethylaminonaphthalene derivatives were prepared according to Fussgänger (1902); 1-aminonaphthalene-3:6:8-sulphonate was a technical product recrystallized three times from 10% NaCl solution.

Precision and errors

The fluorescence-excitation spectra and fluorescence spectra could be reproduced to $\pm 1\%$, i.e. the area under the bands of two samples done under similar conditions would differ by 1% on the average.

The precision of the quantum-yield determination is discussed elsewhere. In general, the reproducibility gives $\Delta q = \pm 0.01$ for $q < 0.20$ and $\Delta q = \pm 0.02$ for $0.20 < q < 0.5$.

RESULTS

Fluorescence-excitation spectra

Figs. 4–6 show the fluorescence-excitation spectra of the three aromatic amino acids plotted with the corresponding absorption spectra. Close correspondence was obtained for all three substances, including the fine structural details in phenylalanine.

Fluorescence spectra

They are shown in Fig. 7. In tryptophan and phenylalanine the overlap of emission and absorption bands is negligible and $F(\lambda)$ has been

plotted as giving directly the molecular spectrum. In tyrosine, where appreciable overlap of the bands occurs, for a 10^{-5} M solution in a cell of 5 mm. thickness, $(\beta-1) \simeq 0.1$ and $q(\beta-1) \simeq 0.02$. The values have been corrected for attenuation only and $F(\lambda)/\alpha(\lambda)\bar{l}$ has been plotted as the molecular spectrum. Table 1 gives the position of the maxima of emission and the half-width of the bands. From an inspection of the values it appears that in a mixture of these amino acids it will be possible in many cases to detect the contribution from each of them.

In contrast to the absorption spectrum and the fluorescence-excitation spectrum, the emission spectrum of phenylalanine gives no sign of the presence of fine structure, although the resolution of the monochromator and the band width used in the scanning of the spectrum would certainly be

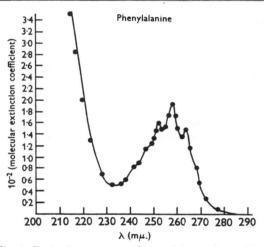

Fig. 4. Excitation spectrum of phenylalanine fluorescence in water. Abscissa: wavelength (mμ.). Ordinate: molecular extinction coefficient. The continuous line is the optical density spectrum; the dots are the values of $\log \Omega$ in equation (5).

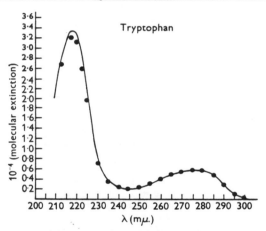

Fig. 6. Excitation spectrum of tryptophan fluorescence in water. Co-ordinates are as in Fig. 4.

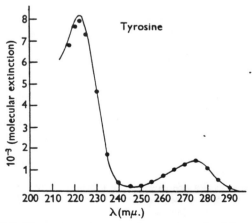

Fig. 5. Excitation spectrum of tyrosine fluorescence in water. Co-ordinates are as in Fig. 4.

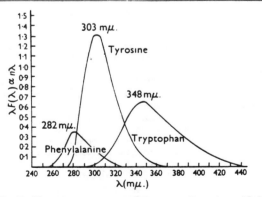

Fig. 7. Fluorescence spectra of the aromatic amino acids in water. Abscissa: wavelength (mμ.). Ordinate: relative number of quanta.

Table 1. *Maxima and band widths of the fluorescence of the aromatic amino acids in neutral aqueous solution*

λ_M is the emission maximum, $\lambda_{\frac{1}{2}}$ and $\lambda_{-\frac{1}{2}}$ are the two values of λ at which the emission becomes one-half of that at the maximum. All values are in mμ.

	λ_M	$\lambda_{\frac{1}{2}}$	$\lambda_{-\frac{1}{2}}$
Phenylalanine	282	298	270
Tyrosine	303	321	287
Tryptophan	348	383	323

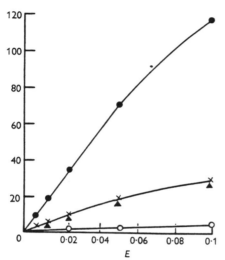

Fig. 8. Quantum yield of the aromatic amino acids in water. Integrating screen: 10^{-2}M solution of sodium 1-dimethylaminonaphthalene-5-sulphonate. Abscissa: optical density of solutions. Ordinate: signal from photomultiplier. ●, Glycogen; ×, tyrosine; ▲, tryptophan; ○, phenylalanine.

Table 2. *Quantum yield of neutral aqueous solutions of the aromatic amino acids*

q_1 is the quantum yield obtained by the use of a concentrated solution of 1-dimethylaminonaphthalene-5-sulphonate as an integrating screen, q_2 is the quantum yield obtained by using the direct response of the photomultiplier and equation 10. The value of $\overline{f(\Delta\lambda)}/f(\lambda_0)$ calculated from the fluorescence spectra and the change of detector response with wavelength is given in the third column. q_r is the yield relative to tyrosine as measured by the area under the emission band corrected for grating transmission. The relative grating transmission factors g calculated are shown in the last column.

	q_1	q_2	$\overline{f(\Delta\lambda)}/f(\lambda_0)$	q_r	g
Phenylalanine	0·045	0·038	0·97	0·23	0·95
Tyrosine	0·21	0·21	0·93	1	0·99
Tryptophan	0·19	0·20	0·76	0·92	0·887

capable of revealing it. This disappearance of fine structure in the fluorescence spectrum indicates the increased interaction between the excited molecular oscillator and the permanent water dipoles nearby. The fine structure of the phenylalanine emission spectrum may conceivably be found in polypeptides and proteins where the water dipoles in the vicinity of the amino acid are replaced by other entities. Fig. 7 also shows that approximately 15 % of the quanta from tryptophan are emitted at wavelengths longer than 395 mμ., appearing as a violet fluorescence to the unaided eye.

Quantum yields

Fig. 8 shows the signal from the detector plotted against the optical density E of the solutions at 2537 Å, for solutions of the three amino acids in water (pH 6·5–7) and for glycogen solutions. The polarizations of the fluorescence for the three amino acids in water are less than 1/100, whereas the polarization of the 2537 Å radiation scattered from glycogens was found to be $0·85 \pm 0·01$. Therefore $3 + p_F/3 + p_s = 0·78$ for the three amino acids. Most of the depolarization of the scattered light from glycogen is probably due to divergence of the exciting beam, which was uncollimated. The values of q calculated are shown in Table 2.

The *relative* efficiencies of the aromatic amino acids have also been measured by integration of the bands obtained in plotting the energy response of the detector against wavelength for constant wavelength band width as shown in Fig. 7. The relative values obtained by integration are in agreement with the absolute values obtained by the previous method, as shown in Table 2.

DISCUSSION

The preceding results show that the fluorescence observed in solutions of the aromatic amino acids is due to the amino acids themselves and not to accidental impurities present in the solutions.

In a recent publication of Shore & Pardee (1956) data are presented on the ultraviolet fluorescence of the aromatic amino acids which are at variance with those presented in this paper. The absolute fluorescence efficiencies quoted by these authors are two to three times lower than those given in Table 1. The relative efficiencies are also different, tyrosine being given as having only two-thirds of the tryptophan fluorescence. Moreover, a variation of the fluorescent efficiency with exciting wavelength, of up to 100 % in one case, is claimed. We believe that these results are due to the method employed by Shore & Pardee for the determination of the fluorescent efficiencies. Their method depends on the comparison of the response of the photomultiplier to the exciting light on one hand and to the fluorescence set up by the absorption of a known fraction of the exciting light on the other, the fluorescent and exciting light being separated by filters. Thus corrections for the geometry, integrated transmission of the filter and integrated response of the photomultiplier have to be calculated for each fluorescent substance. Although by methods neither fully described nor referred to in the paper the first two corrections have been estimated to one significant figure and the third calculated from technical data to the same accuracy, the authors give some of their results to three significant figures. Although the absolute values obtained under these conditions can be taken only as a very rough approximation, their relative figures are explained by the use of a filter with a cut-off at 300 mμ. to separate excitation from emission. Fig. 7 shows that such a filter will intercept about 80 % of the phenylalanine fluorescence, 40 % of the tyrosine and a negligible fraction of the tryptophan fluorescence. An overall transmission factor of 0·6 was used by Shore & Pardee for the three amino acids. The change in fluorescent efficiency with wavelength must be attributed to systematic errors in the method used, since the much more accurate technique of the excitation spectra gives no indication of such changes.

The overlap of the emission spectrum of tyrosine with its own absorption band is of the order of that found in fluorescein and rhodamine B, where transfer of the electronic energy from an excited molecule to a nearby molecule in the ground state is known to occur and to lead to the depolarization of the fluorescence observed in concentrated solutions (Gaviola & Pringsheim, 1924; Pheofilov & Sveshnikoff, 1941). From such depolarization measurements the distance at which probability of transfer equals probability of emission for parallel oscillators has been found for several dyes to be about 30 Å (Weber, 1954). From the number of tyrosine residues and the dimensions of ordinary globular protein molecules it appears that internal transfer of the excitation energy of the tyrosine must occur. Still more probable appears the transfer from phenylalanine to tryptophan or to tyrosine, from tyrosine to tryptophan and from all the aromatic amino acids to the haem group of haem-proteins. This is no doubt the reason for the high quantum yield of the photodissociation of carboxymyoglobin when illuminated by light absorbed by the aromatic residues of the protein (Bücher & Kaspers, 1946). Such possibilities of transfer place definite limitations on the interpretation of the action spectra of photochemical effects in the complex particulate systems of the living cell.

SUMMARY

1. The fluorescence-excitation spectrum, fluorescence spectrum and quantum yield are proposed as criteria to characterize a molecular species in solution as a fluorescent entity.

2. The fluorescence spectra of phenylalanine, tyrosine and tryptophan in neutral water solution are shown to consist of single bands in the ultraviolet with maxima at 282, 303 and 348 mμ. respectively.

3. The fluorescence-excitation spectra determined from 200 to 320 mμ. correspond accurately to the known absorption spectra of the amino acids, showing the constancy of the quantum yield over the range of wavelengths investigated.

4. The quantum yields ($\pm 1 \%$) in neutral water solutions are 4 % for phenylalanine, 21 % for tyrosine and 20 % for tryptophan.

5. The possibility of electronic energy transfer among the aromatic residues in proteins and to the haem in haemoproteins is discussed.

The financial help of the Medical Research Council is gratefully acknowledged. One of us (F.W.J.T.) is indebted to the Agricultural Research Council for a personal grant.

REFERENCES

Birks, J. B. (1954). *Scintillation Counters*. London: Pergamon Press.
Bowen, E. J. (1936). *Proc. Roy. Soc.* A, **154**, 349.
Bowen, E. J. & Sawtell, J. W. (1937). *Trans. Faraday Soc.* **33**, 1425.
Bücher, Th. & Kaspers, J. (1946). *Naturwissenschaften*, **33**, 93.
Conn, G. K. T. & Eaton, G. K. (1954). *J. opt. Soc. Amer.* **44**, 553.
Debye, P. & Edwards, J. O. (1952). *Science*, **116**, 143.
Fellgett, P. (1954). *J. sci. Instrum.* **31**, 217.
Förster, Th. (1951). *Fluorescenz Organischer Verbindungen*, chap. 9. Goettingen: Vandenhoeck and Ruprecht.
Fussgänger, V. (1902). *Ber. dtsch. chem. Ges.* **35**, 976.
Gaviola, E. & Pringsheim, P. (1924). *Z. Phys.* **24**, 24.
Kowalski, I. von (1911). *Phys. Z.* **12**, 956.
Ley, H. & Englehardt, K. von (1910). *Z. phys. Chem.* **74**, 1.
McClure, D. S. (1949). *J. chem. Phys.* **17**, 905.
Marsh, J. K. (1924). *J. chem. Soc.* **125**, 418.
Neporent, B. S. (1947). *Zhur. Fiz. Khim.* **21**, 1111.
Pheofilov, P. & Sveshnikoff, B. J. (1941). *J. Phys., Moscow*, **3**, 493.
Shore, V. G. & Pardee, A. B. (1956). *Arch. Biochem. Biophys.* **60**, 100.
Wavilov, S. I. (1927). *Z. Phys.* **42**, 311.
Weber, G. (1954). *Trans. Faraday Soc.* **50**, 552.
Weber, G. & Teale, F. W. J. (1956). *Trans. Faraday Soc.* (in the Press).
Weissberger, A. (1946). *Physical Methods of Organic Chemistry*, vol. 2, p. 787. New York: Interscience.

Spectral Broadening in Biomolecules

V. Šrajer, K. T. Schomacker, and P. M. Champion

Department of Physics, Northeastern University, Boston, Massachusetts 02115
(Received 22 May 1986)

We have studied the optical line shapes of deoxy myoglobin (Mb) and its ligand-bound form (MbCO). Simulations of the observed line shapes using "transform" and time-correlator techniques show the presence of strong coupling to low-frequency modes in MbCO and substantial *non-Gaussian* inhomogeneous broadening in Mb. The iron-porphyrin disorder parameter in Mb is found to be $\sigma = 0.25$ Å, and the quadratic dependence of the π-π^* electronic excitation energy on the iron coordinates is also determined.

Heme proteins form a class of biological molecules that are involved in a wide variety of fundamental life processes. A large chromophore, known as the heme group, is common to these proteins and plays a key role in the various biological functions (e.g., electron transport, catalysis, oxygen storage, and transport). The iron atom, coordinated at the center of the heme group, generally cycles between two oxidation and/or spin states depending on the biological function being performed. In the case of cytochrome c, the familiar electron-transport protein, two axial ligands along with the four planar porphyrin nitrogens coordinate the central iron atom in a strong-field octahedral arrangement. The iron atom is quite literally "locked" in place in this system and always maintains the low-spin configuration.

In contrast to cytochrome c, the iron atom in the oxgen-storage protein myoglobin maintains its oxidation state at the ferrous level while cycling between the low-spin ($S=0$) ligand bound state and the high-spin ($S=2$) ligand free state. In this study we consider the carbon-monoxide–bound species (MbCO), and the ligand-free "deoxy" state (Mb). This latter state has a five-coordinate iron atom that is displaced by approximately 0.5 Å from the mean heme plane towards the nitrogen atom of the proximal histidine ligand.[1,2]

Previous work on MbCO, involving the kinetics of CO rebinding as a function of temperature,[3] has produced strong experimental evidence for the existence of protein conformational substates. Other spectroscopic probes, involving Mössbauer,[4] x-ray,[5] and far-infrared magnetic resonance,[6] have also displayed evidence of distributions in protein conformation. In the present work we analyze the optical spectra using "transform" and time-correlator techniques[7-10] and then develop a simple model to account for the inhomogeneous broadening found in the deoxy species.

In general, the optical properties of heme proteins are dominated by the broad (~ 2000 cm^{-1}) and intense ($\epsilon \sim 10^5$ mol^{-1} cm^{-1}) near-ultraviolet (~ 400 nm) "Soret" transition. The Soret band is derived from allowed π-π^* electronic excitations of the porphyrin ring and is asymmetrically broadened by the linear (Franck-Condon) coupling of numerous heme normal modes. The situation is quite analogous to the phonon broadening of optical transitions associated with localized defects in crystals (color centers) and much of the original theory is based on early studies of this classic problem.[7,11]

We have recently demonstrated[10] a strategy for the analysis of optical line shapes that is strongly interdependent on the measurement of absolute resonance Raman-scattering cross sections. When the absolute Raman cross sections are known, a simple Kramers-Kronig transform procedure along with the assumption of linear harmonic coupling allows the direct recovery of the electron-nuclear coupling strengths, $\{S_i\}$, of all modes coupled to the resonant excitation. The Raman spectrum also yields the mode frequencies, $\{\bar{\nu}_i\}$, so that the Franck-Condon broadening can be calculated explicitly by use of time-domain expressions that are exact at all temperatures.[8] The absorption cross section at frequency $\bar{\nu}$ is given by

$$\sigma_A(\bar{\nu}) = C_2 \bar{\nu} \, \text{Im} \Phi(\bar{\nu}), \tag{1}$$

with

$$\Phi(\bar{\nu}) = i \int_0^\infty dt \exp(i\bar{\nu}t - \Gamma t)\eta(t) \tag{2}$$

and the time correlator

$$\eta(t) = \exp(-i\bar{\nu}_0 t)\exp\left\{ -\sum_i S_i[(2\langle n_i \rangle + 1)(1 - \cos\bar{\nu}_i t) + i\sin\bar{\nu}_i t]\right\}, \tag{3}$$

where $\bar{\nu}_0$ is the 0-0 transition frequency (zero-phonon line) and $\{\langle n_i \rangle\}$ are the Bose-Einstein factors. In Eq. (2),

Γ describes the damping of the Soret excitation by electronic nonradiative population decay.

Two additional sources of line broadening are also explored in the analysis. First, we consider a bath of low-frequency modes (such as chromophore librations and/or bulk protein motions) that is directly coupled to the resonant excitation and yet is not observed in the resonance Raman spectrum. The fact that the frequency dependence of the resonance Raman-scattering cross section is described by two amplitudes having opposite phase,

$$\sigma_R(\tilde{\nu}, \tilde{\nu}_1) = C_1 \tilde{\nu}(\tilde{\nu} - \tilde{\nu}_1)^3 (\langle n_i \rangle + 1) S_i |\Phi(\tilde{\nu}) - \Phi(\tilde{\nu} - \tilde{\nu}_i)|^2, \tag{4}$$

means that weakly coupled low-frequency modes may not be directly observed, as a result of cancellation of the amplitudes as the Raman mode frequency $\tilde{\nu}_i \to 0$.[9] In order to approximate the presence of the bath, we use an Einstein approximation with an average frequency $\tilde{\nu}_b$, and integrated bath coupling S_b. The presence or absence of the bath is exposed experimentally by careful measurement of the absorption line shape as a function of temperature.[10]

The second additional source of broadening is inhomogeneous effects that can arise from different protein conformations and environments that are imposed on the heme group. Such broadening is usually modeled as a statistical (i.e., Gaussian) distribution of energy levels.

In Fig. 1 we display the Soret transition of the six-coordinate MbCO complex. The high-resolution spectrum ($\times 0.1$) underlying the broad curves is obtained from Eqs. (1)–(3), with $\Gamma = 10$ cm^{-1} for the excited-state damping factor. With the exception of $(\tilde{\nu}_b, S_b)$, the other parameters $\{\tilde{\nu}_i, S_i\}$ are extracted from the resonance Raman spectra (measured on an absolute scale), by use of Eq. (4) and

$$\Phi(\tilde{\nu}) = \frac{1}{\pi} \int_{-\infty}^{\infty} \frac{\epsilon(\tilde{\nu}')}{\tilde{\nu}'(\tilde{\nu}' - \tilde{\nu})} d\tilde{\nu}' + i \frac{\epsilon(\tilde{\nu})}{\tilde{\nu}}, \tag{5}$$

where $\epsilon(\tilde{\nu})$ is the absorption cross section. The broad solid lines in Fig. 1 are the theoretical results obtained by increasing Γ and including low-frequency coupling. The data are artificially scattered in order to allow visu-

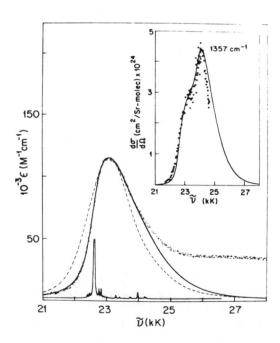

FIG. 2. The experimental Soret band (lower) of deoxy Mb at 290 K is shown with an unsuccessful attempt (dashed line) to fit the data using $\Gamma = 300$ cm^{-1} and a Gaussian inhomogeneous broadening 450 cm^{-1}. A successful fit is generated with $\tilde{\nu}_0 = 22\,630$ cm^{-1}, $\Gamma = 285$ cm^{-1}, and a non-Gaussian inhomogeneous distribution function with $z_0 = 0.5$ Å, $\sigma_z = 0.25$ Å, and $b_z = 2570$ cm^{-1}/Å^2 (solid line). The coupling strengths are again found from the transform analysis and refined to account for the inhomogeneous distribution. When the same distribution is used to calculate the Raman excitation profile of the 1357-cm^{-1} mode (inset) a distinct improvement over the simple transformed absorption band (Ref. 12) is evident.

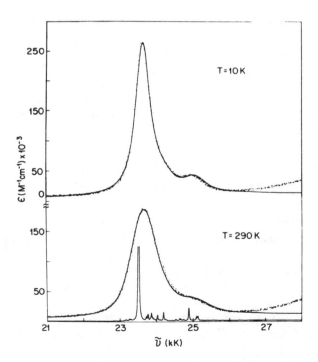

FIG. 1. The experimental Soret bands of MbCO at 10 and 290 K are shown with the solid curves calculated by use of the correlator theory 1 kK (kilokayser) $= 10^3$ cm^{-1}. The fitting parameters Γ and $\tilde{\nu}_0$ are found to be 217 and 23 530 cm^{-1}, respectively. The major reduction in the linewidth observed at low temperature is accounted for by coupling to a low-frequency bath having $\tilde{\nu}_b S_b = 120$ cm^{-1} with $\tilde{\nu}_b < 50$ cm^{-1}.

al differentiation of the theoretical and experimental curves.

It is important to notice in Fig. 1 that a major reduction in the Soret-band linewidth takes place at low temperature. The magnitude of this effect in MbCO is much more pronounced than found for cytochrome c and cannot be explained simply by thermal depopulation of the observed low-frequency modes. Thus, the Soret transition of MbCO must involve coupling to a previously unobserved low-frequency subspace. The parameters leading to the best fits of the MbCO spectra are given in the figure caption. It should be noted that the low-frequency mode coupling in MbCO is at least an order of magnitude larger than found for ferrocytochrome c.[10]

Our attempts to fit the observed asymmetric broadening of the deoxy Mb spectrum, using these same techniques, proved unsuccessful. When the broadening parameters are increased to account for the overall width of the spectrum, a severe mismatch in the low-energy region is inevitable (see dashed line in Fig. 2). These results have led us to reexamine the nature of inhomogeneous broadening in heme protein optical spectra.

We suggest that stochastic environmental perturbations due to the surroundings of the heme may *not* be primarily responsible for the modulation of the π-π^* electronic excitation energy. Rather, it may be disorder in the position and orientation of the central iron atom with respect to the heme plane that determines the π-π^* energy distribution. Since the iron d electrons interact at close range with the π electrons of the porphyrin ring, we might expect this to be a dominant

source of inhomogeneous broadening. In the six coordinate compounds the iron-porphyrin distribution is narrow as found for both cytochrome c and MbCO. In the five-coordinate deoxy complex it is likely that the iron atom is more loosely bound and the relative mean square displacement in its z coordinate (position of the iron atom along the normal to the plane) may be substantial. Additional distributions in the orientational space, as defined by the vector connecting the iron and histidine nitrogen, can also be imagined. Such distributions, may involve the "tilt" of the polar angle (θ) and/or the "rotation" of the azimuthal angle (ϕ) with respect to the heme normal and the porphyrin nitrogen atoms. The azimuthal degree of freedom is of particular interest, since it has been implicated as a source of numerous Raman observations[12] and a wide range of angles ($\phi \sim 0°$–$20°$) is clearly accessible.[1,2]

Given that all three coordinates (z, θ, ϕ) may have importance, we write a general expression for the π-π^* excitation energy:

$$E(Q) \cong E_0 + bQ^2, \tag{6}$$

where Q represents a generalized iron coordinate (z, θ, or ϕ), E_0 is the zero-order π-π^* excitation energy, and b is a constant describing how the coordinates of the iron atom affect the first-order π-π^* excitation energy, $E(Q)$. Only even powers of Q are retained in the expansion because of the approximate reflection symmetry that exists between the iron atom and the porphyrin ring [$E(Q) \cong E(-Q)$]. If we now assume a Gaussian distribution in coordinate space, we find, using Eq. (6), a *non-Gaussian* distribution of electronic energy levels:

$$P(E) = \frac{[(E - E_0)2\pi b]^{-1/2}}{2\sigma} \left[\exp\left(- \frac{[(E - E_0)^{1/2} - Q_0\sqrt{b}\,]^2}{2b\sigma^2} \right) + \exp\left(- \frac{[(E - E_0)^{1/2} + Q_0\sqrt{b}\,]^2}{2b\sigma^2} \right) \right], \tag{7}$$

where Q_0 is the mean coordinate position (e.g., $z_0 = 0.5$ Å) and σ is the disorder parameter of the Gaussian distribution.

When Eq. (7) is employed in the fitting procedure, excellent agreement with the absorption line shape is obtained, as can be seen in the lower portion of Fig. 2 (a separate transition, the "N" band at 27×10^3 cm^{-2} is not fitted with this theory). The effects of the inhomogeneous broadening are also reflected in the Raman excitation profile (REP). When the ensemble average used for the absorption band is included in the REP calculation, good agreement with the data is obtained (Fig. 2, inset). This result should be contrasted with the calculations in Ref. 12 which do not include inhomogeneous broadening and are flawed by a 200-cm^{-1} red shift of the theoretical REP maximum. This is particularly compelling evidence for the non-Gaussian

energy distribution, since all other attempts to fit the REP using Gaussian distributions, non-Condon effects, etc., have been unsatisfactory.

We have listed the fitting parameters in terms of the iron atom displacement coordinate (z), since this is the most commonly considered degree of freedom in these systems. We hold $z_0 = 0.5$ Å to accommodate the x-ray results and let σ_z and b_z vary to achieve the best fits. Alternatively, one can interpret the results in terms of the orientational degrees of freedom (θ, ϕ). For example, if we again fix the mean coordinate position using the x-ray results, $\phi_0 \cong 20°$, then the fitting parameters become $\sigma_\phi \cong 10°$, $b_\phi \cong 1.44$ cm^{-1}/deg^2.

In z space alone the disorder parameter, $\sigma_z = 0.25$ Å, is somewhat larger than might be expected from EXAFS (extended x-ray absorption fine structure) estimates. However, the value for σ is in excellent agreement with the Mössbauer and x-ray crystallography studies.[5] However, since EXAFS is not generally sensitive to the angular degrees of freedom, we believe that the value found for σ may be a reasonable estimate of the overall disorder (displacement and orientational) of the iron atom with respect to the porphyrin ring.

The parameter b is a measure of how the π-π^* energy gap responds to changes in the iron atom coordinates. For example, the value $b_z = 2570$ cm^{-1}/Å^2 indicates that a ~ 25-cm^{-1} (~ 0.5 nm) blue shift of the Soret band accompanies a 0.1-Å movement of the iron atom away from the heme plane. This value of b is in remarkable agreement with the early calculations of Hopfield[13] which yield

$$\left.\frac{dV(z)}{dz}\right|_{z_0} = 2bz_0 = +0.4 \text{ eV/Å} \cong 3200 \text{ cm}^{-1}/\text{Å},$$

where $V(z)$ is the energy of the optical transition as a function of the iron coordinate, and $z_0 = 0.5$ Å.

It has been known for some time that changes in hemoglobin structure, from the low-affinity "T" to the high-affinity "R" conformation, result in alterations of the Soret band position and shape. These spectral shifts are extremely difficult to quantify, as a result of the presence of four heme groups, not all of which necessarily undergo the same spectral changes during the $T \rightarrow R$ transition.[14] At this stage we only comment that the overall magnitude of the Soret-band red shift accompanying the $T \rightarrow R$ transition is consistent with our value for b if a ~ 0.05–0.1-Å motion of the iron atoms toward the heme plane takes place in the α chains only.[14]

In summary, we have demonstrated how optical line-shape analysis of MbCO as a function of temperature indicates little or no inhomogeneous broadening but does suggest relatively strong coupling to low-frequency modes. The behavior observed for MbCO stands in marked contrast to cytochrome c, which has strong covalent attachments between the heme group and the protein macromolecule. Thus, the low-frequency coupling in MbCO may involve torsions or librations of the unrestrained heme chromophore as well as motions of the CO ligand on a predissociative potential surface prior to photodissociation. The present study has also demonstrated that significant non-Gaussian inhomogeneous broadening is able to explain the deoxy Mb spectra at 290 K. Such broadening may be characteristic of the loosely bound iron atom found in the five-coordinate heme complexes.

The rather large value found for the disorder parameter in the deoxy species is in accord with the nonexponential rebinding kinetics described by Austin *et al.*,[3] provided that one recognizes that the relative iron-porphyrin coordinates can affect the potential energy surfaces relevant to the ligand binding process. A more detailed theory is needed in order to relate quantitatively the distributions of the π electron energy levels discussed here with the distributions in rebinding barrier heights described previously.[3] Preliminary work along this line is encouraging and may yield information about the magnitude of the forces involved in returning the iron atom to the heme center as the ligand binds.

This work is supported by the National Science Foundation (Grant No. 84-17712) and the National Institutes of Health (Grants No. AM-35090 and No. AM-01405).

[1]T. Takano, J. Mol. Biol. **110**, 569 (1977).

[2]S. E. V. Phillips, J. Mol. Biol. **142**, 531 (1980).

[3]R. H. Austin, K. Beeson, L. Eisenstein, H. Frauenfelder, I. C. Gunsalus, and V. P. Marshall, Phys. Rev. Lett. **32**, 403 (1974).

[4]K. Spartalian, G. Lang, and T. Yonetani, Biochim. Biophys. Acta. **428**, 281 (1976).

[5]H. Fraenfelder, G. Petsko, and D. Tsernoglou, Nature (London) **280**, 558 (1979).

[6]P. M. Champion and A. J. Sievers, J. Chem. Phys. **72**, 1569 (1980).

[7]V. Hiznyakov and I. Tehver, Phys. Status Solidi **21**, 755 (1967).

[8]C. K. Chan and J. B. Page, Chem. Phys. Lett. **104**, 609 (1984).

[9]K. T. Schomacker, O. Bangcharoenpaurpong, and P. M. Champion, J. Chem. Phys. **80**, 4701 (1984).

[10]K. T. Schomacker and P. M. Champion, J. Chem. Phys. **84**, 5314 (1986).

[11]M. Lax, J. Chem. Phys. 11, 1752 (1952).

[12]O. Bangcharoenpaurpong, K. T. Schomacker, and P. M. Champion, J. Am. Chem. Soc. **106**, 5688 (1984).

[13]J. J. Hopfield, J. Mol. Biol. 77, 207 (1973).

[14]J. S. Olson, Proc. Natl. Acad. Sci. U.S.A. **73**, 1140 (1976).

Platinized Chloroplasts: A Novel Photocatalytic Material

Abstract. *Colloidal platinum was prepared and precipitated directly onto photosynthetic thylakoid membranes from aqueous solution and entrapped on fiberglass filter paper. This composition of matter was capable of sustained simultaneous photoevolution of hydrogen and oxygen when irradiated at any wavelength in the chlorophyll absorption spectrum. Experimental data support the interpretation that part of the platinum metal catalyst is precipitated adjacent to the photosystem I reduction site of photosynthesis and that electron transfer occurs across the interface between photosystem I and the catalyst. Photoactivity of the material was dependent on the nature of the ionic species from which the platinum was precipitated. All photoactive samples were prepared from the hexachloroplatinate(IV) ion, whereas samples prepared by precipitation of the tetraammineplatinum(II) ion showed no hydrogen evolution activity and only transient oxygen activity. This system is among the simplest known for photosynthetically splitting water into molecular hydrogen and oxygen.*

Elias Greenbaum
*Chemical Technology Division,
Oak Ridge National Laboratory,
Oak Ridge, Tennessee 37831*

Photosynthesis can be separated into light and dark reactions (*1–2*). The light reactions occur exclusively in the chloroplast membranes, where the overall process is energized. This ability of the chloroplast to serve as the photochemical "factory" of photosynthesis led to the consideration of systems that drive photoreactions other than carbon reduction.

The three-component CFH system of isolated chloroplasts, ferredoxin, and hydrogenase is one such system. In 1961 Arnon *et al.* (*3*) showed that the CFH system could photoproduce molecular hydrogen when cysteine was used as the electron source. In 1973, Benemann *et al.* (*4*) and Krampitz (*5*) presented biochemical evidence for photosystem II (water splitting)–linked hydrogen evolution. Subsequent development of instrumentation allowed the direct observation of simultaneous photoproduction of molecular hydrogen and oxygen by the CFH system (*6*). Gisbey and Hall (*7*) performed energy conversion efficiency measurements of the CFH system, and Krasna (*8*) demonstrated that platinum catalyst on asbestos can substitute for hydrogenase as the hydrogen-evolving catalyst, although it was necessary to also substitute methyl viologen for ferredoxin as the electron relay between chloroplast and catalyst. Reduced ferredoxin will not interact with platinum catalyst. Ochiai *et al.* (*9*) demonstrated that intact cells of the cyanobacterium *Mastigocladus laminosus* functioned as an anodic photoelectrode for use in a conventional electrochemical cell. An authoritative review of photobiological hydrogen production was presented by Weaver *et al.* (*10*).

In the study reported here, colloidal platinum was precipitated directly onto photosynthetic membranes in an aqueous suspension. The resulting chloroplast–colloidal platinum composition was then entrapped on filter paper. As indicated in Fig. 1, this moistened material was capable of sustained simultaneous photoevolution of hydrogen and oxygen when irradiated with visible light. Since no electron relay was added to the system and the overall reactions occurred in an immobilized matrix, it was concluded that the precipitated colloidal platinum directly contacts the reducing end of photosystem I in such a way that electron flow occurs across the biological membrane–metal colloid interface with preservation of charge continuity and catalytic activity. In addition to the special photocatalytic properties of platinized chloroplasts, the specific ionic species used to prepare the material yield information on the physicochemical properties of the photosystem I reduction site on the thylakoid membrane surface. The data in Fig. 1 were obtained by precipitating platinum from the hexachloroplatinate(IV) ion, $[Pt(Cl)_6]^{2-}$. Platinized chloroplasts prepared by precipitating platinum from the tetraammineplatinum(II) ion, $[Pt(NH_3)_4]^{2+}$, resulted in no hydrogen activity and only a transient oxygen gush. As discussed below, the presence of insoluble platinum on the entrapped filter paper composition was determined by x-ray fluorescence analysis.

Type-C chloroplasts were prepared by the procedure of Reeves and Hall (*11*). In this preparation the chloroplast envelope is osmotically ruptured, exposing the thylakoid membranes to the external aqueous medium. A solution of chloroplatinic acid (5.34 mg/ml), neutralized to *p*H 7 with NaOH, was prepared separately. One milliliter of this solution was combined with 5 ml of chloroplast suspension (containing 3 mg of chlorophyll) in Walker's assay medium (*12*). The 6-ml volume was placed in a temperature-controlled, water-jacketed chamber fitted with O-ring connectors to provide a hermetic seal and inlet and outlet ports for hydrogen flow. The mixture was gently stirred and purged with molecular hydrogen in the headspace above the liquid. The temperature of the sample was held at 21°C.

In the platinum precipitation step, it was determined empirically that a hydrogen incubation time of ~30 minutes was needed to obtain photoactive material. Times of 60 to 90 minutes were typically used. After incubation, the reactor chamber was opened to air and the con-

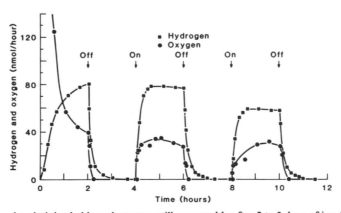

Fig. 1. Simultaneous photoproduction of hydrogen and oxygen by moistened platinized chloroplasts entrapped on filter paper. The ELH projector lamp, providing saturating illumination, was turned on at $T = 0$ and alternated with 2-hour on-off cycles as indicated. The peak oxygen rate was 7 μmol/hour. The area of the disk was 10.2 cm². The activity of

the platinized chloroplasts was still measurable after 2 to 3 days of irradiation cycles at 10 to 20 percent of the initial rates. Greenbaum (*14*) gave a detailed description of the experimental procedure used for measuring the simultaneous photoproduction of hydrogen and oxygen.

tents were filtered onto fiberglass filter paper (AP40, Millipore). The platinum precipitation reaction had a marked effect on filtration properties. Whereas control experiments without hydrogen incubation produced a chloroplast mixture that filtered immediately, the platinized chloroplasts required a considerably longer time—typically 5 to 30 minutes. Also, the platinized chloroplasts were dark green, as opposed to the normal bright green of higher plant chloroplasts. The presence of insoluble platinum on the platinized-chloroplast filter paper composition was identified by x-ray fluorescence analysis after rinsing the filter paper in 2 liters of continuously stirred distilled water for 1 hour. Insoluble platinum was positively identified in this way for platinized chloroplasts prepared by precipitation from $[Pt(Cl)_6]^{2-}$ or $[Pt(NH_3)_4]^{2+}$. As mentioned above, however, only $[Pt(Cl)_6]^{2-}$ yielded photoactive material.

The time profiles of photoactivity are presented in Fig. 1 for three light-dark cycles of 2 hours each. Figure 1 shows that the first irradiation period has a qualitatively different time profile than the subsequent two periods. In the first cycle, the oxygen profile underwent a transient gush (peaking at 7 μmol of O_2 per hour) before settling down to steady state. The hydrogen rate climbed monotonically to steady state. The time required to reach 50 percent of steady state in the first cycle was ~25 minutes, whereas the corresponding time for the second and third cycles was 5 to 8 minutes.

These patterns may be explained as follows. The transient oxygen gush initially represents the filling of reducible pools on the reducing side of photosystem II. Molecular oxygen is capable of oxidizing the components in the electron transport chain linking the two light reactions of photosynthesis, including the plastoquinone pool (13). The platinized chloroplasts were exposed to air during the filtration process plus the time just before insertion in the reaction chamber of the apparatus used for measuring photoactivity (14). Therefore, the initial oxygen gush is believed to represent the filling of the oxidized pool as well as other reducible species in the preparation.

As indicated above, the time for the hydrogen rate to reach steady state was longer for the first interval of irradiation than for the subsequent two intervals, whose time profiles were determined by the response time of the apparatus (15). Since the hydrogen-evolving catalyst in

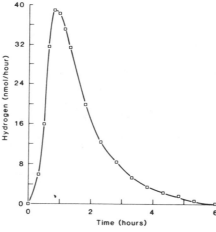

Fig. 2. Control experiment omitting the filtration step in which the platinized chloroplasts are entrapped on filter paper. Hydrogen (and oxygen) produced in the 6-ml aqueous phase is equilibrated with the carrier gas in the headspace and transported downstream to the oxygen and hydrogen sensors downstream. The loss in activity as a function of time is interpreted as the separation of membrane and metal catalyst particle during stirring.

this preparation was elemental platinum, the observed kinetics of the first irradiation interval are interpreted to represent the reduction of the oxide layer formed on the metallic platinum during the period when the platinized chloroplasts are exposed to air, as described above. Since the platinized chloroplast remained anaerobic for the two subsequent irradiations, neither oxygen transients nor hydrogen delays were observed.

A novel aspect of this material is the absence of an added electron relay (such as ferredoxin or methyl viologen) to transport electrons from the reducing end of photosystem I to the hydrogen-evolving catalyst. Two control experiments were performed that emphasize this point. In the first experiment the order of adding chloroplasts and precipitation was reversed. That is, the colloidal platinum was prepared in the absence of chloroplasts (by precipitation from $[Pt(Cl)_6]^{2-}$), and then the chloroplasts were added to the aqueous phase. This mixture was entrapped on fiberglass filter paper. In spite of the physical contact between platinum colloid particles and chloroplasts, no photocatalytic activity was observed, except for a brief oxygen transient.

In the second experiment the filtration step was omitted. The 6-ml aqueous suspension of platinized chloroplasts was stirred in the photoreactor of the assay system with a Teflon stirrer. Hydrogen and oxygen photoproduced in the aqueous phase equilibrated with the carrier

gas in the headspace above the liquid. The time profile for hydrogen evolution is given in Fig. 2. The oxygen profile was virtually the same as this hydrogen profile. As shown in Fig. 2, the hydrogen rate peaked at ~50 minutes and then decayed monotonically with a first half-life of ~110 minutes, a time significantly shorter than the corresponding value for the filter paper–entrapped platinized chloroplasts of Fig. 1. The loss of activity in Fig. 2 was presumably associated with the separation of chloroplast and platinum colloid particles at the photosystem I reducing site.

The only material capable of sustained photoacivity was prepared by precipitating the $[Pt(Cl)_6]^{2-}$ ion in the presence of thylakoid membranes. Platinized chloroplasts prepared by precipitating the $[Pt(NH_3)_4]^{2+}$ ion in the presence of thylakoid membranes were incapable of any hydrogen photoactivity. These results suggest that $[Pt(NH_3)_4]^{2+}$ is repelled (either Coulombically or sterically) by the photosystem I reduction site and that $[Pt(Cl)_6]^{2-}$ has a specific chemical affinity of, perhaps, an ion exchange–like nature for this site. It is known from several types of measurements, such as electrophoretic mobility studies, that thylakoid membrane surfaces bear a net negative electrostatic charge at neutral pH due to carboxyl groups (16). However, the discrete nature of the proteins associated with the membrane prevents this charge from being uniformly distributed across the membrane surface. Photosystem I is capable of donating electrons to negatively charged electron acceptors such as $[Fe(CN)_6]^{3-}$, the classic Hill electron acceptor (17). Moreover, as shown by Zweig and Avron (18) and Kok et al. (19), the methyl viologen (MV^{2+}) cation can be reduced by photosystem I. Chow and Barber (20) have shown that MV^{2+} accumulates in the diffuse layer of the membrane interface. A comparison of the interactions of MV^{2+} and $[Pt(NH_3)_4]^{2+}$ with photosystem I indicates that MV^{2+} can be reduced at a relatively large distance, whereas a smaller distance is required for platinum precipitation and catalytic activity from photosystem I.

In conclusion, a technique has been developed for contacting dispersed platinum catalyst with photosynthetic membranes. It may be possible to extend this technique to other catalysts for dinitrogen and carbon dioxide reduction. Fundamental studies of this nature are clearly of interest in the context of fuel and chemical synthesis from renewable inorganic resources.

References and Notes

1. H. T. Brown and F. Escombe, *Proc. R. Soc. London Ser. B* **76**, 29 (1905).
2. R. Emerson and W. Arnold, *J. Gen. Physiol.* **15**, 391 (1932).
3. D. Arnon, A. Mitsui, A. Paneque, *Science* **134**, 1425 (1961).
4. J. R. Benemann, J. A. Berenson, N. O. Kaplan, M. D. Kamen, *Proc. Natl. Acad. Sci. U.S.A.* **70**, 2317 (1973).
5. L. O. Krampitz, in *An Inquiry into Biological Energy Conversion*, A. Hollaender *et al.*, Eds. (University of Tennessee, Knoxville, 1972), p. 22.
6. E. Greenbaum, *Biotechnol. Bioeng. Symp.* **10**, 1 (1980).
7. P. E. Gisby and D. O. Hall, *Photobiochem. Photobiophys.* **6**, 223 (1983).

8. A. I. Krasna, in *Biological Solar Energy Conversion*, A Mitsui *et al.*, Eds. (Academic Press, New York, 1977), p. 53.
9. H. Ochiai, H. Shibata, Y. Sawa, T. Katoh, *Proc. Natl. Acad. Sci. U.S.A.* **77**, 2442 (1980).
10. P. F. Weaver, S. Lien, M. Seibert, *Sol. Energy* **24**, 3 (1980).
11. S. G. Reeves and D. O. Hall, *Methods Enzymol.* **69**, 85 (1980).
12. D. A. Walker, *ibid.*, p. 94.
13. B. Diner and D. Mauzerall, *Biochim. Biophys. Acta* **305**, 329 (1973).
14. E. Greenbaum, *Photobiochem. Photobiophys.* **8**, 323 (1984).
15. The response time of the apparatus is determined by step functions of hydrogen and oxygen provided by the electrolysis cell and a constant current source.
16. J. Barber, *Biochim. Biophys. Acta* **594**, 253 (1980).

17. A. Trebst, *Methods Enzymol.* **24**, 146 (1972).
18. G. Zweig and M. Avron, *Biochem. Biophys. Res. Commun.* **19**, 397 (1965).
19. B. Kok, H. J. Rurainski. O. V. H. Owens, *Biochim. Biophys. Acta* **109**, 347 (1965).
20. W. S. Chow and J. Barber, *J. Biochem. Biophys. Methods* **3**, 173 (1980).
21. I thank J. P. Eubanks for technical support, D. J. Weaver for secretarial support, and H. W. Dunn for x-ray fluorescence analysis. I also thank O. S. Andersen, W. Arnold, J. Braunstein, G. M. Brown, B. Z. Egan, A. I. Krasna, M. E. Reeves, and J. Woodward for comments and criticism. Supported by the Office of Basic Energy Sciences, Department of Energy, under contract DE-ACO5-840R21400 with Martin Marietta Energy Systems, Inc.

8 July 1985; revised 26 September 1985

Section VI
Photosynthesis

Contents

Introduction

Photosynthesis is the oldest and most reliable method of converting light energy into chemical energy. It is also a field of research that is a natural meeting ground for physicists, chemists, and biologists. In a lucid and elegant discussion, Parson discusses the thermodynamics of the primary reactions of photosynthesis. His work focuses on the proper understanding of the increase in partial molecular free energy of an absorber under illumination and its relation to the nature of the photochemical reactions that can result from excitation of the absorber. A key focus in understanding the physics of photosynthesis has been a study of the early events occurring in photosynthetic antennas following photon absorption. Much of our understanding of the physics of photosynthesis has been derived from studies of the reactions centers of photosynthetic bacteria. In his paper on the physical mechanisms of photosynthesis Clayton focuses on three key issues: the absorption of quanta by light-harvesting pigments and their delivery to the reaction centers; photochemistry of the primary electron transfer reaction; and light-induced creation of electrostatic charge separation, stabilization, and conversion of electrostatic potential energy to energy-rich biomolecules. Feher, Okamura, and Kleinfeld provide an account of electron transfer reactions in bacterial photosynthesis using charge recombination kinetics as a structural probe. Their discussion includes a description of the components of the reaction centers, primary electron donor, and sequential electron acceptors. The triplet state of the primary electron donor in bacterial photosynthesis is formed in bacterial reaction centers when electron transfer to the quinone is blocked. Goldstein, Takiff, and Boxer study the energetics of initial charge separation by monitoring primary electron donor triplet decay rates in very high magnetic fields (up to 135 kG). This information is used to obtain standard free energy and standard enthalpy differences between the radical pair intermediate and the triplet state.

THERMODYNAMICS OF THE PRIMARY REACTIONS OF PHOTOSYNTHESIS

William W. Parson

Department of Biochemistry, University of Washington, Seattle, WA 98195, U.S.A.

(Received 6 January 1978; accepted 15 March 1978)

Abstract—It appears to be widely believed that fundamental thermodynamic limitations prevent the products of the primary electron transfer reactions of photosynthesis from exceeding the reactants in energy by more than about $0.7\,h\nu_0$ where ν_0 is the frequency of the lowest energy absorption band of the reaction center. Specifically, in photosynthetic bacteria, where $h\nu_0 \approx 1.34\,\text{eV}$, it is often said that the midpoint redox potentials of the primary electron donor and acceptor cannot differ by more than about $0.9\,\text{eV}$. This is incorrect. A simple expression is developed for $\Delta\mu$, the increase in the partial molecular free energy of an absorber under illumination. The magnitude of $\Delta\mu$ gives one no information about the nature of the photochemical reactions that can result from excitation of the absorber. It puts no limits on the midpoint redox potentials of the reactants. However, knowing $\Delta\mu$ does allow one to calculate the concentrations of the products if the system comes to equilibrium during illumination.

The thermodynamics of the conversion of light into work has been discussed in depth by Duysens (1958), Ross (1966, 1967, 1975), Ross and Calvin (1967), Ross et al. (1976), Knox (1969, 1978), and Landsberg (1977). A major aim of this work has been to derive expressions for the change in free energy that occurs when a photochemical system is illuminated. Thus, Ross and Calvin (1967) have shown that, if a broadband absorber is exposed to continuous illumination of intensity $I(\nu)$, the partial molecular free energy of the absorber is increased by an amount

$$\Delta\mu = kT \ln \left[\phi_f \textstyle\int I(\nu)\, \sigma\,(\nu)\, d\nu / \int I_b(\nu)\, \sigma(\nu)\, d\nu\right]. \quad (1)$$

Here k is the Boltzmann constant, T the temperature, ϕ_f the fluorescence yield, ν the frequency of the illumination, $\sigma(\nu)$ the absorption cross-section, and $I_b(\nu)$ the black-body radiation intensity at the ambient temperature.* The change in the partial molecular free energy, or chemical potential, of the absorber is the amount that the total free energy of the system is increased, for each molecule of the absorber that is in the excited state. The pressure and temperature of the system are assumed to be constant. Other expressions for $\Delta\mu$ have been derived and have been shown to be essentially equivalent to Eq. 1. For example, Knox (1969) has shown that for a narrowband absorber (one that absorbs only at a single frequency, ν),

$$\Delta\mu \approx (1 - T/T_D)h\nu - kT \ln 1/\phi_f, \quad (2)$$

where T_D is a temperature that characterizes the light source.†

Although this work is almost certainly sound, the complex nature of the final equations may have obscured the simplicity of the problem. Perhaps for this reason, misunderstandings concerning the meaning of $\Delta\mu$ appear to be common among investigators in the field of photosynthesis. One such misconception is the idea that Eqs. like 1 or 2 put limits on the midpoint redox potentials of the electron carriers that participate in the primary reactions of photosynthesis. The origin of this belief is the fact that if one evaluates Eq. 1 or 2, using light with an intensity that is appropriate for the growth of photosynthetic bacteria, and using the absorption characteristics of the bacteria, $\Delta\mu$ turns out to be about 0.7 times $h\nu_0$ (Duysens, 1958; Ross and Calvin, 1967; Knox, 1969). Here ν_0 is the 0–0 frequency of the lowest-energy absorption band of the bacterial reaction center. The value $0.7\,h\nu_0$ is obtained if $\phi_f = 1$; for lower fluorescence yields, $\Delta\mu$ is smaller. Each reaction center that is excited increases the free energy of the system, not by $h\nu_0$, but by less than 70% of this. Since $h\nu_0$ is about $1.34\,\text{eV}$, many investigators evidently have concluded that the primary electron transfer reaction of bacterial photosynthesis can be driven by one photon only if the midpoint redox potentials of the electron-donor and electron-acceptor differ by less than about $0.9\,\text{V}$. This belief frequently has entered into discussions of the mechanism of the photochemical reaction.

As an example, Fong (1976) has argued that "after the usual 70% thermodynamic correction" a single photon does not provide enough free energy to drive an electron from P870 to the initial electron-acceptor in the bacterial reaction center. Fong concludes that the primary electron transfer reaction requires the input of two quanta per electron. Similar reasoning has

---

*For an isotropic absorber, $I_b(\nu) = 8\pi(n\nu/c)^2 \exp(-h\nu/kT)$, where n is the refractive index of the medium, c is the speed of light in vacuo, and h is Planck's constant. Stimulated emission has been neglected.

† T_D is defined by considering the source to be a black body at temperature T_D. The effective temperature of the source depends on the directionality of the absorber and the solid angle subtended by the source (Ross, 1966). The connection between broad-band and narrow-band absorbers is discussed by Ross (1967) and Knox (1969).

been used in discussions of plant photosynthesis. (See, e.g., Knox, 1978; Ke, 1977.)

One can see that this reasoning must be erroneous, by noting that the relationship between $\Delta\mu$ and $h\nu_0$ is not constant (Eq. 1 or 2). Among other things, $\Delta\mu$ depends on the light intensity. With weak light, $\Delta\mu$ approaches zero. The value $0.7\,h\nu_0$ was calculated for a particular light intensity, which actually was quite high relative to the intensities that one must use experimentally in measurements of the quantum yield of photochemistry. Even with strong light, $\Delta\mu$ will be much less than $0.9\,\mathrm{eV}$ if ϕ_f is low, as it is in the reaction centers of photosynthetic bacteria.* If one were to push Fong's (1976) argument to the limit of very low light intensity, where $\Delta\mu$ goes to zero, one would conclude that the quantum requirement for the electron transfer reaction must go to ∞, no matter how small the gap between the midpoint redox potentials of the electron carriers. This conclusion would be quite contrary to experience. Experimental measurements give quantum requirements of essentially 1.0 at very low light intensities (Loach and Sekura, 1968; Wraight and Clayton, 1974).

Uncertainty about the meaning of $\Delta\mu$ can arise if one does not make a clear distinction between concepts (like the molecular energy gap, $h\nu_0$) which describe properties of individual molecules or photons, and concepts (like $\Delta\mu$) which describe the properties of systems containing large numbers of molecules. Although $\Delta\mu$ is an intensive quantity (it does not depend on the absolute number of molecules in the system), it does depend on the composition of the system, i.e. on the ratios of the concentrations of molecules in different molecular states. The same distinction needs to be made between *midpoint* redox potentials, which are molecular properties, and the actual redox potential of a system, which determines the ratio of the concentrations of oxidized and reduced molecules. These distinctions were made clearly by Ross *et al.* (1976), but they frequently seem to be overlooked.

Consider a single molecule (or a single complex of molecules) that can be raised to an excited state, M_2, by the absorption of light (Fig. 1a). The energy of the molecule in state M_2 exceeds that in the ground state (M_1) by $(E_2 - E_1) = h\nu_0$. M_1 and M_2 are quantum mechanical eigenstates of the systems, and their energies do not depend on time or on the light intensity. (One knows, for example, that the wavelength at which a molecule fluoresces does not ordinarily depend on the light intensity.) In between M_1 and M_2 there might be any number of additional eigenstates with various intermediate energies. These could be charge-transfer states, triplet states, excited vibrational substates of M_1, etc. Figure 1a shows one such state, M_3, with energy E_3. Once the molecule is

---

*Zankel *et al.* (1968) and Slooten (1972) found $\phi_f \approx 0.001$ in reaction centers isolated from *Rhodopseudomonas sphaeroides*.

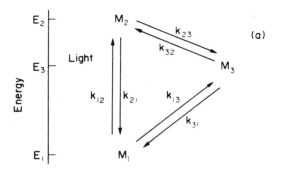

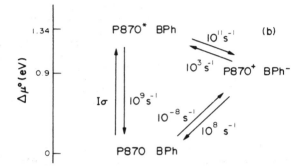

Figure 1. (a). General scheme for a molecule or complex of molecules with ground state M_1, excited state (M_2) reached by absorption of light, and a third state (M_3) with intermediate energy. The k's are rate constants. (b). Approximate rate constants and standard partial molecular free energies $(\Delta\mu^\circ)$ for a particular case, the reaction center of photosynthetic bacteria. I, light intensity; σ, absorption cross-section. The free energies on the vertical scale are expressed relative to the ground state of the system (P870 BPh). See the text for further explanation.

excited to M_2, the path that it takes in returning to M_1 will not depend on the intensity of the light that brought about the excitation. This path can include intermediate states with energies close to (or even above) E_2, even if the light intensity is so low that $\Delta\mu$ is near zero. Neglecting stimulated emission, one could switch off the light after the excitation, without influencing the decay path. The light intensity determines only how frequently the molecule is excited from M_1 to M_2.

Evidently $\Delta\mu$ does not tell one anything about the reactions that are open to the excited molecule. In that case, what is the point of calculating $\Delta\mu$? To clarify this, it may help to consider an expression for $\Delta\mu$ that is simpler than Eq. (1) or (2).

Suppose that we have a large collection of molecules of the same type. The free energy change for the conversion of a molecule from state M_1 to state M_2 is

$$\Delta\mu_{12} = \Delta\mu_{12}^0 + kT \ln \frac{(M_2)}{(M_1)} \qquad (4)$$

where (M_2) and (M_1) are the populations (concentrations or activities) of the two states and $\Delta\mu_{12}^0$ is the *standard* partial molecular free energy change (the standard free energy change per molecule undergoing the conversion$_j$. The standard free energy change is a property of the individual molecules. It is given by

$$\Delta\mu^0_{12} = \Delta\overline{H}^0_{12} - T\Delta\overline{S}^0_{12} \qquad (5)$$

where $\Delta\overline{H}^0_{12}$ and $\Delta\overline{S}^0_{12}$ are the standard changes in partial molecular enthalpy and entropy accompanying the conversion. $\Delta\overline{H}^0_{12}$ consists primarily of the energy difference $(E_2 - E_1)$. If the conversion from state 1 to state 2 causes a volume change $(\Delta\overline{V}^0_{12})$, $\Delta\overline{H}^0_{12}$ also will contain a term $P\Delta\overline{V}^0_{12}$, where P is the pressure. However, this term usually is negligible, and $\Delta\overline{H}^0_{12} = (E_2 - E_1) = h\nu_0$. If M_2 and M_1 have n_2 and n_1 degenerate (isoenergetic) substates, there will be a standard entropy change of $\Delta\overline{S}^0 = k \ln (n_2/n_1)$. The substates of M_1 and M_2 could consist of various vibrational and rotational states, electron spin states, etc. In most cases, the absorption of light probably does not change the number of such substates significantly, and the entropy term also is negligible. This need not always be true, however (Ross, 1975). If $\Delta\overline{S}^0$ is negligible, then $\Delta\mu^0_{12} = h\nu_0$, and

$$\Delta\mu_{12} = h\nu_0 + kT \ln \frac{(M_2)}{(M_1)}. \qquad (6)$$

The free energy change given by Eq. 6 is the same as that described by Eqs. 1 and 2. It represents the maximum amount of useful work that could be obtained from the system, for each molecule of the absorber that returns from state M_2 back to state M_1. From a different perspective, Eq. 6 just expresses the fact that the populations of M_2 and M_1 relax to the Boltzmann equilibrium distribution in the dark, when $\Delta\mu_{12}$ goes to zero.

In order to use Eq. 6, one needs to know the concentration ratio $(M_2)/(M_1)$ that is generated when the system is illuminated. This can be calculated by solving the steady state kinetic equations. For the simple, but rather general scheme illustrated in Fig. 1a,

$$\frac{(M_2)}{(M_1)} = \frac{k_{12}k_{32} + k_{13}k_{32} + k_{31}k_{12}}{k_{21}k_{32} + k_{21}k_{31} + k_{31}k_{23}}. \qquad (7)$$

This expression can be simplified further by dropping unimportant terms, depending on the values of the rate constants. Suppose M_1 is the ground state of the bacterial reaction center, M_2 is the lowest excited singlet state (P870*) of the bacteriochlorophyll complex in the reaction center (P870), and M_3 is the radical pair state (P870$^+$BPh$^-$) in which P870 is oxidized and one of the bacteriopheophytins (BPh) reduced (Fig. 1b). (For reviews on the bacterial reaction center see Parson and Cogdell, 1975; Ke, 1977; Dutton *et al.*, 1977; Blankenship and Parson, 1978.) The excitation rate constant k_{12} is the sum of the rate constants for excitation by the actinic light ($I\sigma$,

where I is the light intensity and σ the absorption cross-section), excitation by the ambient black-body radiation, and spontaneous thermal excitation. Only the first of these ($I\sigma$) is significant if I and σ are measurably different from zero. (The rate constants for thermal excitation and for excitation by the integrated black-body radiation are both on the order of $10^{-16}s^{-1}$.) The electron transfer rate constant k_{23} appears to be on the order of $10^{11}s^{-1}$ (Zankel *et al.*, 1968; Slooten, 1972; Paschenko *et al.*, 1977). The rate constant for reverse electron transfer, k_{32}, is $k_{23}e^{-\Delta\mu^0_{32}/kT} \approx 10^{11}e^{-0.44/0.025} \approx 10^3s^{-1}$. Here $\Delta\mu^0_{32}$ is the standard partial molecular free energy change for the regeneration of P870* by the transfer of an electron from BPh$^-$ back to P870$^+$. The value 0.44 eV comes from redox titrations of the P870 and BPh in the reaction center (Kuntz *et al.*, 1964; Prince *et al.*, 1976) and from $(E_2 - E_1) = 1.34$ eV. This is discussed further below. The decay rate constant k_{21} is the sum of the rate constants for fluorescence, nonradiative decay, and stimulated emission. Under most conditions, stimulated emission is negligible, and k_{21} must be $\leq 0.01 k_{23} \approx 10^9s^{-1}$ to account for the high quantum yield of electron transfer (Wraight and Clayton, 1974). The relaxation rate constant k_{31} is on the order of 10^8s^{-1} (Parson *et al.*, 1975; Cogdell *et al.*, 1976). The rate constant k_{13} for the reverse of the relaxation (spontaneous thermal formation of the radical pair) is essentially zero ($k_{13} = k_{31}e^{-\Delta\mu^0_{13}/kT} \approx 2 \times 10^{-8}s^{-1}$). With these values for the rate constants, the terms $k_{31}k_{12}$ and $k_{31}k_{23}$ predominate heavily in Eq. 7, giving

$$\frac{(M_2)}{(M_1)} \approx \frac{k_{31}k_{12}}{k_{31}k_{23}} = \frac{I\sigma}{k_{23}}. \qquad (8)$$

Combining Eqs. 6 and 8, we have to a very good approximation:

$$\Delta\mu_{12} = h\nu_0 + kT \ln \frac{I\sigma}{k_{23}}. \qquad (9)$$

This equation is useful for $10^{-13}s^{-1} < I\sigma < 10^{11}s^{-1}$, i.e. for virtually any measurable light intensity. If $I\sigma$ is smaller than $10^{-13}s^{-1}$, the approximations made in Eq. 8 break down, and one must return to Eq. 7. Provided that one includes black-body radiation and thermal excitations in k_{12}, k_{32}, and k_{13}, $\Delta\mu_{12}$ will go to zero as I does. Although Eq. 9 has been presented specifically for the bacterial reaction center, it is generally applicable to any system in which the lowest excited singlet state is converted rapidly and efficiently into a metastable state. The introduction of additional metastable states between M_3 and M_1 would not greatly affect the $(M_2)/(M_1)$ ratio or $\Delta\mu_{12}$. For broad-band excitation, one needs only to replace $I\sigma$ by $\int I(\nu)\sigma(\nu)d\nu$.

The absorption cross-section of the bacterial reaction center is approximately $4.9 \times 10^{-16}cm^2$ at 870 nm (Straley *et al.*, 1973).* If a suspension of isolated reaction centers is illuminated continuously with

---

* σ is $3.817 \times 10^{-21} \epsilon$ cm^2, where ϵ is the molar extinction coefficient.

870 nm light having an intensity of 1 nanoeinstein $cm^{-2}s^{-1}$ (6×10^{14} quanta $cm^{-2}s^{-1}$). $I\sigma$ is 0.29 s^{-1}. This gives $(M_2)/(M_1) \approx 2.9 \times 10^{-12}$ and $\Delta\mu_{12} \approx$ 0.67 eV. If a Q-switched laser flash were used for the illumination, providing 1 nanoeinstein cm^{-2} in 20 ns ($I \approx 3 \times 10^{22}$ quanta $cm^{-2}s^{-1}$), $(M_2)/(M_1)$ would be about 1.5×10^{-4} and $\Delta\mu_{12}$ about 1.12 eV. With a ps flash from a mode-locked laser, $(M_2)/(M_1)$ would approach 1, and $\Delta\mu_{12}$ would approach $h\nu_0$. In the simple 3 state scheme of Fig. 1, stimulated emission would no longer be negligible at this point, and Eqs. 8 and 9 would need modification. Equations 8 and 9 can continue to hold in the ps flash regime, if the exciting light initially generates a rapidly relaxing upper vibrational substate of M_2, or a higher electronic state, i.e. if the reaction center is excited at a frequency greater than ν_0. In this case, $(M_2)/(M_1)$ could exceed 1 and $\Delta\mu_{12}$ could exceed $h\nu_0$. Note, however, that M_2 and $h\nu_0$ refer to the state and energy following the relaxation, rather than to the initial excitation.

In intact bacteria or chromatophores, antenna Bchl increases the effective absorption cross-section by a factor of 50 or more, depending on the bacterial species. Assuming that the antenna transfers excitations to the reaction center with high efficiency, the increase in σ will increase $(M_2)/(M_1)$ proportionally, leading to larger values of $\Delta\mu_{12}$. Equation 9 allows one to see clearly the effect of increasing the size of the antenna, whereas Eqs. 1 and 2 may give the incorrect impression that the free energy available at the reaction center does not depend on this parameter. (The effective value of σ will not depend so simply on the size of the antenna if one uses ps excitation. With extremely high light intensities, the migration of excitons to the reaction centers becomes rate-limiting and other quenching processes become important. One must bear in mind also that the efficiency of excitation transfer to the reaction centers will decline even for low light intensities if the size of the antenna is made too great.)

---

* Standard partial molecular free energy differences are needed here, and this is what one obtains by subtracting E_m values. However, E_m values that are measured by separate redox titrations of the two components are not strictly applicable, because they do not allow for interactions between the components. In addition, redox titrations reflect equilibrium states of the system. They may not represent the transient states that actually occur during the electron transfer reactions. The value of 0.9 eV for $\Delta\mu_{13}^0$ is therefore only an approximation. For further discussion of these points, see Case and Parson (1971). Note also that, whereas molecular entropy changes were dismissed as being negligible in the derivation of Eq. 6, this should not be done here. Electron transfer reactions sometimes involve substantial changes in standard partial molecular entropy (Case and Parson, 1971).

Although $\Delta\mu_{12}$ does not provide any information about the mechanism of the primary photochemical reaction, it does allow one to calculate the concentrations of the products, if the system reaches equilibrium in the light. In the bacterial reaction center, the chemical potential of the excited P870 is used to drive the transfer of an electron from P870 to a bacteriopheophytin (BPh). The free energy change for transferring an electron is

$$\Delta\mu_{13} = \Delta\mu_{13}^0 + kT \ln \frac{(P870^+ \, BPh^-)}{(P870 \, BPh)}$$

$$\approx (e)(E_m^{P870} - E_m^{BPh}) + kT \ln \frac{(P870^+ \, BPh^-)}{(P870 \, BPh)}$$

$$\approx 0.9 \, eV + kT \ln \frac{(P870^+ \, BPh^-)}{(P870 \, BPh)}. \qquad (11)$$

Here E_m^{P870} and E_m^{BPh} are the midpoint redox potentials* of P870 and BPh, e is the charge of the electron, and $(P870^+ BPh^-)/(P870 \, BPh)$ is the ratio of concentrations of reaction centers in the charge transfer and ground states. At equilibrium $\Delta\mu_{12} = \Delta\mu_{13}$. Using the values of $\Delta\mu_{12}$ that were calculated above, one can see that a moderately strong continuous light could keep only about 0.1% of the reaction centers in the charge transfer state, but that a Q-switched laser flash can place essentially all of the reaction centers in this state. In either case, any individual reaction center that absorbs a photon is able (and likely) to go to the charge transfer state.

The calculated concentration ratio $(P870^+BPh^-)/$ $(P870 \, BPh)$ usually is not of much significance, because in functioning reaction centers BPh^- is rapidly oxidized by a quinone and $P870^+$ is rapidly reduced by a cytochrome. This keeps $(P870^+$ $BPh^-)/(P870 \, BPh)$ extremely small, and the system never really comes to equilibrium. Analogous ratios for subsequent electron carriers like ubiquinone and the cytochromes are probably of greater interest, if one is concerned with the physiology of the bacterial cells. These components are present in relatively large pools, which turn over more slowly. It is possible that they need to be held predominantly in one redox state or the other, in order for the cells to thrive and to fix CO_2. This could explain the old observation that CO_2 fixation in whole cells has a sigmoidal dependence on light intensity (French, 1937; Wassink et al., 1942). Ross et al. (1976) have used computer modeling to show how the rate of free energy storage in the pools of secondary electron carriers depends on the rate constants of the primary and secondary electron transfer reactions.

Acknowledgements—Supported by NSF grant number RCM 77-13290. I thank R. E. Blankenship, R. S. Knox, D. R. Ort and C. C. Schenck for helpful discussions.

REFERENCES

Blankenship. R. E. and W. W. Parson (1978) *Ann. Rev. Biochem.*, in press.
Case, G. D. and W. W. Parson (1971) *Biochim. Biophys. Acta* **253**, 187–202.
Cogdell, R. J., T. G. Monger and W. W. Parson (1975) *Biochim. Biophys. Acta* **408**, 189–199.
Dutton, P. L., R. C. Prince, D. M. Tiede, K. M. Petty, K. J. Kaufmann, T. L. Netzel and P. M. Rentzepis (1977) *Brookhaven Symp. Biol.* **28**, 213–236.
Duysens, L. N. M. (1958) *Brookhaven Symp. Biol.* **11**, 10–23.
Fong, F. (1976) *J. Am. Chem. Soc.* **98**, 7840–7843.
French, C. S. (1937) *J. Gen. Physiol.* **20**, 711–735.
Ke, B. (1978) *Curr. Top. Bioenerg.* **7**, 76–138.
Knox, R. S. (1969) *Biophys. J.* **9**, 1351–1362.
Knox, R. S. (1978) In *Topics in Photosynthesis* (Edited by J. Barber) Vol. II. pp. 55–97. Elsevier Press, Amsterdam.
Kuntz, I. D., Jr., P. A. Loach and M. Calvin (1964) *Biophys. J.* **4**, 227–249.
Landsberg, P. T. (1977) *Photochem. Photobiol.* **26**, 313–314.
Loach, P. A. and D. Sekura (1968) *Biochemistry* **7**, 2642–2649.
Parson, W. W. and R. J. Cogdell (1975) *Biochim. Biophys. Acta* **416**, 105–149.
Parson, W. W., R. K. Clayton and R. J. Cogdell (1975) *Biochim. Biophys. Acta* **387**, 265–278.
Paschenko, V. Z., A. A. Kononenko, S. P. Protasov, A. B. Rubin, L. B. Rubin and N. Y. Uspenskaya (1977) *Biochim. Biophys. Acta.* **461**, 403–412.
Prince, R. C., J. S. Leigh and P. L. Dutton (1976) *Biochim. Biophys. Acta* **440**, 622–636.
Ross, R. T. (1966) *J. Chem. Phys.* **45**, 1–7.
Ross, R. T. (1967) *J. Chem. Phys.* **46**, 4590–4593.
Ross, R. T. and M. Calvin (1967) *Biophys. J.* **7**, 595–614.
Ross, R. T. (1975) *Photochem. Photobiol.* **21**, 401–406.
Ross, R. T., R. J. Anderson and T.-L. Hsiao (1976) *Photochem. Photobiol.* **24**, 267–278.
Slooten, L. (1972) *Biochim. Biophys. Acta* **256**, 452–466.
Straley, S. C., W. W. Parson, D. C. Mauzerall and R. K. Clayton (1973) *Biochim. Biophys. Acta* **305**, 597–609.
Wassink, E. C., E. Katz and R. Dorrestein (1942) *Enzymologia* **10**, 285–354.
Wraight, C. A. and R. K. Clayton (1974) *Biochim. Biophys. Acta* **333**, 246–260.
Zankel, K. L., D. L. Reed and R. K. Clayton (1968) *Proc. Natl. Acad. Sci. U.S.* **61**, 1243–1249.

Physical Mechanisms in Photosynthesis: Past Elucidations and Current Problems*

RODERICK K. CLAYTON

Division of Biological Sciences and Department of Applied Physics, Cornell University, Ithaca, New York 14850

Photosynthetic tissues are organized functionally into aggregates of light harvesting pigments (mainly chlorophylls, carotenoids, and phycobilins) associated with photochemical reaction centers (1, 2). The pigments absorb light and deliver the energy to the reaction centers, where an oxido-reductive photochemistry ensues. The primary photoproducts, oxidizing and reducing entities, serve as starting points for electron transport that is coupled to phosphorylation. In known cases, the primary photochemical electron donor, at the reaction center, is a chlorophyll (Chl) or bacteriochlorophyll (BChl). In the specialized context of the reaction center this donor is generally designated P (for pigment) followed by a number signifying the peak of the long wave absorption maximum: P700, P870, etc. (3, 4).

This organization defines certain physical problems:

(a) How is energy, absorbed by the light harvesting pigments, delivered to the reaction centers?

(b) What are the details of the photochemical process?

(c) How are the primary photoproducts used effectively and safely, without wasteful recombination or harmful indiscriminate reactions with the surroundings?

We shall consider these questions in turn, especially as related to the photosynthetic bacteria.

ENERGY TRANSFER AND FLUORESCENCE

Fluorescence in photochemical systems

The mechanism of energy transfer has been studied by measurement of the fluorescence emitted by the light harvesting Chl (or BChl) in relation to the chemistry occurring at the reaction centers. The intensity of fluorescence measures the concentration of singlet excited states, or excitation quanta, in the system. These excitations are the direct result of light absorption by the light harvesting pigments. In a steady state† the fluorescence also reflects the rate at which excitation is being quenched by non-fluorescent pathways. The more the "dark" quenching, the less the yield of fluorescence. The dark quenching encompasses dissipation into heat and utilization for photochemistry.

These relationships can be formulated as

$$\phi_f = k_f/(k_f + k_d + k_p) \qquad [1]$$

Abbreviations: Chl, chlorophyll; BChl, bacteriochlorophyll; P, pigment; Cyt, cytochrome; PMS, phenazine methosulfate.

* Presented at the Annual Meeting of the National Academy of Sciences, April 26, 1971, during the Photosynthesis Bicentennial Symposium, Kenneth V. Thimann, Chairman (1971) *Proc. Nat. Acad. Sci. USA* **68**, 2875–2897.

† A quasi-steady state is attained if the time constants for excitation and de-excitation (about 10 nsec) are short compared with the time scale of observation, and compared with the time constants for environmental (e.g., chemical) changes that affect the fluorescence.

where ϕ_f is the quantum yield of fluorescence (quanta emitted/quanta absorbed) and the k's are first-order rate constants for the processes by which singlet excitation quanta become lost: k_f for fluorescence, k_d for radiationless de-excitation (conversion into heat), and k_p for the event that leads eventually to photochemistry. It is usually assumed, for lack of information to the contrary, that k_f and k_d do not vary under changing physiological conditions, but the value of k_p may depend on the functional conditions of the reaction centers.

The migration of energy, and its trapping by photochemical reaction centers

Let us consider first those hypotheses by which a singlet excitation quantum in the light harvesting system is converted locally to some other (metastable) state, and the new state carries the energy to a reaction center. This new state might, for example, be a triplet excited state or an electron–hole pair in the ensemble of Chl molecules. It is not a source of the "prompt" (short lived) fluorescence that characterizes the singlet excited state. In such a model, if the reaction centers were altered so as to be unable to accept and process the energy, the most immediate result would be an increase in the population of metastable states. To a first approximation this would not affect k_p, which in this model is the rate constant for singlet → metastable conversion. The states of the reaction centers would have no effect, or at most a remote and indirect one, on the intensity of fluorescence.

Now suppose instead that the singlet excitation quanta are quenched at the reaction centers, by a process that depends on the state of the reaction center. Specifically, consider a specialized Chl or BChl, P, that can receive an excitation quantum from the light harvesting pigment and can then donate an electron so some acceptor, A:

$$P,A \xrightarrow{h\nu} P^*,A \rightarrow \rightarrow \rightarrow P^+,A^-$$

$$\underbrace{\qquad\qquad\qquad}_{\text{(dark)}}$$

where P^* denotes P in the singlet excited state. The "dark" restoration of the state "P,A" must be completed before the reaction center can perform its function again. Regarding the reaction center as a photochemical trap for singlet excitation quanta, the trap is "open" in the state P,A but becomes "closed" in any of the states $P^+, A; P, A^-$; or P^+, A^-.

When all the traps in a sample are open, k_p has its maximum value and the fluorescence is minimal. When all the traps are closed, $k_p = 0$ and ϕ_f has the maximum value of $k_f(k_f + k_d)$. In a typical experimental observation (5, 6) a sample is illuminated and the yield of fluorescence (from the light harvesting Chl or BChl) is seen to rise as traps become driven into a closed state.

This kind of hypothesis has been borne out by experiments

with photosynthetic bacteria (5–7) and also with green plants in relation to the oxygen-evolving "photosystem II" (8–10).

The states of the reaction centers can be monitored independently, to a limited extent. In cells or subcellular (chromatophore) preparations of some photosynthetic bacteria such as *Rhodopseudomonas spheroides* and *Rhodospirillum rubrum*, aerated and depleted of electron donating substrates, the conversion from P to P^+ can be seen as a light-induced bleaching near 870 nm. The fluorescence rises during illumination, in coordination with this bleaching. The quantitative relationship is as if at any instant, k_p is simply proportional to the fraction of P that has not yet become bleached (5–7). The simplicity of the relationship suggests that the model is correct and that under these conditions, the closing of traps is associated exclusively with the conversion from P to P^+; the state P,A^- does not play an important role. But under a reducing environment the flow of electrons to P^+, causing the reaction $P^+ + e^- \rightarrow P$, may be so rapid that no significant light-induced bleaching of P can be observed. Nevertheless the fluorescence from the light harvesting pigment may rise during illumination (6), suggesting that the traps are becoming closed on the "acceptor" side:

$$P,A \xrightarrow{h\nu} \rightarrow \rightarrow P^+,A^- \xrightarrow[\text{rapid}]{} P,A^-$$

This interpretation cannot be tested directly because of the difficulty of any direct observation of the conversion of A to A^-; we are just beginning to learn how to do this with purified reaction center preparations (see later).

In green plant tissues, most of the fluorescence, and all of the part that varies under physiological conditions, is associated with the oxygen-evolving system II. Although neither the primary photochemical electron donor nor the primary acceptor (usually symbolized Chl_{II} and Q, respectively) has been detected with certainty, an extensive phenomenology has been developed relating these hypothetical entities to fluorescence, oxygen evolution, and electron flow through the reaction center to some ultimate acceptor (Hill reaction) (8–12). Some conclusions that seem appropriate at present are:

(*a*) The yield of fluorescence does change in a way that reflects the states of the traps or reaction centers.

(*b*) The flow of electrons from water (oxygen evolving chemistry) to the "system II" trap is normally so rapid that the trap remains open on the oxidizing side; accumulation of oxidized Chl_{II} is negligible. Closure of the trap, and consequent high fluorescence, is associated with the reduction of Q.

(*c*) Oxidants close to the reaction center, possibly including oxidized Chl_{II}, may accumulate when the flow of electrons from water is impaired: by heating, by washing chloroplasts with Tris buffer, by ultraviolet irradiation, or through manganese deficiency. Electron flow to the reaction center can then be restored by addition of an artificial donor such as hydroquinone.

(*d*) A rapid cycling of electrons from reduced Q to oxidized entities near (and perhaps including) Chl_{II} is possible.

(*e*) Although high fluorescence is associated with the reduction of an entity "Q" (for quencher), that entity is not established conclusively to be the primary electron acceptor. Also there may be two kinds of "Q" with different oxidation–reduction properties.

With respect to green plant system I, for which P700 has been implicated as a primary electron donor, no component of the fluorescence shows variations that can be related sensibly to the chemical state of this system. We shall return to this problem later.

Light absorption, lifetime of the excited state, and yield of fluorescence

There is a useful pair of relationships between the intensity of absorption, the lifetime of the excited state (measured as fluorescence lifetime) and the quantum yield of fluorescence of a pigment. Imagine for the moment that the radiative pathway, fluorescence, is the only mechanism for de-excitation available to an excited molecule. The excited state will then have a mean lifetime τ_0, called the intrinsic lifetime, before it decays by fluorescence. This intrinsic lifetime is governed by the probability (per unit time) of fluorescence; indeed it is the inverse of that probability. On the other hand, the probabilities for absorption and fluorescence, being governed by the same physical considerations, are proportional to each other. Therefore, the probability of absorption is proportional to the inverse of the intrinsic lifetime of the excited state. Absorption probability is measured by the area under the appropriate absorption band; thus

$$1/\tau_0 = \text{Constant} \times (\text{absorption-band area}); \qquad [2]$$

see ref. 13 for the details of this formulation.

Eq. **2** allows a simple computation of τ_0, the lifetime that would prevail if the quantum yield of fluorescence were 100%. But other processes, exemplified by k_d and k_p in Eq. **1**, compete with fluorescence for quenching of the excited state. These other processes shorten the mean lifetime, and reduce the fluorescence yield in the same proportion. Thus, if the yield of fluorescence is ϕ_f and the "actual" mean lifetime of the excited state is τ, we can write

$$\tau = \phi_f \tau_0. \qquad [3]$$

The value of τ_0 computed from the absorption band area is 15 nsec for Chl a and 20 nsec for BChl. For Chl a in ether, the lifetime is 5 nsec and the fluorescence yield is 33% (14), in exact agreement with Eq. **3**. In green plants, the lifetime and yield are about 0.5 nsec and 3%, respectively (15), again in good agreement. In several types of photosynthetic bacteria the yields (16) and lifetimes (17, 18) of BChl fluorescence range from about 2–10% and 0.4–2 nsec, respectively, again in harmony with Eq. **3**.

Summary and anomalies

In summary, there is abundant simple evidence that energy absorbed by light harvesting pigments reaches the reaction centers in the form of singlet excitation quanta, and no compelling evidence to the contrary. This conclusion can be made for green plant system II and for certain photosynthetic bacteria, especially *Rps. spheroides* and *R. rubrum*. In these systems, with photochemical traps fully functioning, the mean lifetime of singlet excitation is usually about 0.5 nsec. This appears to be the time needed for migration of the energy to the traps.

For green plant system I there are several possibilities to account for the absence of any fluorescence that varies with the state of P700:

(*a*) P700 is not the major sink for excitation quanta. This is belied, at least under some conditions, by the high efficiency of light-induced oxidation of P700 (19).

(b) The energy is transferred to the reaction centers, and quenched there, so efficiently that the yield of fluorescence is undetectably small.

(c) The singlet excited state is transmuted by an unknown process independent of the state of the P700. This process might even involve a "true", not yet detected, primary photochemical electron donor that can in turn oxidize P700. In this view, P700 could be a safety device to get rid of excess oxidants when these accumulate.

Because we cannot choose among these alternatives, we cannot pretend to a satisfactory understanding of primary energy transduction in green plant system I. But in the photosynthetic bacteria we can accept, for the present, the simple view that a singlet excitation quantum migrates to a reaction center where it generates an electrically polarized state, precursor of the more stable couple P^+, A^-.

We conclude this part of the discussion by listing three anomalies.

First, evidence exists (20, 21) that at least in *R. rubrum*, there is more than one kind of photochemical system, each with its characteristic reaction center. Whether the variations in fluorescence are related to the conditions at more than one kind of reaction center remains to be settled.

Second, in at least one photosynthetic bacterium, *Ectothiorhodospira Shaposhnikovii*, the component of fluorescence that varies with the states of the reaction centers has a lifetime far less than 0.5 nsec (45). When the traps are all open, the lifetime may be as short as 5×10^{-11} sec, suggesting very rapid migration of excitation quanta to the traps. Perhaps this is the case for green plant system I as well.

Third, we now recognize that the yield of fluorescence associated with green plant system II varies with the electrochemical state of the membrane (thylakoid membrane) that carries the photosynthetic apparatus. Specifically, a quenching of the fluorescence is correlated with the development of a gradient of H^+ concentration across the membrane (22, 23). This may happen in photosynthetic bacteria as well (Sherman, unpublished observations). The mechanism is unknown.

PHOTOCHEMISTRY; CONSERVATION AND UTILIZATION OF THE PHOTOPRODUCTS

Photochemical reaction centers from photosynthetic bacteria

Preparations of photochemical reaction centers made from carotenoidless mutant *Rps. spheroides* (24, 25) are composed of a hydrophobic protein, of molecular weight probably between 60,000 and 140,000, to which BChl and bacteriopheophytin are bound through hydrophobic interactions. There are probably two bacteriopheophytin and three or four BChl molecules attached to each protein molecule (26). The BChl is responsible for absorption bands near 800 and 865 nm; the 865-nm band is identified as P870 in the intact cell. The P870 is bleached reversibly by light; this is identified as a photochemical oxidation of the pigment. No other prosthetic groups, with the possible exception of an iron atom (25), have been found in well purified reaction centers. The identity of the primary electron acceptor remains a mystery; perhaps it is nothing but a locus of electron affinity generated by a special configuration in the protein.

The reaction center protein can be dissociated with sodium dodecyl sulfate into three distinct subunits (27) whose apparent molecular weights based on electrophoresis in polyacrylamide are 19,000, 23,000, and 27,000. If these weights are correct, and if the amount of protein is proportional to the amount of Coomassie Blue stain that it binds in the acrylamide gel, the ratio of the subunits is 1:1:1. This would give a molecular weight of 69,000 for the smallest assemblage of subunits.

Serological experiments (W. R. Sistrom, R. K. Clayton, and R. L. Berzborn, unpublished data) and analysis by acrylamide gel electrophoresis (27) show that this protein makes up about one fifth of the total protein of chromatophores prepared from wildtype *Rps. spheroides*. From such chromatophores this singular protein can be isolated in fairly pure form, but accompanied by the light harvesting pigments, by following the same procedure that yields a reaction center preparation when applied to the carotenoidless mutant strains. The crux of the procedure is to treat the chromatophores with the detergent lauryl dimethyl amine oxide, centrifuge at about $200,000 \times g$ for 2 hr, and discard the pellet. The light harvesting pigments may be bound to the reaction center protein, but they are certainly bound to other proteins in the chromatophore as well.

Sistrom's nonphotosynthetic mutant strain PM-8 of *Rps. spheroides* lacks the reaction center protein. We have shown this both serologically and through analysis by acrylamide gel electrophoresis (27). The light harvesting BChl and carotenoids are bound to other proteins in PM-8; the photochemically active P870 is missing.

The P870 in reaction centers can be oxidized by illumination:

$$P, A \xrightarrow{h\nu} P^+, A^-$$

and the electron can be passed on from A^- to a secondary acceptor such as added ubiquinone or ferricyanide:

$$P^+, A^- + B \rightarrow P^+, A + B^-$$

where B denotes the secondary acceptor. The light-induced state P^+, A may be stable for several seconds, allowing some leisure in attempts to analyze the material while the P870 is in either its reduced or its oxidized form. The chromophores can be extracted from the reaction centers by dilution of the sample with methanol and centrifugation away of the denatured protein. When this is done, the reduced P870 appears as BChl, with an absorption maximum at 770 nm, in the methanolic solution. But extraction of the oxidized form, P^+, yields a bleached form of BChl, presumably $BChl^+$. The bleached form in the methanolic solution can be restored to the unbleached form by addition of ascorbate (S. C. Straley, unpublished data), but the regenerated absorption band is at 780 nm rather than 770 nm. These experiments show that one can prepare a natural photoproduct, oxidized P870, and then study it *in vitro*.

Fluorescence of reaction centers in relation to their chemical activities

The photoproducts in reaction centers can be made to react with external electron donors and acceptors; the possibilities are summarized in Fig. 1.

If the surroundings contain a good donor of electrons such as reduced cytochrome c (Cyt c), the conversion from P^+ to P can be so rapid that the accumulation of neither P^+, A nor P^+, A^- is significant during illumination. The population of reaction centers then shifts from the state P,A to P,A^- during illumination. Alternatively, if a good acceptor (ubiquinone or ferricyanide) is present, the principal states are P,A and P^+,A.

These transformations are attended by variations in the yield of fluorescence emitted by P870 (26, 28, 29) (Note that in these reaction-center preparations, there is no light harvesting BChl to emit fluorescence. Fluorescence from P800 is negligible, and that from BPh can be avoided by exciting at wavelengths greater than 800 nm).

The states P^+,A and P^+,A^- are nonfluorescent because the emitting species, P, is missing. During illumination of reaction centers without added electron donor (upper pathway in Fig. 1) the fluorescence band of P870, centered near 900 nm, disappears along with the absorption band centered near 865 nm. This is shown in Fig. 2, curve a. On the other hand, if an electron donor such as reduced Cyt c or phenazine methosulfate (PMS) is present, the fluorescence rises during illumination, as in curve b. The most direct interpretation of this rise is that the state P,A^- is being formed (lower pathway in Fig. 1). In this state the P870 is more strongly fluorescent because the excited state P^*,A^- cannot be discharged photochemically.

The problem of the photochemical electron acceptor

Following the interpretation of the foregoing paragraph, the fluorescence reveals properties of the primary electron acceptor. The initial fluorescence, f_0 in Fig. 2, can be driven to values approaching f_{max} by chemical reduction, presumably because A is being reduced to A^-. A titration of this effect (29) shows that the system A/A^- has a mid-point potential of -0.05 V, independent of the pH, with the stable oxidized and reduced forms differing by one electron.

The indication that the acceptor can hold just one electron is confirmed by experiments (26) showing that in the absence of secondary electron acceptors, but with an excess of reduced Cyt c present, just one equivalent of Cyt c can be oxidized photochemically for every equivalent of P870:

$$P,A + \text{Cyt} \xrightarrow{h\nu} P^+,A^- + \text{Cyt} \rightarrow P,A^- + \text{Cyt}^+$$

This stoichiometry can be altered by addition of ubiquinone as a secondary electron acceptor; then two additional equivalents of Cyt can be oxidized for every mole of ubiquinone added.

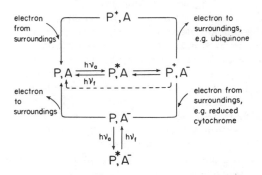

FIG. 1.   A diagram of a model for photochemical oxido-reduction and subsequent electron transport in photochemical reaction centers from *Rhodopseudomonas spheroides*. P denotes P870, P^* is P in the lowest singlet excited state, and P^+ is oxidized P. The hypothetical electron acceptor in the photochemical act is designated A. The surroundings are meant to include endogenous as well as external electron donors and acceptors. Steps involving light absorption and fluorescence are marked $h\nu_a$ and $h\nu_f$.

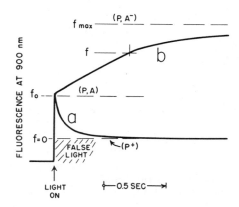

FIG. 2.   Time course of the fluorescence at 900 nm emitted by reaction centers from *Rps. spheroides* during constant illumination. Reaction centers, 1.0 μM, in 0.01 M Tris·HCl buffer (pH 7.5) with 0.1% lauryl dimethyl amine oxide and 5 μM ubiquinone. Exciting light 800 nm; 4.0 mW/cm². *Curve a:* No electron donor added. The fluorescence declined as the emitter, P870, became oxidized (and bleached) to P^+. The "false light" signal with all P870 bleached is ascribed to scattered exciting light and emission from chromophores other than P870. *Curve b:* With 25 μM bovine cytochrome c, added in the reduced form. The P870 was kept predominantly reduced by the cytochrome during illumination, and the fluorescence rose (ultimately to f_{max}) as the photochemistry proceeded. The rise is attributed to reduction of the primary electron acceptor A. The maximum level, associated with "P,A^-", could also be obtained by adding $Na_2S_2O_4$.

lents of Cyt can be oxidized for every mole of ubiquinone added.

The same conclusions have been reached (26) by analysis of the shapes of "fluorescence rise" curves such as curve b in Fig. 2. Such analysis can give information as to the number of quanta needed to drive a sample of reaction centers from the state P,A to P,A^-. In the absence of secondary electron acceptors, about one or perhaps 1.5 quanta are needed for every P870. Independent measurements (26, 30) show that 1–1.5 quanta suffice for the oxidation of one P870 (and by implication, the reduction of one A to A^-). The exact computed value depends on an assumption as to the optical absorption coefficient of P870; this remains somewhat uncertain (26).

There are two other pieces of information that might help to identify the primary acceptor. First, a very broad (several thousand gauss bandwidth) light-induced electron-spin resonance signal has been detected in reaction centers from *Rps. spheroides*, (25) in addition to the familiar narrow one that signals the presence of P^+. The broad signal could be a manifestation of A^-.

Second, we have detected light-induced optical absorption changes (ref. 31, and Straley and Clayton, unpublished data) that might be identified with the interconversion of A and A^-. These absorption changes are perceived when reaction centers are illuminated in the presence of an electron donor such as PMS so as to suppress the appearance of oxidized P870:

$$P,A + \text{PMS} \rightarrow P,A^- + \text{PMS}_{ox}$$

After any changes attributable to the oxidation of PMS have been discounted, we see a residual set of absorption changes that could reflect the conversion of A to A^-. Another approach is to illuminate reaction centers in the presence of mixed ferri- and ferrocyanide. One can detect a slow absorption change after the rapid initial "light on" reaction, and the details of this process suggest that the slow change signifies the reaction

$$P^+,A^- + Fe^{+++} \rightarrow P^+,A + Fe^{++}$$

When the reduction of ferricyanide (Fe^{+++}) to ferrocyanide (Fe^{++}) has been discounted, the net optical change reflects "$A^- \rightarrow A$".

Both kinds of measurement show that the conversion from A to A^- (in our hypothesis) is attended by the following absorption changes: An absorption band appears, centered at 455 nm. There are bathychromic shifts ("red-shifts") of bands near 300, 530, and 760 nm.

The bands near 530 and 760 nm can be identified with bacteriopheophytin, and the shifts of these bands might be due to local electric fields acting on the bacteriopheophytin molecules. Such fields could arise as a result of the photochemical separation of electric charge in the reaction centers. The 800- and 865-nm bands sometimes show small blue-shifts that accompany the red-shifts near 530 and 760 nm; these shifts are distinct from the much larger changes (blue-shift at 800 nm and bleaching at 865 nm) that signal the oxidation of P870. All of these effects may become useful in sensing and mapping the movements of electrons in the reaction centers. At present they simply illustrate the need for caution in interpreting the absorption changes associated with the reaction "$A \rightarrow A^-$".

Our ignorance concerning the primary photochemical electron acceptor in reaction center preparations from Rps. spheroides is representative of our lack of knowledge about this entity in general. Mid-point potentials have been estimated from the way that the redox potential affects the fluorescence of light harvesting Chl or BChl, by use of the assumption that k_p (Eq. 1) becomes zero when the acceptor becomes reduced. Such experiments suggest that in several types of photosynthetic bacteria (32), the acceptor is a "one-electron" agent with a mid-point potential in the range -0.05 to -0.16 V, independent of the pH. Our results with reaction centers are consistent with these measurements for Rps. spheroides. Similar conclusions have been drawn (33, 34) from the ability of reaction center or chromatophore preparations to mediate the photochemical reduction of added substances of various mid-point potentials (methylene blue, indigo sulfonate dyes, etc.).

Chloroplasts and other preparations from green plants are able to effect the photochemical reduction of viologen dyes and other difficult-to-reduce substances, to extents that signify a reducing potential of about -0.6 V for photosystem I (35). In this ability to form very strong reductants photochemically, green plant system I differs strikingly from any of the bacterial systems thus far characterized.

The fluorescence of Chl associated with green plant photosystem II varies with redox potential in a way that suggests two acceptors, or "quenchers", of mid-point potentials about -0.03 and -0.3 V at pH 7 (36). These show a pH dependence of 60 mV per pH unit and a "one-electron" redox titration curve.

We do not know what any of these presumed acceptors are.

Knowledge and speculation about the photochemical act

Returning to the behavior of reaction centers prepared from Rps. spheroides, the quantum yield of fluorescence of "P,A"

if 4×10^{-4}. Therefore, from Eqs. 2 and 3, the lifetime of the excited state P^*,A is 8 psec (28). The most likely fate of this excited state is one that leads to P^+,A^- since the quantum efficiency for the photochemistry is 70% or greater. We can conclude that the first step in the conversion of singlet excitation energy to chemical potential is an event with a half-time of about 8 psec. This event could be the displacement of an electron from P toward A, forming a charge-transfer state that settles into a more stable configuration (P^+,A^-) through nuclear rearrangements. We have no basis for more detailed speculation at present.

Secondary oxidations and reductions; conservation of energy and prevention of undesirable reactions

The earliest electron transfer events following the primary photochemical process have been delineated in photosynthetic bacteria, especially by Parson (37, 38) using laser flash excitation and measurement with time resolution better than 1 μsec. In extracts of Chromatium, the primary formation of "P^+,A^-" is followed, with a time constant of about 1 μsec at room temperature, by the oxidation of one or another Cyt of the "c" type, C553 or C555. Then, with a characteristic time of about 50 μsec, the reaction center returns to the state P,A (in Parson's terminology, P,X) by discharging the electron from A^- to a secondary acceptor. The principal secondary acceptor is probably ubiquinone; this substance abounds in the pigmented membranes of all photosynthetic bacteria and is carried over into some kinds of reaction center preparations (39).

The foregoing time constants vary widely among different kinds of photosynthetic bacteria: the half-time for Cyt oxidation ranges from 0.3 to 10 μsec (40). In every case these secondary electron transfer events are rapid enough to consolidate the primary separation of oxidizing and reducing entities. The most direct return of electrons from A^- to P^+, involving no other recognized electron carriers, has a half time of 20-30 msec at temperatures from 1.3 to about 200°K, in chromatophore preparations and in reaction centers (41-43). At room temperature, this recombination appears to have a half-time of about 60 msec (43). In contrast, the transfer of electrons from Cyt to P^+ and from A^- to ubiquinone requires far less than 1 msec at room temperature in cells and chromatophores of photosynthetic bacteria. Thus, the secondary, energy-conserving electron transfers compete easily against any wasteful direct return of electrons from A^- to P^+.

The secondary electron transfers that occur in living cells can be mimicked in reaction center preparations, by addition purified mammalian Cyt c and ubiquinone to these preparations. The Cyt c appears to be bound electrostatically, and can transfer electrons to P^+ with half-time about 25 μsec at room temperature (44). The ubiquinone is bound, apparently by hydrophobic interactions (43), and can accept electrons from A^- with half-time far less than 20 msec. The reaction centers can therefore be used to reconstruct certain activities of the living cell, and these models show decisively how the primary conversion of light quanta to chemical potential can be consolidated by subsequent oxidations and reductions.

In green plant system II the primary oxidizing entity, perhaps oxidized Chl, is neutralized quickly by electrons from the chemical system that mediates the evolution of oxygen from water. When this process is arrested, as by washing chloroplasts with Tris buffer, the supply of electrons from water can be replaced by electrons from artificial donors such as hydroquinone. But if no such donor is added, the accumulation of strong oxidizing entities is soon made evident by oxidative damage to a variety of components of the system (12). Light harvesting Chl, carotenoid pigments, and presumably many other substances, such as Cyt, succumb to indiscriminate photochemical oxidation under these circumstances. This damage can be halted by addition of a donor of electrons to the oxidizing side of system II. These experiments show how, at least in one photosynthetic system, the rapidity of secondary electron transfer events not only conserves and channels the energy, but also prevents the primary photoproducts from reacting with their surroundings in an uncoordinated and damaging way.

SUMMARY

By examining selected systems and preparations from photosynthetic materials, we have obtained partial insights into the physical mechanisms of photosynthesis. Certain traditional problems have been laid to rest, at least in principle, but many details remain to be understood. Foremost among the questions now outstanding are:

(a) How does the biochemistry of oxygen evolution work?

(b) What are the components of a reaction center for green plant photosystem II, and what is the primary electron acceptor in each of the various photosynthetic systems?

(c) How many distinct kinds of photochemical system can be found among the photosynthetic bacteria? How closely does any of them resemble green plant system I?

(d) What are the detailed steps that intervene between the singlet excited state of Chl or BChl and the appearance of primary oxidizing and reducing entities?

(e) What is unique about Chl and BChl as sensitizers of photochemical oxido-reduction? Is the primacy of these pigments in photosynthesis merely an evolutionary happenstance?

NOTE ADDED IN PROOF

P. A. Loach (verbal communication) has discovered a new light-induced electron-spin resonance signal in subcellular preparations from photosynthetic bacteria after treatment to remove iron. The signal has $g = 2.005$ and 7 gauss bandwidth. Its response to electron donors and acceptors suggests that it is a property of the primary electron acceptor, partner to the oxidation of P870. M. Okamuro, J. McElroy, and G. Feher (verbal communication) confirm that this signal is exhibited by reaction center preparations treated to remove iron. The broad signal described earlier by Feher (25) could be due to an interaction between the primary electron acceptor and an iron atom.

1. Clayton, R. K. (1971) *Advan. Chem. Phys.* **19**, 353.
2. Clayton, R. K. (1971) "Light and Living Matter" (McGraw-Hill, New York), Vol. II, pp. 1–66.
3. Kok, B. (1961) *Biochim. Biophys. Acta* **48**, 527.
4. Clayton, R. K. (1966) *Photochem. Photobiol.* **5**, 669.
5. Vredenberg, W. J. & Duysens, L. N. M. (1963) *Nature* **197**, 355.
6. Clayton, R. K. (1966) *Photochem. Photobiol.* **5**, 807.
7. Clayton, R. K. (1967) *J. Theor. Biol.* **14**, 173.
8. Joliot, P. (1965) *Biochim. Biophys. Acta* **102**, 116, and 135.
9. Kok, B., Malkin, S., Owens, O. & Forbush, B. (1967) *Brookhaven Symp. Biol.* **19**, 446.
10. Duysens, L. N. M. & Sweers, H. E. (1963) in "Studies in Microalgae and Photosynthetic Bacteria": special issue, *Plant Cell Physiol.* (Univ. of Tokyo Press, Japan) p. 353.
11. Clayton, R. K. (1969) *Biophys. J.* **9**, 60.
12. Yamashita, T. & Butler, W. L. (1969) *Plant Physiol.* **44**, 435, and 1342.
13. Strickler, S. J. & Berg, R. A. (1962) *J. Chem. Phys.* **37**, 814.
14. Rabinowitch, E. (1957) *J. Phys. Chem.* **61**, 870.
15. Latimer, P., Bannister, T. T. & Rabinowitch, E. (1956) *Science* **124**, 585.
16. Wang, R. T. & Clayton, R. K. (1971) *Photochem. Photobiol.* **13**, 215.
17. Rubin, A. B. & Osnitskaya, L. K. (1963) *Mikrobiologiya* **32**, 200.
18. Govindjee, Hammond, J. H. & Merkelo, H. (1971) "Lifetime of the Excited State of Bacteriochlorophyll in Photosynthetic Bacteria", *Biophys. J.*, in press.
19. Kok, B. (1963) in "Photosynthetic Mechanisms of Green Plants", ed. Kok, B. & Jagendorf, A. T. (Nat. Acad. Sci.–Nat. Res. Council, Washington, D.C.), Publ. No. 1145, p. 35.
20. Sybesma, C. & Fowler, C. F. (1968) *Proc. Nat. Acad. Sci. USA* **61**, 1343–1348.
21. Sybesma, C. (1969) in "Progress in Photosynthesis Research", ed. Metzner, H. (Int. Union Biol. Sci., Tübingen), Vol. II, p. 1091.
22. Wraight, C. A. & Crofts, A. R. (1970) *Eur. J. Biochem.* **17**, 319.
23. Cohen, W. S. & Sherman, L. A. (1971) *FEBS Lett.* **16**, 319.
24. Clayton, R. K. & Wang, R. T. (1971) in "Methods in Enzymology", ed. Colowick, S. P. & Kaplan, N. (guest editor San Pietro, A.) (Academic Press, New York), Vol. 23, p. 696.
25. Feher, G. (1971), *Photochem. Photobiol.*, **14**, 373.
26. Clayton, R. K., Fleming, H. & Szuts, E. Z. (1971) "Photochemical Electron Transport in Photosynthetic Reaction Centers: II. Interaction with External Electron Donors and Acceptors, and a Re-evaluation of Some Spectroscopic Data", *Biophys. J.*, in press.
27. Clayton, R. K. & Haselkorn, R. (1971), *Biol. Bull.*, **141**, 381.
28. Zankel, K. L., Reed, D. W. & Clayton, R. K. (1968) *Proc. Nat. Acad. Sci. USA* **61**, 1243–1249.
29. Reed, D. W., Zankel, K. L. & Clayton, R. K. (1969) *Proc. Nat. Acad. Sci. USA* **63**, 42–46.
30. Bolton, J. R., Clayton, R. K. & Reed, D. W. (1969) *Photochem. Photobiol.* **9**, 209.
31. Clayton, R. K. & Straley, S. C. (1970) *Biochem. Biophys. Res. Commun.* **39**, 1114.
32. Cramer, W. A. (1969) *Biochim. Biophys. Acta* **189**, 54.
33. Nicolson, G. L. & Clayton, R. K. (1969) *Photochem. Photobiol.* **9**, 395.
34. Loach, P. A. (1966) *Biochemistry* **5**, 592.
35. Kok, B. (1966) in "Currents in Photosynthesis" ed. Thomas, J. B. & Goedheer, J. C. (Ad. Donker, Rotterdam), p. 383.
36. Butler, W. L., Cramer, W. A. & Yamashita, T. (1969) *Biophys. J.* **9**, A-28.
37. Parson, W. W. (1968) *Biochim. Biophys. Acta* **153**, 248; (1969) **189**, 384, and 397.
38. Parson, W. W. & Case, G. D. (1970) *Biochim. Biophys. Acta* **205**, 232.
39. Reed, D. W. & Clayton, R. K. (1968) *Biochem. Biophys. Res. Commun.* **30**, 471.
40. Kihara, T. & Chance, B. (1969) *Biochim. Biophys. Acta* **189**, 116.
41. Arnold, W. & Clayton, R. K. (1960) *Proc. Nat. Acad. Sci. USA* **46**, 769–776.
42. McElroy, J., Feher, G. & Mauzerall, D. (1969) *Biochim. Biophys. Acta* **172**, 180.
43. Clayton, R. K. & Yau, H. F. (1971) "Photochemical Electron Transport in Photosynthetic Reaction Centers: I. Kinetics of the Oxidation and Reduction of P870 As Affected by External Factors", *Biophys. J.*, in press.
44. Ke, B., Chaney, T. H. & Reed, D. W. (1970) *Biochim. Biophys. Acta* **216**, 373.
45. Borisov, A. Y. & Godick, V. I. (1970) *Biochim. Biophys. Acta* **223**, 441.

Electron Transfer Reactions in Bacterial Photosynthesis: Charge Recombination Kinetics as a Structure Probe

G. FEHER [a], M. OKAMURA [a], and D. KLEINFELD [b]

Introduction

During this conference the question of relevance to biology of the systems that were being investigated came up on several occasions. We are in the happy position of not having to defend our system on that score. Photosynthesis is essential to life; it is the source of energy of the entire living world.

Since this is the first talk on photosynthesis and reaction centers at this meeting, we shall start with a brief introduction to the subject. Photosynthesis deals with the conversion of light into chemical energy that is used by the organism to produce energy-rich compounds. The primary process of photosynthesis involves a charge separation, i.e., the formation of oxidized and reduced molecules. In photosynthetic bacteria this process occurs in a protein pigment complex called the reaction center (RC). The RC is composed of three polypeptide subunits called L, M, and H and a number of co-factors associated with the electron transfer chain. These are four bacteriochlorophylls (BChl), two bacteriopheophytines (i.e., a BChl without the central Mg), two ubiquinones (UQ-10) and one high-spin non-heme iron (Fe^{2+}) (for a review, see ref. 1).

Light induces a charge separation with an electron leaving the donor D, a specialized bacteriochlorophyll dimer, and passing via an intermediate acceptor, I, to the primary and secondary quinone acceptors, Q_A and Q_B, respectively (see Fig. 1); (for a review, see ref. 2). The remarkable thing about photosynthesis is that the quantum yield is close to unity. The high yield occurs because the forward reactions are 10^2–10^3 faster than the (energetically wasteful) charge recombinations reactions (see Fig. 1). We shall be hearing a great deal during this meeting about electron transfer reactions. We shall not discuss in detail the underlying theory here, but will mainly use one of the conclusions that seems to have been universally accepted by both theorists and experimentalists; namely, that *the kinetics of the electron transfer reactions are extremely sensitive to the spatial configuration of the charge separated species.* This enables us to use the kinetics as a

[a] Department of Physics B-019, University of California, San Diego, La Jolla, CA 92093.
[b] AT&T Bell Laboratories, Murray Hill, N.J. 07974.

Reprinted from *Protein Structure: Molecular and Electronic Reactivity* (Springer-Verlag, New York, 1987), pp. 399–421; © Springer-Verlag, Inc.

structural probe. Although each of the transfer reactants can serve as such a probe we shall focus mainly on the recombination kinetics between D^+ and Q_A^- and D^+ and Q_B^-, characterized by τ_{AD} and τ_{BD}, respectively. We shall discuss the following topics:

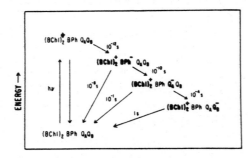

Figure 1 Schematic representation of the electron-transfer reactions in reaction centers from photosynthetic bacteria. After the absorption of a photon, the electron transfers through a series of reactants that are stabilized against charge recombination for progressively longer periods of time. Charged donor-acceptor species are in bold face. Transfer times are given for room temperature and are rounded to the nearest power of 10.

1. Do isolated RCs have the same structure as RCs *in vivo?*

2. The effect of removing the H-subunit on the charge recombination kinetics.

3. Conformational changes associated with the charge separation process.

4. The temperature dependence of the recombination kinetics.

5. The effect of electric fields on the recombination kinetics.

    a) Externally applied fields

    b) Fields due to intrinsic charges.

The first four topics deal with the relative spatial arrangement of the reactants; the fifth topic deals with the electronic level structure of the reactants.

1. Do isolated RCs have the same structure as RCs *in vivo* ?

Historically, the first complex that was called a reaction center was isolated by Reed and Clayton (3) and had a molecular weight of over one million. When we isolated a much smaller reaction center having a molecular weight of $\sim 10^5$, which resembled the "modern reaction center", we were astonished to find great opposition from several quarters when these findings were presented at the Gatlinburg Conference in 1970 (4). Some people just could not believe that such a small unit could perform the marvelous primary process of photosynthesis. They claimed that life had slipped through our fingers during the purification process and that these reaction centers probably bear little resemblance to what happens *in vivo*. Although this attitude has vanished by now, there still remain some lingering questions concerning the extent to which the isolated reaction centers have the same structures as RCs *in vivo*. We have, therefore, resurrected a table from an old piece of work by J. McElroy *et al.* (5) in which that question was addressed by measuring the charge recombination kinetics described by the scheme:

$$DQ_A \underset{\tau_{AD}}{\overset{h\nu}{\rightleftharpoons}} D^+ Q_A^-$$

(1)

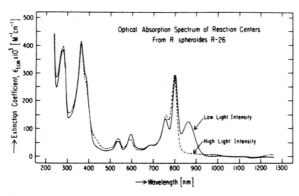

Figure 2 *Absorption spectrum of RCs from R. sphaeroides R-26 obtained under conditions of low light intensity (——) and with strong cross illumination (-----). The ordinate was normalized to the extinction coefficient at 802 nm, i.e., ($\epsilon^{802} = 2.88 \times 10^5 M^{-1} cm^{1-}$(6)). From Ref. 4.*

The state $D^+Q_A^-$ was formed by a short pulse of high intensity light. The charge recombination kinetics were monitored optically via changes in the absorption spectrum caused by the presence of D^+ (see Fig. 2). To prevent the electrons from leaving Q_A^-, these experiments were performed at 77 K. The results showed that whole cells, chromatophores and isolated reaction centers have, within experimental error, the same recombination time τ_{AD} * (see Table I). When reaction centers were

error, the same recombination time τ_{AD} * (see Table I).

TABLE I. OPTICAL DECAY KINETIC OF WHOLE CELLS, CHROMATO-PHORES, AND REACTION CENTERS OF *R. SPHAEROIDES*, R-26

Whole cells and chromatophores were suspended in 50% glycerol, 0.05 M Tris, pH 8.0, to give A_{800nm} of approx. 0.1 (1 mm path). The presence of glycerol did not affect the kinetic behavior. It was used to insure the formation of a transparent glassy matrix of low temperatures. Reaction centers, unless otherwise specified, were suspended in the same buffer to give A_{800nm} of approx. 0.2 (1 mm path) at room temperature. Reaction centers which were exposed to 6 M urea for 4 h at 20°C in the dark had an absorbance A_{800nm} of approx. 1.0 (1 mm path). The decay of the optical change at 795 nm was monitored at a sample temperature of 80°K. The wavelength of the actinic illumination was 900 nm. The measuring beam intensity was approx. 5 $\mu W/cm^2$.

Preparation	1/e decay time	Treatment
Whole cells	30 ± 3 ms	No detergent
Chromatophores	32 ± 3 ms	No detergent
Reaction centers	29 ± 2ms	0.1% lauryl dimethyl-amineoxide
Reaction centers	52%, 28 ms, 48%, 93 ms	6 M urea, 0.1% lauryl dimethylamineoxide, $t = 4$ h
Reaction centers	60%, 260 ms, 40%, 1.8 s	0.1% sodium dodecyl-sulfate, 0.02% lauryl dimethylamineoxide

From: J. McElroy, D. Mauzerall, G. Feher (1974) Biochim. Biophys. Acta 333, 261-277.

treated with strong reagents (e.g., urea or sodium dodecyl sulfate) the decay kinetics changed significantly, indicating a structural change (see Table I). These experiments show that the structural integrity, at least with respect to the donor-acceptor complex, is preserved in isolated reaction centers.

2. The effect of removing the H-subunit

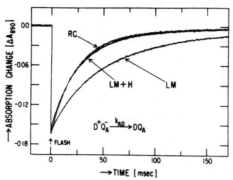

Figure 3 *Kinetics of charge recombination between* Q_A^- *and* D^+ *(k_{AD}) at cryogenic temperatures (77K). The formation and decay of the charge-separated state ($D^+Q_A^-$) was monitored at 890 nm following an actinic flash. From Ref. 7.*

All the prosthetic groups associated with the electron transfer processes are bound to the L and M subunits. The question concerning the role of the H subunit, therefore, naturally arises. R. Debus in our laboratory was able to isolate the LM-pigment-complex and the H-subunit and to subsequently reconstitute LM and H to reform RCs (7). When the charge recombination time, τ_{AD}, was measured at low temperatures, as discussed before, the results shown in Fig. 3 were obtained. The recombination times in LM were about a factor of two slower than those found in RCs. The recombination time between Q_A^- and D^+ is believed to be critically sensitive to changes in the distance between D and Q_A (5,8,9), a point that we shall discuss in more detail in the next section. Since τ_{AD} changes only by a factor of $\sim$ 2 (see Fig. 3), the relative configuration of D and Q_A is affected only to a relatively small extent by the removal of H. Thus, H does not play a major role in this charge recombination step. Incidently, note that the kinetics in the reconstituted LMH are practically identical to those in RCs. This shows that the change in the LM complex was not due to an irreversible denaturing effect accompanying the isolation procedure.

The charge recombination kinetics at room temperature in RCs containing *two quinones* exhibited a more dramatic change when H was removed. The kinetic properties of this system are described by:

$$DQ_AQ_B \xrightleftharpoons[\tau_{AD}]{h\nu} D^+Q_A^-Q_B \xrightleftharpoons[\tau_{BA}]{\tau_{AB}} D^+Q_AQ_B^- \qquad (2)$$

$$\xleftarrow{\hspace{3cm}}$$
$$\tau_{BD}$$

In RCs or reconstituted LMH complexes the recombination time τ_{BD} was $\sim$ 1s, indicative of the characteristic time of recombination between Q_B^- and D^+ (see Fig. 4). In LM, on the other hand, the recombination time was ten times shorter, as expected from the recombination between Q_A^- and D^+ (see Fig. 4). This suggests that the electron transfer from Q_A^- to Q_B was impaired. Indeed, independent experiments have shown that the electron transfer time, τ_{AB}, is about three orders of magnitude longer in LM than in RCs (7). This, of course, is a large effect that would likely be detrimental to the physiological well-being of the bacterium. From a structural point of view, this means that the distances (or angles) between Q_A and Q_B have been significantly changed upon removal of H. Another effect shown in Fig. 4 is the lack of recovery of the absorbance change in LM. This presumably is due to a loss of an electron to exogenous acceptors and may again be a consequence of the opening up of the structure.

3. Conformational changes associated with the charge separation process

There exists some evidence for bulk structural changes in the charge separated state. It comes from the calorimetric study of Arata and Parson (10,11) who found that during charge separation the volume of the RC-solvent system decreased. In a different set of experiments, Noks et al. (12) found that incubation of chromatophores with the cross-linker glutaraldehyde affected the electron transfer kinetics only if incubation was performed in the presence of light. We addressed the question of a conformational change during charge separation by analyzing the charge recombination kinetics in samples prepared under different conditions (13).

We start by describing experiments performed on RCs containing only *one quinone*, i.e., we focus on the charge recombination between D^+ and Q_A^- (see Eq. 1). Two sets of samples were prepared. In one, RCs were cooled to cryogenic temperature under illumination, i.e., in the charge separated state. Thus any possible light-induced structural changes may be trapped when RC conformations are immobilized at low temperatures. The second sample was cooled to cryogenic temperature in the dark. The results of the kinetics of charge recombinations in the two samples is shown in Fig. 5a. There is a significant difference in the recombination time τ_{AD} between the two samples, i.e.,

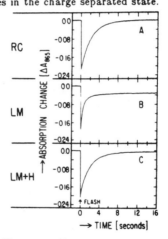

Figure 4 *Charge recombination between Q_A^- or Q_B^- and D^+ in RCs, LM, and reconstituted LMH at 4°C. The lack of complete recovery in the LM subunit is attributed to a loss of the electron from the quinone acceptors. From Ref. 7.*

$$\tau_{AD}^{light} = 120 \text{ ms}, \qquad \tau_{AD}^{dark} = 25 \text{ ms}.‡$$

‡ The value of k_{AD} for UQ differs in recent RC preparations by $\sim 15\%$ from that quoted earlier (see **Table I**) The origin of this discrepancy is not understood, it may be due to a changed binding site

In Fig. 5b the change of absorbence is plotted logarithmically; a single exponential recombination process should give a straight line on this plot. We see that RCs cooled under illumination have not only a longer recombination time, but their kinetics are much more non-exponential than those for RCs cooled in the dark. For comparison we show also the recombination kinetics at room temperature; in this case a good exponential recovery is observed.

Qualitatively, we attribute the non-exponential behavior to a distribution of structural states. Evidence that such distributions may exist in proteins comes from the detailed work of Austin, Frauenfelder and collaborators (14,15), and Woodbury and Parson (16). Furthermore, we see that RCs cooled under illumination deviate much more from exponentiality, i.e. they will have a broader distribution of conformational states. Their recombination time is also longer, indicating that the average distance between D^+ and Q_A^- has increased during illumination.

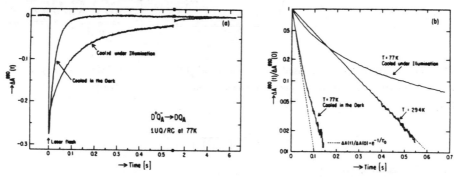

Figure 5 *Electron Donor Recovery Kinetics for 1UQ/RC following a laser flash. (a) Kinetics at 77K for RCs cooled in the dark and under illumination (b) semilog plot of the kinetics shown in part (a) together with kinetics obtained at room temperature. Dashed lines represent fits to an exponential function, with $\tau_0 = 22$ ms for RCs cooled to 77 K in the dark and $\tau_0 = 132$ ms for RCs at 294 K. Note the large deviation from an exponential of the kinetics in RCs cooled under illumination. From Ref. 13.*

To treat this problem quantitatively, we parameterize the recombination kinetics in terms of the D^+–Q_A^- electron transfer distance, r_{AD}. If all the donor acceptor pairs had identical separation distances, the observed absorption change would be given by a single exponential

$$\Delta A(t) \, / \, \Delta A(0) = e^{-t/\tau(r_{AD})} \tag{3}$$

where $\tau(r_{AD})$ is the characteristic recombination time, given by:

$$\tau(r_{AD}) = \tau_0 e^{-r_{AD}/r_0} \tag{4}$$

the value of $r_0 \simeq 1\text{Å}$ (17,18). If r_{AD} varies between different donor acceptor pairs, Eq. 3 is no longer valid and $\Delta A(t)$ is described by a normalized distribution function of distances $D(r)$, i.e.:

$$\Delta A(t) \,/\, \Delta A(0) = \int_{0}^{\infty} D(r)e^{-t/\tau(r)}dr \qquad (5)$$

To solve for the distribution function D(r), we fit the data with an analytic function given by:

$$\Delta A(t) \,/\, \Delta A(0) = [1 + t \,/\, (n\tau_0)]^{-n} \qquad (6)$$

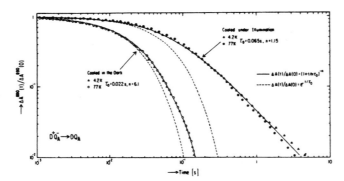

Figure 6 *Log-log plot of the donor recovery kinetics at 4.2 and 77 K in 1UQ/RC samples cooled in the dark and cooled under illumination. Dashed lines represent fits of the initial slopes of the data to an exponential; solid lines are fits to a power law (Eq. 6). The values of parameters τ_0 and n are given in the figure. Note that τ_0 is the same for both functions. From Ref. 19.*

This function is the same as used by Austin *et al* (14); it fits our data well as shown in Fig. 6. Equating this to the expression given by Eq. 5 one can solve for the distribution D(r), i.e.,

$$r_0 D(r) = \frac{1}{\Gamma(n)} \left[n\frac{\tau_0}{\tau(r)} \right]^n e^{-n[\tau_0/\tau(r)]} \qquad (7)$$

where $\Gamma(n)$ is the gamma function.

The result of the calculation of the distribution function is shown in Fig. 7. It bears out our qualitative discussion given before, i.e.:

1.) the average electron transfer distance in RCs cooled under illumination is larger than the average distance in RCs cooled in the dark,

2.) the width of the distribution in RCs cooled under illumination is two and a half times larger than in RCs cooled in the dark.

The shift and width of the distribution is of the order of 1Å, which is similar to the root mean square displacements determined from crystallographic studies on other proteins (19,20) and from model calculations (see, e.g., Refs. 21 and 22). The recombination kinetics were essentially temperature independent between 4.2 and 77 K (see Fig. 6). This indicates that the distribution remained constant with temperature, on the time scale, τ_{AD}, of the measurement.

It is interesting to speculate whether these light-induced changes have a physiological function analogous to those produced by allosteric changes in other proteins (for a review, see Ref. 23). Perhaps the structural changes accompanying the charge separation process act to inhibit the wasteful recombination reactions by stabilizing the charge separated states. Such 'stabilization processes have also been discussed by Warshel (24), and Woodbury and Parson (16).

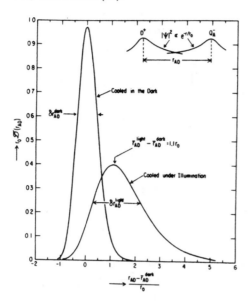

Figure 7 *Calculated distributions (see Eq. 7) of the electron transfer between D^+ and Q_A^- in $1UQ/RC$ samples cooled in the dark and under illumination. This distribution describes the nonexponential decay kinetics of $D^+Q_A^-$, shown in Figs. 5 and 6. The experimental parameters, n and r_0, given in Fig. 6, were used together with Eqs. 4 and 7 to calculate the distributions. Insert shows the exponential decrease of the wave functions that leads to Eq. 4. Note that RCs cooled under illumination have a larger average electron-transfer distance as well as a larger spread in distances than RCs cooled in the dark. From Ref. 19.*

We now turn to the more complicated question of the recombination kinetics from the *secondary quinone*, Q_B, described by the scheme given by Eq. 2. The electron transfer time τ_{AB} is approximately 10^{-4} sec at room temperature but becomes unobservably long at 77 K. If we want to study, therefore, the recombination kinetics of $D^+Q_AQ_B^-$ at low temperature this state has to be trapped at 77 K by cooling RCs under illumination. This was done and the result of the recombination kinetics are shown Fig. 8. The solid line represents a theoretical fit to the same function as was used for the one quinone case (see Eq. 6). However, in this case one observes two features that are distinctly different from those observed in RCs containing one quinone. The recombination time is highly temperature dependent and the spread in characteristic times is very large. (Note the logarithmic scale of the abscissa). If we extrapolate the 18 K data to longer times the recombination is less than half complete even after 10^7s (1 year!).

The observed temperature dependence of the two quinone systems can be explained by the model of Agmon and Hopfield (25). Due to the dynamics of protein motion the RC passes through a number of structural states. The most favorable states for rapid recombination are those for which the distances between Q_B^- and D^+ are small. As the temperature is raised the probability that transitions to these favorable states occur is increased, thereby reducing the recombination time τ_{BD}. For RCs with one quinone the recombination time τ_{AD} is many orders of magnitude shorter than τ_{BD}. Consequently there is no opportunity to sample the different conformational states within the time τ_{AD}. This gives rise to an effective static distribution of distances between Q_A^- and D^+, resulting in temperature independent kinetics as observed in RCs with one quinone.

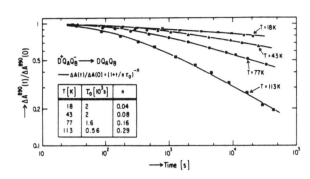

Figure 8 *Log-log plot of the donor recovery kinetics at different temperatures in 2UQ/RC samples cooled under illumination. Data were normalized to the maximum absorption change, $\Delta A^{890}(0)$, found by extrapolating the measured absorption changes back to zero time (more data were acquired at short times than are shown). The maximum absorption level [i.e., $A^{890}(\infty)$], which served as the base line for the absorption changes, was determined by warming and recooling the sample in the dark. Typically, $\Delta A^{890}(0)$ was 80% of $A^{890}(\infty)$. The parameters τ_o and n were found from fitting the data to Eq. 6 and are tabulated in the insert. From Ref. 13.*

The above model also explains the exponential behavior found for τ_{AD} at room temperature. If the transitions between structural states occur in a time that is much shorter than τ_{AD}, the individual states will not be expressed and a single, average, τ_{AD} will be observed. This situation is analogous to motional narrowing in magnetic resonance.

An interesting finding that we will not discuss in detail here is that the electron transfer time from Q_A^- to Q_B is at least 8 orders of magnitude shorter in RCs that have been illuminated while being cooled as compared to those cooled in the dark (13). It is difficult to see how such a large change can be produced by a light-induced conformational change. It is more likely that a proton that associates with Q_B^- at room temperature (26) remains trapped in the vicinity of Q_B^- after the RCs are cooled. This proton cannot associate with RCs at low temperatures. Thus, RCs cooled in the dark will remain unprotonated upon illumination.

4. The temperature dependence of the recombination kinetics

We next discuss the temperature dependence of the charge recombination rate $D^+Q_A^- \rightarrow DQ_A$ (see Eq.1). We shall inquire whether the experimental results can be explained by present theories of electron transfer or whether major contributions are due to temperature dependent structural changes, e.g., thermal expansion.

Measurements were made using RCs containing one quinone, i.e., 1UQ/RC. All samples were cooled to cryogenic temperature in the dark; the value of k_{AD} was stable with time at each temperature and was completely reversible as the temperature was cycled. To insure the presence of an optically transparent sample at all temperatures, RCs were incorporated into a thin film of polyvinyl alcohol (PVA) [27].

The temperature dependence of k_{AD} is shown in Fig. 9a. The recombination rate was essentially temperature independent at low temperature, as discussed earlier (see Fig.

For RCs cooled under illumination, a different behavior of the kinetics was observed Above ~ 90 K, k_{AD}^{light} changed with time heading, toward the value of k_{AD}^{dark} Apparently, the structural changes that had been trapped during illumination were annealing out at $T > 90$ K

6). As the temperature was increased from ~ 90 K to 300 K, k_{AD} decreased by a factor of ~ 6 † (see also Ref. 28-31). The results are similar to the temperature dependence of k_{AD} observed by Loach *et al.* [30] and Mar *et al.* [31] with RCs from *R. rubrum*. These temperature dependences are rather unusual; the rate k_{AD} *decreases* with *increasing* temperature. This is in contrast to the usual behaviour of thermally activated processes.

Can we understand the observed temperature dependence of k_{AD} in terms of the electron transfer theories of Hopfield [8] and Jortner [9,17]? In these theories, the electronic transition $D^+Q_A^- \rightarrow DQ_A$ is coupled to a vibrational mode(s) in the protein. The decrease in electronic energy during the charge recombination is compensated for by an equal increase in the energy of the vibrational mode coupled to the reaction. Thus, energy is conserved during the transition.

The theoretically predicted charge recombination rate can be expressed in a compact form under the approximation that both the electron donor and the acceptor are coupled to the *same*, single, vibrational mode. The recombination rate, valid for any temperature, T, is given by [9,17]:

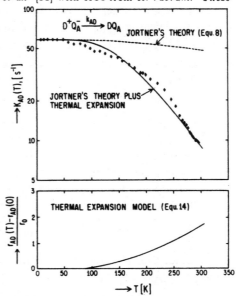

Figure 9 a.) *Temperature dependence of k_{AD} for 1UQ/RCs embedded in a 0.1mm polyvinyl alcohol film (A^{800}=1.2). Dashed line represents Jortner's theory (Eq. 8) with $\Delta E_{redox}=E_{nuc}$ and T_o=500K. Solid line represents the best fit of the expansion model (Eq.13,14) plus Jortner's theory to the experimental data. b.) Temperature dependence of the distance between D^+ and Q_A^-, r_{AD}; computed from the thermal expansion model (Eq.14).*

$$k_{AD} = \frac{(2\pi)^2}{h}|M|^2\frac{1}{k_BT_o}\left[\frac{\upsilon+1}{\upsilon}\right]^{p/2} e^{-s(2\upsilon+1)} I_p\left[2s\sqrt{\upsilon(\upsilon+1)}\right] \qquad (8a)$$

where

$$s = \frac{E_{nuc}}{k_BT_o} \qquad p = \frac{\Delta E_{redox}}{k_BT_o} \qquad \upsilon = \frac{1}{e^{T_o/T}-1} \qquad (8b)$$

The overlap integral M connects the electronic states of $D^+Q_A^-$ and DQ_A, T_o is the characteristic temperature of the vibrational mode (i.e., $k_BT_o = \hbar\omega$), E_{nuc} is the energy required to rearrange the nuclear positions concomitant with the electron transfer, ΔE_{redox} is the

† The values for k_{AD} obtained with RCs in PVA are ~ 50% larger than those obtained with RCs in glycerol. This difference may be caused by the 1 Molar salt concentration in the dried PVA film.

difference in free energy between the D^+/D and Q_A/Q_A^- redox couples (i.e. the energy difference between $D^+Q_A^-$ and the ground state), $I_p(x)$ refers to the modified Bessel function of order p, k_B is Boltzman's constant and h is Plank's constant.

The temperature dependence of the predicted rate (Eqs. 8a and 8b) simplifies considerably in the limit of either high or low temperature. The recombinaton rate is expected to follow an activated temperature dependence when there is sufficient thermal energy available to excite the vibrational mode coupled to the reaction $D^+Q_A^- \rightarrow DQ_A$. In this limit, i.e., for $T >> T_o$, the rate is given by [8,9,17]:

$$\text{For } T>>T_o; \quad k_{AD} = \frac{(2\pi)^2}{h}|M|^2 \frac{1}{\sqrt{4\pi E_{nuc}k_B T}} e^{\frac{-(\Delta E_{redox}-E_{nuc})^2}{4E_{nuc}k_B T}} \tag{9}$$

At low temperature, i.e., for $T << T_o$, the vibrational mode coupled to the reaction remains in the ground state and thus the recombination rate is independent of temperature. In this limit, the rate is given by a Poisson distribution for the rearrangement energy, i.e. [9,17]:

$$\text{For } T<<T_o; \quad k_{AD} = \frac{(2\pi)^2}{h}|M|^2 \frac{1}{k_B T_o} s^p \frac{e^{-s}}{p!} \tag{10}$$

Note that for large p, the term p! in Eq.10 makes the low temperature limit of k_{AD} very sensitive to changes in the redox energy difference, E_{redox}, between $D^+Q_A^-$ and DQ_A. This limit applies to RCs, where $E_{redox} \sim 500$ meV [2] and $k_B T_o$ for proteins typically lies in the range 10 - 100 meV ($\sim$100 - 1000K).

The theoretical model we discussed predicts, in general, for temperatures near or above T_o an *increase* in the recombination rate with *increasing* temperature. This is in contradiction to the experimentally observed temperature dependence of k_{AD} (see Fig. 9a). A hypothesis often suggested (see, e.g., Refs.8,9,17) to circumvent this inconsistency between experiment and theory is that the redox energy difference between $D^+Q_A^-$ and DQ_A equals the nuclear rearrangement energy, i.e., $\Delta E_{redox} = E_{nuc}$. For this special condition, the theory predicts that k_{AD} is constant for $T<<T_o$ (see Eq. 10); and that k_{AD} *decreases* with *increasing* temperature for temperatures near of above T_o. For $T >> T_o$ and $p! >> 1$, one obtains from Eqs.9 and 10 (with p! approximated by $\sqrt{2\pi p} \, p^p e^{-p}$) for the temperature dependence of k_{AD}:

$$\text{For } T>>T_o; \, p!>>1: \quad k_{AD} = k_{AD}(0) \sqrt{\frac{T_o}{2T}} \tag{11}$$

where $k_{AD}(0)$ is the low temperature limit of the recombination rate (see Eq. 10).

Since the observed recombination rate (see Fig. 9a) does not correspond to the high temperature limit we tried to fit the data with the general expression given by Eq. 8. We took the room temperature value of $\Delta E_{redox} = 500$ meV[2] and equated it to E_{nuc} (i.e., p

= s), with $k_{AD}(0) = 58$ s^{-1}. To estimate T_o, we plotted k_{AD} *versus* T/T_o and found that T_o corresponds to ~ 5 times the temperature at which k_{AD} changes from a temperature independent to a temperature dependent value. We see from Fig. 9a that this transition occurs at ~ 100 K, i.e., the characteristic temperature is

$$T_o \sim 500 \text{ K} \tag{12}$$

Using the above values of T_o, ΔE_{redox}, E_{nuc} and $k_{AD}(0)$, the predicted temperature dependence of k_{AD} (Eq. 8) disagrees with the observed behaviour (Fig. 9a). Changing the value of E_{nuc} away from the value of E_{redox} increases the disagreement even further. For redox energies outside the range of 400 meV $> E_{nuc} > 700$ meV (with $E_{redox} = 500$ meV), theory predicts a change in the *sign* of the temperature dependence, in accord with a thermally activated process. To reconcile the disagreement between experiment and theory, Sarai (32) and Kakitani *et al.* (33,34) modified the theories of Hopfield (8) and Jortner (9,17) by including a multiplicity of vibrational modes, as opposed to a single mode[‡]. An alternate mechanism to account for the observed temperature dependence of k_{AD} is the thermal expansion of the protein (29,35,36). Thermal expansion will cause the donor-acceptor distance, r_{AD}, to increase with increasing temperature. This in turn will decrease the value of the overlap integral M, thereby reducing k_{AD} (see Eq. 4). To estimate the magnitude of this effect we write

$$|M(T)|^2 = |M(0)|^2 \, e^{-|r_{AD}(T)-r_{AD}(0)|/r_o} \tag{13}$$

where r_o is the same scaling factor as used in Eq. 4. The change in lattice spacing for a simple solid with an anharmonic interatomic potential (see e.g. ref.37) is given by

$$\frac{r_{AD}(T)-r_{AD}(0)}{r_o} = \left[\frac{\gamma}{r_o}\right] \frac{T_o}{2} \left[\coth\frac{T_o}{2T} - 1\right] \tag{14}$$

where γ/r_o is an adjustable parameter related to the linear expansion coefficient, β, by $\beta = \gamma/r_{AD}(0)$. Assuming that the characteristic temperatures associated with the vibrational mode coupled to the thermal expansion is the same as that coupled to the electron transfer (i.e., $T_o = 500$ K) we fitted the thermal expansion model (Eqs. 13, 14) to the observed data with a value of

$$\gamma/r_o = 1.4 \times 10^{-2} \text{ K}^{-1} \tag{15}$$

This value corresponds to a thermal expansion coefficient that is an order of magnitude larger than that determined for a protein (38). It should be noted, however, that the relevant number is not the *average* expansion coefficient but the change in a *particular* distance, namely r_{AD} with temperature. This can be an order of magnitude larger that the change given by the average expansion coefficient (38).

[‡] Now that the three dimensional structure of RCs is being determined (Deisenhofer *et al.* (1984) *J. Mol. Biol.* **180**, 385 and Michel *et al.*, these proceedings), there is hope that one will be able to correlate the vibrational mode(s) with specific bonds in the vicinity of the primary reactants.

The temperature dependence of r_{AD} found from the expansion model is shown in Fig. 9b. The change in r_{AD} between low temperature and room temperature is $\Delta r_{AD} = 1.7 r_o \sim 1-2$ Å. This is the same as found for some atomic positions in a protein (38). Thus, the expansion mechanism seems to provide a possible mechanism for the temperature dependence of k_{AD}, although it certainly does not constitute a proof.

Before leaving this topic let us briefly discuss the assumption that $\Delta E_{redox} = E_{nuc}$. At first glance it seems odd that nature should have picked this equality since it maximizes the rate (k_{AD}) of a physiologically undesirable reaction. However, this constraint may be a consequence of maximizing the transfer rate for the forward reaction $D^+I^-Q_A \rightarrow D^+IQ_A^-$, where I is the intermediate acceptor (2). The redox energy difference between the I/I^- and Q_A/Q_A^- redox couples is approximately the same as that between the Q_A/Q_A^- and D^+/D couples (2).

The temperature dependence of k_{AD} in *R. rubrum* (30,31) was found to be similar to that observed in *R. sphaeroides* (Fig. 9a). However, Mar *et al.* (31) found a much weaker temperature dependence of k_{AD} in RCs from *Ectothiorhodospira* sp. The simplest explanation of this result is that in this bacterial species $\Delta E_{redox} \neq E_{nuc}$, although the alternate explanation that the temperature dependence of Δr_{AD} has been reduced cannot be excluded.

Can we test experimentally the applicability of the electron transfer theories to RCs, and in particular, whether our assumption $\Delta E_{redox} = E_{nuc}$ is justified? Eq. 8 predicts a parabolic-like dependence of k_{AD} verus ΔE_{redox} with k_{AD} peaking at $\Delta E_{redox} = E_{nuc}$. Thus the most direct test would be to vary ΔE_{redox} and to establish whether k_{AD} exhibits the expected parabolic dependence. ΔE_{redox} was changed by substituting quinones with different redox potentials for the native ubiquinone (39).

The low temperature (77K) values of k_{AD} are plotted together with the theortical curve (Eq.10) in Fig.10 with the assumption that $\Delta E_{redox}(UQ) = E_{nuc}$. The redox potentials of the quinones were taken from ref. (40); the accepted value ΔE_{redox} for UQ is 520 meV (11). The general parabolic feature of the theory are seen to be borne out by the experimental data. However, it should be kept in mind that the redox potentials used were obtained for quinones in dimethylformamide at *room* temperature and are likely to deviate from the values found in situ (41) at low temperatures. Furthermore, substitution of quinones may change other parameters (e.g. r_{AD}) besides ΔE_{redox} that affect k_{AD}. Consequently a quantitative agreement of the experimental results with theory cannot be expected at this point; the rather good agreement shown in Fig. 10 seems to us better than one has the right to expect.

An extensive and systematic set of substitution experiments have been performed by

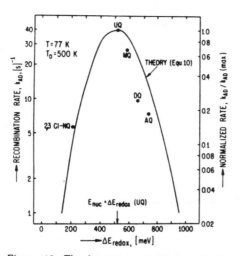

Figure 10 *The low temperature recombination rate,* k_{AD}, *as a function of* ΔE_{redox}. *Solid line represents Jortner's theory (Eq.10) with* $T_0 = 500K$ *and* $E_{nuc} = \Delta E_{redox} = 520\ meV$. *Dots represent experimental points obtained with RCs in which the native ubiquinone (UQ) was substituted with different quinones. The values of* k_{AD} *for menadione (MQ), duroquinone (DQ) and anthraquinone (AQ) were taken form ref. (39), the value for 2,3, dichloronaphtoquinone (2, 3Cl-NaQ) represents a new measurement. The values of the redox potentials were taken from ref. (40).*

Gunner *et al.* (42). They reported that at room temperature the value of k_{AD} was essentially constant over a large range (~ 0.7 eV) of redox potentials. These results seemed to be in disagreement with electron transfer theories as was pointed out during this conference. However, more recently, these authors concluded (43) that their room temperature measurements on the halogenated benzoquinones did not represent a direct recombination process but a transfer of electrons via excess quinones in solutions. Measurements of k_{AD} at low temperatures would eliminate this problem and would provide additional important data to compare with theory. An alternate way of changing ΔE_{redox} is to apply an external electric field across the reaction center as discussed next.

5. The effect of electric fields on the recombination kinetics

Before discussing the effect of an electric field on the direct recombination rate, k_{AD}, we shall consider the case in which the electric field changes the observed recombination via an indirect pathway. Thus, we shall discuss now how the kinetics can be used to probe *electronic energy levels* rather than *conformational changes*.

Fig. 11 shows the electron transfer reactions that we will be concerned with (44) The state $D^+IQ_A^-$ can decay via a direct or indirect pathway, as indicated. The observed decay rate of this state, k_{obs}, is in general a combination of the direct and indirect pathways. Assuming that the rates k_{AI} and k_{IA} are fast in comparison to k_{ID} and k_{AD}, the states $D^+I^-Q_A$ and $D^+IQ_A^-$ can be considered to be in equilibrium and the observed decay rate is given by:

$$k_{obs} = k_{INDIRECT} + k_{DIRECT} = k_{ID}\alpha + k_{AD}(1-\alpha) \qquad (16)$$

where α is the fraction of RCs in the thermally excited state $D^+I^-Q_A$ (see Fig.10). For $\alpha << 1$, the condition that prevails in RCs, Eq. 16 becomes

$$k_{obs} \simeq \alpha k_{ID} + k_{AD} = k_{ID}e^{-\Delta G^0/k_bT} + k_{AD} \qquad (17)$$

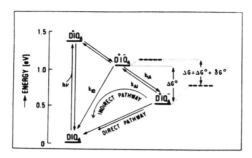

Figure 11 *Simplified energy level scheme showing electron transfer (arrows) in reaction centers of R. sphaeroides. The state $D^+IQ_A^-$ can decay either via the direct pathway (with rate k_{AD}) or via the intermediate state $D^+I^-Q_A$, depending on the value of the energy difference, $\Delta G°$ (Eq. 17). An electric field changes $\Delta G°$ by $\delta G°$ and affects, therefore, the recombination rate via the indirect pathway (Eq. 18). The change in energy levels, is illustrated for a direction of the electric field that reduces $\Delta G°$. When the field is reversed, the energies of the two states are lowered and $\Delta G°$ is increased. From Ref. 44.*

Which of the two pathways predominates depends critically on the value of the energy difference $\Delta G°$. Substituting the measured values of k_{ID} and k_{AD} into Eq. 17 one can show that the two pathways will contribute equally, i.e., $k_{DIRECT} = k_{INDIRECT}$, for $\Delta G° = 400$ meV. The energy gap, $\Delta G°$, in RCs containing the native ubiquinone (UQ) as the primary acceptor has been determined to be 500 - 600 meV (40,45) while for anthraquinone (AQ) $\Delta G° = 340$ meV (40,44,46). Since one of these numbers is larger and the other smaller than the critical value of 400 meV, the direct pathway predominates for UQ whereas the indirect pathway predominates for AQ.

a.) Externally applied fields.

Effect on the indirect pathway: We shall first focus on RCs that have anthraquinone as the primary acceptor; in this case the observed (indirect) recombination rate will be given by the first term of Eq. 17, i.e.:

$$k_{obs} = k_{ID}e^{-\Delta G°/k_bT} \tag{18}$$

Let us now consider the effect of an electric field on the energy levels. The two states $D^+I^-Q_A$ and $D^+IQ_A^-$ will be effected to a different extent since the magnitudes of their dipoles along the electric field are different. This is indicated by the dashed lines in Fig. 11. The energy difference between the two states has been changed by an amount $\delta G°$ producing a change in the recombination rate given by

$$k_{obs} = k_{obs}^°e^{-\delta G°/k_bT} \tag{19}$$

where $k_{obs}^°$ is the recombination rate in the absence of an electric field.

How is an electric field applied across the RCs? A. Gopher in our laboratory incorporated RCs into a lipid bilayer that separates two aqueous compartments (44,47). A voltage was applied across the bilayer and the current produced by the charge recombination following a pulse of light was measured. Fig. 12 shows the results of such an experiment performed on RCs containing AQ. The top panels show the current measured after the light is turned off, i.e. during the charge recombination process. The areas under the curves represent the total transferred charge and are, therefore, equal in all three panels. Consequently, the amplitude increases as the recombination time becomes shorter. In the lower panel the experimental data are plotted logarithmically. The data were fitted with a straight line given by

$$\tau = 1/k_{obs} = 8.5 \times 10^{-3} e^{-V/0.175} \text{ s} \qquad (20)$$

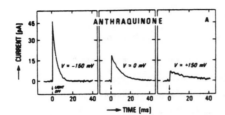

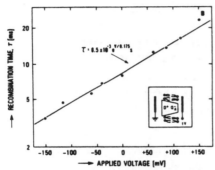

Figure 12 *The effect of an applied electric field on the kinetics of charge recombination in RCs with anthraquinone as the primary acceptor incorporated in a planar bilayer. A.) Time course of current after light pulse is turned off in the absence (V=0) and presence (V=±155mV) of an electric potential. B.) Dependence of the charge recombination rate on the applied voltage. Solid line represents least square fit to the data and obeys the relation $\tau = 8.5 \times 10^{-3} e^{V/0.175}$s. Inset shows the polarity of the voltage with respect to the functionally oriented population of RCs. The polarity of the output signal was inverted by an amplifier. From Ref. 44.*

We see that an e-fold change in the recombination time, τ, results when the applied voltage across the membrane is 175 mV. If I and Q_A were to span the entire membrane one would expect an e-fold change for 25 mV (i.e., $k_b T/q$), where q is the charge of the electron. The fact that we need a seven times larger voltage means that the component of the distance between I and Q_A along the normal of the membrane is only 1/7th of the width of the membrane.

It is interesting to speculate whether the effect of an electric field on the recombination kinetics has any physiological significance. We know that the outside of chromatophores is negative with respect to the inside and that the RCs are oriented in the membrane with the donors pointing towards the inside. Thus the membrane potential created during charge separation *in vivo* decreases ΔG°, thereby decreasing the quantum efficiency at high light intensity. Thus, nature may have build in a negative feedback to prevent detrimental effects at high light intensities.

Effect on the direct pathway: Let us now consider the case of RCs containing UQ. Since now ΔG° is larger than the critical value of 400 meV, the direct pathway predominates. Fig. 13 shows that the recombination kinetics remained uneffected within experimental error ($\pm 5\%$) over the range of applied voltages (± 150meV). How do we reconcile this result with electron transfer theories? If we assume again that $\Delta E_{redox} = E_{nuc}$, then k_{AD} is relatively insensitive to changes in ΔE_{redox} (i.e., $[dk_{AD}/d(\Delta E_{redox}-E_{nuc})]=0$). Under these conditions, Eq.8 predicts that for a 10% change in k_{AD} one needs a change in

$(\Delta E_{redox} - E_{nuc})$ of $\sim$ 200 meV (see Fig.10). Although we have applied 300 meV across

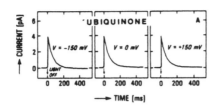

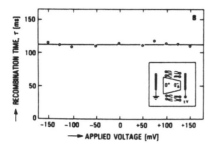

Figure 13 *The effect of an applied electric field on the kinetics of charge recombination in RCs with ubiquinone (UQ-10) as the primary acceptor. Compare the results with those shown in Fig. 12. (Note the difference in time scales and the logarithmic ordinate). From Ref. 44.*

the membrane, the effective voltage across the $D^+Q_A^-$ dipole is reduced by the ratio of the projection of the distance between D^+ and Q_A^- along the field to the thickness of the membrane. This will reduce the effective voltage below the required 200 meV. Thus, the experimental results are compatible with theory if $\Delta E_{redox} - E_{nuc}$ is close to zero.

We plan to repeat these experiments and determine k_{AD} with higher precision (48). We will also attempt to apply higher (pulsed) voltages to the membrane. If $\Delta E_{redox} - E_{nuc}=0$, the change in k_{AD} should be approximately independent of the direction of the electric field (note the near symetry of the theoretical curve in Fig.10). If $\Delta E_{redox} - E_{nuc} \neq 0$, k_{AD} should pass through a maximum for one field direction when the voltage across $D^+Q_A^-$ equals $(\Delta E_{redox} - E_{nuc})/q$.

Z. Popovic *et al.* reported at this conference experiments in which they applied considerably larger fields to RCs embedded in monolayers deposited on a substrate (see also ref. 49). Changes in k_{AD} have been observed although the analysis of the data is complicated by the fact that the RCs are randomly oriented and the decay has to be deconvoluted into a number of exponentials.

b) Fields due to intrinsic charges.

Instead of applying an electric field from an external source, we can also explore the effect of electric fields produced by charges associated with the protein. In particular, we can investigate the protonation of the quinones, a problem that has so far not been solved satisfactorily (see, e.g., Refs. 50-52).

The rationale of the experiment is as follows: The proton produces an electric field, thereby shifting the energy levels of $D^+I^-Q_A$ and $D^+IQ_A^-$ as described previously (see Fig. 11). This produces a change in k_{obs} given by the relation (in analogy to Eq. 19):

$$k_{obs}^{H^+} = k_{obs}^{o} \, e^{-\delta G^o/k_b T} \tag{21}$$

where $k_{obs}^{H^+}$ and k_{obs}^{o} are the recombination rates in the presence and absence of a proton and δG^o is the energy shift caused by the binding of the proton.

As the pH is varied, k_{obs} should change in accordance with the pK value for the pro-

$$k_{obs} = \frac{k_{obs}^{o} + 10^{(pK-pH)}k_{obs}^{H^{+}}}{1 + 10^{(pK-pH)}} \qquad (22)$$

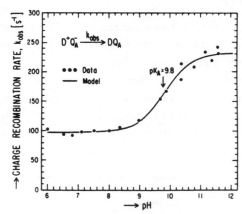

Figure 14 *The pH dependence of the charge recombination rate* k_{obs}. *The solid line (Model) was calculated using Eq. 22 with* $pK_A = 9.8$, $k_{obs}^{H^{+}} = 97 s^{-1}$ *and* $k_{obs}^{o} = 230 s^{-1}$. *From Ref. 59.*

Fig. 14 shows the pH dependence of k_{obs} in RCs containing AQ (53). The solid line represents a theoretical fit (Eq. 22) with $k_{obs}^{o} = 230$ s^{-1}, $k_{obs}^{H^{+}} = 97$ s^{-1} and pK $= 9.8$. The value of pK is in agreement with that found from redox titrations (54-56) and electron transfer measurements (26).

The interaction energy δG^{o} obtained from Eq. 21 is 22 meV. From this value one can make a rough estimate of the location of the proton binding site relative to Q_A^-. Assuming that the interaction of the proton with both Q_A^- and I$^-$ is electrostatic in origin, one calculates a distance that is *larger* than $\sim$ 5 Å. Additional experiments are in progress to obtain this distance from ENDOR experiments on Q_A^- in RCs (57,58).

Summary:

We have shown how electron transfer reactions can be used to probe the spatial and electronic structure of photosynthetic reaction centers. Both "static" strcutural changes (e.g., produced by removal of the H subunit) and "dynamic" changes (e.g., produced by illumination) as well as the effect of an electric field on the energy levels were investigated. Several findings (e.g., the temperature dependence of the recombination kientics and the lack of dependence of an electric field on the recombination kinetics) can be reconciled with present theories of electron transfer reactions by assuming that the difference in redox energy, ΔE_{redox} is approximately equal to the reorganization energy, E_{nuc}. Additional experiments were suggested to investigate the validity of this assumption. The temperature dependence of the recombination kinetics was explained by a thermal expansion model. Although we have focused in this work only on a particular charge recombination reaction, the approach should be applicable to other electron transfer reaction as well.

Acknowledgement:

We gratefully acknowledge the contributions of the many students, post-docs and collaborators whose work was cited in this review. The work from our laboratory was supported by grants from the NIH (GM-13191) and NSF (DMB 82-02811).

NOTE ADDED IN PROOF

On the Determination of the Characteristic Temperature, T_o.

Bixon and Jortner (59) have tried to fit the temperature dependence of k_{AD} (see Fig. 9) with a characteristic frequency $h\omega = 100$ cm^{-1} (i.e. $T_o \simeq 140$K), which is considerably lower than the one we used (Eq. 12). Although their fit at temperatures below 200K is good, they fail to fit the temperature dependence between 200 and 300K. There is, of course, no justification (except simplicity) to fit the entire temperature dependence with <u>one</u> value of T_o, since the vibrations involved in the electron transfer are likely to be different from those associated with the expansion. In the absence of information about the characteristic temperatures of either set of vibrations, we had opted in Fig. 9 for the simple approach of fitting the entire temperature range with a single temperature, T_o. We have now been able to determine the characteristic temperature of one of the vibrations that we believe plays a role in the electron transfer and thus fit the observed temperature dependence of k_{AD} in a more logical way.

The vibrations in question are those of the hydrogens bonded to the two oxygens of the primary acceptor, Q_A (57,58) (see insert in Fig. 15). We have determined the temperature dependence of the O--H bond length by measuring the hyperfine interaction of the proton with the unpaired spin on Q_A^- (60). This interaction has been shown to be dipolar (58,61), i.e., it is proportional to r^{-3}, where r is the O-H bond length. Thus, for small changes, Δr, in the bond length, the change in the hyperfine coupling, ΔA, is given by

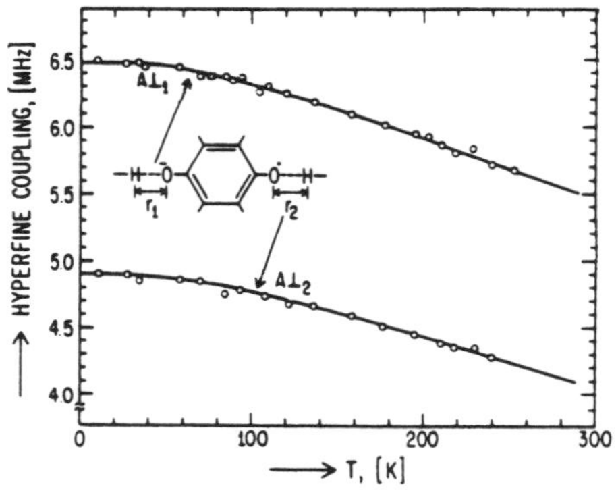

Figure 15 Temperature dependence of the perpendicular components of the hyperfine couplings $(A\perp)$ of the exchangeable protons on Q_A^- (57,58). Solid line represents the theoretical fit (Eq. 24) with $T_o = = 200K$ and $\gamma/r_o = 4.0 \times 10^{-4}$ K^{-1} and 4.8×10^{-4} K^{-1} for $A\perp_1$ and $A\perp_2$, respectively.

$$\frac{\Delta A}{A} \simeq -3 \frac{\Delta r}{r} = -3 \left(\frac{\Delta r}{r_o}\right)\left(\frac{r_o}{r}\right) \quad (23)$$

Substituting Eq. 14 for $\Delta r/r_o$ yields

$$\frac{\Delta A_{1,2}}{A_{1,2}} = -\frac{3}{2} \frac{\gamma_{1,2}}{r_{1,2}} T_o \left[\coth \frac{T_o}{2T} - 1\right] \quad (24)$$

where the subscripts 1 and 2 refer to the two protons. The temperature dependence of the perpendicular component of the hyperfine interaction, $A\perp$, of both protons is shown in Fig. 15. The solid line represents a fit of the data to Eq. 24 with $T_o = 200$K, $\gamma_1/r_o = 4.0 \times 10^{-4}$ K^{-1}, $r_1 = 1.55$Å, $\gamma_2/r_o = 4.8 \times 10^{-4}$ K^{-1}, $r_2 = 1.71$Å. Thus, the characteristic temperature is closer to the value favored by Bixon and Jortner (59). From the sensitivity of the fit to T_o we estimate a possible error in T_o of $\pm$ 50K.

We next have to show that the O-H vibration is involved in the electron transfer reaction under discussion (Eq. 1, Fig. 9). The evidence comes from the isotope effect, i.e., the observed change in k_{AD} when the protons were substituted with deuterons (62). The experimentally determined value of k_{AD} increased at 300K by 6% upon deuteration. A simple theoretical argument showed that in the <u>low temperature limit</u> a 20% effect is expected (62). At $T = 300K$ and $T_o = 200K$ Eq. 8 predicts an order of magnitude smaller isotope effect. Notwithstanding the lack of quantitative agreement between the observed and predicted isotope effect, we take the qualitative agreement as evidence that the hydrogen bonding protons associated with Q_A provide a vibrational mode that is important in the electron transfer reaction.

Having determined the characteristic temperature, T_o^{ET}, of the vibrations coupled to the electron transfer, we leave the other characteristic temperature, T_o^{exp}, associated with the expansion as well as the expansion coefficient γ/r_o as free parameters to fit the observed temperature dependence of k_{AD} with the expression (obtained from Eq. 8 and 14)

$$k_{AD}(T) = k_{AD}(0)\tanh^{\frac{1}{2}}\left(\frac{T_o^{ET}}{2T}\right)e^{-\frac{\gamma T_o^{exp}}{2r_o}\left(\coth\frac{T_o^{exp}}{2T}-1\right)} \qquad (25)$$

where $\tanh^{\frac{1}{2}}\left(\dfrac{T_o^{ET}}{2T}\right)$ is the strong coupling limit (s >> 1) of Eq. 8a. It is plotted in Fig. 16 (dashed line) for $T_o^{ET} = 200K$. A fit of Eq. 25 to the experimental data (dots) with $T_o^{ET} = 200K$, $T_o^{exp} = 1000K$ and $\gamma/r_o = 0.036$ is shown by the solid line in Fig. 16. Although the fit is very good, the high value of γ/r_o is cause for concern. The expansion mechanism invoked probably represents an oversimplification of the situation; other mechanisms may contribute to the temperature dependence of k_{AD}. Clayton for instance, showed that k_{AD} in dehydrated RCs at $T = 300K$ has a value that is similar to the one

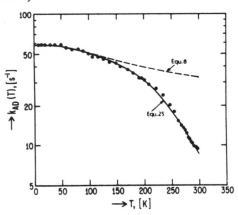

Figure 16 Temperature dependence of k_{AD}. Experimental data (dots) same as in Fig. 9. Solid line represents the best fit of eq. 25 with $T_o^{ET} = 200K$, $T_o^{exp} = 1000$, $\gamma/r_o = 0.036$. Dashed line represents Yortner's theory (eq. 8) with $\Delta E_{redox} = E_{max}$ and $T_o^{ET} = 200K$.

observed at cryogenic temperatures (63). Similarly, we have found that k_{AD} of RCs in PVA films that have been thoroughly dehydrated by prolonged pumping fitted the dashed line of Fig. 16 (Eq. 8) rather than the solid line (Eq. 25) (64). Thus, the water of hydration must play an important role in the electron transfer. Clayton suggested that the observed temperature dependence is due to a phase transition of the bound water (63). The orientational polarizability of the water dipoles may represent another mechanism. It should also be noted that the distance between Q_A^- and D^+ is rather large ($\sim 20\text{Å}$) and consequently encompasses a large amount of protein structure. Any temperature dependent characteristic of the intervening space (e.g. a change in the conformation of the protein backbone) may contribute to the observed temperature dependence. X-ray structure analyses of RCs from *R. sphaeroides* performed at different temperatures should shed some light on this question (65,66).

References

1. Feher, G. and Okamura, M. Y.; in *The Photosynthetic Bacteria*, R. K. Clayton and W. R. Sistrom, eds. (Plenum Press, N.Y.), chapter 19, pp. 349-388 (1978).

2. Parson, W. W. amd Ke, B. (1982) In *Photosynthesis: Energy Conversion by Plants and Bacteria* (Govindjee, ed.), pp. 331-385, Academic Press, New York.

3. Reed D. W., and Clayton, R. K. (1968), *Biochim. Biophys. Res. Commun.* **30**, 471-475.

4. Feher, G. (1971) *Photochem. & Photobiol.* **14**, #3 (Supplement #1), 373-388.

5. McElroy, J. D., Mauzerall, D. C., and Feher, G. (1974) *Biochim. Biophys. Acta*, **333**, 261-277.

6. Straley, S. C., Parson, W. W., Mauzerall, D. C., and Clayton, R. K. (1973) *Biochim. Biophys. Acta* **305**, 597-609.

7. Debus, R. J., Feher, G. and Okamura, M. Y. (1985) *Biochemistry* **24**, 2488-2500.

8. Hopfield, J. J. (1974) *Proc. Natl. Acad. Sci. USA* **71**, 3640-3644.

9. Jortner, J. (1976) *J. Chem. Phys.* **64**, 4860-4867.

10. Arata, H. and Parson, W. W. (1981) *Biochim. Biophys. Acta* **636**, 70-81.

11. Arata, H. and Parson, W. W. (1981) *Biochim. Biophys. Acta* **638**, 201-209.

12. Noks, P. P., Lukashev, E. P., Kononenko, A. A., Venediktov, P. S., and Rubin, A. B. (1977) *Mol. Biol. (Moscow)* **11**, 1090-1099.

13. Kleinfeld, D., Okamura, M. Y., and Feher, G. (1984) *Biochemistry* **23(24)**, 5780-5786.

14. Austin, R. H., Beeson, K. W., Eisenstein, L., Frauenfelder, H., and Gunsalus, I. C. (1975) *Biochemistry* **14**, 5355-5373.

15. Frauenfelder, H. (1978) *Methods Enzymol.* **54**, 506-532.

16. Woodbury, N. W. T. and Parson, W. W. (1984) *Biochim. Biophys. Acta* (in press).

17. Jortner, J. (1980) *J. Am. Chem. Soc.* **102**, 6676-6686.

18. Redi, M., and Hopfield, J. J. (1980) *J. Chem. Phys.* **72**, 6651-6660.

19. Frauenfelder, H., Petsko, G. A., and Tsernoglou, D. (1979) *Nature (London)* **280**, 558-563.

20. Artymiuk, P. J., Blake, C. C. F., Grace, D. E. P., Patley, S. J., Phillips, D. C., and Sternberg, M. J. E. (1979) *Nature (London)* **280**, 563-568.

21. Karplus, M, and McCammon, J. A. (1981) *CRC Crit. Rev. Biochem.* **9**, 293-349.

22. Levitt, M., Sander, C., and Stern, P. S. (1985) *J. Mol. Biol.* **181**, 123-447.

23. Huber, R., and Bennett, W. S. (1983) *Biopolymers* **183**, 261-679.

24. Warshel, A. (1980) *Proc. Natl. Acad. Sci. USA* **77**, 3105-3109.

25. Agmon, N. and Hopfield, J. J. (1983) *J. Chem. Phys.* **78**, 6947-6959; **80**, 592 (erratum).

26. Kleinfeld, D., Okamura, M. Y., and Feher, G. (1984) *Biochim. Biophys. Acta* **766**, 126-140.

27. Eisenberger, P., Okamura, M. Y., and Feher, G. (1983) *Biophys. J.* **37**, 523-538.

28. Parson, W. W. (1967) *Biochim. Biophys. Acta* **153**, 248-259.

29. Hsi, E. S. P. and Bolton, J. R. (1974) *Biochim. Biophys. Acta* **347**, 126-133.

30. Loach, R. A., Kung, M., and Hales, B. J. (1975) *Ann. NY Acad. Sci,* **244**, 297-319.

31. Mar, T., Vadeboncoeur, C., and Gingras, G. (1983) *Biochim. Biophys. Acta* **724**, 317-322.

32. Sarai, A. (1980) *Biochim. Biophys. Acta* **589**, 71-83.

33. Kakitani, T. and Kakitani, H. (1981) *Biochim. Biophys. Acta* **635**, 498-514.

34. Kakitani, T. and Mataga, N. (1985) *J. Phys. Chem.* **89**, 8-10.

35. Hales, B. J. (1976) *Biophys. J.* **16**, 471-480.

36. Hopfield, J. J. (1976) *Biophys. J.* **16**, 1239-1240.

37. Feynman, R. P. (1972) *Statistical Mechanics: A set of Lectures,* pp. 53-55, W. A. Benjamin, Reading, PA.

38. Ringe, D., Kuriyan, J., Petsko, G. A., Karplus, M., Frauenfelder, H., Tilton, R. F., and Kuntz, I. D. (1984) *Am. Crystallographic Assoc. Transactions* **20**, 109-122.

39. Okamura, M. Y., Isaacson, R. A., and Feher, G. (1975) *Proc. Natl. Acad. Sci.* **72** #9, 3491-3495.

40. Prince, R. C., Gunner, M. R., Dutton, P. L. in *Function of Quinones in Energy Conserving Systems,* B. L. Trumpower, ed., Academic Press, Inc., New York., pp. 29-33 (1982).

41. Recently, N. W. Woodbury, W. W. Parson, M. R. Gunner, R. C. Prince, and P. L. Dutton (1986) *Biochim. Biophys. Acta* **851**, 6-22, have determined the redox potentials of different quinones *in situ* from the quantum yield of delayed fluorescence. They found significant deviations from the values determined in dimethylformamide. Unfortunately, their data were also obtained at room temperature and are, therefore, not strictly applicable to the data of Fig.10. We are indebted to N. W. Woodbury and W. W. Parson for making their data available to us prior to publication.

42. Gunner, M. R., Tiede, D. M., Prince, R. C., and Dutton, P. L. in *Function of Quinones in Energy Conserving Systems,* B. L. Trumpower, ed., Academic Press, Inc., New York., pp. 265-269 (1982).

43. Gunner, M. R. (private communication).

44. Gopher, A., Blatt, Y., Schönfeld, M., Okamura, M. Y., and Feher, G. (1985) *Biophys. J.* **48**, 311-320.

45. Arata, H., and Parson, W. W. (1981) *Biochim. Biophys. Acta* **638**, 201-209.

46. Gunner, M. R. Y., Liang, Y., Nagus, D. K., Hochstrasser, R. M., and Dutton, P. L. (1982) *Biophys. J. (Abstracts)* **37**, 226a.

47. Schönfeld, M., Montal, M., and Feher, G. (1979) *Proc. Natl. Acad. Sci. USA* **76**, 6351-6355.

48. We have recently been able to measure changes in k_{AD} with a precision of $\sim 0.5\%$ [M. Y. Okamura and G. Feher (1986) *Biophys. J. (Abstracts)* **49**, 587a].

49. Popovic, Z. D., Kovacs, G. J., Vincett, P. S., and Dutton, P. L. (1985) *Chem. Phys. Letters* **116**, 405-410.

50. Crofts, A. R. and Wraight, C. A. (1983) *Biochim. Biophys. Acta* **726**, 149-185.

51. Maroti, P. and Wraight, C. (1985) *Biophys. J. (Abstracts)* **47**, 5a.

52. Kleinfeld, D., Okamura, M. Y., and Feher, G. (1985) *Biochim. Biophys. Acta* **809**, 291-310.

53. Kleinfeld, D., Okamura, M. Y., and Feher, G. (1985) *Biophys. J.* **48**, 849-852.

54. Prince, R. C. and Dutton, P. L. (1976) *Arch. Biochem. Biophys.* **172**, 329-334.

55. Rutherford, A. W. and Evans, M. C. W. (1980) *FEBS Lett.* **110**, 257-261.

56. Wraight, C. A. (1981) *Isr. J. Chem.* **21**, 348-354.

57. Lubitz, W., Abresch, E. C., Debus, R. J., Isaacson, R. A., Okamura, M. Y., and Feher, G. (1985) *Biochim. Biophys. Acta* **808**, 464-469.

58. Feher, G., Isaacson, R. A., Okamura, M. Y., and Lubitz, W. (1985) *Antennas and Reaction Centers of Photosynthetic Bacteria: Structure, Interaction and Dynamics* (M. Michel-Beyerle, ed.) Springer-Verlag, Berlin, pp. 174-189.

59. Bixon, M., and Jortner, J. (1986) *J. Phys. Chem.* **90**, 3795-3800.

60. Feher, G., Isaacson, R. A., Okamura, M. Y. and Lubitz, W. (1986), unpublished results.

61. O'Malley, P. J., Chandreshekar, T. K. and Babcock, G. T., *Antennas and Reaction Centers of Photosynthetic Bacteria: Structure, Interaction and Dynamics* (M. Michel-Beyerle, ed.) Springer Verlag, pp. 339-344.

62. Okamura, M. Y. and Feher, G., *Proc. Natl. Acad. Sci. USA* (1986) **83**, 8152-8157.

63. R. K. Clayton (1978). *Biochim. Biophys. Acta* 504, 255-264.

64. Arno, T. R., McPherson, P. H., Feher, G. (1986), unpublished results.

65. Allen, J. P., Feher, G., Yeates, T. O. and Rees, D. C. (1986) presented at the VIIth International Congress on Photosynthesis, Brown University, *Proceedings*, Martinus Nijhoff/W. Junk (in press).

66. Allen, J. P., Feher, G., Yeates, T. O., Rees, D. C., Deisenhofer, J., Michel, H and Huber, R. *Proc. Natl. Acad. Sci.* (1986) **83**, 8589-8593.

Acknowledgements

Table 1 and Figure 2 (on page 401) are reprinted from Feher, G. (1971) *Photochem. & Photobiol.* 14(3, Suppl. 1), 373–388, with the permission of Elsevier/North Holland Biomedical Press, Amsterdam, and with the permission of Pergammon Press, Elmsford, New York, respectively.

Figures 5, 6, 7, and 8 (on pages 404, 405, and 407, respectively) are printed from Kleinfield, D., Okamura, M.Y., and Feher, G. (1984) *Biochemistry* 23(24), 5780–5786, with the permission of the American Chemical Society, Washington, D.C.

Figures 11, 12, and 13 (on pages 413, 414, and 415, respectively) are reprinted from Gopher, A., Blatt Y., Schönfeld, M., Okamura, M.Y., and Feher, G. (1985) *Biophys. J.* 48, 311–320, with the permission of The Rockefeller University Press, New York, New York

Figure 14 (on page 416) is reprinted from Kleinfield, D., Okamura, M.Y., and Feher, G. (1985) *Biophys. J.* 48, 849–852, with the permission of The Rockefeller University Press, New York, New York.

Energetics of initial charge separation in bacterial photosynthesis: the triplet decay rate in very high magnetic fields

Richard A. Goldstein, Larry Takiff and Steven G. Boxer

Department of Chemistry, Stanford University, Stanford, CA (U.S.A.)

(Received 21 September 1987)
(Revised manuscript received 16 March 1988)

Key words: Photosynthesis; Electron transfer; Reaction center; Magnetic field effect

The triplet state of the primary electron donor is formed in bacterial reaction centers when electron transfer to the quinone is blocked. The relative quantum yield of formation and the rate of decay of the triplet state were measured in quinone-depleted reaction centers from *Rhodobacter sphaeroides* as a function of temperature and magnetic field up to 135 kG. By analyzing these dependences we obtain a standard free-energy difference at room temperature between the radical pair intermediate and the triplet state of 1370 ± 30 cm^{-1} (0.170 ± 0.004 eV), and a standard enthalpy difference of 1450 ± 70 cm^{-1} (0.180 ± 0.009 eV). These results differ substantially from those obtained previously using lower fields. This difference is ascribed primarily to effects of nuclear spin polarization. In combination with the known energies of the excited singlet and triplet states of the primary electron donor, we calculate that the standard free-energy change for the initial charge separation reaction is 2120 cm^{-1} (0.263 eV) and that the entropy change is small.

Introduction

The initial electron transfer reaction in bacterial photosynthesis occurs between the photoexcited singlet state of the electron donor (denoted ^1P) and the electron acceptor (denoted I) within about 3 ps at room temperature [1,2], and about twice as fast at cryogenic temperatures [1,3]. Extensive spectroscopic studies [4] and analyses of the X-ray structures of reaction centers (RCs) from the photosynthetic bacteria *Rhodopseudomonas viridis* [5] and *Rhodobacter sphaeroides* [6–9] have identified the electron donor as a bacteriochlorophyll dimer,

often called the special pair. The precise identity of the initial electron acceptor is less certain, but it is believed to be a bacteriopheophytin monomer [10]. The resulting radical pair state, P^+I^-, decays in about 200 ps as the electron moves from I^- to a quinone (denoted Q) [11,12]. When this latter reaction is blocked by removal or prior reduction of the quinone, the radical pair decays by charge recombination to re-form ^1PI, the ground state, or ^3PI, the excited triplet state of the donor (see Fig. 1).

The driving force for the initial electron transfer reaction is a question of considerable interest. Extensive theoretical studies and experimental tests with model systems [13] demonstrate a strong relationship between rate and driving force in charge transfer reactions. It is widely believed that studying this relationship can provide insight into the mechanism by which photosynthetic systems optimize the efficiency of forward reactions and minimize back reactions. In contrast to most model

---

Abbreviations: I, primary electron acceptor; P, primary electron donor; Q, quinone; RC, reaction center; TT, 10 mM Tris/0.1% Triton X-100 (pH 8.0) buffer.

Correspondence: S.G. Boxer, Department of Chemistry, Stanford University, Stanford, CA 94305, U.S.A.

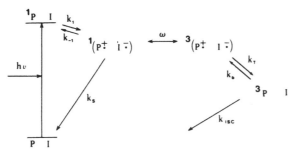

Fig. 1. Reaction scheme used to analyze reaction dynamics in RCs in which electron transfer to the quinone has been blocked.

electron transfer systems, where the driving force can be estimated accurately by measurements of the excitation energy and the reversible one-electron reduction potentials of the donor and acceptor, the energetics of the initial charge separation in RCs must be estimated by indirect methods. In this paper we consider the energy difference between the ^{3}PI and P$^+$I$^-$ states which, when combined with the energies of the ^{1}PI and ^{3}PI states, gives an estimate for the energetics of the initial charge separation reaction, ^{1}PI $\rightarrow$ P$^+$I$^-$.

As shown in Fig. 1, in blocked RCs the radical pair is initially formed from ^{1}PI in a singlet spin configuration, denoted ^{1}(P$^+$I$^-$), with rate constant k_1. ^{1}(P$^+$I$^-$) can decay to the ground state, PI (rate constant k_S), reform ^{1}PI (rate constant k_{-1}), or undergo coherent electron spin evolution to form the triplet spin configuration of the radical pair ^{3}(P$^+$I$^-$). ^{3}(P$^+$I$^-$) either decays by recombination to generate the excited triplet state of P, ^{3}PI (rate constant, k_T), or it evolves back to ^{1}(P$^+$I$^-$). ^{3}PI can decay by intersystem crossing (rate constant, k_{isc}) or reform ^{3}(P$^+$I$^-$) (rate constant, k_b). The rate of interconversion between the singlet and triplet radical pair states is dependent on external magnetic field strength. At zero magnetic field, the nuclear hyperfine interaction causes interconversion of the singlet radical pair state and all three triplet radical pair states. When the magnetic field is increased from 0 to about 1 kG, the rate of interconversion decreases because of loss of the near degeneracy of the singlet and two of the triplet radical pair states. As the field is further increased the rate of interconversion increases as the g-factor difference between P$^+$ and I$^-$ starts to become the dominant mechanism for interconver-

sion [14]. Because of the dependence of the singlet-triplet mixing rate on applied magnetic field, the quantum yield of formation and the decay kinetics of the states represented in Fig. 1 are affected by magnetic field strength [15].

We have shown that in quinone-depleted RCs from *Rb. sphaeroides* the observed rate of ^{3}PI decay (k_{obs}) depends on the strength of an applied magnetic field [16]. We interpreted this result to mean that ^{3}PI decay at room temperature proceeds largely through thermal reformation of ^{3}(P$^+$I$^-$), spin evolution to ^{1}(P$^+$I$^-$), and decay through the k_S path. Using the reaction scheme shown in Fig. 1 and making a series of assumptions about the temperature and magnetic field dependence of the rate constants and the rates of nuclear and electron spin-lattice relaxation (see below), we derived the expression:

$$k_{obs} = k_{isc} + \tfrac{1}{3}k_S \Phi_{^3PI}\, e^{-\Delta G^0 \beta} \qquad (1)$$

where $\Phi_{^3PI}$ is the magnetic-field dependent quantum yield of ^{3}PI formation, ΔG^0 is the standard free energy difference between the ^{3}PI and ^{3}(P$^+$I$^-$) states, and $\beta = 1/kT$, where k is the Boltzmann constant and T is the absolute temperature. In the original derivation [16], entropy changes between the ^{3}(P$^+$I$^-$) and ^{3}PI states were considered negligible and the free-energy difference was equated with the standard enthalpy difference ΔH^0; this assumption is not required in the present analysis.

Although Eqn. 1 is qualitatively followed, detailed measurements between 0 and 50 kG revealed discrepancies with the predictions of Eqn. 1 [16]: at room temperature, k_{obs} was observed not to vary linearly with $\Phi_{^3PI}$. As has been explained in detail elsewhere [17], this deviation is likely due to nuclear spin polarization at low fields where the singlet-triplet mixing is dominated by nuclear hyperfine interactions. The presence of nuclear spin polarization effects can especially complicate the analysis of the temperature dependence of k_{obs}, since the nuclear spin relaxation rate is likely to be temperature dependent.

At very high magnetic fields, singlet-triplet mixing is dominated by the difference in precession rates of the two unpaired electrons (Δg effect) rather than by nuclear hyperfine interactions, and the reaction dynamics are insensitive to the effects of nuclear spin polarization. For this reason we have investigated the magnetic field and temperature dependence of k_{obs} for applied fields up to 135 kG. Our previous analysis indicated that nuclear spin polarization effects should not be significant at this field [17]. Our results demonstrate that k_{obs} does vary linearly with $\Phi_{^3PI}$ within experimental error at room temperature in the high field regime (magnetic field greater than 20 kG). The analysis of k_{obs} at high field gives a larger calculated value for the $^3(P^+I^-) - ^3PI$ energy difference than obtained from earlier analyses using lower field data.

Experimental

Quinone-depleted RCs from *Rb. sphaeroides* R-26 mutant were prepared by standard procedures [18,19] and were suspended in aqueous buffer (10 mM Tris/0.1% (v/v) Triton X-100 (pH 8.0); TT buffer) or viscous buffer (glycerol/TT buffer (2:1, v/v) or ethylene glycol/TT buffer (2:1, v/v)). The sample was placed in a 3.3 mm pathlength variable-temperature cell (temperature control ±1°C, measured by a platinum resistance thermometer) with a total absorbance in the cell of approx. 0.3 at 860 nm at room temperature. The sample cell was placed in a split-bore superconducting magnet (magnetic field accuracy ±1 kG) at the Francis Bitter National Magnet Laboratory. Zero-field values were obtained with the magnet quenched (residual magnetic field of less than 1 G).

The relative concentration of ^3PI was monitored by observing the change in ground-state absorption at 868 nm following subsaturating 8 ns excitation flashes (532 nm, 10 Hz, less than 1.0 mJ/cm² per pulse). Data were taken with the polarization of the probe beam parallel, perpendicular, and at the magic angle to the magnetic field direction. The data reported represent measurements with magic angle polarization unless otherwise indicated. The observed absorbance change was linear with excitation power and independent of probe light intensity. Relative quantum yields and decay rates of ^3PI were calculated by fitting the time evolution of the absorbance change to a single exponential plus a small baseline and to a gaussian distribution of exponentials plus a

small baseline [20]. Triplet quantum yield anisotropies were calculated as described elsewhere [21].

Results

The values of k_{obs} obtained by averaging the decay rates over distributions of exponentials were generally slightly higher (approx. 10%) than those obtained by fits to single exponentials. The second moment of the distribution in k_{obs} was, at all fields and temperatures, approx. 30% of the average value. The fits to a distribution were significantly better and resulted in a smaller error in resulting calculated parameters; the values of k_{obs} thus obtained were used in the following analysis. Both sets of k_{obs} values yielded essentially identical values for ΔH^0 and ΔG^0. This heterogeneity in observed decay rates will be discussed further below.

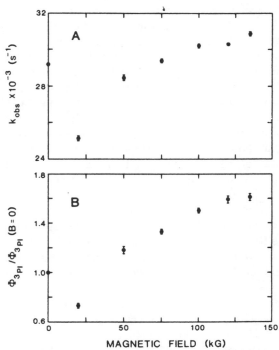

Fig. 2. Dependence of (A) k_{obs} and (B) $\Phi_{^3PI}/\Phi_{^3PI}(B=0)$ on magnetic field between 0 and 135 kG at 15°C for quinone-depleted RCs in aqueous buffer, with probe beam polarized at the magic angle with respect to the applied magnetic field. k_{obs} is the observed rate of ^3PI decay, $\Phi_{^3PI}$ is the quantum yield of ^3PI formation, and $\Phi_{^3PI}(B=0)$ is the ^3PI quantum yield at zero applied field.

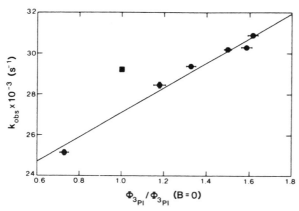

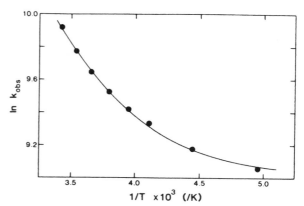

Fig. 3. k_{obs} plotted as a function of relative triplet yield at 15°C from 0 G to 135 kG. The sample was in aqueous buffer. The data from 20 kG to 135 kG (●) are fit to a straight line with intercept $21.09 \cdot 10^3$ s^{-1} and slope $6.02 \cdot 10^3$ s^{-1}; the datum at 0 G (■) deviates significantly from this line.

Fig. 4. Temperature dependence of k_{obs} at 135 kG for RCs in viscous buffer. Experimental errors were smaller than the size of the symbols. The probe beam polarization was at the magic angle with respect to the magnetic field direction. The solid line is a fit to an Arrhenius expression plus constant with a low-temperature asymptote of $8.2 \cdot 10^3$ s^{-1}, a slope of 1450 cm^{-1}, and a pre-exponential of $15 \cdot 10^6$ s^{-1}.

Values of k_{obs} and the quantum yield of ^{3}PI normalized to the yield at zero field as a function of magnetic field strength at room temperature in TT buffer are presented in Fig. 2. At higher magnetic fields, k_{obs} parallels $\Phi_{^3PI}$ as a function of applied field, in agreement with Eqn. 1. This is shown clearly in Fig. 3, where the observed triplet decay rate is plotted vs. relative triplet yield. The high field data (≥ 20 kG) were fit to a straight line with y-intercept $= (21.09 \pm 0.07) \cdot 10^3$ s^{-1} and slope $= (6.02 \pm 0.06) \cdot 10^3$ s^{-1}. The datum at zero field deviates significantly from the fit. Similar data for an ethylene glycol sample obtained at 20°C at fields up to 37.5 kG were also fit to a straight line with y-intercept $(10.60 \pm 0.10) \cdot 10^3$ s^{-1} and slope $(5.70 \pm 0.14) \cdot 10^3$ s^{-1} (data not shown). The data at zero field again deviated from the fit. The temperature dependence of k_{obs}, measured with a glycerol buffer sample at an applied field of 135 kG, is shown in Fig. 4; at room temperature, $k_{obs} = (20.3 \pm 0.1) \cdot 10^3$ s^{-1}. The data were fit to an Arrhenius expression plus a constant with a low-temperature asymptote of $(8.2 \pm 0.2) \cdot 10^3$ s^{-1}, a slope of 1450 ± 70 cm^{-1}, and a pre-exponential of $(15 \pm 5) \cdot 10^6$ s^{-1}. The large difference in ^{3}PI decay rates between samples in aqueous and viscous solution has been noted previously [16]. k_{obs} as a function of temperature at 135 kG with the probe beam polarization parallel and perpendicular to the magnetic field direction

is compared with magic angle data in Fig. 5. Note the large anisotropy in k_{obs} at intermediate temperatures. The anisotropy of the ^{3}PI yield was measured at -20°C in viscous buffer. The anisotropy was independent of magnetic field within experimental error above 50 kG, with a value at 135 kG of -0.10 ± 0.01 (data not shown).

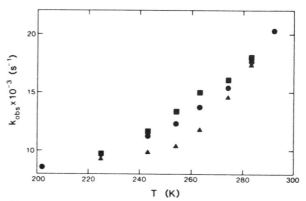

Fig. 5. Temperature dependence of k_{obs} at 135 kG for RCs in viscous buffer, as observed with the probe light polarization perpendicular (■), parallel (▲), and at the magic angle (●) to the magnetic field. Experimental errors were smaller than the size of the symbols. Data at 202 K and 292 K are not resolvable within the experimental error. Note that the anisotropy in k_{obs} goes to zero at low temperatures where ^{3}PI decay is via the magnetic field independent k_{isc} path, and at higher temperatures as the RC rotational correlation time becomes short relative to the ^{3}PI lifetime.

Discussion

Approaches to data analysis

Eqn. 1 provides two methods for obtaining the $^3(P^{\pm}I^{\mp}) - {}^3PI$ energy difference. If k_{obs} as a function of relative ^3PI quantum yield is fit to a straight line, the y intercept is equal to k_{isc}, while the slope is $\frac{1}{3}k_S\Phi_{^3PI}(B=0)\,e^{-\Delta G^0\beta}$, where $\Phi_{^3PI}(B=0)$ is the quantum yield of ^3PI at zero applied field (the field strength is symbolized with a B instead of H to avoid confusion with the enthalpy). k_S and $\Phi_{^3PI}(B=0)$ have been measured previously; substitution allows calculation of ΔG^0.

Alternatively, Eqn. 1 can be rewritten as an Arrhenius-type expression in the form

$$k_{obs} = k_{isc} + \alpha\, e^{-\Delta H^0\beta} \qquad (2)$$

where

$$\alpha = \tfrac{1}{3}k_S\Phi_{^3PI}\,e^{\Delta S^0/k} \qquad (3)$$

and ΔS^0 is the standard entropy difference between $^3(P^{\pm}I^{\mp})$ and ^3PI. If α and k_{isc} are assumed to be temperature independent, then fitting the temperature dependence of k_{obs} to Eqn. 2 yields ΔH^0; the small temperature dependence of α and k_{isc} are discussed further below. Comparison of the derived values for ΔH^0 and ΔG^0 yields ΔS^0.

Assumptions behind Eqn. 1

A series of assumptions is involved in the derivation of Eqn. 1. Three assumptions in particular bear close scrutiny. The first assumption is that the scheme in Fig. 1 is correct. The consequences of possible modifications of the reaction scheme are discussed briefly below and will be discussed in much greater detail elsewhere in the analysis of the ^3PI quantum yield at very high fields. The second assumption is that nuclear spin-lattice relaxation in ^3PI is fast relative to k_{obs}, so the nuclear spins in ^3PI are always at equilibrium. We believe that this assumption is not valid at low fields, and that this explains the discrepancy noted in the introduction. A third assumption is that the RC rotational correlation time is short relative to

the ^3PI lifetime so that anisotropic interactions can be neglected. We have shown previously that RCs rotate rapidly relative to k_{obs} in aqueous solution at room temperature [21], but this is not the case in viscous solutions at low temperature.

Nuclear spin polarization

A quantitative analysis of the effect of nuclear spin polarization on the magnetic field dependence of k_{obs} has been presented earlier [17]. The hyperfine-induced singlet-triplet mixing in the $P^{\pm}I^{\mp}$ state at low field, which leads to formation of ^3PI, enriches ^3PI in nuclear spin states which generate the largest hyperfine fields (nuclear spin polarization). Subsequent decay of ^3PI, again via the singlet-triplet mixing process, will then be faster than if the nuclear spins in ^3PI were at thermal equilibrium. The result is a higher value of k_{obs} than predicted by Eqn. 1, at temperatures at which the radical-ion pair is thermally accessible from ^3PI.

The degree of nuclear spin polarization will change with temperature due to the temperature dependence of the nuclear spin relaxation rate. As the temperature of the sample is lowered, the nuclear spin relaxation rate of ^3PI will decrease; as a result, the assumption of fast spin relaxation will become worse at lower temperatures, increasing the degree of nuclear spin polarization. In addition, nuclear spin polarization effects will go away at low temperature as the quantum yield of ^3PI approaches unity [23] and the probability of the triplet state being formed becomes independent of nuclear spin state. It is not clear which of these two effects will dominate. Since the spin polarization increases the value of k_{obs} above that expected for no spin polarization, if this polarization increases as the temperature is lowered, k_{obs} will decrease more slowly with decreasing temperature than predicted by Eqn. 2. If k_{obs} is fit to Eqn. 2, the fit value of ΔH^0 will be lower than the actual enthalpy difference between $^3(P^{\pm}I^{\mp})$ and ^3PI. Conversely, if spin polarization decreases with temperature, the effect will increase the fit value of ΔH^0. In order to distinguish the slope of the Arrhenius plot from the true ΔH^0, we denote the experimental slope ΔH^0_{app} or apparent enthalpy change. As mentioned above, the nuclear spin polarization effects can be avoided by performing the measurements in the high field regime.

Anisotropic interactions

The spin evolution of P^+I^- is influenced by anisotropic nuclear hyperfine, electron–electron dipole–dipole and electronic spin–orbit (g-tensor) interactions [21]. As a result, the rate of singlet-triplet mixing for a particular RC depends upon its orientation with respect to the external magnetic field. In samples with distributions of orientations, ^{3}PI is formed preferentially in RCs with orientations that favor singlet-triplet mixing. If this orientation is preserved during the decay of the ^{3}PI state, i.e., the rotational correlation time is long relative to the ^{3}PI lifetime, k_{obs} will be larger than predicted by Eqn. 1. This effect will vanish at lower temperatures where the ^{3}PI quantum yield approaches unity and the ^{3}PI decay proceeds only through k_{isc}.

The rotational correlation time can be monitored by observing the ^{3}PI quantum yield anisotropy decay. The assumption of rapid RC rotation is valid at room temperature in both aqueous and viscous buffer, but becomes invalid in viscous buffer as the temperature of the sample decreases below 270 K. A breakdown in this assumption will cause an increase in k_{obs} above the value calculated from Eqn. 1 in a manner similar to the consequences of slow nuclear spin relaxation. As discussed above for the case of nuclear spin polarization, the temperature dependence of the anisotropy will cause the analysis of the temperature dependence of k_{obs} to underestimate ΔH^0. Unfortunately, these effects cannot easily be avoided by going to higher magnetic field. At infinite field the singlet-triplet mixing is so rapid for all RC orientations that $\Phi_{^3PI}$ and k_{obs} should become independent of field strength and orientation. However, we have found that $\Phi_{^3PI}$ is still anisotropic at 135 kG in viscous buffer (anisotropy $= -0.10 \pm 0.01$), and the data in Fig. 5 show that observed values of k_{obs} at 135 kG are likewise dependent on orientation.

Analysis of the room-temperature data: determination of ΔG^0

The assumption of rapid rotation is valid for the data represented in Fig. 3. The fact that a straight line fits the data, including at very high field where nuclear spin polarization effects are not important, indicates that spin polarization effects are not significant down to 20 kG at room temperature in aqueous buffer. The high value of k_{obs} at zero magnetic field is consistent with nuclear spin polarization effects at this low field. Since the nuclear spin relaxation is likely mediated by RC rotation, nuclear spin polarization effects may still be present at higher fields at lower temperature or in viscous solution. The slope of the fit line ((6.02 ± 0.06) $\cdot 10^3$ s^{-1}) is equal to $\frac{1}{3} k_S$ $\Phi_{^3PI}(B = 0)$ e$^{-\Delta G^0 \beta}$; this number can be combined with values of k_S ((4.9 ± 0.4) $\cdot 10^7$ s^{-1}) and $\Phi_{^3PI}(B = 0)$ (0.32 ± 0.04) obtained from independent measurements [24] to yield $\Delta G^0 = 1370 \pm 30$ cm^{-1}. The y-intercept of the fit line is equal to k_{isc}: $k_{isc} = (21.09 \pm 0.07) \cdot 10^3$ s^{-1} for an aqueous sample at 15°C.

ΔG^0 was calculated from the viscous sample data in the same manner as for the aqueous sample data. At room temperature the assumption of rapid RC rotation is valid (Fig. 5). While nuclear spin polarization effects may be present at fields up to 37.5 kG, the fact that all of the non-zero field values of k_{obs} as a function of $\Phi_{^3PI}$ fit a straight line suggests that such effects are small. Assuming that k_S and $\Phi_{^3PI}(B = 0)$ are independent of sample viscosity, the slope of this fit line ((5.70 ± 0.14) $\cdot 10^3$ s^{-1}) gives a value of $\Delta G^0 = 1390 \pm 30$ cm^{-1}, in close agreement with the value of $\Delta G^0 = 1370 \pm 30$ cm^{-1} calculated with the aqueous sample data. The fact that the two values of ΔG^0 closely agree indicates that the radical pair reaction energetics do not depend upon the viscosity of the sample; the difference in the observed ^{3}PI decay rates between aqueous and viscous samples can be explained by the difference in k_{isc}: k_{isc}(aqueous sample) $= (21.09 \pm 0.07) \cdot 10^3$ s^{-1}, k_{isc}(viscous sample) $= (10.60 \pm 0.10) \cdot 10^3$ s^{-1}. Because the calculation of ΔG^0 using the viscous sample data requires assumptions not required of the calculation using the aqueous sample data, the latter will be used below.

Analysis of the temperature dependence of k_{obs}: estimation of ΔH^0

Assuming for the moment that only the exponential term in Eqn. 2 is temperature dependent, the slope of the Arrhenius curve of k_{obs} at any magnetic field should yield the enthalpy difference between $^3(P^{\dagger}I^{\mp})$ and ^3PI. Prior analysis of data obtained at zero and moderate fields (less than 50 kG) using this assumption yielded a calculated enthalpy difference of $\Delta H^0 = 950 \pm 50$ cm^{-1} [16]. As discussed above, this result could be affected by nuclear spin polarization effects. These effects can be avoided by measuring k_{obs} at very high magnetic field where the Δg effect dominates singlet-triplet mixing in the radical pair state. Our analysis predicts that the value of ΔH^0_{app} obtained at high field may be different from that derived at zero field. This prediction is born out by the results presented in Fig. 4, where the data were fit to Eqn. 2 using a least-squares fit. The value obtained for k_{isc} obtained from the low temperature asymptote at 135 kG is $(8.2 \pm 0.2) \cdot 10^3$ s^{-1}, in good agreement with the value obtained previously at low field with a viscous sample ($7.0 \cdot 10^3$ s^{-1}) [16]. The fit value of ΔH^0 is 1450 ± 70 cm^{-1}, significantly higher than the value of 950 ± 50 cm^{-1} obtained at lower fields [16]. Because of the effects of nuclear spin polarization, the value of ΔH^0_{app} obtained at 135 kG is likely most representative of the actual $^3(P^{\dagger}I^{\mp}) - ^3PI$ enthalpy difference.

In the preceding analysis of the temperature dependence of k_{obs}, it was assumed that k_S, $\Phi_{^3PI}$, and k_{isc} are temperature independent. It is known that k_S and $\Phi_{^3PI}$ depend on temperature, k_S decreasing by about a factor of 4 as the temperature is lowered to 200 K [25], while $\Phi_{^3PI}$ increases by about a factor of 3 [26]. The two contrasting temperature dependences at least partially offset each other, since the two terms enter Eqn. 1 as a product. If we separate out the temperature dependence of α and write:

$$\alpha = \zeta(B)\, e^{-\Lambda\beta} \qquad (4)$$

Eqn. 2 then becomes

$$k_{obs} = k_{isc} + \zeta(B)\, e^{-(\Delta H^0 + \Lambda)\beta} \qquad (5)$$

Schenck et al. [18] measured $\Phi_{^3PI}$ as a function of magnetic field and temperature in RCs where electron transfer from $I^{\mp}$ to the quinone had been blocked by prior reduction of the quinone. k_S can

be obtained from our analysis of the radical pair lifetime data [24]. The product of k_S and $\Phi_{^3PI}$ can be fit to an Arrhenius curve with $\Lambda = 120$ cm^{-1} at zero applied field over the temperature range of interest. While this measurement is for a different sample at a different applied field, it indicates the likely magnitude of Λ. Since the slope of the Arrhenius curve is equal to the sum of ΔH^0 and Λ, a value of $\Lambda = 120$ cm^{-1} would decrease the obtained value of ΔH^0_{app} to 1330 cm^{-1}. Because $\Phi_{^3PI}$ depends in a complex way on temperature and magnetic field, Λ will in principle be magnetic field dependent. Separating this effect from the nuclear spin polarization effect would require more knowledge of the temperature dependence of $\Phi_{^3PI}$ as a function of magnetic field.

A temperature dependence to k_{isc} in Eqn. 5 would introduce another source of error in the calculation of ΔH^0. In fact, the value of k_{isc} at room temperature in viscous buffer calculated from the magnetic field dependence of k_{obs} ($(10.60 \pm 0.10) \cdot 10^3$ s^{-1}) does differ slightly from the low temperature asymptote of Fig. 4 ($(8.2 \pm 0.2) \cdot 10^3$ s^{-1}). We can model this weak temperature dependence of k_{isc} with an Arrhenius expression with activation energy $\Delta H^0_{isc} = 20 \pm 10$ cm^{-1}. This value is consistent with the slight temperature dependence of k_{isc} observed previously at temperatures below 200 K [16]. Accounting for the temperature dependence of k_{isc} in this way causes a small increase in ΔH^0_{app} of 100 cm^{-1}.

The presence of anisotropic interactions will cause the temperature dependence of k_{obs} to underestimate ΔH^0 due to the temperature dependence of the rotational correlation time. Since the change from rapid to slow RC rotation relative to k_{obs} occurs over the narrow temperature range from 290 K to 270 K in glycerol/water mixtures, the fact that k_{obs} fits an Arrhenius expression over the whole temperature range of the experiment indicates that anisotropic interactions do not cause a significant perturbation.

The factors discussed above introduce only minor and somewhat offsetting corrections into the value of ΔH^0 derived from the data in Fig. 4. Because these corrections are small and the exact values are unknown, we will use the uncorrected value of ΔH^0_{app} in further discussion. The similarity between the value of ΔH^0 obtained from the slope of the Arrhenius plot at 135 kG and the values of ΔG^0 obtained by substitution into Eqn. 1 indicates that the entropy change between $^3(P^+I^-)$ and ^3PI is small.

Heterogeneity of ^3PI decay rates

As discussed above, the ^3PI decay process exhibits a measurable heterogeneity in decay rates. The fits to a gaussian distribution of exponentials showed a reduced χ^2 (χ^2 divided by the number of degrees of freedom) significantly lower (average value = 1.0) than the reduced χ^2 for fits to single exponentials (average value = 1.4; with approx. 62 degrees of freedom, this represents a change of about 25 standard deviations.) In contrast, simulated single exponential decays with normally distributed noise showed best fits to sharp distributions of exponential decay rates (second moment less than 0.07 of the average value), insignificantly better than fits to single exponentials. A distribution of decay rates is expected due to the distribution of orientations and nuclear spin states in the sample; under conditions when the RC orientation or nuclear spin state is preserved during the lifetime of ^3PI, RCs with orientations and nuclear spin states that favor singlet-triplet mixing will have a faster rate of ^3PI decay. The fact that the second moment of the distribution is an almost constant fraction of the average value, even at low temperatures where the radical pair state is thermally inaccessible and decay is only through k_{isc}, in aqueous solution where RC rotation is rapid, and at high field where nuclear polarization is not important, indicates that there are other sources of heterogeneity in the sample large enough to obscure any temperature or magnetic field dependent heterogeneity. There is precedence for non-exponential decay kinetics in these systems: a distribution in reaction rates for $P^+IQ^- \rightarrow PIQ$ has been observed previously in quinone-containing RCs at low temperatures [27]; this distribution becomes wider when the RCs are frozen under illumination [28].

Initial charge separation energetics

The energy of the ^1PI state is known quite accurately from absorption and fluorescence spectra to be 11 220 cm^{-1} above the ground state [29]. The energy of the ^3PI state has been recently

measured by observing the phosphorescence from this state; based on reasonable assumptions about the ^3P Stokes shift, it is calculated to be 7730 cm^{-1} above the ground state [30]. There is little information available about the entropy changes involved in these transitions; however, since as observed above, the re-formation of the dipolar radical pair state from ^3PI involves only a small entropy change, it is reasonable to assume entropy changes accompanying changes in the electronic state of P would be small. There would be a slight reduction of the free energy of ^3PI of $kT \ln 3$ relative to its energy, due to the spin multiplicity of the triplet state. This would reduce the free energy of ^3PI at room temperature to 7500 cm^{-1} above the ground state. The same spin multiplicity factor would increase the free energy of $^1(P^+I^-)$ to 230 cm^{-1} above $^3(P^+I^-)$. Using the aqueous sample value of $\Delta G^0 = 1370 \pm 30$ cm^{-1}, these factors give a free energy for $^1(P^+I^-)$ of 9100 cm^{-1} above the ground state at room temperature. Combined with the free energy of ^1PI obtained from the absorption and emission measurements, this results in a free energy difference of 2120 cm^{-1} (0.263 eV) for the initial $^1PI \rightarrow {}^1(P^+I^-)$ electron transfer step ($\Delta G^0_{^1PI \rightarrow {}^1(P^+I^-)}$).

Likewise, combining the energies of the ^1PI and ^3PI states with the estimate of ΔH^0 obtained here of 1450 ± 70 cm^{-1} yields an enthalpy difference of 2040 cm^{-1} (0.253 eV) between ^1PI and $^1(P^+I^-)$ ($\Delta H^0_{^1PI \rightarrow {}^1(P^+I^-)}$). Comparisons of this value with $\Delta G^0_{^1PI \rightarrow {}^1(P^+I^-)} = 2120$ cm^{-1} derived above indicates only a small change in entropy accompagnies the initial charge transfer step.

Comparison with other methods

Two other methods have been used to determine the $^1PI - {}^1(P^+I^-)$ energy difference: measurements of the ambient redox potential at which reactions are blocked by prior reduction or oxidation of reactants and measurements of delayed fluorescence due to back electron transfer from $^1(P^+I^-)$ to ^1PI. P can be oxidized in the dark with a midpoint redox potential of approx. + 450 mV vs. NHE [31]. The reduction potential of I in *Rb. sphaeroides* RCs has not been measured to the best of our knowledge. The reduction potential of bacteriopheophytin *a* in organic solvents is -550 mV vs. NHE [32]. Combining these numbers with the energy of the ^{1}PI state above the ground state (11 220 cm^{-1} = 1.39 eV) gives a value of $\Delta G^0_{PI \to {}^1(P^+I^-)}$ of 0.39 eV, or 3150 cm^{-1}. This type of analysis suffers from three problems. First, the redox potentials of P and I are affected by the charge of the other; the free energy of P$^+$I, as measured by redox titration, may not be simply related to the free energy of P$^+$I$^-$. Second, the redox potential of the bacteriopheophytin may be affected by the protein environment. Third, the redox states probed by redox titration are equilibrium states of the RC. It is not clear how these states compare to the transient states formed during RC photochemistry.

The principle of the delayed fluorescence measurements is that the initial amplitude of the delayed fluorescence relative to the amplitude of the prompt fluorescence is a measure of the relative concentrations of the first intermediate state in the reaction scheme and the fluorescing state, assumed to be the $^1(P^+I^-)$ state and the ^{1}PI state, respectively. With the reasonable assumption that the ^{1}P and $^1(P^+I^-)$ states are in thermal equilibrium several hundred ps after excitation, the equilibrium constant for the ^{1}PI $\rightleftarrows$ $^1(P^+I^-)$ reaction can be obtained giving $\Delta G^0_{PI \to {}^1(P^+I^-)}$. Woodbury and Parson [33] obtained the result that for Q$^-$ containing *Rb. sphaeroides* RCs $\Delta G^0_{PI \to {}^1(P^+I^-)}$ equals 1370 cm^{-1} at room temperature and decreases to 400 cm^{-1} at 100 K. Similar results were also obtained for quinone-depleted RCs [34]. Possible reasons for the discrepancy between the delayed fluorescence result and the conclusions of our work are discussed below.

Effect of modification of the reaction scheme

As discussed above, a major assumption of this analysis is that the reaction scheme illustrated in Fig. 1 is correct. A variety of modifications have been proposed by a number of authors. For instance, Woodbury and Parson [33], in their analysis of the delayed fluorescence measurements, explained their observation of non-exponential decay kinetics by proposing that the initial state formed by decay of ^{1}PI relaxes by hundreds of cm^{-1} through a series of intermediate states during the first few ns. There are other indications of the existence of intermediate states en route to formation of $^1(P^+I^-)$. Results from Reaction Yield

Detected Magnetic Resonance (RYDMR) experiments and the magnetic field dependence of $\Phi_{{}^3PI}$ indicate that the state that undergoes singlet-triplet mixing cannot be within 1370 cm^{-1} of ^{1}PI at room temperature. These experiments are sensitive to dephasing processes that interrupt the singlet-triplet spin evolution in the radical pair state, such as reformation of ^{1}PI, and can be used to put limits on k_{-1}. The width of the RYDMR resonance indicates that $k_S + k_T + k_{-1} < (0.4-7.0) \cdot 10^8$ s^{-1} [22,35]. This width is relatively insensitive to temperature [22]. Assuming that $\Delta G^0_{PI \to {}^1(P^+I^-)}$ is mostly enthalpic, k_{-1} should show a strong temperature dependence. The temperature independence of the RYDMR linewidth then argues that k_{-1} must be relatively small compared with k_S and k_T. A value of $k_{-1} < 1.0 \cdot 10^8$ s^{-1} combined with the observed value of $k_1 = 3.0 \cdot 10^{11}$ s^{-1} [1-3] requires that $\Delta G^0_{PI \to {}^1(P^+I^-)}$ at room temperature be greater than 1700 cm^{-1}. Analysis of the magnetic field dependence of the ^{3}PI quantum yield gives similar results. Comparison of these results with the results from the delayed fluorescence measurements indicate that the state formed within a few ps is not the state that undergoes singlet-triplet mixing, but rather is an intermediate state en route to the formation of $^1(P^+I^-)$. A difference between the radical pair state probed by the delayed fluorescence measurement on the ns time scale and the radical pair state repopulated by ^{3}PI on the μs time scale could explain the discrepancy between the values of $\Delta G^0_{PI \to {}^1(P^+I^-)} = 1370$ cm^{-1} calculated by Woodbury and Parson and 2120 cm^{-1} calculated here. The effect of intermediate states in the initial electron transfer process on the reaction dynamics has important consequences for the reaction mechanism and the analysis of the magnetic field dependence of $\Phi_{{}^3PI}$ (Goldstein, R.A. and Boxer, S.G., unpublished results).

Schenck et al. [18] and Chidsey et al. [24] proposed that a fraction of $^3(P^+I^-)$ might decay to the ground state without forming ^{3}PI. This path would lead to a temperature-dependent, but field-independent component to k_{obs}. As we will discuss elsewhere in detail, no set of rate constants can be found that match experimental values of k_{obs}, $\Phi_{{}^3PI}$, the RYDMR linewidth, and the radical pair lifetime with such a branching in the decay of

$^3(P^+I^-)$. Moreover, the experimental observations that led Schenck et al. [18] and Chidsey et al. [24] to propose such a branching path can be explained by the existence of an intermediate state en route to $^1(P^+I^-)$ formation (Goldstein, R.A. and Boxer, S.G., unpublished results).

Conclusion

We have obtained information on the energy difference between the radical pair and ^{3}PI states in quinone-depleted *Rb. sphaeroides* RCs by three methods: the magnetic field dependence of the triplet decay rate and the quantum yield of ^{3}PI formation in an aqueous sample at room temperature and moderate to very high field ($\Delta G^0 = 1370 \pm 30$ cm^{-1}); the magnetic field dependence in a viscous sample at room temperature and moderate to high field ($\Delta G^0 = 1390 \pm 30$ cm^{-1}); and the temperature dependence of k_{obs} at 135 kG in a viscous sample ($\Delta H^0 = 1450 \pm 70$ cm^{-1}). Combining the results of the first and third methods with values for the energies of the ^{1}PI and ^{3}PI states obtained from absorption and emission measurements indicates that the initial electron-transfer reaction, ^{1}PI $\rightarrow$ $^1(P^+I^-)$ has a driving force of approx. 2120 cm^{-1} (0.263 eV) with little change in entropy. This conclusion is in contrast to those based on previous measurements of delayed fluorescence [33]. Because we can find no obvious fault with the delayed fluorescence measurements, we are forced to conclude that the radical pair intermediate being monitored during the decay of the triplet state is different from that being sampled in the delayed fluorescence measurements. The present experiment cannot distinguish whether the difference is due to a shift in the energy of the P^+I^- ion-pair state with time (called nuclear relaxation in Ref. 33, not to be confused with nuclear spin relaxation), or whether the states are actually chemically distinct species. We note that the energy of the radical pair obtained in our measurements is consistent with the energy of the intermediate needed to explain the temperature dependence of the decay of the P^+Q^- state in RCs substituted with different quinones [34].

Acknowledgements

We wish to thank Larry Rubin and Bruce Brandt of the Francis Bitter National Magnet Laboratory for their gracious hospitality and assistance during our visit to their facility. We thank Dr. Neal Woodbury for extensive discussions of delayed fluorescence measurements. This work was supported in part by a grant from the National Science Foundation (DMB8607799) and a Presidential Young Investigator Award to S.G.B.

References

1 Woodbury, N.W., Becker, M., Middendorf, D. and Parson, W.W. (1985) Biochemistry 24, 7516–7521.
2 Martin, J.-L., Breton, J., Hoff, A.J., Migus, A. and Antonetti, A. (1986) Proc. Natl. Acad. Sci. USA 83, 957–961.
3 Martin, J.-L., Fleming, G.R. and Breton, J. (1988) Biophys. J. 53, 66a.
4 Norris, J.R., Scheer, H. and Katz, J.J. (1975) Ann. NY Acad. Sci. 244, 260–281.
5 Deisenhofer, J., Epp, O., Miki, K., Huber, R. and Michel, H. (1984) J. Mol. Biol. 180, 385–398.
6 Chang, C.-H., Tiede, D., Tang, J., Smith, U., Norris, J. and Schiffer, M. (1986) FEBS Lett. 205, 82–86.
7 Allen, J.P., Feher, G., Yeates, T.O., Rees, D.C., Deisenhofer, J., Michel, H. and Huber, R. (1986) Proc. Natl. Acad. Sci. USA 83, 8589–8593.
8 Allen, J.P., Feher, G., Yeates, T.O., Komiya, H. and Rees, D.C. (1987) Proc. Natl. Acad. Sci. USA 84, 5730–5734.
9 Allen, J.P., Feher, G., Yeates, T.O., Komiya, H. and Rees, D.C. (1987) Proc. Natl. Acad. Sci. USA 84, 6162–6166.
10 Fajer, J., Brune, D.C., Davis, M.S., Forman, A. and Spaulding, L.D. (1975) Proc. Natl. Acad. Sci. USA 72, 4946–4960.
11 Rockley, M.R., Windsor, M.W., Cogdell, R.J. and Parson, W.W. (1975) Proc. Natl. Acad. Sci. USA 72, 2251–2255.
12 Kaufmann, K.J., Dutton, P.L., Netzel, T.L., Leigh, J.S. and Rentzepis, P.M. (1975) Science 188, 1301–1304.
13 Marcus, R.A. and Sutin, N. (1985) Biochim. Biophys. Acta 811, 265–322.
14 Boxer, S.G., Chidsey, C.E.D. and Roelofs, M.G. (1982) J. Am. Chem. Soc. 104, 1452–1454.
15 Boxer, S.G., Chidsey, C.E.D. and Roelofs, M.G. (1983) Annu. Rev. Phys. Chem. 34, 389–417.
16 Chidsey, C.E.D., Takiff, L., Goldstein, R.A. and Boxer, S.G. (1985) Proc. Natl. Acad. Sci. USA 82, 6850–6854.
17 Goldstein, R.A. and Boxer, S.G. (1987) Biophys. J. 51, 937–946.
18 Schenck, C.C., Blankenship, R.E. and Parson, W.W. (1982) Biochim. Biophys. Acta 680, 44–59.
19 Okamura, M.Y., Issacson, R.A. and Feher, G. (1975) Proc. Natl. Acad. Sci. USA 72, 3491–3495.

20 Lakowicz, J.R., Cherek, H., Gryczynski, I., Joshi, N. and Johnson, M.L. (1987) Biophys. Chem. 28, 35–50.

21 Boxer, S.G., Chidsey, C.E.D. and Roelofs, M.G. (1982) Proc. Natl. Acad. Sci. USA 79, 4632–4636.

22 Moehl, K.W., Lous, E.J. and Hoff, A.J. (1985) Chem. Phys. Lett. 121, 22–27.

23 Parson, W.W., Clayton, R.K. and Cogdell, R.J. (1975) Biochim. Biophys. Acta 387, 265–278.

24 Chidsey, C.E.D., Kirmaier, C., Holten, D. and Boxer, S.G. (1984) Biochim. Biophys. Acta 766, 424–437.

25 Budil, D.E., Kolaczkowski, S.V. and Norris, J.R. (1987) in Progress in Photosynthesis Research (Biggins, J., ed.), Vol. I, pp. 25–27, Martinus Nijhoff, Dordrecht.

26 Schenck, C.C., Mathis, P. and Lutz, M. (1984) Photochem. Photobiol. 39, 407–417.

27 Morrison, L.E. and Loach, P.A. (1978) Photochem. Photobiol. 27, 751–757.

28 Kleinfeld, D., Okamura, M.Y. and Feher, G. (1984) Biochem. 23, 5780–5786.

29 Woodbury, N.W. and Parson, W.W. (1986) Biochim. Biophys. Acta 850, 197–210.

30 Takiff, L. and Boxer, S.G. (1988) Biochim. Biophys. Acta 932, 325–334.

31 Dutton, P.L. and Jackson, J.B. (1972) Eur. J. Biochem. 30, 495–510.

32 Fajer, J., Brune, D.C., Davis, M.S., Forman, A. and Spaulding, L.D. (1975) Proc. Natl. Acad. Sci. USA 72, 4956–4960.

33 Woodbury, N.W. and Parson, W.W. (1984) Biochim. Biophys. Acta 767, 345–361.

34 Woodbury, N.W., Parson, W.W., Gunner, M.R., Prince, R.C. and Dutton, P.L. (1986) Biochim. Biophys. Acta 851, 6–22.

35 Norris, J.R., Bowman, M.K., Budil, D.E., Tang, J., Wraight, C.A. and Closs, G.L. (1982) Proc. Natl. Acad. Sci. USA 79, 5532–5536.

Author Index